P9-EEG-573

ENVIRONMENTAL MANAGEMENT, SUSTAINABLE DEVELOPMENT AND HUMAN HEALTH

Environmental Management, Sustainable Development and Human Health

Editors

Eddie N. Laboy-Nieves & Fred C. Schaffner
University of Turabo, Gurabo, Puerto Rico, USA

Ahmad H. Abdelhadi
New York Institute of Technology, Amman, Jordan

Mattheus F. A. Goosen
Alfaisal University, Riyadh, KSA

CRC Press is an imprint of the
Taylor & Francis Group, an **informa** business
A BALKEMA BOOK

Cover photo credits: Front cover (above): Ricardo Ríos-Menéndez, front cover (below) and back cover: Eddie N. Laboy-Nieves.

http://ciemades.org

CRC Press/Balkema is an imprint of the Taylor & Francis Group, an informa business

Typeset by Charon Tec Ltd (A Macmillan Company), Chennai, India
Printed and bound in Great Britain by Cromwell Press Ltd, Towbridge, Wiltshire.

Published by: CRC Press/Balkema
P.O. Box 447, 2300 AK Leiden, The Netherlands
e-mail: Pub.NL@taylorandfrancis.com
www.crcpress.com – www.taylorandfrancis.co.uk – www.balkema.nl

Library of Congress Cataloging-in-Publication Data

Environmental management, sustainable development, and human health / edited by Eddie N. Laboy-Nieves …[et al.].
p. ; cm.
Includes bibliographical references and index.
ISBN 978–0–415–46963–0 (hbk. : alk. paper)
1. Environmental health. 2. Environmental management–Health aspects.
3. Sustainable development–Health aspects. 4. Economic development–Health aspects.
I. Laboy, Eddie Nelson. II. Title. [DNLM: 1. Environmental Health.
2. Conservation of Natural Resources. 3. Developing Countries.
4. Ecology. 5. Environmental Medicine. 6. International Cooperation.
WA 30.5 E607 2009]
RA565.E516 2009
362.196'98–dc22

2008039742

ISBN: 978-0-415-46963-0 (hbk)
ISBN: 978-0-203-88125-5 (e-book)

Contents

Ecosystems and Environmental Health

Materials Recycling and Water Treatment

Preface

Access to safe drinking water and sanitation is a basic necessity. However, such access is highly variable around the world and in particular in Africa, Asia and South America. Much progress still remains to be made in infrastructure improvements and poverty reduction. A recent World Bank report, for example, noted that more than 100 million people in the Latin American region alone lack access to potable water and adequate sanitation systems.

Compounding the issue of water availability is contamination of water supplies and the lack of wastewater treatment facilities. This affects human health. The presence of lead as well as excess agricultural fertilizer in the environment, for example, constitutes an ecological and human health concern.

Environmental education plays a major role in achieving changes in attitudes that contribute to environmental awareness in society. The average person is usually somewhat informed about environmental problems. However, not many fully understand the basis of these problems or how to deal with them. One example of rising public awareness is the increasing use of recycled materials, particularly in road systems, paper making and plastics manufacturing. Environmental education is especially critical for young people who will be future decision makers.

All disciplines and all segments of society will need to work together, proactively, in seeking long-term solutions to environmental problems. The recent establishment of an International Center for Environmental Studies and Sustainable Development (http://ciemades.suagm.edu/) is one example of this new approach. This book project is one of the initiatives of this Center.

Sustainable development is now considered by many organizations and their stakeholders as being the model to follow. An increasing number of companies currently act and communicate based on their triple performance (i.e. economic, environmental and social). As educators and scientists we can follow this example by keeping in mind the bigger picture when it comes doing our research and educating our young people. This will help to improve the overall health of the society in which we live.

This main aim of this book is to assess some of the major environmental problems facing the developing as well as developed regions of the world. Solutions are suggested. It includes several papers on environmental education projects that will help to raise awareness in young people about the interrelationships between pollution, the environment and society.

This edited book provides a broad coverage of recent advances in environmental management and sustainable development as it relates to human health. It starts with an overview chapter on the nature of environmental management and sustainable development. This is followed by well over 30 chapters from experts around the world.

Considerable research has been reported on environmental management and sustainable development over the past three decades. International conferences on Sustainable Development were held in Johannesburg in 2004 and Helsinki in 2005. Numerous books on environmental management and sustainable development are available; Environmental Education and Advocacy by Johnson, E.A., 2005; Land, Water and Development: Sustainable Management of River Basin Systems, Newson, M., 1997; Essentials of Medical Geology: Impacts of Medical Geology on Public Health, Selinus, O., 2004; Corporate Environmental Management, Darabis, J., 2007; Handbook of Sustainable Development, Atkinson, G., 2007. Since many of these publications are highly specific, a book is now needed that combines the state-of-the-art knowledge in environmental management and sustainable development and in how these areas relate to improving the overall health of society.

The intended audience for this book includes researchers, practicing engineers, decision makers, environmentalists, medical researchers, contractors, postgraduate and undergraduate students and others working in environmental management and sustainable development. The authors hope that the information provided in this book would help to promote a better knowledge on environmental management and sustainable development and contribute toward the utilization of sustainable technology and environmental friendly practices by society.

The views expressed in the chapters of this book are those of the authors and not necessarily those of their respective institutions. The authors hope that this book will contribute to the advancement in research in sustainable development and help decision makers, business people, and engineers in mounting practical solutions to environmental problems.

Eddie N. Laboy-Nieves
University of Turabo

Fred C. Schaffner
University of Turabo

Ahmad H. Abdelhadi
New York Institute of Technology

Mattheus (Theo) F. A. Goosen
Alfaisal University
2008

List of Authors

Adrián Bonilla-Petriciolet
Instituto Tecnológico de Aguascalientes, México. E-mail: abonilla@ita.mx

Agustín A. Irizarry-Rivera
University of Puerto Rico, Mayagüez. E-mail: agustin@ece.uprm.edu

Ahmad H. Abdelhadi
New York Institute of Technology, P.O. Box 840878, Amman 11184, Jordan.
E-mail: aabdel06@nyit.edu

Aldo Cróquer
Laboratorio de Comunidades Marinas, Universidad Simon Bolivar Apdo. 8900,
Caracas 1080-A, Venezuela. E-mail: acroquer@usb.ve

Anke Arnaud
Embry-Riddle Aeronautical University, Florida, U.S.A. E-mail: arnauda@erau.edu

Arnoldo José Gabaldón
Universidad Simón Bolívar, Caracas, Venezuela. E-mail: arnoldojgabaldon@gmail.com

Axelle Durimel
Université des Antilles et de la Guyane, Guadeloupe. E-mail: adurimel@univ-ag.fr

Carl Thodesen
Clemson University, SC, USA. E-mail: kcdoc@clemson.edu

Chaker Ncibi
High Institute of Agronomy, Chott Meriem. E-mail: nmchaker@yahoo.fr

Dawna L. Rhoades
Embry-Riddle Aeronautical University, Florida, U.S.A. E-mail: rhoadesd@erau.edu

Dimitrios Angelidis
University of Macedonia, Greece. E-mail: d.angelidis@hotmail.com

Eddie N. Laboy-Nieves
Universidad del Turabo, School of Science and Technology, Box 3030, Gurabo,
Puerto Rico 00778-3030. E-mail: elaboy@suagm.edu

Efraín O'Neill-Carrillo
University of Puerto Rico, Mayagüez. E-mail: oneill@ece.uprm.edu

Elizabeth Reynel-Avila
Universidad de Guanajuato, México. E-mail: hreynel@quijote.ugto.mx

Esam Qnais
The Hashemite University, Zarqa, Jordan. E-mail; esam_11@hotmail.com

Evens Emmanuel
Laboratoire de Qualité de l'Eau et de l'Environnement, Université Quisqueya, BP 796 Port-au-Prince, Haïti. E-mail: evemm1@yahoo.fr. Tel: +509 3423 4269/ Fax: +509 2221 4211

F. Nikoi Hammond
School of Engineering and the Built Environment, University of Wolverhampton, UK. E-mail: f.n.hammond2@wlv.ac.uk, Tel: +00 44 1902 322179, Fax: +00 44 1902 322179

Fouad H. Beseiso
Former Governor of Palestine Monetary Authority, E-mail: beseisof@yahoo.com

Francisco Martínez-González
Universidad de Guanajuato, México. E-mail: fmarti@quijote.ugto.mx

Fred C. Schaffner
Office of Science and Technology, Puerto Rico Economic Development Company (PRIDCO) Present address: Universidad del Turabo, School of Science and Technology, Gurabo, Puerto Rico. E-mail: fschaffner@suagm.edu

Fuad A. Abdulla
New York Institute of Technology, Amman, Jordan. E-mail: fabdulla@nyit.edu.jo

Guadalupe de la Rosa
Universidad de Guanajuato, México. E-mail: delarosa@quijote.ugto.mx

Gustavo Cruz-Jiménez
Universidad de Guanajuato, México. E-mail: cruzg@quijote.ugto.mx

H. Okamura
Water Re-use Promotion Center, Japan

Hans Werner Gottinger
STRATEC, Munich, Germany. E-mail: hg528@bingo-ev.de

Hilal Al-Hinai
Department of Mechanical and Industrial Engineering, Sultan Qaboos University, Oman. E-mail: sulaiman@unical.it

Irene Cano-Aguilera
Universidad de Guanajuato, México. E-mail: irene@quijote.ugto.mx

Issa Shehabat
Faculty of Information Technology, Philadelphia University, Amman, Jordan

Jacqueline Taylor Basker
New York Institute of Technology, Amman, Jordan. E-mail: jtaylorbasker@gmail.com

Joseph Somevi
School of Engineering and the Built Environment, University of Wolverhampton, UK. E-mail: f.n.hammond2@wlv.ac.uk, Tel: +00 44 1902 322179, Fax: +00 44 1902 322179

José A. Colucci-Ríos
University of Puerto Rico, Mayagüez. E-mail: jcolucci@uprm.edu

José Raúl Pérez Durán
Advisor on Water Resources, Instituto Nacional de Recursos Hidráulicos (INDRHI) – National Institute for Water Resources, Dominican Republic; EMR (Engineering, Management and Risk) Group, E-mail: jraulperezd@yahoo.com

Jung-Wan Lee
Kazakh-British Technical University, Republic of Kazakhstan.
E-mail: Jwlee1119@yahoo.com

K. Alamar
Dept of Economics, Arul Anandar College, Karumathur, India 625 514

Katerina Lyroudi
University of Macedonia, Greece. E-mail: lyroudi@uom.gr, chatzig@uom.gr

Ketty Balthazard-Accou
Laboratoire de Qualité de l'Eau et de l'Environnement, Université Quisqueya, BP 796 Port-au-Prince, Haïti. E-mail: evemm1@yahoo.fr. Tel: +509 3423 4269/ Fax: +509 2221 4211

Khaldoun Shatanawi
Clemson University, SC, USA. E-mail: kcdoc@clemson.edu

Khulood Abu Maria
College of Information Technology, Arab Academy of Business and Financial Sciences, Amman, Jordan

Louise van Scheers
School of Business Management, University of South Africa, E-mail: vscheml@unisa.ac.za

Luis E. Galván Rico
Universidad Simón Bolívar, Departamento de Tecnología de Servicios, Caracas, Venezuela. E-mail: lgalvan@usb.ve

Mago William Maila
Department of Teacher Education, University of South Africa, PO Box 392, UNISA 0003, TSHWANE, Republic of South Africa. E-mail: mailamw@unisa.ac.za, Tel: 012 429 4395, Fax: 012 429 4909

Marisol Aguilera M.
Universidad Simón Bolívar, Caracas, Venezuela. E-mail: maguiler@usb.ve

Mattheus F. A. Goosen
Alfaisal University, Riyadh, Kingdom of Saudi Arabia. E-mail: mgoosen@alfaisal.edu

Moh'd Mahmoud Ajlouni
Dept. of Banking & Finance, Yarmouk University, Jordan. E-mail: ajlouni4@yahoo.co.uk

Mohammed H. Abu-Dieyeh
The Hashemite University, Zarqa, Jordan. E-mail: dandelion@hu.edu.jo

N. Murali
Dept of Economics, Arul Anandar College, Karumathur, India 625 514
E-mail: nmuralirbs@yahoo.co.in

Naim Ajlouni
Princess Ghazi College of Information Technology, VP, Al-Balqa
Applied Science University, Jordan

Nkobi Moleele
University of Botswana

Osnick Joseph
Laboratoire de Qualité de l'Eau et de l'Environnement, Université Quisqueya,
BP 796 Port-au-Prince, Haïti. E-mail: evemm1@yahoo.fr. Tel: +509 3423 4269,
Fax: +509 2221 4211

Raed Abu Zitar
Faculty of Engineering and Computer Science New York Institute of Technology
(NYIT) Amman, Jordan. E-mail: rzitar@nyit.edu.jo

Rosa E. Reyes Gil
Universidad Simón Bolívar, Departamento de Biología de Organismos,
Caracas, Venezuela. E-mail: rereyes@usb.ve

Sandro Altenor
Université des Antilles et de la Guyane Guadeloupe, Université Quisqueya,
Haïti. E-mail: sandalt@yahoo.com

Sarra Gaspard
Université des Antilles et de la Guyane, Guadeloupe. E-mail: sgaspard@univ-ag.fr

Serji Amirkhanian
Clemson University, SC, USA. E-mail: kcdoc@clemson.edu

Sherrie L. Baver
The City College and The Graduate Centre-City University of New York,
E-mail sbaver@gc.cuny.edu

Shyam. Sablani
Biological Systems Engineering, Washington State University, Pullman, Washington,
USA. E-mail: ssablani@wsu.edu

Siham El-Kafafi
Manukau Institute of Technology, Manukau Business School, Auckland, New Zealand.
E-mail:siham.elkafafi@manukau.ac.nz

Simon W. Tai
Bang College of Business, KIMEP, Republic of Kazakhstan. E-mail: Tai@kimep.kz

Sulaiman K. S. Al-Obaidani
Department of Mechanical and Industrial Engineering, Sultan Qaboos University,
Oman. E-mail: sulaiman@unical.it

Tidimane Ntsabane
University of Botswana, E-mail: ntsabane@mopipi.ub.bw

Y. Taniguchi
Water Re-use Promotion Center, Japan

About the Editors

EDDIE N. LABOY-NIEVES has a Ph.D. in Ecology and works as Associate Professor in the School of Science and Technology, *Universidad del Turabo* (http://ut.pr), Puerto Rico. He has nearly 20 years teaching undergraduate and graduate level courses related to environmental sciences. He worked as Manager of the Jobos Bay National Estuarine Research Reserve (Puerto Rico). His research interests are focused on environmental characterization of mangroves, coral reefs and seagrass beds, and the ecological aspects of shallow water sea cucumbers. Dr. Laboy-Nieves has authored, co-authored, edited and peer-reviewed many publications. He is the Puerto Rico President of the International Center for Environmental and Sustainable Development Studies (http://ciemades.org), the organization that promoted the publishing of this volume. He serves as Scientific Advisor for many community, national, and international organizations.

FRED C. SCHAFFNER is Associate Dean for Graduate Studies and Research and Director of the Doctoral Program in Environmental Sciences in the School of Science and Technology at *Universidad del Turabo* (www.suagm.edu/utdoctoral/doct_env_sc.html). He has a Ph.D. in Biology from the University of Miami (Florida). He spent two years as a post doc at the National Audubon Society Research Department (Florida). In Puerto Rico, he has worked as Scientific Consultant for the Puerto Rico Industrial Development Company, the Department of Natural and Environmental Resources, the US Fish and Wildlife Service, and has held academic and executive positions at the University of Puerto Rico. He has served as Consultant to numerous community and environmental groups, President of the Special Commission for the San Juan Ecological Corridor, and member of the Board of Directors of the Caguas Botanic and Cultural Garden, in Puerto Rico. Dr. Schaffner has written over fifty scientific articles, white papers and technical reports.

AHMAD H. ABDELHADI is Assistant Professor in the College of Arts and Sciences of the New York Institute of Technology, Amman Campus (www.nyit.edu), a non-profit independent private institution of higher education with campuses in the United States and abroad. He obtained his doctoral degree in Galactic Dynamics from Clemson University, South Carolina, in 2003. He worked at James Madison University in Virginia before moving to the New York Institute of Technology as an Associate Chair for the College of Arts and Sciences and later Associate Campus Dean in the Amman Campus. His work on pre-solar silicon carbide grains and computational methods received support from the NASA Origin of Solar System Program. His scientific interests include computational methods in galactic dynamics, galactic chemical evolution, and chaotic systems. Dr. Abdelhadi has lectured in many conferences and published number of papers.

MATTHEUS (THEO) F. A. GOOSEN is Associate Vice President for Research at Alfaisal University (www.alfaisal.edu), a new private non-profit institution in Riyadh, KSA. Previously he held the position of Campus Dean (CAO) at the New York Institute of Technology in Amman, Jordan. Dr. Goosen has also held academic dean positions at the Universidad del Turabo in Puerto Rico, USA, and at the Sultan Qaboos University in Muscat, Oman. He obtained his doctoral degree in Chemical/Biomedical Engineering from the University of Toronto, Canada, in 1981. After graduation he spent three years as a post doc at Connaught Laboratories in Toronto and then ten years at Queen's University in Kingston. He has been on the Board of Directors of two companies. Dr. Goosen has published extensively with over 150 papers, book chapters, books, and patents.

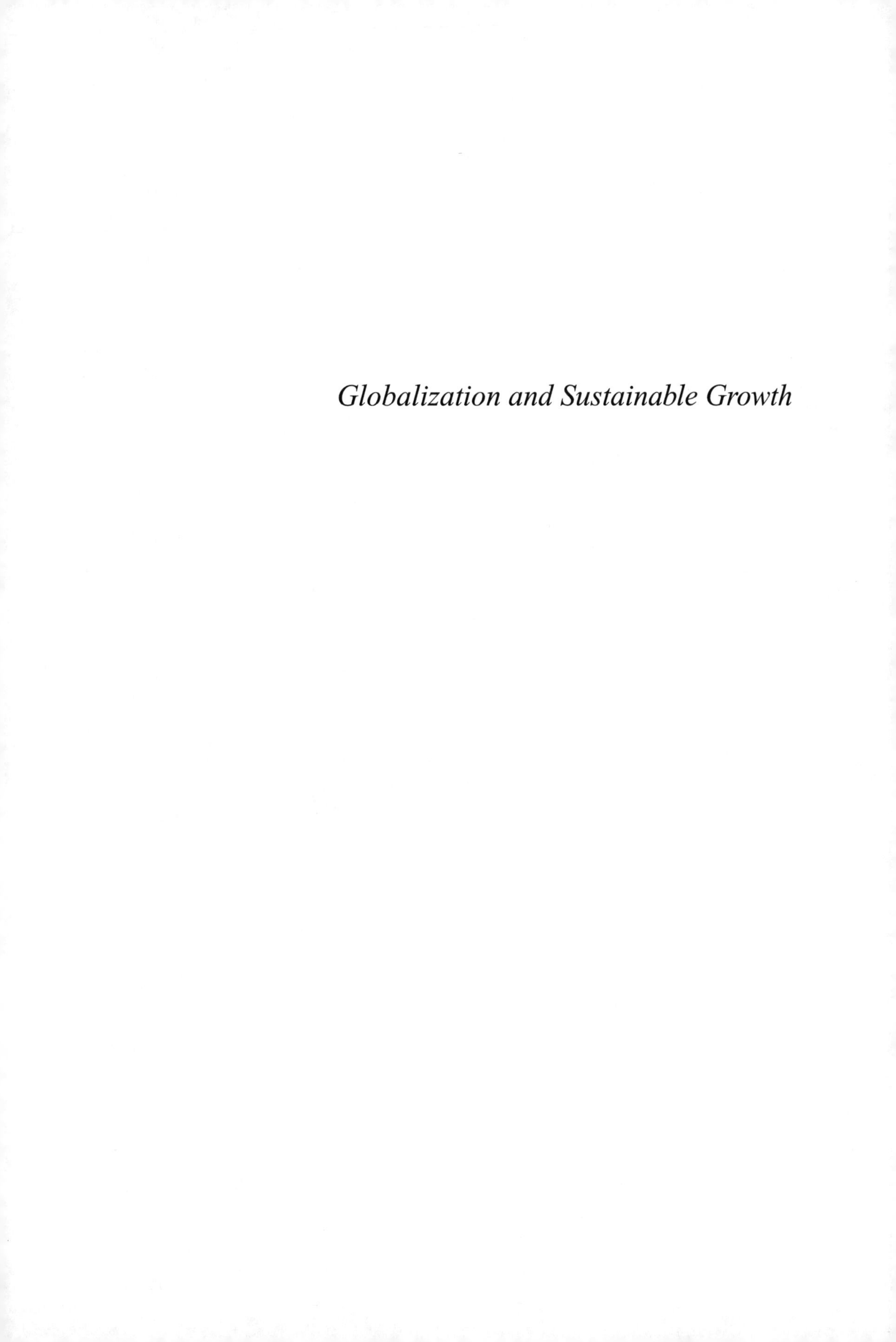

Globalization and Sustainable Growth

The Environment, Sustainable Development and Human Wellbeing: An Overview

Mattheus F.A. Goosen
Alfaisal University, Riyadh, Kingdom of Saudi Arabia

Eddie N. Laboy-Nieves & Fred C. Schaffner
Universidad del Turabo, Gurabo, Puerto Rico

Ahmad H. Abdelhadi
New York Institute of Technology (NYIT), Amman, Jordan

SUMMARY: The aim of this chapter is to provide an overview of the multidisciplinary interrelationships between human health and the environment. Coordinated approaches are necessary for solving the major environmental and sustainability problems facing the developing as well as developed regions of the world. Specific emphasis was placed on globalization and sustainable growth, bioethics and poverty, organizational performance and sustainability, environmental management and individual progress, human and ecosystem health, and water resources and recycling.

1 INTRODUCTION

The interrelationship between economic progress, environmental management and individual wellbeing is a complicated process, affecting both the quality and sustainability of the society in which we live (Abdulla et al., 2009, Maila, 2006). There is a growing realization by the general public as well as practicing engineers, decision makers, environmentalists, and medical researchers, that these three areas are interconnected.

In the past, the world's ecosystems were able to absorb the ecological damage resulting from extensive industrialization and development. However, with the rapid increases in global population and industrialization, as well as enhanced demands on natural resources such as fresh water supplies, the earth is no longer able to sustain a healthy and balanced ecosystem (Misra 2000, Laboy-Nieves, 2009). A coordinated approach is required to solve environmental problems.

Sustainable development is now considered by many organizations and their stakeholders as being the model to follow. For instance, an increasing number of companies currently acts and communicates based on their triple performance in economic, environmental and social areas (Misra 2000, Goosen et al., 2004 and 2009). As educators and scientists, there is a need to follow this example by keeping in mind the larger view to help improve the overall health of the society in which we live.

The aim of this chapter is to provide a brief overview of how sustainable development, environmental management and human health are interconnected, with an emphasis on globalization, bioethics and poverty, organizational performance and sustainability, environmental management and human progress, ecosystem health, and water resources and recycling.

2 GLOBALIZATION AND SUSTAINABLE GROWTH

Alamar and Murali (2009) noted that for sustainable development to be meaningful, overconsumption has to be brought under control. In addition, in a free market economy, the private sector may not bother to conserve nature. For the sake of profit, it may destroy forests, overuse mineral resources, or pollute air and water. This sector may not take into account social costs or benefits (Misra, 2000). Today there is an obsession with economic expansion. Growth should be defined not only in terms of the financial side but also in terms of societal and cultural parameters. The application of technology for the pursuit of profit has resulted in the overexploitation and the excessive utilization of natural resources.

The expansion of industries and domestic markets has drastically altered, for example, the atmospheric concentrations of numerous trace gases (Sharma, 2007). This has distorted nature's auto-balancing mechanism. As a result, the world is now faced with several environmental problems including acid rain, melting of glacial ice, large scale evaporation of water in the tropics, and an increase in cloudiness at higher altitudes. Developing countries, and notably the least developed, are expected to be the most vulnerable to the impacts of global climate change, although their current contribution to the problem is minimal.

Weather change can also have a significant impact on health through vector-borne diseases because of changes in the survival and reproduction rates of the carriers, the intensity and temporal pattern of vector activity, and the life cycle of pathogens within the vectors. Food production, availability and security, fresh water supply, forest biodiversity, coastal settlements, and fishing will also be adversely affected.

Extreme poverty still affects the lives of one out of every five persons in the developing world (Sharma, 2007). Soil degradation from erosion and poor irrigation practices continue to harm agricultural lands, jeopardizing production. Sharma (2007) noted that without a transition to more resource-efficient and less polluting farming methods, it will be difficult to meet world food needs without increasing the environmental burden that stems from intensive agriculture.

Sustainable development has three components: social, economic and ecological (Misra, 2000). Nevertheless these are not always compatible. Sustainability requires a rare balance between these three sets of goals. Once ecological sustainability has been achieved, then it is possible to attain economic sustainability. If this condition is maintained then social stability can be attained.

For sustainable development to be meaningful, overconsumption has to be brought under control. Industrial expansion in the present day has to be made within the carrying capacity of the planet. Human beings are at the centre of concerns for sustainable development. They are entitled to a healthy and productive life in harmony with nature.

Baver (2009) focused on the emerging global concept of democratic environmental governance using Mexico as the case study area. The goal of the research was to examine how several of the country's new national institutions, procedures, and capacities were affecting environmental performance as well as public perceptions of governmental legitimacy. It was reported that Mexico's position as a leader in environmental governance in the developing world did not emanate from the government itself; rather it was derived from a combination of pressures from civil society activists and international organizations and institutions (Rodríguez Bribiesca, 2007). Baver (2009) noted that struggle for democratic governance in this policy area may have contributed to a more accountable and transparent political system.

Three pillars are integral to the concept of democratic environmental governance: citizen access to environmental information held by public authorities; public participation in environmental decision-making; and citizens' effective access to justice. The United States was a global leader in these three pillars through the mid-1980s, having legislation addressing all three issues. However, by the 1990s, various regional and international governmental and non-governmental organizations have bundled the three ideas into one reform package and have promoted it around the world (Speth and Haas, 2006).

Perhaps the key lessons learned from the Mexican case, is that the process of economic globalization can promote modern environmental governance. However, only a highly mobilized movement can monitor these reforms and force them to function effectively.

3 BIOETHICS, ENVIRONMENTAL POLLUTION AND POVERTY

Education for sustainable development should be obligatory for all young people, as this represents the primary vehicle available for catalyzing the cultural changes required for survival (Ferrer and

Alvarez, 2003). Bioethics ought to serve as a platform for education about sustainability. Gabaldón and Aguilera (2009) reported that it is essential to apply a framework of values and principles to guide the conduct of decision-makers, scientists, and technologists. Furthermore, the advances of international environmental jurisprudence also lend a valuable arsenal of bioethical principles to the guidance of world development.

Gabaldón and Aguilera (2009) noted that bioethics is an interdisciplinary development resulting from philosophy, the health sciences, law and social sciences. Any discipline that deals with the study of human beings has had something to contribute to the ethics of life. Its scope also includes the public and sanitary policies, the tasks of all professions, and especially any research on humans or the flora and fauna that constitute ecosystems. This wider view of bioethics goes hand in hand with the concepts of sustainable development.

In its development, bioethics, seen as a branch of philosophy concerned with the study of the morality of human works (Andorno, 1998), incorporated principles that belonged to medical practice, to jurisprudence and to political tradition, as well as to the precepts of human rights, the development of professional ethics, and from religion (Gracia, 1989; Abellán Salort, 2006).

In a related work, Hammond and Somevi (2009) reported on poverty, urban land, and sustainable development using Africa as a case study. This continent is unique when it comes to poverty and sustainable growth. Some 41 per cent of inhabitants of African countries are extremely poor and are surviving on less than one dollar a day (Stern, 2006). At one extreme are the international requirements on governments to rehabilitate their economies in keeping with the United Nations Millennium Development Goals (MDGs). The MDGs set such targets, for example, as halving the proportion of people living on less than US $1 a day by 2015.

The sources of poverty are due to inefficient allocation of natural resources, capital, labour and time, underutilization of resources, and inequitable distribution of incomes (Hammond and Somevi, 2009). Liberal and moderate economists argued that as individuals and economic entities seek to maximise their own self interest and profits, they unwittingly promote the efficient allocation as well as optimal use of resources.

Sustainable development has economic, social, environmental and ethical dimensions. From an economic perspective, a resource is efficiently allocated and optimally utilized if it is put to the use that generates the highest possible returns (Hammond and Somevi, 2009). In free enterprise economies, the first two causes of poverty and economic malaise may be overcome by setting in place a broad legal as well as effective and quality infrastructural framework within which individuals and actors could pursue their self interest. The required legal framework may consist of laws and security rules that promote sound investment, and access to financial, labour and land transactions. To be really effective, the law must remove all manner of discrimination with respect to opportunities regarding education, investments, health services, land resources, capital and employment.

Socially, equitable distribution of the incomes on the other hand may be achieved through mechanisms and social benefits like income supports, unemployment benefits and housing, free or discounted education, health and transportation services. From an environmental perspective, economic and social goals have to be pursued in ways that will cause the least damage to the quality of the environment and limit the exhaustion of irreplaceable resources (Stern, 2006).

The bioethics of sustainable development has raised the spectre of possible exhaustion of vital resources at the expense of future generations. The way to resolve this dilemma is to pursue poverty alleviation methods that adequately reward hard work, initiatives, efforts and productive use of resources within the broad framework of the rule of law but at the same time penalises decision makers with the costs of reckless resource utilisation. The later is not so much to ensure that something is left for the future generation, but more because careless and inattentive use of resources threatens the very survival of the present generation itself.

4 ORGANIZATIONAL PERFORMANCE AND SUSTAINABILITY

Ajlouni (2009) considered the environment as a fifth component of a Balanced Scorecard (BSC) for improving the performance of an organization. The BSC was first proposed by Kaplan and Norton

(2001) as a means for deriving the performance of a system. It was then developed as a managerial mechanism for translating and implementing corporate strategy (Kaplan and Norton, 2001). The BSC was originally a framework of organizational performance measures across four perspectives or aspects: financial, customer, internal business processes, and learning and growth (FCIL). The BSC system has been implemented by numerous for-profit and non-profit businesses. Hence, incorporating and implementing environmental objectives into an organization's strategy should lead to economic sustainability. Top management, however, needs to develop a group environmental strategy, and then communicate it to their employees.

As more businesses around the globe incorporate ecological activities and performance in their financial statements and reports; and as more stakeholders require new and better ways to communicate green issues; top management will become more aware that environmental concerns need to be addressed and tackled within their organizations (Kaplan and Norton, 2001). However, environmental reporting should not be the end of the story. The BSC can help by identifying environmental problems, by targeting key areas for top management attention and reaction, by providing support for needed improvements in current systems, and by formulating an applicable and attainable environmental performance.

In a related study to that of Ajlouni (2009), Lyroudi and Angelidis (2009) examined banking productivity. Financial institutions can be considered as the corner stone of any economic system. Banking institutions are of special interest, for example, to the European Union. The second banking co-ordination directive, which was adopted in 1989, created a regime for regulating all functions of such institutions in the European Union (Siems and Clark, 1997 and Rogers and Sinkey, 1999). This new framework forced banks to adopt new products such as loan commitments, letters of credit, securities underwriting, insurance and derivatives. These are also known as off-balance sheet (OBS) activities. Stable economics systems will help to produce stable societies.

In contrast to the work of Lyroudi and Angelidis (2009), Maila (2009) considered environmentally sustainable economies. Some scholars perceive sustainable development as a North-South; South-South and North-North environmental discourse (Ndlovu-Gatsheni, 2006). Often, institutions find themselves in the wrong corner or part of the globe, especially when they defend sustainable development positions that compromise the lives of poor and marginalized communities, and unsustainable economic policies in developing countries regarding the utilization and care of natural resources.

Maila (2006) noted that development and sustainability must be anchored in action processes that allow participants to have choices in the how and why of doing things. Good quality progress can only occur when ordinary people benefit from it. While big business is dedicated to reap huge profits within a short period of time in their ventures, poor people, on the other hand, are concerned with putting food on the table for their families. It is not only financial expansion and ecological protection that are important for environmental sustainability; human development is also critical. Sen (1999) argued that development has gone from the growth of output per capita to the expansion of human values. He saw development as a process that expands the real freedoms that people enjoy.

Sound ecological sustainability and growth can never be continual if threats to the environment and human well being persist unabated worldwide. Organizational decisions should be made collectively, through discussions. This will allow for the advancement of both economic and human growth as pillars of environmentally sustainable economies.

5 ENVIRONMENTAL MANAGEMENT AND HUMAN PROGRESS

El-Khafafi (2009) assessed the sustainable management tool known as Triple Bottom Line Reporting. (TBL). This instrument has become more important and widespread in recent years. Organisations have been evaluated not only for their performances on the bottom line, but also for their behaviour as worldwide corporate citizens. Global corporate scandals of high profile organisations like Enron Corp., WorldCom Inc. and Arthur Anderson LLP rocked stakeholder confidence, shamed business leaders and led to a higher scrutiny of organisations' integrity. Accordingly, a movement towards corporate responsibility reporting arose to drive transparency in the environment and social arena. In 2004, 1700 corporations filed responsibility reports, up from virtually none in the early 1990s. Some of those corporations are using these reports as a way to push their sustainability commitments further (Assadourian, 2006). However, the success of these techniques depends on the commitment, skill and character of the people implementing them.

How do we make sustainability a commitment rather than just a compliance cost? Building sustainable firms and organisations requires a commitment to people's development. Staff development programs can only be successful if organisations have a clear sense of their place in (Assadourian, 2006). Furthermore, this requires a strong self-identity and an understanding of stakeholder expectations. The development of such understanding with the organisation and community facilitates the building of trust and integrity which are fundamental to sustainability. In pursuing this agenda, there is a strong case for organisations to be proactive in developing their people by helping them analyse problem cases, analyse forces impacting on the organisation and society, become familiar with core ethical literature, develop solutions to ethical problems and revitalise the organisation.

Emotions are an essential part of human life. They influence how we think, adapt, learn, behave, and communicate with each other. As El-Nasr et al. (1998, 2000) noted that the question is not whether intelligent machines can have any emotions, but whether machines can be intelligent without emotions. In El-Nasr's et al. work (2000) neurological evidence was provided which proved that emotions do in fact play an important and active role in the human decision-making process. The interaction between the emotional process and the cognitive process may explain why humans excel at making decisions based on incomplete information; acting on our gut-feelings.

Maria and Zitar (2009) performed a study on modelling of artificial emotions. Two models were built for agent-based systems; one was supported with artificial emotions and the other one without. Both were used in solving a benchmark problem: to have a clean and healthy environment for an orphanage house. The study showed that systems with a proper model of emotions could perform better than systems without emotions. The authors went on to explain that artificial agents can be used as a testing ground for theories about natural emotions in animals and humans. This provides a synthetic approach that is complementary to the analytic study of natural systems.

6 ECOSYSTEMS AND ENVIRONMENTAL HEALTH

Mankind is living in ecosystems that can be recognized at many different levels, ranging from, for example, a small forest to the entire globe (Abdulla et al., 2009; Townsend et al., 2003). Natural ecological systems are dynamically stabilized based on balanced inputs and outputs. All ecological structures are controlled by the same processes including natural and anthropogenic (human caused) turbulence. Yet, with an increasing population, human activities create more disturbances and add additional unbalanced situations to ecosystems.

Humans are among the most successful living things on earth. The present rate of growth of the global population is unsustainably high. This stresses our future by further energy demands; a greater drain on non-renewable resources and extra strain on renewable resources. The response of disturbed ecosystems, in consequence, includes local, regional and/or global problems to human health and the environment. The work of Abdulla et al. (2009) assessed atmospheric pollution, its consequences on humans and the environment, and perspectives toward its control. They also assessed pesticides, their persistence in the environment, the direct effects on human health, and the toxicological aspects of water pollution; and food borne diseases. Understanding of these issues may help scientists, decision makers and the public to make positive modifications in their behaviour that may add to the development of a sustainable environment.

In related studies, Laboy-Nieves (2009) and Cruz-Báez and Boswell (1997) reported that for centuries, peripheral urban developments have affected the evolution of estuaries (i.e. intertidal terrestrial zones) to a point that today some ecosystems exhibit a mosaic of areas that remain almost pristine while others are degraded. The aquatic and terrestrial zones and the populations that inhabit them are tightly linked, showing complex interactions. Laboy-Nieves (2008) emphasized man made factors such as high demographic density, urban sprawling, poor waste management and social indolence with respect to the environment. In his case study area of Jobos Bay, Puerto Rico, Laboy-Nieves (2009) reported that the Bay is a very dynamic ecosystem, where its natural history had been sculpted by physical, biological and anthropological factors. Jobos Bay is a natural laboratory for examining mangroves and upland forests, submerged communities and anthropogenic influences.

A related aquatic environmental health study was presented by Emmanuel et al. (2009). The continual discharge of chemical substances in aquatic ecosystems can bring about changes in the structure and functioning of the biotic community, (i.e. on biotic integrity) (Karr, 1991). As a function of their bioavailability, the pollutants present in effluents cause a large number of harmful effects on the biodiversity of aquatic environments (Forbes and Forbes, 1994).

The main substances involved in chemical pollution phenomena are heavy metals, organic compounds, especially organohalogenated substances, detergents-surfacatnts, pesticides, Polycyclic Aromatic Hydrocarbons (PAHs) and Polychlorobiphenyls (PCBs), nitrates and phosphates and drug residues (Emmanuel et al. 2009). Among the main effects of pollutants on aquatic organisms, are severe pathologies, behavioural problems, and species migration and disappearance.

Given that the toxicity of a substance depends on its available concentration, Emmanuel et al. (2002) not only confirmed the existence of dangerous substances in urban effluents, but also explained the contribution of these pollutant matrices to the loss of genetic, specific and ecosystem biodiversity of aquatic organisms.

7 MATERIALS RECYCLING AND WATER TREATMENT

The field of waste disposal is relatively new (Amirkhanian et al., 2009). It was only in 1965 in the USA, for example, that the first federal legislation was enacted to directly approach the waste problem (Ruiz, 1993). As such, the country is only now starting to come to terms with the amount of waste it produces.

In their feasibility study on the use of waste materials in highway production, Amirkhanian and Manugian (1994) identified the following waste materials with specific applications for the highway construction industry: bottom ash, compost, construction debris, fly ash, plastics, reclaimed asphalt pavement, shingle scarps, slag, sludge and tires. Recycled products have thus emerged as a viable alternative to virgin materials in the highway construction sector.

As natural resources start to dwindle and landfill space gets filled up, the importance of proper waste reduction and management systems increases. The use of waste by-products as a replacement for virgin materials could provide relief for some of the burden associated with disposal and may provide a cost effective construction product exhibiting all the properties of virgin products. The concept of utilizing reclaimed materials as a source for construction is particularly relevant in developing countries. Economical and ecological solutions are particularly important in situations where growing populations are coupled with finite economic and natural resources.

Sustainable fresh water resources are also a critical area. Goosen et al. (2004 and 2009) explained that fiscal development and population growth has put increasing pressure on the world's limited fresh water resources. In order to lessen this problem, desalination processes have been developed to obtain fresh water from the earth's vast supply of seawater. A major concern, however, is the location of a significant fraction of the world's desalination capacity in coastal areas of oil producing countries, such as in the Arabian Gulf (Al-Sajwani, 1998). Oil spills could have catastrophic effects on seawater desalination capacity in these regions.

The two most successful commercial water desalination techniques involve thermal and membrane separation methods (Al Obeidani et al., 2008). Gaining a better understanding, for example, of membrane desalination of oil contaminated seawater is a major challenge facing both scientists as well as plant operators in many parts of the world.

Improvement of living conditions, development of agriculture, tourism and many other important economic activities, are strongly dependent on the sustainability of water resources (Perez Duran, 2009). The importance of water is being increasingly highlighted as we gradually approach the limits of availability. While the world population has tripled during the twentieth Century, water consumption has increased by a factor of seven. Although access to water is considered a basic part for human life and a fundamental right, one billion of the world's people do not have such access and 2.6 billion do not have proper sanitation conditions.

In developed countries, diseases related to poor water quality and inappropriate wastewater disposal account for illness and loss of productivity equivalent to 2% of GDP (UNDP, 2006). Inversely, poor countries which invest in having better access to water and supply and sanitation, have a better economy.

Perez Duran (2009) reported that problems and concerns of water resources in the Dominican Republic consisted of microbial contamination of river and coastal water, overexploitation of aquifers, poor quality of water for aquatic life in rivers passing through cities, and fast paced reservoir sedimentation. He also noted health problems due to the lack of sanitation services, insufficient wastewater infrastructure and irresponsible industrial and municipal discharge of solid and liquid wastes into rivers.

8 CONCLUDING REMARKS

Education for sustainable growth should be obligatory for all young people, as this represents the primary vehicle available for catalyzing the cultural changes necessary for continued subsistence. Bioethics in particular should to serve as the main stage for instruction about sustainability.

In business and economics, an increasing number of companies have started to act and to communicate based on their triple performance in the areas of economics, and environmental and social factors. Building sustainable firms and organisations also requires a commitment to people's development. In addition, decisions should be made collectively, through negotiations.

Poverty continues to be a major challenge facing mankind in many parts of the globe. This has been attributed to inefficient allocation of natural resources, capital, labour and time; underutilization of resources and inequitable distribution of incomes. While development opens up and advances economies, creates new wealth, and ushers many people to a richer lifestyle, millions are forced to struggle to make meaning of the darker side of development that is not environmentally sustainable.

Adequate and dependable water resources are a major issue facing many people. In developed countries, diseases related to poor water quality and inappropriate wastewater disposal account for illness and loss of productivity. On a positive note, poor countries which invest in having better access to water resources and sanitation have better economies.

Finally, the relationship between economic development, environmental management and human health is a complicated process, affecting both the quality and sustainability of the society in which we live. There is a rising comprehension by the general public as well as practicing engineers, decision makers, environmentalists, and medical researchers, that these three areas are interrelated. A synchronized approach is necessary to solving the major environmental and sustainability problems facing the developing as well as developed regions of the world.

REFERENCES

Abdulla, F. A., Abu-Dieyeh, M. H. and Qnais, E. 2009. Human Activities and Ecosystem Health. in: E. N. Laboy-Nieves, F. Schaffner, A. Abdelhadi and M. F. A. Goosen eds. *Environmental Management, Sustainable Development and Human Health*, Taylor & Frances Publ, London, Chapter 28.

Abellán Salort, J. C. 2006. *Bioética, Autonomía y Libertad*. Fundación Universitaria Española, Madrid, España.

Ajlouni, M. M. 2009. The Environment as a Fifth Component of a Balanced Scorecard for Improving Organizational Performance. in: E. N. Laboy-Nieves, F. Schaffner, A. Abdelhadi and M. F. A. Goosen eds. *Environmental Management, Sustainable Development and Human Health*, Taylor & Frances Publ, London, Chapter 10.

Alamar, K. and Murali, N. 2009. Globalization, the Environment and Sustainable Development. in: E. N. Laboy-Nieves, F. Schaffner, A. Abdelhadi and M. F. A. Goosen eds. *Environmental Management, Sustainable Development and Human Health*, Taylor & Frances Publ, London, Chapter 2.

Al Obeidani, S., Al Hinai, H., Goosen, M. F. A., Sablani, S., Taniguchi, Y. and Okamura, H. 2008. Membrane Fouling and Cleaning in Treatment of Contaminated Water: A Critical Review. *Desalination*, (revised paper submitted July 2007)

Al-Sajwani, T. M. 1998. The Desalination Plants of Oman: Past, Present and Future. *Desalination*, 120, 53–59

Amirkhanian, S. N. and Manugian, D. M. 1994. *A Feasibility Study of the Use of Waste Materials in Highway Conctruction.* Columbia, SC: South Carolina Department of Transportation/Federal Highway Administration

Amirkhanian, S., Thodesen, C. and Shatanawi, K. 2009. Utilization of Solid Waste Materials in Highway Construction. in: E. N. Laboy-Nieves, F. Schaffner, A. Abdelhadi and M. F. A. Goosen eds. *Environmental Management, Sustainable Development and Human Health*, Taylor & Frances Publ, London, Chapter 32.

Andorno, R. 1998. *Bioética y Dignidad de la Persona*. Editorial Tecnoc (Grupo Abaya, S. A.). Madrid. España.

Assadourian, E. (March–April 2006). Next Steps for the Business Community. *World Watch: Vision for a Sustainable World.* 19:2, pp. 16–20.

Baver, S. L. 2009. Path to Democratic Environmental Governance in Latin America: A Case Study from Mexico. in: E. N. Laboy-Nieves, F. Schaffner, A. Abdelhadi and M. F. A. Goosen eds. *Environmental Management, Sustainable Development and Human Health*, Taylor & Frances Publ, London, Chapter 4.

Cruz-Báez, A. D., and Boswell, T. D. 1997. *Atlas of Puerto Rico*. The Cuban American National Council. Miami, Florida.

Diez, J. 2006. *Political Change and Environmental Policymaking in Mexico*. New York: Routledge.

El-Kafafi, S. 2009. Building a Sustainable Mindset. in: E. N. Laboy-Nieves, F. Schaffner, A. Abdelhadi and M. F. A. Goosen eds. *Environmental Management, Sustainable Development and Human Health*, Taylor & Frances Publ, London, Chapter 18.

El-Nasr, M. S., Ioerger, T. R. and Yen, J. 1998. Learning and Emotional Intelligence in Agents. *Proceedings of AAAI Fall Symposium*.

El-Nasr, M. S., Yen, J. and Ioerger, T. R. 2000. FLAME – A Fuzzy Logic Adaptive Model of Emotions. *Autonomous Agents and Multi-agent Systems*. 3, 219–257.

Emmanuel, E., Balthazard-Accou, K. and Joseph, O. 2009. Impact of Urban Wastewater on Biodiversity of Aquatic Ecosystems. in: E.N. Laboy-Nieves, F. Schaffner, A. Abdelhadi and M.F.A. Goosen eds. *Environmental Management, Sustainable Development and Human Health*, Taylor & Frances Publ, London, Chapter 30.

Emmanuel, E., Perrodin, Y., Keck G., Blanchard, J.-M. and Vermande, P. 2002. Effects of Hospital Wastewater on Aquatic Ecosystem. *Proceedings of the XXVIII Congreso Interamericano de Ingenieria Sanitaria y Ambiental. Cancun, México*, 27–31 de octubre. CDROM

Ferrer, J. J. and Álvarez, J. C. 2003. *Para Fundamentar la Bioética*. Universidad Pontificia Comillas y Editorial Desclée De Brouwer, S. A. Bilbao, España.

Forbes V. E. and Forbes T. L. 1994. *Ecotoxicology in Theory and Practice*. New York: Chapman and Hall, 220 p.

Gabaldón, A. J. and Aguilera M. 2009. Bioethical Dimensions of Sustainable Development. in: E. N. Laboy-Nieves, F. Schaffner, A. Abdelhadi and M. F. A. Goosen eds. *Environmental Management, Sustainable Development and Human Health*, Taylor & Frances Publ, London, Chapter 6.

Goosen, M. F. A., Al-Obeidani, S. K. S., Al-Hinai, H., Sablani, S., Taniguchi, Y. and Okamura, H. 2009. Membrane Fouling and Cleaning in Treatment of Contaminated Water. in: E. N. Laboy-Nieves, F. Schaffner, A. Abdelhadi and M. F. A. Goosen eds. *Environmental Management, Sustainable Development and Human Health*, Taylor & Frances Publ, London, Chapter 36.

Goosen M. F. A., Sablani S. S., Al-Hinai H., Al-Obeidani S., Al-Belushi R. and Jackson D. 2004. Fouling of Reverse Osmosis and Ultrafiltration Membranes: A Critical Review. *Separation Science and Technology*, 39 (10) 2261–2298.

Gracia, G. D. 1989. *Fundamentos de Bioética*. Eudema, Madrid, España

Hammond, F. N. and Somevi, J. 2009. Poverty, Urban Land and Africa's Sustainable Development Controversy. in: E. N. Laboy-Nieves, F. Schaffner, A. Abdelhadi and M. F. A. Goosen eds. *Environmental Management, Sustainable Development and Human Health*, Taylor & Frances Publ, London, Chapter 9.

Kaplan, R. S. and Norton, D. P. 2001. *The Strategy-Focused Organization*. Boston: Harvard Business School Press.

Karr J. R. 1991. Biological Integrity: A Long-Neglected Aspect of Water Resource Management. *Ecol. Appl.*, 1:66–84.

Laboy-Nieves, E. N. 2009. Environmental Profile & Management Issues in an Estuarine Ecosystem: A Case Study from Jobos Bay, Puerto Rico. in: E. N. Laboy-Nieves, F. Schaffner, A. Abdelhadi and M. F. A. Goosen eds. *Environmental Management, Sustainable Development and Human Health*, Taylor & Frances Publ, London, Chapter 29.

Laboy-Nieves, E. N. 2008. Ética y Sustentabilidad Ambiental en Puerto Rico. *Memorias del Primer Foro Nacional del Agua. INAPA*. Dominican Republic. ISBN 978-9945-406-80-1: 63–76.

Lyroudi, K. and Angelidis, D. 2009. Examining Banking Productivity Across Countries Considering Off-Balance Sheet Activities. in: E. N. Laboy-Nieves, F. Schaffner, A. Abdelhadi and M. F. A. Goosen eds. *Environmental Management, Sustainable Development and Human Health*, Taylor & Frances Publ, London, Chapter 13.

Maila, M. W. 2006. Grounding Sustainable Development in Praxis-Furthering Accountability and Collaboration. A. Ahmad ed *World Sustainable Development Outlook 2006: A Global and Local Resources in Achieving Sustainable Development*. Switzerland: Inderscience. pp. 1–11.

Maila, M. W. 2009. Environmentally Sustainable Economies: A Panacea for Sound Development. in: E. N. Laboy-Nieves, F. Schaffner, A. Abdelhadi and M. F. A. Goosen eds. *Environmental Management, Sustainable Development and Human Health*, Taylor & Frances Publ, London, Chapter 13.

Maria, K. A. and Zitar, R. A. 2009. Modeling of Artificial Emotions and its Application Towards a Healthy Environment. in: E. N. Laboy-Nieves, F. Schaffner, A. Abdelhadi and M. F. A. Goosen eds. *Environmental Management, Sustainable Development and Human Health*, Taylor & Frances Publ, London, Chapter 20.

Misra, B. 2000. New Economic Policy and Economic Development, *IASSI Quarterly*, Vol. 18, No. 4, June, p. 20.

Ndlovu-Gatsheni, S. 2006. Gods of Development, Demons of Underdevelopment and Western Salvation: A Critique of Development Discourse as a Sequel of CODESRIA and OSSREA, in *OSSREA. A Bulletin*, June, Vol. 2: 1–19.

Perez Duran, J. R. 2009. Assessment of Sustainability in Water Resources Management: A Case Study from the Dominican Republic. in: E. N. Laboy-Nieves, F. Schaffner, A. Abdelhadi and M. F. A. Goosen eds. *Environmental Management, Sustainable Development and Human Health*, Taylor & Frances Publ, London, Chapter 37.

Rogers, K. and Sinkey, J. 1999. An Analysis of Non-traditional Activities at U.S. Commercial Banks. *Review of Financial Economics*, 8: pp. 25–39.

Rodríguez Bribiesca, P. 2007. Access to Environmental Information. In: J. Fox (Ed.), *Mexico's Right- to- Know Reforms: Civil Society Perspectives*. Washington, DC: Woodrow Wilson Center. 260–266.

Ruiz, J. 1993. Recycling Overview and Growth. in H. Lund, *The McGraw-Hill Recycling Handbook*. New York, NY: McGraw-Hill Inc.

Sen, A. 1999. *Development as Freedom*. Harvard: Belknap Press.

Sharma, S. N. 2007. Climate Proofing, *The Economic Times*, Chennai, December, 30, p.13.

Siems, T. F. and Clark, T. A. 1997. Rethinking Bank Efficiency and Regulation: How Off-Balance Sheet Activities Make a Difference. *Financial Industry Studies*, Federal Reserve Bank of Dallas: pp. 1–11.

Speth, J. G. and Haas, P. 2006. *Global Environmental Governance*. Washington, DC: Island Press.

Stern, N. 2006. *The Economics of Climate Change: the Stern Review*. UK Government Report, October.

Townsend, C. R., Begon, M. and Harper, J. L. 2003. *Essentials of Ecology*. Blackwell Publishing. MA, USA. 530 pp.

United Nations Development Program (UNDP) 2006. *Report on Human Development, 2006, Beyond Scarcity, Power, Poverty and the World Water Crisis,* Summary. pp. 5–41.

Globalization, the Environment and Sustainable Development

K. Alamar & N. Murali
Department of Economics, Arul Anandar College, Karumathur, India

SUMMARY: The aim of this chapter was to provide a brief review of how globalization, the environment and sustainable development are interrelated. Globalization enables the free flow of goods, capital and technology and thus it becomes a motivating force for nations to develop themselves. For sustainable development to be meaningful over consumption has to be brought under control. The principles of equity, conservation orientation and renewability have to be reestablished. Controlling over-consumption and dealing with the question of ownership are crucial. The earth functions in a systematic way and any unsystematic intervention will cause irreparable damage not only for the development but also for the existence of mankind. Thus, industrial development today has to be done within the carrying capacity of the planet. The chapter discusses how human beings are at the centre of concerns for sustainable development.

1 INTRODUCTION

Globalization is but a modern version of theory of comparative cost advantage which was propagated by the classical economists to provide a theoretical foundation for the unrestricted flow of goods from developed to less developed countries at the time of colonies (Naidu, 2005). International trade benefits nations which enter into trade relations. The same arguments have been put forth by the advocates of globalization who promote an export-led pattern of growth to replace the import substitution trade policies as well as capital and technology flow.

Globalization enables free flow of goods, capital and technology and thus it becomes a motivating force for nations to develop themselves and creates a more gainful environment in the world scenario (Misra, 2000). This logic states that low growth economies get capital and technology from the developed countries to build their wealth and the latter countries get markets for their products. The two Bretton Woods institutions the World Bank and the International Monetary Fund have been playing an active role in promoting market capitalism across the globe through the imposition of the Structural Adjustment Programs (WCED, 2007; Sharma, 2007). The historical evidences revealed that all these trade, capital and technology flows helped the developed nations at the cost of exploitation of natural resources in developing countries. The world has been converted practically into one big market which is determining the fate of many countries.

Globalization is indeed a new dramatic creed. It uses the media to create a sense of lack. Here we are driven to earn more in order to acquire more, yet can never reach a point of contentment. The world in general and developing countries in particular have experienced unprecedented progress since the initiation of the globalization process. Life expectancy in developing countries has risen by more than 20 years, infant mortality rates have been halved and primary school enrolment rates have doubled (IUCN/UNEP/WWF, 1980). Food production and consumption have increased around 20 per cent faster than population growth. Improvements in income levels, health and educational attainment have sometimes closed the gap with industrialized countries. Advances have been made in the spread of democratic, participatory governance and there have been forward leaps in technology and communications. New means of communication support opportunities for mutual learning about national development processes and for joint action over global challenges.

Notwithstanding this remarkable progress, there are also negative trends. These include: economic disparity and poverty; the impact of resources in industrialized countries contributing to climate change; and environmental deterioration and pollution including the impacts of intensive farming, depletion of natural resources and loss of forests, other habitats and biodiversity.

The process of globalization has had adverse repercussions on ecology and the environment and social distortions even in the field of consumption fuelled by aggressive advertising, and slashing of import duties which results in dumping of goods (Naidu, 2005). It has also encouraged the extraction of raw materials in many countries thus contributing to resource depletion and degradation. Under fiscal pressure, the developing countries sell of more and more of their public reserves such as forests, mining resources and maritime assets, thus affecting the inter-generational equity which is essential for sustainable development. Nations have agreed, through processes such as the 1992 Earth Summit, that development should be sustainable. Therefore economic growth and environmental preservation are no longer opposing objectives (Naidu, 2005).

The aim of this chapter was to provide a brief review of how globalization, the environment and sustainable development are interrelated.

2 THE MARKET AND THE ENVIRONMENT

In a free market economy, the private sector may not bother to conserve nature. For the sake of profit, it may, for example, destroy forests, over use mineral resources; and pollute air, water. This sector may not take into account social cost or benefit. The producer may charge a price which covers only internalized production rates without taking into account the cost of pollution, destruction of natural resources or adverse impact on health (Misra, 2000). Today's world is obsessed with economic growth. There is a relentless race for expansion. Every country wants to keep growing faster than the rest. Little thought is spared to its relevance or to the cost. Growth should be defined not only in terms of economics but also in terms of other parameters like societal, cultural and spiritual. Marketers thrive on making people wish for things they never wanted. The application of technology for the pursuit of profit has resulted in the over-exploitation and the excessive utilization of natural resources.

3 ENVIRONMENTAL CHALLENGES

Global warming, environment degradation, ozone depletion and deforestation are caused primarily by the market mode of production (Sharma, 2007). The latter has resulted in rapid increases in the consumption of hydrocarbons. In addition, the expansion of industries and domestic markets has drastically altered the atmospheric concentrations of numerous trace gases. This has distorted nature's auto-balancing mechanism. As a result, the world is now faced with several environmental problems including acid rain, melting of glacial, ice, large scale evaporation of water in the tropics, and an increase in cloudiness at higher altitudes (Sharma, 2007). Thus there is a conflict between the production of economic goods and environmental inputs used in the production of such goods and services.

Environmental degradation is expected to result in shifts of climatic zones, changes in the productivity of ecosystems and species composition, and an increase in extreme weather events. This will have substantial impacts on human health and the Liability of natural resource management in agriculture, forestry and fisheries, with serious implications for all countries. Developing countries, and notably the least developed, are expected to be the most vulnerable to the impacts of global climate change, although their current contribution to the problem is minimal (Sharma, 2007).

The Human Development Report, 2007–08 prepared by the UNDP, identified two degree Celsius as the threshold, above which damages to the global climate will be irreversible. The world needs to alter its level of carbon emissions in the next decade, and to start living within the carbon budget of 14.5 gigatonnes of CO_2 per annum for the remaining years of the 21st century.

3.1 *Natural resource deterioration*

Environmental deterioration continues to increase with serious depletion of natural resources, including soil erosion and loss of forests and fish stocks (IUCN/UNEP/WWF, 1980). Deforestation has reduced the extent and condition of world forests. Some 65 million hectares of woodland were lost between 1990 and 1995 (Sharma, 2007).

Fragile aquatic environments such as coral reefs and freshwater wetlands are under considerable threat from land-based pollution, destructive fishing techniques and dam construction, as well as climate change. It is estimated that almost 60 per cent of the world's reefs and 34 per cent of all fish species may be at risk from human activities. Current patterns of production and consumption and global climate change, raise questions about the continued capacity of the Earth's natural resource base to feed and sustain a growing and increasing urbanized population, and to provided sinks for wastes. As a result of environmental degradation the biodiversity of the earth's ecosystems and the availability of removable natural resources have declined by 33 per cent over the last 30 years while the demands on these resources have doubled.

3.2 *Global energy use*

Since 1971, global energy use has increased by nearly 70 per cent and is projected to continue to increase by over 2 per cent per year over the next 15 years despite the fact that 2 billion people are still largely unconnected to the fossil fuel-based economy (IUCN/UNEP/WWF, 1980). It will raise green house gas emissions by 50 per cent over current levels, unless serious efforts are made to increase energy efficiency and reduce reliance on fossil fuels.

Although there has been considerable growth and technical progress in the use of renewable energy sources such as wind, solar, geothermal, and hydro-electricity, public infrastructure and the convenience of fossil fuels and their low prices seriously inhibit any large-scale switch to the use of clean energy sources in the foreseeable future. In India, for example, over five million people still don't have access to power. The first priority will be to provide energy to everyone in the country. Developed countries should help in giving environment-friendly technology so that they can restrict total carbon emissions.

3.3 *Global warming*

Global warming is caused primarily by the burning of coal, oil and gas (IUCN/UNEP/WWF, 1980; WCED, 1987). Real solution to the problem would include lifestyle changes. This is something that goes against the consumer culture. If the build-up of carbon dioxide continues till the end of the 21st century the earth will be hotter than at any other time in the last two million years. The world is already facing increasing sea intrusions, floods, storms, droughts, heat wavers, disease transmissions and environmental refugees.

Climate change can also have a significant impact on health through vector-borne diseases because of changes in the survival and reproduction rates of the carriers, the intensity and temporal pattern of vector activity and the life cycle of pathogens within the vectors (IUCN/UNEP/WWF, 1980; Misra, 2000). Food production and food security, fresh water supply, forest biodiversity, coastal settlements, and fishing will be adversely affected. Unfortunately, the burden of climate change will fall disproportionately on poor communities. Countries such as India, for example, should adopt strategies for sustainable development irrespective of the climate change debate. It involves economic growth, social equity and environmental sustainability. Developing countries are more vulnerable to climate change than industrialized countries and hence have a greater stake in the success of global climate negotiations and strategies for green house gas stabilization.

3.4 *Deforestation*

Deforestation is a threat to the economy, quality of life and future of the environment. Main causes of deforestation are sudden increases of human and livestock population, increased requirement of timber and fuel wood, expansion of cropland and enhanced grazing (IUCN/UNEP/WWF, 1980; Sharma, 2007). Another cause of forest degradation is construction of roads along the mountains. Ideally one third or 33 per cent of land of a country must be covered by forest. Deforestation has intensified soil accentuation, floods, drought and loss of precious wild life. India is losing about 1.5 million hectares of forest cover each year. Nearly one per cent of the land surface of the country is turning barren every year due to deforestation. In the Himalayan range, rainfall has declined three to four per cent due to deforestation.

3.5 *Ozone depletion*

The maximum ozone concentration of 0.5 ppm occurs between at an altitude of 20 to 35 km. This layer is called the ozone layer. The presence of ozone (O_3) is an essential necessity for life on earth. Stratospheric ozone layer absorbs dangerous VVB rays of the sun and thus protects the Earth's surface from these high-energy radiations. Over the past few decades the O_3 layer has thinned out because of man-made pollutants which catalyze the dissociation of O_3 at a very fast rate (IUCN/UNEP/WWF, 1980). Major pollutants responsible for depletion of ozone are chlorofluro Carbons (CFCs), nitrogen oxides, hydro-carbons and oxides of chlorine and bromine.

4 ENVIRONMENT AND HUMAN DEVELOPMENT

Development in its true sense is an enabler of human freedom and well- being, rather than mere enhancement of inanimate objects of convenience. It is inseparable from environmental, ecological concerns such as clean air, water, epidemic free surroundings and preservation of all life forms.

While over production of material goods results in environmental degradation by way of emitting green house gases (GHGs), environmental problems are also generated due to lack of awareness. Society can tolerate environmental degradation for a while for the sake of economic development, then as income rises people's basic needs are met and the demand for clean environment should emerge.

Access to education also plays a vital role in acquiring knowledge about environment. As early as in 1972, then Prime Minister of India, Mrs. Indira Gandhi, while addressing the UN Conference on Human Environment in Stockholm, noted that poverty is the biggest polluter (IUCN/UNEP/WWF, 1980; Subramanian, 2008). The environment cannot be improved under the conditions of poverty. Hence environment cannot be divorced from development.

5 DIMENSIONS OF POVERTY

Even in these prosperous times, extreme poverty still ravages the lives of one out of every five persons in the developing world (IUCN/UNEP/WWF, 1980; Subramanian, 2008). The social ills associated with poverty are on the rise in many countries. Although world food production is still rising, several trends will make it more challenging to feed a growing world population. Soil degradation from erosion and poor irrigation practices continues to harm agricultural lands, jeopardizing production. Without a transition to more resource-efficient and less polluting farming methods, it will be difficult to meet world food needs without increasing the environmental burden that stems from intensive agriculture. HIV-AIDS and malaria are serious diseases that erode both productive capacity and the social fabric of nations.

6 ENVIRONMENTAL COST OF ECONOMIC GROWTH

Ecologists and environmentalists believe that one of the principal reasons for the existence of environmental problems stems from the emphasis on growth by the industrialized nations

(IUCN/UNEP/WWF, 1980; Subramanian, 2008). They point out that economic growth has been made possible only at the expense of the environment. Ecologists postulate that growth rates were so high, because of the fantastic increase in population and the demands of society. Increased production and consumption had unscrupulously released wastes and pollutants into the environment without consideration of their effects. Fast growth resulted in the destruction of the environment, the impairment in the quality of elemental environmental services, the deterioration of air quality and the contamination of seas, rivers, and lakes.

Production is possible only at the expense of environmental quality. Continued growth of output and population will eventually lead to extinction of many species of flora and fauna. The available resources on this planet are finite. The assimilative capacity of the environment is limited.

6.1 *Zero growth and environmental quality*

Zero growth does not in any way influence environmental quality (WCED, 1987). This growth level does not change the actual stage of economic activities, thus environmental pollution remains fixed. Zero-growth does not even prevent a further deterioration of environmental quality; pollutants are still released into the surroundings. Ecological quality then can decline in spite of zero economic growth. The issue is not whether growth intensifies the conflict between the supply of goods and environmental quality, but rather the way growth has taken place previously. The structure of national output and the relationship between pollution-intensive and environmentally favorable production causes this conflict of objectives.

6.2 *High level growth and environmental quality*

High levels of industrial development will allow the economies to devote additional resources to tackle pollution problems without sacrificing economic growth (WCED, 1987). With continued growth, additional resources may be used in achieving the desired levels of environmental quality. Expansion can be complementary to environmental protection. Anti-pollution devices and technology are part and parcel of the GNP (Gross National Product). Hence, both pollution and its reduction form part of the function of level of income. Environmentalists have taken a very pessimistic view based on Malthusian doctrine of over population and depletion of resources. Economists, on the other hand, have taken an optimistic view, based on the ability of free-market economies to cushion the effects of impending shortages, through the responses of produces and consumers to the various incentives and disincentives in the pricing system and rising prices.

The free use of the environment is the basic reason for its degradation. If the zero price tag of environmental resources use is abolished there will be an incentive to use factors in abatement of pollution, so that production will be reduced. The emission tax, for example, will cause a sectoral reallocation which disfavors the pollution-intensive sector and thus reduces emissions leading to improvement in environmental quality. Industries will have the incentive to accumulate capital in abatement and to develop new abatement technologies.

7 MANAGEMENT OF NATURAL RESOURCES

Environmentalists tend to analyse the economic factors, especially the power of market forces to stimulate ma's ingenuity and inventiveness in solving problems of shortages. A brief look into the economic history of mankind will reveal how man has ingeniously overcome problems and shortages. In 17th century England, the forest resources were unable to meet the requirements of iron making furnaces and construction of ships, as wood was primarily the fuel for industries and ship-building (IUCN/UNEP/WWF, 1980; Subramanian, 2008). The shortage of firewood for industries and the shortage of timber for ships raised hue and cry in the country. But soon, in the 18th century, coke was used in the furnaces and in the 19th century timber gave way to iron in the ships.

In all fields of production, the resource base has changed considerably during the course of this century, due to shortage of traditional resources (IUCN/UNEP/WWF, 1980; Subramanian, 2008). This has happened phenomenally after the Second World War which gave extraordinary impetus for the progress of science and technology. At the same time, economists, on their part, tended to under-estimate the difficulties in responding quickly to a change in market forces. In this context, we have to take into account a new factor that has become a phenomenon after the Second World War. Right from fifties onwards in this century, the pace of economic growth throughout the world has been incredible. The turn over of natural resources has been doubling every 15 or 20 years.

The environment could be generally kept clean by the expenditure of not more than two per cent of GWP (Gross World Product) on anti-pollution measures (WCED, 1987). The earth's natural resources are very large, compared with man's consumption. The so called shortages suffered at present come mainly from our technological and economic limitations. Energy is the key to the solution of most of man's resources problems. Given sufficient-energy, together with the technology and industry needed to apply it food and materials can always be supplied and needed. Minerals could be extracted from low grade deposits and water desalinated. Even the area of good farmland need no longer be a limit now. Agriculture can if required, reinforce solar energy with fuel energy. Apart from human skills, the primary requirements are energy, technology and money; technology to develop such uses of energy and money to turn the technology into new industries.

8 SUSTAINABLE DEVELOPMENT

According to the 1987 Brundtland Report, sustainable development can meet the requirements of the present without compromising the ability of the future generation to meet their own needs (WCED, 1987). Sustainable development has three components: social, economic and ecological. There is a strong interrelation between economic and environmental and social goals and yet these are not always compatible. Sustainability requires a rare balance between these three set of goals. Sustainable development can only occur once ecological sustainability has first been secured. Once the latter has been achieved, it is possible to attain economic sustainability. If this condition is maintained in the long-run then social sustainability can be attained.

8.1 *Measures taken for achieving sustainable development*

Realizing environment trends led to the 1972 UN Conference on the Human Environment in Stockholm which in turn, led to the creation of UNEP and IIED. Since then, worldwide acceptance of the importance of environmental issues has grown enormously. The world conservation strategy (IUCN/UNEP/WWF, 1980) and subsequently, the report of the World Commission on Environment and Development-the Brundtland Commission (WECD, 1987) were developed in response to increasingly informed analyses of the links between environment and development. The world conservation strategy emphasized the need to 'mainstream' environment and conservation values and concerns into development processes.

The Report of the Brundtland Commission emphasized the social and economic dimensions of sustainability, revealing links between, for example, poverty and environmental degradation. At the heart of the concept is the belief that social, economic and environmental objectives should be complementary and interdependent in the development process. Sustainable development requires policy changes in many sectors and coherence between them. It entails balancing the economic, social and environmental objectives of society the three pillars of sustainable development integrating them where ever possible, through mutually supportive policies and practices and making trade-offs where it is not.

There is a direct correlation between economic growth and energy consumption (WCED, 1987). Since global warming has emerged as a serious environmental threat, the case for other sources of power generation that do not emit green house gases assumes importance. Promotion of hydropower and non-conventional sources of energy such as wind power, solar energy and bio-fuel need thrust.

9 OPTIMUM LEVEL OF PRODUCTION

There are fiscal limits to growth since the economy can only exist because of the resources and services provided by the ecosphere and its ecosystems (Naidu, 2005). Due to the operation of the law of diminishing returns there are rising marginal costs with every additional development project. A point is reached where the cost of growth exceeds the benefit. The depletion of nature cannot go on forever; whatever the level of technological sophistication some minimum level of natural capital stock is necessary to sustain economic activity and more importantly life on earth. Economic growth is only beneficial to the society as long as the marginal benefit of further quantitative development is greater than marginal cost. Beyond this point further development of the ecosphere is counter productive and ecologically unsustainable, in which case the economic activity should be stopped at this optimum level. The above argument is evident from the Japanese experience. The economies of Japan are not splurging beyond their already high consumption levels. We should realize that Japan has perhaps reached that desirable state of balance, which should be what every country should aspire to (Subramanian, 2008).

10 CONCLUDING REMARKS

It is ordinary people from different countries, collaborating with one another, who can ultimately bring about the social change needed to prevent the environmental calamity that looms ahead. For sustainable development to be meaningful over consumption has to be brought under control. The principles of equity, conservation orientation and renewability have to be reestablished. Controlling over-consumption and dealing with the question of ownership are crucial. The earth functions in a systematic way and any unsystematic intervention will cause irreparable damage not only for the development but also for the existence of mankind. Thus, industrial development today has to be done within the carrying capacity of the planet. Human beings are at the centre of concerns for sustainable development. They are entitled to a healthy and productive life in harmony with nature.

REFERENCES

IUCN/UNEP/WWF, 1980. World Conservation Strategy IUCN, Gland, Switzerland.

Misra, B., 2000. New Economic Policy and Economic Development, IASSI Quarterly, Vol. 18, No. 4, June, p. 20.

Naidu, K., 2005. Impact of Economic Reforms on Environment Protection, in: "Environmental protection and Development" – Emerging Issues, Reforms and Strategies, S.B. Verma and S.K. Singh, Eds, Deep & Deep Publications Pvt. Ltd. New Delhi, p. 8.

Sharma, S.A.N., 2007. Climate Proofing, The Economic Times, Chennai, December, 30, p. 13.

Subramanian, B.B., 2008, Is Growth always good? The Hindu, Jan. 13, Vol. 131, No. 2, p. 15.

WCED (World Commission on Environment and development) 1987, Our Common Future, New York, Oxford University press, p. 43.

Sustainable Urban Growth for European Cities

Hans Werner Gottinger
STRATEC, Munich, Germany

SUMMARY: This chapter seeks to explain urban growth by looking at the effects of traditional, geographical externality and socio-political variables on productivity, capital and labour growth, respectively, using data sets constructed at the city-industry level for a cross section of European cities. The conjecture that geographic externality and socio-political factors all vary significantly with aggregate growth (though in very specific ways) is empirically explored. For example, the size of a city (a measure of the degree of urbanisation) is uncorrelated with output growth, positively correlated with labour growth, and negatively correlated with capital growth. No one extant theory of growth accounts simultaneously for all the observed phenomena. More importantly, significant interactions are found between the variables that have not been addressed within the theoretical literature that drive the impact of these sources on growth. The paper concludes that a better understanding of the empirical relationships between various growth sources is needed and that, to be complete, theories of urban growth need to respect and understand the importance of such interactions.

1 INTRODUCTION

Theories of economic growth highlight the forces considered important for characterising relative aggregate growth. In addition to the more traditional adjustment factors such as changes in relative prices, the capital-labour ratio and productivity, more recent economic theories have emphasised the importance of externalities in driving economic growth. In particular, externalities may arise from the geographic proximity of production centers (Krugman, 2007). Empirical analysis attempting to understand the role of any of these factors in growth must control for the effect of other externality-based sources in addition to the more traditional adjustment factors.

In this chapter issues were addressed in analysing relative aggregate (city based) growth for a cross section of European cities. A methodology was employed by which the patterns of correlation can be described between individual growth sources and measures of aggregate growth. An analysis was made of how potential interactions between individual sources alter the observed correlations. Since externalities can be considered explicity, both negative and positive, i.e. those that are likely to constrain the city's growth as environmental pressures, demographic changes and congestion, as compared to those that promote them though positive network effects of scale and scope. Making this tradeoff would enable to strive for 'sustainable'growth, as growth internalised by externalities (Gottinger, 1998).

The methodology employed was to categorise the sources that are predicted to vary with growth and to establish empirical proxies for these variables. Then, in order to consider properly the effects of externality-based variables, an explicit distinction was made in the empirical analysis between productivity, labour and capital growth. In so doing, it was possible to identify whether particular externality-based variables influence aggregate growth through specific pathways rather than foccussing on output per capita only.

In departing from much of the conventional growth literature the city was chosen as the relevant unit of observation. Furthermore, data was used from the postwar period in Europe from 1960 to 2000, which was marked by significant change and intercity variation among cities, to estimate a multiple equation growth system.

This chapter focused on sustainable growth of contemporary large European cities from four major EU members: Italy, Germany, France and UK. A study was made of factors of growth in a multiequation econometric (regression) model, subject to sensitivity analysis. The cities covered were: Naples, Turin, and Milan in Italy, as indicated by South; Cologne, Hannover and Munich as Central in Germany, Bordeaux, Lyon and Marseille in France (West), and Manchester, Bristol and Leeds in the UK as North. This initial first choice was made foremost on consideration of the need for a geographical balance and of comparable size where those cities created regional centers. Also accessibility of sufficient data on a time frame of 40 years was a consideration. In a first examination census data were taken for every ten years so that with a selection of 12 cities 48 observations were made. Additional census years were incorporated to improve forecasting ability. No single theory appeared to explain, in a consistent way, the observed growth in the period.

2 GROWTH IN CITIES: METHODOLOGY AND APPROACH

A first important feature of our approach was to focus on cities per se rather than as derivative features of the economies of each member state. Metropolitan regions are a natural geographic unit for economic analysis as regional economic life is primarily organised around city centers. In choosing this focus we followed a group of researchers who have argued that economic growth can be better understood by examining cities. Earlier, Jacobs (1969, 1984) and Porter (1990) both presented strong historical cases for the appropriateness of studying cities as opposed to nations in exploring the sources of growth. This emphasis has also emerged in more formal empirical studies. For example, Glaeser et al (1992) focused on the role of dynamic externalities, including measures of competition, urbanisation and diversity, in explaining the growth of city-industry labour demand. Other studies have examined aggregate regional or city growth (Blanchard and Katz, 1992, Henderson, 1994, Barro and Sala-i-Martin, 2003, Henderson and Thisse, 2004), further in a cross-atlantic comparative framework by Savitch and Kantor (2004).

The starting point of our investigation was two fundamental pathways by which aggregate growth can occur: factor accumulation (increases in the levels of capital and labour operating within a region) and productivity growth (increases in the production capabilities of given levels of capital and labour). These pathways can be written down in the following form:

$$g_{Q,t} = \phi_c(g_{A,t}, g_{K,t}, g_{L,t}) \tag{1}$$

or

$$\ln[Q_{c,t+1}/Q_{c,t}] = \phi_c\{\ln[A_{c,t+1}/A_{c,t}], \ln[K_{c,t+1}/K_{c,t}],\ \ln[L_{c,t+1}/L_{c,t}]\} \tag{2}$$

where $Q_{c,t}$, $A_{c,t}$, $K_{c,t}$ and $L_{c,t}$ denote the levels of output, productivity, capital and labour in region c, at time t, and ϕ_c is a nondecreasing function of the growth rates in capital, labour and productivity. Differences in growth rates will therefore be a function of the magnitudes of factor accumulation and productivity growth rates in each region. That is,

$$g_{Q,1} - g_{Q,2} = \phi_1(g_{A,1},\ g_{K,1},\ g_{L,1}) - \phi_2(g_{A,2},\ g_{K,2},\ g_{L,2}) \tag{3}$$

The growth rates of capital, labour and productivity growth may themselves be functions of economic variables. Identifying those variables which affect factor accumulation and productivity growth is the primary step. Relative labour and capital growth are functions of the incentives for movement that face workers and investors and, also, of the incentives to supply more of those factors. A first item to recognise is that labour and capital growth is interdependent.

Labour and capital may be technological complements in production and the incentives facing decision-makers may be highly correlated (e.g., a region which experiences a positive shock may become more attractive to both workers and investors even if they are technological substitutes in production). Also, prices will induce movements of capital and labour to the highest rents and wages, respectively. The relative factor mix (the capital-labour ratio) might reflect the relative marginal productivities of capital and labour, and thus the incentives for factor movements. Finally, a variety of production externalities may operate at the industry, regional, or economy-wide level that affect incentives influencing the location of factors (agglomeration economies) as well as their qualities (social externalities). The growth of labour or capital, then, is potentially dependent on input prices, the factor mix, and the presence of particular externalities.

After controlling for input growth, the remaining share of output growth can be attributed to productivity growth. Productivity growth arises from technical and organisational change as well as the exploitation of economies of scale or scope. A principal means by which regional productivity growth may occur is through the adoption of more productive techniques from other regions. With ongoing interaction among regions the implication is that there is a tendency for productivity growth rates across regions to converge. Correspondingly, there may be a tendency toward 'adaptive efficiency' in economic institutions (North, 2004). Adoption may also occur within regions and within regional industries. Furthermore, local knowledge spillovers can result in a positive relationship between the growth rate of productivity and these local externalities.

3 EMPIRICAL RELATIONS IN URBAN GROWTH

In modeling the sources of urban growth one would like to include as many theoretical correlative predictions as possible under an empirical framework. To this end it is useful, by reviewing the literature, to group the phenomena we want to address into three categories: traditional economic factors, geographic production externalities, and other external factors.

3.1 *Traditional economic factors*

From its early origins the neoclassical growth model has provided sharp predictions on the effects that relative factor prices, factor utilisation and productivity levels have on relative growth rates (Solow, 1957; King and Rebelo, 1993). Factor prices and factor accumulation should be positively correlated, as higher wages and rents attract more labour and capital. Relative factor utilisation should engender opposing effects on relative labour and capital growth. Capital-intensive cities will induce more labour inflows than less capital-intensive cities. The primary reason for this is that the marginal productivity of labour is higher in a capital-intensive area. Conversely, cities which are highly labour-intensive will attract more new capital. Finally, there is a tendency for productivity growth rates across regions to converge (Barro and Sala-i-Martin, 2003, Rhode and Toniolo, 2006). The principal empirical implication of the traditional convergence hypothesis is that the relative level of productivity is negatively correlated with the growth rate in productivity.

These traditional economic variables are easily represented by city level aggregate measures. The average prices of labour and capital and the aggregate capital-labour ratio can proxy for the factor prices and the factor utilisation level which affect individual incentives for movement and accumulation of capital and labour. Also, measures of productivity, such as output per worker, provide empirical representations for productivity differentials. These aggregate measures are employed in empirical work.

3.2 *Geographic production externalities*

Additional determinants of the rate of input and productivity growth emerge when one considers the potential role of externalities that resemble 'networks' (Gottinger, 2003). Relative urban growth can be tied to production externalities that arise from geographical proximity and a variety of approaches have been used to characterise them. These varied approaches have been compressed into three general categories: urbanisation (URB), localisation (LOC), and specialisation (SPEC).

3.2.1 *Economies of urbanisation*

Size and breadth of urban regions have been thought to generate potentially important externalities that affect the growth of both inputs and productivity. Economic theories based on such ideas predict positive correlations between relative growth rates and relative measures of city size. For example, the decision to immigrate to a city may be affected by expected fluctuations in the employment environment. It has been widely recognised that larger urban areas are more attractive to immigrants because they offer a reduced risk of unemployment during downturns. A large local economy may also stabilise the demand for intermediate goods, thereby providing additional investment incentives for suppliers of these goods. Further, others have argued that the size of an urban region may affect productivity in one of two ways. First, the potential for local interindustry knowledge spillovers may cause the rate of technological change and adoption to be positively related to the size of an urban region. Secondly, a dense economy may reduce production costs directly through the proximity of producers to inputs and a large market. All externality-based effects that result from city size can be classified as urbanisation economies.

As with the traditional variables, there exist obvious aggregate city-level variables which measure a city's degree of urbanisation. In fact, various measures of aggregate economic activity in our estimations were utilised. These included total population, total levels of output and employment, and past population growth (to capture expected local demand) with a particular city.

3.2.2 *Economies of localisation*

Localisation is the degree to which an industry's economic activity takes place in one or a small number of geographical areas. Under this notion, some high technology industry has been linked to externalities which operate at the city-industry level. It has been long thought that productivity growth in a city-industry is positively correlated with the degree of localisation in that city industry. As the share of an industry's employment in a particular location increases, intraindustry knowledge spillovers also increase (Porter, 1990). Since local intraindustry knowledge spillovers are a source of industry productivity growth (Beckmann, 1999), localisation thus promotes such growth. Factor input growth is also predicted to be positively related with localisation. By lowering the unemployment risk of specialised workers and reducing employer monopsony power over wages, localised industries foster labour force growth. (Krugman, 2007). Furthermore, specialised capital, as supplied by nontradable intermediate input and service providers, is attracted to localised industries because of the advantageous bargaining position resulting from numerous potential outside opportunities. Each of these effects can be classified as an 'economy of localisation'.

Localisation's impact upon aggregate city level variables differs qualitatively from that of urbanisation or the traditional variables in that its force upon the economy results from its effects upon those industries which are localised. As the number of city-industries which are localised increases, and as the relative size of those industries increases, the aggregate growth benefits to localisation will increase. To capture this empirically, it is useful to distinguish those city-industries which are localised from those that are not. This is accomplished by determining some threshold share of national employment a city-industry would need to employ to be considered localised. For example, if the threshold share for industry x is 10 percent of employment nationally, industry x would need to employ more than 10 percent of national employment in its industry to be considered localised. A city's measure of localisation, then, would be the share of its employment contained in localised industries.

3.2.3 *Economies of specialisation*

This type of economy refers to the degree to which a city's output is dominated by a single or a number of closely related sectors. Specialisation, a city-level concept, differs from localisation in that it deals directly with a city's sectoral composition. The largest ties to factor growth come through its effects on risk. Specialised cities are especially vulnerable to business cycles – particularly on the downside. When a negative shock occurs in a dominant city-industry, these cities do not have the industrial diversity to absorb labour and capital losses in the dominant industry.

During downturns specialised cities may be subject to particularly high rates of capital and labour outflows and relative productivity decreases (Jacobs, 1984) In addition, the level (as opposed to the growth) of productivity has been positively linked to specialisation. As the division of labour between cities increases, productivity can be increased through trade. Such cities may have higher levels of productivity due to their exploitation of comparative advantage. Moreover, these cities will likely have public infrastructure tailored to serve dominant industries, decreasing production costs and increasing the level of productivity (Arthur, 1988). As these are levels rather than growth effects, though, they suggest no relation between specialisation and productivity growth given our growth equation framwork. We classify the effects on growth which result directly from a city's sectoral composition as economies of specialisation.

Empirical measures of specialisation must capture the degree to which a city is concentrated in a small number of sectors. A slightly modified Herfindahl index, widely used in the industrial organisation literature to measure the level of concentration of a particular industry, is well suited to measure the industrial concentration of a particular city. The Herfindahl index for urban specialisation utilised in the empirical work is given by:

$$\mathrm{SPEC}_{c,t} = \Sigma_i^I (L_{i,t}/L_{c,t}) \tag{4}$$

where I is the total number of industries. It is useful to note that we can measure specialisation with respect to inputs or outputs (i.e., the specialisation of output, employment, or the capital stock). To measure the specialisation of output or the capital stock, the city-share of each industry's employment in the specialisation equation would be replaced by the city-share of each industry's output or the industry's capital stock, respectively.

3.3 *Other external effects*

There are a number of other external effects that can potentially influence growth. Here the focus is on the role of political and social variables in growth. Government expenditure and taxation produce important shifts of incentives for factor input movement through tax policy and the like, (Barro, 1990), and the level and growth of productivity through the provision of public goods and the coordination of legal norms and standards. The precise direction of these shifts is dependent on the nature of the policy provision in question. The most negative ones could be described by transportation congestion, various sorts of pollution, and degradation effects through overbuilding and degenerated land use. Such effects can be parametrised as critical levels for an aggregate (SOC). It is difficult to attain aggregate empirical proxies for many of these variables or, in some cases, to even predict their aggregate impact.

Educational expenditures (or achievement) are potentially empirical proxies for human capital accumulation. Unfortunately, educational expenditure information on a city basis is not available over this time period for quite a number of cities. Competition, in contrast, is difficult to characterise empirically, even at the industry level. Government policy proxies, on the other hand, are readily available. City-level data for government expenditures, revenues, and tax rates, can be obtained so the impact of these policies can be evaluated statistically.

4 ECONOMETRIC FRAMEWORK

This brief review and empirical classification of the urban growth debate serves as the motivation for the econometric-statistical framework. The potential role of each of these variables on both input and productivity growth was emphasized. An attempt was made to be as inclusive and as precise as possible in the construction of empirical measures for the underlying effects in the study.

The empirical framework assumes that for a particular city, the increase in output is a nondecreasing function of the growth rates of each of the determinants of total output in each period (1). Additionally, each of the variables on the right-hand side of the growth Equation (1) is potentially governed by the other economic variables which have been identified earlier. In other words, the growth system can be summarized by foccusing on the following four equation system for each:

$$g_{Q(t)} = \phi_c(g_{A(t)}, g_{K(t)}, g_{L(t)}) \tag{5}$$

$$g_{K(t)} = G_K(g_{A(t)}, g_{L(t)}, K(t)/L(t), SPEC(t), LOC(t), URB(t), SOC(t), \text{interest rate, others}, \varepsilon_{K(t)}) \quad (6)$$

$$g_{L(t)} = G_L(g_{K(t)}, g_{A(t)}, K(t)/L(t), SPEC(t), LOC(t), URB(t), SOC(t), \text{wages, others}, \varepsilon_{L(t)}) \quad (7)$$

$$g_{A(t)} = G_A(Q(t)/L(t), SPEC(t), LOC(t), URB(t), SOC(t), \text{scale economies, others}, \varepsilon_{A(t)}) \quad (8)$$

Unfortunately, even if the precise functional form for each of these equations (5)–(8) was known, this system is underidentified due to the absence of a direct measure of productivity or productivity growth. However, the system can be rewritten as a structural model of city growth in the absence of direct information on productivity growth:

$$g_{Q(t)} = \phi_c(g_{K(t)}, g_{L(t)}, Q(t)/L(t), SPEC(t), LOC(t), URB(t), SOC(t), \text{scale economies, others}, \varepsilon_{A(t)}) \quad (9)$$

$$g_{K(t)} = G_K(g_{A(t)}, g_{L(t)}, K(t)/L(t), SPEC(t), LOC(t), URB(t), SOC(t), \text{interest rate, others}, \varepsilon_{K(t)}) \quad (10)$$

$$g_{L(t)} = G_L(g_{K(t)}, g_{A(t)}, K(t)/L(t), SPEC(t), LOC(t), URB(t), SOC(t), \text{wages, others}, \varepsilon_{L(t)}) \quad (11)$$

The main task is then to adjust the data set in the context of this three equation system (9)–(11). Within this system we are interested in identifying the most salient and robust multivariate correlations which exist in the data. From this identification, initial evaluations can be made of individual theories of growth. This will provide insights into the creation of a unified framework for understanding and discussing growth. In consideration of the small sample size, each of the equations can be estimated separately. This increases the precision, allowing for a sharper examination of the main sources of variation. In addition, to make the exposition of the econometric work a little easier, the presentation was restricted to growth measures which utilise first differences of the log of output, capital, and labour as dependent variables.

The above equation system will proceed along a broad Ordinary Least Squares (OLS) specification. There will be a number of alternate regression specifications, including both instrumental variable and reduced form estimates to explore important potential regularities more closely. These additional tests also serve to demonstrate the degree to which qualitative results concerning the conditional correlation are robust to numerous functional forms.

5 EMPIRICAL EVIDENCE

The relevant observables and proxies for the variables discussed in the previous section were constructed from datasets of national as well as European industrial statistics, covering the period 1960 to 2000. The statistics for the U.K., Germany, France and Italy, allowed for a compilation of a breakdown of manufacturing inputs and outputs by city industry. The number of operating firms was obtained and the recalculated Euro value of output for every city-industry was included. It was thus possible to compute value-added per city industry. The aggregate statistics for the manufacturing sector for each year included the levels of capital, employment, total labour income, and value added. From this information various aggregate city statistics were constructed for the levels and growth rates of value-added, capital, employment and wages. The city output, employment and capital-labour ratios were also computed. Additionally, measures of geographic externality variables were constructed, as described previously. Finally, city-level data were gathered

Table 1. OLS Estimates from exhaustive linear specifications of the Output, Capital and Labour Growth Equations (standard errors in parentheses).

	Dependent Variables (log)		
	Output Growth	Capital Growth	Labour Growth
North	**3.2985**	**3.4415**	**−3.5141**
	(0.6845)	(0.8248)	(0.75551)
Central	**3.3344**	**3.4756**	**−3.5251**
	(0.6905)	(0.8300)	(0.7595)
South	**3.1856**	**3.5236**	**−3.3864**
	(0.6919)	(0.8187)	(0.7613)
West	**3.3286**	**3.4680**	**−3.4677**
	(0.7018)	(0.8362)	(0.7697)
LG(log)	**0.7878**	**0.8820**	
	(0.0857)	(0.0623)	
CG(log)	**0.1894**		**0.7927**
	(0.0850)		(0.0650)
OLR(log)	**−0.8552**		
	(0.1614)		
CLR(log)		**−0.5483**	**0.4623**
		(0.0782)	(0.0798)
RelW(log)	0.2060	0.0560	0.1009
	(0.1667)	(0.1202)	(0.1135)
LOC.	0.0529	**0.2691**	**−2838**
	(0.1294)	(0.1387)	(0.1306)
SPEC.	**−0.5088**	**−0.4626**	0.4027
	(0.2262)	(0.2388)	(0.2275)
Pop.(log)	0.0199	**−0.0586**	**0.0576**
	(0.0242)	(0.0275)	(0.0265)
PopG(log)	0.1200	**0.2164**	−0.0944
	(0.0752)	(0.0790)	(0.0783)
GovEx(log)	0.0164	0.0611	**−0.0888**
	(0.0410)	(0.0453)	(0.0422)
Immigr(log)	0.0721	**0.1331**	**−0.1005**
	(0.0453)	(0.0505)	(0.0488)
Ttax	0.0274	0.0033	0.0215
	(0.0250)	(0.0281)	(0.0265)
$AdjR^2$	0.857351	0.856247	0.842605

Boldface significance at 5 pc. level

on population (Pop), population growth (PopG), government expenditures (GovEx) and taxation rates, and the population share which was foreign born (Immigr).

All regression results referred to are summarised in four tables (Tables 1–4) which show the results of regressions for the three Equations (9)–(11) using various combinations of right-hand side variables. We begin by presenting a broad Ordinary Least Squares (OLS) specification.

The task is to explore, first in a preliminary way, the three equation system. Each equation is estimated separately to increase the precision. So that the general structure of conditional correlation is clear, OLS estimates are presented initially from exhaustive linear specifications of the output, capital, and labour growth equations (Table 1). Not surprisingly, these rich specifications account for much of the growth variance in the sample (adjusted $R^2 > 0.84$ in each equation). However, only a few variables in each equation are statistically significant. This is partially due to high correlations among regressors, as discussed above (Table 2), suggesting that approximately the same amount of variance could be captured by a drastically reduced set of regressors.

Table 2. Covariance Matrix.

	Outp.Gth.	Cap.Gth.	Lab.Gth.	OLR	CLR	Local.
Outp.Gth.	1.000					
Cap.Gth.	0.8642	1.000				
Lab.Gth.	0.8941	0.8638	1.000			
OLR	0.1966	0.4290	0.4316	1.000		
CLR	−0.0845	−0.1863	0.0609	0.3416	1.000	
LOC.	−0.2769	−0.2594	−0.3245	−0.1155	0.0506	1.000
Pop.	−0.0524	−0.0656	−0.1098	0.0722	−0.0881	0.5000
SPEC.	−0.1701	−0.1942	−0.1884	−0.3560	0.0721	0.5018
PopG	0.2196	0.3653	0.2300	0.3228	0.0021	−0.0396
Wage	0.1949	0.3671	0.3779	0.8644	0.2461	−0.1052
Immigr.	−0.0156	−0.0452	−0.0622	0.1822	0.3179	0.3340
Gov.Ex.	−0.2435	−0.2201	−0.2834	−0.0543	−0.0767	0.4238
	Pop	**Spec.**	**PopG**	**Wage**	**Immigr.**	**Gov.Exp.**
Outp.Gth.						
Cap.Gth.						
Lab.Gth.						
OLR						
CLR						
LOC						
Pop	1.000					
Spec.	−0.1471	1.000				
PopG	−0.0448	0.0264	1.000			
Wage	0.0874	−0.3420	0.2575	1.000		
Immigr.	0.2296	0.0984	0.1796	0.2236	1.000	
Gov.Ex.	0.4139	0.0624	−0.1008	0.0753	0.2403	1.000

In examining the output growth equation (Table 1), the most striking finding is the strong positive significance of capital and labour growth. Further, output per worker (a measure of labour productivity) is negatively correlated with output growth. Finally, the only other economic variable significantly correlated with output growth is the specialisation measure, which has a negative coefficient. The localisation and urbanisation measures are not significantly correlated with output growth. The output growth equation can be summarized by noting that the traditional factors enter with the expected signs, while those variables associated with geographic production externalities do not demonstrate a strong statistical correlation with output growth. The output growth equation can be summarized by noting that the traditional factors enter with the expected signs, while those variables associated with geographic production externalities do not demonstrate a strong statistical correlation with output growth.

As in the output growth equation, the highly positive simple correlation coefficient between capital and labour growth is borne out in both input growth regressions (Table 1). Additionally, the capital-labour ratio comes in as expected in both equations.

Further, localisation and urbanisation are significant in both input growth equations, though not always with the expected sign. The specialisation measure is correlated with capital growth, and, as in the output equation, its coefficient is negative. Finally, the share of immigrants in a city is significantly related to both capital and labour growth rates, while government expenditure only enters the labour growth equation and with a negative coefficient.

Earlier the potential and previously unexplored interrelatedness of different externalities was noted. Each of the externality measures was correlated with each other (Table 2). To see the effect of this upon the estimation, Table 3 re-estimates each of the preferred OLS regressions including

Table 3. Exclusion of Externality Based Variables (standard errors in parentheses).

	Output Growth	Capital Growth	Labour Growth
North	**2.9102**	**4.5702**	**−3.9684**
	(0.5624)	(0.6596)	(0.7657)
Central	**2.9100**	**4.5816**	**−3.9756**
	(0.5689)	(0.6633)	(0.7860)
South	**2.8409**	**4.8341**	**−3.9979**
	(0.5641)	(0.6688)	(0.7860)
West	**2.8966**	**4.5907**	**−3.9151**
	(0.5826)	(0.6618)	(0.7785)
LG(log)	**0.7371**	**0.8981**	
	(0.0920)	(0.0673)	
CG(log)	**0.2767**		**0.7684**
	(0.0819)		(0.0591)
CLR(log)		**−0.5555**	**0.4785**
		(0.763)	(0.0782)
RelW			0.0350
			(0.1074)
OLR(log)	**−0.5233**		
	(0.0884)		
LOC.	−0.1448	**0.0951**	**−0.1655**
	(0.0957)	(0.1073)	(0.0993)
Spec.	**−0.4874**	**−0.5138**	0.3095
	(0.2287)	(0.2180)	(0.2158)
Pop.(log)	0.0332	**−0.0547**	**0.0381**
	(0.0232)	(0.0260)	(0.0224)
PopG(log)	0.1113	**0.1868**	
	(0.0763)	(0.0778)	
Immigr.(log)		**0.1672**	**−0.1045**
		(0.0482)	(0.0486)
GovEx.			−0.0672
			(0.0324)
AdjR^2	0.844363	0.849948	0.841473

Boldface significance at 5 pc. level

the localisation, specialisation, and urbanisation variables in turn. The most salient insight in Table 3 is that exclusion of one of the measures leads to different patterns of correlation than had been observed earlier. For example, by excluding specialisation from the output growth equation, a positive correlation between output growth and urbanisation was observed. In other words, a theoretical model which suggested inclusion of the localisation and urbanisation variables, but which ignored specialisation, would lead to different conclusions concerning the validity of the theory than the regression which does include specialisation.

The interrelatedness and the importance of interactions was demonstrated directly by including interaction terms in the regressions. Table 4 shows the results of regression runs that include two reinforcing interaction terms, such as specialisation jointly with (∗) localisation and urbanisation jointly with (∗) localisation. The small sample size does not allow us to fully disentangle interaction effects with the direct effects of geographic proximity variables. Due to this, we do not focus on the implications of particular coefficients. Nor was there an emphasis on how the introduction of these interactions alters the magnitudes and precision of the estimates. Instead, it was noted that in both the capital and labour Equations (10) and (11) no coefficients changed sign while the interaction term specialisation with localisation appears to be at least as important a covariate of capital and labour growth as others. In contrast, and conforming to an earlier finding, interaction terms are not significant in the productivity equation.

Table 4. Interaction Effects (standard errrors in parentheses).

	Dependent Variables (log)		
	Output Growth	Capital Growth	Labour Growth
North	**3.9036**	**4.8093**	**−4.0948**
	(0.7164)	(0.6677)	(0.8336)
Central	**3. 9209**	**4.8394**	**−4.1244**
	(0.7311)	(0.6715)	(0.8360)
South	**3.8806**	**5.1107**	**−4.1668**
	(0.6919)	(0.6787)	(0.8543)
West	**3.9064**	**4.8440**	**−4.0807**
	(0.7367)	(0.6680)	(0.8449)
LG(log)	**0.2423**	**0.8939**	
	(0.0848)	(0.0652)	
CG(log)	**0.2423**		**0.7777**
	(0.0848)		(0.0576)
OLR(log)	**−0.6309**		
	(0.1013)		
CLR(log)		**−0.5506**	**0.4704**
		(0.0732)	(0.0757)
RelW(log)			0.0553
			(0.1107)
LOC	−0.0978	**1.4672**	**−1.8807**
	(0.8918)	(0.9305)	(0.8712)
SPEC	**−0.6060**	**−0.1913**	−0.0145
	(0.2262)	(0.2913)	(0.2868)
Pop.(log)	0.0349	**−0.0501**	**0.0472**
	(0.0262)	(0.0290)	(0.0280)
PopG(log)	0.1114	**−0.0501**	−0.0944
	(0.0772)	(0.0290)	(0.0783)
GovEx(log)			**−0.0708**
			(0.0315)
Immigr(log)		**0.1718**	**−0.1164**
		(0.0484)	(0.0486)
SPEC*LOC.	0.3192	−1.0937	**1.1970**
	(0.5224)	(0.5596)	(0.5347)
URB*LOC.	0.0069	−0.0798	0.1141
	(0.0680)	(0.0716)	(0.0672)
$AdjR^2$	0.871770	0.884089	0.851824

Boldface significance at 5 pc. level

6 CONCLUDING REMARKS

The empirical results in the Appendix support the perspective that geographic externalities together with their negative correlates must be considered in a dynamic setting of city based growth paying attention to the interactions between different variables. The importance of omitting individual variables was demonstrated as well as interacting variables. This is in contrast to current theoretical literature on such externalities, which indicates variables as operating in isolation from other forces. The one-variable theory, by construction, cannot account for the role of externalities in urban growth. This is also, of course, of significance for urban policy analysis.

A broader estimation approach is recommended, using three equations to characterise the growth process rather than just a single equation, as in conventional approaches. This allows for a better understanding of the systemic nature of aggregate growth. Hence, for example, the impact of particular variables differs according to whether one considers productivity, capital, or labour growth. In addition, city size, a proxy for urbanisation, can be found to be positively correlated with labour growth, negatively correlated with capital growth, and uncorrelated with productivity growth. A significant implication of this is that, by solely estimating a single equation, whether it be productivity, capital, or labour, in trying to characterise aggregate growth, many interesting and significant effects are not recognised. If the common procedure is employed of examining only productivity output per capita, the important relationship would be missed between city size and input factor growth.

Another significant finding involves the ties between capital growth and each of the externality-based variables, in particular, the potential role of the SOC – type externalities on city size and growth. This needs further exploration.

In closing, analyzing statistical results, reinforced by sensitivity analysis, provides a better framework for policy analysis and public policy in the urban arena. With a significantly enlarged database the model clearly has the potential of forming the core of a decision support system for urban policy.

REFERENCES

Arthur, W.B. 1988. Urban Systems and Historical Path Dependence, in *Cities and their Vital Systems,* Ed. by J.H. Ausubel and R. Herman, Washington: National Academy Press

Barro, R.J. and X. Sala-i-Martin 2003. *Economic Growth*, Cambridge: MIT Press

Barro, R.J. 1990. Government Spending in a Simple Model of Economic Growth, *Journal of Political Economy*, 95, S103–S125

Beckmann, M. 1999. *Lectures on Location Theory*, Berlin: Springer

Blanchard, O.J. and L.F. Katz 1992. Regional Evolutions, *Brookings Paper in Economic Activity*, (1), pp. 1–75

Glaeser, E.L., H.D. Kallal, J.A. Scheinkn, and A. Shleifer 1992. Growth in Cities, *Journal of Political Economy* 100(6), pp. 1126–1152

Gottinger, H.W. 1998. Optimal and Sustainable Turnpikes, *Review of Economics*, 25, pp. 35–49

Gottinger, H.W. 2003. *Economies of Network Industries*, London: Routledge

Henderson, J.V. 1994. Externalities and Industrial Development, mimeo, Brown University

Henderson, J.V. and J.F. Thisse (eds.) 2004. *Handbook of Regional and Urban Economics*, Vol. 4, Amsterdam: Elsevier

Jacobs, J. 1969. *The Economy of Cities*, New York: Random House

Jacobs, J. 1984. *Cities and the Wealth of Nations: Principles of Economic Life.* New York: Random House

King, R.G. and S. Rebelo 1993. Transitional Dynamics and Economic Growth in the Basic Neoclassical Model, *American Economic Review*, 83, pp. 908–931

Krugman, P. 2007. *Development, Geography and Economic Theory*, Cambridge: MIT Press

North, D.C. 2004. *Understanding the Process of Economic Change*, Princeton: Princeton Univ. Press

Porter, M.E. 1990. *The Competitive Advantage of Nations*, New York: Free Press

Rhode, P.W. and G. Toniolo (eds.) 2006. *The Global Economy in the 1990s, A Long-Run Perspective*, Cambridge: Cambridge Univ. Press

Savitch, H.V. and P. Kantor 2004. *Cities in the International Marketplace: The Political Economy of Urban Development in North America and Western Europe*, Princeton: Princeton Univ. Press

Solow, R. 1957. Technical Change and the Aggregate Production Function, *Review of Economics and Statistics*, 30, pp. 312–320

Path to Democratic Environmental Governance in Latin America: A Case Study from Mexico

Sherrie L. Baver
The City College and The Graduate Centre – City University of New York

SUMMARY: This chapter focuses on the emerging concept of democratic environmental governance as it is being implemented in Mexico. The notion of this type governance derives from the three access pillars elaborated in Principle 10 of the Rio Declaration of 1992; access to environmental information, access to participation in environmental decision-making, and access to justice in environmental matters. Mexico serves as a case study of a Latin American nation relatively advanced in implementing the Principle 10 access reforms. Key events in the country's history of procedural environmental reforms are the 1988 and 1996 environmental laws, the environmental side agreement in the 1994 NAFTA Accord, the 2002 Access to Public Information Law, and the 2006 publication of the country's first Pollution Release and Transfer Registry. The main conclusion is that both a mobilized civil society and strong institutions are needed to promote adequate environmental protection in Mexico and elsewhere.

1 INTRODUCTION

This chapter focuses on the emerging global concept of democratic environmental governance as it is being implemented in Mexico. Environmental governance reforms are a subset of more general governance reforms; as such, they can offer a lens through which to analyze the process of democratic deepening. Specifically, the goal of this study is to examine how several of Mexico's new national institutions, procedures, and capacities are affecting environmental performance as well as public perceptions of governmental legitimacy. Two hypotheses guide this research effort. First, Mexico's position as a leader in environmental governance in the developing world did not emanate from the government itself; rather it derived from a combination of pressures from civil society activists (both domestic and transnational) and international organizations and institutions. Second, the struggle for democratic governance in this one policy area may contribute to a more accountable and transparent political system more generally.

2 THE NOTION OF DEMOCRATIC ENVIRONMENTAL GOVERNANCE

Three pillars are integral to the concept of democratic environmental governance. They are: citizen access to environmental information held by public authorities, including data on hazardous substances and activities in their communities; public participation in environmental decision-making; and citizens' effective access to justice. The United States was a global leader in these three pillars through the mid-1980s, having legislation addressing all three issues. However, by the 1990s, various regional and international governmental and non-governmental organizations have bundled the three ideas into one reform package and have promoted it around the world.

The first multilateral statement of what has come to be called "democratic environmental governance" is the 1972 Stockholm Declaration approved at the United Nations Conference on the Human Environment. Twenty years later, the three basic pillars became known to the world as "Principle 10" of the Rio Declaration, approved at the 1992 United Nations Conference on Environment and Development. Western Europe has been particularly active in promoting these reforms beginning in 1973 with the region's First Environmental Action Program. The European Union became much more proactive after the institutional reforms related to the Maastrict Treaty went into effect in 1993 (Weale, 2003). Most important for Europe has been the 1998 Aarhus Convention, which embodies all three pillars; all EU members are required to adopt the Convention and the EU has been working on diffusing the Aarhus norms beyond the European region. The United Nations Economic Commission for Europe (UNECE) and other international organizations such as the United Nations Institute for Training and Research (UNITAR) are trying to promote this reform package worldwide (Speth and Haas, 2006). In sum, Principle 10 and Aarhus are unlike traditional environmental agreements that target specific environmental problems. These reforms are less tangible but possibly more promising – to invite more voices, including the historically excluded, to participate in decision-making and to provide adequate information to those participants.

2.1 *Pillar I: The right to environmental information*

The right to environmental information at a global level grew out of the spectacular accidents involving toxic and hazardous materials in places such as Seveso, Italy (1976), Bhopal, India (1984), and Chernobyl, Ukraine (1986). A developing worldwide fundamental sense of justice demanded that local citizens needed information about, as well as the right to participate in the risks they were being asked to bear. This sense of environmental justice, with its three pillars, was formally articulated as Principle 10 of the Rio Declaration of 1992.

The first task in translating Pillar I of Principle 10 into concrete institutions and procedures requires that countries adopt a Freedom of Information Act (FOIA). To create the constitutional and statutory provisions of a FOIA, officials must resolve several issues such as: what information is considered confidential and what delivery mechanisms does the government use to provide information to the public? Governments must ensure that the information is systematic, understandable to average citizens, and produced on a predictable timetable. Two final practices considered essential to environmental information access. The first involves developing hazard and eco-labeling schemes on foods and other products to aid environmentally-conscious consumers (UNITAR, 2004). A second subset of FOIA laws are typically referred to "Community Right t to Know" laws and refer to Pollutant Release and Transfer Registries (PRTRs).

Pollutant registries are based on the U.S. Environmental Protection Agency's Toxic Release Inventory developed in the 1986 Emergency Preparedness and Community Right to Know Act and the 2003 Kiev Protocol of the Aarhus Convention.This involves a government establishing procedures and models to disseminate pollution emissions data from private sources. The idea is to bridge the information gap between corporations and governments on one hand and citizens on the other. While the primary goal of Pollutant Registries is to force polluters to reduce their emissions, they can also indirectly obligate governments to be more accountable, and to regulate industrial emissions more stringently. For Pollutant Registries to work as intended, however, a strong civil society must be present to act as both lobbyist and monitor (Pacheco Vega, 2007). At present, Mexico is the only Latin American country with a PRTR, although Peru is in the process of developing one.

2.2 *Pillar II: Public participation in environmental decision-making*

Pillar II of Principle 10 refers to citizen participation in specific environmental impact assessments as well as a role in more general environmental policymaking and implementation. Not only is citizen participation essential to a concept of justice but also numerous studies in environmental decision-making show that citizen input makes for more effective, sustainable policies that garner citizen support rather than opposition (e.g. Beierle and Cayford, 2002). The OAS's *Inter-American Strategy for Participation* in 2001 provides more specific benefits of citizen participation in the region such as: 1) encouraging consensus for alternative solutions; 2) creating trust between participants; 3) building bridges between actors who often do not otherwise communicate, both within civil society and between government and civil society; and 4) serving as a first step toward more formal partnership among participants.

At a minimum, governments must allow citizens to participate in Environmental Impact Assessments (EIAs). For activists in civil society and international governmental organizations, however, Principle 10's participation requirement goes beyond simple input into Impact Assessments. Participation also requires giving citizens adequate notification plus the right to comment with adequate time frames on assessments, draft laws and draft regulations. Citizens must see draft documents early in the decision-making process; and some stakeholders (e.g. the private sector) may not have privileged access to draft documents. Officials must establish opportunities and mechanisms for participation at various levels of decision-making (e.g. state and municipal levels) and must institute channels for the results of deliberations to reach the highest relevant official body for review and endorsement. Finally, citizens should have the right to propose new legislation or policies.

2.3 *Pillar III: Access to justice in environmental matters*

Pillar III of Principle 10 requires that a wide range of individuals and groups have legal standing before courts in order to participate formally in environmental cases. Furthermore, citizens who have not received an adequate response to their requests for environmental information or have felt their participation in decision-making has been circumscribed must have access to an administrative (e.g. ombudsperson) or judicial review procedure. Finally, governments must provide access to environmental justice at a reasonable cost; there should be no financial barriers to litigation. It seems that globally, Pillar III reforms have been the more difficult to implement than participation requirements (Rose-Ackerman and Halpaap, 2001).

3 DEMOCRATIC ENVIRONMENTAL GOVERNANCE IN LATIN AMERICA

In Latin America, the formal bases for democratic environmental governance begin with the "Additional Protocol of the American Convention of Human Rights of 1988." After adoption of Principal 10 by attendees at the 1992 United Nations Conference on Environment and Development in Rio, Latin America nations began to move on governance reforms throughout the decade. Representatives at the 1994 U.N. Conference on Sustainable Development of Small Island Developing States in Barbados affirmed the importance of Principle 10 reforms, especially citizen participation in decision-making, as did representatives at the Central American Ecological Summit for Sustainable Development in Managua and the Miami Summit of the Americas also in 1994. At the 1996 Santa Cruz (Bolivia) Summit on Sustainable Development, representatives authorized the Organization of American States to devise a regional participation strategy that would be implemented by the thirty-four OAS governments working jointly with civil society. The 1998 Santiago Summit of the Americas also highlighted the role of public participation as well as the value of public sector – civil society partnerships. Against this backdrop of several regional and international conferences, the OAS's Unit for Sustainable Development and Environment undertook a three-year project between to promote Principle 10 reforms in the region. The resulting guidance document was *The Inter-American Strategy for the Promotion of Public Participation in Decision – Making for Sustainable Development*, published in 2001.

Part of the research for the *Inter-American Strategy* involved producing an inventory of environmental laws in the region requiring government or citizen action. The inventory consisted of 296 laws from twelve countries. The inventory showed that fewer than 43% of the laws allowed some form of participation, but that there was a strong trend toward including more citizen participation in more recent legislation. Before 1990, few laws included any requirement for citizen participation, but by 2001, most legislation included access to information, process, or justice. Process was emphasized most often, justice second, and information third. Finally the provisions in the 296 laws defining legal standing, remedies, or the right to appeal were not clear. The authors of the strategy contrasted their findings with the Aarhus Convention applicable to the European Union, noting that Aarhus contained a lengthy annex of activities in which governments must permit public participation.

The *Inter-American Strategy* report concludes it is important that practically all of the countries in the region develop the suitable capabilities and conditions for extending the coverage and reach of current access to information tools, socializing the public to use the tools, and to expand access to justice. Since the report's authors had been mandated to focus especially on Pillar II, participation, they made several recommendations. They were especially interested in bolstering the Inter-American Committee on Sustainable Development, a subcommittee of the OAS's "Inter-American Council for Integral Development." They also advocated for additional funding for each country's National Sustainable Development Council. These councils would have clear statutory authority and the power to convoke other agencies. The expectation was that National Sustainable Development Councils would be used for alternative means of dispute resolution. A final *Inter-American Strategy* recommendation to broaden public participation was to form local sustainable development councils that, among other responsibilities, would have the right to review and influence public budgets (ISP, 2001).

A guiding hypothesis of this study is that international and domestic forces have promoted legal frameworks and institutions that are a part of the quest for democratic environmental governance, but the rights inherent in democratic governance cannot be said to exist, in fact, without practice. Typically this means monitoring by civil society, especially domestic and transnational NGOS. Networking efforts have been important to promote Principle 10 reforms and two worldwide efforts, with a significant Latin American presence, have been facilitated the Washington-based World Resources Institute (WRI). The two initiatives are The Access Initiative (made up solely of NGOs) and The Partnership of Principle 10, composed of NGOs, national governments, and International Organizations.

Latin America has two Access Initiatives: IA-Mexico and IA-Latin America. The Mexican initiative began with four NGOs in 2001: the *Centro Mexicano de Derecho Ambiental(CEMDA), Comunicación y Educación, Ciudadana Presencia,* and *Cultura Ecológica.* IA-Latin America (comprising NGOs in Mexico and Chile) worked with the WRI for over a year to research the three access principles in 10 LA countries and three Mexican States (WRI, 2002; Carillo Fuentes, 2007). The findings in the World Resources Institute's "Closing the Gap" were similar to those reported by the Inter-American Strategy (2001); they also found that Latin American NGOs were especially concerned with increasing participation of marginalized groups such as indigenous communities.

The Partnership for Principle 10 is a global group of civil society organizations, governments, and international organizations that have networked to promote a culture that values the three PP10 pillars of democratic environmental governance. The PP10 represents a "Type II" effort agreed to at the World Summit on Sustainable Development in Johannesburg in 2002. In 2007, the global partnership had eight government partners with three from Latin America (Bolivia, Chile, and Mexico), four international organizations, (World Conservation Union, United Nations Development Program, United Nations Environmental Program and the World Bank), and thirty-three civil society groups. (The Latin American civil society groups came from Venezuela, Mexico, Chile, Bolivia, Argentina, Ecuador, Colombia, and Paraguay.

4 DEMOCRATIC ENVIRONMENTAL GOVERNANCE IN MEXICO

Mexico provides a rich analytical case for environmental governance for at least two reasons. It is an emerging democracy, which, along with the democratically consolidated Chile and Costa Rica, has moved farthest in terms of environmental management reforms in Latin America. Second, Mexico is a highly industrialized economy with a wide range of environmental challenges (Mumme, 1998). Succinctly stated, Mexico's key environmental problems are a scarcity of hazardous waste disposal facilities; natural fresh water resources are scarce and polluted in the north, inaccessible and poor quality in center and extreme southeast; raw sewage and industrial effluents polluting rivers in urban areas; deforestation; widespread erosion; desertification; deteriorating agricultural lands; serious air and water pollution in the national capital and urban centers along the U.S.-Mexican border . . . the government considers lack of clean water and deforestation national security issues (CIA World Factbook, 2007).

Thus, both factors, the "youth" of Mexico's democracy as well as the breadth of its environmental challenges have mobilized the country's civil society. Environmental organizations have partnered with transnational activists and International Organizations, to push both the procedural environmental democracy agenda as well as policy reform in specific environmental sectors.

5 INSTITUTIONAL INITIATIVES FOR THE ENVIRONMENT

The focus of this discussion is the role of the government and civil society groups in promoting democratic environmental governance reforms in Mexico. The goal is to understand how Mexico has become an environmental leader, at least in terms of legislation, within the developing world and at the same time to understand how environmental governance reforms in a newly democratizing nation such as Mexican add to that country's overall quality of democracy.

The Mexican government drew up its first modern environmental legislation in preparation for the 1972 United Nations Conference on Human Settlements in Stockholm, but with virtually no citizen participation (Mumme, 1998). In the mid-1980s, Mexico experienced at least two catastrophic events, which mobilized the incipient Mexican environmental movement. The first event, in 1984, involved a PEMEX gas plant explosion at San Juan Ixhuatepec outside of Mexico City. The accident killed over 500 people and dovetailed with the global movement for communities' "Right to Know" the risks they are being asked to bear (Pillar I). The second event was the Mexico City earthquake of 1985. The government's wholly inadequate response served as a catalyst to energize the country's NGO sector. In that crisis, Mexican NGOs began viewing themselves as "green brigades that were aiming to assist the swollen ranks of squatters living in shantytowns around Mexico City" (Diez, 2006; Rodríguez Bribiesca, 2007). In addition and significantly, through 1993, "nine of the fifty-five most serious industrial accidents in the world . . . occurred in Mexico" (Albert-Palacios, 2004).

Since Mexico remained an authoritarian political system in the 1980s, President Miguel de la Madrid (1982–1988) made only minor reforms that encouraged formation of anti-pollution advocacy groups and offered new institutional channels for petition and protest through the Cabinet-level environmental agency, *Secretaría de Desarrollo Urbano y Ecología*. Occasionally, key groups were invited to participate in government forums on environmental and urban planning themes (Mumme, 1998). While Raul Salinas (1988–1994) was even less prone than his predecessor to concern himself with the environment or public participation, his government was forced to become more active in these areas because of the ongoing North American free Trade Agreement (NAFTA) negotiations. One of the early steps in the process of readying Mexico for free trade with the United States and Canada was the 1988 reform of the country's environmental law. According to Carruthers (2007), similar to many other Latin American countries, Mexico largely "imported" an environmental policy architecture modeled along U.S. lines, starting with the 1988 *Ley General de Equilibrio Ecológico y Proteccíon Ambiental (LGEEPA)*.

5.1 *The 1988 environmental law reforms*

In 1988, to coincide with the Salinas presidential campaign, the Mexican government reformed the country's key environmental law first passed in 1982. It enhanced the Environmental Agency's authority giving it greater coordinating power as well as greater powers of enforcement. It also mandated Environmental Impact Assessments (EIAs) for all federal public works and various private sector projects involving potentially polluting industries, mining operations, and tourism developments.

The Mexican government originally incorporated Impact Assessments s into the legal system in 1982, but they were not considered a serious assessment tool until 1988. Also, legislators showed increased attention to participatory designs in 1988 EIA reform as well (Palerm and Aceves, 2004). The 1988 Impact Assessment reforms provided numerous avenues to broaden participation in general, to make the system more accountable, and, more specifically, to include participation from indigenous communities (approximately 9% of the population) as well as indigenous knowledge (Palerm and Aceves, 2004). Thus, participation in the EIA process represented a step towards the kind of public participation envisioned in Pillar II of Principle 10.

However, the current Mexican Environmental Impact Assessment process is considered "closed" in that there are predetermined projects that require assessments and the content of the assessment is strictly regulated. Although the system is rigid, a 1999 OECD report, suggests this is a better approach for consolidating democracies since it allows less room for corruption. Still, Mexico's Environmental Ministry, now called Secretaría de Medio Ambiente y Recursos Naturales (SEMARNAT) could do more to approach international best practices. Palerm and Aceves (2004) suggest that: translators are always available for indigenous communities; public hearings are mandatory for projects and not left to the discretion of the Ministry; officials publicize hearing notices more widely than is now the case; and SEMARNAT must justify the adoption or non-adoption of public input in their final decisions. The result of a transparent and accountable EIA system would aid in overall democratic consolidation. In short, "in the Current EIA system, opportunities for participant are very limited when compared to best practice principles".

5.2 *North American Free Trade Agreement (NAFTA) 1994*

The 1994 NAFTA agreement remains controversial on many accounts not least of which has been its effect on the Mexican environment. The scholarly jury is still out on whether or not NAFTA has turned Mexico into an industrial pollution haven. What seems clear, however, is that because of NAFTA, and the NGOs that monitor the agreement, activists have been rather successful in gaining procedural environmental reforms in Mexico as well as some degree of environmental cleanup (Mumme, 1998; Diez, 2006) (this reference needs to be cleared). For example, in the early 1990s because of the transnational activists monitoring NAFTA negotiations, Mexican President Salinas was forced to pay significant attention to the northern border with United States (with its large concentration of assembly plants) and help to create the Integrated Border Environmental Program. The program focused on improving sewage treatment in seven sister cities along the U.S.-Mexican boundary.

Due to the success of environmental activists and the institutions they created, the NAFTA agreement set an institutional precedent for future free trade agreements. While the Commission on Environmental Cooperation (CEC) is the best know of the NAFTA environmental institutions, other NAFTA institutions deserving mention and deserving of closer analysis are the Border Environmental Cooperation Commission and the North American Development Bank to fund needed border infrastructure projects.

In turning specifically to the three-nation Commission on Environmental Cooperation, Speth and Haas (2006) characterize it as "almost unprecedented in international relations because it gives NGOs and other non-state actors the right to formally challenge the legitimacy of governments' actions." The Commission has representatives from the national environmental agencies of Canada, the United States, and Mexico it has supranational authority to invoke sanctions, including trade sanctions if member countries fail to enforce their own domestic laws. These new institutions open avenues to environmental groups to lodge complaints against any of the three NAFTA members and provide a new tool for citizens, especially in Mexico with its weaker regulatory capacity, to hold their government and the private sector accountable in cases of environmental abuse.

At least one case, the 1996–1997 Cozumel case provides an example in which the "Article 14-15 process" of the CEC worked as intended, at least by transnational environmental activists. While the Mexican government planned a large development in Cozumel involving a pier, shopping mall and golf course, environmentalists demanded a natural protected area and alleged that the government had not followed the required EIA process for the proposed development. Since at that time, local environmental NGOs did not have standing in Mexican courts (reflecting the lack of the third pillar of Principle 10), environmentalists used the NAFTA process instead. This was the first case to go through the entire Article 14-15 process. The decision resulted in creating a protected area and terminal port but not the much larger development originally proposed. The case also raised public awareness and put pressure on Mexico's Federal Attorney for Environmental Protection to enforce laws and regulations more effectively (Alanis Ortega, 2002).

Although several long-term studies of the Commission on Environmental Cooperation are now in process regarding its overall effectiveness, given the reality of a free trade agreement among the United States, Canada, and Mexico, scholars agree that, at a minimum, it is better to have the Commission than not (Kirton and McLaren, 2002; Mumme, 2007; Pacheco-Vega, 2007). Still, as a preliminary conclusion, few see the CEC as a rousing success. Alanis Ortega (2002) concluded that after the Cozumel case, Mexico and Canada wanted to limit petitions sent to the CEC and Speth and Haas note "although several dozen petition have been filed, few of the challenges have produced results."

5.3 *The 1996 environmental law reform (LGEEPA)*

As Mexicans demanded numerous democratizing reforms in the 1990s, President Ernest Zedillo (1994–2000) moved the country ahead in the process. He also gave increased prominence to the environmental agenda; this parallels what is also true in U.S. environmental policymaking; presidential leadership along with the choice of environmental chief is the key variable determining national policy success or failure. Diez (2006) documents the numerous accomplishments of Zedillo's Environmental Minister, Julia Carabias; she can be seen as instrumental in bringing Principle 10 access pillars to Mexican environmental governance. As a first step, to reforming the environmental law, Environmental Minister Carabias created the National Consultative Council for Sustainable Development along with four regional consultative councils in 1995.

Specifically, what led to the 1996 environmental law reforms? As the Zedillo administration democratized, Mexican NGOs raised two general complaints about the 1988 law. The first problem was the overall lack of government responsiveness to citizen complaints; and second was the perception that the government was handing out only symbolic sanctions to violators. The 1988 environmental law contained provisions for public participation in various environmental activities such as land use planning, drafting official norms, and environmental impact evaluations (a Pillar II reform) and a Public Complaint and the Appeals Review (a Pillar III access to justice reform). These activities plus information access, were strengthened in the 1996 law. For example, the 1996 legislation dedicates its fifth chapter to "Social Participation and Environmental Information."

This Pillar I reform was also formalized in the 1996 law. The reforms contained both substantive and procedural mandates regarding environmental information. The mandate was to establish a formal right for Mexicans to have access to environmental information (Rodríguez Bribiesca, 2007), while the substantive mandate was for the government to create a National Environmental and Natural Resources Information System specifically to publish and disseminate existing registries and databases. Most NGOs were pleased with the outcome of the reforms; and given this broad increase in Principle 10 access rights during the Zedillo years, environmentalists were able to win several major victories.

5.4 *The 2002 transparency and access to information law (LFTAIPG)*

The Pillar I Right to Environmental Information granted in Mexico's 1996 environmental law was solidly strengthened with passage of Mexico's 2002 Access to Information Law (*Ley Federal de Acceso y Transparencia de la Información Pública Gubernamental*. The Right to Government Information was first mentioned in Mexico's 1977 constitutional reforms under President Lopez Portillo but had no teeth until 2002. The idea of a meaningful information access law evolved between 1994 and 2000 as part of the program to modernize Mexico's public administration. A key goal emerging from this process was "*rendición de cuentas*" or governmental accountability, responsibility, and citizens' rights to information and evaluation (Bareda Vidal, 2000).

It is reasonable to suggest that the winning of the Right to Environmental Information granted in 1996 paved the way for the more general Freedom of Information Act in 2002. The main institutional innovation of the 2002 law was creation of Federal Institute for Access to Information (IFAI), which has been hailed as "an exceptional tool in the slow transition to democracy and the slow opening of the Mexican government's he culture of secrecy" (Hofbauer, 2007). All public information requests go through the IFAI, and requests for environmental information are among the top five issues of concern from Mexican requestors. The Institute may also be considered a Pillar III, access to justice reform in that it has an appeal process if citizens are denied the public information they are requesting. At present, though, it is not clear how objective the IFAI is in handling citizen appeals. While in March 2008, this researcher found strong support for IFAI among two leading Mexican ENGOs another researcher tested the IFAI appeals process and found that the Institute unquestioningly privileged industry's concern to keep information secret rather than considering the public's Right-to-Know (Pulido Jiménez, 2007).

In terms of requests for environmental information, the Access to Information law is an improvement on the 1996 environmental law. In the 1996 Law, requestors had to present official identification and a reason in writing for requesting the information. Furthermore, the 1996 law contained the threatening provision that requesters would be held liable for damages if information were used inappropriately, such as the sharing of trade secrets from a private firm. In 1996, the assumption was that the government had the right to deny information. By 2002, the presumption was that the information existed and it was the government's responsibility to provide it.

A central Pillar I environmental governance tool begun through NAFTA but reinforced by the 2002 Access to Information Law, was the creation of a mandatory Pollutant Release and Transfer Registry (in Spanish, *Registro de Emisiones y Transferencia de Contaminantes or RETC*). Because of NAFTA, Mexico's PRTR would have to be comparable to the published databases in Canada and the United States. Before this change, PRTR reporting in Mexico was entirely voluntary (Ranger 2004). However, the implementing regulations for the Pollutant Registry only began in mid-2005. The data must be publicly available and disaggregated by business location, chemical substance, and the means by which the pollutant is being released or transferred. The first report became available in August 2006. However, one analyst of Mexico's Registry argues that senior officials in Mexico's Environmental Ministry have little interest in enforcing reporting requirements (Pacheco Vega, 2007). Another, possibly less confrontational, Registry analyst noted that although the Federal Prosecutor for Environmental Protection can levy fines or other administrative sanctions for a company's failure to comply with PRTR reporting, it usually doesn't simply because of a lack of capacity. Thus Mexico is only beginning to implement the PRTR piece of Pillar I "Access to Information" and Pillar III "Access to Environmental Justice" procedures.

While the Mexican government allots minimal funding to implementing the Access to Information Law, it represents a major step forward in the process of democratization. LFTAIPG has helped begin to promote a culture of participation among citizens, a culture of transparency among bureaucrats, and a broader interpretation of who can have access to environmental information. A requestor no longer must prove a vested or legal interest (Carillo Fuentes, 2007).

The construction of any right is a long-term process, including Principle 10 environmental access rights of information, participation, and justice; and Mexico is a Latin American leader in this process. The Right to Environmental Information, Pillar I, was included in the country's 1988 and 1996 environmental laws. However, it became a much more substantive right with the 2002 Access to Information Law, which also created a federal institute to manage citizen requests for all categories of public information. In addition, in part because of NAFTA obligations, Mexico is one of the only Latin American countries with a Pollutant Release and Transfer Registry. This Right to Information, like all rights, can only be guaranteed by constant use by the public as well as the Mexican bureaucracy. The country is beginning to create a culture of transparency, which refers to changes in the beliefs, practices, and expectations, embedded in both the state and society, about the public's "Right to Know" (Fox and Haight, 2007). In fact, Mexican expectations about information and transparency are no becoming so embedded that in 2007, the government set minimum national standards for transparency in the public sector.

Due to the closely linked concept of democracy, Pillar II, the Right to Citizen Participation, seems, at least formally, to be the easiest to implement. This has been the case in Mexico. Resolute citizens, especially in environmental NGOs have been able to participate in environmental decision-making with some ease at least since the reform of the environmental law in 1996. Still, some citizens; especially the indigenous groups in Mexico are *de facto* excluded from participation. This is a clear issue of environmental injustice to which the government must attend to make this right meaningful for all Mexicans (Palerm and Esteves, 2004; Diez, 2006).

Finally Pillar III, Access to Justice, may be the hardest to implement, probably due to a lack of awareness and training in environmental law among judges and their staffs. It is also true that relatively few political systems have autonomous judiciaries that can make law apart from politicians whose main task, especially in developing countries, is to promote economic growth. For instance Mumme (1998) noted that environmental organizations have virtually ignored Mexican courts. Not a single case has been successfully brought by an environmental advocacy organization against the government or the private sector. This is slowly beginning to change but the judiciary, on its own, is unlikely to become a major arena of environmental policy reform.

6 CONCLUDING REMARKS

Mexico has made impressive strides in the past decade in terms of overall democratization, as well as democratic environmental governance and now is a leader in the developing world in terms of Principle 10 environmental access reforms. With the general process of democratization plus the fear of added environmental stress coming from increased economic activity with NAFTA, Mexican civil society organizations along with transnational activists have successfully demanded formal rights and institutions for environmental protection. Perhaps the key lessons learned from the Mexican case, is that the process of economic globalization can bring some ideas and institutions that can promote modern environmental governance. However, only a highly mobilized environmental movement can monitor these Principle 10 access reforms and force them to function effectively.

REFERENCES

Alanis Ortega, G. (2002) "Public Participation within NAFTA's Environmental Agreement: The Mexican Experience." In J. Kirton and V. McLaren (Eds.), *Linking Trade, Environment, and Social Cohesion*, pp. 193–187. Burlington, VT: Ashgate Press.

Albert-Palacios, L. (2004) "Emergencias Químicas y Salud Pública en Mexico." *Revista de la Facultad de Salud Pública y Nutricíon* 5(4): 4.

Barreda Vidal, P. (2007) "Ley de Acceso a la Información Pública: Reto Ciudadano de los Mexicanos." Paper presented at the Latin American Studies Association Meeting, Montreal, September 5–8. Retrieved from libertadenaccion@yahoo.com.

Beierle, T. and Cayford, J. (2002) *Democracy in Practice: Public Participation in Environmental Decisions*. Washington, DC: Resources for the Future Press.

Carrillo Fuentes, J. C. (2007) "The 'Access Initiative' in Mexico . . . and Latin America." In J. Fox, et al. (Eds.), *Mexico's Right to Know Reforms: Civil Society Perspectives*, pp. 267–269. Washington, DC: Woodrow Wilson Center.

Carruthers, D. V. (2007) "Environmental Justice and the Politics of Energy on the U.S.-Mexican Border." *Environmental Politics*. Vol. 16, 3: 394–413.

Central Intelligence Agency. (2007) *World Factbook*. www.cia.gov/library/publications/the-world-factbook/geos/mx.html.

Diez, J. (2006) *Political Change and Environmental Policymaking in Mexico*. New York: Routledge.

Hofbauer, H. (2007) "Preface." In J. Fox et al. (Eds.), *Mexico's Right-to-Know Reforms: Civil Society Perspectives*, pp. 13–16. Washington DC: Woodrow Wilson Center.

Kirton, J. and McLaren, V. (Eds.). (2002) *Linking Trade, Environment, and Social Cohesion*. Burlington, VT: Ashgate.

Latin American Public Opinion Project (LAPOP). (2006) "The Political Culture of Democracy in Mexico." http://www.vanderbilt.edu/lapop. Accessed 2/28/08.

Magaloni, B. (2005) "The Demise of Mexico's One-Party Dominant Regime: Elite Choices and the Masses in the Establishment of Democracy." In F. Hagopian and S. P. Mainwaring (Eds.), *The Third Wave of*

Democratization in Latin America: Advances and Setbacks, pp. 121–148. New York: Cambridge University Press.

Mumme, S. (2007) "CEC in North America: Lessons for Export?" Paper presented at the Latin American Studies Association Meeting, Montreal, September 5–8.

Mumme, S. (1998) "Environmental Policy and Politics in Mexico." In U. Desai (Ed.), *Ecological Policy and Politics in Developing Countries: Economic Growth, Democracy, and the Environment*, pp. 183–204. Albany: SUNY Press.

Organization of American States (OAS). (2001) *Inter-American Strategy for the Promotion of Public Participation in Decision-Making for Sustainable Development ISP*. Washington, DC: Unit for Sustainable Development and Environment.

Organization for Economic Cooperation and Development (OECD). (1999) *Environment in the Transition to a Market Economy: Progress in Central and Eastern Europe and the New Independent States*. Paris.

Pacheco Vega, R. (2007) "Mexico's Pollutant Release Registry: Taking Stock, Looking Ahead." In J. Fox, et al. (Eds.), *Mexico's Right to Know Reforms: Civil Society Perspectives*, pp. 270–274. Washington, DC: Woodrow Wilson Center.

Palerm, J. and Aceves, C. (2004) "Environmental Impact Assessment in Mexico: an analysis from a 'consolidating democracy' perspective." *Impact Assessment and Project Appraisal* 22: 99–108.

Partnership for Principle 10 (PP10). (2007) Fourth Committee of Whole Meeting, Mexico Cityat www.pp10.org. Accessed 12/14/07

Pulído Jimenez, M. (2007) "Optimizing Transparency and the Challenge Facing the IFAI: Making the Principle of Maximum Possible Disclosure Effective in Environmental Issues." In J. Fox, et al. (Eds.), *Mexico's Right-to-Know Reforms: Civil Society Perspectives*, pp. 265–266, Washington, DC: Woodrow Wilson Center.

Ranger, E. M. (2004) "Pollutant Registry Enacted." *Business Mexico*, Vol. 14, no. 8: 15.

Rodríguez Bribiesca, P. (2007) "Access to Environmental Information." In J. Fox, et al. (Eds.), *Mexico's Right-to-Know Reforms: Civil Society Perspectives*, 260–266. Washington, DC: Woodrow Wilson Center.

Rose-Ackerman, S. and Halpaap, A. (2001) "The Aarhus Convention and the Politics of Process: The Political Economy of Procedural Environmental Rights." Draft Paper for *The Law and Economics of Environmental Policy: A Symposium*, Faculty of Laws, University College London, Sept. 5–7.

Speth, J. G. and Haas, P. (2006) *Global Environmental Governance*. Washington, DC: Island Press.

United Nations Economic Commission for Europe (UNECE). (2003) Kiev Protocol on Pollutant Release and Transfer Registers to the Convention on Access to Information, Public Participation in Decision-making, and Access to Justice in Environmental Matters (Aarhus Convention). New York: United Nations.

United Nations Institute for Training and Research. (2004) *Guidance Document: Creating a National Profile*. Geneva: UNITAR.

Weale, A. (Ed.). (2003) *Environmental Governance in Europe: An Ever Closer Ecological Union? (2nd ed.)* Oxford: Oxford University Press.

World Resources Institute (WRI). (2002) *Closing the Gap: Information, Participation, and Justice in Decision-Making for the Environment*. Washington, DC.

Environmental Issues, Challenges and Initiatives in Southern Africa: A Case Study from Botswana

Tidimane Ntsabane & Nkobi Moleele
University of Botswana, Botswana

SUMMARY: This chapter, using a historical approach, will look at the major environmental issues, challenges and initiatives faced by southern Africa and argue that there is now consensus that environmental conservation is essential to attain sustainable development. It will argue that there is a need to find solutions which are not only technical but also socio-economic as the environment does not exist as a separate sphere from human actions, ambitions and needs. For the continent to catch up with the developed world there must be some intensive exploitation of resources. Human populations on the continent and indeed globally, now pose the greatest threat to the environment than ever before. The unsustainable consumption patterns of natural resources and the ever-increasing population are putting ever-increasing stress on the land, water and ecological systems.

1 INTRODUCTION

African societies have through their pre-colonial history maintained an intimate relationship with nature, showing a greater sensitivity to the workings of natural ecosystems (Oakes, 1998). Their simple technologies and small population sizes were such that they had limited impact on the environment. Communities and families in these societies had through the decades been teaching children and adults the value of conserving, protecting and sustaining their resources. The system was ecologically balanced, as it was not losing products faster than these could be regenerated. This living with nature resulted in a greater sensitivity to the workings of natural ecosystems.

For most of colonial and post-colonial Africa, however, the predominant thinking at some point was that environmental responsibility and development were competing objectives. For the continent to catch up with the developed world there must be some intensive exploitation of resources. Human populations on the continent and indeed globally, now pose the greatest threat to the environment than ever before. The unsustainable consumption patterns of natural resources and the ever-increasing population are putting ever-increasing stress on the land, water and ecological systems (Oakes, 1998).

It is clear from the preceding discussion that there are serious regional and global environmental problems which require attention and action. There is also consensus that they can no longer be reduced to rich versus poor or north versus south issues but are similar to both and global in nature. From erosion through pollution to the depletion of some resources, the evidence is now in – that many of our human activities are currently reducing the long term ability of the natural environment to provide goods and services as well as adversely affecting current and future human health and well-being.

This chapter, using a historical approach, will look at the major environmental issues, challenges and initiatives faced by southern Africa and argue that there is now consensus that environmental conservation is essential to attain sustainable development. It will argue that there is a need to find solutions which are not only technical but also socio-economic as the environment does not exist as a separate sphere from human actions, ambitions and needs.

2 CONTEXT OF BOTSWANA'S ENVIRONMENTAL PROBLEMS

2.1 *Land area and population*

Botswana is a landlocked country in the interior of the southern African sub-region. It shares borders in the north with Zambia, in the northeast with Zimbabwe, in the east and south with South Africa, and in the northwest with Namibia. The country has a land area of 582,000 km^2 and a population of 1,326,796 (CSO, 1991). The population density, according to the 1991 census, was 2.3 people per km^2. This low density however conceals regional, district and rural-urban variations. The western part of the country (Chobe, Ngamiland, Kgalagadi and Gantsi) for example, was very sparsely populated with 14–17 persons per 100 km^2. Localities with relatively good soils and water resources covered only one third of the country and tended to have a higher density than those with poorer resources. While on the other hand smaller districts such as Barolong and South East have a higher population density of around 31 persons per km^2. Urban areas have the highest density per square km.

Botswana is one of those countries with a fast growing populace. As population statistics from the three post-independence census and preliminary figures from the recently completed 2001 census show, the de facto population in 1971 was 596,944 growing to 941,027 in 1981, 1,326,796 in 1991 and 1,678,891 in 2001. These figures imply growth rates of 4.7 percent between 1971 and 1981 and 3.5 percent between 1981 and 1991 and 2.4 percent between 1991 and 2001. These expansion rates are still relatively high compared to the annual averages of 2.8 percent for Africa, 2.0 percent for the less developed countries (LDCs) and the world's average of 1.7 percent (CSO, 1991).

The 1991 census showed further that 60 percent of the population was aged less than 30 years of age. It is thus a youthful population with implications not only for demands on education, training, health, housing, and job creation but also for massive increases in the use of environmental resources on a finite geographical environmental area.

There are however major challenges facing the country's social sector. Principal is the challenge of HIV/AIDS. In 1998 it was estimated that approximately 17 percent of the general population were infected with HIV. Infant mortality rates, according to the 2000 Multiple Indicators Survey, has now increased to more than 57 per 1,000 live births from around 37 per 1,000, while the under-five mortality rate rose to 77 per 1,000 live births in 2000, from around 48 per 1,000 live births in the previous decade (CSO, 1991). The World Health Organization estimated that life expectancy at birth has dropped from 67 years to 50 years. Projections show that unless some drastic measures are taken soon to stop this pandemic, the country's life expectancy will continue to drop.

2.2 *Land ecosystem*

Environmentally, Botswana is a country of contrasts. On the one hand 80 percent of the country is a vast tableland (flatland), covered by Kalahari sand beds, with no surface water, except for short-lived small pans. On the other hand there is the Okavango Delta, situated in the northwest with an estimated area of between 10,000 and 18,000 km^2 (Botswana Society, 1976). The Okavango River forms the delta and as a result of the deep sand layer, the river's speed and volume is drastically reduced such that the water spills out to form a complex system of channels, ridges, swamps and pools of the delta. It has since time memorial not only been home to a wide variety of both flora and fauna but also a source of surface water for both domestic and agricultural use, food in the form of fish and wild edible plants, wildlife, recreation, transport and building materials.

2.3 *Climate and resource base*

Botswana has a highly fragile resource base as a result of the harsh climatic conditions. Like all semi-arid areas, Botswana is subject to highly variable rainfall over space and time. Severe droughts are frequent and population fluctuations prevent plants and herbivores from developing closely coupled interactions. Ecosystems seldom reach a climatically determined equilibrium point (Langeni, 1999).

Average rainfall varies from 200 mm in the southwest to about 650 mm in the northwest in the Okavango-Chobe system. Day temperatures are normally high and as a result evapo-transpiration exceeds rainfall especially in the sandveld. Daily rates of open-water evaporation may reach 7.5 mm (Magole, 1998; Atlhopheng et al., 1998). Drought is endemic and occurs with quasi 10–12 year periodicity.

The livestock sector is one of the major industries in Botswana. Cattle outnumber people two to one. Grazing and watering of these large herds of cattle is a major land use activity. The consequence of these land use activities have led to overgrazing, range degradation, competition with wildlife resources, soil erosion and an overall imbalance in the ecosystem (Cantrel, 1992).

Over the years government investment and subsidy policies have been heavily biased towards livestock. There has been a relative decline in arable agriculture's share of GDP (Gross Domestic Product). The decline in arable agriculture is fairly significant as this sector affects a majority of the rural population, which engages in small subsistence farming and petty commodity production. The result has been persistent poverty, unemployment and migration to urban areas in search of jobs and a better life.

Mining is the dominant sector in Botswana's economy contributing not less than 80 percent to the total GDP. The key mineral is diamonds with small-scale mining of copper, nickel and coal. There are, however, inherent weaknesses in both the mineral led economic growth and the nature of mining in the case study country. Mining in Botswana is capital, not labor intensive, so its contribution to employment creation is relatively small. Secondly, the phenomenal economic growth based on mining has limits unless new mines can be continually developed. Mining commodities like diamonds are susceptible to changes in the world market.

3 BOTSWANA'S ENVIRONMENTAL PROBLEMS AND CHALLENGES

The unsustainable consumption patterns, the country's fragile natural resource base and the ever-increasing population are putting ever-increasing stress on the land, water and the ecological system. The consequences of these have been land degradation, air pollution, water pollution, threat to biodiversity and the population problem. We now turn to some of the major environmental issues and challenges.

3.1 *Water resources*

Water is a basic human need and in Botswana as is in much of southern Africa, it is a scarce resource with rainfall spread across seasons and years creating temporal and spatial disparities. High evaporation rates exceed the total rainfall by a factor of about four. Ground water resources from past wetter climates are limited and hence uncertain surface sources are the only leeway for the future. Physical characteristics (e.g. turbidity, color, temperature, taste and odor); chemical properties (e.g. pH, hardness and dissolved oxygen); and readily available water to meet demand, are critical issues in water supply.

Various environments receive different quantities of water from the hydrological cycle with some places having a deficit and others having a surplus. There is an increasing demand against a finite water resource. Sources of water (i.e. rivers, lakes, springs, dams, reservoirs, aquifers and rainwater catchments systems) are affected by pollution. Indiscriminate and inappropriate methods of water disposal and some shallow aquifers are a primary concern in maintaining high water quality in Botswana. Urban encroachment and other competition in land use may adversely affect watersheds (Botswana Government, 1990; Botswana Conservation Strategy Agency, 1990; Mosothwane, 1998).

Botswana has a very limited recharge (0–5%) or 3 mm or less per year. Water is being mined and there is need for close monitoring of the extraction of underground water. If the rate of replenishment is exceeded by the extraction rate, then alternative sites or sources have to be sought. Surface water depends on climate system which is uncertain, low, erratic, varying spatially and temporary and evaporation exceeds precipitation.

Low yield from dams due to unpredictable inflows and large evaporation losses from shallow reservoirs is also a problem. Water is secured under very difficult and expensive conditions that it would be very unfair to compare the cost of safe water in Botswana with the cost of water in other countries where in some cases all you have to do is stick a pump in the constantly flowing river.

Extraction from aquifers, streams, lakes, deltas and other sources at a faster rate than replenishment could be perceived as degradation. National demand has shown marked increases in just 30 years. By 2020, Botswana National Water Master Plan projects that demand will equal supply. Water shortage is a general Southern African problem reaching desperation with Namibia's unilateral extraction of water from the Okavango.

Water is a cornerstone of both health and infrastructural development. The cost of water resources is likely to frustrate this. Long term hopes are on surface water since underground systems are from past wetter climates. Surface water, however, depends on rainfall that is uncertain due to current droughts. Drought reduces both aquifer recharge and river flow levels. High evaporation rates make a mockery of expensive water schemes in the form of dams. It is crucial that the meager fresh water resources be protected from pollution as a future investment as no future money would be used in cleaning the environment.

Water is a finite resource and there is need for frugality due to acute water shortages and other problems. There is also an increasing pressure on water resources due to increasing population, development and urbanization. The use of a pricing system has been meant to ease monitoring water scarcity but reduction in consumption is limited and non-market water meant for livestock, wildlife and the mining sector is not costed but significant.

The co-existence of subsidies and price increases gives consumers a confusing signal. Relatively small price increases tend to have minimal effect on resource use patterns (Botswana Government, 1999, 1993; Colclough and McCarthy, 1990). Government subsidies have led to a low water use efficiency and extra demand. Subsidies are only useful for meeting the basic needs of the poorest and encouraging water saving appliances and technologies.

3.2 *Rangeland degradation*

The primary contributor to range degradation is the livestock industry which constitutes one of Botswana's major industries. Cattle outnumber people two to one. Grazing these large herds is a major land use activity. The consequence of this overuse has led to overgrazing, range degradation, soil erosion and an overall imbalance in the ecosystem. The traditional communal land tenure system has also become a contributory factor to the overgrazing problem. Most people graze their animals on communally owned land where there is no limit on the number of animals to graze. Livestock in most cases easily exceeds the carrying capacity of the land leading to overgrazing and soil erosion. Efforts at bettering range management through for example fencing and stocking rates have been met with some controversy. They do not also address the question of dual rights of commercial farms grazing in communal areas.

3.3 *Vegetation and wood resources*

The high dependency of the rural majority on wood as a source of firewood, building materials, fuel wood and medicinal purposes has led to the over-exploitation of timber resources. This has led to increased demand by individuals and industry for the resource and has often led to reductions in their sustainable yield.

Forests provide many products to the inhabitants of Botswana. However, the natural forest is finite, especially in the context of the country's semi-arid environment. Past intensive harvesting and other ecological factors have led to some species such as Pterocarpus angolensis (Mukwa) being so heavily exploited that its survival is now severely threatened (Atlhopeng et al., 1998; Langeni, 1999; Mordi, 1991; Mosesane, 1998).

The rapid rise in deforestation has been accompanied by increases in the human population. This increase is associated with an expansion of cropland and increasing demands for forest products

(Letamo and Totolo, 1997; Ngwane, 1998; Totolo and Douglas, 1998). Population growth impacts on deforestation are also a function of the distribution of the population; the latter tends to be higher around highly populated areas.

3.3.1 *Natural regeneration*

This is a mechanism used by plants to maximize the potential for successful seedling establishment in specific natural environments: can include vegetative expansion; persistent seed or spore bank or persistent juveniles. The rate at which forests are harvested and the harsh climatic conditions of Botswana make regeneration negligible (Atlhopeng et al., 1998; Mosesane, 1998; Magole, 1998; Ngwane, 1998).

3.3.2 *Exotic tree plantations*

In reacting to the deforestation problem, the Botswana Government has encouraged planting of woodlands since 1972. The woodlots were mainly Eucalyptus species, which were selected because it was believed then that these grow fast, are well adapted to arid conditions and are generally do well. The exotic forest estates or plantations were created to supply future domestic needs for timber; fuel wood and other forest produce in most parts of Botswana (Atlhopeng et al., 1998; Langeni, 1999; Mordi, 1991; Mosesane, 1998; Magole, 1998; Ngwane, 1998). Problems with Eucalyptus plantations were encountered. The plantations were not stable as natural ecosystems. The outbreaks of pest diseases were found to be catastrophic.

Eucalyptus plantations are very wasteful i.e. water consumption is very high and drops in the ground water table have been reported for places with such woodlots. Suppression of other ground vegetation due to Eucalyptus' high competitive ability for groundwater has been documented. Eucalyptus plantations are also poor in controlling soil erosion, because its litter does not decompose fast due to bactericidal substances in the leaves. This results in decreases in the water holding capacity of plantation soils (Mordi, 1991; Mosesane, 1998; Magole, 1998; Ngwane, 1998). Woodlots also tend to produce fewer products, poorer fuel woods and fodder sources are sometimes inferior in durability compared to indigenous species.

3.3.3 *Indigenous woodland management*

As a result of problems experienced with the woodlots, performance of exotics was compared with that of indigenous species to assess the possible advantages of using local species. As a result of such studies, forestry programs now tend to concentrate on management of existing indigenous woodlands rather than on the development of exotic woodlots and tree plantations alone (Atlhopeng et al., 1998; Langeni, 1999). Mixed plantations of indigenous and exotic species are being pushed forward as they are much more diverse and offer more benefits as they produce diversified and non-timber products, lower the risk of disease and pest outbreak, protect the soil and improve fertility, provide shade and protection to valuable under-storey plants and offer better habitat for wildlife. The local species are often preferred as the species are well known by the local people and the resource may already be partly established.

Managing indigenous species through in situ conservation has proven to be problematic. Firstly in situ conservation is limited by the fact that large samples may be required to maintain genetic diversity (e.g. 5,000 to 20,000 plants to maintain variation in a species) (Atlhopeng et al., 1998; Langeni, 1999; Mordi, 1991; Mosesane, 1998; Magole, 1998; Ngwane, 1998). This means that the land requirements will be high especially in the case of forest trees. Furthermore, plants preserved in natural conditions are exposed to pests, pathogens and fires, which are unpredictable and can be quite dangerous.

Alternative strategies, taking into account the foregoing problems, have been developed by organizations such as the Forestry Association Botswana (FAB) and the Forestry Unit and the National Tree Seed Centre of the Ministry of Agriculture. One of these strategies involves the development of laboratory based storage systems such as seed stores or in vitro cultures.

3.4 *Depletion of veld products*

Broadly, veld products are range resources that include wildlife and plants such as wild fruits and roots, and mineralized and clay soils. For the purposes of this study, fuel wood and wildlife are excluded in this definition, as these components are treated on their own. Of primary concern here are wild fruits, herbs, roots and insects that are used for consumption and or medicinal purposes (Atlhopheng et al., 1998).

Just like wood resources, veld products are utilized on a subsistence and commercial basis. Subsistence harvesting is, however, the more prevalent of the two. It is the threat of over-exploitation of veld products that led the Government to include it in the list of the major environmental issues in Botswana (Atlhopheng et al., 1998).

Over-exploitation and depletion is related to the commercialization of veld-products such as phane and grapple plant, poverty, land use competition (e.g. expansion/encroachment of villages, arable lands and grazing pressure), and prolonged droughts that result in environmental degradation. The use of inappropriate harvesting methods for the grapple plant was found to be a factor that led to depletion around settlements in the western part of the case study region.

Atlhopheng et al., (1998) have also argued that access to veld products is a major factor that should be considered as contributing to veld product depletion. Veld product harvesting is currently not recognized or considered as a distinctive land use activity that deserves to be preserved. This therefore implies that the veld product areas can always lose their status to more recognized land uses such as arable farming, livestock grazing, settlements etc. Inappropriate use is therefore often worsened by local communities loss of control over local resource management.

3.5 *Wildlife resources*

Various types of conflicts characterize the situation over protected areas by the local people. This is attributed to the drastic increases in populations in the African countries (Rakgoasi, 1999; Letamo and Totolo, 1997), as in the case of Botswana. Most people live in rural areas and only a small portion in the urban centers. The rural people derive their livelihood from land; agriculture and livestock. Agriculture requires productive land, but as populations increase at high rates as experienced in most African countries, fertile land becomes scarce. Increases in rural populations put a lot of pressure on protected areas. Throughout Africa, people considered the protected areas (wildlife areas) outside the agricultural periphery as potential land that could be tribalised. In the case of Botswana, freehold land has been tribalised in the northeast and southeast of the country. Conflicts between the rural population and the governments over protected areas have been recorded in this country as well as in Kenya and Zimbabwe (Letamo and Totolo, 1997; Mordi, 1991; Molutsi, 1988).

The other problem facing African protected areas is an increasing demand for land by poor masses. Poverty, an increasing demand for land and the desire to improve people's standard of living are the three main issues causing many problems of conservation in the African countries and in developing countries in general (Langeni, 1999; Molutsi, 1988).

The revenue from protected areas was hardly of a direct benefit to the rural populations surrounding these areas as it was directed almost entirely to the urban centers. The local population is not allowed to hunt in the protected areas, but the wildlife can move out of these areas and damage the land that remains to the people. This has further hardened the people's attitudes towards the protected areas, and has forced the rural people to engage in the illegal activity of killing the animals themselves i.e. poaching for meat and to assist organized criminals who are only interested in the financial gain from selling products of wild animals. The other type of poaching occurs when African traditional healers require the parts of certain animals for medicinal purposes. This can place such animals under threat as, for example, has happened to the pangolin in Zimbabwe Letamo and Totolo, 1997; Mordi, 1991; Molutsi, 1988.

One other problem that is now being discovered in Africa is that when some of the protected areas were established, little attention was paid to the needs of migratory species. This is especially evident in Botswana in the Central Kalahari and in Tanzania. Animals sometimes wander across park boundaries into private land. This is one reason which strengthens the suggestions that management of national parks and protected areas should not only concentrate within the park boundaries, but must include the surroundings. In some places wrong harvesting methods of wildlife have led to concentrations of game, which in turn destroyed the restricted important habitats.

4 EVOLUTION OF NATURAL RESOURCE CONSERVATION STRATEGIES

This part of the paper discusses the various attempts at managing the major environmental issues in Botswana that were discussed in the previous section. These have also been identified and discussed in the country's National Conservation Strategy (Atlhopheng et al., 1998). It was found befitting to consider the evolution of the country's natural resource conservation strategies in three historic phases: traditional management strategies (very obvious in the pre-Colonial period); management strategies during the Colonial period; and the post-independence period. To understand the causes of contemporary problems, the historical origins of those problems must be traced. The past is the key to the present and studying the past also shows what worked and what failed, hence avoiding the possibility of repeating those errors.

4.1 *Traditional natural resource conservation strategies in the pre-protectorate period (before 1885)*

People lived in social groups that identified themselves as tribes (e.g. Bangwato, Bakgatla). The chief headed the tribe, and under him were subdivisions led by the designate of the chief (Botswana Society, 1976; Langeni, 1999).

The management strategies of natural resources during this period were embedded into the cultural and administrative set-ups of the tribes. The chief effected natural resources management strategies through tribal land divisions. The natural resource base (trees, pastures, water, wildlife) was under the tribal land. The chief in consultation with the elders and headmen appointed land overseers for different sections of the tribal land (Botswana Society, 1976; Langeni, 1999). The role played by the overseer involved monitoring the status of the grazing pastures and general availability of other resources (e.g. wildlife and water) within the tribal land. In the presence of a problem, the chief would convene a tribal meeting to discuss and resolve the problem. Solutions included removal of livestock from affected areas for recovery purposes. For instance, cattle posts were moved to the sandveld areas (western part of the country) from the Bangwato cattle posts on the hardveld areas (eastern part of the country) to reduce grazing pressure on natural resources. Below follows a few examples of strategies employed by tribes in their management systems.

4.1.1 *Conservation strategies for plant and water resources*

In the traditional natural resources management strategies, tree resources were associated with myths and taboos. Such myths and taboos played a major role in the conservation of rare and unique species (e.g. Adansonia digitata-fig tree or mowana). Chiefs' decrees/restrictions also played a major role in the conservation of tree resources. For instance Spirosytachies africans was an important tree species used as timber and goat browsing by the Batswapong people. As a result of the chief's decree, a stand of this species was set-aside in Pilikwe and no harvesting was allowed to take place until recently. Chief's representatives were responsible for seeing that the decrees/restriction were followed or complied with. The Barolong chiefs are also known to have regulated the cutting of brushwood through permits and overseers (Botswana Society, 1976; Langeni, 1999; Botswana Government, 1990; Mosesane, 1998).

4.1.2 *Management strategies for wildlife resources*

Tribesmen could hunt small game wherever they pleased, but there were however, qualifications required concerning the age group of hunters. Big game (e.g. buffalo, giraffe, gemsbok, and zebra) were protected by the chief's decree/restriction and could only be killed with the chief's permission.

The Kgori bustard (a bird) was only killed and utilized by the chief, so as to avoid mass killing as the bird bred poorly (Mordi, 1991; Botswana Society, 1976; Langeni, 1999; Botswana Government, 1990).

The oldest form of customary law that emphasized sustainable use of natural resources is reflected in totemic laws of tribes. The tribes in Botswana (among other things) distinguished themselves from each other by their totems, which in many cases were animal species (Botswana Society, 1976; Langeni, 1999; Botswana Government, 1990; Mosesane, 1998). Tribes could not kill or touch their respective totems, and this placed some protection on a range of animals. The present authors believe that during the pre-Protectorate Period (before 1885) such practices were major components of traditional natural resources management.

Methods of killing and storage during this period also contributed to the success of the management strategies employed. The methods of killing were not destructive as compared to today. Lack of refrigeration facilities set limits to what could be killed by each family at a time. It is well documented in the literature that large game was only killed in the winter through the chief's permission. Meat would be slowly dried for use through the course of the year when hunting was still restricted. This kind of system therefore allowed for breeding to go undisturbed, which is normally between September to February. The only people who could hunt throughout the year were the Basarwa (Bushmen) because their lives were so intricately tied to the wildlife Mordi, 1991; Botswana Society, 1976; Langeni, 1999; Botswana Government, 1990.

4.2 *Natural resource Conservation Strategies during the protectorate period 1885–1966: A period of gradual power loss by the chiefs over natural resources management*

In 1885 the British proclaimed a protectorate over Bechuanaland, after three local chiefs went to London to seek protection from the Boers/Afrikaners in South Africa and the Germans in South West Africa. It is therefore important to make an analysis of how the Protectorate administration impacted upon the traditional natural resources management strategies that were embedded within the cultures of the existing tribes.

Before the protectorate period, Tswana customs and traditions emphasized a communal use of resources. However, white settlers during the protectorate period seemed eager to change the societies from the traditional communal attitude to the individualist capitalist control of the resources especially land (Botswana Government, 1990; Botswana Conservation Strategy Agency, 1990). The settlers believed that individual control was essential given the range of ecological constraints affecting land use in the existing set-up. The sections that follow illustrate how the Protectorate Government interfered with the traditional strategies of natural resources management.

4.2.1 *Early days of the protectorate and conservation*

The first commissioner of Bechuanaland Protectorate in 1885 was advised by the British Government not to interfere with administration of tribes headed by chiefs (Langeni, 1999; Botswana Government, 1990; Botswana Conservation Strategy Agency, 1990). However, in 1891 the High Commissioner was ordered to ensure administration of justice, raising of revenue, and generally order and good government of all persons. This was supposed to apply to Europeans only while Chiefs and their tribes were to be entirely left alone. During this period tribes continued practicing most of the conservation strategies they had before the protectorate days. In 1910 the British Government decided to include everyone (both Europeans and tribes) under the administration of the Protectorate under its commissioner. New land institutions were established. To some extent, this diminished the traditional leaders' powers even though the leaders (chiefs) still retained autonomous rule in their tribal areas.

The institutions were detached from the traditional communal ways of land ownership (Langeni, 1999; Botswana Government, 1990). They evolved from tribal or communal systems during the early days of the Protectorate land and included the introduction of different land tenure systems: Freehold land was land taken over by the white settlers when they got in the country, which was initially communally owned. This became private land owned by white settlers. Crown land had few inhabitants or was not occupied and there was little done in a way to developing such land during the protectorate period. Crown land was later transferred to Government ownership and renamed state lands at the end of the protectorate period. Most of this land is used for parks, game reserves and forest reserves. At the end of the protectorate period, 48.9% of land in Botswana was tribal or communal land; 46.7% crown lands and 4.4% freehold land. This shows that in actual fact a small percentage of land was taken over as private land during the protectorate period. However, it must be noted that although small in percentage this was the most productive land in the country.

With time, conflicts of interests related to natural resources management, especially wildlife erupted in the early 1890's between the Protectorate Administration and the tribes. For instance Police Officer Major Grey of the Bechuanaland Protectorate complained through the High Commissioner about the large numbers of animals killed by the Ngwato Tribe (Langeni, 1999; Botswana Government, 1990; Botswana Conservation Strategy Agency, 1990). Statistics put forward as part of the evidence included more than 600 heads of large game killed by the locals between March 1893 and March 1894. The current paper argues that the arrival of the Europeans in South Africa and South West Africa (Namibia) prompted the locals to kill more wildlife for sale in exchange for other valuable goods like alcohol and jewellery. The Ngwato tribesmen had also seen a lot of game being destroyed by the Boer trekkers on their way to Damaraland. This brought about a situation where the chiefs were left in a dilemma. On the one hand they were discouraging the locals from indiscriminating killing of wildlife, whilst the foreigners could just do it and go unpunished.

The Protectorate Administrators viewed the uncontrolled killings as irresponsibility on the side of the tribesmen. They tried to pursue corrective measures through Chiefs Sebele of the Bakwena, Bathoen of Bangwaketse and Linchwe of Bakgatla on the subject in 1894. In general the chiefs and their people were against any new laws on game killing coming from the Protectorate Administrators.

Although not supportive of some activities of the tribes in relation to natural resources management, the Protectorate Administrators realized the need to work hand in hand with the chiefs. This is evident from the nature of the chief's decrees (melao) passed before and during the Protectorate Administration.

These decrees were responses to growing hunting pressure on wildlife resources from both local and commercial hunters in specific regions of the country. It was evident (Table 1) that commercial hunters were in the country long before the Protectorate Administration. Due to the increased commercial value of ivory, the Bangwaketse and Bangwato Chiefs passed in 1815 and 1878, respectively, decrees (melao) aimed at controlling the activities of commercial hunters. In some parts of the country, missionary-influenced natural resource conservation strategies were introduced by chiefs. For example, in Kweneng the Chief passed a decree in 1856 prohibiting hunting on Sundays. The excessive exploitation of game by European commercial hunters led to further restrictions during the Protectorate Administration. This undesirable exploitation was the main reason why the Chiefs were so willing to cooperate with the Protectorate Administration in matters of game protection (as evidenced by the series of decrees passed by chiefs during the Protectorate Administration).

The heightening pressure for change by the whites in the traditional conservation strategies was further reflected in new laws passed by the chiefs. The latter began to make special laws about the use of their land and resources, which did not seem necessary before then.

4.2.2 *Middle days of the protectorate and conservation*

The middle days of the Protectorate saw a change of policy from not interfering with the Native Administration to laws that would regulate natural resources conservation on both freehold and communal land. The Chiefs powers as custodian of all natural resources were being gradually eroded. For instance in 1895, the Protectorate Administration through the British South Africa Company instructed Chiefs that they shall have/enjoy hunting rights only on condition that they agreed to observe a close season (Langeni, 1999; Botswana Government, 1990; Botswana Conservation Strategy Agency, 1990). Following this, a series of game laws that had been previously enacted in South Africa were applied to the Protectorate. Examples of such laws are summarized in Table 2.

Table 1. The decrees passed by chiefs as means of natural resources management before and during the Protectorate. Some decrees were a direct influence from the White Settlers and Missionaries.

Year	Decree	Tribe
1815	Ivory the property of the chief	Bangwaketse
1856	Hunting prohibited on Sundays	Bakwena
1877	Capture of young ostriches prohibited	Bangwato
1878	Hunting by European Commercial prohibited, sport hunters permitted on personal application to the chief	Bangwato
1892	Hunting of giraffe and other big game prohibited without permission of the chief	Bakwena
1892	Hunting of ostrich prohibited, but chief gave permission to hunt cock ostriches	Bangwaketse
1893	Hunting of elephant, giraffe, eland and other big game prohibited without special permission of the chief	Bangwaketse
1895	Hunting of giraffe, eland and other big game prohibited without permission of the chief	Bangwato
1898	The use of deadfalls, staked pits and trapping on roads prohibited. Hoofed game to be caught only with jackal (iron) traps	Bangwaketse
1910	Hunting of elephant prohibited without permission of the chief. Hunting of giraffe, buffalo, eland, rhinoceros and hippopotamus prohibited	Batawana
1913	Immigrants obliged to obey the Chief's laws concerning the destruction of game. The killing of white storks and secretary birds prohibited. Hyrax and guinea-fowl are totally protected on Serowe Hill	Bangwato
1920	Elephants to be hunted only with permission of the Chief, and one tusk to be given as tribute	Batawana
1926	Hunting of big game east of the railway line prohibited. Setting traps in other peoples' fields prohibited	Bakgatla
1936	Sale of lion and leopard skins to traders prohibited	Bakwena
1937	Hunting of giraffe and other Royal Game prohibited without the permission of the Chief	Batawana

Generally speaking, during the middle days of the protectorate era, contact with Europeans led to greater skill in the use of land and its resources. For example, underground water supplies were extensively tapped through boreholes.

4.3 *Introduction of novel implements and new techniques improved cultivation methods*

New methods of combating agricultural pests and stock diseases were introduced. Better means of transport and communication notably the railway line was built in 1896–7 (Langeni, 1999; Botswana Government, 1990; Botswana Conservation Strategy Agency, 1990). This gave the natives greater freedom of movement from the restrictions formally imposed by their surroundings. Development of the transport facilities brought about a wider range of goods so that the natives had a choice not to entirely depend on their immediate environment for survival.

About this time, the habit of seeking work outside the territory especially in the mining areas of South Africa began. Standards of living were on the rise. The introduction of horses, guns and iron traps caused extensive destruction of game. Imported goods such as blankets were slowly replacing traditional handicraft products, and metal goods such as sewing machines were becoming very important to the locals.

All these advances were achieved at the expense of overcrowding and over-utilization of the communal areas. The European influence had allowed the natives to acquire the skills and tools to more effectively challenge the limitations of their environments.

Table 2. Game laws passed during the Protectorate and Post-Protectorate Periods as means by the Protectorate Administrators to regulate the hunting and killing of wildlife.

Year	Game law passed during and after the Protectorate Administration	Summary of major functions
1891	The Game Law Amendment Act of 1886	
1891	The Ostrich Export Duty Act of 1884	Imposed a tax on the export of ostriches and their products
1893	1893 Proclamation of the 19th September (Appendix II)	The first Protectorate Proclamation concerning game, but did not repeal the Game Law Amendment Act
1904	The Large Game Preservation Proclamation No. 22	It repealed the 1893 Proclamation
1907	1907 Proclamation No. 2	Instituted Section 11 of the Act of 1886 passed in South Africa, empowering the High Commissioner to protect species in specified areas for up to three years at a time
1907	1907 Proclamation No. 39	Repealed the Ostrich Export Duty Act of 1886 passed in South Africa (making unlawful the export of any ostrich or ostrich egg)
1911	1911 Proclamation No. 42	Instituted a licence to trade in game products. Exemption was made for landowners trading in game products derived from their own land and for tribesmen in tribal areas
1914	1914 Proclamation No. 44	Plumage Birds Protection and Preservation, and it made possible it an offence to trade in, export the plumage of any wild birds
1924	1924 High Commissioner's Notice No. 20	Protected large game for a period of I year in all Crown Land north of the Molopo River.
1925	1925 Proclamation No. 17, the Bechuanaland Protectorate Game Proclamation	It repealed the Act of 1886 and its eleven subsequent active Proclamations (excluding the Plumage Birds Protection and Preservation Proclamation), and made some amendments to existing laws on game
1929	1929 Proclamation No. 48	It placed the burden of proof on the accused to prove that any game in his possession was not hunted in contravention of the law
1930	1930 Proclamation No. 27	Introduced the forfeiture of firearms and ammunition found in the possession of an accused at the time of commission of an offence of unlawful hunting
1932	1932 High Commissioner's Notice No. 53	Instituted an important new protected area, protecting both large and small game for a period of 3 years in an estimated 15,550 km^2 in Chobe District. The area is part of the present Chobe National Park.
1934	1934 Proclamation No. 74 Native Administration Proclamation	Empowered the Native Administration to issue through the chiefs any order thought desirable for the protection and preservation of game
1940	1940 Proclamation No. 19, the Bechuanaland Protectorate Game Proclamation	It repealed Proclamation No. 17 of 1925 and its amendments and introduced new provisions in line with the 1933 London Convention
1940	1940 High Commissioner's Notice No. 42	Extended the area protected under the 1930 Proclamation No. 27 to include the 'whole of the Kgalagadi District'. For the first time native residents could be issued permits to hunt and kill game in reasonable quantities for food in this area.
1940	High Commissioners Notice No. 107	The first Game Reserve was established under Proclamation 19 of 1940 along the Nossop River

(*Continued*)

Table 2. (Continued)

Year	Game law passed during and after the Protectorate Administration	Summary of major functions
1950	1950 High Commissioner's Notice No. 228	Brought into force the Laws of Bechuanaland, game becoming Chapter 114. This law consolidated the amendments enacted since 1940.
1960	1960 High Commissioners Notice No. 65	Established the Chobe Game Reserve, protection of the area (under the 1932 High Commissioner's Notice No. 53) having lapsed since 1943.
1961	1961 Proclamation No. 22, the Fauna Conservation Proclamation	Further and better provision of the conservation and control of the wild animal life and to give effect to the International Convention of 1933 as amended for the protection of the fauna and flora of African continent
1967	1967 Act No. 47, the Fauna Conservation (Amendment) Act	Retained most of Proclamation No. 22 and introduced some principal amendments
1967	1967 Act No. 48, the National Parks Act	Introduced for the establishment of National Parks and for the preservation of wild animal and fish life, vegetation and objects of scientific interest, and to provide for the control and management of such Parks. Chobe was declared a National Park by Act.
1967	1967 Statutory Instrument No. 64	Promulgated the first Tribal Territory hunting regulations for the Bangwato Tribe hunting on the Tribal Territory
1968	1968 Statutory Instrument No. 4	Announced the first Controlled Hunting Areas in the Kweneng District
1968	1968 Statutory Instrument No. 13	Exempted Remote Area Dwellers from the Batawana Tribal Territory Hunting Regulations (S. I. No. 65 of 1967)
1968	Statutory Instrument No. 23	Provided regulations for hunting in Controlled Hunting Areas on Tribal land. These did not apply to member of a tribe hunting in the Tribal Territory of such a Tribe

4.3.1 *Last days of the protectorate and conservation*

In accordance with the limitations of their environment, the pre-protectorate Tswana were primarily huntsmen and pastoralists. The products of cultivation were sparse, undependable and seldom marketable. Arable farming was therefore of secondary importance within their economy (Langeni, 1999; Botswana Government, 1990; Botswana Conservation Strategy Agency, 1990).

The introduction of irrigation techniques and ploughs during the protectorate period formed a potential change in the importance of their traditional ways by making arable farming of primary importance. This was to be expected because simple traditional ways of hunting could not exploit the resources in the same way as in a system where animal husbandry, agriculture and industrial, and mining civilization were important land use components.

Large-scale firearms were introduced in the country in the 1800's, together with increasing penetration of merchant capital from outside. Initially this penetration was in search of cattle but soon the interest focused on the region's resources of ivory and ostrich feathers. This had a lot of consequences for the Tswana's natural fauna. For instance, the firearms and ammunitions led to a decline in availability of game meat as a major source of food.

As more of the pastoral land was cultivated, settlements spread out further in search of pasture. However, the coming of Europeans did not have the same effects upon the Tswana agriculture as it had in other parts of Africa where the native's acquired new crops like cocoa, coffee, and tea on a very large scale for the market. In Bechuanaland, the people still relied mainly upon their traditional products although their yields especially maize improved.

4.4 *The period after independence (after 1966)*

During the protectorate days, population increases for both human and livestock resulted in a diverse array of environmental issues in comparison to the pre-colonial period. As a result of the complex nature of the issues, management strategies were then compartmentalized to focus on specific problems within Ministries and Government Departments. Below follows an account of characteristic features typical of conservation issues for most African countries at independence. Much environmental conservation has been the domain of the state or the government in a relationship of a strong state versus a weak civil society. The state, relatively wealthy and better organized, directed both the content and direction of environmental conservation. Through control of state institutions it has had a relative monopoly over environmental conservation.

4.4.1 *Historical Background of Protected Areas (in the African context)*

The 1960's saw most of the African countries gaining independence. The new governments were inheritors of imperial institutions such as administrative structures, education and others such as natural resources conservation. Of all these institutions, conservation was the least understood by the new governments and the least they were prepared to take over (Langeni, 1999; Botswana Government, 1990). The new Government inherited the protected areas for wildlife and forest reserves that were established during the Protectorate/Colonial days in response to pressure and interest of the Europeans/white settlers. There had been little consultation with the local people in setting up these protected areas. The local people had a negative attitude towards these protected areas as they were set up at the cost of displacing their traditional practices. Thus at independence, most African local people were grateful thinking that the protected areas would once again be opened up and they would be allowed to hunt the old way (Langeni, 1999; Botswana Government, 1990; Botswana Conservation Strategy Agency, 1990).

Most African people preferred wild animal meat to that of livestock. However, the game laws passed for the protected areas during the Protectorate days and inherited by the new Government, restricted local traditional hunting. The people could not understand this situation. All this hardened the local attitudes towards the conservation efforts. In the African context, the protected areas had no support of the former nationalists, many of whom were now rulers, or from the local populations. But despite this indifference, many African states emerging into independence during the 1960's were fast to adopt the preservation of indigenous fauna and flora as a matter of prestige rather than anything else (Langeni, 1999).

It is against this background that the following sketch of the main factors operating in the present situation should be read. Mistakes in management were bound to happen and they did. The new governments in Africa had no trained personnel in this field of protected areas i.e. there were no personnel who had both professional commitment and emotional understanding of what they were expected to do. Thus, while new personnel were being trained, the actual management was being carried out by inexperienced people to whom conservation meant anti-poaching or poaching.

Poaching and illegal trading in game trophies and plant products became epidemic across the continent (Langeni, 1999; Botswana Government, 1990). Worse still, high-ranking officials did all these anti-conservation acts rather than the local people, for whom there would have at least been the excuse of being displaced from their land and restricted for hunting. These practices continued until some animal and plant species were in danger of extinction and it was at this point, during the early 1980's, that the international community put pressure on African governments to correct the situation.

5 CONCLUDING REMARKS

For most of colonial and post-colonial Africa the predominant thinking at some point was that environmental responsibility and development were competing objectives. For the continent to catch up with the developed world there must be some intensive exploitation of resources. Human populations on the continent and indeed globally, now pose the greatest threat to the environment than ever before. The unsustainable consumption patterns of natural resources and the ever-increasing population are putting ever-increasing stress on the land, water and ecological systems. There are serious regional and global environmental problems which require attention and action. There is also consensus that they can no longer be reduced to rich versus poor or north versus south issues but are similar to both and global in nature. From erosion through pollution to the depletion of some resources, the evidence is now in that many of our human activities are currently reducing the long term ability of the natural environment to provide goods and services as well as adversely affecting current and future human health and well-being.

REFERENCES

Atlhopheng, J. et al, 1998, Environmental Issues in Botswana: A Handbook. Lighthouse.

Botswana Government, 1999, Annual Government Economic Report. Government Printer, Gaborone.

Botswana Government, 1990, National Conservation Strategy. Gaborone: Government Printer.

Botswana Government, 1993, Population and Development Issues in Botswana: a National Report for the International Conference on Population and Development.

Botswana Conservation Strategy Agency, 1990, National Policy on Natural Resources Conservation and Development. Gaborone: Government Printer.

Botswana Society, 1976, Proceedings of the Symposium on the Okavango Delta and its Future Utilisation. Gaborone: National Museum.

Central Statistics Office, Population Census Reports for 1971, 1981 and 1991. Ministry of Finance and Development Planning.

Cantrell, M. 1992, Environmental Education in Botswana: A National Planning Workshop.

Colclough and McCarthy 1980, The Political Economy of Botswana: A Study of growth and Distribution. Oxford University Press.

Langeni, T.T. 1999, 'History, Geography and Economy' in Gaisie, S.K. and Majelantle, R.G. Demography of Botswana. Mmegi Publishing House, Gaborone. Pp. 23–32.

Letamo, G. and Totolo, O. 1997 Population Growth and Environment. A Paper presented at a seminar on 'Democracy in Botswana' organised by the Department of Demography University of Botswana. Gaborone Sun 9th–11th December 1997.

Magole, I.L. 1998, 'Tree Planting As a form of Addressing Desertification' in Garret, J. et al. Proceedings of the Environmental Education Association of Southern Africa 1998 Annual Conference: EE in the 21st Century: From Rhetoric to Action. Gaborone, Botswana.

Mordi, A.R. 1991, Attitudes Towards Wildlife in Botswana. New York: Garland Publishing.

Molutsi, P. 1988, 'The State, the Environment and Peasant Consciousness' in Review of African Political Economy 42, Pp. 40–47.

Mosesane, N.E., 1998, 'Botswana National Botanical Garden; A Tool for Environmental Rehabilitation' in Garret, J. et al. Proceedings of the Environmental Educational Association of Southern Africa 1998 Annual Conference: EE in the 21st Century: From Rhetoric to Action.

Mosothwane, M. 1998, 'An investigation of Children's Understanding of Environmental Problems' in Garret, J. et al. Proceedings of the Environmental Educational Association of Southern Africa 1998 Annual Conference: EE in the 21st Century: From Rhetoric to Action. Pp. 175–188.

Ngwane, M. 1998, 'Losing Indigenous Plants and Indigenous Knowledge: Teachers Role In Moving From Rhetoric to Action' in Garret, J. et al. Proceedings of the Environmental Educational Association of Southern Africa 1998 Annual Conference: EE in the 21st Century: From Rhetoric to Action. Pp. 200–210.

Oakes, D. 1998, 'Environmental Education in the 21st Century: The Need for Something Radically New' in Garret, J. et al. Proceedings of the Environmental Educational Association of Southern Africa 1998 Annual Conference: EE in the 21st Century: From Rhetoric to Action. Pp. 292–300.

Rakgoasi, S.D. 1999, 'Population Growth and Composition' in Gaisie, S.K. and Majelantle, R.G. Demography of Botswana. Mmegi Publishing House, Gaborone. Pp. 33–76.

Totolo, O. and Douglas, C. 1998 'Environmental Conservation Education and Action' in Garret, J. et al. Proceedings of the Environmental Educational Association of Southern Africa 1998 Annual Conference: EE in the 21st Century: From Rhetoric to Action. Pp. 233–237.

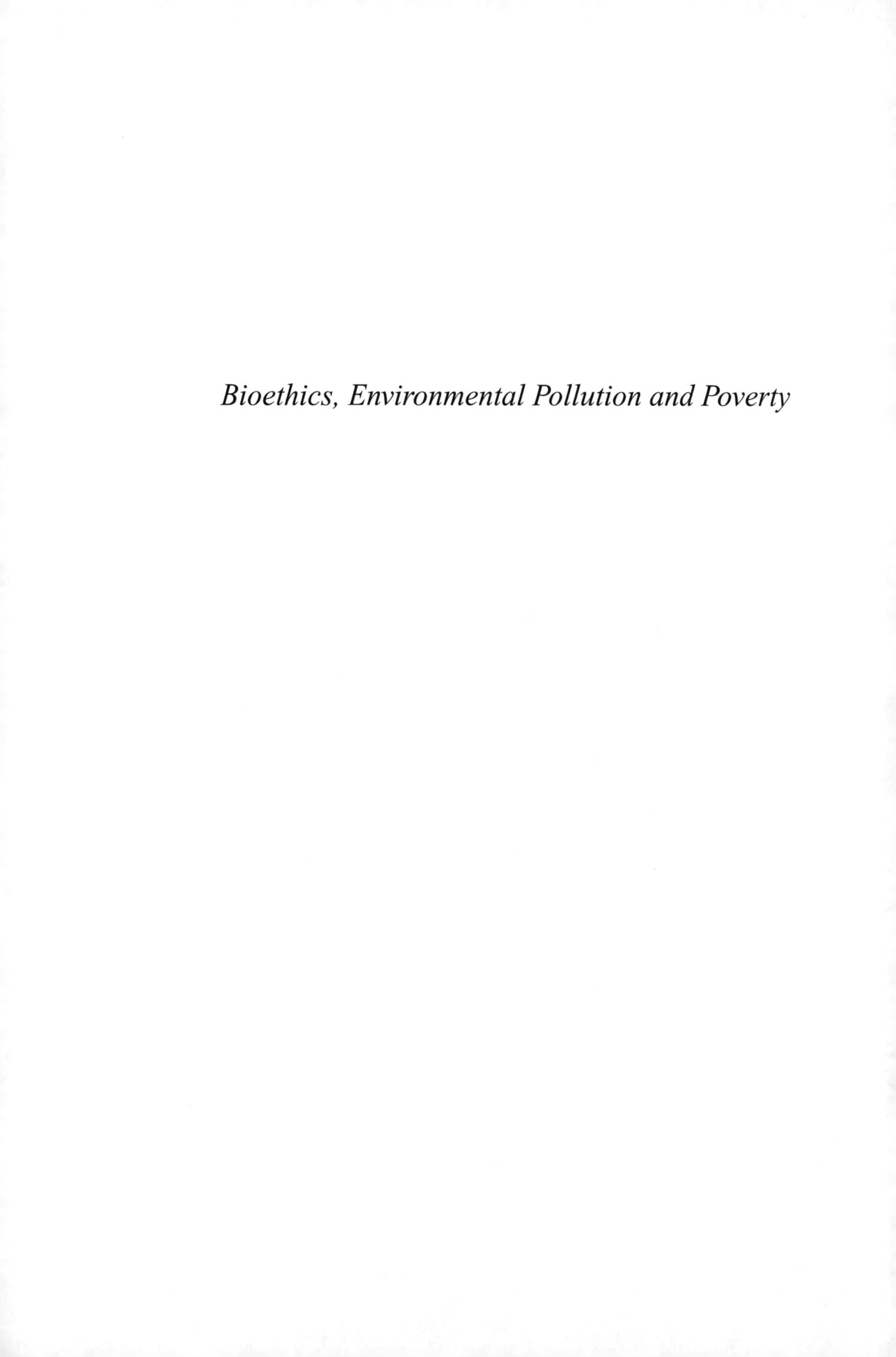

Bioethics, Environmental Pollution and Poverty

Bioethical Dimensions of Sustainable Development

Arnoldo José Gabaldón
Universidad Simón Bolívar, Caracas, Venezuela

Marisol Aguilera M.
Universidad Simón Bolívar, Caracas, Venezuela

SUMMARY: Bioethics is a new subject of study that joins biological knowledge with the awareness of systems related to human values. The historical evolution of the term will be described, starting from the initial scope mainly restricted to the biomedical field, until its widening to enclose the features and problems of life of all species including humans and their relationships with ecosystems. Global bioethics, as it has also been named according to this broader meaning, is linked to the concept of sustainable development. Thus, the latter acquires a bioethical dimension indispensable for the fulfillment of its own purpose. The objective of this chapter is to contribute to the understanding of the relationships between the two important concepts that concern all humankind, i.e. bioethics and sustainable development. In order to appreciate how the former may contribute to the latter some particular situations will be analyzed in three separate ethical dimensions. One, concerning the international relations chiefly regulated by the instruments of international law; another one which concerns the intertemporal relations among human beings, and a third one regarding the relationships among diverse types of living beings. As concluding remarks, some reflections are stated to emphasize the importance of guiding sustainable development with solid ethical values and principles. Also, we consider the role(s) of these in the advancement of science and technology as linked to the lives of humans and all other species. Finally, the role played by bioethics is pointed out, stressing the need for the establishment of education regarding sustainability and the promotion of a culture of harmony.

1 INTRODUCTION

The last century was profuse in discoveries, innovations and advancements in the field of technology. This led to numerous reflections and opinions about the moral limits of humankind's technological capability. In the 70s, two Dutch clinical scientists, Van Rensselaer Potter (an oncologist and biochemist, 1911–2001) and André Hellegers (a human embryology physiologist, 1925–1979) concurred in the need to incorporate ethical principles into the biomedical field as well as into other fields of scientific and technological research. This led to the term bioethics (from the greek <u>Bios</u> = life, and <u>Ethikós</u> = ethics) coined by Potter (1970, 1971) and used in several of his articles. Bioethics has appeared as a new study matter that links biological knowledge with the knowledge of human value systems; the bio to indicate biological knowledge, the science about living systems, and ethics to indicate knowledge about values systems (Ciccone, 2006).

In Potter's line of thought, one can perceive preoccupation about the challenges imposed by knowledge and its applications: The human genus urgently requires wisdom to guide actions, a wise use of knowledge for the welfare and the future of the human condition, a science for survival, that is, bioethics which has the fundamental requisite of promoting the quality of life. It was believed that this science for survival ought to be constructed upon the biological sciences albeit reaching beyond its traditional boundaries to include the basal elements of the social and anthropological sciences, with particular reference to philosophy in its strict sense of love of wisdom.

Today we have diverse definitions of bioethics that have surpassed the field of biomedicine, such as that expressed by the philosopher Camps who noted that bioethics is an interdisciplinary reflection developing from the concourse of several disciplines: philosophy, the health sciences, law and social sciences (Ferrer and Alvarez, 2003). Any discipline that deals with the study of human beings has had something to contribute to the ethics of life. Also, it has to be pointed out that, at present, a widening of the concept of bioethics can be recognized, based on the Potterian ideals, and which has been named global bioethics (Ferrer and Alvarez, 2003). Hence, it may be stated that its scope also includes the public and sanitary policies, the tasks of all professions, and especially any research on humans or the flora and fauna that constitute ecosystems. This wider view of bioethics goes hand in hand with the concepts of sustainable development, which will be considered again later on.

In its development, bioethics, seen as a branch of philosophy concerned with the study of the morality of human works (Andorno, 1998), incorporated principles that belonged to medical practice (beneficence and no-maleficence, to jurisprudence (autonomy) and to political tradition (justice), as well as to the precepts of human rights, the development of professional ethics, and from Christianity and the Catholic Church, among others (Gracia, 1989; Abellán Salort, 2006).

Currently, bioethics encompasses several guiding principles (Código de Bioética y Bioseguridad, 2003):

- Autonomy, which is the right of each person to decide over the matters that affect her or his life, directly or indirectly.
- Beneficence, which consist in doing good and avoid inflicting harm or damage.
- No-maleficence (*primun non nocere*), which specifically means not to harm and avoid behaviors that may harm or damage oneself, other living beings or the environment.
- Justice, as the rights to equity or equality of opportunities, giving each one her or his rightful share.
- Responsibility, understood as the commitment with life by assuming behaviors that favor it, and
- Precautionary principle, which pertains to the adoption of an excess of safety in case of doubt and/or lack of information or scientific understanding about the environmental consequences: the lesser the knowledge the greater any caution(s)

According to the United Nations, Rio Declaration (1992a, 1992b), in order to protect the environment, the precautionary approach should be widely applied by States according to their capabilities. Where there are threats of serious or irreversible damage, lack of full scientific certainty should not be used as a reason for postponing cost-effective measures to prevent environmental degradation (United Nations, 1948, 1966, 1992).

The progress of bioethics has allowed it to overcome the principles, and today it is accepted that the most convenient way is to adopt an attitude of reflection with regard to any human conduct that may affect the environment, biodiversity (including humans), as well as the scientific-technological activities and products, factories, agro-industry, health and coexistence in general, among other aspects (Código de Bioética y Bioseguridad, 2003). This view allows the establishment of value-judgments in order to decide as a whole and through dialogue, if a given behavior matches to what should be or not.

By the end of 1983, the United Nations General Assembly, concerned about the intense process of ecological degradation and the increase in poverty around the world, decided to form the World Commission on the Environment and Development. The general purpose of this Commission was to formulate a Global Agenda for Change, within the context of the critical situations that were occurring at the interface between the environment and development. Specifically, the purpose was to propose a long-range environmental strategy to accomplish sustainable development. This Commission, presided over by Gro Harlem Brundtland, Prime Minister of Norway, presented its Final Report in 1987, under the title Our Common Future (The World Commission on Environment and Development, 1987). It was the product of a profound search that collected the opinions of Heads of State, experts from the academic world, civil servants and employees, young people, inhabitants of slums, farmers, factory-owners, and other exponents of the civil society of many countries. The Report was comprised of three parts: concerns, challenges and common efforts. It is important to note that among the concerns that were highlighted included poverty, the deepening of the gap between rich and poor countries, the survival of biological life, economic growth as well as that of the population, and the world's energy crisis (Aguilera, 1992).

From an analysis of the problems emerged the necessity of a form of development that could carry the adjective of sustained or sustainable, defined as development that meets the needs of the present without compromising the ability of future generations to meet their own needs.

Among other lines of thought, in the context of the multiple relationships between human progress and the state of the environment, it becomes evident that the evolution of humankind and its interaction with the natural physical surroundings has not advanced in step with scientific and technological developments. This inequality forces us to think about the causes of such separation if we wish to attain sustainable development as defined above. There are diverse causes that have led to and maintain that gap. One that can be immediately identified is the lack of political willingness and of ethical principles regarding the significance of biological life, in the leaders that control development. This can be ascribed in general, to the lack of understanding of the meaning of development that qualifies as sustainable, and to the scarcity of education regarding this concept.

The goal of this chapter is to contribute to the understanding of the link between the two important concepts that pertain to all of humanity, one which refers to facts and reality (sustainable development), and another that ponders about facts and values and establishes judgments within the area of knowledge which has become known as bioethics. Sustainable development, in view of all its inherent changes including individual and collective behaviors, is not viable if a wide ethical base is not attained within societies. The implementation of bioethics appears to be the best way to achieve such a purpose.

How can bioethics contribute to sustainable development? In order to answer this question, we shall adopt the levels of analysis pointed out by Marcos (2001), for the detection of problems that usually emerge and the taking of decisions to solve them. This author suggested placing those levels in three new ethical dimensions: one that considers the supranational relations, another pertaining to human generations separated in time (the temporal concept within sustainable development) and the last one regarding relationships between different types of living beings.

2 FRAMEWORK OF SUPRANATIONAL RELATIONS

Contemporary society, every day more interconnected, contains institutions and events that have brought about a series of statements and agreements that led to international commitments of various dates and scopes, which in turn act as guidelines for a better society. In this supranational context, the leading document is the Universal Declaration of Human Rights (UDHR) proclaimed by the United Nations General Assembly in Paris on December 1948. This is regarded as the global ethics code par excellence. In its 30 Articles, it refers to the rights of human beings, of which stand out the rights of life, freedom, equality, work, social security and assistance (health, housing and public services) and education. International Covenant on Economic, Social and Cultural Rights and the International Covenant on Civil and Political Rights have constantly upheld these rights, both since 1966.

The Convention on Biological Diversity (United Nations, 1992a) was approved during the United Nations Conference on Environment and Development in Rio de Janeiro, Brazil. Its general objectives were the preservation of biological diversity, the sustainable use of its components and a fair and equitable sharing of the benefits derived from the use of genetic resources. The document stressed as priorities economic and social development, the eradication of poverty and the sustainable use of biological resources and diversity in order to satisfy the needs of food, and health of all the world's population.

The United Nations Framework Convention on Climate Change also approved during the Rio de Janeiro Conference, the stabilization of greenhouse gas concentrations in the atmosphere at a level that would prevent dangerous anthropogenic interference with the climate system. Such a level should be achieved within a time frame sufficient to allow ecosystems to adapt naturally to climate change, to ensure that food production is not threatened and to enable economic development to proceed in a sustainable manner. Indeed, the Convention identified as adverse effects the changes in the physical or biotic environments resulting from climatic changes that cause significant negative effects on the composition, the capability for recovery or the productivity of natural ecosystems or those subjected to controlled use, or on the functioning of socioeconomic systems, or health and human welfare.

The 21st century began with two vigorous proposals. The first one was the Earth Charter (ECI, 2000), which was proposed by the Earth Council, presided by Maurice Strong, and the International Green Cross, led by Mikhail Gorbachov. It was aimed at expressing the fundamental principles for sustainable development. The Earth Charter, which was supported by UNESCO as an educational instrument, contained a synthesis of values, principles and aspirations regarding the respect and care of the community of life, ecological integrity, social and economic justice, and democracy, non-violence and peace. This is a universal code of conduct based on fundamental values and principles, a declaration of interdependence and responsibility, and a participatory process involving civil society to reflect on their decisions.

The second proposal, sponsored by the United Nations, was named the Millennium Declaration (United Nations, 2000). It reiterated the goals concerning eradication of poverty, the increase of development, the decrease of diseases, and the reduction of injustice, inequality, terrorism and delinquency, and the protection of the environment. The leaders of the world agreed on these objectives.

The General Conference of UNESCO (2005) adopted the Universal Declaration on Bioethics and Human Rights which dealt with ethical matters related to medicine, the life sciences and connected technologies applied to human beings, taking into account their social, legal and environmental dimensions. This Declaration offered a frame of principles and procedures for the formulation of policies, legislation and ethical codes. It must be pointed out that it refered to social responsibility (Article 14), and it particularly stressed that the progress of science and technology should promote the well-being of all people and the human species, favoring especially access to high-quality medical attention, to essential medicines, food, and adequate water supplies. Likewise, it emphasized the preservation and protection of the environment, the biosphere and biodiversity.

It is evident that these ideas and reflections expressed in the Declaration and the participation of many people, have contributed to spreading concern about the deficiencies and problems confronted by the world's societies. In particular there is a need for actions to achieve respect for the right to survival of all species, the rights of the human population to equity in the distribution of benefits derived from development regarding food, clean water, health, education and housing, as well as the benefits from science and technology and the prevention of their social or ecological risks.

International guidelines have been reviewed pertaining both to bioethics and sustainable development. They have contributed to some of the improvements that can be seen worldwide, as well as to the necessary thinking demanded by society at present. However, if we analyze the Global Environmental Outlook (GEO, 2000) of the United Nations Environment Program (UNEP, 2000), developed by some 850 people from diverse places of the world, it becomes clear that the main environmental problems show high diversity and magnitude. Many of them have become aggravated instead of being diminished over the last few decades. The GEO report lists 36 problems that go from greater to lesser importance according to the percentage of interviewed people who referred to them. The first 10 problems listed are climate change, scarcity of fresh water, deforestation and desertification, freshwater pollution, a deficient public administration, loss of biodiversity, the growth and movements of human populations, changes in social values, disposal of wastes and air pollution.

Analyses of all the problems as a whole allowed for several recommendations: I) The need to understand the interactions and consequences of worldwide and interregional processes, as well as to discern whether the new environmental policies and the budgets assigned for the environment attain the desired results; II) Establish the fundamental causes of environmental problems that should be confronted, as many of them are not strictly of environmental origin, for instance the massive consumption of resources, which renders it necessary to modify the values of society in order to diminish such consumption; III) To use an integrated criterion that allows the linking of environmental matters to the prevailing ways of thought; and IV) Mobilizing efforts for solving problems, as a cooperative effort, is indispensable, involving individuals, non-governmental organizations, local and national governments, and international organizations and institutions.

It suffices to examine the Indices of Environmental Performance for 2008 published by the Universities ofYale and Columbia to infer that the environmental radiography of the world depicted by GEO 2000 is not very different from what exists today and that their recommendations are still valid (http//epi.yale.edu).

3 THE WELFARE OF FUTURE GENERATIONS

A central objective of sustainable development lies in satisfying the needs and requirements of future generations, which implies the commitment to maintain the equilibria and potentials of the diverse ecosystems on the planet. This is the second level of analysis, according to Marcos (2001). For some authors, this contains the precise ecological meaning of any form of development that may be regarded as sustainable. Beyond the requirements of the present time, with its character of immediacy from the dialectics of conventional development, humans have a bioethical commitment with future generations, which we must attend to by taking the necessary steps to offer them a better quality of life and not the opposite, as is regrettably happening in many places around the world.

The decisions inherent to development between the present and the future cannot be governed solely by considerations of profit, as is often the case under prevailing economic systems. Ecosystems constitute living complexes that do not know economics and for which it is very difficult to judge their economic value. Hence, it becomes necessary to impose a bioethical viewpoint on the patterns of preferences and public use of ecosystems, so that any decisions taken focus on the welfare of future generations. Likewise, it becomes essential to plan or program on short- or long-term basis, also including the ecological aspects.

4 OUR RELATIONSHIP WITH OTHERS

The third level of analysis is represented by the interactions among living beings. We believe that this is a domain in which efforts should be emphasized. The relationships of humans with their surroundings and especially with other living beings have generated diverse lines of thought that may be cataloged within what today is regarded as environmental ethics. Marcos (2001) has grouped the different viewpoints of the human-nature relationship into: anthropocentrism; ecological viewpoints (biocentrism, ecocentrism, earth ethics and deep ecology); social ecology and ecofeminism, and humanism (utilitarianism, ethics of responsibility, Christian ethics and Aristotelian environmental ethics). We do not pretend to place ourselves in any of these lines of thought; on the contrary, discarding the more extreme positions (anthropocentrism, biocentrism, ecofeminism and utilitarianism) we propose to rescue the best, in our judgment, from some of them as further criteria for the elaboration of a scale of values that guide human conduct. Within line of ecological thought, we hold the position that all living beings possess a moral status and therefore they deserve consideration and respect. Likewise, we strongly support the attention to ecosystems, that complexity of structures and functions that contains so many direct and indirect values.

From social ecology, we back the need for changes that include as primary reference human rights and the socio-ecological relationships, as postulated by sustainable development. From humanism, we support as fundamental the capability to meditate about ethical principles and values. This is why the study of Potter (1971) serves as a guide in this debate: There are two cultures -science and humanities- that seem to be unable to speak to each other. If this were the reason for the uncertain future of humanity, then possibly a bridge could be built to the future constructing the discipline of bioethics as a link between the two cultures. The ethical values cannot be separated from the biological facts.

How can we contribute in a practical way to build bridges between the sciences and the humanities? There are two ways, one that has to do with every individual person, and another that refers to the social collective. Regarding the first one, our daily life should be expanded consciously keeping in mind to ponder about facts, commitments, dialogues and debates, respect for the plurality of thought and the necessity of knowledge. Complementarily, as part of any community, good use should be made of the scenarios for reflection and for decision making, in different spaces. For some people this possibility arises from the participation in independent bioethical committees, all multidisciplinary and pluralistic.

Freedom represents the fundamental right of all humankind. The democratic system of government can be viewed as one that has the best conditions to ensure freedom; that it is highly convenient to intensify international technical and financial cooperation as an expression of global justice.

Sustainable development requires the building of a widely shared ethical doctrine that is ethics based on a commitment to life and the welfare of humankind and all other species that constitute the biosphere (Gabaldón, 2006). Such a doctrine represents its bioethical dimension, and development thus qualified relies on the fundamental pillars of science and technology to attain its goals. However, management of those two branches poses many questions of ethical character, above all whenever the life of human beings becomes threatened.

5 CONCLUDING REMARKS

It is essential to apply a frame of values and principles that guide the conduct of decision-makers, scientists, and technologists. Bioethics, besides orienting and setting boundaries to the multiple decisions and imponderables implicit in developments that may affect life on the planet, ought to serve as a platform for education about sustainability. Education for sustainable development should be obligatory for all young people, as these represent the primary vehicle available to catalyze the cultural changes required for a bioethics that grants a privilege to survival.

The advances of international environmental jurisprudence also lend a valuable arsenal of bioethical principles to the guidance of world development. Therefore, the developed and developing countries are morally committed in their construction, refinement and application. Finally, we offer a comment about the importance of peace. There is no worse aggression against life than the violence unleashed by warfare, either domestic or international. Thus, we all are obligated, from the ethical perspective of sustainability, to promote a culture of peace.

REFERENCES

Abellán Salort, J.C. 2006. *Bioética, Autonomía y Libertad*. Fundación Universitaria Española, Madrid, España.

Aguilera, M. 1992. Ambiente, Desarrollo y Reforma del Estado. In *Ciencia y Tecnología en Venezuela*. Ediciones de la COPRE. Caracas, Venezuela. Vol. 12:269–280.

Andorno, R. 1998. *Bioética y Dignidad de la Persona*. Editorial Tecnoc (Grupo Abaya, S. A.). Madrid. España.

Ciccone, L. 2006. *Bioética. Historia. Principios. Cuestiones*. Ediciones Palabra (2nd. Edition), Madrid, España.

Código de Bioética y Bioseguridad. 2003. Fondo Nacional de Ciencia y Tecnología (Fonacit), Caracas, Venezuela.

ECI. 2000. The Earth Charter.

Ferrer, J.J. y Álvarez, J.C. 2003. *Para Fundamentar la Bioética*. Universidad Pontificia Comillas y Editorial Desclée De Brouwer, S. A. Bilbao, España.

Gabaldón, A.J. 2006. *Desarrollo Sustentable. La salida de América Latina*. Grijalbo, Caracas, Venezuela.

Gracia G., D. 1989. *Fundamentos de Bioética*. Eudema, Madrid, España.

Marcos, A. 2001. *Ética Ambiental*. Secretariado de Publicaciones e Intercambio Editorial, Universidad de Valladolid. Valladolid, España.

Potter, V. R. 1970. Bioethics: the science of survival. *Perspectives in Biology and Medicine*, 14:120–153. University of Wisconsin, USA.

Potter, V. R. 1971. *Bioethics: Bridge to the Future*. Editorial Prentice-Hall. Engelwood Cliffs, USA.

The World Commission on Environment and Development. 1987. *Our Common Future*. Oxford University Press, Oxford-New York.
UNEP. 2000. Global Environmental Outlook (GEO 2000). Earthscan. Earthscan Publications Ltd, London.
UNESCO. 2005. Universal Declaration on Bioethics and Human Rights.
United Nations. 1948. Universal Declaration of Human Rights.
United Nations. 1966. International Covenant on Economic, Social and Cultural Rights.
United Nations. 1992a. Convention on Biological Diversity.
United Nations. 1992b. Framework Convention on Climate Change.
United Nations. 2000. Millennium Declaration.

The Impact of Global Warming and Human Activities on Coral Reefs: How Much Are We Willing to Pay?

Aldo Cróquer
Laboratorio de Comunidades Marinas, Universidad Simon Bolivar Apdo. 8900, Caracas 1080-A, Venezuela

SUMMARY: Over the past few centuries, human population has expanded to reach an unprecedented size. It is believed that industrialization is rapidly changing the climate system on a planetary scale with serious ecological consequences. Coral reefs are particularly vulnerable to global warming as temperature and other climate-related variables are important factors controlling the distribution, structure and function of these ecosystems. In view of the capital importance of these reefs for human societies, we will review how global warming is affecting coral reef ecosystem function and resilience. We are particularly interested in analyzing what has changed in our societies compared to ancient human groups, how these changes are jeopardizing the future of coral reefs and to what extent coral reefs will be able to respond to climate change. The controversial issue of artificial reef restoration will also be discussed in the context of imminent coral reef decline. The chapter shall conclude by describing and listing the wide variety of economic goods and services that coral reefs provide to human societies, and discussing why we cannot afford to loose coral reefs and what we should do to prevent and/or slow down coral reef decline.

1 INTRODUCTION

Human population has expanded over the past few centuries, reaching an unprecedented size during our times. Earth's population grew about 10-fold from 600 million people in 1700 to 9.3 billion in 2007 (Cohen, 2003; United States Census Bureau, 2008); by 2050, 2 to 4 billion more people are likely to inhabit the planet (Cohen, 2003). The rapid advance and expansion of technology in the field of medicine (e.g. discovery of antibiotics and development of surgery and therapeutic treatments) and agriculture (e.g. development of systems for mass production of food and other goods) has improved our quality of life; increasing the mean life expectancy for humans and therefore the rate of population growth.

Paradoxically, the quest for a better quality of life for humans and the concomitant rise of industrialized economies designed to satisfy the increasing demand for goods and services, has produced a myriad of environmental impacts on both terrestrial and aquatic ecosystems, jeopardizing the quality of life for future human societies (Beznosov and Suzdaleva, 2004). More important is the fact that industrialization is rapidly altering the climate system, water supply and other biogeochemical cycles on a planetary scale (Becker *et al.*, 1999; Vorosmarty *et al.*, 2000).

Climate change (i.e., any change in climate over time, whether due to natural variability or as a result of human activity, [IPCC 2007]) is a global problem related with rapid human expansion. During the past few decades, the public debate about the effect of greenhouse gas emissions on climate and the concomitant cascade effects recorded in different ecosystems have gained special attention (Harley *et al.*, 2006; Vorosmarty *et al.*, 2008). Amongst all ecosystems affected by climate change, coral reefs are particularly vulnerable, as temperature and other climate-related variables are important factors controlling the distribution, structure and function of these ecosystems.

Coral reefs are important for humans as they provide societies with countless ecological good and services (Moberg and Folke, 1999). In view of the importance of coral reefs for human societies, we will review how global warming is affecting coral reef ecosystem function and resilience, i.e., the capacity of an ecosystem to maintain and recover its function in the face of human disturbance (Bellwood *et al.*, 2004; Nyström, 2006). Of particular interest is how our societies have changed compared to ancient human groups, how these changes are threatening the future of coral reefs and to what extent coral reefs will be able to respond to climate change. The controversial issue of artificial reef restoration will also be discussed in the context of imminent coral reef decline. The chapter shall conclude by describing and listing the wide variety of economic goods and services that coral reefs provide to human societies, arguing why we cannot afford to loose coral reefs and discussing what we should do to prevent and/or slow down coral reef decline.

2 CLIMATE CHANGE AND HUMAN SOCIETIES: WHAT IS NEW?

Paleoclimate analyses of pollen (Andreev and Klimanov, 2000; Tarasov *et al.*, 2005; Wu *et al.*, 2007), ice (IPCC, 2007), sediment (Battarbee *et al.*, 2002; Drescher-Schneider *et al.*, 2007) and coral cores (Linsley *et al.*, 2004) performed at different geographical locations indicate that our planet has been characterized by a dynamic climate system (Kuznetsov, 2005). Climate change is a normal process and may occur abruptly or gradually when triggering factors drive the climate towards particular thresholds (Alley *et al.*, 2003).

Climate change can be triggered by the natural variation of orbital forces (Alley *et al.*, 2003). Attention has recently focused on the possibility of solar forcing contributing to abrupt climate change (e.g. moderate climate oscillations during the Holocene, such as the Little Ice Age) (Bond *et al.*, 2001). Other forces acting at smaller scales such as amplifiers and sources of persistence are abundant in the climate system and can produce large changes with minimal forcing (Alley *et al.*, 2003). Persistence of climate patterns arises from the wind-driven circulation of the oceans, stratospheric circulation and related chemistry (Hartmann *et al.*, 2000). Whereas triggers, amplifiers, and sources of persistence are easily identified, globalizers that spread anomalies across the whole planet are less obvious (Alley *et al.*, 2003). While solar forcing is recognized as a force capable of producing global climate change, the rapid increase of greenhouse gases that began with the industrial revolution and has continued to reach unprecedented levels during our times, is for many scientists the primary cause of recent global warming (e.g. Hoegh-Guldberg *et al.*, 2007; IPCC, 2007).

Several factors are relevant in determining whether a society prospers or collapses, the combination of environmental impacts, climate change and how each society responds to its particular set of environmental problems being particularly important (Diamond, 2005). Contrary to popular belief, many ancient human societies overexploited their natural resources and produced dramatic environmental impacts (Gillson and Willis 2004; Diamond, 2005). Environmental and ecological disasters produced by ancient people have led to the collapse of entire civilizations (e.g. the Mayan empire, the Anasazi and the Pacific Islanders of Pascua, Pitcairn and Henderson), while well-planned natural resource exploitation has allowed other ancient societies to continue prospering even until our times (e.g. New Guinea Highlanders, Tikopia, Tokugawa and Japan) (Diamond, 2005).

Modern societies tend to be larger, grow faster and have more advanced technology than ancient ones, making it possible to produce environmental changes at larger spatial scales (Diamond, 2005). Furthermore, contemporary human societies have enormous evolutionary consequences and can greatly accelerate evolutionary change in different organisms, especially in pathogens, agricultural pests, commensals, and species that are commercially hunted (Palumby, 2001). Thus, identifying what our modern societies have in common with those that collapsed in the past is an important step in avoiding the same mistakes and taking corrective measures in time. In this regard, modern societies have great advantages over ancient ones; we now have the ability to learn from the past, we have the capacity to create new environmentally friendly technology, we have the capability to communicate on a global basis, and, since 1962, we have been developing environmental awareness, seeding this issue inside the collective mind (Diamond, 2005). Today, we also have pro-environment NGOs, environmentally-focused political parties and an increasing number of individuals working towards conservation. All these are good reasons to be hopeful for the future of our planet.

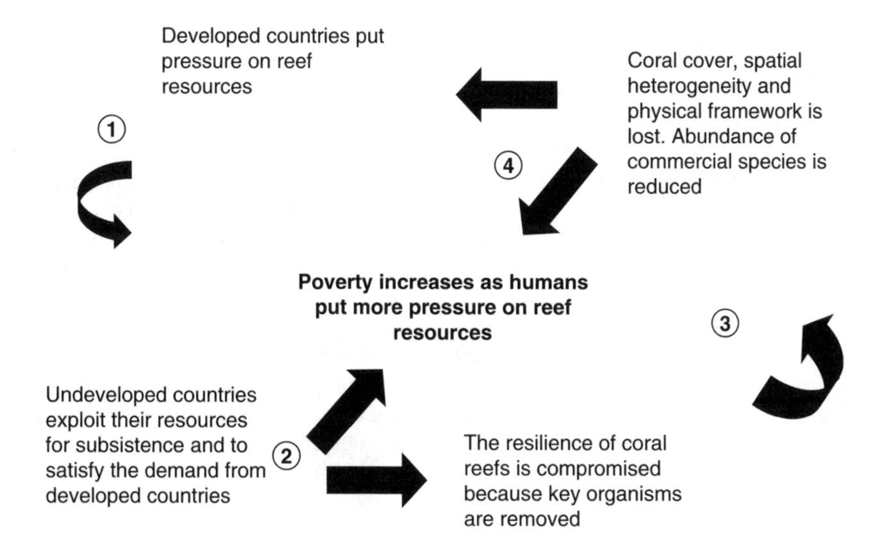

Figure 1. The negative spiral between the excessive pressure of reef resources, coral reef decline and poverty in undeveloped countries.

In short, climate changes have occurred throughout Earth's history when the Earth system had been forced across thresholds; however, the current rate of change is unprecedented. As for humans, two major characteristics have changed in our contemporaneous societies. One is the problem of large-scale overpopulation which is a consequence of an ever-increasing rate of human population growth. The other is the rapid development and expansion of technology, which combined with the problem of overpopulation results in the overexploitation of ecosystems at larger spatial scales due to the increasing demand of natural resources and more efficient modes of production (e.g. Harte, 2007).

Coral reefs are particularly sensitive to these influences, as they mainly occur in tropical latitudes where human populations are rapidly expanding; both developing and developed countries have increased their demand on coral reef resources for different purposes. People from poor countries sell their reef resources to developed countries enhancing reef decline; this in turn, creates more poverty and dependence on large economies (Figure 1).

3 A PLANET WITH A CHANGING CLIMATE: WHAT ARE THE CONSEQUENCES FOR CORAL REEF ECOSYSTEMS?

Global warming is unequivocal, as is now evident from observations of global increases in average air and ocean temperatures, widespread melting of snow and ice, and the rising of mean global sea level (IPCC, 2007). These include changes in Arctic temperatures and ice glaciers, widespread changes in precipitation, ocean salinity, and wind patterns, and an increase in extreme weather conditions including droughts, heavy precipitation, heat waves and tropical cyclones (IPCC, 2007).

Coral reefs are complex ecosystems largely affected by meteorological variables, ocean chemistry and physical oceanographic processes which are susceptible to climate change (Hoegh-Guldberg, 1999). Despite these ecosystems occupy less than 0.09% of the ocean's floor (Spalding and Grenfell, 1997); they harvest by far the greatest number phyla per hectare (Birkeland, 1997) and approximately a quarter of all fish species (Spalding and Grenfell, 1997). This high diversity results in a wide variety of functional groups (i.e. different organisms that perform the same ecosystem function regardless of their taxonomic affinities) that provide coral reefs with functional redundancy (i.e. the capacity of one species to compensate for the loss of another), and therefore, with high resilience (Hughes *et al.*, 2003; Bellwood *et al.*, 2004; Nyström, 2006). Functional groups and redundancy are important for resilience because organisms complement each other while keeping the structure and function of the ecosystem relatively stable over large temporal scales (Pandolfi and Jackson, 2006). Nevertheless, despite of their high functional redundancy coral reefs have been shown to change towards non-desired alternative stages relatively easy (Bellwood *et al.*, 2003).

Thus, although coral reefs can thrive with disturbances at different spatial and temporal scales but their resilience is limited specially when stressing factors are intense and occur to fast. The ability of coral reefs to recover from disturbances depends in part on the type of disturbance (i.e., frequency and magnitude, Connell, 1997) and on life history traits of organisms subjected to disturbances (Hughes and Tanner, 2000). The rapid change of earth's climate is likely to affect coral reef ecosystem function and resilience in at least 5 ways: (1) by decreasing reef accretion and enhancing bioerosion (Kleypas *et al.*, 1999; Hoegh-Guldberg *et al.*, 2007), (2) by narrowing physiological tolerance of key organisms (i.e. coral bleaching) (Hoegh-Guldberg 1999), (3) by promoting disease epizootics that deplete populations of corals and other reef organisms (Harvell *et al.*, 1999, 2002), (4) by altering dispersion capabilities and connectivity of reef organisms (Ayre *et al.*, 2004) and (5) by increasing the frequency and intensity of hurricanes and cyclones (Goldenberg *et al.*, 2001), forces capable of destroying the reef framework relatively easy.

4 GLOBAL WARMING AND BIOEROSION

Current levels of CO_2 (i.e. 380 ppm) in the atmosphere exceed the highest values recorded over the past 740,000 if not 20 million years (Hoegh-Guldberg *et al.*, 2007). During the 20th century, increasing atmospheric CO_2 has depleted seawater carbonate concentrations by $\sim$30 mmol kg^{-1} and seawater acidity by 0.1 pH unit (IPCC 2007). Approximately 25% (2.2 Pg C year^{-1}) of the CO_2 currently produced by all anthropogenic sources (9.1 Pg C year^{-1}) enters the ocean (Canadell *et al.*, 2007), where it reacts with water to produce carbonic acid (Fig. 2). Carbonic acid dissociates to form bicarbonate ions and protons, which in turn react with carbonate ions to produce more bicarbonate ions, reducing the availability of carbonate to biological systems (Figure 2).

Decreasing carbonate-ion concentrations reduces the rate of calcification of marine organisms such as reef-building corals, ultimately favoring erosion and decreasing the accretion of the reef framework (Hoegh-Guldberg *et al.*, 2007). Reef accretion may also be affected by the rapid sea level rise ($\sim$17 cm over the 20th century) that results from global warming (Hoegh-Guldberg *et al.*, 2007) because calcification depends in part on temperature (Marshal and Clode, 2004) and light availability (Barnes and Crossland, 1980; Yap and Gomez, 1984); it is unclear whether corals will keep up with the current rise of sea level (Hoegh-Guldberg *et al.*, 2007).

5 GLOBAL WARMING AND PHYSIOLOGICAL TOLERANCE OF CORALS

Current levels of CO_2 are also producing a rapid increase ($\sim$0.74°C over the 20th century) of the average temperature of the oceans, which is affecting the physiology of corals. Temperature anomalies challenge the stability of the mutualistic symbiosis between corals and their unicellular algae (zooxanthellae) leading to bleaching (i.e. expulsion of zooxanthellae from coral tissues and breakdown of the coral-algae symbiosis). During the past few decades, bleaching events, caused by more prolonged and severe sea surface temperature anomalies (e.g. El Nino Southern Oscillation), have become a serious threat to coral reef health world wide.

A variety of mechanisms are involved in coral bleaching: *in-situ* degradation of zooxanthellae followed by expulsion; exocytocis (i.e. zooxantellae move to apex of the cell to be released); apoptosis (i.e. programmed cell death); necrosis (i.e. cell death) and; cell host detachment (zooxantellae are expelled together with the animal cell during extreme environmental stress) (Lesser, 2004). Coral bleaching produces a variety of negative consequences for coral reefs at different levels of organization. Zooxanthellae are important for several fundamental processes in corals

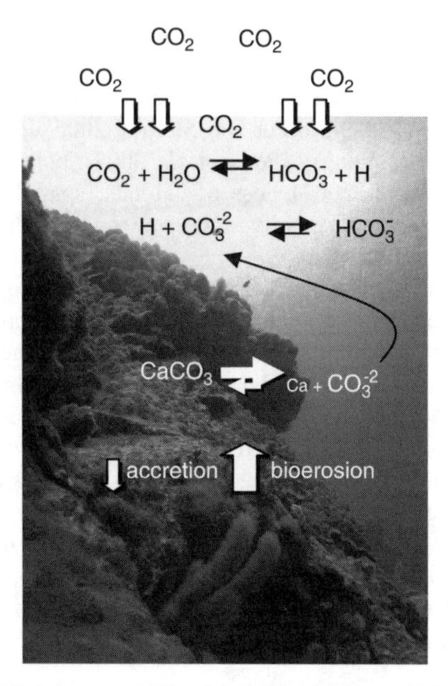

Figure 2. Effect of ocean acidification on calcification, reef accretion and bioerosion (modified from Hoegh-Guldberg *et al.*, 2007).

(Muscatine, 1990). Thus the loss of zooxanthellae due to bleaching may reduce calcification and growth (Mendez and Woodley, 2002), compromise coral's energy budget (Anthony *et al.*, 2007), decrease reproductive output (Szmant and Grassman, 1990; Ward *et al.*, 2000; Baird and Marshall, 2002; Mendez and Woodley, 2002), increase susceptibility to disease (Muller *et al.*, 2008); and in severe cases lead to death (Lesser, 2004). At the population level, populations of corals which are highly vulnerable to bleaching (e.g. branching corals such as acroporids among others) might be reduced to critical numbers which would decrease the number of larvae available for recruitment (Hoegh-Guldberg, 1999). Coral bleaching also has significant impacts at the community and ecosystem levels [i.e. loss of coral cover, habitats, spatial heterogeneity and fish abundance (Brown, 1997; Roessig *et al.*, 2004; McClanahan *et al.*, 2007; Smith *et al.*, 2008); and may promote phase shifts from coral-dominated to algal dominated states (Hoegh-Guldberg, 1999)].

6 GLOBAL WARMING AND DISEASES OF REEF ORGANISMS

A cursory look at the geography and history of infectious diseases shows that they are prevalent in places where the weather is hot and wet (e.g. WHO, 2005). Climate change is most likely to affect free-living, intermediate, or vector stages of pathogens infecting humans and terrestrial animals (Harvell *et al.*, 2002; Derraik and Slaney, 2007); because rising temperatures affects vector distribution, parasite development, and transmission rates (Kovats *et al.*, 2001). Many vector-transmitted diseases are climate limited because parasites cannot complete development before the vectors die (Harvell *et al.*, 2002). If temperatures increase, parasites might expand their range and may reach more hosts. Thus, in a scenario of global warming, one would expect to see infectious diseases spread into new regions and perhaps intensify (McMichael *et al.*, 2001).

In marine ecosystems, compelling evidence demonstrates a link between changing ocean temperatures and pathogens, including vectors of human diseases such as cholera (Colwell, 1996; Pascual *et al.*, 2000) and emerging coral pathogens (Harvell *et al.*, 2002). Coral diseases have been highlighted as one of the primary causes of recent coral reef decline (Weil, 2004). While the Caribbean is viewed as a hot-spot for coral diseases because of its relatively small size and the large number of emergent diseases and epizootic events recorded over the last three decades (Weil, 2004), coral diseases appear to be increasing in other bioregions as well (Willis *et al.*, 2004). Coral diseases have been shown to produce multiple negative effects on coral reef ecosystems [e.g. rapid tissue mortality (Nughes, 2002; Borger and Steiner, 2005; Bruckner and Bruckner, 2006; Ainsworth *et al.*, 2007), loss of fecundity (Petes *et al.*, 2003), population depletion of important reef-builders (Edmunds and Elahi, 2007), loss of habitats, spatial heterogeneity and biodiversity (Aronson and Precht, 2001), phase shifts and species replacements (Aronson *et al.*, 1998, 2003, 2004)].

The rapid expansion of coral diseases has been linked to local anthropogenic disturbances such as nutrient enrichment (Bruno, 2003; Voss and Richardson, 2006) and human pollution (Kaczmarsky *et al.*, 2005), but more importantly, to ocean warming and/or other factors associated with climate change (Ben-Haim and Rosenberg, 2002; Harvell *et al.*, 2002). Growth rates of marine bacteria (Shiah and Ducklow, 1994) and fungi (Holmquist *et al.*, 1983) are positively correlated with temperature. Coral pathogens are known to grow well at temperatures close to the host optima, which suggest that they would increase in warmer seas (Harvell *et al.*, 1999, 2002). The ethological agents of bacterial bleaching (i.e., *Vibrio shiloi* and *Vibrio coralliitycus*) are temperature sensitive (Kushmaru *et al.*, 1998; Ben-Haim *et al.*, 1999, 2003) and the epizootiology/physiology of other diseases such as black band (Richardson and Kuta, 2003; Rodriguez and Croquer, in press) and yellow blotch (Cervino *et al.*, 2004) are controlled in part by temperature. Heat-induced viruses could also be involved in temperature-induced coral bleaching (Wilson *et al.*, 2005; Davy *et al.*, 2006).

Compelling evidence indicates that corals are normally associated with a diverse microbiota (Klaus *et al.*, 2005; Johnston and Rower, in press). Coral-associated bacteria can be divided into four functional groups: (a) bacteria with possible roles in coral nutrition, (b) pathogenic bacteria, (c) bacteria which can act as probiont, aiding the growth of beneficial bacteria but limiting the growth of pathogenic forms (e.g. production of antibiotics or chemical defense [Kim *et al.*, 2000; Geffen and Rosenberg, 2005; Ritchie, 2006] and physical barriers [Kim *et al.*, 1997]), and (d) purely commensal bacteria with no impact on the other three groups (Klaus *et al.*, 2005). Changes in the environment (e.g. temperature) can alter the structure and function of normal microbiota, facilitating the invasion of pathogenic microorganisms and making the host more prone to disease (Reshef *et al.*, 2006; Ritchie, 2006). Thus, under a global warming scenario, stress on corals is expected to increase and the structure and function of normal microbiota is expected to change, thereby making corals more susceptible to disease (Fig. 3). Among marine fungi, optimum temperatures for growth coincide with thermal stress and bleaching for many corals (Holmquist *et al.*, 1983), leading to likely co-occurrence of bleaching and fungal infection.

In the Great Barrier Reef (GBR), major outbreaks of white syndrome (i.e. a coral disease that affects a wide variety of hosts) only occur on reefs with high coral cover after warm years, being absent on cooler and low-cover reefs (Bruno *et al.*, 2007). Another climatic anomaly hypothesized to initiate coral disease is transport of aeolian dust from Saharan Africa (mediated by a shift in the North Atlantic Oscillation) to the Caribbean (Shinn *et al.*, 2000).

7 GLOBAL WARMING AND CONNECTIVITY

In the ocean, large-scale dispersal and replenishment by larvae is a key process underlying biological changes associated with global warming (Ayre and Hughes, 2004). Populations of corals at high or isolated latitudes (e.g. oceanic reefs of the Pacific Ocean) may be more vulnerable to climate change because they are typically at the margins of geographical ranges, and are likely to be small and isolated (Hughes *et al.*, 2002) and also because larval connections between these reefs occur along latitudinal or thermal gradients. Long-distance dispersal by corals to geographically isolated reefs has been found to be very rare (Ayre and Hughes, 2004). Consequently, it has been predicted that localized extinctions of isolated populations (e.g. because of oil spills or thermally induced bleaching) will have persistent impacts over very long periods (Ayre and Hughes, 2004). Coral

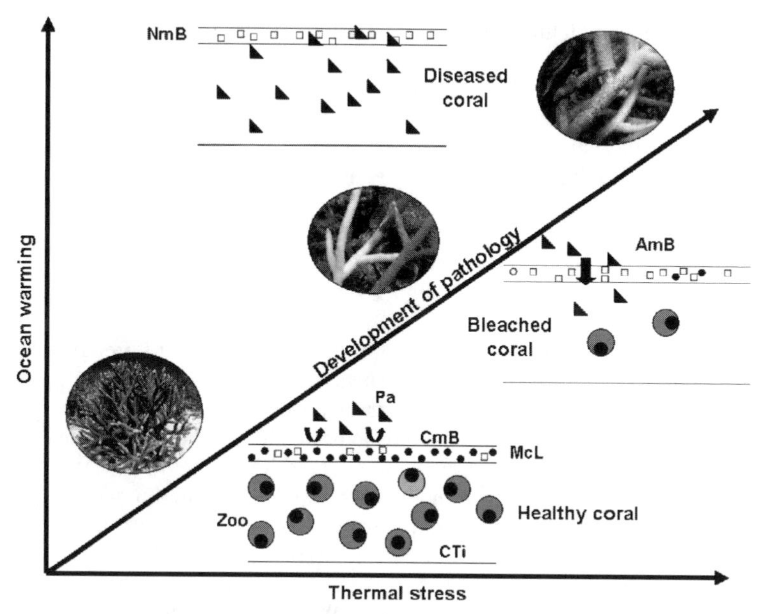

Figure 3. Expected effect of ocean warming and the increase of thermal stress on the development of coral pathologies. Zoo = zooxanthellae, CTi = coral tissue, McL = mucus layer, CmB = coral microbiota, Pa = Pathogen, AmB = abnormal microbiota NmB = new microbiota different to normal coral microbiota.

bleaching may produce massive mortalities, reducing connectivity due to a significant depletion of larval supply and a concomitant failure of recruitment (Hoegh-Guldberg, 1999). In consequence, the exchange of allelic variation across reefs could be compromised. Thus, reef organisms like corals with limited genetic diversity are predicted to have limited capacity to respond to environmental change (Ayre and Hughes, 2004).

8 GLOBAL WARMING AND HURRICANES

Global warming is also linked with the distortion of natural climate phenomena (e.g. hurricanes and cyclones) which are often implicated in the extensive loss of reef habitats over short periods of time (Goldenberg *et al.*, 2001). The oceans are the primary energy source for tropical cyclones. Localized sea surface temperatures (SSTs) play a direct role in providing moist to power incipient tropical cyclones (Saunders *et al.*, 1997). Local SST greater than 26.5°C is necessary for tropical cyclone development (Goldenberg *et al.*, 2001), and higher SST can increase overall activity (Saunders *et al.*, 1997). Multidecadal variations in major hurricane activity have been attributed to changes in the SST structure in the Atlantic (Gray, 1990; Landsea *et al.*, 1999) because tropical North Atlantic SSTs correlate positively with major hurricane activity. Thus, under a scenario of global warming, the frequency and intensity of hurricanes and cyclones are expected to increase with uncertain impacts for coral reefs ecosystems.

In short, in the next 50 to 100 years, atmospheric carbon dioxide concentration is expected to exceed 500 parts per million and global temperatures to rise by at least 2°C (Hoegh-Guldberg *et al.*, 2007). The negative impacts of these changes on coral reefs will most likely include a decrease in accretion rates, an increase in erosion rates, and decreases in coral cover and diversity, as few taxa are expected to adapt or acclimate to new environmental conditions (Hoegh-Guldberg *et al.*, 2007). Climate change will also exacerbate local stresses causing further decline in water quality and overexploitation of key species, driving reefs increasingly toward the tipping point for functional collapse (Hoegh-Guldberg *et al.*, 2007). Coral bleaching and disease epizootic events are also expected to increase in warmer seas, and both might act synergistically with bleaching making corals more susceptible to disease or vice versa (Fig. 3). Three major questions arise from these scenarios: Have coral reefs suffered these sorts of changes in the past? Have they recovered from previous large-scale disturbances and would they adapt to the scenario above described? What would the consequences for human societies be if coral reefs were to collapse?

9 CORAL REEF DECLINE AND RECOVERY: WHAT CAN WE LEARN FROM THE PAST?

Bioerms have been dominated by a wide variety of organisms capable of producing calcium carbonate (i.e. cyanobacteria, sponges, polychaetes, bivalves and scleractinian corals among others) throughout most of the Phanerozoic. Reef-building organisms have evolved under the presence of different natural disturbances. Geological records indicate that coral reefs have undergone significant changes in community structure throughout the earth's history (Wood, 1998). Major shifts have been triggered by sea level, climate and biogeochemical changes, but more importantly due to changes in the trophic demands of reef-building organisms and the disturbances they have faced (Wood, 1998). On a geological scale, massive extinctions of certain taxa have been followed by the evolution of other organisms capable of producing new reef structures with different community structures and/or along different geographical ranges (Wood, 1998; Pandolfi *et al.*, 2006). Knowing whether these events are natural or coincident with recent human impact is vital for management (Pandolfi *et al.*, 2006).

Modern reefs are formed as the direct or indirect result of organic activity, developing because of the aggregation of sessile epibenthic marine organisms, with the resultant higher rate of *in-situ* carbonate production than in surrounding sediments (Wood, 1998). Two major sets of evolutionary changes determined the characteristic structure, function and stability of modern coral reefs: (1) the coral-algal symbiosis and (2) the rise of predation, herbivory and bioerosion (Wood, 1998). These features are important because the dynamic nature of symbiotic combinations in corals may have allowed them to persist through hundreds of millions of years of rapid, and sometimes extreme, environmental change (Buddemeier, 1997) and also because modern reef building corals have anti-predatory characteristics, including rapid regeneration from partial mortality, an important life history trait necessary for recovering from disturbances (Jackson, 1983).

Like the ancient, modern Caribbean reef-coral fauna has demonstrated remarkable stability over the past 1.5 Million years (Jackson *et al.*, 1996; Jackson and Johnson, 2000; Pandolfi and Jackson, 2001). Coral communities have persisted over the same temporal scale in the Caribbean (Pandolfi and Jackson, 2006; Aronson *et al.*, 1998, 2004; Greenstein *et al.*, 1998) and other bioregions (e.g. Pandolfi, 1996).

While pre-industrial human societies impacted coral reefs (e.g. Jackson 1997), modern societies have produced more rapid changes on coral reef ecosystem function on a global scale (Wilkinson, 1999). For example, in the Caribbean modern reefs differ from Pleistocene communities partly because of the lack and/or a significant decrease in the abundance of acroporids and other structural species (Pandolfi and Jackson, 2006). The significant reduction of Caribbean acroporids has been linked to a myriad of factors linked to human activities: e.g., overfishing (Hughes, 1994; Jackson *et al.*, 2001), coral (Patterson *et al.*, 2002) and herbivore (Lessios *et al.*, 1984) epizootics as well as land use, soil erosion and modification of river catchments (Lewis *et al.*, 2007). These differences between modern and ancient coral reefs can all be at least partly attributed to overpopulation (Wilkinson, 1999), overexploitation of reef resources (Wilson *et al.*, 2006), global warming or a combination of the three (Buddemeier *et al.*, 2004).

While coral reef decline is easy to describe, coral reef recovery is more difficult to understand; in part because coral decline occurs much faster (i.e., from months to years) than recovery (from decades to centuries or even more). Therefore, understanding the factors controlling the ecological processes and mechanisms behind recovery is complicated. Recovery of depleted populations depends on successful recruitment (i.e., the incorporation into a population of new individuals which will eventually become reproductive) (Hughes and Tanner, 2000), the survivorship of new settlers (Lirman, 2000) and the exchange of propagules among populations, i.e., connectivity (Cowen *et al.*, 2006). The majority of organisms inhabiting coral reefs have microscopic planktonic stages in their life cycles which are released into the water column. Thus, improving our understanding of connectivity, recruitment and survivorship is imperative for understanding and addressing the role of global warming in coral reef recovery.

In summary, if we look into the past it is clear that coral reef structure has changed throughout the evolutionary history of our planet, with relatively frequent episodes of large-scale extinctions, species turnovers and replacement of dominant taxa and changes in their geographic distribution. Such events have primarily been linked with natural climate and sea level changes. Despite changing conditions, coral reef communities have prevailed and survived, which indicates that they can thrive by adapting or acclimating to new environmental conditions. Nevertheless, over the past millennia, reef communities remained relatively stable until recent massive mortalities of corals due to overfishing, pollution and climatic distortions (i.e. increase in the intensity and frequency of hurricanes) changed the structure and function of modern reefs.

Addressing whether or not coral reefs will recover from the series of threats that global warming is producing is uncertain for two reasons: (1) current environmental changes seem to be occurring much faster than in the past and (2) recovery is a complicated issue which involves the interaction of factors, processes and mechanisms operating at different levels of organization at different temporal and spatial scales.

10 ARTIFICIAL REEF RESTORATION: A VIABLE TOOL FOR RECOVERY FROM CLIMATE CHANGE?

Maragos (1974) and Birkeland *et al.* (1979) were among the first to demonstrate the ability of coral transplants to rehabilitate reefs damaged by elevated eutrophication (Hawaii) and by heated effluents produced by a power station (Guam), respectively. The immediate rationale for coral transplantation is to increase live coral cover and possibly diversity in a damaged reef area (Guzman, 1991; Epstein *et al.*, 2001; Rinkevich, 2005), while circumventing the initial larval recruitment stage, which can be prolonged (Riegl and Luke, 1998; Fox, 2004). However, this approach is perhaps useful to promote long-term recovery from disturbances that operates at small to medium spatial scales; but little useful to deal with the problem of large-scale degradation of corals as only few coral species tolerate transplants and/or adverse environmental conditions.

Providing reefs with optimal conditions (e.g. through improvement of water quality, management of fisheries, control and planning of coastal development, etc) should be the priority for avoiding coral reef decline. Prevention of ecological catastrophes is cheaper and easier than any attempt at artificial restoration. Thus, good science for management should be the priority for the prevention, slowing down and or/mitigation of coral reef decline rather than artificial restoration which is extremely expensive and in some cases infective in achieving recovery at large spatial scales.

Achieving the goal of reef management and promoting natural recovery requires a change in the behavior of our societies, at both local and global scales. To do this, scientists, politicians, stakeholders and the public need to join efforts towards the same objective. We need to share a common interest: the conservation of coral reefs and their ecological and economical values for future generations.

11 THE VALUE OF CORAL REEFS FOR HUMAN SOCIETIES: CAN WE AFFORD TO LOSE THEM?

Coral reefs are important for human societies world-wide as they supply vast numbers of people with ecological goods and services (Moberg and Folke, 1999) (Fig. 4). Almost a third of the world's marine fish species are found on coral reefs (McAllister, 1991) and the catch from reef

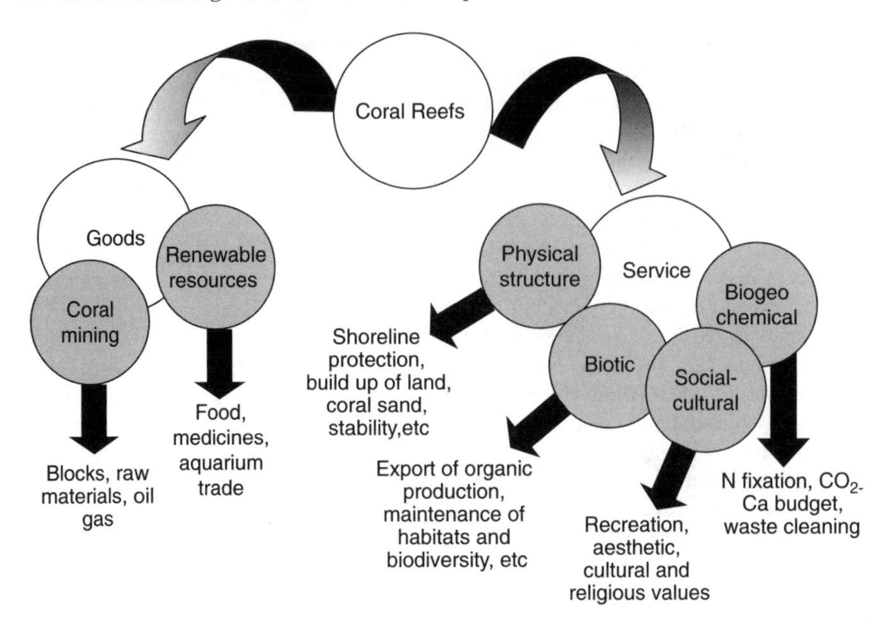

Figure 4. Ecological good and services that coral reefs provide to human societies (modified from Moberg and Folke 1999).

areas constitutes around 10% of the fish consumed by humans (Smith, 1978) and in some parts of the Indo-Pacific region, the reef fishery constitutes up to 25% of the total fish catch (Cesar, 1996).

Coral reefs provide ecological goods to over 100 million people in more than 100 countries that have coastlines with coral reefs (Craik *et al.*, 1990; Birkeland, 1997; Moberg and Folke, 1999; Salvat, 1992; Hoegh-Guldberg *et al.*, 2007) (Fig. 4). The pharmaceutical industry has discovered potentially useful substances with anticancer, AIDS inhibiting, antimicrobial, antiinflammatory and anticoagulating properties among the seaweeds, sponges, mollusks, corals and sea anemones that inhabit coral reefs (Sorokin, 1993; Carte, 1996; Birkeland, 1997). Also coral skeletons have proven to be promising in bone graft operations (Spurgeon, 1992).

The marine aquarium market is also a multi-millionaire industry in many countries (Wood, 1985). In the Maldives and Panama, coral blocks, rubble and sands serve as the main construction materials with approximately 20,000 m^3 of corals mined every year (Cesar, 1996; Guzman *et al.*, 2003). Lime, produced by grinding coral skeletons, is used as a pH regulator in agriculture (Cesar 1996), and in some regions coral debris is also collected and crushed to be used as fertilizer (Kühlmann, 1988). Oils and gas are thought to exist in large quantities below living reefs (Moberg and Folke, 1999) (Figure 4).

Coral reefs also provide human societies with a wide range of ecological services (i.e., physical structure, biotic, biogeochemical and socio-cultural) (Moberg and Folke, 1999). As for physical structure, coral reefs prevent erosion of the shoreline by protecting it from currents, waves and storms. In Indonesia, between US$ 820,000–1,000,000 per km of coastline has been lost as a consequence of coral destruction (Cesar, 1996). In tropical nations and in the Indian and Pacific oceans, large human populations are situated on islands built by coral reefs (Stoddart, 1973). Bioeroders, such as algae, sponges, polychaetes, crustaceans, sea urchins, and fishes are important in producing the reef sediments (rubble, sand, silt, and clay) that form sandy beaches with high value for tourism (Trudgill, 1983; Moberg and Folke, 1999). Furthermore, coral reefs promote changes in the physical environmental that are important for the establishment of other ecosystems such as seagrass beds (Richmond 1993) (Figure 4).

Biotic services can be found both within and between ecosystems (Moberg and Folke, 1999). In the first case, coral reefs function as important spawning, nursery, breeding and feeding areas for a multitude of organisms of commercial value (Fig. 4). In the second case, organisms such as fish and urchins migrate back and forth from reefs to adjacent areas; therefore representing a link between coral reefs and nearby ecosystems (Moberg and Folke, 1999). Coral reefs support the pelagic food web by exporting the excess of organic production such as mucus, wax esters, and dissolved organic matter as well as bacterioplankton, phyto and zooplankton (Sorokin, 1990). This net flow to surrounding waters enhances the productivity of local planktonic communities and consequently also supports local fisheries (Sorokin, 1990).

From the biogeochemical point of view, coral reefs function as nitrogen fixers in nutrient poor environments (Sorokin, 1993). The nitrogen fixing ability is not only of local importance to the reef system itself but also to the productivity of the adjacent pelagic communities due to the release of excess nitrogen fixed in the reefs (D'Elia and Wiebe, 1990; Sorokin, 1990). Reefs appear to act as sinks for carbon dioxide over geological scales and are net sources of carbon dioxide in time scales relevant for humans (Gattuso *et al.*, 1996; Hallock, 1997). Biochemical processes on coral reefs play a significant role in the world's calcium balance (Kühlmann, 1988) as they precipitate approximately half of the 1.2×10^{13} mol of calcium delivered to the sea each year (Smith 1978). Coral reefs can transform, detoxify, and sequester wastes released by humans, thus providing a cleaning service (Moberg and Folke, 1999).

The recreational value of coral reefs is perhaps the most important socio-cultural service they provide (Dixon *et al.*, 1993; Pendleton, 1995; Cesar, 1996). The financial value of tourism in the Great Barrier Reef World Heritage Area was estimated to be about AUS$ 682 million per year (Driml, 1994). In 1990, Caribbean tourism earned US$ 8.9 billion and employed over 350 000 people (Dixon *et al.*, 1993). Another important and often forgotten service of reefs is their support of cultural and spiritual values (Moberg and Folke, 1999). For instance religious rituals have developed around reefs in southern Kenya, where traditional management with the primary purpose of appeasing spirits has also served to regulate fish stocks (McClanahan *et al.*, 1996).

In summary, the problem of the world-wide coral reef decline is a complex issue with serious ecological (e.g. loss of species diversity) biogeochemical (e.g. alteration of carbon and nitrogen cycles) and socioeconomic (e.g. loss of fish stocks, potential collapse of local and/or global economies, increase of poverty, etc) consequences for human societies which rely one way or another on coral reefs. Thus, managing the problem is not only a scientific but also a social and a political issue. Human societies cannot afford to loose coral reefs, simply because without these important ecosystem our quality of life would be severely compromised.

What can we do: is there any hope for coral reefs?

Coral reefs are confronting a global crisis today (Hughes *et al.*, 2004) and are expected to collapse in the next 50–100 years (Hoegh-Guldberg *et al.*, 2007). This fact calls for a change of behavior in our societies the major challenge being encouraging people to be proactive towards the problem. Scientists can no longer limit their work to the production and publishing of data to support predictions and hypotheses. Instead, we need to compile the evidence and make it digestible for broader audiences, so that all the parties involved in coral reef conservation (i.e. local people, politicians, developers, environmentalists) may participate in and contribute to the creation and implementation of specific solutions. While it is true that coral reef conservation is a complicated issue because it involves many countries with different social, economic and political systems, our modern world enables us to communicate much faster and also to learn from the past. Today it is pointless to ask whether or not coral reefs are declining; it is worthless to be trapped in the same argument. Instead, we need to ask what we can do for coral reefs.

Several things can be done globally and locally in order to slow down the coral reef crisis. On a global basis it is imperative to (1) reduce greenhouse gas emissions, especially in developed countries, (2) create a balance between development and environmental change, especially in coastal zones, (3) reduce population growth, especially in developing countries, (4) reduce the demand on reef resources by developing countries and (4) improve management of coral reef resources in both developed and undeveloped countries. On a local scale, it is imperative to educate local people and create environmental awareness of the importance of coral reefs for their own quality of life. To do this, a multidisciplinary approach which integrates biological, social, and economic and political sciences to evaluate the ecological reality and the social history of each particular community is required. I do believe that our modern societies have the tools (if not the political will) to implement the necessary adjustments to stop or at least slow down the coral reef crisis in time.

12 CONCLUDING REMARKS

Global warming is now producing drastic changes in the structure and function of coral reefs. Projections based on expected levels of CO_2 for the next few decades, indicate that greenhouse emissions will deteriorate coral reefs to a critical point in less than 50 years. If this happens, there will be serious consequences for human society. The key to preserving coral reefs for future generations is asking ourselves what we can do about to reduce the problems that they face. Human societies may either argue that there is nothing we can do or take action to avoid the predicted demise of coral reefs. We believe that the second option is the proper answer to the crisis. Society has the opportunity to prevent or at least slow down coral reef decline. The scientific community has been successful in detecting the causes behind the crisis; the next step should be to take corrective measures. If scientists, politicians and people from developed and undeveloped countries join efforts in a common cause, coral reefs should have a chance. These ecosystems have an extraordinary capacity to recover from disturbances both natural and anthropogenic.

REFERENCES

Ainsworth, T.D., Kvennefors, E.C., Blackall, L.L., Fine, M., and Hoegh-Guldberg, O. 2007. Disease and cell death in white syndrome of Acroporid corals on the Great Barrier Reef. Mar. Biol. 151: 19–29.

Alley, R.B., Marotzke, J., Nordhaus, D., Overpeck, J.T., Peteet, D.M., Pielke Jr, A., Pierrehumbert, R.T., Rhines, P.B., Stocker, T.F., Talley, L.D., and Wallace, J.M. 2003. Abrupt Climate Change. Science 299: 2005–2010.

Andreev, A.A., and Klimanov, V.A. 2000. Quantitative Holocene climatic reconstruction from Arctic Russia. J Paleolim 24: 81–91.

Anthony, K.R.N., Connolly, S.R., and Hoegh-Guldberg, O. 2007. Bleaching, energetics, and coral mortality risk: effects of temperature, light and sediment regime. Limnol. Oceanogr. 52: 716–726.

Aronson, R.B., Precht, W.F., and Macintyre, I.G. 1998. Extrinsic control of species replacement on a Holocene reef in Belize: the role of coral disease. Coral Reefs 17: 223–230.

Aronson, R.B., and Precht, W.F. 2001. White-band diseases and the changing face of Caribbean coral reefs. Hydrobiologia 460: 25–38.

Aronson, R.B., Bruno, J.F., Precht, W.F., Glynn, P.W., Harvell, D.C., Kaufman, F., Rogers, G.S., Shinn, E.A., and Valentine, J.F. 2003. Cause of coral reef degradation. Science 302: 1502.

Aronson, R.B., Macintyre, I.G., Wapnick, C.M., and O'Neill, M.W. 2004. Phase shifts, alternatives stages and the unprecedented convergence of two reef systems. Ecology 1876–1891.

Ayre, D.J., and Hughes, T.P. 2004. Climate change, genotypic diversity and the gene flow in reef building corals. Ecol. Lett. 7: 273–278.

Baird, A.H., and Marshall, P.A. 2002. Mortality, growth and reproduction in scleractinian corals following bleaching on the Great Barrier Reef. Mar. Ecol. Prog. Ser. 237: 133–141.

Barnes, D.J., and Crossland, C.J. 1980. Diurnal and seasonal variations in the growth of a staghorn coral measured by time-lapse photography. Limnol. Oceanogr. 25:1113–1117.

Battarbee, R.W., Grytnes, J.A., Thompson, R., Appleby, P.G., Catalan, J., Korhola, A., Birks, H.J.B., Heegaard, E., and Lami, A. 2002. Comparing palaeolimnological and instrumental evidence of climate change for remote mountain lakes over the last 200 years. J. Paleolim. 28: 161–179.

Becker, A., Wenzel, V., Krysanova, V., and Lahmer, W. 1999. Regional analysis of global change impacts: Concepts, tools and first results. Environ. Mod. Asses. 243–257.

Bellwood, D.R., Hoey, A.S., and Choat, J. 2003. Limited functional redundancy in high diversity systems: resilience and ecosystem function in coral reefs. Ecol. Lett. 6: 281–285.

Bellwood, D.R., Hughes, T.P., Folke, C., and Nystrom, M. 2004. Confronting the coral reef crisis. Nature 429: 827–833.

Ben-Haim, Y., Banim, E., Kushmaro, A., Loya, Y., and Rosenberg, E. 1999. Inhibition of photosynthesis and bleaching of zooxanthellae by the coral pathogen *Vibrio shiloi*. Environ. Micro. 1: 223–229.

Ben-Haim, Y., Zicherman-Keren, M., and Rosenberg, E. 2003. Temperature-Regulated Bleaching and Lysis of the Coral *Pocillopora damicornis* by the Novel Pathogen *Vibrio coralliilyticus.* App. Environ. Micro. 69: 4236–4242.

Ben-Haim, Y., and Rosenberg, E. 2002. A novel *Vibrio* sp pathogen of the coral *Pocillopora damicorni*s. Mar. Biol. 141: 47–55.

Beznosov, V.N., and Suzdaleva, A.L. 2004. Potential Changes in Aquatic Biota in the Period of Global Climate Warming. Water Resour. 4: 459–464.

Birkerland, C. 1997. Life and death of coral reef. Ed. Birkeland, C., Chapman and Hall, N.Y., U.S.A. 520 pp.

Birkeland, C., Randall, R.H., and Grim, G. 1979. Three methods of coral transplantation for the purpose of reestablishing coral community in the thermal effluent area of the Tanguisson Power Plant. University of Guam, Marine Laboratory Technical Report 60.

Bond, G., Kromer, B., Beer, J., Muscheler, R., Evans, M.N., Showers, W., Hoffmann, S., Lotti-Bond, R., Hajdas, I., and Bonani, G. 2001. Persistent solar influence of North Atlantic climate during the Holocene. Science 294: 2130.

Borger, J.L., and Steiner, S.C.C. 2005. The spatial and temporal dynamics of coral diseases in Dominica, West Indies. Bull. Mar. Sci. 77: 137–154.

Brown, B.E. 1997. Bleaching: causes and consequences. Coral Reefs 16: 129–138.

Bruckner, A.W., and Bruckner, R. 2006. Consequences of YBD on *Montastraea annularis* (species complex) populations on remote reefs off Mona Island, Puerto Rico. Dis. Aqua. Org. 69: 67–73.

Bruno, J.F., Peters, L.E., Harvell, C.D., and Hettinger, A. 2003. Nutrient enrichment can increase the severity of coral diseases. Ecol. Lett. 6: 1056–1061.

Bruno, J.F., Selig, E.R., Casey, K.S., Page, C.A, and Willis, B.L. 2007. Thermal stress and coral cover as drivers of coral disease outbreaks. PLoS One 6: 001–008.

Buddemeier, R.W. 1997. Making light work of adaptation. Nature 388: 229–30.

Buddemeier, R.W., Kleypas, J.A., and Aronson, R.B. 2004. Coral reefs and global climate change. Report available at: www.pewclimate.org.

Canadell, J.G., Le Que re, C., Paupach, M.R., Field, C.B., Buitenhuis, E.T., Ciaisf, P., Conway, T.J., Gillet, N.P., Houghton, R.A., and Marlandi, G. 2007. Contributions to accelerating atmospheric CO2 growth from economic activity, carbon intensity, and efficiency of natural sinks. Proc. Natl. Acad. Sci. USA 104: 18866–18870.

Carte, B.K. 1996. Biomedical potential of marine natural products. BioScience 46: 271–286.

Cervino, J., Hayes, L.H., Shawn, W., Polson, S.C., Goreau, T.J., Martinez, R.J., and Smith, G.W. 2004. Relationship of Vibrio species infection and elevated temperatures to yellow blotch/band disease in Caribbean corals. App. Environ. Micro. 70: 6855–6864.

Cesar, H. 1996. Economic Analysis of Indonesian Coral Reefs. The World Bank.

Cohen, J.E. 2003. Human population: the next half century. Science 302: 1172–1175.

Colwell, R.R. 1996. Global climate and infectious disease: the Cholera paradigm. Science 274: 2025–2031.

Connell, J.H. 1997. Disturbance and recovery of coral assemblages. Coral Reefs 16: 101–113.

Cowen, R.K., Paris, C.B., and Srinivasan, A. 2006. Scaling of connectivity in marine populations. Science 311: 522–527.

Craik, W., Kenchington, R., and Kelleher, G. 1990. Coral-Reef Management. In: Dubinsky, Z. (Ed.), Ecosystems of the World 25: Coral Reefs. Elsevier, New York, pp. 453–467.

Davy, S.K., Burchett, S.G., Dale, A.L., Davies, P., Davy, J.E., Muncke, C., Hoegh-Guldberg, O., and Wilson, W.H. 2006. Viruses: agents of coral disease? Dis. Aq. Org. 69: 101–110.

D'Elia, C.F., and Wiebe, W.J. 1990. Biochemical nutrient cycles in coral reef ecosystems. In: Dubinsky, Z (Ed.), Ecosystems of the World 25: Coral Reefs. Elsevier, New York, pp. 49–74.

Derraik, J.G.B., and Slaney, D. 2007. Anthropogenic Environmental Change, Mosquito-borne Diseases and Human Health in New Zealand. EcoHealth 4: 72–81.

Diamond, J. 2005. Collapse: How societies choose to fall or succeed. Penguin, New York pp. 574.

Dixon, J.A., Scura, L.F., and van't Hof, T. 1993. Meeting ecological and economic goals: marine parks in the Caribbean. Ambio 22: 117–125.

Drescher-Schneider, R., Beaulieu, J.L., Magny, M., Walter-Simonnet, A.V., Bossuet, G., Millet, L., Brugiapaglia, E., and Drescher, A. 2007. Vegetation history, climate and human impact over the last 15,000 years at Lago dell'Accesa (Tuscany, Central Italy). Veget. Hist. Archeobot. 16: 279–299.

Driml, S. 1994. Protection for Profit-Economic and Financial Values of the Great Barrier Reef World Heritage Area and Other Protected Areas. Great Barrier Reef Marine Park Authority. Research Publication No. 35, Townsville, Australia, pp. 83.

Edmunds, P.J., and Elahi, R. 2007. The demographics of a 15-year decline in cover of the Caribbean reef coral Montastraea annularis. Ecol. Mongr. 77: 3–18.
Epstein, N., Bak, R.P.M., and Rinkevich, B. 2001. Strategies for gardening denuded coral reef areas: the applicability of using different types of coral material for reef restoration. Restor. Ecol. 9: 432–442.
Fox, H.E. 2004. Coral recruitment in blasted and unblasted sites in Indonesia: assessing rehabilitation potential. Mar. Ecol. Prog. Ser. 269: 131–139.
Gattuso, J.P., Frankignoulle, M., Smith, S.V., Ware, J.R., and Wollast, R. 1996. Coral reefs and carbon dioxide (technical comment). Science 271: 1298.
Geffen, Y., and Rosenberg, E. 2005. Stress-induced rapid release of antibacterials by scleractinian corals. Mar. Biol. 146: 931–935.
Gillson, L., and Willis, K.J. 2004. 'As Earth's testimonies tell': wilderness conservation in a changing world. Ecol. Lett. 7: 990–998.
Goldenberg, S.B., Landsea, G.W., Metas-Nunez, A.M., and Gray, W.M. 2001. The Recent Increase in Atlantic Hurricane Activity: Causes and Implications. Science 293: 474–479.
Gray, W.M. 1990. Strong association between West African rainfall and US landfall of intense hurricanes. Science 249: 1251–1256.
Greenstein, B.J., Curran, H.A., and Pandolfi, J.M. 1998. Shifting ecological baselines and the demise of *Acropora cervicornis* in the western North Atlantic and Caribbean Province: a Pleistocene perspective: Coral Reefs 17: 249–262.
Guzman, H.M. 1991. Restoration of coral reefs in Pacific Costa Rica. Conservation Biol 5: 189–195.
Guzman, H.M., Guevara, C., and Castillo, A. 2003. Natural disturbances and mining of Panamanian coral reefs by indigenous people. Conser. Biol. 17: 1–7.
Hallock, P. 1997. Reefs and reef limestones in Earth history. In: Birkeland, C., (Ed.), Life and Death of Coral Reefs. Chapman and Hall, New York, pp. 13–42.
Harley, C.D.G., Hughes, A.R., Hultgren, K.M., Miner, B.G., Sorte, C.J.B., Thornber, C.S., Rodriguez, L.F., Tomanek, L., and Williams, S.M. 2006. The impacts of climate change in coastal marine systems. Ecol Lett 9: 228–241.
Harte, J. 2007. Human population as a dynamic factor in environmental degradation. Pop. Environ. 28: 223–236.
Hartmann, D.L., Wallace, J.M., Limpasuvan, V.D.W., Thompson, J., and Holton, J.R. 2000. Can ozone depletion and global warming interact to produce rapid climate change? Proc. Natl. Acad. Sci. USA 97: 1412–1417.
Harvell, C.D., Kim, K., Burkholder, J.M., Colwell, R.R. Epstein PR, Grimes, D.J., Hofmann, E.E., Lipp, E.K., Osterhaus, A.D.M.E., Overstreet, A.M., Porter, J.W., Smith, G.W., and Vasta, G.R. 1999. Emerging marine diseases-climate links and anthropogenic factors. Science 285: 1505–1510.
Harvell, C.D., Mitchell, C.E., Ward, J.R., Altizer, S., Dobson, A.P., Ostfeld, R.S., and Samuel, M.D. 2002. Climate warming and disease risk for terrestrial and marine biota. Science 296: 2158–2162.
Hoegh-Guldberg, O. 1999. Climate change, coral bleaching and the future of the world's coral reefs. Mar. Freshw. Res. 50: 839–866.
Hoegh-Guldberg, O., Mumby, P.J., Hooten, A.J., Steneck, R.S., Greenfield, P., Gomez, E., Harvell, C.D., Sale, P.F., Edwards, A.J., Caldeira, K., Knowlton, N., Eakin, C.M., Iglesias-Prieto, R., Muthiga, N., Bradbury, R.H., Dubi, A., and Hatziolos, M.E. 2007. Coral Reefs Under Rapid Climate Change and Ocean Acidification. Science 318: 1737–1742.
Holmquist, G.U., Walker, H.W., and Stahr, H.M. 1983. Influence of temperature, pH, water activity and antifungal agents on growth of *Aspergillus flavus* and *A. parasiticus*. J. food. Sci. 8: 778–782.
Hughes, T.P. 1994. Catastrophes, phase shifts, and large-scale degradation of a Caribbean coral reef. Science 265: 1547–1551.
Hughes, T.P., and Tanner, J.E. 2000. Recruitment failure, life histories, and long-term decline of Caribbean corals. Ecology 81: 2250–2263.
Hughes, T.P., Bellwood, D., and Connolly, S. 2002. Biodiversity hotspots, centers of endemicity, and the conservation of coral reefs. Ecol. Lett 5: 775–784.
Hughes, T.P., Baird, A.H., Bellwood, D.R., Card, M., Connolly, S.R., Folke, C., Grosberg, R., Hoegh-Guldberg, O., Jackson, J.B.C., Kleypas, J., Lough, J.M., Marshal, P., Nyström, M., Palumby, S.R., Pandolfi, J.M., Rosen, B., and Roughgarden, J. 2003. Climate change, human impacts and the resilience of coral reefs. Science 301: 929–933.
IPPC. 2007. Climate Change: The Scientific Basis. The Contribution of Working Group I to the Third Assessment Report of the Intergovernmental Panel on Climate Change, J.T. Houghton *et al.*, Eds. Cambridge Univ. Press, New York, 2007.

Jackson, J.B.C. 1983. Biological determinants of present and past sessile animal distributions. In Biotic Interactions in Recent and Fossil Benthic Communities, (eds) Tevesz, M., and McCall, P.W., New York, Plenum. 837 pp.

Jackson, J.B.C., Budd, A.F., and Pandolfi, J.M. 1996. The shifting balance of natural communities? In: Jablouski, D., Erwin, D.H., and Lipps, J.H. (eds) Evolutionary paleobiology, Universiy of Chicago Press, Chicago, pp. 89–122.

Jackson, J.B.C. 1997. Reefs since Colombus. Coral Reefs 16: 23–32.

Jackson, J.B.C., Kirby, M.X., Berger, W.H., Bjorndal, K.A., Botsford, L.W., Bourque, B.J., Bradbury, R.H., Cooke, R., Erlandson, J., Estes., J.A., Hughes, T.P., Kidwell, S., Lange, C.B., Lenihan, H.S, Pandolfi, J.M., Peterson, C.H., Steneck, R.S., Tegner, M.J., and Warner, R. 2001. Historical overfishing and the recent collapse of coastal ecosystems. Science. 293: 629–638.

Johnston, I.S., and Rohwer, F. in press. Microbial landscapes on the outer tissue surfaces f the reef-building coral *Porites compressa*. Coral Reefs.

Kaczmarsky, L.T., Draud, M., and Williams, E.H. 2005. Is there a relationship between between proximity to sewage effluent and the prevalence of coral disease? Car. J. Sci. 41: 124–137.

Kim, K., Harvell, C.D., and Smith, G.W. 1997. Mechanisms of sea fan resistance to a fungal epidemic. I. Role of sclerites. Am Zool 37:132A.

Kim, K., Kim, P.D., Alker, A.P., and Harvell, C.D. 2000. Chemical resistance of gorgonian corals against fungal infections. Mar. Biol. 137: 393–401.

Klaus, J.S., Frias-Lopez, J., Bonheyo, G.T., Heikoop, J.M., and Fouke, B.W. 2005. Bacterial communities inhabiting the healthy tissues of two Caribbean reef corals: interspecific and spatial variation. Coral Reefs 24: 129–137.

Kleypas, J.A., Buddemeier, R.W., Archer, D., Gatusso, J.P., Langdon, C., and Opdyke, B.N. 1999. Geochemical consequences of increased atmospheric carbon dioxide on coral reefs. Science 284: 118–120.

Kovats, R.S., Campbell-Lendrum, D.H., McMichael, A.J., Woodward, A., and Cox, J. St H. 2001. Philos. Trans. R. Soc. London Ser. B 356: 1057–1068.

Kühlmann, D.H.H. 1988. The sensitivity of coral reefs to environmental pollution. Ambio 17: 13–21.

Kushmaro, A., Rosenberg, E., Fine, M., Ben Haim, Y., and Loya, Y. 1998. Effect of temperature on bleaching of the coral *Oculina patagonica* by Vibrio AK-1. Mar. Ecol. Prog. Ser. 171: 131–137.

Kuznetsov, V.G. 2005. Atmospheric Carbon Dioxide and Climate: Historical–Geological Aspects. Lithol. Min. Resour. 40: 320–331.

Landsea, C.W., Pielke, Jr. R.A., Mestas-Nunez, A.M., and Knaff, J.A. 1999. Climatic Change 42: 89–129.

Lesser, M.P. 2004. Experimental biology of coral reef ecosystems. J. Exp. Mar. Biol. Ecol. 300: 217–252.

Lessios, H.A., Robertson, D.R., and Cubit, J.D. 1984. Spread of Diadema mass mortality through the Caribbean. Science 226: 335–337.

Lewis, S.E., Shields, G.A., Kamber, B.S., and Lough, J.M. 2007. A multi-trace element coral record of land-use changes in the Burdekin river catchment, NE Australia. Palaeo 246: 471–487.

Linsley, B.K., Wellington, G.M., Schrag, D.P., Ren, L., Salinger, M.J., and Tudhope, A.W. 2004. Geochemical evidence from corals for changes in the amplitude and spatial pattern of South Pacific inter-decadal climate variability over the last 300 years. Clim. Dyn. 22: 1–11.

Lirman, D. 2000. Fragmentation in the branching coral *Acropora palmata* (Lamarck): growth, survivorship, and reproduction of colonies and fragments. J. Exp. Mar. Biol. Ecol. 251: 41–57.

Maragos, J.E. 1974. Coral transplantation: a method to create, preserve and manage coral reefs. Sea Grant Advising Report SEA-GRANT-AR.74-03-COR-MAR-14. University of Hawaii. Honolulu.

Marshall, A.T., and Clode, P. 2004. Calcification rate and the effect of temperature in a zooxanthellate and an azooxanthellate scleractinian reef coral. Coral Reefs 23: 218–224.

McAllister, D.E. 1991. What is the status of the world's coral reef fishes? Sea Wind 5: 14–18.

McClanahan, T.R., Rubens, J., Glaesel, H., and Kiambo, R. 1996. The Diani-Kinondo coral reefs, fisheries, and traditional management. Coral Reef Conservation Project Report, The Wildlife Conservation Society, Mombasa, Kenya, pp. 28.

McClanahan, T.R., Ateweberhan, M., Muhando, C.A., Maina, J., and Mohammed, M.S. 2004. Effects of climate and seawater temperature variation on coral bleaching and mortality. Ecol. Mongr. 77: 503–525.

McMichael, A.A., Githeko, R., Carcavallo, A.P., Gubler, D., Haines, A., Kovats, R.S., Martens, P., Patz, J., and Sasaki, A. 2001. 'Human health'. In McCarthy, J.J., Canziani, O.F., Leary, N.A., Dokken, D.J., and White, K.S. (eds), Climate Change 2001: Impacts, Adaptation, and Vulnerability – Contribution of Working Group II to the Third Assessment Report of the Intergovernmental Panel on Climate Change, Cambridge: Cambridge University Press.

Mendez, J.M., and Woodley, J.D. 2002. Effect of the 1995–1996 bleaching event on polyp tissue depth, growth, reproduction and skeletal band formation in *Montastraea annularis*. Mar. Ecol. Prog. Ser. 235: 93–102.

Moberg, F., and Folke, C. 1999. Ecological goods and services of coral reef ecosystems. Ecol. Econ. 215–233.

Muller, E.M., Rogers, C.S., Spitzack, A.S., and van Woesik, R. 2008. Bleaching increases likelihood of disease on *Acropora palmata* (Lamarck) in Hawksnest Bay, St John, US Virgin Islands. Coral Reefs 27: 191–197.

Muscatine, L. 1990. The role of symbiotic algae in carbon and energy flux in reef corals. Coral Reefs 25: 1–29.

Nugues, M.M. 2002. Impact of a coral disease outbreak on coral communities in St Lucia: what and how much has been lost? Mar. Ecol. Prog. Ser. 229: 61–71.

Nyström, M. 2006. Redundancy and Response Diversity of Functional Groups: Implications for the Resilience of Coral Reefs. Ambio 35: 30–35.

Palumby, S.R. 2001. Humans as the world's greatest evolutionary force. Science. 293: 1786–1790.

Pandolfi, J.M. 1996. Limited membership in Pleistocene reef coral assemblages from the Huon Peninsula, Papua New Guinea: constancy during global change. Paleobiology 22: 152–76.

Pandolfi, J.M., and Jackson, J.B.C. 2001. Community structure of Pleistocene coral reefs of Curacao, Netherlands Antilles. Ecol. Monogr. 71: 49–67.

Pandolfi, J.M., and Jackson J.B.C. 2006. Ecological persistence interrupted in Caribbean coral reefs. Ecol. Lett. 9: 818–826.

Pandolfi, J.M., Tudhope, A.W., Burr, G., Chappell, J., Edinger, E., Frey, M., Steneck, R., Sharma, C., Yeates, A., Jennions, M., Lescinsky, H., and Newton, A. 2006. Mass mortality following disturbance in Holocene coral reefs from Papua New Guinea. Geology 34: 949–952.

Patterson, K.L., Porter, J.W., Ritchie, K.B., Polson, S.W., Mueller, E., Peters, E.C., Santavy, D.L., and Smith, G.W. 2002. The etiology of white pox a lethal disease of the Caribbean elkhorn coral *Acropora palmata*. Proc. Natl. Acad. Sci. 99: 8725–8730.

Pascual, M., Rodo, X., Elner, S.P., Colwell, R., and Bouma, M.J. 2000. Science 289: 1766–1769.

Pendleton, L.H. 1995. Valuing coral reef protection. Ocean Coast. Manag. 26: 119–131.

Petes, L., Harvell, C.D., Peters, E., Webb, M., and Mullen, K. 2003. Pathogens compromise reproduction and induce melanization in Caribbean sea fans. Mar. Ecol. Prog. Ser. 264: 167–171.

Reshef, L., Koren, O., Loya, Y., Ziber-Rosenberg, I., and Rosenberg, E. 2006. The coral probiotic hypothesis. Environ. Micro. 8: 2068–2073.

Richardson, L.L., and Kuta, K.G. 2003. Ecological physiology of the black band disease cyanobacterium *Phormidium corallyticum*. FEMS Micro. Ecol. 43: 287–298.

Richmond, R.H. 1993. Coral reefs: present problems and future concerns resulting from anthropogenic disturbance. Am. Zool. 33: 524–536.

Riegl, B., and Luke, K.E. 1998. Ecological parameters of dynamited reefs in the northern Red Sea and their relevance to reef rehabilitation. Mar. Pollut. Bull. 37: 488–498.

Rinkevich, B. 2005. Conservation of coral reefs through active restoration measures: recent approaches and last decade progress. Environ. Sci. Technol. 39: 4333–4342.

Ritchie, K.B. 2006. Regulation of microbial populations by coral surface mucus and mucus-associated bacteria. Mar. Ecol. Prog. Ser. 322: 1–14.

Rodriguez, S., and Cróquer, A. in press. Dynamics of Black Band Disease in a *Diploria strigosa* population subjected to annual upwelling on the northeastern coast of Venezuela. Coral Reefs.

Roessig, J.M., Woodley, C.M., Cech, J.J., and Hansen, L.J. 2004. Effects of global climate change on marine and estuarine fishes and fisheries. Rev. Fish. Biol. Fisher. 14: 251–275.

Salvat, B. 1992. Coral reefs: a challenging ecosystem for human societies. Glob. Environ. Change 2: 12–18.

Saunders, M., and Harris, A. 1997. Statistical evidence links exceptional 1995 Atlantic hurricane season to record ocean warming. Geophys. Res. Lett. 24: 1255–1258.

Shiah, F.K., and Ducklow, H.W. 1994. Temperature and substrate regulation of bacterial abundance, production and specific growth rate in Chesapeake bay, USA. Mar. Ecol. Prog. Ser. 103: 297–308.

Shinn, E.A., Smith, G.W., Prospero, J.M., Betzer, P., Hayes, M.L., Garrison, V., and Barber, R.T. 2000. African dust and the demise of Caribbean coral reefs. Geophys. Res. Lett. 27: 3029–3032.

Smith, S.V. 1978. Coral-reef area and the contribution of reefs to processes and resources of the world's oceans. Nature 273: 225–226.

Smith, L.D., Glimour, J.P., and Heyward, A.J. 2008. Resilience of coral communities on an isolated system of reefs following catastrophic mass bleaching. Coral Reefs 27: 197–205.

Sorokin, Yu I. 1990. Aspects of trophic relations, productivity and energy balance in reef ecosystems. In: Dubinsky, Z. (Ed), Ecosystems of the World 25: Coral Reefs. Elsevier, New York, pp. 401–410.

Sorokin, Yu I. 1993. Coral Reef Ecology. Ecological Studies 102. Springer Verlag, Berlin, pp. 4–28.

Spalding, M.D., and Grenfell A.M. 1997. New estimates of global and regional coral reef areas. Coral Reefs 16, 225–230.

Spurgeon, J.P.G. 1992. The economic valuation of coral reefs. Mar. Pollut. Bull. 4: 529–536.

Stoddart, D.R. 1973. Coral reefs of the Indian Ocean. In: Jones, O.A., and Endean, R. (Eds.), Biology and geology of coral reefs, vol. 1. Academic Press, New York, pp. 51–92

Szmant-Grassman, A.M. 1990. The effects of prolonged 'bleaching', on the tissue biomass and reproduction of the reef coral *Montastrea annularis*. Coral Reefs 8: 217–24.

Tarasov, P., Granoszewski, W., Bezrukova, E., Brewer, S., Nita, M., Abzaeva, A., and Oberhansli, H. 2005. Quantitative reconstruction of the last interglacial vegetation and climate based on the pollen record from Lake Baikal, Russia. Clim. Dyn. 25: 625–637.

Trudgill, S.T. 1983. Measurements of rates of erosion of reefs and reef limestones. In: Barnes, D.J. (Ed.), Perspectives on coral reefs. Brian Clouston, A.C.T., Australia, pp. 256–262.

United Nations Population Division, World Population Prospects: the 2008 Revision, *Highlights* (online database). ESA/P/WP.180, revised 21 February 2008, Available at: http://esa.un.org/unpp/

Vorosmarty, C.J., Green, P., Salisbury, J., and Lammers, R.B. 2008. Global Water Resources: Vulnerability from Climate Change and Population Growth. Science 289: 284–289.

Voss, J.D., and Richardson, L.L. 2006. Nutrient enrichment enhances black band disease infections in corals. Coral Reefs 25: 269–576.

Ward, S., Harrison, P.L., and Hoegh-Guldberg, O. 2000. Coral bleaching reduces reproduction on scleractinian corals and increases susceptibility for future stress. Proc. 9th Int. Coral Reef Symp. Bali 2: 1123–1128.

Weil, E. 2004. Coral reef diseases in the wider Caribbean. In: Rosenberg, E., and Loya, Y. (eds) Coral Health and Disease Springer-Verlag, pp. 35–67.

Winkler, R., Antonius, A., and Renegar, D.A. 2004. The skeleton eroding band disease on coral reefs of Aqaba, Red Sea. Mar. Ecol. 25: 129–144.

WHO (2005) Global Health Atlas, World Health Organization. Available at: gamapserver.who.int/mapLibrary/Files/Maps/global_cases.jpg

Wilkinson, C.R. 1999. Global and local threats to coral reef functioning and existence: review and predictions. Mar. Fresh. Res. 50: 867–878.

Willis, B.L., Page, C.A., and Dindsdale, E.A. 2004. Coral disease in the Great Barrier Reef. In: Rosenberg, E., and Loya, Y. (eds) Coral Health and Disease. Springer-Verlag, New York pp. 69–104.

Wilson, W.H., Dale, A.L., Davy, J.E., and Davy, S.K. 2005. An enemy within? Observations of virus-like particles in reef corals. Coral Reefs 24: 145–148.

Wilson, S.K., Graham, S.J., Pratchett, M.S., Jones, G.P., and Polunin, N.V.C. 2006. Multiple disturbances and the global degradation of coral reefs: are reef fishes at risk or resilient? Global Change 12: 2220–2234.

Wood, E.M. 1985. Exploitation of coral reef fishes for the aquarium fish trade. Marine Conservation Society, Rosson-Wye, UK.

Wood, R. 1998. The ecological evolution of reefs. Ann. Rev. Ecol. Sys. 29: 179–2006.

Wu, H., Guiot, J., Brewer, S., and Guo, Z. 2007. Climatic changes in Eurasia and Africa at the last glacial maximum and mid-Holocene: reconstruction from pollen data using inverse vegetation modeling. Clim. Dyn. 29: 211–229.

Yap, H.T., and Gomez, E.D. 1984. Growth of *Acropora pulchra* 2. Responses of natural and transplanted colonies to temperature and day length. Mar. Biol. 81: 209–215.

Preventing Environmental Pollution through Monitoring, Clean Technologies, Education, Economics and Management

Rosa E. Reyes Gil
Universidad Simón Bolívar, Departamento de Biología de Organismos. Caracas, Venezuela

Luis E. Galván Rico
Universidad Simón Bolívar, Departamento de Tecnología de Servicios. Caracas, Venezuela

SUMMARY: Concerns about environmental problems became evident in the mid-twentieth century, as a consequence of pollution brought about by rapid industrial development. In order to combat this problem, industries initially opted to install equipment at the end of the production process. The latter; however, was expensive, thus increasing the end-cost of the products. More recently, companies have started adopting measures to prevent pollution from the beginning of the manufacturing processes. This chapter describes the main tools used to prevent environmental pollution; development of preventive monitoring programs using early-warning signals, the design and application of clean technologies, the formation of a pro-environmental education in citizens, the evaluation of costs and values of natural resources, as well as ascribing costs to pollution, and the use of Environmental Management as a business strategy for the best exploitation of resources and a minimal generation of wastes.

1 INTRODUCTION

The problems of contamination and pollution started with the Industrial Revolution, through the widespread use of steam engines in the mid-eighteenth century (Hunt & Johnson, 1998). The urban and industrial explosions of the 18th and 19th centuries led to an increased damage to the environment. The more severe problems arose when the emissions and discharges of the industrial processes surpassed the self-purification capacity of the receiving bodies (i.e. water, air and soil), and generated human health problems together with alterations of the ecosystems

It should be pointed out that the world's industrialization has not been equitable, as the countries which are nowadays called developed started the process much earlier than the developing ones. The Industrial Revolution began in Europe, chiefly in England. The factories brought about new job opportunities that were well paid and this attracted a high migration of populations from rural to urban areas. Cities like Manchester, London and Munich grew rapidly. Soon various health and water-pollution problems were detected. As the magnitude of these problems rose, diverse solutions were proposed: at first, sewage systems were built, later followed by biological treatment with activated sludge and finally, the disinfection of waters by means of chlorine, a tertiary treatment. Air and soil pollution was no less severe due to the burning of coal that began to affect the health of the inhabitants of the cities (Hunt & Johnson, 1998; Villegas *et al.*, 2004).

By the end of the 19th century other industrialized countries such as Germany, Canada, France, Holland and the United States, among others, became aware that the receiving bodies of the various discharges, wastes and emissions began to loose their capability for self purification, due to saturation; this disturbed the dynamic equilibria. The flora and fauna began to perceive the effects of environmental degradation. Thus, the first agencies for environmental protection were created with the main purpose of starting and endorsing laws for the control or mitigation of environmental impacts (Miller, 1994).

The industries responded to the legal pressures by introducing alternatives for the reduction of wastes at the final processes of production (end of pipe solutions), such as dust collectors, filters and other similar methods. However, this equipment tended to be rather expensive. Therefore, it became necessary to transfer a percentage of their cost to the price of the final products in order to avoid financial losses (Freeman, 1998; Bifani, 1999; Kiely, 1999).

The developing countries did not have to comply with any environmental legislation in the mid-20th century. Many did not use mitigation processes for the wastes of their products, which would have increased their production costs. This led to an unfair competitive advantage, in which their own markets had less access to those cheaper products. Thus, the developed countries had to seek other solutions. They found them in strategies to optimize industrial processes and improve the life-cycles of products, all of which avoided the generation of wastes and contaminants right from the beginning of the manufacturing process and reduced energy expenditure, thereby increasing the economic benefits (Kiely, 1999; Guédez *et al.*, 2003; Villegas *et al.*, 2004).

The industrialization of the underdeveloped countries started at the end of the 19th and the beginning of the 20th centuries. It had the same consequences that had earlier occurred in the developed countries. However, relevant laws and regulations came into effect only towards the end of the 20th century. The industries adopted the same initial control measures already tried by the developed countries, that is, the use of systems at the end of processes and trying not to modify these. Such equipment usually had to be imported from elsewhere and increased the final product price. Today, the industries of these countries are being forced to adopt preventive measures of pollution control from the beginning of the manufacturing process (Reyes *et al.*, 2002; Guédez *et al.*, 2003; Villegas *et al.*, 2005).

Prevention of pollution before it occurs avoids risks that could affect not only the members of any given community but also the workers involved in the management of the contaminants (Nebel and Wright, 1999; Glyn and Heinke, 1999). An important benefit of pollution prevention is that it includes a solution to economic aspects. The minimization or elimination of wastes altogether leads to savings in costs of materials, ending up in increased productivity.

The careful examination of the manufacturing process required for a successful prevention of undesirable effluents, emissions or wastes may bring about a variety of collateral benefits as well as significant improvements such as less expenditure of water and energy, or a higher quality end-product (Freeman, 1998; Field, 1995). Preventive methods or systems may also lead to greater savings related to fines or clean-up. Often the most important economic saving is a consequence of the reduction of future legal responsabilities and costs caused by the effective control of polluting agents (Reyes *et al.*, 2005).

This chapter describes the principal tools used to prevent environmental pollution; the development of preventive monitoring programs using early-warning signals, the design and application of clean technologies, the formation of pro-environmental awareness in the citizens through Environmental Education Programs, the economic valuation of environmental resources as well as ascribing costs to pollution and feasible prevention tools, and Environmental Management as a strategy for the successful exploitation of natural resources with minimum generation of wastes or residues.

2 EARLY-WARNING SIGNALS

The need to assess the impact of pollution of chemical compounds on environmental quality has led to the development of biological markers (Livingstone, 1993; Reyes, 1999; Capó, 2002). In this regard, the deterioration of environmental health appears to be caused by the synergistic mode of action of a complex mixture of pollutants. On one side, conventional tools for environmental monitoring can evaluate the levels of pollutants and the environmental health status but not the relationships between both. The use of organism responses, on the other hand, may relate cause and effect (Bucheli and Fent, 1995; Capó, 2002).

It is widely accepted that the effects of toxic substances on an ecosystem are initiated by a biochemical reaction. This first response occurs at the lowest levels of biological organization.

It is highly specific and reversible. Subsequently, while the impact rises, a sequence of alterations takes place at more complex levels, leading to disturbances of vital functions and the death of the organism(s). In this case, the effects may appear at the level of entire populations and eventually in the ecosystem (Bucheli and Fent, 1995; Depledge *et al.,* 1995).

At the higher organizational levels the sensitivity, specificity and accuracy tend to decrease while the ecological relevance increases (Adams *et al.*, 1989). The initial biological responses are known as Molecular Biomarkers. They render biologically relevant and complete information about the potential impact of toxic contaminants on the health of organisms. The advantage of their use is that they give early signs of the damage caused by the pollutant; thus, they are quite suitable as alarm signals of the presence of chemical contaminants (Livingstone, 1993; Melacon, 1995).

Among the specific biomarkers widely reported in the available literature there is the synthesis of metalothioneins in response to heavy metals (Roesijadi, 1992; García and Reyes, 1996, 1998, 2001; Reyes, 1999; Salazar and Reyes, 2000; Reyes *et al.,* 2001; Pernía *et al.*, 2008), and the increased activity of monooxygenase P450 in the presence of pollution by polyaromatic hydrocarbons (PAHs) (Livingstone, 1993; Pothuluri *et al.*, 1998; Li *et al.,* 2001; Mastandrea *et al.*, 2005; Bozo *et al.*, 2007). The proven capabilities of microbial communities to utilize few groups of PAH compounds as sources of energy and carbon, have allowed their use as bioremediation tools for spills of crude oil or other organic compounds (Lizarraga-Partida *et al.*, 1991; Martínez-Alonso and Gaju, 2005). The interactions between bacteria and pollutants have been described identifying the sites where those take place, either at intracellular or extracellular levels, or on the bacterial surface (Suárez and Reyes, 2002). Increased populations of coliforms are currently used as bioindicators of organic-matter pollution in diverse water bodies (APHA, 1998; Fujioka, 2002; Caruso *at al.*, 2002; Bozo *et al.*, 2007).

The inclusion of Biomarkers as effective tools has been proposed for Ecological Risk Assessments (ERA) (Neuberger-Cywiak, 2004). These are relevant for the management of natural resources as they allow evaluating temporal and spatial tendencies of the quality or health status of the environment (Suter, 2006). That assessment permits (EPA, 1990, 1998): the implementation of *a priori* actions before environmental degradation becomes irreversible; the determination of the effects of introduction of contaminants to the environment; the identification of the sources of pollution; and the development of criteria for environmental quality with regulatory purposes.

The use of biomarkers is thus proposed for ERA in order to obtain reliable data on the damages inflicted by a contaminant on an organism, population or ecosystem. Such values of potential damages can then be used to propose preventive or mitigation measures to avoid or stop them (INE, 2006). Nonetheless, the use of biomarkers may be limited or restricted by the following factors: changes of the environmental variables, specificity, dose-response values, understanding of the mechanism(s) involved, and lack of technical expertise or adequate instruments to obtain and interpret the results (Livingstone, 1993; Bucheli and Fent, 1995).

Another important restriction to the use of biomarkers is that they might change with seasonal or reproductive conditions, or with other variables of the ambient (Depledge *et al.,* 1995; Knap *et al.,* 2002). The specificity of a biomarker has been strongly related to diverse groups of pollutants such as heavy metals and organic compounds, whereas this does not appear to be the case for other contaminants. Dose-response values may differ between field and laboratory measurements. This has to be kept in mind if the emphasis is placed on the biomarker as a biological measure of the levels of pollution instead of its use as an integrator of the effects of a mixture of contaminants on the organism. Finally, understanding the underlying mechanisms of action is a requisite for the interpretation of results and the effective application of biomarkers (Melacon, 1995), an area that obviously needs further research.

3 CLEAN TECHNOLOGIES

The choice of clean technologies to replace the older more contaminant ones as a prevention tool is becoming more widespread every day. Industries can adopt either of two positions regarding pollution, knowing that if they contaminate they will have to face legislation and the public in general. Those two postures are Decontamination and Anticontamination; the formers refers to the installation of systems or equipment that control or eliminate the polluting agents by means of external treatments such as filters, cleaners, purifiers, recycling plants, incinerators and controlled dumps, among others. The latter option seeks to intervene in the processes applying clean technologies, so as to avoid the generation of contaminants, attempting to include internal treatments to the processes in order to eliminate the necessity to further handle any wastes (Aldy *et al.*, 1998; Seoánez, 1998; Kiely, 1999; Marco and Reyes, 2003).

Clean technologies comprise the application of strategies that include methods such as recycling, substitution or replacement, recovery and re-evaluation. These technologies may be simple as in the case of a single change in a procedure, or they may be highly sophisticated (Seoánez, 1998; Kiely, 1999; Marco and Reyes, 2003). For example, a simple change would be to add urban residues compost to agricultural soils to convert the residual organic matter into fertilizer to improve soil quality. A more sophisticated procedure is to improve and diversify the cultivable plant varieties that are tolerant to herbicides as well as to pests (biotechnology).

Among the advantages of the application of clean technologies that stand out include: better business results based on less pollution produced; the applied procedures are efficient, reliable and permanent, with high incidence on productivity and hence, good financial returns; payments of fines or other expenses can be saved, and prevention which implies an evident economic advantage compared to the costs of decontamination (Seoánez, 1998).

In an ideal form or a real model, diverse clean technologies may be integrated, aimed at reducing both the generation of wastes and the negative impacts on natural resources; likewise, the subsidy of energy costs to produce and use fertilizers and biocides may be decreased by the use of alternative recycling methods. At the same time, the energy flows or fluxes within the system should be maximized. To achieve this, various techniques may be applied to reinforce the integration; diversification, nutrient recycling, atmospheric nitrogen fixation, the use of local or alternative energy sources, fractioning of plant biomass, high-yield crops, multi-purpose trees, the use and control of wastewaters, as well as the overall management of pests, rotation and association of crops and minimal soil turnover (Escobar *et al.*, 1998; Nieto *et al.*, 2002).

Corchete (1986) proposed methods that may be regarded as clean technologies applied to agriculture, such as collection and classification of inorganic residues to be later sold or reused or conveniently destroyed; also, removal of the corpses of dead animals, chiefly cattle, to places where carrion-eater organisms could dispose of them; further on, to bury in deep trenches the weeds and other vegetal residues, and to use weeds and stubble as an alternative source of agroenergy. Among the views envisaged by Altieri (1998, 1999) for the improvement of productivity and resiliency of agroecological systems there could be mentioned the diverse designs of polycultures, diversified systems of weed or cover crops, among others. Due to the success of clean technologies, presently a large number of commercial produce is being regarded as organic and its use and consumption is widening as more farmers join agroecological processes. The following few examples may be mentioned: earthworm *hummus* (a fertilizer), Tricobiol (a fungicide), Nemabiol (a nematocide), Biograss (which is insecticide), among others (Rigby and Cáseres, 2001; Anzola, 2002).

Framed within the use of clean technologies is the concept of Design for the Environment (DftE), proposed in 1992 to incorporate environmental awareness in the development of new products. The DftE concept is a systematic regard of the design function to environmental concerns, as well as health and safety along the whole life-cycle of the process and the product. In short, DftE means the design of safe and eco-efficient products. Its purpose is to include prevention in the design phase of processes prior to product specifications or the building of industrial plants, so that eco-efficiency reaches its highest magnitude. It would be of little use to correct pollution problems when the basic process parameters have already been established, as their rectification is always more expensive (Fiksel, 1997). It is also noteworthy to point out that DftE is based on a complete knowledge of the life-cycle of the manufactured products, which ensures that innovations always lead to reductions of pollution and of undesirable residues at one or more stages of the cycle, while other aspects of costs, yields and profits are also satisfied (Kiely, 1999).

4 ENVIRONMENTAL EDUCATION

The need to educate the citizens about the features and functions of the environment has always been recurrent, through diverse approaches centered on natural resources. However, this was not recognized until recently as Environmental Education, as this definition spread only in the 60s of the 20th century. It has been stated that the term emerged when humanity started to question its role in the conservation or degradation of its (Pellegrini, 2002). Within the context of this chapter, Environmental Education is shown as a prevention tool of contamination, starting from the fact that sensibilization and environmental awareness or knowledge by the citizens should lead to appropriate environmental attitudes or conducts. Especially, the environmental education of university students is highly important as it is necessary to supply these future professionals the required tools to understand and generate changes in their personal behaviour, as well as of the policies of their future working places, in relation to the environment.

It has to be borne in mind that in a chain of co-responsability, the Universities play a key role because they form and shape the future members of governments, who will be creating new legislation in environmental matters and will also be orienting public policies. Furthermore, the members of technical and consulting committees and professionals responsible for the manufacturing or construction of new products usually hold a university degree (Domingo *et al.*, 2004). In this context, and as a first step, it becomes fundamental to transmit to university students the basic aspects of sustainability and related environmental concepts. Today this is absent from most study programs (Pellegrini and Reyes, 2001; Reyes *et al.*, 2006, Galvan and Reyes, 1996).

Activities aimed at the incorporation of environmental themes into academic programs should be reinforced, either introducing the subject into existing courses or creating new ones on environment and sustainable development on an obligatory basis. A feasible proposal would be to initiate a project for environmental formation and awareness in all universities, based on actions to define policies, respond to environmental problems and educate communities in these aspects. Such actions should rely on the establishment of an institutional environmental policy, the creation of new courses stressing this theme, and the continuous formation of teachers and professors (see Pellegrini *et al.*, 2007 a,b).

Furthermore, environmental education for adults could be tested on mine-worker populations from developing countries. For instance, a case of study could be the Imataca Forest Reserve, in southern Venezuela, which is a rainy tropical forest of the Amazonia. This Reserve has been exploited by artisanal procedures for many years to obtain alluvial gold, using methods that have aggressively damaged the environment, such as the release of large amounts of mercury and the deforestation of extensive jungle areas (Reyes *et al.*, 2003). An Environmental Education Program can be envisaged, to sensibilize and train the miners to restore the areas disturbed by the mining activities and to warn them about the unhealthy effects of mercury. The results of the Program could generate a reference standard for other places or countries, and they could serve as a starting point for other educational programs concerning the environment and directed at adults in other socioeconomic contexts as rough as artisanal gold mining.

5 ENVIRONMENTAL ECONOMICS

The worldwide environmental problems caused by pollution may be regarded as economic variables. Thus they should be analyzed in economic terms. Various methods for the economical assessment of environmental quality represent powerful tools for the prevention and control of pollution (Azqueta, 1995; Azqueta and Pérez, 1996). These tools allow us to: rely on real-economy arguments that justify the use of clean technologies, evaluate the economic magnitude of an environmental offense or damage, and develop a "culture of prevention" based on costs as incorporated into the cost-benefit analyses of any investment project (Pearce, 1976; Pearce and Turner, 1995; Azqueta, 2002).

When examining aspects concerning the economy factors of pollution, that is, setting permissible levels or values both for the diverse economic agents involved and the society in general, and the methods proposed to achieve this goal (Reyes *et al.*, 2005), contamination might be defined as a negative external offshoot of any production activity, and its associated costs must be taken up (i.e. internalized) by the economic agent that causes it. Hence, pollution always implies a cost for the economic agent that is affected, as long as this is not compensated by the causing agent. From this point of view, the overall social least-level of pollution is determined by the intercept of the curve of marginal cost of the affected agent with that of marginal benefit of the causal one (Romero, 1993, 1997).

The optimal acceptable level of contamination may be estimated by means of various methods. The ones that stand out include tax method considered by Pigou (1920), and the free negotiation between parts as described by Coase (1960). Other environmental policies in use for the establishment of optimum least-levels of pollution are: the setting of environmental standards that fix the maximum level of pollution permitted to the contaminating agent, the granting of subsidies to decrease pollution, in order to stimulate the contaminating agent to introduce clean technologies, and the issue of payable pollution permits, among others. These policies, together with prior research on the prices and costs of environmental goods (Reyes and Galván, 1999) and the value of unique ecosystems (Galván and Reyes, 1999), represent a starting point to explore the viability of these tools for the prevention and control of environmental contamination and pollution, and the appropriate management of natural resources.

6 ENVIRONMENTAL MANAGEMENT

Among the main strategies proposed at high enterprise levels to prevent or impede pollution and its sequels are the Environmental Management Systems (EMS). These systems constitute a way that an organization or industry can use to systematically identify and solve its environmental impacts (Cascio *et al.*, 1996). The EMS are closely related to Quality Management Systems, which are oriented towards a systematic and cyclic process of continuous improvement. Thus, EMS are built into the organizational structure of the enterprise or industry. They are focused on the control of processes that might lead to damage to the environment, and to minimize possible impacts while improving the yield of those processes. Likewise, they identify policies, procedures and resources to comply and maintain an effective environmental management of the company (Clements, 1997; Hunt and Johnson, 1998).

The EMS are mechanisms to ensure a cyclic process of continuous improvements, starting with the planification of a desirable result (i.e., a better environmental role), followed by the carrying out of a plan, checking if this works and, finally, rectifying and improving the planned actions based on the observations emerging from the checking process (Roberts and Robinson, 1999; Reyes *et al.,* 2002). The usage of EMS is one of the most frequently applied tools of industries seeking to improve their environmental image whilst raising their economical goals, as such systems are centered on a sustainable development under an eco-efficient scheme (Villegas *et al.*, 2005). Also, the EMS is at the base of most of the environmental normatives specified by the International Standards Organization (ISO) in 1996. They may be certified under one of such norms (ISO 14001), conferring a competitive advantage to any given industry when placing its product(s) at the international markets (ISO, 1996 a,b).

In view of these comments, it is interesting to revise the efforts in environmental programs carried out worldwide by oil companies (Guédez *et al.*, 2003). This analysis reveals that only those companies firmly compromised with their own environmental policies have successfully implemented and gained from the application of the proposed plans for environmental preservation.

The support of the top-level directors and managers appears to be essential to reach all the proposed goals, together with the internalization of the overall purposes by the operational units and their various Departments. This all-round compromise is evident in companies such as Shell, Statoil, BP, and Exxon-Mobil. Chevron-Texaco and Unocal are catching up by establishing integrated systems in their operations; in some cases, they have already published fruitful results that resulted in international certificates. This situation is also shared by the state-owned companies Pemex, Petrobras and PDVSA, although the latter is still planning and/or integrating the system.

It is possible follow the steps of diverse energy companies that have embraced these Integrated Systems as a positive way to achieve a solid image of environmental compromise, investing large amounts to obtain tangible changes. The picture that emerges is that the development of a certifiable system is not enough unless it is regarded as an essential part of the bussiness, which under strict controls and continuous improvements leads to a better productivity and high competitivity.

Further results resembling those observed in PDVSA were found by examining 100 small and medium companies from Venezuela (Villegas *et al.,* 2005). Most of them are still planning and developing EMS, even though transnational companies in Venezuela and Latin America have imposed a strong pressure on the smaller companies to implement clean technologies and EMS. Presently, research is being carried out to diagnose the medium and large companies from Latin America with respect to their involvement in environmental matters and the incorporation of environmental management attitudes within their organizational structure. In this same context, results of studies regarding industrial strategies for environmental preservation in Brasil, in which various owners and executives of diverse companies were interviewed, indicated that both groups perceive as remote the returns of projects centered on environmental conservation (Clemente *et al.,* 2005). Executives were less reticent and more proactive than owners with regard to environmental investments, and the former perceived more easily the strategic value of preservation projects. Nevertheless, both groups from the most contaminating companies were more involved in resolving legal complaints and disputes than in recognizing the strategic importance of investing in favour of the environment. In contrast, the owners and staff of less-contaminating companies envisaged such investments as profitable and strategically convenient.

7 CONCLUDING REMARKS

Efforts aimed at minimizing the wastes and effluents generated by modern industrial and urban processes have led to various preventive strategies cleanup procedures. At the ecological level, the early detection of signals that may predict possible future irreversible harm or disruption may be of great help to prevent or impede such damages. Recent proposals in this regard include the use of biomarkers in environmental impact studies and in ecological risk studies as preventive strategies. An interesting prevention tool against pollution is the environmental education of citizens. Likewise, at an industrial level the preventive procedures are linked to the application of clean technologies along production processes, the maintenance protocols and even to the generation of final products more friendly to the environment. The economic valuation of environmental goods and benefits, and the estimation of costs of pollution may represent economic tools that could be helpful to environmental management. Also, the incorporation of environmental variables into the organizational structure of companies or industries represents an important advancement on the way to minimize wastes by preventing contaminating processes.

REFERENCES

Adams, S., Shepard, K., Greeley, M., Jimenez, B., Ryon, M., Shugart, L. and McCarty, J. 1989. The use of bioindicators for assessing the effects of pollutant stress on fish. Mar. Env. Res. 28: 459–464.

Aldy, J., Hrubovcak, J. and Vasavada, U. 1998. The role of technology in sustaining agriculture and the environment. Ecological Economics 26: 81–96.

Altieri, M. 1998. Modern Agriculture: Ecological impacts and the possibilities for truly sustainable farming. Division of Insect Biology, University of California, Berkeley, USA. http://www.internet.agroecology in action.htm

Altieri, M. 1999. Multifunctional Dimensions of Ecologically-based Agriculture in Latin America. Department of Environmental Science Policy and Management University of California, Berkeley, USA.http://www.internet.agroecology in action.htm

Anzola, L. 2002. Índice Agropecuario. Edición 27. Estado Aragua, Venezuela. 90 p.

APHA. 1998. Standard methods for the examination of water and wastewater. Washington, DC. Am. Public Health Assoc. 234 pp.
Azqueta, D. 1995. Valoración Económica de la Calidad Ambiental. McGraw-Hill Interamericana, S.A. España. 299 pp.
Azqueta, D. and Pérez, L. 1996. Gestión de Espacios Naturales. La Demanda de Servicios Recreativos. McGraw-Hill Interamericana S.A. España. 237 pp.
Azqueta, D. 2002. Introducción a la Economía Ambiental. Mc.Graw-Hill Interamericana, S.A. España. 420 pp.
Bifani, P. 1999. Medio Ambiente y Desarrollo Sostenible. IEPALA Editorial. España, 593 pp.
Bozo, L., Fernández, M., López, M., Reyes, R. and Suárez, P. 2007. Biomarcadores de la contaminación química en comunidades microbianas. Interciencia 32(1): 8–13.
Bucheli, T. and Fent, K. 1995. Induction of cytochrome P 450 as a biomarker for environmental contamination. Crit. Rev. Env. Sci & Tech. 25(3): 201–268.
Capó, M. 2002. Principios de Ecotoxicología. McGraw Hill/Interamericana de España S.A. Madrid. 314 p.
Cascio, J., Woodside, G. and Mitchell, P. 1997. Guía ISO-14000. Las Nuevas Normas Internacionales para la Administración Ambiental. Editorial McGraw-Hill Interamericana S.A. México. 224 pp.
Caruso, G., Crisafi, E. and Mancuso, M. 2002. Development of an enzyme assay for rapid assessment of *Escherichia coli* in seawaters. Journal of Applied Microbiology 93: 548–556.
Clements, R. 1997. Guía Completa de las Normas ISO 14000. Ediciones Gestión 2000. España. 285 pp.
Clemente, A., Souza, A., Galván, L. and Reyes, R. 2005. Estrategias empresariales para la conservación ambiental en el Sector Industrial. Universidad, Ciencia y Tecnología, (UCT) 9(33): 3–9.
Coase, R. 1960. Problems of Social Cost. J. Law and Economics 1: 1–44.
Corchete, S. 1986. El Agricultor ante la conservación y mejora del Medio Ambiente. Hojas Divulgadoras. N° 13/86 HD. Ministerio de Agricultura, Pesca y Alimentación. Madrid. 24 p.
Depledge, M., Aagaard, A. and Gyorkos, P. 1995. Assessment of trace metal toxicity using molecular, physiological and behavioral biomarkers. Mar. Pollut. Bull. 21: 19–27.
Domingo, J., Gámiz, J. y H. Martínez. 2004. Los planes de medio ambiente en las universidades catalanas. Ideas Brillantes. Espacio de reflexión y comunicación en Desarrollo Sostenible.
EPA. 1990. Risk Assessment Methodologies: comparing EPA and status approaches. No. 570990012, 122 pp.
EPA. 1998. Guidelines for Ecological Risks Assessment. EPA/630/R-95/002F. 171 pp.
Escobar, A., Messa, H., Ruiz-Silvera, C. and Rodríguez, J. 1998. Proyecto de Establecimiento y evaluación de un Modelo Físico de Agricultura Tropical Sostenible. Taller Internacional: Agricultura Tropical Sostenible. Experiencias y desafíos para el tercer milenio. Fundación Polar. CIARA. Caracas, Venezuela. 65–72 pp.
Field, B. 1995. Economía Ambiental. Una Introducción. Editorial McGraw-Hill Interamericana S.A. Colombia. 587 pp.
Fiksel, J. 1997. Ingeniería de Diseño Medioambiental. DFE. Desarrollo Integral de Productos y Procesos Ecoeficientes. Editorial McGraw-Hill. España. 512 pp.
Freeman, H. 1998. Manual de Prevención de la Contaminación Industrial. Mc Graw-Hill Interamericana S.A. Ciudad de México. 943 p.
Fujioka, R. 2002. Indicators of marine recreational water quality. pp: 234–243. En: Hurst C.J., Knudsen G.R., McInerney M.J., Stetzenbach L.D. y Walter M.V. Manual of Environmental Microbiology. 2nd Edition. ASM Press, Washington D.C.
Galván, L. and Reyes,R. 1999. Asignación de precio a los ecosistemas como bienes ambientales únicos. Interciencia 24(1): 14–16.
García, E. and Reyes, R. 1996. Bioconcentration of mercury in *Acetabularia calyculus*: Evidence of a polypeptide in whole cells and anucleated cells. Tox. Env. Chem. 55: 11–18.
García, E. and Reyes, R. 1998. Induction of mercury-binding peptide in whole cells and anucleated cells of *Acetabularia calyculus*. Tox. Env. Chem. 67: 189–196.
García, E. and Reyes, R. 2001. Synthesis pattern of an Hg-binding protein in *Acetabularia calyculus* during short-term exposure to mercury. Bull. Environ. Contam. Tox. 66(3): 357–364.
Glynn, J. and Heinke, G. 1999. Ingeniería Ambiental. 2° Edición. Editorial Prentice Hall Hispanoamericana S.A. México. 778 pp.
Guédez, C., de Armas, D., Reyes, R. and Galván, L. 2003. Los sistemas de gestión ambiental en la industria petrolera internacional. Interciencia 28(9): 528–533.
Hunt, D. and Johnson, C. 1998. Sistemas de Gestión Medioambiental. Editorial McGraw-Hill Interamericana. Colombia. 318 pp.
Instituto Nacional de Ecología (INE). 2006. Capitulo 4: La evaluación de riesgo ecológico. México. http://www.ine.gob.mx/ueajei/publicaciones/libros/400/cap4.html

ISO (International Standard Organization). 1996a. Environmental Management Systems. Specification with guidance for use ISO 14001. Switzerland.

ISO (International Standard Organization). 1996b. Environmental Management Systems. General Guidelines on Principles Systems and Supporting Techniques (ISO 14004). Switzerland.

Kiely, G. 1999. Ingeniería Ambiental. McGraw-Hill. Madrid, España. 1.331 pp.

Knap, A., Dewailly, E., Furgal, C., Galvin, J., Baden, D., Bowen, R., Depledge, M., Duguay, L., Fleming, L., Ford, T., Moser, F., Owen, R., Suk, W. and Unluata, U. 2002. Indicators of ocean health and human health: developing a research and monitoring framework. Environ. Health Perspect. 110: 839–845.

Li, Q., Ogawa, J., Schmid, R. and Shimizu, S. 2001. Engineering cytochrome P450 BM-3 for oxidation of polycyclic aromatic hydrocarbons. Appl. Environ. Microbiol. 67: 5735–5739.

Livingstone, D. 1993. Biotechnology and pollution monitoring: use of molecular biomarkers in the aquatic environment. J. Chem. Tech. Biotechnol. 57: 195–211.

Lizarraga-Partida, M.L., Izquierdo-Vicuna, F. and Wong-Chang, I. 1991. Marine bacteria on the Campeche Bank Oil Field. Mar. Pollut. Bull. 22: 401–405.

Marco, O. and Reyes, R. 2003. Tecnologías limpias aplicadas a la agricultura. Interciencia 28(5): 252–258.

Martínez-Alonso, M. and Gaju, N. 2005. El papel de los tapetes microbianos en la biorrecuperación de zonas litorales sometidas a la contaminación por vertidos de petróleo. Ecosistemas. Revista técnica y científica de ecología y medio ambiente. pp 1–12. http://www.revistaecosistemas.net

Mastandrea, C., Chichizola, C., Luduena, B., Sánchez, H., Álvarez, H. and Gutiérrez, A. 2005. Hidrocarburos aromáticos policíclicos. Riesgos para la salud y marcadores biológicos. Acta Bioquím. Clín. Latinoam. 39: 27–36.

Melacon, M. 1995. Bioindicators used in aquatic and terrestrial monitoring. Handbook of Ecotoxicology. Lewis Publisher Press Inc. Boca Raton. USA. 755 pp.

Miller, G. 1994. Ecología y Medio Ambiente. Grupo Editorial Iberoamérica S.A. de C.V. México. 867 pp.

Nebel, B. and Wright, R. 1999. Ciencias Ambientales. Editorial Prentice Hall Hispanoamericana S.A. México. 698 pp.

Neuberger-Cywiak, L. 2004. La ecotoxicología como herramientas en las Evaluaciones de Riesgo Ecológico y en la Evaluaciones de Impacto Ambiental. 6th. Congreso Ibérico y 3rd. Iberoamericano de Contaminación y Toxicología Ambiental (CICTA-2005). Cádiz, España.

Nieto, A., Murillo, B., Troyo, E., Larinaga, J. and Garcia, L. 2002. El uso de compostas como alternativa ecológica para la producción sostenible del chile (*Capsicum annuum* L.) en zonas áridas. Interciencia 27: 417–421.

Pearce. D. 1976. Economía Ambiental. Fondo de Cultura Económica. México. 258 pp.

Pearce, D. and Turner, R. 1995. Economía de los Recursos Naturales y del Medio Ambiente. Celeste Ediciones. España. 448 pp.

Pellegrini, N. 2002. Environment in the National Park System of Venezuela. Environ. Educ. Res. 8(4): 463–473.

Pellegrini, N, and y Reyes, R. 2001. Los mapas conceptuales como herramientas didácticas en la educación científica. Interciencia 26: 144–149.

Pellegrini, N., Reyes, R., Martín, A., Aguilera, M. and Pulido, M. 2007a. La Dimensión Ambiental en la Universidad Simón Bolívar. Universidad, Ciencia y Tecnología 11(42): 45–50.

Pellegrini, N., Reyes, R. and Pulido, M. 2007b. Programa de Interpretación Ambiental en la Universidad Simón Bolívar: sus recursos, su cultura y su historia. Educere 11(39): 605–611.

Pernía, B., De Sousa, A., Reyes, R. and Castrillo, M. 2008. Biomarcadores de la contaminación por cadmio en las plantas. Interciencia 33(2): 112–119.

Pigou. A. 1920. The Economics of Welfare. McMillan, S.A. UK. 298 pp.

Pothuluri, J., Sutherland, J., Freeman, J. and Cerniglia, C. 1998. Fungal biotransformation of 6-nitrochrysene. Appl. Environ. Microbiol. 4: 3106–3109.

Reyes, R. 1999. Las metalotioninas como biomarcadores de la contaminación ambiental por metales pesados en organismos acuáticos. Interciencia 24: 366–371.

Reyes, R. and Galván, L. 1999. Asignación de precios a los bienes ambientales como instrumento de gestión tecnológica. Universidad, Ciencia y Tecnología 3(9): 3–7.

Reyes, R., Salazar, R. and García, E. 2001. Incorporation of cadmium by *Acetabularia calyculus*. Bull. Environ. Contam. Tox. 67: 749–755.

Reyes, R., Galván, L., Guédez, C. and de Armas, D. 2002. La Gerencia Ambiental en el sistema productivo venezolano. Universidad, Ciencia y Tecnología 6(23): 155–159.

Reyes, R., Pellegrini, N. and Farah, D. 2003. La educación ambiental para la pequeña minería aurífera en la reserva forestal Imataca, Estado Bolívar, Venezuela. Universidad, Ciencia y Tecnología 7(28): 262–266.

Reyes, R., Galván, L. and Aguiar, M. 2005. El precio de la contaminación como herramienta económica e instrumento de política ambiental. Interciencia 30(7): 436–441.
Reyes, R., De Sousa, A. and Petersen, J. 2006. La prevención de la contaminación industrial como asignatura para la formación ambiental universitaria. Universidad, Ciencia y Tecnología 10(40): 198–204.
Roberts, H. and Robinson, G. 1999. ISO 14001 EMS. Manual de Sistema de Gestión Medioambiental. Editorial Paraninfo. España. 425 pp.
Roesijadi, G. 1992. Metallothionein in metal regulation and toxicity in aquatic animals. Aq. Tox. 22: 81–114.
Romero, C. 1993. Economía Ambiental. Aspectos Básicos. Revista de Occidente 149: 25–39.
Romero, C. 1997. Economía de los Recursos Ambientales y Naturales. Alianza Editorial. España. 214 pp.
Rugby, D. and Cáceres, D. 2001. Organic farming and the sustainability agricultural systems. Agricultural Systems 68: 21–40.
Salazar, R. and Reyes, R. 2000. Efectos tóxicos y mecanismos de tolerancia al cadmio en los seres vivos. Universidad, Ciencia y Tecnología 13: 17–22.
Seoánez, M. 1998. Medio Ambiente y Desarrollo. Manual de gestión de los Recursos en Función del Medio Ambiente. Mundi-Prensa. Madrid. 592 pp.
Suárez, P. and Reyes, R. 2002. La incorporación de metales pesados en las bacterias y su importancia para el ambiente. Interciencia 27: 160–164.
Suter, G. 2006. Ecological Risk Assessment. CRC Presss. 643 pp.
Villegas, A., Reyes, R. and Galván, L. 2004. Problemática Ambiental en Venezuela y el Mundo. Universidad, Ciencia y Tecnología 8(30): 117–125.
Villegas, A., Reyes, R. and Galván, L. 2005. Gestión ambiental bajo ISO 14001 en Venezuela. Universidad, Ciencia y Tecnología 10(34): 63–69.

Poverty, Urban Land and Africa's Sustainable Development Controversy

F. Nikoi Hammond
School of Engineering and the Built Environment, University of Wolverhampton, UK

Joseph Somevi
Aberdeenshire Council, UK

SUMMARY: The concept of sustainable development has raised the spectre of possible exhaustion of vital resources at costs to posterity. This has called for measures to impose discipline on the way resources are used. At the same time, Africa has the added obligation to half the number of its inhabitant living under extreme poverty by 2015. These concurrent goals, from the urban land use perspective, appear at odds. The pursuit of poverty requires intensive and extensive resource usage and yet sustainable development, when taken to the extreme may be opposed to such use. How can a delicate balance be achieved in this conundrum in an impoverished region like Africa?

1 INTRODUCTION

This chapter, which presents a focus on Africa, argues that whilst quantitative restrictions in the use of urban resources may be popular, its effects on the sustainable development and poverty alleviation of Africa remains to be fully explored. The point advanced here is that efficiency and intensity of urban resource usage holds the key to the sustainable development and poverty alleviation on this continent.

African is unique when it comes to poverty and sustainable development. Like other developing regions, this continent is caught up in a world of two ideals. Both are issued from the vistas of the global co-operation in international development and poverty alleviation. This is owed primarily to a multitude of problems (e.g. hardship, misery, ignorance, disease and deprivation) confronting the continent as well as the rapidly widening disparity in material standard of living between industrialised and the so called poor countries. Currently, some 41 per cent of inhabitants of African countries are extremely poor and are surviving on the equivalent of 0.82 dollars a day (i.e. less than one dollar a day). At one extreme are the international requirements on African governments to rehabilitate their economies in keeping with the United Nations Millennium Development Goals (MDGs) as agreed in 2000. The MDG set such targets as halving the proportion of people living on less than US $1 a day by 2015. To deal adequately with the poverty problem, it is important for the causes to be identified.

The causes of poverty are threefold: inefficient allocation of natural resources, capital, labour and time; underutilization of resources and/or inequitable distribution of incomes. Liberal and moderate economists, following primarily after Adam Smith (1776), argue that as individuals and economic entities seek to maximise their own self interest and profits, they unwittingly also promote the efficient allocation as well as optimal use of resources. Consequently, a requisite for reaching the targets set out in the MDGs is the intensive, efficient and effective exploitation and utilization of the continent's resources without damaging the environment. This is consistent with the concept of Sustainable Development in the 1987 report of the United Nation's World Commission on Environment and Development as fleshed out later in the United Nations Agenda 21 and recently energised by Nicholas Stern (2006). in a report for the UK government.

Sustainable development has economic, social, environmental and ethical dimensions. From an economic perspective, a resource is efficiently allocated and optimally utilized if it is put to the use that generates the highest possible returns. In free enterprise economies, the first two causes of poverty and economic malaise may be overcome by setting in place a broad legal as well as effective and quality infrastructural framework within which individuals and actors could pursue their self interest. The required legal framework may consist of, at a minimum, laws and security rules that promote sound investment, access to financial, labour and land transactions. The law must be non-discriminatory in principle and practice. Rather, it must remove all manner of discrimination with respect to opportunities regarding education, investments, health services, land resources, capital and employment.

Socially, equitable distribution of the incomes on the other hand may be achieved through mechanisms and social benefits like income supports, unemployment benefits and housing, free or discounted education, health and transportation services. From an environmental perspective, economic and social goals have to be pursued in ways that will cause the least damage to the quality of the environment and limit the exhaustion of irreplaceable resources. Before delving into the way these may be achieved, it is useful to ponder the meaning of poverty.

2 WHO ARE THE POOR?

Like many other socio-economic phenomena, poverty is capable of bearing a variety of meanings. By behavioural patterns and the extent to which people can consume goods and services, Schreider *et al.* (1971) defined poverty as the inadequate consumption of goods (such as food and clothing) and services (such as medical care and recreation). Looking at the cause of poverty, Charles Booth writing in 1899 classified the impoverished into two main categories; poor and very poor. The poor, according to Booth, are those whose means may be insufficient but are barely sufficient for decent independent life. The very poor, a term that has now been replaced by the World Bank and the United Nations with the term extremely poor, are, according to Booth, those whose means are insufficient for decent independent life according to the usual standard of life in a given country. The poor, Booth contends live under a struggle to obtain the necessaries of life and make both ends meet, while the very poor or extremely poor live in a state of chronic want.

Seebohm Rowntree (1901) following Booth, looked at the poor in terms of family units rather than individuals and classified poverty into primary poverty and secondary poverty. According to Rowntree, families whose total earnings are insufficient to obtain the minimum necessaries for the maintenance of merely physical efficiency fall under primary poverty. On the other hand, families whose total earnings would be sufficient for the maintenance of merely physical efficiency were it not that some portion of them is absorbed by other expenditure, fall under secondary poverty. Rowntree's primary poverty corresponds in essence to Booth's very poor and secondary poverty corresponds to the poor.

According to these classifications the dividing line between the poor and the non-poor is the ability to afford decent life or physical efficiency. There cannot be common worldwide criteria for determining the poverty status of a person; a poor person in one country may be an extremely poor or even a well-off person in another country. This presents difficulties in international comparative analysis of poverty as no common measure or basis can be applied. Yet without comparative analysis it is impossible to ascertain how well countries are doing in striving to rid themselves of the poor.

3 HOW IS POVERTY MEASURED?

Rowntree (1901) provided what arguably remains the most reliable statistical measure of poverty. His basic argument was that to achieve physical efficiency or, in the words of Booth decent life, a person must be able to afford the minimum calories needed to keep them working efficiently. At the time of Rowntree, men required 3,500 calories; women and children required less. The maintenance of physical efficiency required decent shelter and household sundries such as clothing and fuel. Thus by computing the costs of attaining the requisite calories using the minimum price of food required to achieve 3,500 calories and adding the result to the rent payable of decent shelter and the costs of household sundries, Rowntree established in essence the cash amount required to achieve this. This amount became known as the poverty line. Any body earning less than this amount that is, falling below the poverty line was to be regarded as very poor or falling under secondary poverty.

Since 1980, the cut-off line (poverty line) for the extremely poor (otherwise termed very poor or under primary poverty) employed by the World Bank and generally accepted by development economists for the developing world is one US dollar a day. The cut-off line for the poor or those under secondary poverty is two dollars a day. Accordingly, those earning below one dollar a day across the developing world regardless of their country of residence are considered equally extremely poor whilst those earning below two dollars a day are regarded as equally poor. The trouble is, since the 1980 (almost three decades ago) this cut-off line has not changed though prices of food, sundries, rents and those things required for decent life have gone up over the years. Consequently, current poverty statistics on which floods of development economic analysis are conducted may be presenting overly optimistic views of the proportion of people that are extremely poor or even poor across the developing world than the actual truth. What then is the state of poverty across the developing world and sub-Saharan Africa?

4 POVERTY IN THE DEVELOPING WORLD

Poverty exists everywhere, in both developed and developing countries. Then again, this is not a recent phenomenon. It is the scale of poverty in the developing world that gives cause for greater concern. As at 1981, the number of people living below the extreme scarcity line of one dollar a day or less stood at about 1.5 billion globally. More than half of these people were in East Asia and more than a fourth in South Asia. Of the total only, 164 million were at the time in sub-Saharan Africa (Sachs, 2005).

Since 2000, the international development community has made an explicit commitment to eliminate extreme poverty across the world, a commitment enunciated by the United Nations through the Millennium Development Goals (MDGs). The MDG set the year 2015 by which the number of people living on or below the extreme poverty line of less than one dollar a day across the world should be halved.

On the aggregate, the statistics based on the one dollar a day cut-off mark show considerable progress towards achieving this target. For example the number of extremely poor worldwide decreased from 1.5 billion in 1981 to 970 million in 2004. This rate of progress is much faster than the 0.6 per cent annual rate required from 1990 to achieve the target. When the data is examined on regional basis, it becomes clearer that not all regions are making encouraging progress; some regions indeed represent serious drag on worldwide progress.

5 LOSING OUT IN THE RACE TO THE FINISHING LINE

Table 1 reports the proportion of national populations of respective developing regions that earn below the cut-off for extremely poor of one dollar a day. The data was retrieved from the World Bank's online poverty database *POCVAL* (2008) which is currently the most extensive database on poverty across the developing world. The table shows that as of 1981 sub-Saharan Africa (SSA) had the third largest collection of extremely poor people in the developing world. The East Asia and Pacific region (EAP) had over half of its population extremely poor. South Asia (SA) came second with about half of its population extremely poor. By 1999 when the MDG was promulgated sub-Saharan Africa had attained the unenviable position of the world's poorest region with 46.1 per cent of its population classified as extremely poor.

Table 1. Developing Countries: Population Living below US1 a day (Headcount).

	1981	1984	1987	1990	1993	1996	1999	2002	2004
East Asia and Pacific	57.7	39	28.2	29.8	25.2	16.1	15.4	12.3	9.07
Europe and Central Asia	0.7	0.52	0.36	0.47	3.61	4.22	3.6	1.28	0.95
Latin America and the Caribbean	10.8	13.1	12.1	10.2	8.42	8.88	9.6	9.05	8.63
Middle East and North Africa	5.08	3.82	3.11	2.33	2.12	2.23	2.08	1.69	1.47
South Asia	49.6	45.4	45.1	43.1	36.9	36.1	35	33.4	31.1
Sub-Saharan Africa	42.4	46.3	47.5	46.8	45.7	48	46.1	42.6	41.1

Source: POCVAL, World Bank (2007)

Table 2. Proportion of Population Surviving on Less than US1 a day (Headcount).

	1981	1984	1987	1990	1993	1996	1999	2002	2004
East Asia and Pacific	84.80	76.79	68.53	69.72	65.04	52.50	49.23	41.68	35.99
Europe and Central Asia	4.62	3.99	3.09	4.31	16.48	17.94	18.29	12.94	9.83
Latin America and the Caribbean	28.45	32.24	29.61	26.25	24.09	25.29	25.23	24.72	22.14
Middle East and North Africa	29.16	25.59	24.31	21.69	22.03	23.51	23.61	21.09	19.70
South Asia	88.52	86.95	86.59	85.63	82.20	82.12	80.44	79.70	77.20
Sub-Saharan Africa	74.49	76.88	77.34	76.99	76.15	76.79	76.05	73.81	71.96

Source: POCVAL, World Bank (2007)

Since then a lot has changed. Presently, SSA is no longer the region with the third largest extremely poor inhabitants in the developing world; it is now the region with the most extremely poor inhabitants. It is clear from the table that with the exception of SSA and SA most developing regions are set to achieving the target by 2015. At this rate, it will take the SSA in particular another 41 years from 2004 to eradicate extreme poverty.

It is important to draw attention once again to the fact that these statistics may be presenting an overly optimistic view of the performance of these regions as the cut-off line has not taken into account price changes that have occurred since 1993. Besides, it is unclear from the statistics what caused the reduction in the number of extremely poor in the respective regions. For, extreme poverty may be reduced by lifting or empowering people to work their way out of poverty. But this is not the only way. It can also be reduced through deaths of those in that category arising from natural causes but also from conflicts, malnutrition, hardships of all sorts, poor sanitation and so forth. In sub-Saharan Africa where conflicts are rife, life expectancy very low and ever reducing, child mortality very high, birth related maternal deaths very high, it cannot be surprising that a considerable proportion of the reduction in extreme poverty arise from deaths rather than actual economic uplifts.

6 THE FORGOTTEN FACET OF POVERTY

In the haste to meet the UN Millennium Development Goals (MDGs) target of halving poverty by 2015, countries are losing sight of the fact that the poverty question is indeed far more profound and encompasses not just the extremely poor (or those under primary poverty) but also the poor (or those under secondary poverty and earning below two dollars a day). Thus eradicating extreme poverty completely can not by itself resolve the problem.

Table 2 presents data on the proportion of national population that survive on less than two dollars US a day. This data is represented graphically in Figure 1. As the data show, even whereas less than one per cent of the population of Europe and Central Asia, for example, are extremely poor, as much as 10 per cent remain poor. Indeed with the exception of Europe and Central Asia

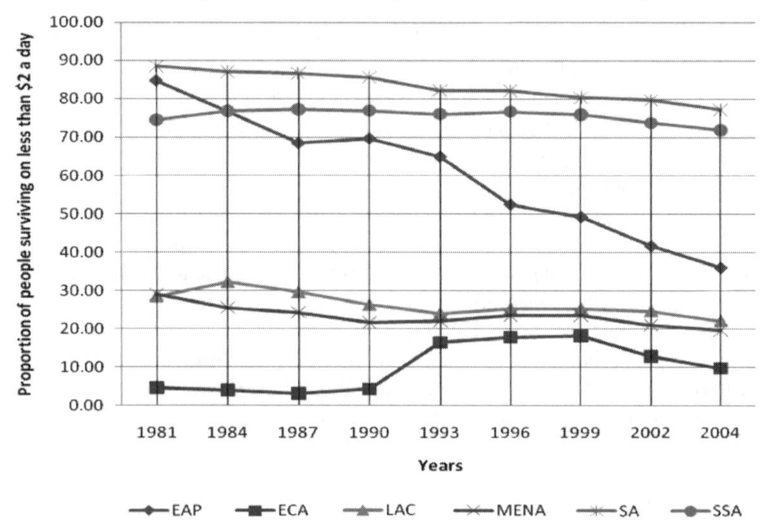

Figure 1. Proportion of the Population Surviving on less than US1 a day (Headcount). See Table 2 for the country/region code.

with less than 10 percent poor, 20% of the population of all the developing regions are poor with South Asia having the highest proportion (77.2%), followed by sub-Saharan Africa (71.96%). This shows that even when the MDG is achieved a great number of the population of the developing world would still be poor (under secondary poverty).

Remarkably, whilst Europe and Central Asia have, for instance, managed to reduce the proportion of their population that are statistically extremely poor, the number that are poor have actually increased between 1981 and 2004. Likewise, South Asia and sub-Saharan Africa made very little progress in reducing its proportion of poor people. The only region that made very rapid progress in reducing the size of its statistically poor population is East Asia and Pacific. This may be indicating that, rather than confronting the poverty question as a whole, the exclusive concentration on the extremely poor has left the proportion of the poor only slightly improved, leaving the quest for poverty alleviation very far from achieved.

Poverty, in common with most other social and economic phenomena that affects quality of life, has to be experienced first-hand to be correctly understood. For instance, it is true that over 70 percent of the population of Africa are known to survive on less than US$2 a day. Nonetheless, such a statistic by itself does a poor job in telling the severity of the hardships, gloom, melancholy, the joint suffering of the mind and body, deprivation, indecency of the living conditions, miseries, dejection, dullness and hopelessness in which the vast majority of the population of sub-Saharan Africa are trapped. It is instead the qualitative aspects of poverty that separates Africa from the rest of the world.

7 RICH CONTINENT, POOR PEOPLE

The way a problem is defined ultimately dictates the solutions prescribed for it; wrong problem definition is likely to lead to wrong solutions and this can lead to waste of time and resources that could yield better returns elsewhere. Firstly the classification of countries as rich or poor on the basis of the proportion of its population that survive on less than US$1.00 or US$2.00 a day is seriously flawed. Nations with equal proportions of people surviving on less than US$1.00 or US$2.00 are equally poor even though one may have large reservoir of untapped vital resources than the other. To be poor is essentially to be in lack of the resources or means needed to attain quality material standard of living.

A country is truly poor when it actually lacks the resources and means to support the livelihood of a vast majority of its people. Such a country requires resources in the form of external aid, favours, and technical and financial support. Africa with her abundant natural resources such as land, minerals, forest, and water human resources cannot be truly classified as poor. What is true though is that Africa currently harbours the largest proportion of the world's poor. So the continent is rich, the people are poor.

The vast majority of Africans are poor because they have failed to exploit and manage their resources efficiently and effectively. This situation may stem from their lack of capacity, or having been incapacitated and hence unable to transform resources; misuse and wastage of the available productive resources; external constraints and weak and counterproductive institutional context in which the resources operate (Hammond, 2006).

The majority of these problems are internal, a few are external. Removing the external bottlenecks such as trade barriers may go a long way to deal with the issues. However, external aid, debt relieve and support can do no more than kick start growth or provide partial, not sustainable relieve.

8 COMBATING POVERTY: CLOSING THE GAP

The point has already been made that a third cause of poverty is inequity in the distribution of incomes in society. How can this be resolved? There are two main ways to address the challenge. Firstly there is the need for a thoughtful program to remove all discriminatory customary or formal arrangements, rules and regulations that impede equal access to education, employment, investment, capital, and land markets. This must be accompanied by programs to reduce time and monetary constraints (directly and indirectly) to investment opportunities. This will enable individuals to work their way out of poverty. The second component is to institute a program of direct income redistribution. The aim here is to close the poverty gap directly. The gap for each individual is the difference between the earning of that individual and the poverty line.

Table 3 reports the statistics on the poverty gap using the two dollars a day cut-off line. The gap indices reported in the table show the mean gap below the poverty line as a proportion of the poverty line. Thus taking the 2004 poverty gap index for sub-Saharan Africa, the actual poverty gap, the short fall is 38.09 per cent of the poverty line which in this case is two dollars. This gives approximately US$ 0.76 poverty gap (i.e. 38.09% X $2.00). This amount multiplied by the number of people below the poverty line gives the total costs required to eliminate poverty with income redistribution. But this later method is not without complications particularly when applied in the developing world where the economies are themselves highly impecunious. Firstly the amount required is not simply the average shortfall multiplied by the number below the poverty line. Additional cost would have to be incurred in setting up and running the institutions to be responsible for the redistribution of income. This obviously would require big bureaucracies.

Big bureaucracies come with big costs. Moreover, there is the added cost of targeting. To ensure that the right persons receive the right amount of the redistributed income, a way has to be found by which the right people would be identified. This requires complete and regularly updated national

Table 3. The Poverty Gap Indices.

	1981	**1984**	**1987**	**1990**	**1993**	**1996**	**1999**	**2002**	**2004**
East Asia and Pacific	47.28	36.59	29.69	30.66	27.45	19.56	18.54	15.26	12.39
Europe and Central Asia	1.37	1.12	0.82	1.13	5.28	5.83	5.63	3.30	2.41
Latin America and the Caribbean	11.61	13.58	12.57	10.93	9.50	9.92	10.42	9.98	8.87
Middle East and North Africa	8.87	7.45	6.78	5.76	5.77	6.19	6.11	5.28	4.85
South Asia	44.73	42.31	41.84	40.59	36.79	36.38	35.41	34.49	32.75
Sub-Saharan Africa	39.05	41.79	42.54	42.15	41.51	42.96	41.77	39.26	38.09

database or register of all citizens and their income details. Apart from the associated huge expense in setting up and keeping such a register up to date, the risk of wrong targeting could end up inflating the register with the wrong people thereby transferring incomes to wrong persons, leaving a vast number of the poor still poor. With all these costs and risks, it is legitimate to ask if poor countries are truly ready for income redistribution approach to poverty reduction.

9 COMBATING THE CAUSES OF POVERTY IN AFRICA

To offer measured solutions to the problem, it is important to understand the extent to which poverty in Africa has resulted from: inefficient resource allocation, underutilization of resources and/or inequitable distribution of incomes. Since the 1940s, economists have sought to address these questions in the context of Africa and the rest of the developing world. This led to the development of a body of economic theory, collectively called development economics. These theories provided analytical tools for diagnosing and prescribing economic solutions that are chiefly suitable to the poverty and economic problems faced by the developing world as a whole. With support from the international development community and national governments these theories were put into practice in many African countries particular during the mass post colonisation decades of the 50s and 60s. The consequences that resulted from these theories surprised many economists.

The anticipated development and growth never materialised. Instead some countries experienced reverse development. Poverty exacerbated in many of these countries. After a long period of despondency, contemporary economists now claim to have learned a great deal during the past few years about how countries develop and what roadblocks can stand in the way (Sachs, 2005). This claim could be taken to mean that modern day development economists are better able to answer the above questions correctly than their predecessors.

It is curious that having learned a great deal, the solutions offered by contemporary economists to the problems of Africa bear a striking resemblance to those framed by economists of the 1940s and 1950s that failed and were discarded amidst severe criticism (Easterly, 2005).

Both earlier and current mainstream development economists argue that large-scale investments in strategic sectors of African economies are the prime stimulant required to kick start development and eradicate poverty across the region. This prescription was described in the late 40s and 50s as the big push development theory. Recent discussions in the literature show that the earlier theories failed largely because they reposed too much confidence in the state as the prime investor without necessarily taking care of the risk and threats that bad governance presents to the success of the theory.

Availability of statistics on corruption particularly from Transparency International, however imperfect they may be, have provided a solid basis for measuring the level of threats posed by the prospects of bad governance across respective countries (Sachs, 2005). The basic thesis of modern development economists can thus be put as: massive investments within a framework of appropriate and less obnoxious legal system, quality infrastructure and public services in economies with low risk of bad governance is most likely to kick start self sustained growth and reduce poverty considerably. The trouble is, using the level of corruption as a proxy for bad governance actually conceals a great deal of bad governance issues embodied in misallocation of resources, to build presidential mansions and buy presidential jets for instance, that are not necessarily due to theft, bribery and other corrupt practices. Misallocation of state resources may as well be more insidious than corruption. Perhaps another index on misallocations of resources will help improve perception of the undermining effects of bad governance in these countries.

10 SUSTAINABLE DEVELOPMENT

Sustainable development, a 'not-easily-defined' term, is perceived as 'confused, vague, inherently self-contradictory and reflecting ambivalent goals' (Toman *et al.*, 1997). The challenges that human progress and development face and the limits of resources to meet the needs of current and future generations of mankind and other life forms lie at the heart of the sustainable development debate. While the modern debate about sustainable development gathered serious global momentum after the publication of the Brundtland Commission Report (WCED, 1987), environmentalism – philosophical thought and concerns about the environment, limits to resources and their impact on the environment dates back to antiquity (Mebratu, 1996). It seeks to safeguard man's survival, health and well being within the context of man-nature relationships. It also addresses the question as to whether humans are separate from nature; and therefore should exploit and dominate nature for human benefit; or whether humans are part of nature, and therefore must respect, protect and live in harmony with it regardless of its values to man (Pepper, 1997; Dryzek, 1997; McManus, 1996; Dobson, 1996; Mebratu, 1996).

Gladwin and Kennelly (1997) coined the term sustain centrism a normative construct of a distinct paradigm based on a nurturing of the earth and its process. The sustain centric paradigm represents an emergent synthesis and an attempt at higher and deeper integration. Milton (1997) argues from anthropological perspectives that knowledge of the environment, in the non-developed world, is cultural and generated by ways in which that society engages the environment with the physical world and with each other. In parts of the non-industrial world, the environment is seen as giving, reciprocating, protective and reliable; in other parts, it is dangerous and unpredictable. In the developed world people's understanding of the world varies according to how they organise their social relationships (McAllister, 1994).

Over time, sustainable development debates centre around, economic, social and environmental sustainability interlaced with ethical considerations. From an economic perspective, it is the extent to which man-made capital and natural capital are complements or substitutes. The latter can be the overall capital stock or base, critical natural capital, irreversible natural capital or units of significance (Dobson, 1996). The MDG also seeks to ensure environmental sustainability and to integrate the principles of sustainable development into country policies and programs; reverse loss of environmental resources; reduce biodiversity loss, achieving, by 2010, a significant reduction in the rate of loss; halve, by 2015, the proportion of people without sustainable access to safe drinking water and basic sanitation; and by 2020, to have achieved a significant improvement in the lives of at least 100 million slum-dwellers.

The environmental sustainability of human ways of life refers to the ability of the environment to sustain those ways of life. The environmental sustainability of economic activity refers to the continuing ability of the environment to provide the necessary inputs to the economy to enable it to maintain economic welfare (Ekins, 1997). Both of these "sustainabilities" depend on the maintenance of requisite environmental functions. A wide range of environmental functions, include the capacity of natural processes and components to provide goods and services that satisfy human needs and the absorption of wastes from human activities. Natural processes and components are stocks of and flow from natural capital; regulation, carrier, production and information (Dobson, 1996; Ekins, 1997).

Social sustainability refers to a nation's ability to maintain a shared sense of communal purpose to foster connectivity, societal integration and cohesion that relates to the question of culture and values as well as the state of the economy. The sense of identity and social purpose of very many people as well as their income derive in a large part from their employment (Dobson, 1996; Ekins, 1997). Unemployment therefore leads to these characteristics, which in turn probably leads to ill health, mental stress and family breakdown.

Membership of and involvement in a local community contributes to people's sense of identity (Dobson, 1996; Ekins, 1997). The concept of cultural capital and sustainable livelihoods is consistent with social sustainability. It includes the way societies exist from generation to generation, their adaptation to the environment, their self-supporting collective security systems, their interdependence on one another, their ability to withstand stress and shocks are dependent partly on their traditional ecological knowledge or worldviews. According to Turner (1996), in these are a large number of self-regulating regimes governing access to resources in common property and their scope in limiting the level of economic stress on particular ecological systems is clearly very wide.

11 SUSTAINABLE DEVELOPMENT AND THE URBAN LAND USE DILEMMA

The key issues in urban areas include the need to maintain sustainable communities, provide good land use planning, address social deprivation in slums and poorer deprived areas, sort property rights, expand and manage environmental assets and problems (Jones, 1991; EC, 2004). There is also a need to add value to the inheritance of future generations by husbanding resources, use renewable building materials, construct thermally efficient buildings, and adopt sustainable methods of developing infrastructure. The location of work, schools, leisure facilities, shops social and medical services to minimise travel is essential (Cullingworth and Nadin, 2002).

However, because of the stage of poverty at individual and government levels, management of urban environmental resources is weak. For instance, there are threats to biodiversity in many respects: non-native plants displace indigenous plants; urban development results in the loss and fragmentation of urban habitats and species (Pearce *et al.*, 1995). There is a disruption of hydrology, pollution, light pollution, increased background noise, rubbish dumping, increased predation of urban fauna, increased risk of fire, disturbance, trampling, vandalism, and disruption to conservation management.

Other challenges common in urban areas include inefficient management of waste, energy, land, and other environmental resources. There are also barriers to clean air from traffic emissions. In sub-Saharan Africa, there are large urban market of second-hand energy efficient good including cars, vans and trucks which pollute the air. When indoor and outdoor air quality becomes poor, it affects human health, attitudes, productivity, and people's livelihoods reducing long-term economic performance. The lack of response to housing designers to minimise the dependence on air conditioning and energy consumption is contribution to global warming and climate change (Rogers *et al.*, 2008).

Urban land, whether in Africa or anywhere else, is a finite and non-renewable resource. Additionally, the majority of urban land uses such as residential, commercial, industrial and infrastructure developments are, unlike most rural land uses, permanent or semi permanent in nature. These two parameters lead to a situation in which the residue of urban land available for the future generation necessarily diminishes with increasing allocation.

One factor that influences the rate of urban land development is the size of the population. Urban population continues to grow progressively. Actually, the evidence from the relevant statistics is that urban population in Africa is growing rapidly. The significance of this point is that, overtime; the residue of urban land resources within its current boundaries will inescapably and progressively diminish. Urban land will not be able to meet the needs of future generations unless more peri-urban land is brought into development. Improvement in technology may slow down the process of depletion would be unable to stop it. While rapid population growth can have a negative impact on poverty, it could be worse for posterity.

Policy makers, governments, and local authorities seeking to manage resources sustainably have several tools at their control. These include command and control, economic instruments, efficiency and conservations, renewable resources, tradable permits, eco-planning, sequestration and carbon capture (Toman *et al.*, 1994). Command and control could be exercised to legal and regulatory instruments, where, for example, the importation of certain undesirable commodities could be banned. Through good land use planning, it is quite possible to prevent development on certain areas of urban land for the sake of conservations or for other environmentally sound purposes (Grubb, 1991 & 1996).

Economic instruments such as green taxes, rebates, or incentives can be used to dissuade people from trading in certain urban goods or the development of certain areas which may be heavily taxed. It may also encourage the development of certain areas where incentives and rebates are given. Tradable permits system is an emission trading system, which regulates how much pollution can be permitted through the sale of emission credits. Selling some of your credits means you can pollute less; buying more credits means, you can emit more pollution. The overall pollution levels remain the same under the scheme (Tetienberg, 1997).

Efficiency opportunities abound in the urban context. These opportunities may include land use planning to minimise urban sprawl, building design to reduce the need for air conditioning, industrial energy savings through the improvement of the performance of plants and machinery in urban areas (Cullingworth & Nadin, 2002). Eco-planning, or least cost planning (closely related to third party agreement) is a combination of economic and legal arrangements. Under the policy, utility companies, private individuals, and companies are encouraged to invest in energy efficiency for the benefit of the poor or even through the provision of some free services for the less rich. The regulatory arrangements allow them to recoup the cost of their investment by the margin of the efficiency savings made (Worrell *et al.*, 1997).

In order that the carbon dioxide emitted through the consumption of fuels, trees can serve are sinks for the absorption. Allowing more trees to stand and planning more trees can sequester carbon dioxide emission resulting from urbanisation and rapid motorisation. Through technology, it is becoming possible for carbon dioxide pumped into the atmosphere be captured under ground, and particularly under oil fields.

In sub-Saharan African environmental assets are not limited to urban land for development. They include non-depletable renewable resources such as wind, hydro, solar, geothermal and tidal energy depending upon which resources are available in a particular urban area (Karekezi & Mackenzie, 1993). Depletable renewable resources available include solid wood, charcoal, municipal waste and briquettes. Other depletable resources include liquid ethanol, and methanol; or biogas anaerobic digesters, and gasifier-produced gases (Walsh *et al.*, 1997).

Unlike urban land, and wood resources, other potential renewable resources require substantial investment to harness them. With the level of poverty identified in sub-Saharan Africa, it is difficult for the very poor to invest into all the urban resources available. Policies such as emissions trading have not yet developed; land use planning lags behind development; and if green taxes are to be imposed of the very poor, it can put pressure on the limited resources available to them, and exacerbate their plight. The scope for sustainable policy using efficiency is not without its challenges. The key challenge that arises is whether markets are developed and regulations enforceable to encourage firms to invest in efficiency. Any rigid application of strict policy instruments to deal with environmental problems and which requirements have high financial implications within the urban context can have graver effects on the very poor. While the technology is important and while, the poor may lack technical knowledge and the financial resources for investment; their traditional ecological knowledge is better suited to support their local habitats.

Second, the discussion of poverty in terms of hardships, gloom, melancholy, suffering of the mind and body, deprivation, indecency, miseries, dejection, dullness and hopelessness raises psychosocial issues. This means that greater weight would have to be given to the pursuit of social goals as a means to ecological goals. Bartlett (1986) argues that under certain circumstances, economic sustainability, like social sustainability can be in conflict with environmental sustainability by ignoring environmental implications of actions. Environmental sustainability should not and could not fully supplant economic sustainability; any more than environmental sustainability should or could replace social sustainability. Nevertheless, in light of the level of poverty discussed in this paper, it would appear that issues of social sustainability in terms of poverty alleviation or economic sustainability – wealth creations – should have a priority over other forms of sustainability, which over the long run must be subordinate to it. Without a doubt, what is required is a delicate balance between the needs of the present generation and the future. That also means a reasonable compromise to ensure a balance between the concurrent pursuit of poverty alleviation and sustainable development.

Third, although there are constraint to urban land availability of urban land for development, availability of vast environmental resources provide the scope for developing green business with the support and governments and external agencies as equal partners to development. Business can be developed around renewable resources solar energy. For instance, instead of environmental hazards that waste causes, investment can be made into waste as an input to energy. Any viable green business development cannot be divorced from strong institutions and secured property rights. Since the land tenure of every society provides the setting within which rationality in urban land uses take place, this presupposes that the solution must be adapted the peculiar tenure system of Africa. In order words the tenure system should be capable of ensuring that extraneous land uses are avoided whilst permitting relevant and necessary land uses. Because of the grave misunderstanding of the African land tenure system which has resulted in the low opinion that many policy experts have for it; it is crucial in arguing for it to be relied on in the fight against poverty and for the pursuit of sustainable development to labour an evaluation of the debate surrounding the land tenure systems of Africa.

Fourth, the role of forestry resources to sequester carbon means that developing countries have global assets. The effects of carbon dioxide are global, and so the sequestration of carbon dioxide is a global asset. Getting the rich nations to see the conservation of these resources as global assets and therefore paying economic rents for these assets in perpetuity. Use and non-use values of these assets can be determined through methods such as hedonic approaches, contingent valuation, and surrogate markets (Pearce & Turner, 1990). Channelling the income into the development of rural areas and creation of jobs in these areas will not only save the planet from the warning effects of climate change. It will reverse rural-urban migration thereby stabilising pressure and poverty in urban areas.

Fifth, population growth can cause environmental degradation and increase, and put pressure on the limited resources of the urban poor. It could be argued that rapid population would necessitate the need to slow down the rate of urban land usage by individuals through quantitative restrictions or using economic or legal instruments. For, so long as urban population continue to grow, the progressive reduction in the residue of urban lands cannot be totally abated. Actually any measure that imposes quantitative restrictions on the size of urban land usage by the extant population has the danger of, given that the population growth rate of the society remains the same, concentrating too many people on a given unit of land. Beyond certain point, the reasonable bearing capacity of usable urban land units is bound to be exceeded. This could result in the creation of slums, hardships, health and environmental problems for the present generation, a situation that is clearly unsustainable. On the other hand, it is a known economic fact that population increase per se does not have a significant impact on the environment. It is population growth, combined with increased consumption and affluence that causes environmental degradation (Ehrlich, 1968; Williams *et al.*, 1987).

Sixth, promoting the goals of sustainability development raises some moral questions. For instance, the question that often arises is whether there is a moral justification for asking people like the poor in Africa, faced with extreme hardship and misery, to protect the environment for the yet unborn generation. The question is asked as to why government policy instruments like planning, or green belts constrain development. In a society like Africa plagued with a swarm of socio-economic problems, what merits are there in being so concerned with the welfare of those yet unborn when those living presently cannot make ends meet even at the current rate of resource use.

This difficulty can be illustrated by supposing that A, a citizen of an African country lives on less than $1 dollar a day and needs Q quantities of a resource R to work his way out of poverty. If the total stock of R in that country is exactly Q, the real dilemma that ought to be resolved is, what proportion of Q should A be allowed to utilize to improve his or her livelihood? And it is here that poverty alleviation clashes with sustainable development ideals. From the poverty alleviation standpoint, A has to utilize the entire stock of R if that is the only way he or she can make his or her way out of poverty. This means that, if A is allowed to utilise the entire existing stock of R as that is what he or she requires, there would be none left for the future generations. So, from the sustainable development standpoint, A should be allowed only a fraction of Q and reserve sufficient amount for the future generation. This means also that A would be unable to cross the $1 a day mark as expected under the MDGs.

A real life example of such quantitative restriction on resource utilization and consumption is the current carbon emission quota allocations to all countries of the world as a way of sustainably managing climate change. This certainly neat mechanism is hardly a pragmatic solution to the urban land use conundrum. It is for instance impractical to set maximum limits on how much urban land space should be developed globally so that each nation is then apportioned a land use allowance in such a way that if they are unable to fully utilize they can trade the remainder with nations that wish to exceed their limit for cash. Of course, the boundaries of urban land could always be pushed. But that is only possible because rural lands are reduced proportionately. This will eventually lead to shortage of rural lands, forcing farmers to employ high yielding technologies such as fertilizers to produce the same quantity of agricultural products as before. These have their own environmental consequences.

In many sectors the sustainable development concept is quite fluid. For, to be able to make provision in the allocation of resources for the future generation, the decision maker must have a clear appreciation of the percentage of Q that should be reserved for future generations. And by what formula this is to be determined. To answer this question adequately, there is the need, firstly to know with fair certainty the size of the future generation in question, their needs and the sort of resources they will require for their needs. This will dictate the type and quantity of resources they will require to meet this need and hence the provisions to make for them. It may seem to be a false confidence to anticipate that a person in *A's* position will readily accept to give up the consumption of a proportion of the resource R that he needs and continue to languish in hardship, misery, disease and so forth just for the sake of the so called future generation. For, *a fortiori*, as Adam Smith argues, "before we can feel much for others, we must in some measure be at ease ourselves. If our own misery pinches us very severely, we have no leisure to attend to that of our neighbour: and all savages are too much occupied with their own wants and necessities, to give much attention to those of another person" (Smith, 1759).

Poverty alleviation may be seen as the foremost challenge of the region, and pursuing poverty alleviation through social and economic sustainability may be the top priority. To ignore the environment in development will result in the legacy of unbalanced development including traffic congestion, polluted air, deteriorating infrastructure and buildings, urban sprawl, social exclusion, insecurity and criminality (EC, 2004). This will adversely affect human health, attitudes, productivity, life expectancy and people's livelihoods reducing long-term economic performance. Urban sustainability issues such as energy and waste management, mobility and transport, a air quality, housing, cultural heritage, tourism, land use planning, redevelopment and regeneration, and social cohesion can be compatible with poverty alleviation if properly managed (EC, 2004).

12 CONCLUDING REMARKS

The concept of sustainable development has raised the spectre of possible exhaustion of vital resources at the expense of future generations. The way to resolve this dilemma is to pursue poverty alleviation methods that adequately reward hard work, initiatives, efforts and productive use of resources within the broad framework of the rule of law but at the same time penalises decision markers with the costs of their reckless resource utilisation. The later is not so much to ensure that something is left for the future generation, but more because careless and inattentive use of resources threatens the very survival of the present generation itself. By alleviating poverty through cautious uses of resources, as people pursue improvements in their own life they will unwittingly also be preserving resources for future generations. Thus with the right conditions, this could serve to prevent extraneous use of urban lands and balance the interest of current and future generations.

REFERENCES

Bartlett, R.V. (1986), Ecological rationality: Reason and environmental policy. *Environmental Ethics*, 8(3), 221–239.

Booth, C. 1899, "Old Age Pensions and the Aged Poor: A proposal", London, Macmillan.

Cullingworth, B. and Nadin, V. (2002), *Town & Country Planning in the UK*. London: Routledge.

Dobson, A. (1996), 'Environment Sustainabilities: An Analysis and a Typology,' in *Environmental Politics*, 5(3).
Dryzek, J.S. (1997), *The Politics of the Earth. Environmental Discourses*, Oxford.
Easterly, W. 2005, National policies and economic growth: A reappraisal, The handbook of economic growth, vol. 1, part 1, pp. 1015–1059.
EC, 2004, "EU Research for sustainable urban development and land use", European Commission, Brussels.
Ehrlich, P.R. (1968) *The Population Bomb*, Balentine Books, New York.
Ekins, P. (1997), 'Sustainability as a Basis of Environmental Policy,' in *Environment, Equity and Welfare Economics in Sustainability and Global Environmental Policy. New Perspectives*, Edward Edgar, Chiltenham, UK and Lyme, USA.
Gladwin, T.N. and Kennelly, J.J. (1997), 'Business Environment: Philosophical Perspectives' in Bansal, P. and Howard, E. (eds.), *Business and the Natural Environment*, Butterworth-Heinemann, Oxford.
Grubb, M. (1991), *Energy Policies and the Greenhouse Effects*, Vol. 1, Royal Institute of International Affairs, Dartmouth *Edition*, Duke University Press, Durham and London.
Grubb, M. (1996), *Energy Policies and the Greenhouse Effect*, Dartmouth [for] The Royal Institute of International Affairs, Aldershot.
Hammond, F. (2006), The Economic Impacts of sub-Saharan African Urban Real Estate Policies, in School of Engineering and the Built Environment. University of Wolverhampton, Wolverhampton.
Jones, D.W. (1991), 'How Urbanization Affects Energy Use in Developing Countries,' in *Energy Policy*, September 1991.
Karekezi, S. and Mackenzie G.A. (eds.), (1993), *Energy Options for Africa-Environmentally Sustainable Alternatives*, Zed Books, London.
McAllister, I. (1994), 'Dimensions of Environmentalism: Public Opinion, Political Activism and Party Support in Australia,' in *Environmental Politics,* 3(1), 22–42.
McManus, P. (1996), 'Contested Terrain: Politics, Stories, and Discourses of Sustainability,' in *Environmental Politics*, 5(1).
Mebratu, D. (1996), 'Sustainability and Sustainable Development: Historical and Conceptual Review,' in *Environmental Impact Assessment Review*, 12(6).
Milton, K. (1997), 'Different Lives, Different Worlds: Anthropology and Environmental Discourse,' in Bansal, P. and Howard, E. (eds.), *Business and the Natural Environment*, Butterworth-Heinemann, Oxford.
Pearce, D. and Turner, K. (1990), *Economics of Natural Resources and the Environment*, Harvester Wheatsheaf, London, New York.
Pearce, D.W., Markandya, A. and Barbier, E.B. (1995), *Blueprint for a Green Economy*, Earthscan, London.
Pepper, D. (1997), *The Roots of Modern Environmentalism*, Routledge, London.
POCVAL. (2008), World Bank poverty database, World Bank.
Rogers, P.P., Jalal, K.F. and Boyd, J.A. (2008), *An Introduction to Sustainable Development*. London: Earthscan.
Rowntree, B.S. (1901), *Poverty, the study of Town Life*, London, MacMillan.
Sachs, J.D. 2005, "*The End of Poverty: Economic Possibilities for Our Time*", New York, Penguin Press.
Shreiber, A.F., Gatons, P.K. and Clemmer, R.B. (1971), *Economics of Urban Problems: An Introduction*, Boston, Houghton-Miffin.
Shrestha, R.M. *et al.* (1998), 'Environmental and Electricity Planning implications of Carbon Tax and Technological Constraints in a Developing Country,' in *Energy Policy*, 26(7).
Stern, N. (2006), *The Economics of Climate Change: the Stern Review*. UK Government Report, October 2006.
Tetienberg, T.H. (1997), 'Economic Instruments for Environmental Regulation,' in Lewis, O. and Unwin, T. (eds.), *Environmental Management. Readings and Case Studies*, Blackwell Publishers, Oxford.
Toman, M.A., Pezzey, J. and Krautkraemer, J. (1997), 'Environmental Policy,' in *Environment, Equity and Welfare Economics in Sustainability and Global Environmental Policy. New Perspectives*, Edward Edgar, Cheltenham, UK and Lyme, USA.
Turner, R.K. (1996), 'Sustainability: Principles and Practice,' in Turner, R.K. (1993), (ed.), *Sustainable Environmental Economics and Management: Principles and Practice*, Belhaven Press, London and New York.
UN Millennium Development Goals. (2008). www.un.org/millenniumgoals
Walsh, M., Perlack, R., Turhollow, A., TorreUgarte, D., Becker, D.R., Graham, R., Slinsky, S. and Ray, D. (1997), *Evolution of the Fuel Ethanol Industry: Feedstock Availability and Price. Biofuels Feedstock Development Program*, Oak Ridge National Laboratory, Oak Ridge, TN.
Williams, R.H. Larson, E.D. and Ross, M.H. (1987), 'Materials, Affluence, and Industrial Energy Use,' in *Annual Review of Energy*, 12, 99–144.

World Commission on Environment and Development (WCED), (1987), *Our Common Future* Oxford, University Press.
Worrell, E., Levine, M.D., Price, L.K., Martin, N.C., van den Broek, R. and Blok, K. (1997), *Potential and Policy Implications of Energy and Material Efficiency Improvement*, United Nations, New York.
Smith, Adam (1977-02-15), *An Inquiry into the Nature and Causes of the Wealth of Nations*, University Of Chicago Press.
Smith, Adam (1982), *The Theory of Moral Sentiments, (ed.), D.D. Raphael and A.L. Macfie, vol. I of the Glasgow Edition of the Works and Correspondence of Adam Smith*, Liberty Fund.

Organizational Performance and Sustainability

The Environment as a Fifth Component of a Balanced Scorecard for Improving Organizational Performance

Moh'd Mahmoud Ajlouni
Department of Banking & Finance, Yarmouk University, Jordan

SUMMARY: The purpose of this study was to incorporate an environmental aspect into the balanced scorecard (BSC). It provided a way of linking environmental objectives with sustainable development and business strategy. It filled a gap by linking environmental themes with other outlooks, as well as formulating an integrated environmental Scorecard. It showed how to incorporate an environmental viewpoint into the BSC organizational performance measures framework by providing a new way of linking green objectives with sustainable development in business strategy.

1 INTRODUCTION

The Balanced Scorecard (BSC) was first proposed by Kaplan and Norton (1992) as a means for deriving the performance of a system. It was then developed as a managerial mechanism for translating and implementing corporate strategy (Kaplan and Norton, 1996). Recently, it was enhanced further as a method for transforming a corporation into strategy-focused business (Kaplan and Norton, 2001). The Balanced Scorecard is a framework of organizational performance measures across four perspectives or aspects: financial, customer, internal business processes, and learning and growth (FCIL). It recognizes that there might be other sides than these four points of view.

The purpose of the current study was to incorporate an environmental viewpoint or aspect into the balanced scorecard (eBSC). In this chapter, environmental or green indicates health, safety, impacts, natural resources scarcity, pollution, and other related issues, for staff, workplace, products, community, and other stakeholders. This study also provides a way of linking environmental objectives with sustainable development and business strategy. The Scorecard is a quality and performance improvement system that has been implemented by numerous for-profit and non-profit businesses. Hence, incorporating and implementing the environmental objectives into an organization's strategy will lead to micro and macro economic sustainability.

For most associations, the environmental issues of reducing pollution, utilizing available energy sources efficiently and/or reducing environmental risks can provide the necessary linkage with an organization's strategy. However, there is a scarcity of reports on the linkage between environmental themes and customer perspective, internal business processes and/or employees and systems. This might be due to a lack of an environmental perspective in most quality management models, such as Total Quality Management (TQM), Competency Management, Kolb's Learning Cycle, Six Sigma, and the BSC.

The current study fills the gap by linking environmental themes with other outlooks, as well as formulating an integrated environmental Scorecard. Background literature is provided on the effects of the environment on sustainable development. The micro level of environmental effects on corporations is introduced by explaining the reasons for companies to go green and how they can do it. Applicability of the Balanced Scorecard in various business sectors and in different kinds of organizations is explained. Formulating the green perspective and building a totally environmental scorecard is discussed in the last section.

2 SUSTAINABLE DEVELOPMENT AND THE ENVIRONMENT

Sustainable development necessitates meeting the requirements and wants of the present generation without jeopardizing the capacity of future generations. This cannot be brought about by governmental policy alone or by individual companies. It has to be considered by society as a whole as a guiding principle in political and economic decisions and in consumption and production. International organizations have recognized the need for developing strategies to help governments and businesses in achieving this type of growth.

2.1 *Institutional concerns*

Governments and international organizations are discovering the importance of effective approaches for dealing with the problems of the environment and development. They assign high importance to promotion and maintenance of health and safety, and sustainable environment in all their activities. There is an increase, for example, in new environmentally related legislations.

The development potential of economies is increasingly threatened by environmental deterioration. It has obvious effects not only on human health and well-being but also on socio-economic growth. Environmental deterioration, such as water, air and soil pollution, imposes extra costs on households, business and industry.

The European Union (EU) adopted a sustainable development strategy in 2001 and enlarged it in 2006 (European Commission, 2007). The aim of the strategy is to identify and develop actions to enable the EU to achieve a continuous long-term improvement of quality of life through the creation of sustainable communities able to manage and use resources efficiently, able to tap the ecological and social innovation potential of the economy and, in the end, able to ensure prosperity, environmental protection and social cohesion. The strategy sets general objectives and concrete actions for seven key priority challenges. These are: climate change and clean energy, sustainable transport, sustainable consumption and production, conservation and management of natural resources, public health, social inclusion, demography and migration, and global poverty and sustainable development challenges (European Commission on Sustainable Development).

The International Chamber of Commerce (ICC) developed a business charter for sustainable development (ICC, 2007). Their charter promotes environmental policy in several areas:

2.1.1 *Corporate priority*

Environmental Management should become one of the highest priorities for all corporations demonstrated by the creation of policies, programmes, procedures and practices for achieving environmentally sound operations for the company. The executive has authority for directing the company and therefore is the route of all future changes in that direction.

2.1.2 *Integrated management*

The full integration of environmental elements into the normal business operation of the company is essential to ensure that the management system functions as a coherent entity. This integrated approach is further demonstrated by amalgamation of the management systems for environment, quality, customer satisfaction, health and safety of staff.

2.1.3 *Process of improvement*

To use the company organization, customer expectations, community expectations, legal regulations, development of tools and new techniques in order to continually improve the environmental performance of the organization.

2.1.4 *Staff education and training*

To provide awareness ability and motivation to employees in order to conduct their activities in an environmentally responsible manner.

2.1.5 *Prior assessment*
To evaluate the potential environmental impacts of the company's business before embarking on new projects, products or activities, and to consider these effects when ceasing production or closing facilities.

2.1.6 *Products and services*
To consider the life cycle of products and services to ensure that they have no undue environmental impact and are fit for their intended use, are efficient in their consumption of resources, can be re-cycled, re-used and disposed of safely.

2.1.7 *Facilities and operations*
A stronger emphasis can be placed on the promotion of renewable resources, minimization of pollution and generation of waste and a safe and responsible husbandry of the land resource.

2.1.8 *Contractors and suppliers*
These are an extension, but integrated, part of the company's operations and products. The organization can encourage contractors and suppliers to also adopt sound environmental principles combined with a system of continuous improvement ensuring that the suppliers become consistent with the aims of the organization itself. This also encourages the wider adoption of environmental improvement practices within suppliers and sub-contractors.

2.1.9 *Openness to concerns*
To foster an open exchange of information and communication between the organization and its stake holders, employees, suppliers, society, the public, stockholders.

2.1.10 *Compliance and reporting*
This is the measurement of ecological performance. Following definition of the company's environmental aspects and impacts, measurements can be performed to demonstrate stability or improvement and confirming compliance with company objectives and legal requirements. This information can also be used to periodically provide information to the stake holders.

2.2 *Linking sustainable development with environment*

The contemporary interest in the relationship between sustainable development and environmental quality is a rational expansion of a debate about the consequences of economic growth. In 1971, The Report to the Club of Rome (Meadows et al., 1972) drew the shocking conclusion that the continued growth of the world economic system is not sustainable because of the exhaustion of critical environmental resources. The global response was that rising prices of scarce environmental resources would induce the production and use of the less limited reserve alternatives. Market-driven feedback mechanisms would keep the economy away from any resource-precipice. This resulted in switching the focus away from the predicament of resource depletion towards the pollution dilemma (Ansuategi et al., 1998).

Thirty years after Club of Rome argument, it has been recognized that the most important environmental effects are external to the market-driven feedback mechanisms. The environmental Kuznets curve (EKC) proposed that higher incomes will induce the proposition for spending on improving the quality of the surroundings. The results of EKC literature are not firmed and the conclusion is that EKC is not a panacea for environmental issues (Stern et al., 1996). For the measurement of sustainable development see Pearce et al. (1998) and Nijkamp and Ouwersloot (1998). Empirical literature has focused on pollution issues rather than resources depletion (Ansuategi et al., 1998). Linkage and the perception of resource scarcity were discussed by Opschoor (1998).

A similar picture can be seen at the micro level, (*i.e.* corporate). As evidence of environmental issues and challenges continue to grow, it is likely that a firm's performance and valuation will be raised by financial institutions and markets as well as by shareholders and investors. There is still an on-going and long-running debate on whether firms should incorporate environmental policies into their tactical and strategic decisions.

The classical economic approach towards the surroundings is largely based on making a profit. This has contributed to a mass of environmental problems, including global warming, pollution, the extinction of some animal species, deforestation and drought. The major environmental concerns that have occurred as a result of economic development are currently gaining attention. They have also generated a heightened awareness of the environment. As a result, businesses are seeking to operate in environmentally friendly ways. While the movement has taken awhile to generate momentum, the stance seems vivid as worry about the surroundings continues to escalate. This trend has created new terms in the business world, such as green investing; green financing (Thompson and Cowton, 2004); green bonds; green manufacturing; green delivery; and green information technology (IT).

Green investing involves allocating money to projects that actively promote environmental responsibility. The concept of green investing is an outcome of the socially responsible investing movement. Socially responsible investors often seek to avoid investing in companies that produce products such as alcohol, tobacco and firearms. Green investors seek to invest in companies that protect the environment. Green bonds, on the other hand, have had a major boost from an amendment to the America Jobs Creation Act of 2004. The amendment was officially titled the Brownfield's Demonstration Program for Qualified Green Building and Sustainable Design Projects. It was designed to provide funding, in the form of $2 billion worth of AAA-rated bonds issued by the United States Treasury, to finance environmentally friendly development. The objective was to reclaim contaminated industrial and commercial land (brown fields), and to encourage energy conservation and the use of renewable energy sources.

3 ORGANIZATIONS AND THE ENVIRONMENT

3.1 *Why go green?*

In order to achieve environmental objectives, two sorts of incentives are required. Local or governmental policy should reward a company for making environmental improvements in the form of tax incentives, reduction in regulatory costs or any other financial benefits. Examples include extending permits issued under clean air and water regulations for companies reducing releases of regulated pollutants; providing project financing or low interest loans for encouraging voluntary cleanup and redevelopment projects; improving tax treatment of the use of innovative remedial and control equipment technologies; and providing incentives for those companies improving the environmental relationships with their suppliers, customers and contractors (GEMI, 2004a, 2004b, 2001, & 1999).

There are several incentives for corporations to do business in a green way. These must match company needs. It has been argued that 50%–90% of a firm's market value is influenced by intangibles like environmental aspects, 35% of institutional investors' portfolio allocation decision is based on intangibles like ecological performance, and 81% of global 500 executives rate environmental issues among top ten driving value in their businesses (GEMI, 2004a and 2004b). Figure (1) shows the relative performance of the 2006 global 100 companies back-tested against the MSCI World Index during the period 2000–2005 (GEMI, None).

There are many advantages to improving business environmental performance. Smeets and Wetering (1999) reported on three indicators; to supply information on environmental problems, allowing policy-makers to prioritizing issues; to supply policy development and optimize the assignment of resources to addressing priority issues; and to effectively monitor the effects of policy responses. Examples include, improved process efficiency and cost savings as a result of increasing the level of pollution prevention; having subsidies, tax exemptions and extension of permit issued under clean air and water regulations as a result of reducing releases of regulated pollutants;

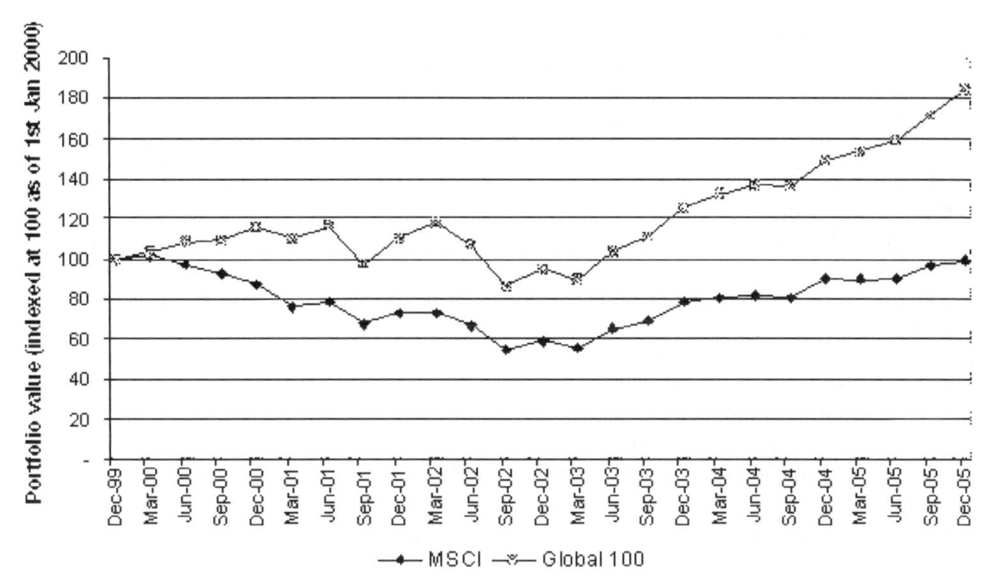

Figure 1. The Relative Performance of Global 100 Companies against the MSCI World Index during the period 2000–2005.
Source: GEMI (None b).

the use of green bonds to fund environmentally friendly projects may give tax-exempt income to investors; green products score points with consumers; save energy costs, avoiding regulation penalties, disruptions lawsuit expenses and compensations; finding new avenues for business; having good public relations, hence, attracting new customers and opening new markets. In the IT businesses, for example, benefits include energy-efficient IT is high performance IT; lower exposure to energy prices; and cost-savings (Ambrosio, 2007; GEMI, 1999).

3.2 *How to go green*

A common focus of many innovative and effective ideas and approaches being worked out nowadays is a central concern with the actual process of environmental planning and management (EPM). The framework of EPM comprises four inter-related aspects. These are (Sustainable Cities Programme, None): Identification and prioritization of environmental issues and stakeholders involvement; Formulating environmental management strategies; Formulating and implementing environmental action plans; and Institutionalizing environmental planning and management.

3.2.1 *Environmental management systems*

Environmental management systems (EMS) have emerged to assist organizations to meet their responsibilities. Since 1990, the Global Environmental Management Initiative (GEMI) has created tools and provided strategies to help businesses foster global environmental, healthy and safety excellence and economic success. Based in the USA, GEMI currently has 40 member companies from 22 business sectors having annual sales of US$1 trillion worldwide. The UK was among the first to have its own EMS standard. It has had an EMS standard, coded BS7750, since 1992, but this ceased to be relevant upon the introduction of ISO 14001. The European Union has had the Eco-Management and Audit Scheme (EEMAS) since 1993. It is intended to provide recognition for those companies who have established a program of environmental action designed to protect and to continuously improve their environmental performance.

ISO14001 was issued in 1996 and presents a series of standards of specification, guidance and advice on a wide range of environmental issues such as auditing, performance evaluation, labeling, life-cycle assessment, and environmental aspects in product standards. Other countries such as the USA, Canada and Japan, have introduced eco-labeling programs. There is a similarity in the components of these different national and international standards. The most common features of these EMS include the following: An objective understanding of environmental aspects and their impacts; The need for an environmental policy, clarifying the environmental principles promoted by the company; Objective and targets defining the environmental goals and the path towards achieving them; Defining how the objectives and targets are to be realized; Internal audits to ensure effectiveness and compliance; and management review of the system to ensure that it continues to be suitable and effective for the organization and its aims.

Environmental management systems have many benefits. Successful green management will evaluate all opportunities for cost savings. The most common benefits derive from a review of resource/ energy utilization and its efficiency, forcing full consideration of alternative energy sources and their cost effectiveness. The other primary element will be minimization of waste and result and cost of disposal.

Many companies have addressed the management system for customer requirements related to quality and ISO 9000. The range and diversity of customer needs and expectations is constantly growing with many customers increasing preference for use of suppliers and sub-contractors who can demonstrate that they are good environmental citizens. No customer would want to risk a tarnished reputation (or non-compliance to legislation) from the poor environmental performance of their suppliers and sub-contractors. The safest option for the customer is to use suppliers and sub-contractors who can demonstrate their positive environmental performance.

The ability to demonstrate a responsible environmental attitude can dramatically improve the image of the corporation fostering better relations with the company's stake holders. Even more importantly, adverse publicity about the organizations environmental performance is always highly damaging.

The scope and severity of environmental legislation is ever increasing. A management system that ensures recognition of the requirements and compliance with them will ensure that fines are avoided and staff is not imprisoned in addition to avoidance of the publicity that inevitably follows an environmental prosecution.

Investors are increasingly moving to green portfolios. The financial performance of these portfolios has been good in comparison to more traditional investment. In seeking additional investment for the organization it is sensible to ensure the widest scope and this is only aided by a demonstrably sound environmental performance.

Insurance companies are fully aware of the risk to their policies from poor environmental performance of the insured. Companies with a sound and effective environmental management system are able to demonstrate that they pose less risk to the insurance company and create a negotiating tool for lower premiums. Some insurance companies now require an environmental audit of the company prior to agreeing cover.

Finally, all companies seeking growth obviously want their product and services attractive to a widest possible market. Poor environmental performance will encourage many potential customers to decide not to buy from the company; good environmental performance will ensure continuation of the widest possible market.

3.3 *Relationship between environmental issues and a firm's performance*

There have been a number of studies attempting to investigate the relationship between environmental issues and firm performance (Table 1). Spicer (1978) reported that better pollution control results in a higher profitability and lower stock risk. While Chen and Metcalf (1980) and Hamilton (1995) reported that pollution influences stock performance. In addition, Konar and Cohen (2001) concluded that firms depositing fewer toxins and those confronted with fewer environmental lawsuits tend to have higher valuation. King and Lenox (2002) found a positive relationship between waste prevention and firm value. Moreover, Klassen and Mclaughlin (1996) showed that stock price increased following positive environmental information. White (1996) stated that green portfolios

Table 1. Relationship between a Firm's Environmental and Financial Performance.

Ref	Country	Relationship	Variables	Method-	Results
Spicer (1978)	USA	Pollution Control & Firm Performance	Profitability & Stock Beta (Risk)	Regression/Correlation	Better pollution control, higher profitability and lower stock risk.
Chen and Metcalf (1980)	USA	Pollution Control & Stock Performance	Stock Price	Regression/Correlation	Pollution control is NOT rewarded with higher stock performance.
Mahapatra (1984)	USA	Pollution Control & Stock Performance	Stock Price	Regression/Correlation	Pollution control is NOT rewarded with higher stock performance.
Freedman and Jaggi (1988)	USA	Environmental Pollution Disclosure and Firm Profitability	Financial Ratios	Regression/Correlation	Little evidence on the relationship between environmental pollution disclosure and financial performance of the firms.
Hamilton (1995)	USA	Environmental Pollution & Stock Price	Stock Return	Event Study	Pollution influences stock performance.
Klassen and McLaughlin (1996)	USA	Environmental Information & Stock Price	Stock Return	Event Study	Stock price increased following positive environmental information, but less decrease in response to negative news.
White (1996)	USA	Green Portfolios & Stock Performance	Risk-Adjusted Returns of Portfolios	Capital Asset Pricing Model (CAPM)	Green portfolios have significant positive risk-adjusted returns, while not-green ones have not.
Cohen et al. (1997)	USA	Investing in Environmental Leader Companies	Environmental Firm Characteristics	CAPM	NO rewards or penalties for investing in environmental leader companies.
Russo and Fouts (1997)	USA	Environmental Performance and Firm Profitability	ROA	Regression/Correlation	Environmental performance is positively correlated with firm profitability.
Dowell et al. (2000)	USA	US International Firms Using Higher Environmental Standards & Firm Value	Market Valuation	Tobin's Q	US international firms using higher environmental have higher firm valuation than peers using less standards.
Konar and Cohen (2001)	USA	Depositing Small Amount of Toxic & Firm Value	Market Valuation	Tobin's Q	Firms depositing fewer Toxic and those confronted with fewer environmental lawsuits tend to have higher valuation.
Thomas (2001)	UK	Environmental & Stock Performance	Risk-Adjusted Returns of Portfolios	Regression/Correlation	Moderate relationship between environmental & financial performance.

(Continued)

Table 1. Continued

Ref	Country	Relationship	Variables	Method-	Results
Ziegler et al. (2002)	Europe	Environmental & Stock Performance	Risk-Adjusted Returns of Portfolios	Regression/Correlation	Moderate positive relationship between environmental & financial performance.
King & Lenox (2002)	USA	Waste Prevention and Future Firm Value	Market Valuation	Tobin's Q	Positive Relationship between waste prevention and firm value.
Orlitzky et al. (2003)	All Over the World	Corporate Environmental and Financial Performance	Financial Ratios and Market Value	Meta Analysis	Environmental responsibility is likely to pay off. Environmental performance is highly correlated with corporate financial performance measured by financial ratios than that with market-based indicators.
Derwall et al. (2004)	USA	Eco-Efficiency & Stock Performance	Risk-Adjusted Returns of Portfolios	Enhanced Perform. Evaluation Techniques	Eco-efficient companies provide anomalously positive equity returns relative to their less-eco efficient peers.

have significant positive risk-adjusted returns. Furthermore, Russo and Fouts (1997) and Orlitzky et al. (2003) showed that ecological performance is positively correlated with firm profitability. Also, Derwall et al. (2004) demonstrated that eco-efficient companies provide anomalously positive equity returns relative to their less-eco efficient peers; while Dowell et al. (2000) illustrated those US international firms using higher environmental standards have higher firm valuation than peers using fewer standards.

4 THE BALANCED SCORECARD

The Balanced Scorecard (BSC) presents a network of financial measures of past performance and driver measures of future performance. It has a multi-generic measure that views organizational performance from four perspectives. During the last 15 years, the BSC has been adopted by various organizations in the world ranging from for-profit and not-for-profit, to public and private. In fact, the BSC has been adopted by almost all *Fortune 1000* organizations, and has been considered as one of the 75 most influential ideas in the 20th century (Niven, 2002). The BSC has been applied in companies from different sectors: manufacturing, construction, transportation, statistics bureau, military sales, schooling and higher education, and healthcare organizations (Table 2).

5 CONSTRUCTING AN ENVIRONMENTAL POINT OF VIEW FOR THE BALANCED SCORECARD

Six steps can be suggested as a guideline for constructing a fifth angle for a Balance Scorecard; the environmental point of view (Figure 2). These steps identify and describe ways in which organizations can effectively plan and implement their environment strategy, translating that approach into measurable actions and performance.

5.1 *Step one: Plan the environmental project*

The first step in constructing and implementing an environmental side for the Balanced Scorecard (i.e. BSC or eBSC) is to plan for it. The plan or project should include six components (Niven, 2002):

- Develop clear objectives of the environmental perspective in the BSC or eBSC project, by reviewing its rationale. This involves determining the key environmental priorities of the organization and making sure to addressing them, then embedding them in the management systems, so that they become the cornerstone of executive analysis, support and decision-making.
- Determine the starting point and location, *i.e.* managerial unit, of the eBSC project within the organization, based on criteria such as strategy, need, scope, resources, data, participant acceptance, and sponsorship.
- Make top management committed to sponsoring the scheme, because otherwise it will fail. Convincing methods should be used. These include identifying the senior manager who fits the profile, demonstrating other business success in pursuing the project, using key statistics on the implementations of the venture by competitors, taking a proactive step against the warning signs of performance measurement problems, and educating about the eBSC itself as a management system and a performance tool.
- Build an effective team to carry out the eBSC plan, by choosing those who work together using a common approach and common purpose for translating strategy into performance measures, having mutual accountability and commitment to the project. In addition, the roles and responsibilities of each member of the team should be identified, followed by training.
- Assemble a development plan for the scheme to guide the work of the team, by outlining the most critical tasks and tracking them. The plan should be accepted from the beginning by the team and top management, and be based upon the prevailing culture of the organization.

Table 2. Examples of Applications of BSC

Ref	Country	Application	Reasons for applying the BSC
Walton et al. (2004)	USA	Foreign Military Sales Center	Developing and executing international agreements by linking strategic initiatives to the BSC.
Elefalk (2001)	Sweden	Police Service	Improving the analysis, planning, and management and follow-up of the work performed.
Bonfim et al. (2004)	Brazil	Construction	Integrating synchronously management and production efforts and actions.
Mohamed (2003)	Australia	Construction	Translating safety policy into a clear set of goals across the BSC perspectives.
Phillips (2004)	USA	Public Transit System	Adapting the BSC for appraising the performance of the public transit system
Poli and Scheraga (2003)	USA	Motor Carrier	Assessing the performance of Less-Than-Truck motor carrier in meeting the expectations of their customers with regards to quality.
Dickinson and Tam (2004)	Australia	Bureau of Statistics	Measuring clients servicing for its four clients segments: key clients, professionals' regular users, specialists, and sole users.
Voytek et al. (2004)	USA	Manufacturing	Developing performance metrics for science and technology programs.
Nielsen and Sorensen (2004)	Denmark	Manufacturing	Surveying the motives for implementing the BSC.
Karathons and Karathons (2005)	USA	Education	Describing how an education performance excellence program has adopted the BSC.
Kettunen (2004)	Finland	Regional Higher Education	Analyzing the role of regional development using the BSC.
Storey (2002)	UK	School Management	Examining the potential and the limitation of the BSC in the school context.
Chang and Chow (1999)		Accounting Education	Examining the BSC as a potential tool for supporting change and continuous improvement.
Bailey et al. (1999)	USA	Business School	Explaining the BSC approach to stimulating and sustaining continuous improvement.
O'Neil, Jr. et al. (1999)	USA	Academic	Designing and implementing the BSC.
Huang et al. (2004)	Taiwan	Emergency Department	Using the BSC to improve the performance.
Sugarman & Watkins (2004)	UK	Clinical Processes	Setting out how performance indicators can be developed to capture clinical processes.
Inamdar and Kaplan (2002)	USA	Healthcare Providers	Surveying executives who implementing the BSC about the BSC roles, motivations, barriers and benefits.
Meliones et al. (2001)	USA	Children's Hospital	Applying the BSC.
Holt (2001)	USA	Army Medical Department	Developing an activity-based management system by using the BSC.
Pink et al. (2001)	USA	Hospital System	Creating the BSC.
Oliveria (2001)	USA	Healthcare Management	Assessing the BSC as an integrated approach to performance evaluation.
Weber (2001)	USA	Healthcare Corporations	Advising corporations on improving performance by formulating the BSC.
Voelker et al. (2001)	USA	Healthcare Organizations	Describing the application of the BSC.
Curtright et al. (2000)	USA	Outpatient Services in Mayo Clinic	Developing a performance measurement by applying the BSC.
Griffith and King (2000)	USA	Healthcare Organizations	Advising management on how to become a championship by constructing the BSC.
Hageman et al. (1999)	USA	Community Health Partnership	Focusing on the benefit of applying the BSC as a collaborations system.
MacStravic (1999)	USA	Healthcare Organizations	Advising organizations on building the BSC.

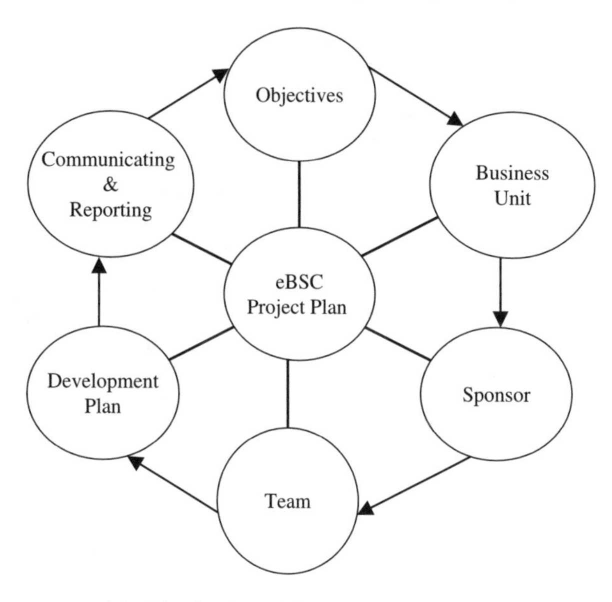

Figure 2. The Components of the Plan for the eBSC Project.

- Erect strategies for communicating and reporting the project. These should include building awareness at all levels of the business, provide education on key environmental concepts to all staff, and generate engagement, commitment and enthusiasm of key stakeholders, and disclose and communicate results without delay.

5.2 *Step two: Define the environmental mission, vision and values*

The mission statement should clarify the business identity and the core environmental purpose of the organization. It should answer questions such as why does it exist? What is it? What does it do for the environment? What is its primary role? What is its ultimate environmental goal? Who are the stakeholders? In addition, the mission statement should be long-term, inspire changes, and be easily understood (Rampersad, 2003).

The vision statement is a futuristic term that is linked to a time horizon, strategic environmental objectives and core values of the organization. It has to reflect what the group intends to become and what its ultimate environmental ambition is. It should answer questions such as where does the association go from here, and what does it want to be.

The core values of the business are the fundamentals that guide it. They are about the behavior of top management, and employees in terms of integrity, respect, excellence and communication. These are the inner values that bind an organization together. Top management should not only create environmental values, but should also reflect them in their work and speeches. The core green values of an organization should become part of its culture.

5.3 *Step three: Specify strategic environmental success factors*

The French used the *tableau de bord* as a dashboard for key success indicators. It was designed to assist staff to steer the group towards its objectives by identifying measurable key achievement factors (Lebas, 1994).

Mintzberg et al. (1998) stated that strategic success factors are related to core competence and can be identified from an organization's vision. These factors are the ones that the business has to

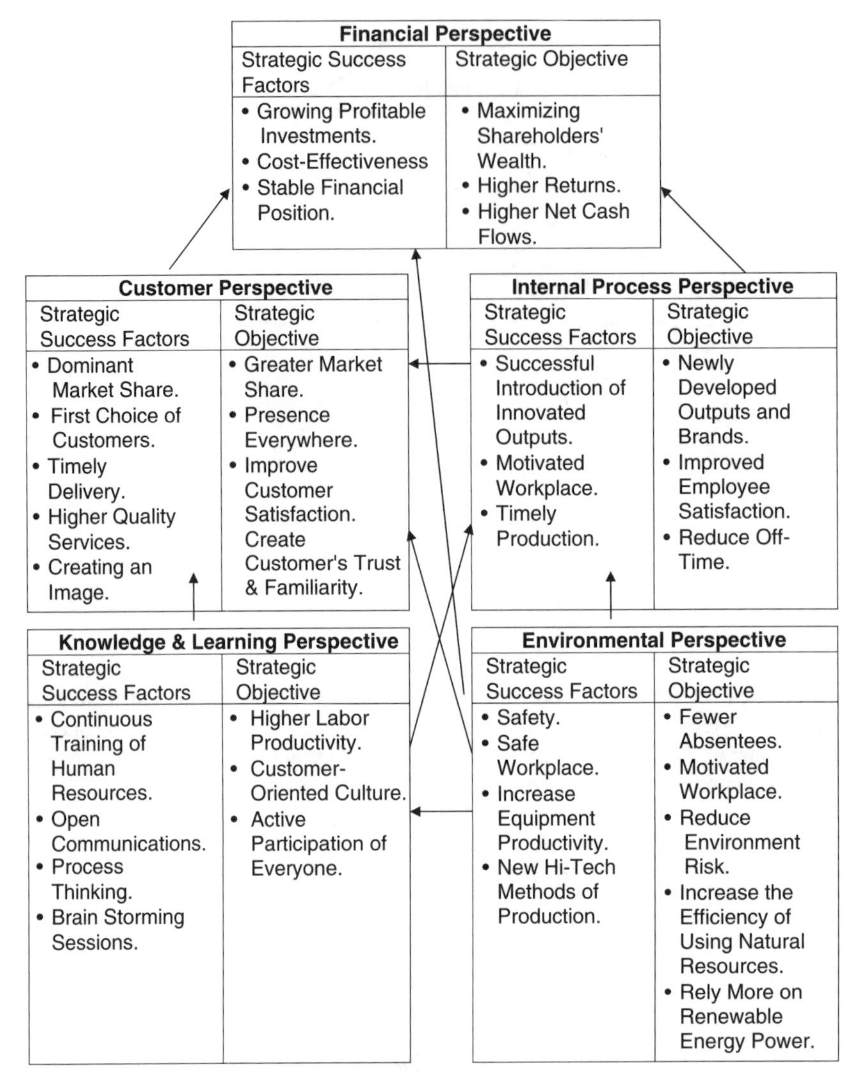

Figure 3. General Examples of Strategic Success Factors and Related Strategic Objectives to Each Perspective of the BSC.

excel in order to achieve its environmental objectives. They should identify the processes and units that are most critical for achieving environmental goals. These aspects are not quantitative ones. Instead, implementing them will produce quantitative measures.

Strategic success factors are a qualitative means for achieving an organization's tactical objectives within the five FCILE perspectives of the BSC (Figure 3). It can be seen from the figure that when the ultimate financial perspective is the maximization of shareholders' wealth, the planned achievement issue is the continuous profitability and growing business. In other words, in order to achieve the shareholders' objective, the organization business has to be profitable and expanding. Likewise, for the products or services to be everywhere, they have to be the first choice of the customers. Also, to improve customer satisfaction, the product or service should be timely delivered with high quality in-sales and post-sales services. Moreover, to provide motivation, the workplace itself, for example, should be safe.

5.4 *Step four: Setting environmental strategic objectives*

5.4.1 *Approaches to environmental objectives*

A major issue in setting ecological objectives is the question of how the environment is identified as a normative orientation for organizations. Sometimes, it refers to the need to maintain natural and environmental capital. In other cases, it emphasizes the need to ensuring a continuous flow of inputs from a specific environmental or natural capital stock.

Three approaches, borrowed from organization theory, can categorize the numerous ways for setting environmental objectives (Locke, and Latham, 1984, and Miner, 1988). These are briefly explained as follows:

The environmental goal approach may take a variety of forms. Specifying the aims may serve as a driver for an organization and direct its efforts toward a particular type of endeavor. However, green goals might be considered as constraints on decisions. Top management should consider not only the ecological constraints of their tactical decisions, but also the implications for the effectiveness and quality of their business processes. Otherwise, the company value will become less.

An organization is a set of various inter-related business units, surrounded by a variety of constraints. Thus, it has multiple goals. Different production units have different objectives depending on what their role is in the organization, and what is required from each. There is, at least, one aim for each business unit, and a set of goals to overcome the surrounding constraints and obstacles. The purpose might be financial, production, sales, development and/or supporting services. However, there is an environmental aspect in each business unit. Thus, there is an environmental aim along with each unit's goal. Each of these environmental targets should have a metric measure, so that one can judge whether the organization has reached it. It has to be explicit and communicated among the business unit staff and stakeholders, including top management, internal clients of the unit as well as the external customers and related parties, if any. In fact, making environmental goals specific, measurable, communicated and public can increase the staff motivational impact and contribute to a more precise evaluation of organizational overall environmental performance.

The environmental systems approach considers a business as environmentally effective. It measures the organization's ability to meet its internal challenges and external constraints. Output goal indexes include not only the profit but also flexibility, adaptability, allocation of resources, utilization of available resources, coordination and flow of information between business units, and dealing with stress and difference (Strasser et al., 1981). Its emphasis on system maintenance and survival remain unchanged cross-sectional, (*i.e.* from one organization to another), and cross organization, (*i.e.* from time to time within the association). This feature distinguishes the systems approach from the goal approach The latter may differ between organizations and changes from time to time.

The environmental stakeholder approach considers that the business functions as a tool or means to carry out certain purposes. The functions might those of the top management, owners/shareholders, employees, suppliers, customers, creditors, competitors, media, local community and/or unions. Those stakeholders want different functions from the organization, thus, having different aims. This approach encompasses the concept of goals, but gives it a second priority. The judgment here is about which stakeholders' goal defines the organization's effectiveness.

The development of this approach is in response to a growing concern of the social responsibility of the organizations, usually referred to as corporate social responsibility (CSR). Hence, managements have come under influence of their stakeholders in planning policy and strategy (Freeman, 1984). However, Zammuto (1982) concluded that the various and different stakeholders purposes create a problem of unrelated goals in an organization. Therefore, measurement outcomes of those goals might lead to different rating of performance effectiveness of the group. However, in the context of surroundings, this conclusion is not valid. Regardless of whose stakeholder's goal is the organization's priority or is the performance effectiveness definer. All environmental purposes and goals are related, and their measures can be summed up within the business, and averaged among all organizations. Either way, environmental objectives should be based on scientific principles, be measurable, be predictable, be relevant, be flexible, be manageable, and be compatible.

Boisvert et al. (1996) identified several practical indicators; The structure and dynamic behavior of system concerned should be represented; Aims should be constructed on a spatial and temporal scale relevant to natural, economic and social phenomena; Objectives should be quantifiable, legible and transparently suitable to decision-making; They should include deliverable scopes; Incremental values should be specified; and goals should be predictable.

Every organization is different. Objectives will be the result of cost-reduction, business sustainability, regulatory requirements, and social responsibility promises. Environmental goals vary depending on the business sector of the organization. Examples in the industrial sector include: reducing dependence on fossil fuels, using non-polluting technology, employing longer-life energy-efficient computer sets, optimum utilization of equipment, efficient utilization of resources, minimizing waste, maximizing ash utilization, improving pollution control standards, monitoring air, water land and noise parameters and minimizing adverse consequences of industrial activities on the environment. For instance, Toyota and other big companies are already selling hybrid cars, and most major auto manufacturers are racing to develop the next generation of fuel technology, with ideas ranging from engines that deliver dramatically increased gas mileage to hydrogen fuel cells that forego gasoline altogether.

5.4.2 *Translating environmental objectives into actions*

Achieving an organization's environmental objectives starts with an essential commitment from the top management. The green aims will act as a bridge from strategy to related performance measures (Figure 4). The figure translates the environmental objectives framework into actions, through the production process of converting the organization's inputs, *i.e.* raw materials, or data, into output, *i.e.* goods, services or software.

The first step in creating an environmental perspective in the BSC (or eBSC) is to set a plan. An effective ecological viewpoint in the eBSC stresses the need to identify and specify the organization's green aims. As with the traditional BSC, this takes the form of a top-down process. That is, top management sets the environmental objectives, discloses them within the organization, and gets feedback about not only its attainability, but also its acceptance from their staff. This process involves defining the action plan for implementation and the metric-measures for monitoring. However, it is important to relate environmental objectives and measures with financial goals in order to maintain or increase a company's value.

Setting an environmental plan starts with answering key questions. What are the most important environmental issues? To what extent does ecological performance need improvement? Who are the major stakeholders? What are their environmental needs? What are the priorities for each stakeholder? Answering these questions will lead to developing specific performance goals and action plans.

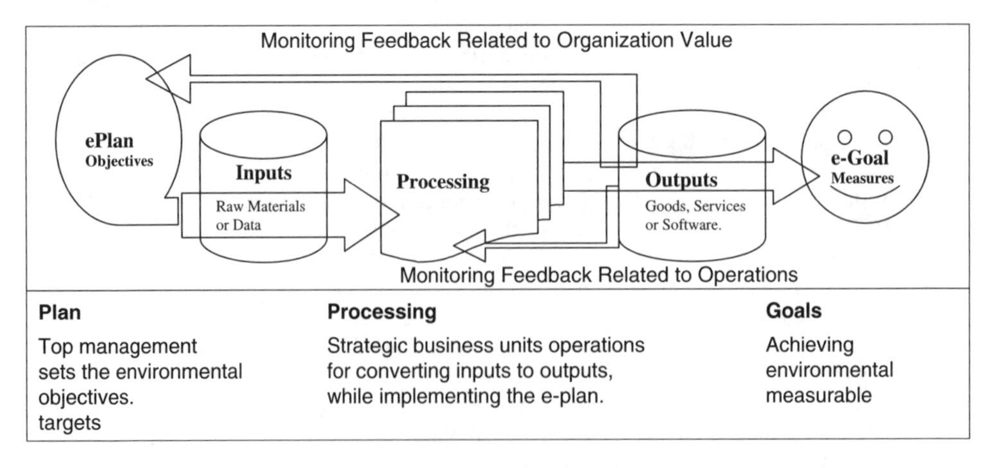

Figure 4. A Framework for Translating Environmental Objectives into Actions.

Processing and implementation are typical functions for any organization through which it can achieve its operational purposes by converting or processing raw materials or data inputs into goods, services or even software outputs. Starbucks, Nike, General Electric and McDonald's are a few examples of companies that are far ahead of their competitors in terms of the way they operate. Many of these companies track and measure their impact on global warming, have set goals for the reduction of energy use and provide easy-to-access information about their efforts to address environmental issues.

Monitoring is an essential measure to judge progress towards stated environmental objectives and performance. In fact, monitoring can be seen as an in-process environmental measure which is valuable for tracking progress and identifying areas for further improvement. While production is a continuous process, a monitoring system will generate feedback. In the later case, the environmental plan may need to be modified, altered or even changed.

Overall results should be compared with the environmental plan objectives. A number of options become available to improve performance. These include, revise existing plan and goals, adjust the implementation systems, and/or reconsider the environmental measures. Since change is the only constant, then no matter how the group is good environmentally, it can always do better. Performance should be communicated within the organization as well as to its stakeholders. For environmental disclosures by companies, see for example, Campell (2004), Rahaman et al. (2003), Al-Kadash (2004), Williams (1999), Fekri et al. (1996) and Niskala and Pretes (1995). The eBSC provides a mechanism whereby business unit processing can identify areas in which changes or improvements in the implementation systems and environmental objectives are needed.

5.5 *Step five: Developing performance measures*

A performance measurement system serves as a useful tool for monitoring quality and performance with the BSC. This should be derived from the organizational strategy and related to the tactical success factors to communicate and help implement the broad objectives (Kaplan and Norton, 1996). A performance gauge is a metric judgment on the functioning of a business unit. It is the banner that measures the progress of achieving a strategic objective. A target is a quantitative object of a performance measure. Performance measures and targets provide management with timely flags on the progress of a strategic objective. Each strategic objective should have no more than three performance measures (Rampersad, 2003).

Performance measures should comply with the following SMART criteria: They must be specifically expressed so that they can influence behavior; They must be stated in such a way that they can measure the objective; They must be realistic, realizable, feasible and acceptable; They must be related to firm results; They must be time-constrained so that their realization may be followed through time. Examples of environmental performance indicators include: number of recordable illnesses/injuries, number of lost workday cases, number of hazardous waste generated, quantity of toxic chemicals released, number of notices of violation, type/volume of non-regulated material recycled, type/volume of non-regulated material disposed, amount of fines and penalties, number/type of reportable releases, permitted air emissions, amount/type of fuel used, amount of water used, total amount of environment, health and safety operating costs, number of regulatory inspections, ozone depleting substance use, and total amount of annual environment, health and safety capital costs. Other examples include the proportion of people without access to improved water supply and sanitation, as economic analysis supports the notion that investment in water supply and sanitation would generate three to seven-fold returns (Hutton and Haller, 2004), pH concentration in the water treatment plants, the ash left behind after combustion of coal, reduction in land requirements for main plant, capacity addition in old plants, and reduction in water requirement through recycling and reuse of water (National Association for Environmental Management (1996), and Segnestam, (2002)).

In many cases conflict exists on how to give some performance metrics higher importance than others when analyzing the overall performance system. Youngblood and Collins (2003) addressed

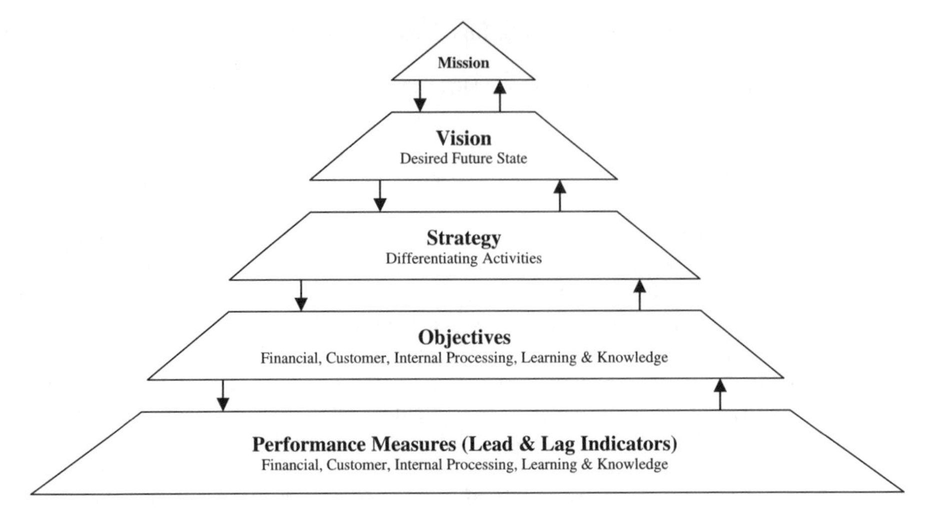

Figure 5. Development of the eBSC Construction Process.

this issue using multi-utility theory to identify different utility functions. They provided a means for a situation-dependent utility function to be incorporated into performance measurement systems.

The traditional BSC indicates that there are three kinds of performance measures (Niven, 2002): Lagging indicators are evaluation scores representing the consequences of actions previously taken. They measure outputs and they are historically in nature and do not reflect current activities, and normally easy to identify and capture. Leading indicators are performance drivers leading to the results achieved in the lagging indicators. They are in-process measures of performance, and they are predictive in nature and allow the organization to make adjustments based on outcomes. They are new measures with no history and normally difficult to identify and capture. However, there is another measure or indicator of performance. Benchmarking measures progress against goals or against other competitors or average industry performance, or against regulatory standards or business codes.

Each perspective of the BSC should have its own strategic objective(s) and a mix of lead and lag performance measures (Figure 5). An environmental perspective in the eBSC must retain emphasis on outcomes. Thus, all perspectives, including the environmental one, should contribute to the strategic objectives of the financial viewpoint (Figures 7 and 8). A lagging indicator of reducing the cost of good sold can be driven by leading indicators such a increasing the productivity of the equipments used, improving the efficient use of energy, and reducing spelled materials. Table 3 summarizes the main features of both types of indicators.

5.6 *Step six: Developing cause-and-effect linkages*

Building an ecological point of view should motivate organizations to link their environmental objectives to business strategy. The aims should be related to financial goals and measures. The selected green procedures should be part of a linked cause-and-effect relationship that contributes to improving financial performance. The importance of making these relationships explicit is to monitor, test and validate them. The measure that does not contribute to strategic organizational performance should be left out of the eBSC. In addition, possible gaps in the eBSC objectives or missing measures for achieving the desired performance cannot be identified without analyzing the cause-and-effect relationships.

The chain of cause-and-effect interactions is a handy tool for communicating the BSC to lower organizational levels (Niven, 2002). Mapping these cause-and-effect relationships is a quantitative summary of all indicators. It should start from the deriver indicators to lagging ones for each and

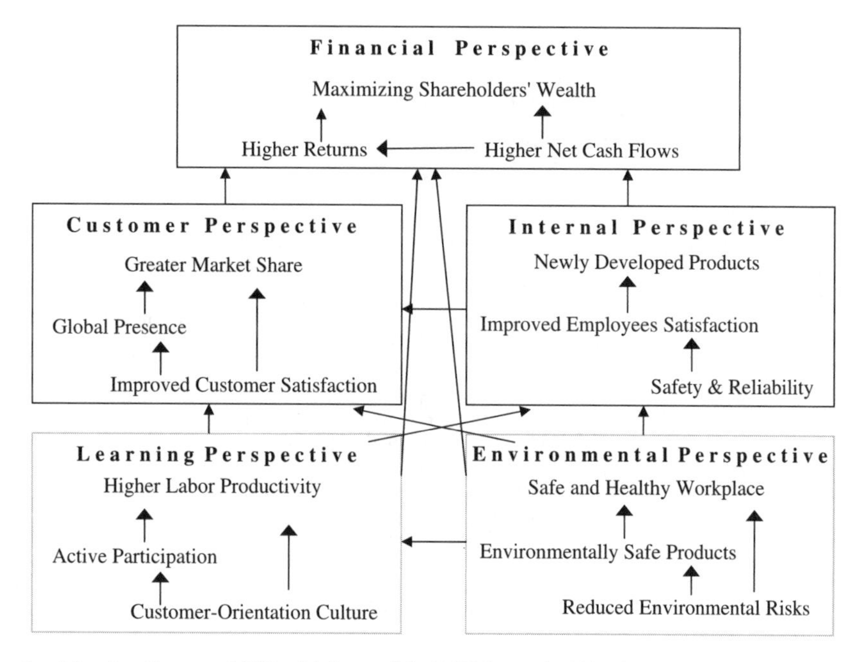

Figure 6. Mapping Cause-and-Effect Linkage of the BSC Strategic Objectives.

every perspective's objectives in the BSC. Figure 6 gives an example of three strategic objectives for each perspective. In the environmental perspective, for instance, achieving the first strategic objective, reducing environmental risk, will contribute to attaining the second objective, environmentally safe products; as well as the third objective, safe and healthy workplace within the same perspective. Also, it will induce employees to actively participate in the improving performance, in the learning and knowledge perspective. In addition, achieving environmentally safe products will enhance the identification process for new products and brands, in the internal process perspective, on one hand; and improve customer satisfaction. Lastly, safe and healthy workplace will induce employee productivity, in the learning and knowledge perspective, reducing injuries, hence, absenteeism and provides a safety and reliability business process as well as improving employees' satisfaction, in the internal process perspective.

5.7 *Step seven: Incorporating environment perspective in the balance scorecard*

Two methods can be proposed to incorporate environmental themes in the BSC. The integral method indicates that environmental objectives should be included in the customers, internal business processes and employees and systems objectives and measures. This BSC is called eBSC. The green goals are represented by the aggregate environmental measures derived from the points of view of customers, internal business processes and employees. This implies that there is no independent environmental viewpoint. However, the overall ecological performance should be reflected in the financial outlook. It has a cause-and-effect relationship with the views of the customers, internal business processes and employees (Figure 7).

The discrete method proposes that various ecological aims and procedures should be included in the environmental outlook. This indicates that the latter lies between the business processes and learning and growth perspectives, and has a cause-and-effect relationship with each of them. This scorecard is called 5-Perspectives-BSC to differentiate it from the earlier eBSC. The outputs provided to the customers are environmentally friendly, thus having an indirect cause-and-effect. This will lead to satisfying the stakeholders, which in turn is reflected in a good reputation of the organization in the marketplace. Hence, more demand on its output (Figure 8).

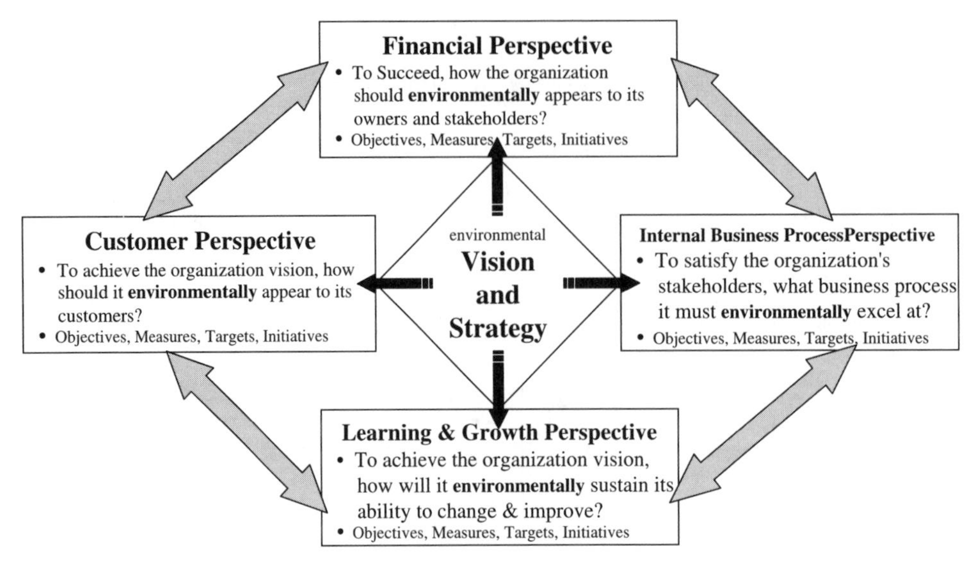

Figure 7. The Environmental Components (Objectives, Measures, Target and Initiatives) in the BSC Perspectives (FCIL) that Translates an Organization Environment Strategy into Action Adapted from Kaplan and Norton (1996).

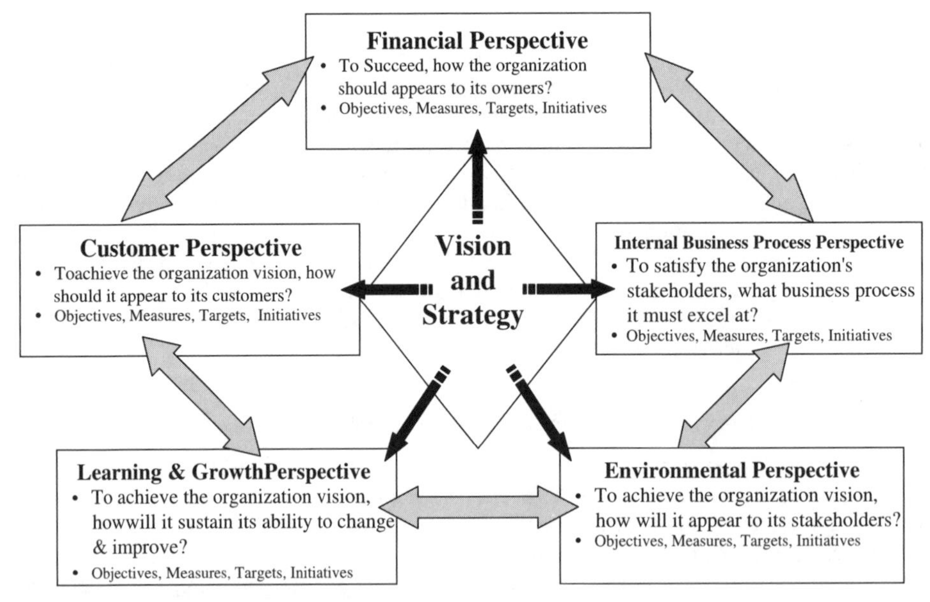

Figure 8. The Environmental Perspective in the BSC that Translates an Organization Environment Strategy into Action. Adapted from Kaplan and Norton (1996).

Both the 5-Perspectives-BSC and eBSC emphasize that environmental aims must be part of the information system communicated not only between the employees but also to the organization's stakeholders. Front-line employees must understand the ecological consequences of their actions; senior executives must understand the drivers of long-term environmental sustainability and improvement. External stakeholders must reconsider the organization value and performance effectiveness in the light of its green objectives.

Table 3. Summary of the Main Features of Lagging and Leading Indicators

Feature	Lagging Indicators	Leading Indicators
Type of Measure	Output indicators	Management indicators
Location	End-of-process	In-process
What does it Measure	Past Performance	Future anticipation
Approach	Quantitative	Qualitative, however, sometimes quantitative
Strength	Easy to quantify and preferable	Difficult to quantify
Weakness	Time lag	Hard to build support for use
Example (1)	Number of hazardous waste notices of violation	Percentage of facilities conducting regular self-inspections
Example (2)	Pounds of toxic chemical released to air, water and land	Percentage of facilities conducting self audits

Source: Adapted from GEMI (1998).

6 CONCLUDING REMARKS

This study has shown how to incorporate an environmental viewpoint into the Balanced Scorecard framework (i.e. eBSC) by providing a new way to linking green objectives with sustainable development in business strategy.

The Balanced Scorecard includes the organization's mission, vision, core values, critical success factors, strategic objectives, performance measures, targets and improvement actions. Environmental concerns should be reflected in all of these. Top management needs to develop a group environmental strategy, and then communicate it to their employees.

As more businesses around the globe incorporate ecological activities and performance in their financial statements and reports; and as more stakeholders require new and better ways to communicate green issues; top management will become more aware that environmental concerns need to be addressed and tackled within their organizations. However, environmental reporting should not be the end of the story. The eBSC can help by identifying environmental problems, by targeting key areas for top management attention and reaction, by providing support for needed improvements in current systems, and by formulating an applicable and attainable environmental performance.

REFERENCES

Al-Kadash, Husam (2004), *The Accounting Disclosure of Social and Environmental Activities: A Comparative Study of the Industrial Jordanian Shareholding Companies*, Abhath Al-Yarmouk (Yarmouk Researches, Human and Social Sciences), Vol. 20, No. 18, March, pp. 21–39.

Ambrosio, Johanna (2007), *Green IT: Popularity due to Savings or Morals*?, Computerworld, September 13th.

Ansuategi, Alberto; Barbier, Edward and Perrings, Charles (1998), *The Environmental Kuznets Curve*; in Bergh, Jeroen C.J.M. van den and Hofkes, Marjan (Editors), Theory and Implementation of Economic Models for Sustainable Development, Dordrecht, The Netherlands: Kluwer Academic Publishers, pp. 139–164.

Bailey, Allan R.; Chow, Chee W.; and Haddad, Kamal M. (1999), *Continuous Improvement in Business Education: Insights From the For-Profit Sector and Business School Deans*, Journal of Education for Business, Jan./Feb, pp. 165–180.

Boisvert, V.; Holec, N; Farnoux, C. M. and Vivien, F. D. (1996), *Ecological and Environmental Information for Sustainability*, in Faucheux, S. (Editor), Application of Non-Monetary Procedures of Economic Valuation for Managing a Sustainable Development, C3E, Universite Paris: Pentheon-Sorbonne, pp. 51–80.

Bonfim, Walace Satori; de Lima, Edson Pinheiro; and da Costa, Sergio Eduardo Gouvea (2004), *Process: An Approach to Strategic and Operational Integration*, Journal of Integrated Design and Process Science, Vol. 8, No. 3, Sept. pp. 37–47.

Campbell, David (2004), *A Longitudinal and Cross-sectional Analysis of Environmental Disclosure in UK Companies: A Research Note*, The British Accounting Review, Vol. 36, pp. 107–117.

Chang, O. H.; and Chow, Chee W. (1999), *The Balanced Scorecard: A Potential Tool for Supporting Change and Continuous Improvement in Accounting Education*, Issues in Accounting Education, Vol. 14, pp. 395–412.

Chen, K. H.; and Metcalf, R. W. (1980), *The Relationship Between Pollution Control Record and Financial Indicators Revisited*, The Accounting Review, Vol. 55, No. 1, pp. 168–177.

Cohen, M. A.; Fenn, S. A.; and Konar, S. (1997), Environmental and Financial Performance: Are They Related? Working Paper.

Curtright, J. W.; Stolp-Smith, S.; and Edell, E. (2000), *Strategic Performance Management: Development of a Performance Measurement System at the Mayo Clinic*, Journal of Healthcare Management, Vol. 45, No. 1, pp. 58–68.

Derwall, J.; Guenster, N.; Bauer R.; and Koedijk, K. (2004), *The Eco-Efficiency Premium Puzzle*, Financial Analysts Journal, Vol. 61, No. 2, pp. 51–63.

Dickinson, Teresa; and Tam, Siu-Ming (2004), *Measuring Client Servicing in the Australian Bureau of statistics (ABS): A Balanced Scorecard Approach*, Statistical Journal of the United Nations ECE, Vol. 21, pp. 7–16.

Dowell, G.A.; Hart, S.; and Yeung, B. (2000), *Do Corporate Global Environmental Standards Create or Destroy Market Value?* Management Science, Vol. 46, No. 8, pp. 1059–1074.

Elefalk, Kjell (2001), *The Balanced Scorecard of the Swedish Police Service: 7000 Officers in total Quality Management Project*, Total Quality Management, Vol. 12, No. 7&8, pp. 958–966.

European Commission (2007), European Commission on Sustainable Development: http://ec.europe.eu/sustainable/.

Fikrat, M. Ali; Inclan, Carla; and Petroni, David (1996), *Corporate Environmental Disclosures: Competitive Disclosure Hypothesis Using 1991 Annual Report Data*, The International Journal of Accounting, Vol. 31, No. 2, pp. 175–195.

Freedman, M.; and Jaggi, B. (1988), *An Analysis of the Association between Pollution Disclosure and Economic Performance*, Accounting, Auditing and Accountability, Vol. 1, No. 2, pp. 43–58.

Freeman, R. Edward (1984), Strategic Management: A Stakeholder Approach, Boston: Pitman.

Global Environmental Management Initiative (GEMI) (2004a), Clear Advantage: Building Shareholder Value Working Paper, USA: Global Environmental Management Initiative.

Global Environmental Management Initiative (GEMI) (2004b), Forging New Links: Enhancing Supply Chain Value through Environmental Excellence, Working Paper, USA: GEMI.

Global Environmental Management Initiative (GEMI) (2001), New Paths to Business Value: Strategic Sourcing: Environment, Health and Safety, Working Paper, USA: GEMI.

Global Environmental Management Initiative (GEMI) (1999), Environmental Improvement through Business Incentives, Working Paper, USA: GEMI.

Global Environmental Management Initiative (GEMI) (1998), Measuring Environmental Performance: A Survey of Metrics in Use, Working Paper, USA: GEMI.

Global Environmental Management Initiative (GEMI) (None), Fostering Environmental Prosperity: Multinationals in Developing Countries, Working Paper, USA: Global Environmental Management Initiative.

Griffith, J. R. and King, J. (2000), *Championship Management for Healthcare Organizations*, Journal of Healthcare Management, Vol. 45, No. 1, pp. 17–31.

Hageman, W. M. et al. (1999), *Collaborations That Works*, Health Forum Journal, Vol. 42, No. 5, pp. 46–48.

Hamilton, J. T. (1995), *Pollution as News: Media and Stock Market Reactions to the Toxics Release Inventory Data*, Journal of Environmental Economics and Management, Vol. 28, pp. 98–113.

Holt, T. (2001), *Developing an Activity-Based Management System for the Army Medical Department*, Journal of Healthcare Finance, Vol. 27, No. 3, pp. 41–46.

Huang, Shu-Hsin; Chen, Ping-Ling; Yang, Ming-Chin; Chang, Wen-Yin; and Lee, Haw-Jenn (2004), *Using a Balanced Scorecard to Improve the Performance of an Emergency Department*, Nursing Economics, Vol. 22, No. 3, May/June, pp. 140–146.

Hutton, Guy and Haller, Laurence (2004), The Evaluation of Costs and Benefits of Water and Sanitation Improvement at the Global Level, World Health Organization.

Inamdar, Noorein; and Kaplan, Robert S. (2002), *Applying the Balanced Scorecard in Healthcare Provider Organizations*, Journal of Healthcare Management, Vol. 47, No. 3, May/June, pp. 179–195.

The International Chamber of Commerce (ICC) (2007), *ICC Business Charter for Sustainable Development.*

Kaplan, R.S. and Norton, D. P. (1992), "The Balanced Scorecard: Measures That Derive Performance", Harvard Business Review, January–February, pp. 71–79.

Kaplan, R.S. and Norton, D. P. (1996), "The Balanced Scorecard: Translating Strategy into Action", Boston: Harvard Business School Press.

Kaplan, R.S. and Norton, D. P. (2001), "The Strategy-Focused Organization", Boston: Harvard Business School Press.

Karathanos, Demetrius; and Karathanos, Patricia (2005), *Applying the Balanced Scorecard to Education*, Journal of Education in Business, March/April, pp. 222–230.

Kettunen, Juha (2004), *The Strategic Evaluation of Regional Development in Higher Education*, Assessment and Evaluation in Higher Education, Vol. 29, No. 3, June, pp. 357–368.

King, A.; and Lenox, M. (2002), *Exploring the Locus of Profitable Pollution Reduction*, Management Science, Vol. 48, No. 2, pp. 289–299.

Klassen, R. D.; and Mclaughlin, C.P. (1996), *The Impact of Environmental Management on Firm Performance*, Management Science, Vol. 42, No. 8, pp. 1199–1214.

Konar, S.; and Cohen, M. A. (2001), *Does the Market Value Environmental Performance*? Review of Economics and Statistics, Vo. 83, No. 2, pp. 281–289.

Lebas, M. (1994), *Managerial Accounting in France: Overview of past Tradition and Current Practice*, European Accounting Review, Vol. 3, No. 3, pp. 471–487.

Locke, Edwin A. and Latham, Gary P. (1984), Goal Setting: A Motivational Technique That Works, NJ: Prentice-Hall.

MacStravic, S. (1999), *A Really Balanced Scorecard*, Health Forum Journal, Vol. 42, No. 3, pp. 64–67.

Mahapatra, S. (1984), *Investor Reaction to a Corporate Social Accounting*, Journal of Business, Finance and Accounting, Vol. 11, No. 1, pp. 29–40.

Meadows, D.H.; Meadows, D.L.; Randers, J. and Behrens, W.W. (1972), The Limits to Growth, NY: Universe Book.

Meliones, J. N.; Ballard, R.; Liekweg, R.; and Burton, W. (2001), *No Mission ≠ No Margin: It's That Simple*, Journal of Healthcare Finance, Vol. 27, No. 3, pp. 21–29.

Miner, John B. (1988), Organizational Behavior: Performance and Productivity, NY: Random House.

Mintzberg, H.; Ahlstrand, B. and Lampel, J. (1998), Strategy Safari, New York: The Free Press.

Mohamed, Sherif (2003), *Scorecard Approach to Benchmarking Organizational safety Culture in Construction*, Journal of Construction Engineering and Management, Jan&Feb., Vol. 129, No. 1, pp. 80–88.

National Association for Environmental Management (1996), Performance Measurement of Environment, Health and Safety Management Programs Survey, November, National Association for Environmental Management.

Nielsen, Steen; and Sorensen, Rene (2004), *Motives, Diffusion and utilization of the Balanced Scorecard in Denmark*, International Journal of Accounting, Auditing and Performance Evaluation, Vol. 1, No. 1, pp. 103–124.

Nijkamp, Peter and Ouwersloot, Hans (1998), *Multidimensional Sustainability Analysis: The Flag Model*; in Bergh, Jeroen C.J.M. van den and Hofkes, Marjan (Editors), Theory and Implementation of Economic Models for Sustainable Development, Dordrecht, The Netherlands: Kluwer Academic Publishers, pp. 255–273.

Niskala, Mikael; and Pretes, Michael (1996), *Environmental Reporting in Finland: A Note on the Use of Annual Reports*, Accounting, Organizations and Society, Vol. 20, No. 6, pp. 457–466.

Niven, Paul R. (2002), "Balanced Scorecard Step-By-Step: Maximizing Performance and Maintaining Results", New York: John Wiley & Sons, Inc.

O'Neil, Jr., Harold; Simon, Estela; Diamond, Michael; and Moore, Michael (1999), *Designing and Implementing an Academic Scorecard*, Change, Nov./Dec., pp. 33–40.

Oliveira, J (2001), *The Balanced Scorecard: An Integrative Approach to Performance Evaluation*, Healthcare Financial Management, Vol. 55, No. 5, pp. 42–46.

Opschoor, Hans (1998), *Delinking, Relinking and the Perception of Resource Scarcity*; in Bergh, Jeroen C.J.M. van den and Hofkes, Marjan (Editors), Theory and Implementation of Economic Models for Sustainable Development, Dordrecht, The Netherlands: Kluwer Academic Publishers, pp. 165–172.

Orlitzky, Marc; Schmidt, Frank L. and Rynes, Sara L. (2003), *Corporate Social and Financial Performance: A Meta Analysis*, Organization Studies, Vol. 24, pp. 403–441.

Pearce, David; Atkinson, Giles and Hamilton, Kirk (1998), *The Measurement of Sustainable Development*; in Bergh, Jeroen C.J.M. van den and Hofkes, Marjan (Editors), Theory and Implementation of Economic Models for Sustainable Development, Dordrecht, The Netherlands: Kluwer Academic Publishers, pp. 175–193.

Phillips, Jason Keith (2004), *An Application of the Balanced Scorecard to Public Transit System Performance Assessment*, Transportation Journal, Winter, pp. 26–55.

Pink, G. H.; McKillop, M.; Schraa, E.; Montgomery, C.; and Baker, G.R. (2001), *Creating a Balanced Scorecard for Hospital System*, Journal of Healthcare Finance, Vol. 27, No. 3, pp. 1–20.

Poli, Patricia M.; and Scheraga, Carl A. (2003), *A Balanced Scorecard Framework for Assessing LTL Motor Carrier Quality Performance,* Transportation Quarterly, Vol. 57, No. 3, Summer, pp. 105–130.

Rahaman, Abu Shiraz; Lawrence, Stewart; and Roper, Juliet (2003), *Social and Environmental Reporting at the VRA: Institutionalized Legitimacy or Legitimation Crisis*? Critical Perspectives on Accounting, Vol. 15, pp. 35–56.

Rampersad, Hubert K. (2003), Total Performance Scorecard: Redefining Management to Achieve Performance with Integrity, Netherlands: Butterworth-heinemann.

Russo, M.V. and Fouts, P.A. (1997), *A Resource-Based Perspective on Corporate Environmental Performance and Profitability*, Academy of Management Journal, Vol. 40, No. 3, pp. 534–559.

Segnestam, Lisa (2002), Indicators of Environment and Sustainable Development: Theories and Practical Experience, Environmental Economics Series, Paper No. 89, December, The World Bank.

Simon, H.A. (1976), Administrative Behavior: A Study of Decision Making Process in Administrative Organization, NY: Free Press.

Smeets, E. and Wetering, R. (1999), Environmental Indicators: Typology and Overview, Technical Report No. 25, Copenhagen: European Environment Agency.

Spicer, B.H. (1978), *Investors, Corporate Social Performance and Information Disclosure: An Empirical Study*, The Accounting Review, Vol. 53, No. 4, pp. 781–796.

Stern, D.I; Common, M.S. and Barbier, E.B. (1996), *Economic Growth and Environmental degradation: The Environmental Kuznets Curve and Sustainable Development*, World Development, Vol. 24, pp. 1151–1160.

Storey, Anne (2002), *Performance Management in Schools: Could the Balanced Scorecard Help*? School Leadership and Management, Vol. 22, No. 3, pp. 321–338.

Strasser, Stephen; Eveland, J.D.; Cummins, Gaylord; Deniston, O. Lynn and Romani, John (1981), *Conceptualizing the Goal and System Models of Organizational Effectiveness: Implications for Comparative Evaluation Research*, Journal of Management Studies, Vol. 18, pp. 321–340.

Sugarman, Philip; and Watkins, James (2004), *Balancing the Scorecard: Key Performance Indicators in a Complex Healthcare Setting*, Clinician in Management, Vol. 12, pp. 129–132.

Thomas, A. (2001), *Corporate Environmental Policy and Abnormal Stock Price Returns: An Empirical Investigation*, Business Strategy and the Environment, Vol. 10, pp. 125–134.

Thompson, Paul; and Cowton, Christopher J. (2004), *Bringing the Environment into Bank Lending: Implications for Environmental Reporting*, The British Accounting Review, Vol. 36, pp. 197–218.

Voelker, Kathleen E.; Rakich, Jonathan S.; and French, G. Richard (2001), *The Balanced Scorecard in Healthcare Organizations: A Performance Measurement and Strategic Planning Methodology*, Hospital Topics, Vol. 97, No. 3, Summer, pp. 13–24.

Voytek, Kenneth P.; Lellock, Karen L.; and Schmit, Mark A. (2004), *Developing Performance Metrics for Science and Technology Programs: The Case of the Manufacturing Extension Partnership Program*, Economic Development Quarterly, Vol. 18, No. 2, pp. 174–185.

Walton, Len; McDaniel, Tiyette; and Shyne-Turner, Schneata (2004), *Air ForceSecurity Assistance Center, Foreign Military Sales Center Institutionalizes how it Develops and Executes International Agreements by Linking Strategic Initiatives to the Balanced Scorecard Process*, The DISAM Journal, Spring, pp. 113–120.

Weber, D. O. (2001), *A Better Gauge of Corporate Performance*, Health Forum Journal, Vol. 44, No. 3, pp. 20–24.

White, M.A. (1996), *Corporate Environmental Performance and Shareholder Value*, Working Paper WHI002, McIntire School of Commerce.

Williams, S. Mitchell (1999), *Voluntary Environmental and Social Accounting Disclosure Practices in the Asia-Pacific Region: An International Test of political Economy Theory*, The International Journal of Accounting, Vol. 34, No. 2, pp. 209–238.

Youngblood, Alisha D.; and Collins, Terry R. (2003), *Addressing Balanced Scorecard Trade-off Issues between Performance Metrics Using Multi-Attribute Utility Theory*, Engineering Management Journal, Vol. 15, No. 1, March, pp. 11–17.

Zammuto, Raymond F. (1982), Assessing Organizational Effectiveness, Albany: State University of New York Press.

Ziegler, A.; Rennings, K. and Schroder, M. (2002), *The Effect of Environmental and Social Performance on the Shareholder Value of European Stock Corporations*, Working Paper, Centre for European Economic Research (ZEW).

Creating a Climate for Long-term Ecological and Organizational Sustainability

Anke Arnaud & Dawna L. Rhoades
Embry-Riddle Aeronautical University, Florida, USA

SUMMARY: This chapter proposes that long-term organizational sustainability requires a firm's deep commitment to sustainability and ethics. This commitment needs to be integrated in the organization's fabric, its climate. This climate is characterized by its commitment to ethics (ethical climate) and sustaincentrism (sustaincentric climate). The dimensions of the ethical and sustaincentric climates are described. Finally, suggestions are offered for future research.

1 INTRODUCTION

According to Viederman (1994) sustainability is a participatory process that creates and pursues a vision of community that respects and makes prudent use of all its resources (e.g. natural, human, human-created, social, cultural, and scientific). Sustainability seeks to ensure that present generations attain a high degree of economic security and can realize democracy and popular participation in control of their communities, while maintaining the integrity of the ecological systems upon which all life and all production depend. At the same time they are assuming responsibility for future generations to provide them with the where-with-all for their vision, hoping that they have the wisdom and intelligence to use what is provide in an appropriate manner.

Two decades after the World Commission on Environment and Development (WCED) final report, commonly called the Brundtland report, articulated the most commonly cited definition of sustainability as meeting the need of the present without compromising the ability of future generations to meet their needs (WCED, 1987), Europeans have discussed and debated the concept in an effort to identify its dimensions, to define key indicators and metrics of each dimension, to develop strategies to implement sustainable systems, and to integrate local, national, and regional programs (CSD, 1996; EC, 1992; Stiflung, 2001; ICSU, 2002). As governments strive to develop regulations and policies to address environmental issues and promote the greening of organizations, companies are exploring avenues to integrate ecological sustainability into their systems in order to reduce their carbon footprint and achieve long-term ecological and organizational sustainability.

In the US, mainstream politicians, Fortune 500 companies, venture capital firms, and Wall Street investors are also going green (Kluger, 2007; Stone, 2006). A recent United Press International (UPI) opinion poll has even reported that a majority of the population in the US now believe that there is a connection between human activity and climate change and that the US should cooperate with other governments and world organizations on solutions (Ecochard, 2007). In large part, this new trend is an outgrowth of a stream of films and specials such as *An Inconvenient Truth* and recent natural disasters such as Hurricane Katrina. While the rest of the world would certainly view this trend as positive, there is at least one word missing from the public debate in the US; sustainability. For the most part, green has been presented as either an environmental issue (Hoffman, 2006) or an energy issue (Breslau, 2007). Recent attention to the US addiction to oil has framed the search for non-oil sources of fuel as a national security issue (Bash and Malveaux, 2006; McKinnon and Meckler, 2006; NEPDG, 2001).

At the organizational and individual level, green is frequently touted as a cost reducing issue (Kuchment, 2007; Walsh, 2007a/b) and/or a profitability/competitive advantage issue (Esty & Winston, 2007) or the need to conform to legal requirements and avoid governmental punishment.

In short, U.S. corporations have not developed either a broad, coherent vision of sustainability or a plan to implement it. In fact, a recent essay in Time suggests that citizens in the US are only at the "eco-anxiety stage" where they have been forced "to add environmentalism to their already endless checklist of things to fret about" (Cullen, 2007).

This is not to suggest that the academic community in the US has been silent on the issue of sustainability, although Gladwin *et al.*, (1995), reported that less than .003 % of the abstracts in leading management journals from 1990 to 1994 contained the phrases biosphere, environmental quality, ecosystem or sustainable development, but the situation has been improving. This article appeared in an Academy of Management Review Special Topic Forum on Ecologically Sustainable Organizations and seems to have been a precursor for more scholarly research on organizations, business, and sustainability. Still, Alexander (2007) is correct in noting that US business has not reached the stage where profitability is "seen in terms of how it is affected by, and derived from, maximizing ideal environmental sustainability instead of determining our commitment to environmental sustainability as determined by how it affects profitability."

Researchers have noticed the need to promote ecological and sustainability values, attitudes, and behaviors in their organizations (Jennings and Zandbergen, 1995; Starik and Rands, 1995), yet we are not aware of any work that explicitly defines how these values, attitudes and behaviors should be embedded in the organization's immediate environment, its climate. In this paper, we suggest that efforts need to be made at all levels to instill the concern for ecological sustainability into the organization's fabric. Clearly, there is an interaction between the individual and the organizational level that can act as a positively reinforcing feedback loop such as the kind described by Senge (1990). Yet, in this chapter, we focus on the organizational level where the pursuit of lower costs and competitive advantage will encourage even skeptical market-oriented individuals to pursue formal policies as well as informal systems (organizational climates) of sustainability (Alexander, 2007; Shrivastava, 1995a).

2 BIRTH OF A CONCEPT: SUSTAINABILITY

The concept of sustainable development is in some sense an outgrowth of the observations by Malthus (1933) on the limits of food supply and geometric population growth and similar work by Hardin (1968). In the US, Carson (2002) helped to give birth to an environmental movement whose primary focus until the 1980s was "the reduction of local or regional environmental degradation through pollution of certain environmentally harmful outputs" (Giljum *et al.*, 2005). The focus has shifted since that time to explore problems associated with production and consumption patterns of the developed and developing world. This new focus is closely linked to a second key strand in the growing sustainability movement – post-World War II efforts to foster economic growth and development as a means of reducing poverty. The consensus among many scholars and policy-makers is that this effort has failed. The Millennium Development Goals, espoused by the United Nations in 2002, have not been met and poverty is increasing not decreasing in many regions of the world (Kemp *et al.*, 2005; Sachs, 2007).

Sustainability became the focus of the World Commission on Environment and Development (WCED, 1987) and the later Rio Declaration on Environment and Development in 1992. According to Kemp *et al.*, (2005), subsequent debate and discussion has led to general agreement on the following basics of sustainability: 1) current paths of development are not sustainable, 2) sustainability is about protection and creation, 3) requirements of sustainability are multiple and interconnected, 4) pursuit of sustainability hinges on integration; 5) core requirements and general rules must be accompanied by context specific elaboration, 6) diversity is necessary, 7) surprise is inevitable, 8) transparency and public engagement are key characteristics of decision making for sustainability, 9) explicit rules and processes are needed for decision making about trade-offs and compromises, and 10) the end is open. In short, the implementation of a sustainability agenda requires not only a deeper understanding of the concept but a detailed elaboration of the objectives, tactics, and potential outcomes of the process which must be transparent and inclusive.

3 DIMENSIONS OF SUSTAINABILITY

While sustainable development was initially seen as a matter of environmental concern, it is now viewed as a complex interaction of four dimensions – environmental, economic, social, and institutional. The environmental dimension is generally framed as a matter of insuring that 'the total volume of resources extracted does not overburden the environment, while being sufficient to maintain the functions of society and the overall economy (Giljum *et al.*, 2005). Economic sustainability has been conceptualized as a problem of maintaining a "permanent income for mankind, generated from non-declining capital stocks" (Spangenberg, 2005). Criteria subject to study have included innovation, competitiveness, public debt, aggregate demand, consumption levels, and savings rates (Pearce & Turner, 1991; Pearce & Barbier, 2000; Rennings, 2000; Spangenberg, 2005). Controversies in these areas include the impact of rapidly growing developing nations such as India and China (Diamond, 2005), the failure of regions such as Sub-Saharan Africa to development (Sachs, 2007), and the questions surrounding growth itself and possible inherent limits (Gladwin *et al.*, 1995). Further, Shrivastava (1995b) suggests that an ecocentric view of organizations must consider the risks and externalities of productive activity rather than viewing organizations as "innocent systems of production that produce products desired by consumers."

While a good deal of work has taken place on the first two dimensions, the social and institutional dimensions are still in a formative stage where their importance has been recognized but scope, indicators, and implementation have yet to coalesce. Becker *et al.*, (1999) have suggested that "sustainability describes a topic of research that is basically social, addressing virtually the entire 'process' by which societies manage the material conditions of their reproduction, including the social, economic, political, and cultural principles that guide the distribution of environmental resources." The institutional dimension first joined in the debate after the 1995 UN Commission for Sustainable Development recommended its inclusion and added it to its indicator system for evaluating country progress, but at the time the focus was on organizations not the institutional aspects and practices that shape the outcome of any activity (Pfahl, 2005). According to Keohane *et al.*, (1993), institutions (or more broadly institutionalization) can be defined as "persistent and connected sets of rules and practices that prescribe behavioral roles, constrain activity and shape expectations." Jennings and Zandbergen (1995) explain that an institutional focus is vital to the spread of sustainability as it provides a framework for understanding how values and practices are embedded in organizations' environments, in their climates.

4 THE MISSING LINK

Pressured by world organizations, the European Union, and the undeniable evidence of environmental problems, such as the Greenhouse Effect, the U.S. government has begun to take action. More than 100 independent federal environmental statues, at least seven of which span several thousands of pages (Burtraw & Portney, 1991) and a variety of state level regulations were composed and enacted to promote health and reduce dangerous chemical substance levels in the environment (Orts, 1995; Rosenbaum, 1991). The Clean Air Act, the Clean Water Act, the Emergency Planning and Community Right-to-Know Act, the Pollution Prevention Act, and the Resource Conservation and Recovery Act represent a few examples of these government interventions. While regulations and policies have mitigated many environmental problems and have certainly steered corporate attention to environmental problems and long-term sustainability concerns, it is not enough. Organizations need to support and supplement governmental efforts in order to address these ecological problems (Lee, 1993; Shrivastrava, 1995a/b). However, formal organizational policies and rules prescribing desired values, attitudes, and behaviors with regard to ecological long-term sustainability may lead to compliance in order to prevent, detect, and punish violations of such policies and rules, but they are limited by the need to control and enforce these policies and rules. Support for this argument exists in the ethics literature, which suggests that formalized, rules-based ethics programs are less effective than more informal values-based programs in generating sustainable ethical outcomes (Schminke *et al.*, 2007; Treviño and Weaver, 2001; Treviño *et al.*, 1999).

Values-based approaches, such as the development of a climate that promotes desired values, attitudes and behaviors, encourage self-directed and self-motivated adaptation (by employees) of an organization's behavioral and attitudinal norms. Starik and Rands (1995) explain that organizations need to develop shared ecological sustainability attitudes,values and behavior that are directly built into the fabric of the organization. This is not to say that values-based approaches should replace compliance- and rules-based approaches, but both should be combined and developed to encourage employee discipline and the successful creation of sustainable environments in organizations (Treviño *et al.*, 1999).

This evidence suggests that above and beyond governmental and organizational compliance-based systems, the organization's climate presents a potent force influencing employee values attitudes, and behaviors. In the following pages we will explain the foundation for the development of a climate for long-term ecological and organizational sustainability, the sustaincentric climate, to supplement regulatory intervention. We believe this to be critical in creating truly ecological and sustainable organizations.

Sustaincentric climates represent a subset of the more general concept known as organizational climates. In broad terms, climates include employees' perceptions of "how things are done around here," and include characteristics, which the members of the organization perceive and come to describe in a shared way (Verbeke *et al.*, 1998). More specifically, Payne (1990) defines organizational climate as a molar concept that represents the relatively enduring quality of the work environment reflecting the content and strength of the prevalent values, attitudes, and behaviors of the members of a social system such as an organization, department and work group. Organizational climates reflect employee perceptions of the policies, practices and procedures that the organization rewards, supports and expects, and ultimately drive employees' attitudes and actions. Our focus is the subclimate of the broader organizational work that defines the prevalent values, attitudes, and behaviors with regard to long-term organizational and ecological sustainability, the sustain-centric climate.

5 WHY SUSTAIN-CENTRIC CLIMATE?

A key contributor to the current ecological crisis is that, even until very recently, we viewed and studied organizations as separate from the natural world. Despite the fact that we understand organizations to be open systems that depend and interact with their environments, intellectually, we separated the organization from the sources of life that are absolutely essential to its existence such as the ecosystems, food chain, biochemical and nutrient cycle, and biodiversity. As dominators and owners of earth, we have learned to use nature to our advantage; organizations emerged into high efficiency tools to take more effective control over the environment and its elements. This technocentric worldview is deeply rooted in liberal social theory and the "invisible hand" reasoning that is biased toward human dominion over nature and that some suggest has its roots in Western religions (Capra, 1982; Daly and Cobb, 1994; Dyck and Schroeder, 2005; Orr, 1992). The technocentric worldview views Earth as inert and passive and legitimately exploitable. Ehrenfeld (1981) suggests this view to be arrogantly human centered: the person is separated from the superiority of nature and humans have the right to control and master natural creation for human benefit. Nature is a commodity and has mere instrumental and monetarily quantifiable values (Gladwin *et al.*, 1995). Proponents of this view suggest that environmental dangers are greatly exaggerated and explain that nature is resilient and damage is reversible (Easterbrook, 1995).

This view is so prevalent among management theory and practice, that attention to the nonhuman nature has, until very recently, been absent from the strategic management literature (Hosmer, 1994) and is even limited in the field of business ethics (Hoffman, 1991). A number of researchers have noted the paucity of attention management scholars and educators place on the biophysical world (Gladwin *et al.*, 1995; Starik and Markus, 2000).

The opposing, ecocentric worldview offers a view of the world that is more holistic, integrative, and less arrogant anthropocentric. It draws inspiration from Eastern philosophies, which suggest that we need to conform to the critical order of nature because earth gives life, further referred to as land ethic by Leopold (1949), and rejects human domination over nature (Sessions, 1995) while promoting a new age systems thinking (Capra, 1982). Earth is the nurturing mother of all life, where and interlocking order exists of which humans are but one strand. Humans should not intervene with Mother Nature because they represent but a strand of nature's interlocking nature. Ecocentrists explain that everything is connected to everything and humans are ontologically and phylogenetically inseparable from other elements of nature. We do not occupy a privileged place in nature and nonhuman nature should only be used to satisfy vital needs of sustenance. Nature is viewed as volatile and fragile, easily destroyed and we are the culprits who enable and cause irreversible damage in and to nature. Human well-being is secondary to the well-being of earth (Swimme and Berry, 1992).

Both views, technocentrism as well as ecocentrism are limited in practicability and applicability. They are morally monistic because they either view productivity to dominate the natural environment (technocentrism) or concerns for the natural environment to dominate the needs of humankind (ecocentrism) (Dyck and Schroeder, 2005). While humans need to respect and honor their dependence and interlocked existence with nature, they have an ecological need to be predators, prey, competitor and symbiont (Grizzle, 1994). "Some degree of domination of nature by humans is necessary to prevent the domination of humans by nature" (Nash, 1991). Gladwin *et al.*, (1995) argue that technocentrism as a practical view for long-term sustainability fails because it represses the connection between nature and humankind required for long-term sustainability. It fails to deal adequately with intergenerational, intragenerational and interspecies equity. It creates wealth and power for a privileged minority and creates risks and imbalances that threaten the future of the entire human race. Ecocentrism fails as a practical view for long-term sustainability because ecological sustainability is necessary, but not sufficient, for long-term sustainable development. For example, ecological sustainability is unachievable under conditions of social and economic unsustainability. Ecocentrism ignores fundamental relationships bearing upon human security and therefore, ecological integrity. In order to act and prescribe sustainable development in a practical matter, it is important to integrate the technocentric view with the ecocentric view.

The sustaincentric worldview emerges as a product of the ecocentric and technocentric view. Gladwin *et al.*, (1995) explain that sustaincentrism receives its inspiration from claims of universalism of life, and the stewardship arguments at the base of all major religions, ecological economics (Costanza, 1991), traditions of conservationism and scientific resource management (Norton, 1991), and recent theories emphasizing the inherent complexity of nature (Botkin, 1990; Prigogine and Stengers, 1984; Wheatley, 1992). Humans represent organic and ecological elements in the biosphere, but intellectually, they are above the biosphere. Intelligence raises humans to be stewards of life's continuity on earth. Sustaincentrism rejects the moral monism of technocentrism and ecocentrism and embraces moral pluralism. It is our moral obligation to assume responsibility for sustainability, which we have to defend based on our intelligence. Current generations are obligated to maintain the liberties, opportunities, or welfare-generating potentials available to future generations to levels below those enjoyed to date. The global ecosystem is not growing, but materially closed and vulnerable to human interference. It has limited regenerative capacities. We have to strive for and achieve a state of balance between use and regeneration of nature in order to achieve sustainability. For example, waste emission can not exceed the natural assimilation capacity, biodiversity loss can not exceed biodiversity preservation, ecosystem damage can not exceed ecosystem rehabilitation and nonrenewable resource depletion can not exceed renewable resource substitution. A prosperous economy depends on a healthy ecology. Markets are to allocate resources efficiently, but policy instruments and economic incentives are required to place preemptive constraints on the pursuit of purely market criteria bearing upon natural resource use and satisfaction of basic human needs. This view accepts that material and energy growth are bounded by ecological and entropic limits. Human values depend on a healthy ecological, social, and economic context and all living things have value independent of their usefulness to human purpose.

We agree with Gladwin *et al.*, (1995) that long-term sustainability should be grounded in a sustaincentric worldview. It should focus on human development (widening or enlarging the range of people's choices; United Nations Development Programme, 1994) based on the premise that human development occurs over time and space and embraces ecological, social, and economic interdependence. It needs to be guided by intergenerational, intergenerational, and interspecies fairness based on duties of technological, scientific and political care and prevention and demanding safety from chronic threats and protection from harmful disruption.

6 THE ETHICAL FOUNDATION OF A SUSTAIN-CENTRIC CLIMATE

To unleash the vast potential of corporations to resolve ecological problems, researchers and managers must reconceptualize their roles in society (Shrivastava, 1995a). Organizations pursuit of ecological sustainability requires moral development and a deepened sense of moral obligation to the environment Sustainability and sustaincentrism are deeply rooted in ethical decision making. The pursuit of the sustainability of all living things and ultimately organizations, require a strong moral obligation to reduce unethical conduct such as polluting the environment and endangering the survival of any species or engaging in activities that endanger the welfare of humans today or in the future (Howarth, 1992). Organizations need to decide if it is right to benefit from exposing another group, such as customers, suppliers, employees and shareholders, to risk and if it is right to exploit current resources without considering long –term sustainability of the plant and its human and non-human habitants. Environmental decisions are ethical in nature since the failure to properly treat the environment has significant ramifications for life and the well-being of species and ecosystems. Sustaincentric decisions require a focus on ethical constructs such as social justice; avoid harm, and focuses on safety and health. Sustainable development requires an understanding of the relationship between organizations and the natural environment and requires a focus on the decision processes, such as the ethical decision making processes, of organizational participants (Flannery and May, 2000).

Individuals and social systems are constrained by the norms, procedures and resources of the institutions in which they live and work (Tetlock, 1985). Organizational climates can limit or promote the ethical judgments of decision makers (Cohen, 1993; Treviño *et al.*, 1998). Therefore, a sustaincentric climate is, in large part, characterized by its ethical make-up, the ethical climate.

The Psychological Process Model (PPM) presented by Arnaud and Schminke (2007) explains that ethical climates include the four dimensions of collective moral sensitivity, collective moral judgment, collective moral motivation, and collective moral character. Collective Moral Sensitivity involves two factors including the prevalent mode of imagining what alternative actions are possible, and evaluating the consequences of those actions in terms of how they affect others and who would be affected by them. More specifically, collective moral sensitivity includes the prevalent norms of (a) moral awareness and (b) empathy/role-taking that exists in a social system. Cohen *et al.*, (2001) found that students with higher levels of moral awareness had more intention to behave ethically. Also, empathetic individuals have been found to be more committed to following through with helping behaviors (Davis, 1983) and displayed more organizational citizenship behaviors such as altruism (Kidder, 2002). Based on this evidence, higher levels of collective moral sensitivity, including higher levels of norms of moral awareness and norms of empathetic concern, should define more ethical climates and give rise to more ethical behaviors.

Collective Moral Judgment reflects the prevalent form of moral reasoning used to decide which course of action is morally justifiable. More specifically, collective moral judgment is defined as the norms of moral reasoning (CMD) used to judge which course of action is morally right. Peterson identified that in general, EWCs characterized by norms of postconventional and conventional moral reasoning (norms of moral judgment focusing on the benefits of others such as peers and society in general) were negatively related to unethical behaviors such as property deviance, ethnic or sexually harassing remarks, production deviance, and deviant political behavior. EWCs characterized by norms of preconventional moral judgment (norms of moral judgment focusing on the benefit for oneself) were positively associated with employees working on personal matters during company time and taking company property without permission. This evidence suggests that norms of collective moral judgment with a focus on others such as peers and the society should characterize more ethical climates and give rise to more ethical behaviors. Norms of collective moral judgment with a focus on oneself should characterize less ethical climates and negatively affect ethical behaviors of employees.

Collective Moral Motivation involves assessing whether ethical concerns dominate other concerns when determining actions and reflect whether individuals in a social system generally intend to do what is morally right. More specifically, collective moral motivation involves the prevalent values of the social system, and whether moral values, such as honesty, honor, or integrity, are generally prioritized over other values, such as power, dominance, or economics. Evidence exists, linking individuals' emphasis on moral values above other personal values to ethical behaviors. Gaerling (1999) found universalism to be related to prosocial behavior. In addition, individuals who identified benevolence and universalism values as guiding principles in their lives engaged in more socially desirable behaviors such as helping than individuals who identified other values as guiding principles in their lives (Franc *et al.*, 2002). Also, relationships between values related to benevolence and universalism and moral behavior have been identified, as have relationships between values that serve the collective interest and moral behavior (Schwartz, 1992; Schwartz & Bielsky, 1990).

These findings suggest that collective moral motivation characterized by prevalent moral values such as benevolence and universalism should characterize a more ethical climate and should be positively linked to ethical behaviors.

Collective Moral Character involves the norms for implementing a planned course of action characterized by the norms of self-control and assuming responsibility. Collective moral character is defined as the norms for carrying out a planned course of action characterized by the norms of self-control and assuming responsibility. Some support exists at the individual-level linking moral character and self-control (i.e., a strong sense of responsibility and self-control) with ethical action. For example, Krebs (1969) stated that individuals with the ability to self-regulate their actions cheat less than individuals who lack this ability. Gottfredson and Hirschi (1990) demonstrated that crimes were committed by individuals with low self-control. Schwartz (1973) found that individuals who volunteered had higher personal norms for ascribing responsibility to themselves. People high in responsibility denial tended to ignore standard norms and rationalized their behavior by blaming depersonalized others, such as organizations. In addition, Schwartz and David (1976) and Zuckerman *et al.*, (1977) found that individuals who had high personal norms for assuming responsibility were more likely to engage in prosocial behaviors such as helping others and volunteering.

These findings suggest that collective moral character including higher norms of self-control and assuming responsibility characterize more ethical climates and should be positively linked to ethical behavior.

7 DIMENSIONS OF THE SUSTAIN-CENTRIC CLIMATE

The sustaincentric climate is characterized by attitudes, values, and behavioral norms that (a) promote human development and require employees to consider the organization's impact on humankind today and in the future; (b) based in ecological, social, and economic interdependence as well as intergenerational, intragenerational, and interspecies fairness; (c) exhibit technological, scientific and political care; and (d) focus on human and environmental safety from chronic threats and protection from harmful disruption that could be caused by the organization and its members (Gladwin *et al.*, 1995). More specifically, we propose that a sustaincentric climate encompasses three dimensions: collective sustaincentric sensitivity, collective sustaincentric motivation, and collective sustaincentric character.

Collective Sustaincentric Sensitivity includes the prevalent mode of identifying and assessing what can be done to promote ecological and long-term sustainability, and evaluating the consequences of those possible actions in terms of how they affect the organization and its environment (all living beings and nature) today and in the future. Collective sustaincentric sensitivity is characterized by the organization's level of awareness (exhibited by its employees) with regard to how their work and organization affect the ecological and long-term sustainability of the environment and what can be done to revere, reduce, and correct the damage that has been done. It is characterized by the level of systems thinking (Senge, 1990) prevalent in the organization. This creates an awareness that the organization and the natural environment are interconnected and interdependent such that the organization is ecologically, socially, and economically dependent on its environment and affects the ecological, social, and economical balance of the environment. It encompasses the degree of understanding employees exhibit regarding the interconnectedness of daily business operations with the environment and issues such as poverty, population growth, income distribution and distribution of wealth between generations. This further promotes an awareness that economic goals can not be achieved without achieving social and environmental goals such as education and employment opportunity and health for all, equitable access to and distribution of resources, stable populations, and a sustainable natural resource base. In addition, collective sustaincentric sensitivity is characterized by the organization's level of empathy (exhibited by its employees) for the poor, marginalized, and vulnerable segments of society and all living beings and nature.

Organizations with higher levels of sustaincentric sensitivity are generally aware of their systematic interconnectedness and interdependence with the environment. This leads to more rigorous life-cycle analysis of impacts, and a more integrative view of sustainability such that employees will try to save energy in their daily operations such as using energy-saving light fixtures, turning off energy-consuming devises, and recycling and using recycled materials whenever possible. Organizations with high levels of collective sustaincentric sensitivity understand the importance of recycling packaging, organic production upstream in the supply chain, reducing emissions of harmful substances and waste and increasing reliance on distribution systems that save energy (e.g. rail transportation via use of trucks).

Collective Sustaincentric Motivation involves identifying and assessing whether sustaincentric values dominate other values when determining actions and reflect whether organizations (exhibited by employees) generally intend to do what is right with regard to ecological and long-term sustainability. More specifically, sustaincentric motivation involves the prevalent values of the organization, and whether sustaincentric values, such as universalism (understanding, tolerance, and protection for the welfare of all living beings and for nature), benevolence (protecting the welfare of all living beings and nature), respect for the environment (avoiding harm to all living beings and nature and promoting ecological balance), conservationism (peaceful and dignified existence; restraint with regard to use of resources, natural and human made), health (technological, scientific prevention of harm to all living beings and nature today and in the future), fairness (with regard to distribution and property rights, within and between generations, social, political, and economical fairness, interspecies), openness to change, and self-transcendence are generally prioritized over other values such as self-enhancement, tradition driven (resisting change), economic power (prestige and control over others (human, animals and nature) and other organizations), political power (influence on state and federal decisions, policies and procedures to advance company goals), dominance (control over natural and human-made resources), control over nature (nature is used for organization's benefit), organizational expansion (focus on organizational achievement, and shareholder wealth and growth).

A strong sustaincentric climate will be characterized by values that exhibit consideration for ecological concerns to prevent ecological damage and harm to the environment. For example, leaders with a strong focus on environmental concerns understand the need to protect the welfare of all living beings and nature and have a deep respect for the environment. They are generally more open to experiences, service oriented, and altruistic (Egri and Herman, 2000; Kempton *et al.*, 1995; Snow, 1992a/b; Westley, 1997). Also, self-transcendence values are closely related to the goals of the environmental movement (Karp, 1996; Stern *et al.*, 1995), as well as environmental leadership and strategies (Egri and Herman, 2000).

Collective Sustaincentric Character encompasses the norms for implementing actions that promote ecological and long-term sustainability. It is characterized by the general level of responsibility the organization (exhibited by its employees) assumes for the well being of current and future generations, all living being and nature today and in the future. It is defined by the organization's level of ecoefficiency; the production and delivery of goods and services, while reducing the ecological impact and resource intensity to a level compatible with Earth's carrying capacity (Starik and Marcus, 2000). It is defined by the level of support and reward for sustainability oriented innovation and performance, life cycle assessment, design for the environment, product stewardship, and cradle-to-cradle management (Henn and Fava, 1994). Collective sustaincentric character is defined by the level of balance that exists in the organization with regard to the organization's financial, social and ecological performance. Waste emissions should not exceed natural assimilative capacity, harvest rates for renewable resources should not exceed natural regenerative rates and human activities should result in no net loss of generic, species, or ecosystem diversity (Hawken, 1993; Roberts, 1994). Collective sustaincentric character is defined by employees' general willingness to deal with external constituents as well as their commitment to achieving integrated solutions to meet both ecological and economic goals (Egri and Pinfield, 1996).

Environments described by strong collective sustaincentric character empower employees to act on a sense of duty for technological, scientific, and political care and prevention and avoid hazards, prevent harm to the environment and humankind at all cost. They avoid chronic threats and do what is right to achieve ecological balance, and long-term sustainability. They employ individuals with a strong personal commitment to environmental issues and reward that commitment (Portugal and Yukl, 1994). Sustaincentric leaders are less willing to compromise their values in the name of running a business and consider the impact of the organization on the natural environment and environmental support and efforts they generate (Egri and Herman, 2000). They dedicate their work to something meaningful with the desire to promote long-term ecological sustainability. For example, hundreds of 3M employees have initiated and implemented several thousand Pollution Prevention Pays (3P) projects (Starik and Rands, 1995) and Tokyo Electric Power Company has developed closed-loop energy production and recycling systems. National Audubon Society cut its use of energy by 40% through solar architectural design, energy efficient lighting fixtures, conservation-oriented maintenance, and energy use program (Shrivastava, 1995a).

8 CONCLUDING REMARKS

Ecological sustainability can not be attained without and organization's commitment to adopt the vision and objective of sustainability into its strategy and structure. Ecological sustainability needs to become part of the organization's day-to-day activities. It needs to be embedded in the climate of the organization: a sustaincentric climate.

A sustaincentric climate is characterized by its strong commitment to ethics, as exhibited by its ethical climate, because sustainability can only be achieved by promoting a balance between economy, ethics, and ecology. A sustaincentric climate is characterized by the dimensions of collective sensitivity, collective values, and collective character. This model of a climate provides a necessary and important contribution to the study of ecological and long-term sustainability because climates determine how things are done around organizations. It includes the values, attitudes and standards of behaviors that influence employee attitudes and activities and ultimately organizational outcomes and performance. Hence, it is important that organizations, which pursue ecological and long-term sustainability, implant their strategic focus of ecological sustainability deeply into their fabric: their climates. The sustaincentric climate is our recommendation for embedding a strong focus on ecological and long-term sustainability into an organization's make-up.

REFERENCES

Alexander, J. (2007). Environmental sustainability versus profit maximization: Overcoming systemic constraints on implementing normatively preferable alternatives, Journal of Business Ethics, 76, 155–162.

Arnaud, A., and Schminke, M. (2007). Ethical work climate: A weather report and forecast. In Gilliland, S.W., Steiner, D.D., & Scarlicki, D.P. (Eds.), Social issues in management: Managing social and ethical issues in organizations, 5.

Bash, D., and Malveaux, S. (2006). Bush has plan to end oil addiction. CNN.com, 1 February.
Becker, E., Jahn, T., and Steiss, I. (1999). Exploring uncommon ground: sustainability and the social sciences. In E. Becker and T. Jahn (eds.), Sustainability and the Social Sciences: A Cross-disciplinary Approach Integrating Environmental Considerations Into Theoretical Reorientation, 1–22, Zed Books, London.
Botkin, D.B. (1990). Discordant harmonies: A new ecology for the twenty-first century. New York: Oxford University Press.
Breslau, K. (2007). It's all about energy, stupid. Newsweek, (November 1998), 34–37.
Burtraw, D., and Portney, P.R. (1991). Environmental policy in the United States. In D. Helm (Ed.), Economic policy towards the environment, 289–291, Oxford, England: Blackwell.
Capra, F. (1982). The turning point: Science, society and the rising culture. New York: Bantam Book.
Carson, R. (2002). Silent Spring, Special 40th Anniversary Edition with Forward by Edward O Wilson and Linda Lear, New York: Mariner Books/Houghton Mifflin
Cohen, J. R., Pant, L. W., and Sharp, D. J. (2001). An Examination of Differences in Ethical Decision-Making Between Canadian Business Students and Accounting Professionals. Journal of Business Ethics, 30, 319–336.
Cohen, J.R. (1993). Creating and maintaining ethical work climates: Anomie in the workplace and implications for managing change. Business Ethics Quarterly, 3, 343–358.
Costanza, R. (Ed.) (1991). Ecological Economics: The science and management of sustainability. Now York: Columbia University Press.
CSD (Commission for Sustainable Development) (1996) Indicators for Sustainable development, Framework and Methodology, http://www.un.org/esa/sustdev/isd.htm.
Cullen, L.T. (2007). It's inconvenient being green, Time, December 2007, 110–111.
Daly, H.E. and Cobb, J. (1994). For the common good: Redirecting the economy toward community, the environment, and a sustainable future. Boston: Beacon Press.
Davis, M. H. (1983). Measuring Individual Differences in Empathy: Evidence for a Multidimensional Approach. Journal of Personality and Social Psychology, 44, 113–126.
Diamond, J. (2005). Collapse: How Societies choose to Fail or Succeed, New York: Viking Penguin.
Dyck, B., and Schroeder, D. (2005). Management, theory and moral points of view: Toward an alternative to the conventional materialist-individualist ideal-type of management. Journal of Management Studies, 42, 705–735.
Easterbrook, G. (1995). A moment on the earth: The coming age of environmental optimism. New York: Viking.
EC. (1992). Toward Sustainability: A European Community Programme of Policy and Action in Relation to the Environment and Sustainable Development. The European Commission, Brussels.
Ecochard, K. (2007). Analysis: Concern over climate change. United Press International. Available at: http://www.upi.com/Zogby/UPI_Polls/2007/06/25/analysis_concern_over_climate_change/6716/print_view/
Egri, C.P., and Herman, S. (2000). Leadership in the North American environmental sector: Values, leadership styles, and contexts of environmental leaders and their organizations, Academy of Management Review, 43, 571–604.
Egri, C.P., and Pinfield, L.T. (1996). Organizations and the biosphere: Ecologies and environments. In S. Clegg, C. Hardy, W. Nord (Eds.), Handbook of organization studies, 459–483. London: Sage.
Ehrenfeld, D. (1981). The arrogance of humanism. Oxford, England: Oxford University Press.
Esty, D.C., and Winston, A.S. (2007). Green to Gold: How Smart Companies use Environmental Strategy to Innovate, Create Value, and Build Competitive Advantage. New Haven: Yale University Press.
Franc, R., Sakic, V., and Ivicic, I. (2002). Values and value orientation of adolescents: Hierarchy and correlation with attitudes and behaviours. Rdustvena Istrazivanja, 11, 215–238.
Flannery, B.L., and May, D.R. (2000). Environmental ethical decision making in the U.S. metal-finishing industry. Academy of Management Journal, 43, 642–662.
Gaerling, T. (1999). Value priorities, social value orientation and cooperation in social dilemmas. British Journal of Social Psychology, 38, 397–408.
Giljum, S., Hak, T., Hinterberger, F. and Kovanda, J. (2005). Environmental governance in the European Union: Strategies and instruments for absolute decoupling. International Journal for Sustainable Development, 8, 31–46.
Gladwin, T.N. (1992). Building the sustainable corporation: Creating environmental sustainability and competitive advantage. Washington, DC: National Wildlife Federation.
Gladwin, T.N., Kennelly, J.J., and Krause, T. (1995). Shifting paradigms for sustainable development: Implications for management theory and research. Academy of Management Review, 20, 874–907.
Gottfredson, M.R. and Hirschi, T. (1990). A general theory of crime. Stanford University Press.
Grizzle, R.E. (1994). Environmentalism should include human ecological needs. Bioscience, 44, 263–268.

Hardin, G. (1968). The tragedy of the commons. Science, 162, 1243–1248.
Hawken, P. (1993). The ecology of commerce: A declaration of sustainability. New York: Harper Business.
Henn, C.L., and Fava, J.F. (1994). Life-cycle analysis and resource management. In R.V. Kolluru (Ed.), Environmental strategies handbook, 923–970. New York: McGraw Hill.
Hoffman, A.J. (2006). Getting ahead of the curve: Corporate Strategies that address climate change. Pew Center for Global Climate Change, Arlington, VA.
Hoffman, W.M. (1991). Business and environmental ethics. Business Ethics Quarterly, 1, 169–184.
Hosmer, L.T. (1994). Strategic planning as if ethic mattered. Strategic Management Journal, 15, 17–34.
Howarth, R.B. (1992). Intergenerational Justice and the chain of obligation. Environmental Values, 1, 133–140.
ICSU: International Council for Science (Ed.) (2002). Science and Technology for Sustainable Development. Science for Sustainable Development, ICSU, Paris.
Jennings, P.D. and Zandbergen, P.A. (1995). Ecologically sustainable organizations: An institutional approach. Academy of Management Review, 20, 1015–1052.
Karp, D.G. (1996). Values and their effect on pro-environmental behavior. Environment and Behavior, 38, 111–133.
Kemp, R., Parto, S., and Gibson, R.B. (2005). Governance for sustainable development: moving from theory to practice. International Journal for Sustainable Development, 8, 12–30.
Kempton, W., Boster, J.S., and Hartley, J.A. (1995). Environmental values in American culture. Cambridge, MA: MIT Press.
Keohane, R.O., Haas, P.M., and Levy, M.A. (1993). Institutions for the Earth: Sources of Effective International Environmental Protection. Cambridge: MIT Press.
Kidder, D. L. (2002). The influence of gender on the performance of organizational citizenship behavior. Journal of Management, 28, 629–648.
Kluger, J. (2007). What Now?: Our feverish planet badly needs a cure, Time, April 9; 49–59.
Krebs, R.L. (1969). Teacher perceptions of children's moral behavior. Psychology in the Schools, 6, 394–395.
Kuchment, A. (2007). Going green at work, Newsweek, 11 June, 73.
Lee, K. (1993). Compass and gyroscope: Integrating science and politics for the environment. Covello, CA: Island Press.
Leopold, A. (1949). A Sand Country almanac. New York: Oxford University Press.
Malthus, T. (1933). An Essay on Population. New York: Dutton.
MCKinnon, J.D. and Meckler, L. (2006). Bush eschews harsh medicine in treating US oil 'addiction.' The Wall Street Journal, 7 August, A1.
Nash, J.A. (1991). Loving nature: Ecological integrity and Christian responsibility. Nashville, TN: Abingdon Press.
Norton, B.G. (1991). Toward unity among environmentalists. New York: Oxford University Press.
NEPDG-National Energy Policy Development Group. (2001). Reliable, Affordable, and Environmentally Sound Energy for America's Future. Washington, DC: US Government Printing Office.
Orr, D.W. (1992). Ecological literacy: Education and he transition to a postmodern world. Albany: State University of New York Press.
Orts, E.W. (1995). Reflexive environmental law. Northwestern University Law Review, 89, 1–139.
Payne, R. L. (1990). Method in our madness: A reply to Jackofsky and Slocum. Journal of Organizational Behavior, 11, 77–80.
Pearce, D.W. and Turner, R.K. (1991). Economics of Natural Resources and the Environment, Prentice-Hall: Baltimore.
Pearce, D.W. and Barbier, E.B. (2000) Blueprint for a Sustainable Economy. London: Earthscan.
Pfahl, S. (2005). Institutional sustainability. International Journal of Sustainable Development, 8, 80–96.
Portugal, E., and Yukl, G. (1994). Perspectives on environmental leadership. Leadership Quarterly, 5, 271–276.
Prigogine, I., and Stengers, I. (1984). Order out of chaos. New York: Bantam Books.
Roberts, K.H. (1994). Den Naturliga Utmaningen [The natural challenge]. Falun, Sweden: Ekerlids Förlag.
Rosenbaum, W.A. (1991). Environmental politics and policy. Washington, DC: Congressional Quarterly.
Rennings, K. (2000). Redefining innovation-eco-innovation research and the contribution of ecological economics. Ecological Economics, 32, 319–332.
Sachs, J.D. (2007). The End of Poverty: Economic Possibilities for our Time. New York: The Penguin Press.
Schminke, M., Arnaud, A., and Kuenzi, M. (2007). The Power of Ethical Work Climates. Organizational Dynamics, 36, 171–183.
Schwartz, S. H. (1992). Universals in the content and structure of values: Theory and empirical tests in 20 countries. In M. Zanna (Ed.), Advances in experimental social psychology, 25, 1–65. New York: Academic Press.

Schwartz, S. H. (1973). Normative Explanations of Helping Behavior: A Critique, Proposal and Empirical Test. Journal of Experimental Social Psychology, 9, 349–364.
Schwartz, S.H., and Bielsky, W. (1990). Toward a theory of the universal content and structure of values: Extensions and cross-cultural replications. Journal of Personality and Social Psychology, 58, 878–891.
Schwartz, S., and David, A.B. (1976). Responsibility and helping in an emergency – effects of blame, ability and denial of responsibility. Sociometry, 39, 406–415.
Senge, P.M. (1990). The Fifth Discipline: The Art and Practice of the Learning Organization. New York: Doubleday.
Sessions, G. (Ed.). (1995). Deep ecology for the 21st century. Boston: Shambhala.
Shrivastava, P. (1995a). The role of corporations in achieving ecological sustainability. Academy of Management Review, 20, 961–985.
Shrivastava, P. (1995b). Econcentric management for a risk society. Academy of Management Review, 20, 118–137.
Snow, D. (1992a). Greening business: Profiting the corporation and the environment. Cincinnati: Thompson, Executive Press.
Snow, D. (1992b). Voices from the environmental movement: Perspectives for a new era. Washington, DC: Island Press.
Spangenberg, J.H. (2005). Economic sustainability of the economy: concepts and indicators. International Journal of Sustainable Development, 8, 47–64.
Starik, M., and Marcus, A.A. (2000). Introduction to the special research forum on the management of organizations in the natural environment: A field emerging from multiple paths, with many challenges ahead. Academy of Management Journal, 43, 539–546.
Starik, M., and Rands, G.P. (1995). Weaving and integrated web: Multilevel and multisystem perspectives of ecologically sustainable organizations. Academy of Management Review, 20, 908–935.
Stern, P.C., Dietz, T., Kalof, I., and Guagnano, G. (1995). Values, beliefs and pro-environmental action: Attitude formation toward emergent attitude objects. Journal of Applied Social Psychology, 25, 1611–1636.
Stiftung, H. B., (Ed.) (2001). Pathways to Sustainable Future: Results from the Work and Environment Interdisciplinary Project, Dusseldorf, Germany.
Stone, B. (2006). The color of money, Newsweek, 13 November, E10–E14.
Swimme, B., and Berry, T. (1992). The universe story. New York: Harper Collins.
Tetlock, P.E. (1985). Accountability: The neglected social context of judgment and choice. In L.L. Cummings and B.M. Staw (Eds.). Research in organizational behavior, 7, 297–332. Greenwich, CT: JAI Press.
Treviño, L.K., Butterfield, K.D., and McCabe, D.L. (1998). The ethical context in organizations: Influences on employee attitudes and behaviors. Business Ethics Quarterly, 8, 1083–1093.
Treviño, L.K. and Weaver, G.R. (2001). Organizational justice and ethics program "follow-through": Influences on employees' harmful and helpful behavior. Business Ethics Quarterly, 11, 651–671.
Treviño, L.K., Weaver, G.R., Gibson, D.G., and Toffler, B.L. (1999). Managing ethics and legal compliance: What works and what hurts. California Management Review, 41, 131–151.
United Nations Development Programme. (1994). Human development report 1994. New York & Oxford: Oxford University Press.
Verbeke, W., Volgering, M., and Hessels, M. (1998). Exploring the conceptual expansion within the field of organizational behaviour: organizational climate and organizational culture. Journal of Management Studies, 35, 303–329.
Viederman, S. (1994). The economics of sustainability: Challenges. Paper presented at the workshop, The Economics of Sustainability, Fundacao Joaquim Nabuco, Recife, Brazil.
Walsh, B. (2007a). The cost of being clean. Time, 29 October, 54–55.
Walsh, B. (2007b). Bring eco-power to the people, Time, 3 December, 95.
WCED. (1987). World Commission on Environment and Development: Our Common Future. Oxford, England: Oxford University Press.
Westley, F. (1997). "Not on our watch": The biodiversity and global collaboration response. Organization & Environment, 10, 342–360.
Wheatley, M.J. (1992). Leadership and the new science: Learning about organizations from an orderly universe. San Francisco: Berrett-Koehler.
Zuckerman, M., Siegelbaum, H., Williams, R. (1977). Predicting helping – behavior – willingness and ascription of responsibility, Journal of *Applied Social Psychology, 7,* 295–299.

Evolution of Wholesale and Retail Distribution Channels in South Africa

Louise van Scheers
School of Business Management, University of South Africa

SUMMARY: The evolution of wholesaling moves through characteristic phases. The stage in which wholesalers are operating in a specific country is determined by the level of economic development. The growth of the cash and carry trader in the South African distribution conduit from 1960 to 2007 was highlighted. The development of the informal sector and the allotment channels which were created for the spaza shop showed the important role that the cash and carry wholesaler plays in the supply pathway.

1 INTRODUCTION

Environmental trends in a community cause a unique distribution channel structure to form. However, there will also be similarities to supply channels of other countries at the same stage of economic development (Kotler & Armstrong, 2007). A distribution channel is a set of interdependent participants involved in the process of providing a product or service for the consumer. It also provides the utility of place, of having products where the customer wants when the customer wants them. A typical South African distribution channel consists of the manufacturer, wholesaler, retailer, spaza shop and consumer.

The evolution of distribution conduits is closely related to the economic development of countries. The economic phase in which a specific state finds itself reflects on the type of commercial institutions and type of distribution channel that exist. When a nation moves to a higher level of industrialisation, there is a tendency for the phase in which the distribution channel is operating to move up to that of more developed countries like Germany. In this chapter the evolution of distribution outlets in South Africa, the role of channel participants, namely: the wholesalers, retailers, and the informal sector as well as franchising will be outlined. Internet distribution channels also forms part of the new development.

More than 100 million people worldwide can connect to Internet. Traffic is estimated to double every hundred days (www.**ecommerce**times.com/). The multimedia nature of the Internet is suitable for high-impact branding. The Internet has become an integral part of the branding mix for many advertisers. New forms of branding are animated banner ads, sponsor logos, interstitial's, "advertorials," "advertainment," and 3-D visualization. E-commerce was $20 billion USD/y in 1998 and is expected to pass 350 billion USD/y by 2006. Information flows in an instant all over the glob at no cost. Most marketers have now realized that Internet is different from conventional media in several respects. First, Internet serves as a communications channel, a transaction, and distribution conduit. Consumers can get information, make intelligent choices, and finally pay all through the Internet. No other medium can accomplish these functions instantly, without resorting to other means. Second, the Internet is by nature interactive. Users can initiate a shopping process by visiting a website and then clicking on hyper-linked text for more information. Third, it has the capacity for multimedia content. It can carry not only text and graphics but also audio and video content.

1.1 *Evolution of south african distribution channels*

The development of modern distribution channels in South Africa began when importing houses started springing up. The latter were of British origin and were established in principal ports such as

Cape Town and Durban (Black Market Report, 1991). The importing houses were usually set up in one port and did not have branches in other cities. These importers sold a whole variety of products to manufactures and did not specialise in product ranges. The merchandise was mainly imported from Europe. In addition, small independent retail trading stores were set up in the coastal towns and supplied with goods by the importing houses. There were also regular marketing days where mainly agricultural produce was sold, as well as stalls where imported merchandise was sold. In the interior, northward migration led to the establishment of wholesalers in towns like Grahamstown, Kimberley, Graaff-Reinet, Pretoria and Pietermaritzburg. The wholesalers in the interior did not import a great deal and mainly purchased merchandise from the importers at the ports. They formed the second link in the expansion of the distribution path in South Africa. Most of the traders in the interior were affiliated companies of the importers at the principal ports.

After the discovery of gold on the Reef in Johannesburg in 1886, a metropolis sprang up which created a large consumer market. The mining centres on the Witwatersrand were the first competition for the importing houses, since the mining houses imported their own technical equipment. Before the growth of Johannesburg, the importing houses controlled all exports, but after the large-scale development of mining on the Witwatersrand, exports were arranged directly from Johannesburg. The discovery of gold resulted in South Africa's rapid development from an agricultural country to an industrial nation.

The Second World War disrupted the supply of imports from the traditional overseas sources, and forced local industries to produce these goods themselves (www.tigerbrands.co.za, 2005). During this period Johannesburg became the country's main supply hub. As a result of industrialisation, a radical change took place in the economic structure of South Africa. The economy became more diversified after the discovery of gold and the inception of secondary industry issue such as iron and food manufacturing. This ushered in a new phase of development, centring on mining, which affected all spheres of the economy. Economic growth accelerated, and within five years Johannesburg had grown into a large city with thousands of inhabitants. The population of the city increased from 50,000 to 250,000 in the period between 1800 and 1895 (Finance Week, 1991).

The large domestic market was promising from the point of view of consumer products and an efficient transport system. By 1892 there was a direct rail link between Johannesburg and Kimberley. This made it easier to transport products from the ports. Johannesburg became South Africa's main railway junction and many of the importers switched their headquarters to that city. The importers fulfilled the marketing functions of a wholesaler, but with the growth of trade the marketing functions of the importers shrank progressively. Technical developments, such as improvement in transport facilities and communication, favoured the expansion of rural wholesalers and further reduced the importance of the importing houses. National advertising media, such as radio and newspapers, contributed to the dispersion of markets throughout the country. This process is still going on. The South African market has not yet been fully integrated.

By 1967 there were still very few manufacturers that controlled supply paths, because of the lack of a large, country-wide product turnover. At that time most local manufacturers still supplied the wholesalers, but there were already signs of escalation in direct sales to the retail trade. Many producers also established their own wholesale branches. They soon realised that the wholesale function can be performed far more cheaply by an independent, especially where a limited product range or products with a low unit value are marketed.

Fifty years ago an independent wholesaler, WG Brown, gave credit and undertook deliveries to retailers. WG Brown realised that deliveries to concession holders such as the retailer Spar could afford them an opportunity to survive in the distribution channel. During the period from 1952 to 1985 this wholesaler grew by an average real annual rate of 4.9 percent and currently the wholesaler play an important role in the South African distribution (Boyd et. al, 2006)

In the rural areas wholesalers are still very important participants in the delivery conduit, but in the urban areas manufacturers tend to shorten the channel, especially in the case of convenience products. By 1991 popularity of the wholesalers suffered a further blow with the emergence of large retailers who buy directly from manufacturers. In this case the large retail trade takes over the marketing functions, so that the wholesalers no longer have a place in the distribution channel. At this stage there was a tendency for corporate chain store groups to acquire an increasing share of total retail sales. In fact, only two percent of South African retailers are responsible for 68 percent of the total sales. The main groups of retailers are OK Bazaars with 190 stores, Checkers/Shoprite with over 170 and Pick 'n Pay with 100 (Black Market Report, 1991).

For many years wholesalers have occupied a dominant position in the South African distribution channel. At present they are experiencing the same problems as in other countries since the concept of wholesaling is in the process of changing. Various trends influence the traditional wholesaling patterns, such as the movement of population to the urban areas, the decentralisation of commercial institutions, and the improvement of infrastructure and communication. Changes in the surroundings and also the fact that both manufacturers and retailers are operating on a larger scale posed a threat to wholesalers. These alterations have resulted in both groups leaving the wholesalers out of the way (disintermediation). The wholesalers reacted by improving and adjusting their marketing activities, as is the trend in developing countries. The trade established its own retail/self-service outlets, which has led to the emergence of cash and carry retail.

2 DISTRIBUTION OUTLET TRENDS IN SOUTH AFRICA

A delivery trend that is typical of developing countries is the prevalence of informal markets (i.e. street markets); these have become very popular in South African over the last few years. People who lose their employment in the formal sector attempt to make a living in this way. Also many unemployed people form small informal grocery stores called spaza shops, which sell convenience products to the consumer, are to be found in the black urban residential areas.

The influence of modern distribution institutions is to be seen in South Africa such as in hypermarkets and cash and carry retailers (Etzel et. al, 2004). Since 1986 the development of large-scale self-service group of retailers has had a definite effect on the wholesale trade. The big retailers like Pick 'n Pay have integrated the retail function with the wholesale functions and developed their own cash and carry wholesaler (Price Club) in order to gain more control over the wholesale functions. In America the Robinson-Patmans Act prohibited the vertical and horizontal integration of enterprises in order to overcome possible monopolies. But in South Africa there are no regulations forbidding this kind of integration. Interestingly enough, in America the large retailers also attempted to force the full-service retailer out of the distribution channel in the early twenties (Etzel et. al, 2004).

Competition from large retailer groups has come in two ways; firstly through the development of wholesale self-service outlets (cash and carry wholesalers). Metro Cash and Carry wholesaler (Metro) for instance has established 170 shops countrywide that employ this concept. Secondly, they set up retail sales outlets and create manufacturing facilities (ie. forward and backwards vertical integration). The merging of Metro and Score Foods under the Premier Group has meant that the Premier Group, as a manufacturer, can market its own products directly (i.e. shortening the distribution channel). The full service wholesaler therefore forms self-service units that are very popular with small retailers. This corresponds to the trend towards full-service wholesalers as seen in America in the 1920's. The independent retailers used the cash and carry wholesaler to keep the selling price of products low.

The wholesaler in urban areas no longer merely carries out the traditional functions in the delivery outlet. They run big cash and carry outlets and become involved, through controlling companies, in manufacturing and packing and in the importing of products. The activities of cash and carry wholesalers have therefore diversified so that they market to small informal units (like spaza shops), formal retailers, catering businesses, as well as to the consumer. The direct sales to the public by Makro could be regarded as an innovation in the trade in South Africa. The cash and carry wholesaler makes provision for each of the above target markets in his product mix. The modern wholesaler has no intention of being a passive observer while the manufacturer and the retailer attempt to force him out of the distribution channel. Let us now look at the position of wholesalers in the South African distribution channel.

2.1 *Wholesalers*

The wholesale sector in South Africa is often known as the invisible industry because the final consumer does not usually purchase directly from them and therefore has little contact with it. They however played an important role in the development of the South African distribution system.

2.1.1 *The evolution of wholesalers*

The phase in which wholesalers are operating is determined by the economic level of development of a country. In the first phase the wholesaler separated its functions from those of the importing houses. The early wholesalers carried out many different marketing activities, many of which could not be classified as purely wholesaling activities. They carried out importing functions, collected raw materials and were also involved in the retail distribution of products. In the second phase the wholesale role was separated from the retail task. As institutions developed along with economic development in a particular country, the wholesalers delegated activities to other trading institutions such as retailers. In the third stage specialisation of products occurs, along with the integration of the wholesale trade with manufacturing. In the last phase of development new types of wholesalers such as the cash and carry emerge (Wooltru, 1990).

The phase at which the wholesalers were in South Africa in 1963 can be compared to the phase at which their German equivalents were operating in the early 1800s. Distinct and recognisable similarities in allocation outlet progress can be identified between these countries. It is also clear that definite differences caused by the environment occur but that identifiable similarities also crop up. America is the model developed country. Such comparisons reveal that Japan and Germany as developed countries are on the same level of economic development as America.

South African merchants such as Metro South Africa and Massmart have been at the forefront of supplying the formal and informal trade sector through the expansion of cash & carry outlets in several Sub-Saharan countries such as Botswana, Uganda, Kenya and Zimbabwe. Their efforts to supply the vast number of traditional stores have brought in essence modern retailing to these markets, as stores adopt larger product lines of higher quality. Metcash is currently present in eight African countries, including South Africa, with nearly 300 Cash & Carry and wholesale trade centres; It supplies several hundred associated stores, mainly under the Lucky 7 brand, and a huge number of traditional stores, street vendors and other forms of traders. The return on investment from these stores has not only helped the retailer's revenues, but also generated much needed after tax income, which under good governance is being reinvested back into the economy to improve the country's infrastructure (Wooltru, 1990).

Massmart is a managed portfolio of nine wholesale and retail chains, each focused on high volume, low margin, low cost distribution of mainly branded consumer goods for cash, through 157 outlets and two buying associations serving 1 596 independent retailers and wholesalers, in nine countries in Southern Africa (www.massmart.co.za, 2006). The channel strategy followed by Massmart refers to any shared or collaborative activity that enhances profitability beyond that which would otherwise be achieved independently by the chain groups such as Pick 'n Pay. Activities are planned and coordinated through a series of forums focused on the core trading and functional deeds of the group. All costs and benefits of any such activity, including those of the holding company, are allocated to chain groups such as Pick 'n Pay. This conduit continues to be a core element of the Massmart Business Model and an increasing source of profitability.

2.1.2 *The importance of the wholesaler in developing countries*

In growing nations there is an economy based mainly on scarcity of products. Manufacturing activities take place on a small scale, and numerous retailers run many small operations. The wholesaler plays a very important role by usually being in control of the distribution channel. They are responsible for far more marketing activities than in developed countries. They collect products and raw materials, build up contact with markets, keep stocks, break up bulk quantities and also carry the financing burden (Kotler & Amstrong 2007). The retailer and the manufacturer are not able to handle the financing of products. The extent of the influence that the wholesaler has in the distribution channel is usually an indication of the economic stage of development of the country. Japan is an exception, the only developed country where there are still so many wholesalers in the delivery outlet.

Japan has more wholesalers and retailers in relation to its population than any other developed country (Ministry of International Trade and Industry (MITI), Japan). There are virtually twice as many wholesale and retail institutions as in America. This is underlined by the fact that the total volume of wholesale sales was 4.8 times that of retail sales in 2006. There are 1 721 000 retailers as against 429 000 wholesalers (i.e., four retailers for every wholesaler). The products are handled by various kinds of wholesalers (called Tonya), usually by primary, secondary, regional and local wholesalers, before they eventually reach the retailer and consumer.

Many different types of wholesalers exist but from all the different ones the functions of the cash and carry have changed most and therefore its role will be discussed in more detail. The cash and carry wholesaler performs essential marketing functions between the manufacturer and the retailer, and therefore provides a link between the two. The retailer, in turn, is the link between the cash and carry wholesaler and the consumer in the distribution channel. The retailer can therefore be regarded as the primary target market.

2.1.3 *Evolution of the cash & carry wholesaler*

Etzel et. al (2004) emphasised the fact that the cash and carry wholesaler markets a limited range of products with a fast turnover. It concentrates primarily on a limited range of products in order to obtain the competitive advantages of purchasing in bulk.

The cash and carry wholesaler developed in the 1950s and in spite of an impressive start, initially did not make a great contribution to the wholesale sector. In the USA, where the concept originated, in 1954, there were only 922 such businesses with joint annual sales of $300 million. At that stage, the US cash and carry wholesaler sold mainly food products. However, the scope of this market grew dramatically over the years.

In South Africa, the cash and carry wholesaler concept only developed about 30 years later, with the establishment of Makro in the then Pretoria/Witwatersrand/Vereeniging (PWV) area. The country's wholesale sector was at that stage ravaged by increasing competition from the manufacturers and retailers. The manufacturers contended that the cash and carry wholesaler unnecessarily increased costs in the distribution channel, and attempted to perform the wholesale function themselves. The upshot was that the manufacturers eliminated the cash and carry wholesaler from the distribution channel and began marketing directly to the retailer – initially with big success.

The rise of the large retailer further reduced the involvement of the wholesaler. The larger retailers were in a position to perform the wholesale functions themselves. Consequently, the manufacturers own marketing endeavours and the development of large retailing groups led to intense competition. This resulted in the cash and carry wholesaler being increasingly removed from the supply path. Competition between the larger retail and independent retailers also developed. As the large retailer was in a position to market products more cheaply, it also posed a threat to the independent retailer.

The cash and carry wholesaler gradually began to realise that its survival was linked to that of the smaller or independent retailer because the latter was compelled to support the cash and carry wholesaler for its own survival. Owing to the fact that the retailer purchases smaller quantities, it was unable to buy from the traditional full-service wholesaler and therefore was ideally suited to support the cash and carry wholesalers' activities.

2.1.4 *The development of the cash & carry wholesaler over the past 45 years*

During the period 1960 to 1970, there were no cash and carry wholesalers in South Africa. According to Dalrymple and Parsons (2003), the cash and carry wholesaler concept was unknown in the South African distribution channel.

The period 1971 to 1985 saw the introduction of the cash and carry wholesaler. Makro was one of the first in the country, and in 1971, the first Makro was established in the then PWV area by the Dutch group Nederlandse Steenkolen Handelsvereniging (NSHV). In 1987, owing to anti-apartheid pressure on the Dutch enterprise, Makro was taken over by the Wooltru Group.

The second cash and carry wholesaler to be established in South Africa was Metro. This company came into being in 1979 when the former Kliptown Wholesalers started taking over competitive wholesale groups such as Leiserowitz Brothers, Savex and Kirsh. Metro grew so rapidly that in the same year, it was listed on the Johannesburg Securities Exchange.

The period from 1986 to 1990 was characterised mainly by the growth of Makro and Metro as well as the establishment of new cash and carry wholesalers in the country. The latter development increased intra-type competition between the two companies and the other smaller ones. Increased competition resulted in further takeovers which resulted in the race to acquire a greater share of the cash and carry wholesale market.

In 1987, Makro changed owners because the Dutch Group was compelled by political pressure to completely withdraw from South Africa. The group sold its interests to the Wooltru Group. Makro went from strength to strength over the years, and in 1987 already boasted five outlets in densely populated urban areas. The company's unprecedented growth emphasised the importance of the cash and carry wholesaler in the distribution channel (Makro's Method 1991).

Metro also expanded its interests and by 1987 already had more than 250 wholesale outlets. It concentrated on the specific needs of builders as well as the formal retailers and spaza shops in the black residential areas. In March 1987, the company took over the Frasers Group, comprising mainly 41 cash and carry wholesale outlets in Lesotho, for R21 million (i.e. 2.8 million USD). The upshot of this new takeover was that the Metro group had become a formidable competitor, and in 1987 was already the largest group in the country. During this period, the Metro group operated three different kinds of cash and carry wholesalers, namely Metro, Bingo and Trade Centre. Trade Centre has a wider product composition than the other wholesale groups and sells groceries, building products and household fittings, while Bingo markets household fittings only and Metro groceries only. Metro concentrated on the consumer who comprises mainly the C and D income groups. This means that the company was in direct competition with the large retailer Pick 'n Pay. Owing to this rivalry, Pick 'n Pay established its own cash and carry wholesale outlet, Price Club. The first Price Club was opened in October 1986 and concentrated on the black retail market.

In 1983, another cash and carry wholesaler, Success was established by the Shield Trading Corporation. Success grew rapidly, and five years later the company already had seven outlets. In July 1987, the Shield Trading Corporation established another cash and carry wholesaler, namely Shield Cash and Carry. The Shield Trading Corporation then had 150 cash and carry wholesale outlets and marketed to approximately 22 000 retailers.

Finally during the period 1991 to 2007 the industry was characterised by a large number of cash and carry wholesaler outlets under the Metro and Makro brand name which were all subsidiaries of large listed groups. The Wooltru Group controlled the cash and carry wholesalers, Makro and Shield, while the Premier group controlled, Trador, Bingo and Trade Centre. Furthermore, Pick 'n Pay, had its own cash and carry wholesaler outlets, namely Price Club. An additional 25 smaller cash and carry wholesalers were established in South Africa during these years.

Metcash Trading Limited, a subsidiary of the listed company Metro Cash & Carry was in court over VAT fraud (www.massmart.co.za, 2006). The Receiver's two-year investigation into Metro came to a head when more than 50 members of the revenue service removed documents from the company's head office. The main allegations against Metro were doubtful suppliers of goods totalling R632-million (i.e.79 million USD).

The merger of Metro and Score Foods under the Premier Group in 1991 gave rise to the largest cash and carry wholesale industry in South Africa. This group owned about 28 percent of market with a turnover of approximately R6 billion per annum. Metro expanded into 28 counties with 500 outlets.

Metro, Trador and Trade Centre followed a niche marketing strategy. Owing to stiff competition between Metro and Makro, the latter in collaboration with Makro-office, Shield and Drop Inn, formed a new wholesale group known as Massmart. The latter is a competitive cash and carry wholesale group that owns approximately 18 percent of the market in South Africa (Wooltru, 1990).

The traditional South African cash and carry wholesaler has survived for more than twenty years in its present form but is showing signs of being influenced by a continued innovative process.

Factors such as smaller packaging, credit, delivery and consumer marketing, which go against traditional practices, are taking root. Innovation is being spurred on by changes in the target market, specifically as a result of the inclusion of the spaza shop market and the issue of Black Economic Empowerment (BEE).

Massmart has stores in Botswana, Lesotho, Mauritius, Mozambique, Namibia, Swaziland, Tanzania, Uganda, Zambia and Zimbabwe. The total store portfolio comprises seven Game general merchandise discount stores; nine CBW wholesale food and liquor outlets; and two Makro general merchandise, food and liquor stores. The company approved an integrated BEE strategy to develop their transformation scorecard, to rate their current progress, and to develop Group wide plans for implementation. They achieved their self-imposed equity targets for mid and low-level employment grades and registered progress at higher levels. They also created an Enterprise Development Division and an investment company tasked with nurturing and developing small and medium black businesses (www.massmart.co.za).

2.2 *South African retailers*

The retailers in the formal sector are managed independently and own a relatively small share of the total market. There are about 53,644 retailers in the South African formal sector with retail trade sales at current prices for May 2006 of R31 million (4 million USD) (www.massmart.co.za, 2006) . Although the country has a relatively small retail proportion compared to the rest of the world (Table 1), retailers definitely form a stable market and are the cornerstone of the continued existence of the cash and carry wholesaler. The official retailer is the institution that is most prevalent in South Africa. Initially this type operated in the coastal towns. It was only in the densely populated Witwatersrand area where it increased in both size and power to such an extent that it became independent of some of the wholesalers. The growth of the large independent retailer was the cornerstone of the retail revolution in South Africa.

The revolution served as the emancipation of the retailer, the wholesaler and the manufacturer. There was unprecedented growth in this sector during the 1960s, which surprised the manufacturers and the larger retailers. The smaller formal retailers had difficulty adjusting to the rapid and radical changes in the distribution structure. It is this group of retailers that currently constitute the target market of the cash and carry wholesaler. During the period 1971 to 1989, the number of formal retailers rose by an average of 4.9 percent per annum, which consequently increased the marketing potential of the cash and carry wholesaler substantially.

Formal retailers contribute approximately 29 percent to South Africa's gross national product and form a vital target market of the cash and carry wholesaler (www.statssa.gov. 2005). Economic trends such as unemployment and limited job opportunities in the formal sector generate increasingly more retailers. Although the formal retailer generally showed a small increase during the period from 1988 to 1994, the urban retailer grew by 34 percent (Table 2).

Table 1. Retailing: Africa versus the Rest of the World, 2006

Continent	GDP (USD mn)	Share (%)	Retail sales (USD mn)	Share (%)
Africa	769,538	1.9	296,512	3.0
Europe	14,178,063	34.7	2,813,882	28.3
North America	12,743,037	31.3	2,875,465	28.9
South America	1,131,590	2.8	378,441	3.8
Central America	870,859	2.1	284,009	2.9
Asia & Pacific	10,446,679	25.6	3,165,421	31.9
Middle East	637,139	1.6	121,607	1.2
Total	40,776,879	100	9,935,337	100

Source: www.massmart.co.za

Table 2. Africa: Leading retailers and their grocery formats, 2006

Retailer	No. of stores	Retail banner sales (USD mn)	Formats
Metcash	1,5446*	3,272*	Cash & carries, supermarkets & neighbourhood stores, convenience stores, non-food formats
SPAR (S.A)	813	1,829	Superstores, supermarkets & neighbourhood stores
Shoprite	743	4,067	Hypermarkets, supermarkets, discounters, convenience stores, cash & carries
Pick 'n Pay	594	4,288	Hypermarkets, supermarkets, convenience stores, drug stores
Woolworth (RSA)	221	1,744	Department stores, discounters, convenience stores
Massmart	167	2,532	Cash & carries, variety stores, DIY Neighbourhood stores
Uchumi	31	120	Hypermarkets, supermarkets Casino 24 1,384 Hypermarkets, supermarkets, discount stores
Nakumatt	14	126	Hypermarkets
Carrefour	4	905	Hypermarkets, supermarkets

Source: www.massmart.co.za

2.2.1 *The large retailers*

In South Africa, the three main groups of large retailers, Hyperama, Shoprite & Checkers, Spar and Pick 'n Pay are responsible for 63 percent of all food purchases (www.tigerbrands.co.za, 2005). These large organisations purchase directly from the manufacturer and are therefore in direct competition with the cash and carry wholesaler. Increased growth in the retail sector exacerbates this competitive situation. From the definition of a wholesaler (i.e. an enterprise that obtains 50 percent of its gross trade from wholesale sales) one sees that wholesalers, like retailers, also sell to consumers. Their target markets thus overlap and they are involved in a competitive struggle. The spaza shops also buy products from the large retailers. Hence the cash and carry wholesaler cannot lay claim to the total spaza shop target market (Morris, 2006). The spaza shop market increases the competitive situation between the large retailer and the cash and carry wholesaler even further. The spaza shop can be defined as a retailer who sells provisions from his or her home in the black suburbs. In certain respects, the spaza shop is similar to the traditional general retailer who sells products over the counter (The Citizen, 1990).

The distinction between the large retailer and the cash and carry wholesaler is in the process of blurring. Some retailers are even being operated as hypermarkets that perform wholesale functions. This is possibly another reason why the target markets of the cash and carry wholesaler and the large retailer overlap and cause stiff competition. Another reason for the increasing competition is the establishment of larger cash and carry wholesale groups such as the Premier Group. However, the large retailer is also not innocent and endeavours to take over part of the cash and carry wholesale market by sending large trucks with low-priced products into black townships. This new trend is threatening the very survival of the spaza shop.

2.2.2 *The smaller retailer*

The popularity of the small retailer in South Africa is another important consumer characteristic. Limited income and storage facilities, together with a strong preference for fresh food, make it essential for housewives to undertake regular, daily shopping trips. This creates marketing opportunities for neighbourhood shops situated close to the consumer. These conditions call for a distribution structure that incorporates a large number of small, conveniently established retailers supported by a large number of wholesalers.

3 THE INFORMAL SECTOR & DISTRIBUTION CHANNELS CREATED AROUND THEM

Poverty, low literacy levels and extremely high population growth are generic problems experienced by developing countries. South Africa's population is expected to reach the 70 million mark by 2025. In contrast to the relatively rapid population increase, the economic growth rate declined from 7.7 percent in 1965 to 2.5 percent in 2005. As a rule, an increase in population and low levels of economic growth lead to high unemployment levels. The South African formal sector is unable to provide enough job opportunities and the unemployed are thus compelled to create their own job opportunities in the informal sector. The term "informal sector" is used to define an economic phenomenon that is widespread in the rural and urban areas of Third World countries.

A shortage of natural resources is experienced in developing countries. This is because resources are still largely caught up in the potential phase as regards capital, trained management and labour. High and rising levels of unemployment are therefore found as a result of the inadequate and ineffective use of labour. The growing urban population have difficulty in finding employment. The result was the development of the informal sector in the early seventies (Kotler & Armstrong, 2007). The current rate of open unemployment in developing countries on average is 10 to 15 percent of the urban population. Migration from rural areas has resulted in a growth of 5 to 7 percent in the urban population in many African countries and job creation cannot keep pace with the population increase (Schoell, 2006). The informal sector, while creating employment opportunities, can only be considered as a temporary arrangement for people.

A significant number of unemployed people establish their own retail operations in the informal sector and buy from the cash and carry wholesaler. This wholesaler is an important link in the distribution channel between spaza shops and manufacturers. This applies particularly to the informal retailer or spaza shop which, because of population growth, is showing huge expansion opportunities.

3.1 *Informal retailer or Spaza shop*

Spaza shops can be defined as home shops in black suburbs which provide essential products at reasonable prices in places where there are no traditional buying amenities (Van Scheers, 1998). They are part of the informal sector and are not regulated according to the normal, conventional prescribed methods. The right of existence and feasibility of the spaza concept can be ascribed to the convenience shopping trend of black consumers. It is important for the cash and carry wholesaler to determine precisely how large the spaza shop target market is in order to justify its own existence in this distribution channel. The shop can be classified into three categories; in category A, a carport is erected and used as a spaza shop; in category B, the owner fits out and uses a room in his or her house to serve as a spaza shop; and in category C, business is conducted in corrugated iron shack separate from the house.

It is impossible to establish the exact size of the spaza shop market. According to Black Market Report (1991) an estimate by on he East Rand alone, there were about 4,000 of these shops. Scott-Wilson and Mailoane (2006) estimated that there are approximately 20,000 spaza shops in the whole of South Africa. These estimates are probably too conservative. Kotler & Amstrong, 2005 (Pick 'n Pay tills ring the changes 2006) painted a more realistic picture in his estimate of about 66,000 spaza shops in 1995. These estimates are further complicated by the fact that new spaza shops are continually entering the market, while others are going under almost daily.

The joint spaza shop purchasing power of about R16 billion per annum (i.e. 2.2 billion USD/y) is indeed extremely high and therefore constitutes a promising market for the cash and carry wholesaler. The combined retail purchasing clout motivates the cash and carry wholesaler to exploit the spaza shop target market. The wholesaler regards this target market as an important outlet for the black consumer. This access to the black consumer market was previously not possible.

However, the spaza shop may not be an answer to South Africa's unemployment problems and merely offers a temporary solution. The chances of survival of the smaller spaza shops that are not strategically located are slim. On the other hand, the larger spaza shops situated near to throughways definitely have a better chance. New urban areas are continually being developed, and in places where there are no formal retailers, new spaza shops that will purchase from the cash and carry wholesaler, will undoubtedly develop. If one considers the incredible growth of the spaza shop target market and the fact that the wholesaler supplies most of the products, this shop can undoubtedly be regarded as a factor that justifies the existence of the cash and carry wholesaler in the distribution channel.

The reason for the continued success of the cash and carry wholesaler in the South African delivery conduit is twofold. On the one hand, manufacturers are increasingly confronted with rising distribution costs such as delivery and the high risk of credit provision. This means that they tend to foist these marketing functions on to the cash and carry wholesaler who is able to perform them more cheaply. On the other hand, the smaller retailers are experiencing problems purchasing products from the full-service wholesalers because the latter do not sell small consignments. One could say that the cash and carry wholesaler owes its very existence in the distribution channel to the rise of the spaza shop. The increase in the number of spaza shops because of increased population growth, low levels of economic growth and unemployment has possibly led to a shift of emphasis in the cash and carry wholesaler's target market. The escalation of spaza shops encourages the cash and carry wholesaler to exploit this target market. It is also important to note that this target market. alone cannot possibly justify the existence of the cash and carry wholesaler in the distribution channel. This could be because the spaza shop market is still too small. However, there is a possibility that the increase in unemployment in South Africa and the resultant growth of the informal sector will in the future create a greater marketing potential for the cash and carry wholesaler.

Greater involvement in spaza shops, marketing aimed at the consumer and adjustments to the credit and delivery functions appear to be the future trends that can be expected in the wholesale sector. The cash and carry wholesaler is currently undergoing structural changes, is performing increasingly more retail functions of the large retailer and is marketing to the consumer to a greater degree. All of this is impacting on the basic characteristics of the cash and carry wholesaler. The latter is also even willing to give credit and deliver products, and is thus increasingly impinging on the terrain of the full-service wholesaler. Innovation is unavoidable.

4 CONCLUDING REMARKS

This chapter highlighted the growth of the cash and carry trader in the South African distribution channel from 1960 to 2007. The spaza shop can be considered as a phenomenon development which created a reason for the cash and carry wholesaler to maintain its role in the country's distribution channel.

5 GLOSSARY

Backwards integration: Retailers acquire (buy out/take ownership of) their own suppliers.
Cash-and-carry wholesalers: Like service wholesalers, except that the customer must pay cash.
Channel of distribution: Any series of firms or individuals who participate in the flow of products from producer to final user or consumer.
Financing: Provision of the necessary cash and credit to produce, transport, store, promote, sell and buy products.

Forward integration: A manufacturer acquires (buys out or takes ownership of) some wholesalers and retailers in the distribution channel.
Franchise operation: A franchisor develops a good marketing strategy, and the retail franchise holders carry out the strategy in their own units.
Franchisor: The parent company that gives franchisees the right to operate a business according to a prescribed marketing plan and to use an established trademark.
Marketing strategy: The strategy that specifies a target market and a related marketing mix.
Retail trade Retail trade includes the resale (sale without transformation) of new and used goods and products to the general public for household use.
Retailer A retailer is an enterprise deriving more than 50% of its turnover from sales of goods to the general public for household use.
Spaza (shop): A uniquely South African retail establishment that serves the needs and wants of customers in rural areas, informal settlements and townships.
Wholesalers: Firms whose main function is providing wholesaling activities.
Wholesaling: The activities of those persons or establishments that sell to retailers and other merchants, and/or to industrial, institutional and commercial users, but who do not sell in large amounts to final consumers

REFERENCES

Black Market Report. 1991. The role of small business in big business. January: 2–3.
Boyd, HW, Walker, OC & Larréché, JC. 2006. *Marketing management.* 2nd edition. London: Irwin.
Dalrymple, DJ & Parsons, LJ. 2003. *Markeing management strategy and cases.* 6th edition New York: Wiley.
Etzel, MJ, Walker, BJ & Stanton, WJ. 2004. *Marketing management.* New York: McGraw-Hill.
Finance Week, June 1991.
Kotler, P & Armstrong, G. 2007. *Marketing management: analysis, planning and control.* 15th edition. Englewood Cliffs, NJ: Prentice-Hall.
Makro's Method. 1991. *Financial Mail*, May: 31.
Morris, R. 2006. *Markeing to black townships*. Durban: Butterworth.
Schoell, WF. 2006. *Marketing management.* 8 th edition. Boston: Allyn & Bacon.
Van Scheers, ML. 1998. Die plek en funksie van die kontant-afhaal-groothandelaar in die Suid-Afrikaansedistribusiekanaal. DCom-proeskrif, Universiteit van Suid-Afrika, Pretoria.
Wooltru R173,4 m Capex. 1990. *The Cape Times,* October:18.
www.massmart.co.za, 2006
www.tigerbrands.co.za, 2005.

Examining Banking Productivity Across Countries Considering Off-Balance Sheet Activities

Katerina Lyroudi & Dimitrios Angelidis
University of Macedonia, Greece

SUMMARY: This paper examines the impact of off-balance sheet (OBS) activities on the productivity of banking institutions across eleven European countries for the period 1995-2002. The data envelopment analysis (DEA) technique was employed to calculate the Malmquist indices of total factor productivity (TFP) change. This procedure was repeated twice; first the OBS items were omitted and then they were added as the fourth output variable. The results revealed that productivity varies according to both approaches since for some countries it is enhanced while in others it is worsened. However, when the OBS items were not included as an additional variable the predicted TFP indices fit better than the actual TFP indices, since the mean absolute percent error (MAPE) was lower for the majority of the countries. Hence, in order to evaluate banking productivity the model without OBS activities should be preferred.

1 INTRODUCTION

There are different types of productivity, such as land, labor, physical capital and total yield. In general, productivity can be defined as the ratio of goods and services to the factors of production that are utilized for the manufacturing of these goods. Yield can be augmented by increasing the number of final products, by keeping the factors of output stable, or by reducing this factor and keeping the number of final products stable, or by a combination of the two, having a maximum output result with the lowest input resources possible.

It is interesting to examine what factors determine the level of productivity of a country. The present study concentrated on financial institutions since it is preferable to concentrate on a particular sector of the economy. These institutions, due to their importance and magnitude, can be considered as the corner stone of any economic system.

Banking institutions are of special interest to the European Union. The second banking co-ordination directive, which was adopted in 1989, created a regime for regulating all functions of such institutions in the European Union. This new framework along with technological improvements and intensified competition in the financial services industry forced banks to adopt new products and services beyond the conventional deposits and loans. These new products consisted of intermediation and other fee based activities such as loan commitments, letters of credit, securities underwriting, insurance and derivatives. These are also known as off-balance sheet (OBS) activities [Siems and Clark (1997) and Rogers and Sinkey (1999)].

In the literature, some studies have evaluated banking productivity without taking into consideration OBS activities. Others have included them. Ambiguous results have been obtained on whether or not OBS activities enhance productivity. Hence, there is a need for further investigation to gain more insight.

The purpose of this study was to evaluate the banking productivity in eleven European countries, namely: Austria, Belgium, Denmark, France, Germany, Holland, Italy, Luxemburg, Spain, Switzerland and the U.K. Two approaches were used; the first one without considering the off-balance sheet (OBS) activities and the second one by including them in the evaluation process. This was done to determine the better approach was. The results of this study will be useful for policy makers and practitioners as well as academicians, since it will increase our understanding of this pertinent subject.

2 BANKING PRODUCTIVITY

In this chapter, the focus was only on those studies that used OBS activities as an additional factor for checking bank efficiency. Jagtiani, Nathan and Sick (1995) investigated whether failure to incorporate off-balance sheet (OBS) products or activities in estimating bank efficiency might lead to a misspecification problem. Their sample consisted of US commercial banks over the period 1988–1990. They found that OBS products seemed to have little or no significant effect on the scale economies. Furthermore, the authors suggested that the volume of OBS activities had little or zero impact on bank costs.

Jagtiani and Khanthavit (1996) studied the largest 120 banking institutions in the U.S. The final sample contained 91 banks for the period 1984–1991. They used a translog cost function. The authors considered both on- and off- balance sheet products and allowed for bank product mixes to differ. Their paper examined the impact of the risk based capital requirements (RBC), which were approved in July 1988, on bank cost efficiencies. They found that RBC requirements lead to a decrease in the efficiency of large banks. However, it did not reduce the OBS products adopted by these institutions.

Rogers (1998) examined the cost and profit efficiency of U.S. commercial banks first by not incorporating the OBS activities and then by incorporating them. He used the distribution free frontier approach (DFA) and found that when the OBS activities were excluded bank efficiency was understated.

Rogers and Sinkey (1999) examined common features of US commercial banks that were heavily engaged in non-traditional activities in 1993. The empirical results suggested that the more productive banks were the larger ones which faced more competitive interest rate conditions, had fewer core deposits, had more diverse sources of revenue by also having OBS activities and had greater access to financial markets, which reduced their risk.

Stiroh (2000) examined the performance of US banks from 1991 to 1997. The analysis of cost and profit functions suggested that the gains were primarily due to productivity growth and changes in scale economies. He tested productivity growth with traditional and non-traditional output specifications, which were both significant. The author found that failure to account for non-traditional activities like OBS items led to profit efficiency, but not to cost efficiency, which was understated especially for the largest banks.

Clark and Siems (2002) examined the effect of including non traditional or off-balance-sheet activities on the efficiency of US banks. This was measured from three points of view: production cost efficiency, economic cost efficiency and profit efficiency. Their results showed that economic cost and production cost x-efficiency increased when the OBS activities were incorporated. On the other hand, profit x-efficiency was almost unaffected. As proxies for the OBS items the authors used a credit equivalent measure, an asset equivalent measure and a revenue-based measure. The former was not a very good proxy for the OBS activities since there was no x-efficiency when it was included representing the OBS items. X-efficiency in economics, is a concept introduced by Leibenstein (1966) that indicates how effectively a given set of inputs can produce the maximum possible outputs.

There are a few studies that focus on European banks. Rime and Stiroh (2003) examined the performance of Swiss banks from 1996 to 1999. Using a broad definition of bank output they found evidence of large relative cost and profit inefficiencies. A more narrow definition that focused only on traditional activities (excluding OBS activities) led to efficiency estimates that were even lower. They also found evidence of economies of scale for small and mid-size banks, but not for the very large banks. They concluded that OBS activities should be included in the efficiency valuation of Swiss banks.

Tortosa-Ausina (2003) examined the importance of non-traditional activities in measuring cost efficiency of banking institutions in Spain for the period 1986-1997. She chose the data envelopment analysis technique to measure effectiveness. Two bank output definitions were considered; one with traditional output variables and one treating non-traditional activities as an 'additional output'. The sample was composed of Spanish commercial and saving banks. The empirical results revealed that average cost efficiency was improved for the model that included the non-traditional (OBS) activities. However, this result varied over time, between different size and types of financial institutions. Specifically, cost efficiency was more enhanced for saving banks when the OBS items were included in the analysis than for commercial ones.

Casu and Girardone (2004) used the Malmquist total factor productivity index to analyse the effect of the inclusion of OBS activities. Their data included annual information for a balanced panel of over 2000 European banks from France, Germany, Italy, Spain and the UK between 1994 and 2000. They employed the nominal value of banks' OBS items as an output measure in Euros. The authors tested for differences between mean TFP indices when the OBS activities were first excluded and then included. Their results suggested that omitting the non-traditional activities in the definition of bank output led to understatement of productivity levels. The technological change index was higher when OBS items were included.

Goddard, Molyneux and Wilson (2004) investigated the profitability of banking institutions in six European countries, namely: Denmark, France, Germany, Italy, Spain and the U.K., in relation to OBS activities. Their sample included any commercial, saving or co-operative bank from member countries of the European Union in 1992 that had complete and consistent data for the period 1992 to 1998. There were observed differences between banks in different countries regarding the relationship of OBS activities and the banks' profitability. For Germany this relation was significant and negative, while for the U.K. it was significant and positive. For the rest of the sample countries the above relationship was not significant statistically.

3 METHODOLOGY

3.1 *Malmquist Total Factor Productivity (TFP) index*

The data envelopment analysis (DEA) technique was employed to estimate the Malmquist indices of Total Factor Productivity (TFP) change [Malmquist (1953), Charnes, Cooper and Rhodes (1978), Lyroudi and Angelidis (2006)]. DEA measures the relative efficiency of a set of firms among similar decision units that have the same technology (processing procedure) to pursue similar objectives (outputs) by using similar resources (inputs). These authors proposed a model, that had an input orientation and assumed constant returns to scale. The present paper followed the above model.

The Swedish statistician Malmquist (1953) developed the Total Factor Productivity Index (TFP) which measures changes in total output relative to inputs. The Malmquist TFP index is one of the most frequently used methods to evaluate productivity change. Berg, Forsund and Jansen (1992 were the first ones that used it in order to capture the productivity changes in the banking sector. Since then, many banking studies employed the index to assess output of financial institutions.

Following Fare et al. (1994) the Malmquist TFP index calculates the change in productivity between two points by estimating the ratio of the distances of each relative to a common technology. The Malmquist input oriented TFP change index between the base period t and the following period t + 1 is defined as:

$$M(Y_t, X_t, Y_{t+1}, X_{t+1}) = \left[\frac{d_{t+1}(Y_{t+1}, X_{t+1})}{d_t(Y_t, X_t)} * \frac{d_{t+1}(Y_{t+1}, X_{t+1})}{d_{t+1}(Y_{t+1}, X_{t+1})}\right]^{\frac{1}{2}} \quad (1)$$

where, X_t input vector at time t.
X_{t+1} input vector at time t + 1.
Y_t output vector at time t.
Y_{t+1} output vector at time t + 1.
d_t output distance function at time t.
d_{t+1} output distance function at time t + 1.

A value of M greater than unity implies a positive TFP growth from the period t to period t+1. Otherwise, a value of M less than one indicates a TFP decline for that period. Equation (1) is the geometric mean of two TFP indices. The first index is calculated with respect to period t technology, while the second index is calculated with respect to period t + 1 technology.

The productivity change (M) can be decomposed into technical efficiency change (TEC) and technological change (TC). Using symbols for this decomposition, Equation (1) can be written as follows:

$$M(Y_t, X_t, Y_{t+1}, X_{t+1}) = \frac{d_{t+1}(Y_{t+1}, X_{t+1})}{d_t(Y_t, X_t)} \left[\frac{d_t(Y_{t+1}, X_{t+1})}{d_{t+1}(Y_{t+1}, X_{t+1})} * \frac{d_t(Y_t, X_t)}{d_{t+1}(Y_t, X_t)} \right]^{\frac{1}{2}} \tag{2}$$

The first ratio on the LHS in Equation (1) represents the technical efficiency change (TEC) and the second ratio represents the technological change (TC). The technological change captures the improvement or the deterioration in the performance of the best practice financial institutions, while the technical efficiency change reflects the convergence towards, or divergence from the best practice bank by the remaining banks. The Malmquist index gives the advantage to analysts by allowing them to distinguish the shifts in the production frontier (TC) from the movements of the sample banks towards the frontier (TEC).

3.2 *Neural network approach*

There are a variety of studies such as Jagtiani, Nathan and Sick (1995), Jagtiani and Khanthavit (1996) for the U.S. market, Tortosa-Ausina (2003) and Casu and Girardone (2004) for the European markets, that mention the significance of the inclusion of the OBS activities for the evaluation of banks' productivity change. All of the above studies concluded that OBS items tend to overestimate banks' efficiency. It would be interesting to measure which approach provides more trustworthy results in the sense of how close are the estimated figures of productivity to the real ones.

To answer the above question the present study employed the non-parametric approach of neural network systems (NNS). The NNS was preferred instead of the classical Ordinary Least Square as it provided more reliable results [Lyroudi and Angelidis (2006)]. Neural network systems use a set of processing nodes. These processing nodes are interconnected in a network which can identify patterns as it gets exposed to the data. Using back propagation a neural network learns through an iterative procedure. The network is repeatedly shown examples of the data to learn and make adjustments to the weights so that it will fit the model better. This process is repeated thousands of times. In order to perform a neural network analysis the data set has to be split into three subsets; training, test and validation. The neural network learns the problem by using the training set. Then, the test set is used during training to monitor the learning performance. Finally, the validation set is used after training as a final inspection to determine how well the model performs. According to Specht (1991), about 65 percent of the observations are employed for the first set, 15 percent for the second set and the remaining 20 percent are employed for the last.

The predicted values that had been extracted by the last set were compared with the actual values that had been obtained by the DEA program. Then, the mean absolute percent error (MAPE) of the predicted values was computed using the following formula:

$$MAPE = \frac{\sum_{i=1}^{n} \left| \frac{RV_i - PV_i}{RV_i} \right|}{N} * 100, \qquad RV_i \neq 0 \tag{3}$$

where RV_i is the real value of the ith bank's productivity, PV_i is the predicted value of the ith bank's productivity and N is the number of observations. MAPE was computed for each approach by country and by year. First the OBS business items were excluded and then these activities were included. In the next step of the analysis, the values of MAPE obtained by each approach were

Table 1. Number of estimations by country and year.

	1995–96	1996–97	1997–98	1998–99	1999–00	2000–01	2001–02	Sum
AUSTRIA	45	53	107	118	128	136	158	745
BELGIUM	61	68	75	78	89	96	96	563
DENMARK	83	87	88	89	89	97	99	635
FRANCE	298	335	354	374	423	432	438	2,658
GERMANY	696	747	801	868	925	953	995	5,985
HOLLAND	40	44	48	51	55	58	57	353
ITALY	252	280	357	429	522	578	574	2,992
LUXEMBURG	102	111	120	122	126	141	140	862
SPAIN	99	112	132	138	149	166	179	976
SWITZERLAND	177	212	247	272	296	338	388	1,930
UK	94	113	125	154	159	171	175	991

compared to each other. The lower (higher) value of MAPE indicated the higher (lower) fit of the underlying approach. So, the approach that presented the lower value for MAPE was considered the most appropriate.

3.3 *Data and definition of inputs and outputs*

Information was obtained from the banks' balance sheets for the period 1995–2002, using Thomson's Bankscope database. The maximum of available estimations based on all observations were used. Eleven major European markets constituted the sample; Austria, Belgium, Denmark, France, Germany, Holland, Italy, Luxemburg, Spain, Switzerland and the United Kingdom. In order not to calculate a bank twice, three consolidation codes of Bankscope were selected. First, the consolidated statements with an unconsolidated companion, second the consolidated statements with no unconsolidated companion and third the unconsolidated statements with no consolidated companion.

The banks employed in this study are presented in Table 1. It must be noted that the sum of the last column in Table 1 is not necessary the sum of the observations for each year. This is because the DEA program does not always provide figures for all banks and as a result some of them had to be omitted.

The classification of a bank's variable as an input or as an output is an issue related directly to its function explanation. There are three basic approaches: value added, intermediation and user cost. The value added approach considers deposits as outputs, since funds are collected from depositors and there is competition among banks to attract customers (depositors). Berger and Humphrey (1992) modified this approach and considered deposits as both outputs and inputs. Nathan and Neave (1992) followed the intermediation approach, where only banks' assets are thought as outputs, while deposits are regarded as inputs, since banks buy and sell funds acting as intermediaries between borrowers and receivers of funds. Lastly, the user cost approach classifies a variable according to its contribution to bank income as input or output. That means that if the financial return on the assets exceeds the opportunity cost of funds, the banks' assets are considered as outputs.

The value added method was preferred for the present paper, according to Pastor, Perez and Quesada (1997) and Volekova (2004) and because it was believed to be closer to the European banking standards. Therefore, three variables were defined as outputs: total other earning assets, total customer loans and total deposits. On the other hand, three variables were defined as inputs: personnel expenses, other operating expenses and total fixed assets. Other operating expenses were not available for Denmark, France, Germany and Luxemburg. All these figures were considered as their corresponding natural logarithms (ln). This was because the natural logarithms allowed for more precise estimation of TFP indices than the use of the nominal value of variables [Sengupta (2000)].

The off-balance sheet (OBS) items were employed as output variables in the present study. The DEA program was used twice for each approach. Similarly to Casu and Girardone (2004) first the OBS items were omitted for the calculation of TFP indices and then the OBS items were included for the calculation of the total productivity change. Next, the two sets of total productivity change indices were compared in order to determine whether the inclusion of the OBS items had improved the results of the DEA analysis or not.

4 RESULTS AND DISCUSSION

Malmquist total factor productivity (TFP) change indices were calculated and then decomposed into technological change (TC) and technical efficiency change (TEC) using (TFP) = (TC) ∗ (TEC). Based on the results in Table 2, the total productivity without the inclusion of OBS items, increased (M>1) in nine out of the eleven countries for the whole period 1995–2002. The only exceptions were the Netherlands and the U.K. In contrast, when the OBS activities were included in the output variables the total productivity rose (M>1) only for four cases (France, Germany, Holland and Spain). This was a first indication that the exclusion of OBS items leads to higher total productivity change of banking institutions. However, the results were rather ambiguous, since in only seven countries were the TFP indices higher when the OBS items were omitted; Austria, Belgium, Denmark, Germany, Italy, Luxemburg and Switzerland. So, no clear conclusions could be extracted on the question of whether the inclusion of OBS activities enhances or worsens the total productivity indices.

Regarding the components of the TFP indices, the technological change indices were greater when the OBS activities were excluded for Austria, Belgium, Italy, Luxemburg, Spain and the UK. This means that in these countries the best practice banks did not achieve gains by adopting non-traditional banking activities. On the other hand, the best practice banks improved their total productivity by engaging OBS activities in Denmark, France, Germany, Holland and Switzerland. Technical efficiency change indices were greater when OBS activities were included for Austria, France, Italy, Luxemburg, Spain and the UK. Hence, the banks that represented the "catch up" term gained by the adoption of OBS items in these countries. Again, the results by the decomposition of the Malmquist TFP index were rather mixed and no specific conclusions could be extracted for the influence of the OBS items on productivity.

The value ranges between the TFP indices with and without OBS items were small from the minimum 0.001 for France to the maximum 0.016 for Holland. The t-test was employed to examine the statistical significance of differences between the TFP indices that included OBS items and those that excluded such activities. The null and the alternative hypotheses that were under examination stated that:

H_0: The estimated productivity change indices determined by each of the two approaches are expected to be statistically equal to each other.
H_a: The estimated productivity change indices determined by each of the two approaches are expected to be statistically different.

The results in Table 3 indicated that the null hypothesis was rejected for Denmark, France, Germany, Italy, Luxemburg, Spain, Switzerland and the UK for any level of significance and for the 10 percent level of significance for Holland. The null hypothesis could not be rejected for Austria and Belgium. This meant that although there was not a clear direction about the influence of the OBS activities on the productivity measurement on the whole sample, since for some countries the productivity improved by including the OBS activities and for some others the productivity was worsened, the empirical results may differ significantly on an individual country level. We observed that for most countries in our sample, the TFP index obtained by the approach where the OBS items were included gave lower values than the TFP index where the OBS items were excluded. This was in

Table 2. TFP indices and their components by country and year.

Country	Year	Without OBS			With OBS		
		TEC	TC	TFP	TEC	TC	TFP
Austria	1995–1996	0.992	1.305	1.295	0.982	1.036	1.017
	1996–1997	1.007	0.758	0.763	1.023	0.894	0.914
	1997–1998	0.932	1.090	1.016	0.951	1.059	1.007
	1998–1999	1.000	0.974	0.974	0.994	0.982	0.977
	1999–2000	0.969	1.071	1.037	0.936	1.116	1.045
	2000–2001	0.986	1.023	1.009	1.029	0.988	1.016
	2001–2002	1.049	0.908	0.953	1.142	0.831	0.949
Sum	1995–2002	0.959	1.043	1.001	1.010	0.983	0.994
Belgium	1995–1996	0.979	1.040	1.019	0.963	0.991	0.955
	1996–1997	1.002	0.952	0.954	0.981	1.000	0.981
	1997–1998	1.003	1.017	1.019	1.012	1.025	1.038
	1998–1999	1.023	0.944	0.966	1.008	0.949	0.957
	1999–2000	1.010	0.978	0.988	0.977	1.103	1.078
	2000–2001	0.976	1.003	0.979	0.963	0.993	0.956
	2001–2002	0.984	1.034	1.017	1.062	0.896	0.952
Sum	1995–2002	1.002	0.998	1.001	0.999	0.993	0.992
Denmark	1995–1996	1.821	0.528	0.962	0.908	1.011	0.919
	1996–1997	0.889	1.019	0.906	1.170	0.815	0.953
	1997–1998	1.310	0.775	1.015	1.022	0.968	0.989
	1998–1999	0.936	1.026	0.960	1.000	0.993	0.993
	1999–2000	1.242	0.861	1.069	1.011	0.998	1.008
	2000–2001	1.476	0.747	1.103	1.030	1.032	1.064
	2001–2002	0.960	1.061	1.019	0.285	3.491	0.995
Sum	1995–2002	1.144	0.884	1.012	0.362	2.762	0.999
France	1995–1996	0.901	1.316	1.186	1.137	0.900	1.023
	1996–1997	7.026	0.131	0.917	8.321	0.111	0.931
	1997–1998	0.746	1.416	1.057	6.915	0.154	1.062
	1998–1999	0.463	2.100	0.973	1.064	0.925	0.984
	1999–2000	0.853	1.178	1.005	0.240	4.216	1.010
	2000–2001	0.674	1.513	1.020	5.171	0.194	1.058
	2001–2002	1.277	0.793	1.013	2.840	0.359	1.020
Sum	1995–2002	1.191	0.861	1.026	1.306	0.924	1.027
Germany	1995–1996	1.116	0.900	1.005	0.996	1.008	1.004
	1996–1997	0.981	1.028	1.008	1.091	0.926	1.010
	1997–1998	1.012	0.988	1.000	0.492	2.023	0.995
	1998–1999	1.048	0.957	1.003	0.422	2.265	0.957
	1999–2000	1.009	0.995	1.004	1.872	0.543	1.017
	2000–2001	0.985	1.023	1.008	1.440	0.704	1.014
	2001–2002	0.998	1.011	1.009	0.658	1.530	1.006
Sum	1995–2002	1.046	0.961	1.006	0.890	1.126	1.003
Holland	1995–1996	1.032	0.875	0.903	1.012	0.893	0.905
	1996–1997	0.981	1.184	1.161	0.985	1.133	1.116
	1997–1998	0.985	0.936	0.922	0.988	0.950	0.939
	1998–1999	0.982	0.988	0.969	0.971	1.063	1.032
	1999–2000	0.989	1.051	1.039	1.040	0.992	1.032
	2000–2001	1.009	1.001	1.010	1.014	0.992	1.005
	2001–2002	1.017	0.962	0.979	1.038	1.006	1.044
Sum	1995–2002	1.011	0.987	0.998	1.010	1.004	1.014

(*Continued*)

Table 2. (Continued)

		Without OBS			With OBS		
Country	Year	TEC	TC	TFP	TEC	TC	TFP
Italy	1995–1996	0.976	1.031	1.006	1.379	0.726	1.001
	1996–1997	0.943	1.068	1.007	1.193	0.870	1.039
	1997–1998	0.990	1.028	1.018	1.029	0.961	0.989
	1998–1999	1.076	0.948	1.020	0.964	1.022	0.986
	1999–2000	0.904	1.115	1.008	1.062	0.916	0.973
	2000–2001	1.183	0.843	0.997	1.443	0.670	0.967
	2001–2002	0.711	1.406	0.999	0.941	1.085	1.021
Sum	1995–2002	0.798	1.268	1.013	1.114	0.896	0.999
Luxemburg	1995–1996	1.087	0.915	0.995	0.880	1.186	1.043
	1996–1997	0.954	1.044	0.996	0.961	1.032	0.992
	1997–1998	1.019	0.985	1.004	6.470	0.159	1.030
	1998–1999	0.943	1.088	1.027	1.003	0.993	0.996
	1999–2000	1.041	0.976	1.016	1.014	1.046	1.061
	2000–2001	1.088	0.939	1.022	0.700	1.511	1.058
	2001–2002	0.915	1.052	0.963	0.682	1.388	0.947
Sum	1995–2002	0.950	1.055	1.003	2.278	0.438	1.000
Spain	1995–1996	0.999	0.991	0.990	0.989	1.022	1.011
	1996–1997	0.964	1.101	1.061	0.993	1.061	1.054
	1997–1998	0.982	1.020	1.002	0.837	1.237	1.035
	1998–1999	1.093	0.897	0.980	1.269	0.767	0.974
	1999–2000	0.894	1.150	1.028	0.962	1.041	1.002
	2000–2001	0.980	0.991	0.972	0.747	1.309	0.978
	2001–2002	0.730	1.377	1.005	0.830	1.212	1.006
Sum	1995–2002	0.691	1.455	1.006	0.708	1.440	1.019
Switzerland	1995–1996	1.033	0.955	0.987	1.074	0.910	0.977
	1996–1997	0.921	1.092	1.005	1.001	1.006	1.007
	1997–1998	1.149	0.890	1.023	1.125	0.888	0.998
	1998–1999	0.915	1.078	0.986	0.839	1.166	0.978
	1999–2000	1.022	0.956	0.976	0.877	1.086	0.952
	2000–2001	1.014	1.026	1.040	1.001	1.006	1.007
	2001–2002	0.976	1.010	0.986	1.039	0.898	0.933
Sum	1995–2002	1.020	0.984	1.004	0.995	0.997	0.992
UK	1995–1996	1.011	0.975	0.985	0.962	1.032	0.993
	1996–1997	0.955	1.039	0.992	0.850	1.126	0.957
	1997–1998	1.042	0.936	0.975	0.848	1.170	0.993
	1998–1999	1.041	0.983	1.023	1.503	0.711	1.069
	1999–2000	1.002	1.002	1.005	1.021	0.969	0.989
	2000–2001	1.079	0.897	0.968	1.093	0.908	0.993
	2001–2002	1.205	0.807	0.973	1.175	0.859	1.009
Sum	1995–2002	1.257	0.785	0.987	1.359	0.735	0.999

contrast to the work of Casu and Girardone (2004). They examined the period 1994–2000, while we studied the period 1995–2002, which includes some years after the adoption of the single currency of euro. This event could have made a difference, but it was beyond the scope of this study. Another possible reason for the difference in our results could be methodological, since natural logarithms were used and not the nominal values for the variables' estimation.

As statistically significant differences between the TFP indices were observed, it would be quite interesting to examine the correlation of the figures that were obtained with and without OBS

Table 3. T-tests for differences between TFP indices.

Country	T-statistic	P-Value	Country	T-statistic	P-value
Austria	0.2	0.838	Italy	−6.06***	0.000
Belgium	1.25	0.211	Luxemburg	−6.94***	0.000
Denmark	3.98***	0.000	Spain	−3.77***	0.000
France	6.94***	0.000	Switzerland	−5.13***	0.000
Germany	−19.95***	0.000	UK	−5.49***	0.000
Holland	−1.79*	0.074			

Note: *→ statistically significant at the 10% level of significance
***→ statistically significant at the 1% level of significance

Table 4. Correlation analysis.

Country	Pearson correlation coefficient	Country	Pearson correlation coefficient
Austria	0.511	Italy	0.678
Belgium	0.385	Luxemburg	0.509
Denmark	0.356	Spain	0.546
France	0.947	Switzerland	0.236
Germany	0.503	UK	0.343
Holland	0.427		

Note: All values are statistically significant at the 1% level of significance

activities. For this purpose, Table 4 illustrates the Pearson correlation coefficients between the TFP indices obtained by the two approaches for each country. Every correlation coefficient was positive, as expected, and statistically significant. Hence, the two pairs of TFP indices obtained by the two different approaches were moving towards the same direction. The coefficient correlations were positive and statistically significant for every case implying that the differences (spotted by the t-test) between the TFP indices that include OBS items and those that exclude such activities, were owed not due to the results by each approach, but due rather to the relatively limited potential bilateral fluctuations of individual banks' performances.

The addition of OBS items may enhance or worsen the productivity change of a bank. However, it is essential to examine for which option the obtained TFP was more accurate. The acceptance that productivity indices rise or fall when the OBS activities are included is not enough. For that reason, the present study employed the non-parametric method of neural networks to measure how well the selected variables could explain the TFP index. This is a new approach where neural network methodology is employed for the evaluation of indices produced by DEA [Lyroudi and Angelidis (2006) and Angelidis and Lyroudi (2006)].

The neural network approach was applied twice; first, there was inclusion of the OBS items as outputs in the computation of the TFP index and second, there was exclusion of this variable in the TFP index calculation. In each case the dependent variable was the TFP index and the exogenous variables were the input and output variables.

The mean absolute percent error (MAPE) of the predicted values versus the real values of TFP indices are depicted in Table 5. The results showed that the MAPE without OBS items was lower for the banks of nine countries, while the MAPE with OBS items was lower only for the banks of Denmark and Holland. This meant that when the OBS items were not included in the model the predicted TFP indices were closer to the actual TFP indices. Also, the banks in countries for which the variable "other operating expenses" was unavailable presented the highest MAPE. So, this particular variable should be included because it appears that it is very crucial, as long as it is available, for the estimation of productivity change.

Table 5. MAPE of predicted TFP indices by country.

Country	Without OBS	With OBS
Austria	8.49	14.47
Belgium	6.23	8.05
Denmark	184	86.66
France	321.3	430.5
Germany	9.31	406.5
Holland	11.97	9.44
Italy	35.56	52.26
Luxemburg	8.37	305.3
Spain	36.6	47.17
Switzerland	29.15	93.03
UK	52.58	84.49

The findings of the present study were consistent with the results of Jagtiani and Khanthavit (1996) and Tortosa-Austina (2003). In contrast, Casu and Girardone (2004) mentioned that the inclusion of OBS items resulted in an increase of the estimated productivity levels for most of the cases. Special factors for each country, based on macroeconomic variables, or some industry indices, or some bank characteristics could be incorporated in the examination at a future date, in order to determine the differentiating factors.

5 CONCLUDING REMARKS

This study investigated whether the inclusion of OBS items in the calculation of the TFP index of selected banking institutions influences the value of the TFP index. The results were ambiguous, since for some countries the inclusion of the OBS items improved the TFP indices, while for others it did not. The t-statistic test showed the magnitude of the differences of the TFP indices which resulted from the two different approaches. It was revealed that although there was not a clear direction about the influence of the OBS activities on the productivity measurement on the whole sample, the empirical results differed significantly on an individual country level.

The correlation coefficient for every country was positive and statistically significant. This implied that the differences between the TFP indices that included OBS items and those that excluded such activities were owed not to the results for each approach but rather to relatively limited potential bilateral fluctuations of individual banks' performances.

The empirical results suggested that the approach without the inclusion of OBS items presented lower MAPE for the majority of the countries. This means that when the OBS items were not included in the model the predicted TFP indices were closer to the actual TFP indices. Also, the banks in countries for which the variable other "operating expenses" was unavailable presented the highest MAPE.

Future research should focus on the selection of the appropriate variables that will minimize the MAPE, on the inclusion of some macroeconomic variables that possibly differentiate among the countries and on the inclusion of more countries in the sample in order to get better insight.

REFERENCES

Angelidis D. and Lyroudi K. (2006). "Efficiency in the Italian Banking Industry: Data Envelopment Analysis and Neural Networks" International Research Journal of Finance and Economics, Issue 5: pp. 155–165.

Berg S.A., Forsund F. and Jansen E. (1992) "Malmquist Indices of Productivity during the Deregulation of Norwegian Banking, 1980-89" Scandinavian Journal of Economics, 94: pp. 211–28.

Berger A. and Humphrey D. (1992) "Measurement and Efficiency Issues in Commercial Banking" University of Chicago Press.
Casu B. and Girardone C. (2004) "An Analysis of the Relevance of Off-Balance Sheet Items in Explaining Productivity Change in European Banking" Presentation at the 13th Annual Meeting of the European Financial Management Association, Basel, Switzerland, June – July 2004.
Charnes A., Cooper W. and Rhodes E. (1978) "Measuring the Efficiency of Decision Making Units" European journal of Operation Research, 2: pp.429–44.
Clark J. and Siems T. (2002) "X-Efficiency in Banking: Looking beyond the Balance Sheet" Journal of Money, Credit and Banking, 34 (4): pp. 987–1011.
Coelli, Tim (1996), A Guide to DEAP Version 2.1: A Data Envelopment Analysis (Computer Program), Center for Efficiency and Productivity Analysis, Dept. of Econometrics, University of New England, Australia, Working Paper 96/08.
Färe, R., Grosskopf, S. and Lovell, C.A.K., (1994) Production Frontiers, Cambridge University Press, Cambridge.
Goddard J., Molyneux P. and Wilson J. (2004) "The Profitability of European Banks: A Cross-Sectional and Dynamic Panel Analysis", The Manchester School, 72(3), June 2004: pp. 363–381.
Jagtiani J. and Khanthavit A. (1996) "Scale and scope economies at large banks: Including off-balance sheet products and regulatory effects (1984–1991)" Journal of Banking and Finance, 20: pp. 1271–87.
Jagtiani J., Nathan A. and Sick G. (1995) "Scale economies and cost complementarities in commercial banks: On-and off-balance-sheet activities" Journal of Banking and Finance, 19: pp. 1175–89.
Leibenstein, Harvey (1966) "Allocative efficiency vs X-efficiency" The American Economic Review, 56, issue 3: pp. 392–415.
Lyroudi, K. and Angelidis, D., (2006) "Measuring Banking Productivity of the Most Recent European Union Member Countries: A non-Parametric Approach", East-West Journal of Economics and Business, 9(1): pp. 37–57.
Malmquist S. (1953) "Index numbers and indifference surfaces" Trabajos de Estadistica, 4: pp. 209–42.
Nathan A. and Neave E. (1992) "Operating Efficiency of Canadian Banks", Journal of Financial Services Research, 6: pp. 265–276.
Pastor J.M, Perez F. And Quesada J. (1997) "Efficiency Analysis in Banking Firms: An International Comparison", European Journal of Operational Research, 98: pp. 395–407.
Rime B. and Stiroh K. (2003) "The performance of universal banks: Evidence from Switzerland" Journal of Banking and Finance, 27: pp. 2121–50.
Rogers K. (1998) "Nontraditional activities and the efficiency of US commercial banks" Journal of Banking and Finance, 22, issue 4: pp. 467–482.
Rogers K. and Sinkey J. (1999) "An analysis of nontraditional activities at U.S. commercial banks" Review of Financial Economics, 8: pp. 25–39.
Sengupta, J.K., 2000, "Efficiency Analysis be Stochastic Data Envelopment Analysis", Applied Economic Letters, 7, 379–383.
Siems T.F. and Clark T.A. (1997) "Rethinking Bank Efficiency and Regulation: How Off-Balance Sheet Activities Make a Difference" Financial Industry Studies, Federal Reserve Bank of Dallas: pp. 1–11.
Specht D. (1991) "A General Regression Neural Network" IEEE Transactions on Neural Networks, 2–6 November 1991.
Stiroh (2000) "How did bank holding companies prosper in the 1990s?" Journal of Banking and Finance, 24: pp. 1703–45.
Tavares G. (2002) "A Bibliography of Data Envelopment Analysis" Rutcor Research Report, January 2002.
Tortosa-Ausina E. (2003) "Nontraditional Activities and Bank Efficiency Revisited: A Distribution Analysis for Spanish Financial Institutions" Journal of Economics & Business, 55: pp. 371–95.
Volekova D. (2004), "The Bank Efficiency Among European Economical and Political Arrays" Working Paper 2271.

Assessing Advertising Mediums used by Small Businesses in Soweto to Promote Sustainable Development

Louise van Scheers
School of Business Management, University of South Africa

SUMMARY: There has been significant growth in the number of small businesses in Soweto, a black urban residential area in South Africa, as a result of new money being invested in the region. This has resulted in an increase in competition. Qualitative, exploratory analysis of these businesses was done in the form of in-depth interviews with managers and owners. It was found that owners appreciated the need for advertising to differentiate themselves from other companies in the area. Results from this study revealed new ways in which small businesses operating in the changing Sowetan environment can advertise and which marketing mediums are most effective. This information can be employed to increase the chance of success and thus to promote the sustainable development of small companies.

1 INTRODUCTION

Advertising for small businesses in Soweto, a black urban residential area in South Africa, is becoming increasingly important. As infrastructure improves and more funds are being invested in this area, an increased number of small businesses will emerge. If these businesses wish to survive it is imperative that they have effective marketing campaigns in place and the correct means to implement them. It is also important that the company utilises effective low-cost advertising mediums to promote their product/service successfully. With the development of Soweto underway, the township's large population and the increase in tourism to the district, a great potential exists for small businesses to emerge, survive and grow as it becomes a commercial hub. This will result in greater competition and a flexible and destabilised business environment. Knowledge of advertising mediums and the means to successfully implement them is imperative for maintaining a competitive advantage and achieving overall success.

Marketing helps to build business identify and also helps people to identify and remember the value and function of a product or service (Maravilla, 2000 and Du Plessis *et al.* 2006). Small industries will have to consider cost effective advertising mediums due to their size and shortage of start-up finances.

Due to the increasing number of undersized businesses in Soweto, augmented competition occurs, resulting in a greater need for differentiation in order to survive. Promotion provides the means to accomplish this. Therefore it is important for these companies to be able to identify, select and implement effective advertising programmes.

This chapter aims to assess the advertising mediums that small businesses are using in Soweto to promote sustainable development. Specific objectives were to identify advertising mediums which are currently used by small businesses; to determine which promotion standards achieve the most success; and lastly to identify key success factors that makes advertising in this district effective.

2 CASE STUDY AREA

Soweto was established in 1904 as a township to house mainly black labourers. It has grown into the most populated black urban residential area in South Africa (ANON, 2003 a). Soweto stretches across a vast area, 20 kilometres from Johannesburg (Davie, 2001). The township's population is estimated to be between one and three million people (Davie, 2001 and Smith, 2004). Despite its continuing problems of over-crowding, high unemployment and lack of infrastructure, this district is being reinvented as a feasible centre of commercial and cultural activity (ANON, 2003b and Thale, 2003).

The government has made attempts to upgrade the overall standard of living in Soweto by planting trees and providing basic amenities (ANON, 2003c). Thale (2003) explains that a new development called the Baralink project, promises to improve the overall infrastructure of the area and should encourage the growth of a viable business environment. This activity will lead to the emergence of a multitude of small companies in the vicinity. A comprehensive marketing campaign is therefore vital for these new businesses to survive and succeed.

Soweto is fast becoming a commercial hub with an increasing number of small companies opening every year. This has led to amplified competition in the area. Consequently it has become increasingly important for Sowetan businesses to differentiate themselves from their competitors. This can be done effectively through the use of advertising. For that reason it is necessary to be able to identify relevant promotion opportunities.

3 ADVERTISING OBJECTIVES AND MEDIA

Marketing is any form of mass communication about a product or services which is paid for by an organisation or an identified sponsor (Du Plessis *et al.* 2006:31). Advertising helps sell products and helps companies build a product brand. It also builds industry identity as observed by Maravilla (2006). Du Plessis *et al.* (2006) defined media as a communications channel or a group of channels used to convey information, news, entertainment, and advertising message to an audience. Two major publicity media types exist: broadcast and print. Television and radio are broadcast media, while newspapers, magazines, billboards, direct mail and leaflets are examples of print media. (Du Plessis *et al.* 2006). A small company in Soweto would probably need to use different types of promotion compared to a larger company in more developed areas, such as advertising in taxis and in various outdoor mediums.

Russell & Lane (2006) cited that adverting is usually delivered through a form of mass communication. The key differentiating element that thus defines this type of promotion is that it is a paid-for message. The communication is controlled by the advertiser. Marketing is a business activity, using creative techniques to design a persuasive communications in mass media that promote ideas, goods, and services in a manner consistent with the achievement of the advertiser's objectives, the delivery of consumer satisfaction, and the development of social and economic welfare (Cohen, 2006). Dirksen *et al.* (2006), defined advertising as any form of announcement that will be paid for by an identified sponsor, that has been sent through a mass media, and is directed to a specific group of individuals or organisations. Russell *et al.* (2006) also described advertising as a sales tool to bring buyers and sellers together for the exchange of goods and services.

Marketing objectives are important to a company because they offer the firm a standard against which the results of the advertising can be evaluated (Cohen, 2006). Advertising can also be used to support other elements in the marketing mix. This can improve the effectiveness of the firm's total marketing strategy with the ultimate objective of increasing sales (Russel *et al.* 2006). The following lists goals a firm may want to obtain from its promotion campaign (Cohen, 2006: 123):

- Increasing the number of customers
- Increasing total demand
- Attracting non-users
- Maintaining current sources
- Rekindling interest in a mature product
- Rediscovering former uses
- Attracting users of competitor's brands

- Increasing the usage rate
- Increasing the frequency of use
- Reducing the time between purchases
- Marketing a new product, a product modification, or a new package

Mass media such as newspapers, magazines, radio, and television are especially well suited for delivering advertisements (Sissors & Bumba, 2006). Mass media is important because it is able to deliver to large audiences at a relatively low cost (Sissors *et al.*, 2006: 7). They can deliver advertisements to special types of audiences who are attracted to each medium's editorial or programming, and they tend to develop strong loyalties among audiences who return to their favourite medium with a high degree of regularity. Cohen (2006) noted that the largest share of promotion expenditures is allocated to newspapers. The largest expenditures for national advertising, however, occur in television. Direct mail is the third largest medium in terms of advertising expenditures.

Television was introduced into South Africa as a mass medium in 1975 (Du Plessis, 2006). This is the single most powerful medium for a company (Morris, 1992 and 2006) with 70% of people in Soweto watching television regularly.

According to Morris (2006) radio is one of the few mediums that actually reaches the majority of the black population at any one time. Radio has a very cost effective reach in both urban and rural areas by offering businesses the best mix of formal advertising, infomercials and sponsorships (Du Plessis *et al.* 2006. In comparison to television, radio is a relatively low cost medium. Compared to other media, radio has a fairly low cost per thousand listeners (Cohen, 2006). Due of its low cost it can often be used as a supporting medium. Radio can give a business personality through the creation of advertising campaigns that make use of sounds and voices (Pleshette, 2005).

There is a variety of very effective forms of outdoor media that can be used in advertising (Morris, 2006 and Russell *et al.*, 2006). These include using buses, shopping bags, posters, bin panels, bus stop shelters, wall painting, and supermarket trolley panels. Taxis can also be used as a form of outdoor media and are especially useful when wanting to advertise in the townships. Point of purchase displays can also be placed outside retail outlets in the townships (Morris, 2006). Other forms of outdoor media are billboards, inflatables, rolling boards, point of purchase materials, mobile advertising boards, and aerial advertising (Du Plessis *et al.* 2006). .Outdoor advertising is a medium that continues to grow (Du Plessis *et al.* 2006).

4 KEY FACTORS THAT MAKE ADVERTISING SUCCESSFUL

Advertising messages that are novel or unusual are more likely to attract attention and stand out (Foxall *et al.* 2005). Other attention getting stimulus factors that can make marketing more successful are the size of the message, its placement in the environment, the colour of the advertisement, and how vibrant it is. It is important that consumers are exposed to a constant message. This can be done by advertising regularly (Egelhoff, 2004) even when a maximum level of awareness is already achieved. Campaigns do not have to be continuous but should be repeated at carefully regulated periods.

The placement of advertisements, especially outdoors, close to the point of sales, serves as a reminder to consumers (Reiter, 2004). Colour is very important. Bright eye catching colours should be used to attract the customer's attention and differentiate the message from surrounding advertisements (Reiter, 2004). Frequency and repetition is very vital; the more a potential customer see a message in a short period of time the greater the chance of remembering the product or service in the future (Anon, 2004). If a company's adverts are placed all over a selected area, they will be seen repeatedly by potential clients. According to Anon (2004) most people need to be exposed to an advert on the average of thirteen times before remembering and responding to the advertisement. Dupont (2002) stated that repeated viewing of an advertisement increases peoples perception of the promoted product. Through repeated viewing of the product through advertising people recognize the brand and come to like it.

When advertising in a black township, black and white people should be used in the advert, or even blacks only (Morris, 2006). This representation helps the black consumer appreciate that the product is universal and used by all. Situations and humour must be employed that black people can relate to (Morris, 2006). Sporting personalities should be used with care; the reason being biases could result from opposing team supporters. English should also be utilized because most black people understand it. There are so many different black languages that they can become confusing, even for the black consumer.

Mail order is becoming an increasingly popular means of purchasing among black consumers in townships (Morris, 2006). Marketers can make use of catalogues when trying to capture an audience in a township. Graphic or photographic images should be used heavily in print and visual media, accompanied by short copy statement. Using this type of marketing in South Africa is very important due to the high illiteracy levels in the country.

When advertising in a township, the message should be happy, free of negative situations, uncluttered, and reality based (Morris, 2006). Companies should be aware of offensive slogans and scenes that may be contrary to religious, cultural, tribal or even political beliefs. In closing, it can be seen that although there are many different forms of advertising not all such mediums are necessarily suitable due to factors such as cost, specific target markets and illiteracy.

5 METHODOLOGY

Due to limited published studies on the topic, exploratory research was necessary to determine which advertising mediums are used by small businesses in Soweto and which were most effective in that environment. Personal in-depth interviews using open-ended questions were employed to collect data. This method also eliminates the chance of misinterpretation of the questions, which is vital in order to collect reliable data.

The population from which the sample was selected was small businesses in Soweto. Judgement (purposive) sampling was employed. This is a non-probability technique in which the sample that is selected is based on personal judgement about some appropriate characteristic of the test member. Interviewees selected were sixty owners and managers of small companies in the district. This ensured a full range of perspectives.

The study was conducted during the period 9 to 19 October 2007. Owners and managers were interviewed at their relevant businesses. Three open ended questions were used in the survey; information was requested about the advertising mediums currently used in Soweto, the most successful advertising mediums for small businesses, and what are the key factors that make advertising in the district successful.

6 RESULTS AND DISCUSSION

6.1 *Advertising media currently used in Soweto*

Figure 1 shows the advertising mediums used by small businesses in Soweto. The most commonly utilised mediums included: brochures, the Internet, pamphlets, signboards, door-to-door, local newspapers, radio, shopping centre information boards, company cars, posters, directional sign-boards, product displays, word of mouth and wall painting. One respondent found the Internet especially useful to reach external target markets such as tourists as it allows anyone in the world with a computer and an Internet connection to have access to the business's information. This is essential as tourists allow these small companies to expand their consumer base thereby increasing profits. This agrees with studies Anon (2001) and Stirling (2003) that show that the Internet allows a company to target a worldwide audience.

According to one respondent, brochures are a widely used publicity standard utilised by small business in Soweto. The fliers are usually made available at locations that are most likely to

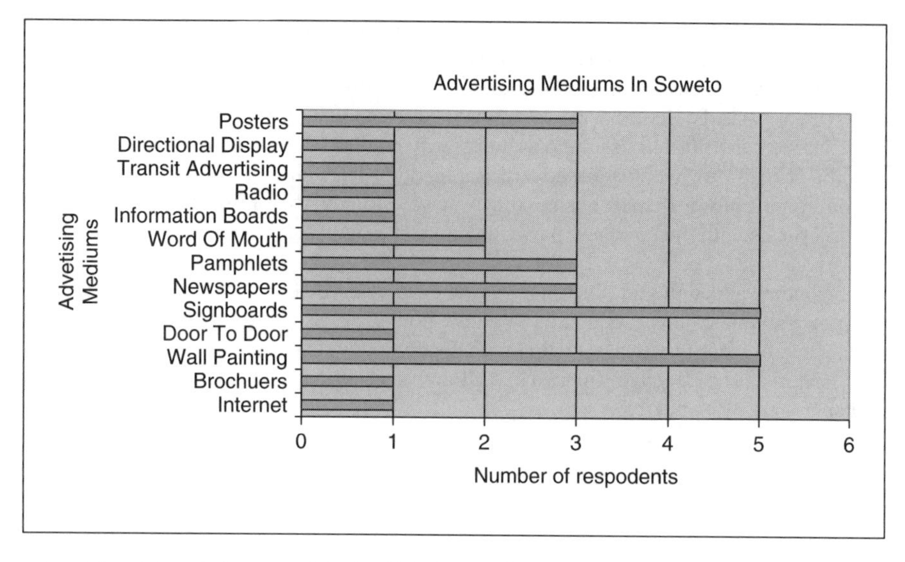

Figure 1. Advertising Mediums used by Small Businesses in Soweto.

be frequented by the target market. This ensures that the correct type of consumer is targeted and money is not wasted. One example given was various tourist sites in Soweto such as the Hector Petersen Memorial. The large number of visitors allowed for a wider consumer base to be targeted.

The survey indicated that a substantial number of small businesses in the case study area make use of wall painting in order to advertise. This was attributed to the fact that this form of promotion is inexpensive and if placed in the right location, can be seen by a large number of people. Some large walls had actually been specifically earmarked for advertising such as, for example, the Bara Advertising Wall which was situated on the Old Potchefstroom Rd. This opinion was echoed by Morris (2006).

Figure 1 illustrates that one respondent found that door-to-door promotions and signboards are also very effective. This was in agreement with Morris (2006), who reported that door-to-door provides an effective way of ensuring that a company's product sample gets to the consumer. The investigation indicated that sign boards are also a reasonably inexpensive advertising medium.

Three of the people surveyed regard community newspapers and pamphlets as useful mediums available to small businesses. However, newspapers are not widely utilised in the district under study due to the cost. This agrees with the results of Du Plessis *et al.* (2006) that pamphlets are cheap to produce and to distribute. They are flexible in design and easy to modify to display the latest specials. The respondents added that the individuals who hand out the pamphlets can be given t shirts or wear display boards over their shoulders displaying the company's details and product. This is an effective way to increase awareness of the companies brand around the townships. In contrast to what was found in our study Morris (2006) observed that newspapers are an effective way of advertising in townships because they enjoy a fairly high penetration rate.

Two of the people surveyed considered word of mouth as an efficient advertising method. Soweto is a densely populated community in which positive information about a business can easily be communicated amongst people. This agrees with the work of Foxall *et al.* (2005) that word of mouth communication is an excellent form of advertising and can be more effective than formal advertising. According to the respondents in our study, for word of mouth to be a successful marketing medium a small company needs to offer a high quality product or service. Satisfied consumers who spread positive information are seen as more credible and are trusted by other potential clients.

Radio, shopping centre information boards, and company cars (Figure 1) are other forms of advertising that are used by small business in Soweto. Morris (2006) stated that radio is one of the most effective forms of advertising in black townships as the majority of people living there listen to it on a regular basis. By contrast, the respondents inferred that only the larger or more successful small businesses can afford to use radio as a publicity medium due to the high cost.

Shopping centres provide a useful advertising medium through their information boards; these provide an opportunity for small businesses in Soweto to advertise fairly cheaply. Here A4 size information sheets with the company name and contact details are placed on the boards and often tear offs are included.

One respondent added that product displays can also be used as an advertising medium and can be set up on the side of busy roads and in active public places; these allow consumers to view the product and to identify and associate it with the company. This agrees with Morris (2006) that product displays can be effectively placed in busy area such as store outlets and community centres. Poster displays are also valuable.

6.2 *The most successful marketing mediums in Soweto*

Due to the large quantity of advertising methods mentioned it was important to find out which of them where viewed by the owners and managers surveyed as the most successful. Advertising is expensive for small businesses and therefore it is very important to find out which mediums are the most effective in a particular environment.

According to Figure 2, twenty-five percent of the respondents believe that wall painting is a very successful advertising medium. This method of promotion is not very expensive and is valuable as small businesses often do not have huge advertising budgets. Wall painting is practical as it gives a company creative freedom and allows advertisements to include bold, bright, eye-catching designs. This sentiment is echoed by Russel *et al.* (2006).

Fifteen percent of the respondents indicated that newspapers provide a useful and effective advertising medium, particularly local, community newspaper. They also indicated that it is an expensive medium and a company has to advertise in a newspaper regularly for it to be successful which increases costs. For this reason newspaper advertising is not suited to most small businesses. The study reported by Du Plessis *et al.* (2006) supports this but still maintains that it is inexpensive when compared with television as an advertising medium.

Twenty five percent of the respondents indicated that signboards placed outside the small business are an important form of advertising for small businesses. Signboards allow for potential clients to identify the small business location, name and service.

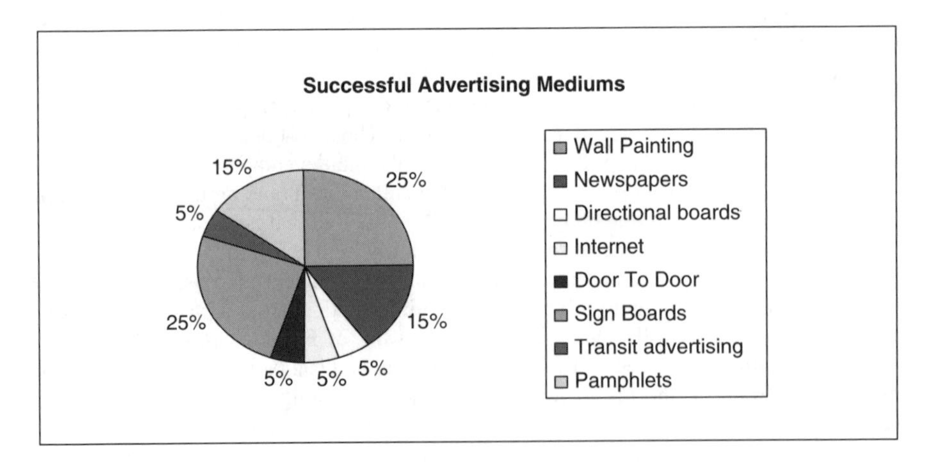

Figure 2. Successful Advertising Mediums used in Soweto.

6.3 *Key success factors for advertising in the case study area*

Figure 3 shows that 28 percent of the respondents stated that in order for outdoor advertising to be effective it needs to be placed in high traffic areas. . This agrees with Morris (2006) who noted that outdoor advertising such as wall painting and posters are only effective when placed busy areas. The owners and managers interviewed added that main roads and shopping centres are recommended high traffic areas especially for wall painting, product displays, posters and directional posters. Businesses should be placed on these main roads where they will be seen by a dense flow of people and will acquire more customers than business which are situated on side streets away from the flow of people and traffic. There are a few chief thoroughfares that lead into Soweto, with the Old Potchefstroom road being one of the key ones were most of the small businesses are situated.

Taxi ranks were also mentioned as another area in which many potential clients gather and would therefore also be a successful location to advertise. The majority of people living in Soweto make use of public transport and therefore a taxi rank is an important place for small business to target an advertising campaign.

According to the respondents, when advertising in high traffic areas there are usually many different businesses advertising in the same area so it is important for a small business to make their advertising stand out from the rest. Eighteen percent of the respondents stated that advertising needs to stand out from the other advertisements that it is surrounded by (Figure 3). This can be done by making them more eye-catching than others in the area. The respondents suggested that this can be done, for example, by using a light background on the wall such as white and then paint the writing in a darker colour such as red.

Figure 3 indicates that 18 percent of the managers and owners surveyed thought that advertising can also be made more effective by the size. This can be achieved by making them large enough to be seen and by using bold print that can be read easily by a passenger in a moving car. The respondents added that when advertising on main roads cars are moving fast and people are generally too busy to notice every poster, therefore advertisements in theses areas should be large, easy to read and eye catching. It is very important that the advertisement is not cluttered with too much information that makes it increasingly difficult for passing potential consumers to read it especially when driving.

Twenty-four percent of the respondents claimed that repetition is essential to the success of a business's advertising efforts (Figure 3). When a potential client sees the advert they may not make use of the business straight away, but through repeated and frequent viewing, when the potential client needs that particular product or service, they will then remember the advertised product and

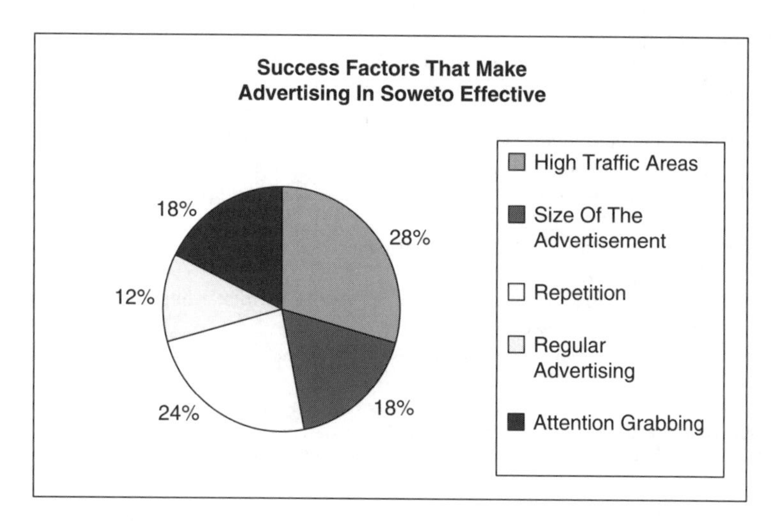

Figure 3. Success Factors that make Advertising in Soweto Effective.

will then buy from that business. Increasing the amount of times people notice the advertisement in a short period of time increases memory retention. The respondents added that repetition is important when using all forms of advertising.

7 CONCLUDING REMARKS

For advertising to be successful it needs to be implemented correctly and key success factors must be identified. According to the survey, the main success factor is the area which is chosen to place the announcement. Advertising should be placed in high traffic areas. Repetition in a short period of time is also very important as it helps a person to remember the advertisement. The size, colour, graphics, and writing of the advertisement all have an effect on the marketing campaign and helps potential clients differentiate the advertisement from surrounding advisements. This is especially necessary in areas which are cluttered with competitor's advertisement.

Based on the outcome of the survey the following recommendations should be considered: in the future new advertising mediums should be looked at due to advertising clutter and limited space. More door-to-door advertising should be implemented since it was observed that the majority of household in the Sowetan area had at least one adult member of the family at home through the majority of the day. With an increase of tourist to the district, the Internet should be considered a more popular form of advertising. It is imperative for small businesses in Soweto to erect sign boards by their businesses building. This provides a dual function; it shows potential clients where the business is located and helps potential clients to differentiate the business from its competitors.

In closing, this study has uncovered new ways in which small business operating in changing environments can advertise and which advertising mediums are effective. This examination will present owners and managers of small businesses operating in changing business environments such as Soweto, with a better understanding of advertising and advertising mediums. This information can be implemented to increase the success and thus to promote the sustainable development of small businesses.

REFERENCES

Anon. 2003 (a) *Advertising objectives.* [Internet] Available from: <http://www.examstutor.com/business/resources/studyroom/marketing/advertising/2-advertisingobjectives.php?style=> [Accessed: 18 Apr 2004].

Anon. 2003 (b) *The making of Soweto.* [Internet] Available from: <http://www.joburg.org.za/soweto/overview.stm> [Accessed: 15 Apr 2004].

Anon. 2003 (c) *What are the advantages of advertising on the Internet?* [Internet] Available from: <http://www.happy-online.co.uk/tutorial/advertising.htm> [Accessed: 10 Oct 2004]

Anon. 2004. *CTN launches living township billboards.* [Internet] Biz-Community. Available from: <http://www.biz-community.com/Article.aspx?c=70&1=196&ai=4796> [Accessed: 6 Oct 2004].

Cohen, D. 2006. *Advertising.* Illinois: Scott, Foresman and Company.

Davie, L. 2001. *Soweto, city of contrasts.* [Internet] Available from: <http://www.joburg.org.za/december/soweto.stm> [Accessed: 13 Apr. 2004].

Dirksen, CJ. Kroeger, A. & Nicosia, FM. 2006. *Advertising: Principles and Management Cases.* 6th edition. Illinois: Richard D. Irwin. INC.

Dupont, L. (2002). *The six effects of repetition.* [Internet] Available from: <http://bellzinc.sympatico.ca/en/content/503046?skin=sli> [Accessed: 14 Oct 2004]

Du Plessis, F. Bothma, N. Jordaan, Y. & van Heerden, N. 2006. *Integrated Marketing Communication.* Claremont: New Africa.

Egelhoff, T. 2004. *How to develop a successful advertising campaign.* [Internet] Available from: <http://www.smalltownmarketing.com/adprogram.html> [Accessed: 16 October 2004]

Foxall, G. Goldsmith, R. & Brown, S. 2005. *Consumer psychology for marketing*, Second edition. London: Thomson.

Maravilla, N. 2006. *The perceived value of advertising.* [Internet] Available from: <http://www.powerhomebiz.com/vol/ad-value.htm> [Accessed: 10 Sep 2004].

Morris, R. 2006. *Marketing to black townships.* Cape Town: Juta & Co.

Pleshette, LA. 2005. *Choosing the right advertising medium for your small business.* [Internet] Available from: <http://www.powerhomebiz.com/vol118/admediums.htm> [Accessed: 19 Sep 2004].

Reiter, J. 2004. *The great outdoors* [Internet] Available from: <http://www.findarticles.com/p/articles/mi_m3301/is_n5_v91/ai_9175677> [Accessed: 18 October 2004]

Russel, JT. & Lane, WR. 2006. *Kleppner's advertising procedure.* 14th edition. New Jersey: Prentice Hall.

Sissors, JZ. & Bumba, L. 2006. *Advertising media planning.* 5th edition. Illinois: NTC Business Books.

Smith, T. 2004. So where to now? *The Property Magazine*. October 2004: 43–52. Johannesburg: Media Nova (Pty) Limited.

Stirling, R. 2003. *Advantages of Internet advertising*. [Internet] Available from: <http://au.geocities.com/comptroller_pboro/advantagesinternetadvertising.htm> [Accessed: 11 Oct 2004]

Thale, T. 2003. *Soweto's gravel roads to become history* [Internet] Available from: <http://www.joburg.org.za/2003/sept/sep11_roads.stm> [Accessed: 13 Apr. 2004].

Environmentally Sustainable Economies: A Panacea for Sound Development

Mago William Maila
Department of Teacher Education, University of South Africa, TSHWANE, Republic of South Africa

SUMMARY: Using a socially critical framework, the aim of this chapter is to argue that environmentally sustainable economies should be perceived as a panacea for sound development supported by quality and relevant education. Although education and environment are perceived as two opposite sides of the same coin, they are also perceived as intertwined and mutually inclusive to each other. The land, the air, the biodiversity and the world's waters have been frequently treated as free and limitless. Although these nature's services are enormous and wealthy they also underline the fact that far too many are becoming limited as a result of abuse, poor management and over-exploitation in the name of development. Of course, management and utilization of natural resources is generally good for sustainable development in general and for the promotion of good governance in particular. However, the downside of good quality and sustainable progress is growth that misuses, mismanages and over-exploits natural resources for quick monetary benefits. There is a need for better environmental education policies and strategies to minimize this phenomenon.

1 INTRODUCTION

Numerous scholars have alluded to the complexity of sustainable development as a concept that is embedded in practice or praxis rather than in rhetoric debates (Ndlovu-Gatsheni 2006; UNEP Educator January-June 2006). Some of these scholars perceive sustainable development as a North-South; South-South and North-North environmental discourse. Often the sustainable trajectory is polarized along these positions, and is not seen as a contextually bound discourse that engages all people in different contexts and ways or experiences, especially by higher education and research institutions. Most of the time these institutions find themselves in the wrong corner or part of the globe, especially when they defend sustainable development positions that compromise the lives of poor and marginalized communities, and unsustainable economic policies in developing countries regarding the utilization and care of natural resources.

The environmentally sustainable economies debate in this paper tries to illuminate some critical issues pertaining to sustainable development as a base for sound development. Notwithstanding there is a claim made by Ndlovu-Gatsheni (2006), that there is presently a deep-seated poverty of imagination characterizing contemporary development debates and a stalemate in thinking and policy. Worldwide initiatives discoursing on development and sustainability (Ahmed 2006; Neefjes 2000) are contradicting this claim. The trajectory of development and sustainability is on course, but the engagement must be anchored in praxis, that is in action processes that allow participants to have choices in the how and why of doing things.

Good development can only be said to be excellent when ordinary people benefit from it. Therefore lively debates about ideas must be grounded in the actual doing of things for ordinary people; poverty stricken people and marginalized people to harvest what can be put on the table. Of course such a view is in contrast to big business agenda of development. Big business is keen to harvest huge profits within short periods of time in their ventures, and poor people are interested in putting food on the table for their families. For both these two groups the benefits are not actually sound; depending on their perspective. Economic benefits are sometimes not worthwhile if they undermine processes that ensure that the environment is cared for and used in a responsible and ethical manner. On the other hand, poor people do not just need economic opportunities or values in their employment; they also desperately need individual benefits which are only possible through human development processes (Maila 2006).

Development that ignores the value of the environment is definitely appalling. The document Ecologically Sustainable Development (Commonwealth of Australia 1990) notes that the task confronting us is to take better care of the environment while ensuring economic growth, both now and in the future, and that both economic growth and environmental protection are critical. But it is not only financial expansion and ecological protection that are important for environmental sustainability, human development is also critical. All of these necessities must ultimately benefit people. Sen (1999) argued that development has gone from the growth of output per capita to the expansion of human values. He saw development as a process that expands the real freedoms that people enjoy.

Development does not denote economic growth only, but that it also means human progress. It means a deliberate process of ensuring that opportunities for people to grow their capabilities, to enjoy choices, are purposively enlarged by development processes. This orientation means that progress is not an activity to its end, but must be guided by a teleological perspective to reality, that says, improvement as a discourse of community of practice is ongoing.

If sustainable growth is narrowly defined as expansion that caters for the present people economically without compromising the ability of the environment to cater for future generations, then such development will persist. Those benefiting from such an orientation will label it as sustainable development (Alkire 2002).

Economies of the world are supported by processes that stockpile and continue to feed the production machines in order that they can continue to churn out products. If not needed then they will be dumped somewhere in, for example, an African country in a guise of cheap service. Whilst development opens up and advances economies, create new wealth, and usher many people to a richer lifestyle, millions are forced to struggle to make meaning of the darker side of development that is not environmentally sustainable.

Using a socially critical framework, the aim of this chapter is to argue that environmentally sustainable economies should be perceived as a panacea for sound development supported by quality and relevant education.

2 MAKING MEANING OF SOCIAL CRITICAL DISCOURSE

Various scholars and researchers' perspectives regarding the critical theory (or critical social theory) differ considerably. It is the aim of this chapter to draw on a few of these perspectives.

There seems to be consensus as to the social evaluation of critical social theory in communal enquiry, namely, that the restrictive and alienating conditions of the status quo are brought to light. This view is on the notion that critical researchers assume that social reality is historically constituted, and that it is produced and reproduced by people. Although people can consciously act to change their social and economic circumstances (forced by environmental conditions), critical researchers recognize that their ability to do so is constrained by various forms of social, cultural and political domination. The main task of critical research is, therefore, seen as being one of social critique, whereby critical research focuses on the oppositions, conflicts and contradictions in contemporary society, and seeks to be emancipatory (i.e. to eliminate the causes of alienation and domination) (http://www.qual.auckland.ac.nz/, November 2007).

2.1 *Sustainable development*

Sustainable development is the insurer of humanity's vast natural resources. Governmental and the broader stakeholder community, use both contextually based and globally based ways of responding to concerns, problems or risks threatening the sustainable use of this natural bounty. Nations through their own development programs and the United Nations (UN) Program for Education for Sustainable Development (ESD), have embraced sustainable development as a viable alternative to environmental crisis.

What sustainable progress means is debatable. The World Commission on Environment and Development (UNESCO 1997) sees sustainable growth as development that meets the needs of the present without compromising the ability of future generations to meet their own needs. Yeld (1991) noted that it refers to improving the quality of human life while living within the carrying capacity of supporting systems. UNESCO (2005) linked education to sustainable development; stating that education for sustainable development is a life-wide learning endeavor which challenges individuals, institutions and societies to view tomorrow as a day that belongs to all of us. The UN and other stakeholders proposed a number of strategies, values and principles for ESD with vested interest in both education for sustainable development and development.

Proposed strategies must be robust and broad in application. For example, it is a fact that some donor nations include their own nationals (home-staff) in projects that received funding from them. This is done as soon as the proposal has been accepted for funding. Is this a morally clean value judgment or is it corruption?

Most donor agencies are the core managers of the very funds, which are sometimes squandered. What is the role of the home-staff personnel in such development projects? This issue is therefore, broad and multifaceted. Many-sided approaches are needed in addressing it. It is argued that all of these strategies, values and principles will not accomplish desired outcomes if human development is not an integral component of education for sustainable development and environmental sustainability.

2.2 *Environmental Sustainability*

Sustainability goes beyond the environmental aspect. It is often defined as a term that relates to the way people think about the world, including forms of personal and social practices that lead to issues that are: ethical, empowered and personally fulfilled individuals; communities built on collaborative engagement, tolerance and equity; social systems and institutions that are participatory, transparent and just; and environmental practices that value and sustain biodiversity and life-supporting ecological systems (UNESCO 2005). It is therefore imperative that programs and strategies for environmental sustainability be endowed with these elements in order to contribute to environmentally sustainable economies.

Palmer (1998) argues that three priorities must be incorporated in economic development programs, namely, the maintenance of essential ecological processes, the sustainable use of natural resources, and the preservation of genetic diversity and the conservation of wild species. Needless to say, these priorities also underscore the role of the environment in economic growth and human development. Palmer (1998) argued that this strategy is anti-development, because it largely emphasizes the conservation of the physical environment; it stresses sustainability in ecological terms and is less interested in economic development.

It is imperative that not only should economic and human development be advanced, but that ecological growth should also be fostered for the people's wellbeing. Environmental sustainability can therefore not be defined in a narrow and shallow perspective that ignores the politico, socio-economic and bio-physical dimensions of human life. It must also be defined in relation to poverty, equity, gender and youth issues, disabilities, and justice issues. Hence the following issues are pertinent to environmental sustainability: environmental ethics; economic ethics; empowered and personally fulfilled individuals; communities built on collaborative engagement, tolerance and equity; social systems and institutions that are participatory, transparent and just; and environmental practices that value and sustain biodiversity and life-supporting ecological systems (UNESCO 2005). Anything less, threatens the very essence of environmentally sustainable growth.

3 THREATS TO SUSTAINABLE DEVELOPMENT AND ENVIRONMENTAL SUSTAINABILITY

Stunt growth policies are excellent on paper but can not be implemented due to bureaucratic procedures within an organization or institution. In countries were developments depend on government legislation only, such policies are definitely a threat to environmental sustainability and human development.

The major contributory factor to unsustainable progress economies is poverty. Amongst many causes of the latter is endemic socio-economic inequality throughout the globe. Ndlovu-Gatsheni (2004) and Nuwagaba (2006) insisted that inequality has a negative impact on growth, leading to externalities such as crime, corruption and other social ills. Hence, such ill-managed socio-economic, political and bio-physical dimensions of inequality perpetuate and exacerbate poverty, making it a vicious trap for millions of people.

Natural disasters can occur without notice. Perhaps in the near future most of the weather related disasters will be sensed and proper evacuation and protection of people can be taken. How as human beings we forecast and minimize the effects of this threat is imperative for a growing economy that is supported by a sustainable environmental.

Most conflicts and wars have their roots in inequalities regarding the distribution of services (e.g. land, sources of water, forests), poor political and justice policies, and religious intolerance. All of these lead to certain groups of people being marginalized, often using religion or language as a determining factor for legitimizing the discriminatory acts. Africa has quite a number of such examples; the marginalized Ndebele people in Zimbabwe, conflict in Durfur, and subtle marginalization tendencies experienced in most if not all of the Southern African Development Community states.

Injustices perpetrated to certain individuals or groups of people are often picked up by global organizations like the World Trade Organization (issues regarding injustices in the Agriculture sector), the International Labor Organization (injustices in the labor market), the World Health Organization (injustices in the health sector with all related fields). The general norm is that most of the injustices are seldom perceived as environmental problems or crises. Hence their resolution is often in piece-meal and not multifaceted.

Blind justice is the tendency of justice state organs to execute judgment in situations where the country wants to be seen as pro-people (good in the face of the people), but either shuns or avoids situations which demand retribution against it. Sightless justice makes it very hard for most communities to have and to lead decent lives for if the state wrongs them; there is no recourse for them. They loose everything they have, be it land or any resources in their possession.

Inferior quality of education is a complex problem which cannot be resolved easily. A vicious cycle of corruption exists in most African countries. It seriously compromises the quality of education. Teachers need to be adequately trained if they are to teach qualitatively. They cannot be in colleges for sufficient years if a country does not have adequate resources to support them while in training, money included. Those who come out of the college poorly trained will pass on the inferior quality of their way of doing things to the learners. So the cycle continues, unless efforts are put in place to stop it.

In development initiatives the local peoples' ways of knowing are often scoffed at and perceived as inferior. Unfortunately, broad-based community research results have shown that indigenous ways of knowing can be used as both indicators to development initiatives and as critical policy directives to sustained growth (Maila 2007). It is hoped that the global community will recognize this flaw and avoid it so that environmentally sustainable economies are embedded in sound development strategies.

4 CAN SOUND ECONOMIES BE ENVIRONMENTALLY SUSTAINABLE?

Economies can be environmentally sustainable but under certain conditions. Different governances worldwide have their economic systems pillared around the socio-cultural, socio-political and socio-economical dimensions of their countries. However, most of these systems are driven by industrialist regimes that are highly controlled by the big financial institutions of the world like the International Monitory Fund and the World Bank. Ndlovu-Gatsheni (2006) sees these international institutions as Western growth prophets, who continue to demand that aid in mainstream development be acknowledged as the salvation for Africa. This section briefly sketches the disparities in power and control of the world economic processes, and the dire need to operationalize these on a leveled playing field for all the nations of the world.

How can leveled terrains or equality on socio-economic, socio-cultural, socio-political and biodiversity issues be achieved on a globally uneven-world terrain that thrives on marginalizing its clientele? For these are dimensions that underscore services and goods grounded in environment for growth that must be sustainable to ensure a quality life for all. Sound development processes are imperative if they are to be viewed as a panacea for environmentally sustainable economies. According to Neefjes (2000: 152) these must include a process in which decisions are made collectively, or indeed through comparatively harmonious negotiation and compromise between many different stakeholders, must find ways of involving social actors in different ways. In such a participatory process, problems are gradually analyzed, in separate groups and together, and solutions arise from research, discussions and various interactions. These processes are not strictly defined and structured, and decisions emerge gradually from them, instead of being taken as fixed and predetermined moments by particular people.

5 FACTORING SOUND DEVELOPMENT IN ENVIRONMENTALLY SUSTAINABLE ECONOMIES

There are probably numerous and diverse ways for good development in order to enhance and improve the quality of human life. The following five factors are not only seen as suitable for the argument of this paper, but are also aptly illuminating on the essentials of sound growth that is guided by environmental sustainability.

Development in its nature depends on raw materials for the production of goods, and raw materials obviously must be sought in the environment (both the natural raw material and human material). It is not sensible to stockpile development activities in one area and neglect others. There are numerous gains when progress is founded on sensible measures rather than on inequality notions of marginalizing people. Cogneau (2004) confirmed that redistribution of income and assets contribute to the advancement of ordinary people, hence reducing poverty and other social ills.

Development needs to be grounded on reliable governance processes. Murali and Alamar (2006) argued that the ultimate aim of public governance is service to the people with honesty, courtesy and efficiency. They further pointed out that transparency, accountability, and electoral reforms, credible law enforcement agencies, sensitizing the people and effective delivery mechanisms, are crucial ingredients of a sound, reliable and sustainable development process. Maila (2006) reported that the sound sustainable economies for development must be embedded in the accountability and collaboration praxis of institutional processes in order to ensure that expansion outcomes are people-centered and not just profit-centered. This factor does not only refer to the fact or practicality of good development processes, but also refers to constructive decisions and actions embarked upon regarding the praxis and implementation of good development practices underpinned by sound economic processes.

Ensuring a 'being positive' process in development calls for the participation of the community in development processes that are community based. Not for 'rubber-stamping' what needs to be done, but for ensuring meaningful engagement and deliberate contributions in what should be accomplished for the positive growth of society.

In order that development is underpinned by sustainability, economies of the world must be guided by environmentally sound practices that focus on socio-cultural, political, bio-physical or ecological dimensions of human services and practices. These dimensions are indisputable and irrefutable when it comes to their contribution in focusing the education system. Hence the emphasis that environmentally good economy is crucial for good development practices and those unassailable dimensions of human endeavor should underscore development strategies.

Positive results or outcomes are eminent-certain because the strategy for development ensures success. The strategy is focused and resolved because it has been researched and piloted, and its clientele or communities of practice agree with its implementation process. Therefore, it is also perceived as compact and useful in economic development and human development.

6 CONCLUDING REMARKS

Environmental sustainability ought to be grounded in social critical theory for the mere fact that economic growth and human growth are underscored by socio-cultural, political, and bio-diversity dimensions in/for human advancement. Critical social theory enables researchers to investigate constraints and emancipatory processes regarding environmentally sustainable economies and sustainable development in diverse contexts. It is worth noting therefore, that social critical research focuses on the oppositions, conflicts and contradictions in contemporary society, and seeks to be emancipatory in its practice. In other words, it seeks to eliminate the causes of alienation and domination in environmental sustainability issues.

Sound ecological sustainability and growth can never be continual if threats persist unabated worldwide. Their impact must be eliminated or minimized. Some strategies for achieving such an outcome have been suggested. Each problem or risk is unique and needs a different solution. It is for that reason that decisions should be made collectively, through negotiations. This will allow for the advancement of economic growth and human growth as pillars of environmentally sustainable economies.

REFERENCES

Ahmed, A. (ed). (2006). *World Sustainable Development Outlook 2006: A Global and Local Resources in achieving Sustainable Development*. Switzerland: Inderscience.

Alkire, S. (2002). *Valuing freedoms: Sen.'s capability approach and poverty reduction*. Oxford: Oxford University Press.

Cogneau, D. (2004). Comment, edited by EUDN, in *Poverty, inequality and growth*. Proceedings of the AFD-EUDN Conference, 2003. pp 113–123.

Ecologically Sustainable Development. (1990). Australia: Commonwealth.

Ndlovu-Gatsheni, S. (2006). Gods of development, demons of underdevelopment and Western Salvation: A Critique of development discourse as a sequel of CODESRIA and OSSREA, in *OSSREA. A Bulletin*, Vol 2: 1–19.

Maila, M. (2007). Indigenous knowledge and sustainable development: investigating the link. *Indilinga: African Journal of Indigenous Knowledge*. Vol. 6(1): pp 76–84.

Maila, M. W. (2006). Grounding Sustainable Development in praxis – furthering accountability and collaboration, edited by A. Ahmad in *World Sustainable Development Outlook 2006: A Global and Local Resources in achieving Sustainable Development*. Switzerland: Inderscience. pp 1–11.

Murali, N., & Alamar, K. 2006. Governance and development, edited by A. Ahmad, in World Sustainable Development Outlook 2006: global and Local Resources in Achieving Sustainable Development. pp. 74-78.

Neefjes, K. 2000. *Environments and livelihoods: Strategies for sustainability*. Oxford: Oxfam. NEPAD Document, October 2001.

Nuwagaba, A. 2006. Towards MDGs Localization and poverty reduction: the case of Uganda, in *OSSREA. A Bulletin*, June 2006, Vol 2: 1–13.

Palmer, J. A. 1998. *Environmental Education in the 21st Century: theory, practice, progress and promise*. London: Routledge.

Sen, A. (1999). *Development as Freedom*. Harvard: Belknap Press.

UNEP Educator, January – June 2006.

UNESCO. 2005. UN Decade of education for Sustainable Development 2005-2014 Draft Implementation Scheme, October, 2005.

UNESCO. 1997. Rio + 5 Conference. UNESCO.

UNESCO. *World Conservation Strategy*. 1980. IUCN.

Yeld, J. 1991. *Caring for the Earth: A Strategy for Sustainable living*. Stellenbosch: WWF-SA. htt://www.qual.auckland.ac.nz/ (November 2007).

Environmental Management and Sustainable Development in the Oil and Gas Industry: A Case Study from Kazakhstan

Jung-Wan Lee
Kazakh-British Technical University, Republic of Kazakhstan

Simon W. Tai
Bang College of Business, KIMEP, Republic of Kazakhstan

SUMMARY: This chapter presents the management practices of the oil and gas industry and the local government's approach to solve the problems of environmental protection and sustainable economic development in the Eurasian country of Kazakhstan. In particular, this study describes innovative approaches to environmental sustainability in this industry. The practices of three major players, TengizChevrOil, Karachaganak Petroleum Operating (KPO) Consortium, and Agip KCO at Kashagan oil field, are presented in order to find out how these companies deal with environmental problems. Finally, governmental efforts are assessed on environment management and sustainable development. A special program was developed by the government with the assistance of the World Bank.

1 INTRODUCTION

Kazakhstan with its abundant supply of accessible mineral and fossil fuel resources is situated in the middle of the Eurasian continent. The country's geographical boundaries extend approximately 3,000 kilometers from east to west and 1,800 kilometers from north to south. Kazakhstan is the ninth largest country in the world with a total area of 2.72 million square kilometers. Peculiarities of terrain and climate contribute to the uneven distribution of surface water in the country. About 30,000 rivers and streams flow on the territory. Small plain rivers and streams of less than 10 meters across account for over 90 percent of the river network (UNDP, 2004). As a result, such streams and small rivers appear during spring floods and usually dry up in summer. Flooding there usually occurs during spring and summer periods. Many regions of Kazakhstan possess significant reserves of fresh (about 40 cubic kilometres) and low salinity ground water (about 21 cubic kilometres), of which approximately 2.6 cubic kilometres is allocated annually for industrial and agricultural needs (UNDP, 2004).

Development of petroleum, natural gas, and mineral extraction has attracted most of the over 40 billion US dollars in foreign investment inflows in Kazakhstan since 1993 and accounts for 57 percent of the nation's industrial output or approximately 13 percent of gross domestic product (Agency on Statistics of the Republic of Kazakhstan, 2007). The country has the second largest uranium, chromium, lead, and zinc reserves, the third largest manganese reserves, the fifth largest copper reserves, and ranks in the top ten for coal, iron, and gold reserves (Embassy of Kazakhstan to the USA & Canada, 2008). Perhaps most significant for economic development of the country, it currently has the 11th biggest proven reserves of oil and natural gas. In total, there are 160 deposits with over 2.7 billion tons of petroleum (International Crisis Group, 2007) (Table 1).

Led mainly by the oil sector, a vigorous economic growth started in 2000 and has continued through 2006, with an average gross domestic production (GDP) growth of 10 percent during the period. Perhaps the most impressive achievement has been the strong increase in gross national savings from 16 to 28 percent of GDP between 1997 and 2006 (The National Bank of Kazakhstan, 2007). Foreign direct investment inflow has averaged over 9 percent of GDP for the 2000–2006 period, although the bulk of these investments are in the oil industry (Table 2).

Table 1. Country overview and oil production of Kazakhstan.

Location	Central Asia
border countries:	China 1,533 km, Kyrgyzstan 1,051 km, Russia 6,846 km, Turkmenistan 379 km, Uzbekistan 2,203 km
Population	15,340,533 (July 2008 estimate) Population growth rate: 0.374% (2008 estimate)
Natural resources:	major deposits of petroleum, natural gas, coal, iron ore, manganese, chrome ore, nickel, cobalt, copper, molybdenum, lead, zinc, bauxite, gold, uranium
Economy – overview:	Kazakhstan, the second largest of the former Soviet republics in territory, possesses enormous fossil fuel reserves and plentiful supplies of other minerals and metals. Kazakhstan's industrial sector rests on the extraction and processing of these natural resources. Kazakhstan enjoyed double-digit growth in 2000–01 – 8% or more per year in 2002–07 – thanks largely to its booming energy sector, but also to economic reform, good harvests, and foreign investment. In the energy sector, the opening of the Caspian Consortium pipeline in 2001, from western Kazakhstan's Tengiz oilfield to the Black Sea, substantially raised export capacity. In 2006 Kazakhstan completed the Atasu-Alashankou portion of an oil pipeline to China that is planned in future construction to extend from the country's Caspian coast eastward to the Chinese border.
Industries:	oil, coal, iron ore, manganese, chromite, lead, zinc, copper, titanium, bauxite, gold, silver, phosphates, sulfur, iron and steel; tractors and other agricultural machinery, electric motors, construction materials
Oil–reserves:	9 billion barrels (1 January 2006 estimate)
Natural gas – reserves:	1.765 trillion cubic meters (1 January 2006 estimate)
Environment – international agreements:	• party to: Air Pollution, Biodiversity, Climate Change, Desertification, Endangered Species, Hazardous Wastes, Ozone Layer Protection, Ship Pollution, Wetlands • signed, but not ratified: Climate Change-Kyoto Protocol

Source: CIA, World Factbook, 2008

Table 2. Main indicators of socio-economic development in Kazakhstan.

Year	1997	1998	1999	2000	2001	2002	2003	2004	2005	2006
GDP[1)]	101.7	98.1	102.7	109.8	113.5	109.8	109.3	109.6	109.7	110.7
GDP per capita, USD	1,446	1,469	1,129	1,229	1,491	1,659	2,068	2,874	3,771	5,292
FDI inflows, million USD	2,107	1,233	1,852	2,781	4,557	4,106	4,624	8,317	6,619	10,437
Production of petroleum[2)]	112.3	100.6	116.1	117.2	113.5	117.9	108.8	115.6	103.4	105.7

[1)] Gross domestic product, as percentage of previous year
[2)] production indices of petroleum includes gas condensate, as percentage of previous year
Source: Agency on Statistics of the Republic of Kazakhstan (2007)

Despite these favorable developments, Kazakhstan's performance, as measured by social indicators, reveals a mixed picture. Income per capita in 2006 reached 5,292 US dollars, but with substantial income inequalities. Preliminary calculations based on the government's minimum subsistence level, the population living on less than 2 US dollars per day still stood at 19% in 2004, access to improved sanitation at 72%, access to an improved water source at 86%, and life expectancy stood at 64 years (World Resources Institute, 2007). Access to safe water was well below the European average, particularly in rural areas where only 6.4 percent of households had access to water supply (44 percent of population lives in rural areas) (UNDP, 2004). The country also faces major environmental challenges related to water availability, water pollution, and a legacy of mismanagement of natural resources, oil and radioactive industries.

The current environmental issues that the country is facing include:

- Radioactive or toxic chemical sites associated with former defence industries and test ranges scattered throughout the country pose health risks for humans and animals;
- Industrial pollution is severe in some cities; because the two main rivers which flowed into the Aral Sea have been diverted for irrigation, it is drying up and leaving behind a harmful layer of chemical pesticides and natural salts; these substances are then picked up by the wind and blown into noxious dust storms;
- Pollution in the Caspian Sea;
- Soil pollution from overuse of agricultural chemicals and salination from poor infrastructure and wasteful irrigation practices.

Environmental protection is a priority in oil extraction operations. Oil extracting, despite its profitable economic effect, has a harmful influence on the environment. For instance, burning torches of natural gas, which is habitual in Kazakhstani oil and gas exploration history, pollute the atmosphere with sulphide. So that in a radius of 250 meters around the whole of flora disappears and in a radius of 3,000 meters trees get dry and shed all the leaves. This is not the only environmentally harmful impact of oil exploration. In the Caspian region the amount of spilled oil comes to 5 million tons. The total area of polluted soil is more than 19 thousand hectares. The topsoil is soaked with oil beginning from a depth of ten centimeter to a depth of ten meters. Near the TengizChevrOil refinery, 10 million tons of sulfur is stored (Akanayeva, 2004), that has a negative impact on the region's ecology. Thus environmental safety remains one of the most crucial directions of governmental concerns as well as hydrocarbon extracting firms. Nowadays many oil operators face the necessity to develop innovation technologies and invest huge funds in R&D in order to mitigate this problem.

Nowadays major oil and gas technologies development focus on:

- Identifying additional and incremental hydrocarbon reserves;
- Accessing reserves cost effectively;
- Sustaining and improving existing production from mature fields;
- Improving environmental performance with technological development.

One of the main challenges governments face is how to develop the country's economy using oil and gas resources, but at the same time maintaining appropriate environmental sustainability. To ensure that the benefits of recent economic growth spread over a larger proportion of the population and to ensure environmental sustainability, the government is looking for ways to improve public investment and to promote national environmental management. Accordingly, the government has to facilitate innovative ecological management and to invest in new technologies.

The aim of the current research was to investigate the practices of environmental management and technological development in the oil and gas industry of Kazakhstan. Three cases of environmental management practices were reviewed. The paper concludes by suggesting policy alternatives for the government toward resolving environmental problems.

2 CHANGES IN THE ENVIRONMENT DUE TO ECONOMIC ACTIVITIES OF THE OIL AND GAS SECTOR IN KAZAKHSTAN

The prospective oil and gas bearing region of Kazakhstan is some 1.7 million square kilometres in area, which is more than 62 percents of the total territory. To date, more than 208 oil and gas deposits have been discovered with oil reserves of 2.2 billion tons, 690 million tons of condensate and about 2 trillion cubic meters gas, excluding Caspian Sea shelf reserves (UNDP, 2004). Oil

Table 3. Emissions of pollutants and environmental management in Kazakhstan.

Year	1997	1998	1999	2000	2001	2002	2003	2004	2005	2006
Investment for environment[1)]	1.0%	1.0%	1.0%	1.0%	2.4%	1.8%	2.0%	2.2%	1.7%	2.3%
Emission of pollutants[2)]	2,368	2,327	2,309	2,429	2,583	2,529	2,884	3,017	2,968	2,921
Pollution incidents[3)]	88	628	25	54	18	27	15	57	59	71

[1)] Investment for environmental protection, as percentage of total investment in fixed capital
[2)] Emissions of harmful pollutants (include sulphur dioxide, nitrogen oxides, and carbon monoxide) into the air from stationary sources, thousand tons
[3)] Number of registered pollution incidents (include water resources and air pollution)
Source: Agency on Statistics of the Republic of Kazakhstan (2007)

explorations have shown that the deposits on the Caspian shore are only a part of a much larger deposit. It is said that 3.5 billion tons of oil and 2.5 trillion cubic meters of gas could be found in that area. Overall Kazakhstan's combined onshore and offshore proven hydrocarbon reserves have been estimated between 9 and 40 billion barrels (Energy Information Administration, 2008).

Oil production holds the promise of providing stable and sustainable development of the economy. In 2006, Kazakhstan was producing approximately 1.426 million barrels (226,700,000 cubic meters) of oil daily and 23.5 billion cubic metres of natural gas annually (British Petroleum, 2007). Production of the Tengiz oil field, discovered in 1979, now accounts for almost 80 percent of the total volume of hydrocarbon production in Atyrau Oblast. Long-term prospective oil production development hopes are high for the Caspian Sea shelf and the sub-salt complex of the Pre-Caspian depression. The country's oil and gas complex includes both its oil refining and gas processing industries. Kazakhstan presently has three functioning oil refineries: in Atyrau, Pavlodar and Shimkent. Their total annual capacity is 18.5 million tons of petroleum, with an average utilization rate of 65.3 percent (UNDP, 2004).

While playing a vital role in the national economy, oil and gas production also is one of the main causes of adverse ecological conditions in production and processing regions. The pollution caused by oil refineries has generated strong public interest in connection with increased production and processing of sulphurous oil and associated sulphur emissions. In addition to oil containing traditional sulphur compounds, the production of oil and condensate containing compounds of active sulphur has been increasing, making their production and processing ecologically hazardous. Of total oil industry pollutants, approximately 75 percent occurs as atmospheric emissions, with about 20 percent water pollution and 5 percent in soil (UNDP, 2004) (Table 3). According to estimates, about 3.5 percent of oil extracted is lost in the field. Some petroleum is lost in oil collection and separation systems and during transportation through pipelines. Losses of petroleum from tanks are also considerable due to their generally poor design.

Despite some recent efforts to improve practices, considerable pollution still occurs during oil field exploitation. During exploration and development of hydrocarbon sites, 70–80 percents of vegetation is destroyed within a radius of 500–800 meters (UNDP, 2004). Atmospheric emissions and oil spills present the largest threats of pollution during oil field exploitation. The main causes of oil spills are: corrosion, defects of construction and erection works, and mechanical damage. Further, there are no reliable emergency spill prevention systems on the main domestic pipelines.

The most acute environmental problems are as follows:

- Development of deep "sub-salt" strata (for example, Karachaganak, Tengiz and other deposits) with a high content of sulphurous gas, sulphur dioxide, carbon sulphide, sulphide, disulphides, seriously impacting the environment and also hazardous to human health;

- Increased volume of technical and technological waste products with associated waters, gases, tailings, and wastes associated from dehydration, oil demineralization at preparation and millions of tons sulphur lumps;
- Crude oil production in the reserve area of the Caspian Sea, where offshore drilling increases the probability of accidents (for example, emissions of carbon sulphide, petroleum) and threatens catastrophic pollution of the sea, seabed and foreshore brushwood as well as poisoning of living organisms in significant areas.

Removal of associated gas in oil production remains a major environmental and sanitary problem. Disposal is only performed in Manghistau Oblast. Overall in Kazakhstan, over 800 million cubic meters of associated gas is burnt in flares annually with millions of tons of pollutants emitted into the atmosphere (UNDP, 2004). Data on activities of the largest oil companies and their impacts on the country's environment are presented next.

3 EVIDENCE OF ENVIRONMENTAL MANAGEMENT IN THE OIL AND GAS SECTOR

3.1 *TengizChevrOil in Tengiz oil field*

The TengizChevrOil (TCO) joint venture developed the Tengiz oil field since 1993 (Figure 1). The major partners in this venture are Chevron (50% ownership), ExxonMobil (25% ownership), the Kazakhstani government through KazMunayGas (20% ownership) and Russian LukArco (5% ownership). TCO was established with the aim to develop the Tengiz oil field, which was located in western Kazakhstan in the territory of Atyrau Oblast. Foreign investments required for this project was up to 20 billion US dollars. Estimated at up to 25 billion barrels ($4\,km^3$) of oil originally in place, Tengiz is the sixth largest oil field in the world; recoverable crude oil reserves have been estimated at 6 to 9 billion barrels (0.9 to $1.4\,km^3$). Figures for predicted volumes of geological reserves are put at 3.133 billion tons. In January 2003, after contentious negotiations with the government of Kazakhstan, the TCO consortium members initiated a 3 billion US dollar expansion project designed to boost production to approximately 450,000 barrels (72,000 m^3) per day by 2006 (TengizChevrOil, 2006).

Though TCO's production volumes have been growing for the past 9 years, the amount of environmental damage has also grown, respectively. The information provided seems controversial: some sources state TengizChevrOil is successful in meeting the environmental norms (TengizChevrOil, 2007); others provide the facts that show clearly the company has problems in managing the issues. The main problem is that the extraction process brings not only oil and gas but also some by-products such as sulfur that have a negative impact on the nature. Since the oil from Tengiz contains up to 17 percent sulfur, an estimated 6 million tons of this material has been stored in the form of large sulfur blocks as of December 2002 (The Guardian, 2002). At the time, about 4,000 tons a day were being added. In 2007, the Kazakhstani Environment Ministry was reported to be considering imposing fines against TCO for alleged breaches in the way the sulfur was stored.

In 1994 bilateral agreement about Maximum Allowable Emission (MAE) was established between the government and TCO participants. It was stipulated that according to the special technology, crude and acid gases containing hydrogen sulphides must be processed. However, during 2004 there were 18 illegal gas flare accidents registered in Tengiz field. The environmental damage constituted more than about 1.9 million US dollars. In 2006 the amount of accidents increased to 98 cases, and as a result more than 290 million cubic meters of natural gas were flared (Martynyuk, 2007). In accident situations gas plants cannot accommodate crude and acid gas, that is why gas is.

Another important source of environmental pollution provided by TCO activity is storage of huge amounts of sulfur derivable from the process of crude oil refinement. Currently about 10 million tons of sulfur are stored in the Tengiz area. The sulfur spoils soil in the area, moreover there were some cases of spontaneous ignition of the sulfur which means that the sulfur include huge amounts of dangerous gas – hydrogen sulphide. Besides, sulfur dust pollutes the Caspian Sea

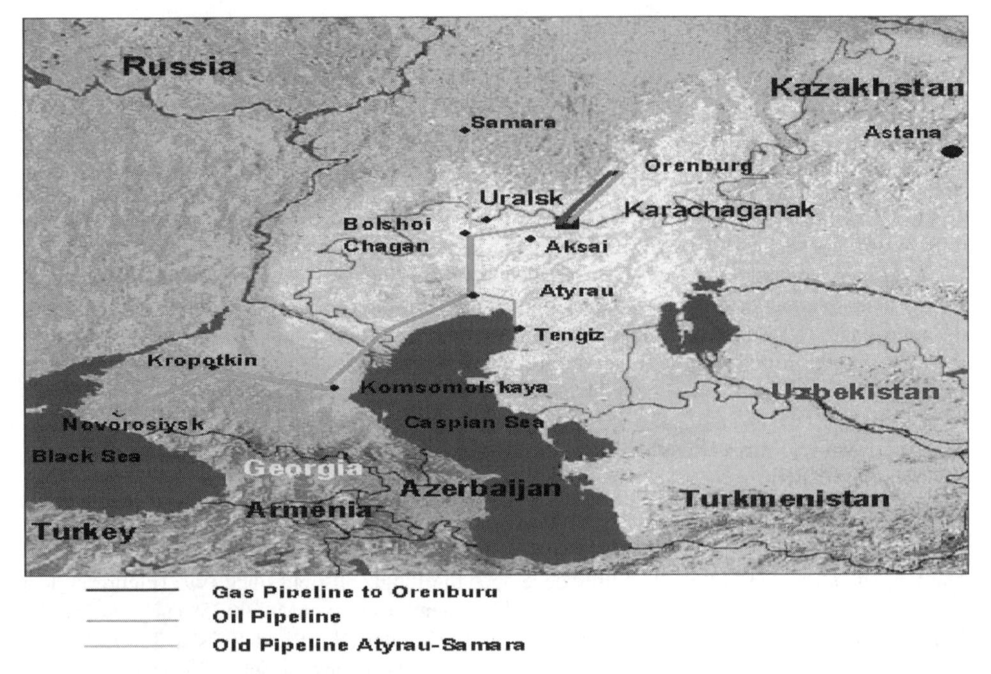

Figure 1. Major oil fields in Kazakhstan.
Source: LUKOIL Oil Company (2008). http://www.lukoil.com/static.asp?id=76

in the form of sulphuric acid, which is very dangerous. The Ministry of Environment Protection of Kazakhstan has been bringing a suit against TCO during several last years. The claims of the suit dealt with environment protection rules violation by TCO and demanded a payment of penalty for violation and improvement of environmental protection technology. Thus TCO faced a serious necessity to improve its environmental protection technologies.

In June 2004 TCO started the realization of Tengiz's crude natural gas re-injection project. The essence of the project was in a technology of natural gas volumes re-injection into oil reservoir. The re-injection was planned to be implemented by special high-power compressors. During the first stage of the project a gas re-injection station was built. After 12–18 months the second stage of the project was started. During this stage crude gas, including hydrogen sulphide touch, was pumped into the oil reservoir. The technology helped to increase TCO productivity by more than 3 million tons of oil per year (Kazakhstan Today Information Agency, 2007).

There are two major advantages of using gas re-injection technology. The first is the ability to maintain high seam pressure during a longer period of time, which, in turn, allows increasing of reservoir output. Another advantage is a possibility to utilize by-products. In particular, there is no necessity to refine natural gas and, therefore, sulfur storage is not necessary. Thus the technology helps to significantly reduce the amount of sulfur waste and, consequently, the amount of dangerous sulfur stored in the field area. Moreover, during last two years TCO considerably increased investments directed to environmental R&D activities. In turn, total investments for environmental protection constituted 1.23 billion US dollars (TengizChevrOil, 2007).

TCO aspires to develop the strategy which will reduce the fines imposed by the Kazakhstan government and to improve the environmental management policies. Still there is no strong incentive for the company to develop its environmental sustainability as the penalties imposed on the production process are simply taken as the additional cost of production which is covered by the revenues of TCO.

3.2 *Agip KCO in Kashagan oil field*

On behalf of seven companies and under the North Caspian Sea PSA (Production Sharing Agreement), Agip KCO (Agip Kazakhstan North Caspian Operating Company N.V.), wholly owned by "Eni" through Agip Caspian Sea B.V., is the single operator of the appraisal and development operations in the Kazakhstan sector of the Caspian Sea. Currently KCO is made up of partners, which include the national oil and gas company KazMunaiGaz (8.33% share), Agip Caspian Sea B.V. (18.52% share), ExxonMobil Kazakhstan Inc. (18.52% share), Shell Kazakhstan Development (18.52% share), Total E&P Kazakhstan (18.52% share), ConocoPhillips (9.26% share), and INPEX North Caspian Sea (8.33% share). The main oil field operated by Agip KCO is in Kashagan field (Agip KCO, 2006).

Kashagan is the largest oil field discovered in the north Caspian Sea area. It is currently estimated that there are 38 billion barrels of oil in place (Agip KCO, 2006). Its development represents one of the greatest current challenges of the petroleum industry given the following characteristics: deep, high-pressure reservoir; high (16–20%) sulfur content with associated production of hydrogen sulfide; shallow waters that range from 3 to 4 meters and freeze from November to March and sea-level fluctuation during the rest of the year; wide temperature variations from −30 to +40C and a very sensitive environment with a variety of internationally protected species of fauna and flora.

As far as Kashagan oil field development is concerned with associated petroleum gas utilization, the company had to implement a range of measures and technologies in this sphere to protect the environment. One of the newest innovation technologies applied to oil extraction process is the usage of special chemical agents based on amino acids salts (Oil and Gas vertical, 2006). The chemical agents were developed for refining gas streams while oil processing activities. The agents are characterized by high stability of ingredients that helps significantly increase effectiveness of gas refining process. New chemical agents surely delete hydrogen sulphide from the gas mixture and, if it is necessary, will remove carbonic acid as well. This helps support the highest level of environment protection standards.

Agents' chemical action is strictly selective. Owing to their physical and chemical properties, new agents became more stable than earlier used chemical compounds. As a result, in conditions of cyclic process, these chemical agents need to be replenished in much smaller volumes. Moreover, this new technology can be used together with other processing technologies. Thus, technology implementation provides opportunity to make gas-refining process more effective and efficient.

One of the largest discoveries of the decade and possibly one of the most challenging, the project faced not only the technical difficulties of extracting the oil in the harsh climate but also the political and geopolitical altercations associated with the region. On 27 September 2007, Kazakhstan parliament approved the law enabling the government to alter or cancel contracts with foreign oil companies if their actions threatens the nation's interests. This resulted in suspending by Kazakhstan government work at the Kashagan development for at least three months due to environmental violations. Therefore, the original date of 2005 of production/extraction date has been postponed to 2010 (Kazakhstan Today, 2007).

3.3 *Karachaganak Petroleum Operating (KPO) in Karachaganak oil field*

The Karachaganak oil and gas condensate field in the west Kazakhstan oblast is one of the world largest fields. It occupies an area of 280 square kilometers, and the field's reserves are evaluated at 1.2 billion tons of oil and condensate, and over 1.35 trillion cubic meters of gas. Discovered in 1979, the field began delivering small amounts of gas and condensate to the Orenburg processing plant in Russia via pipeline in 1985. Today, KPO is carrying out its operations at the field according to a 40-year Production Sharing Agreement (PSA) signed by the partners of the international consortium. At present, Karachaganak is one of the largest investment projects in the country, with over 4 billon US dollars in foreign investment. The consortium also includes BG Group (32.5% share), Italy's Eni, (32.5% share), the Chevron (20% share) and Russia's LUKOIL (15% share) (KPO, 2007a).

KPO manages Health, Safety and Environment (HSE) issues at every stage of field development and operations. Key issues include the followings. Biodiversity is essential to the environmental health of the planet and is emerging as a critical business issue. In 1992, the United Nations Convention on Biological Diversity met and its principles became international law in 1993. Each of the 137 signatory countries, including Kazakhstan, the United Kingdom, Italy and the United States was required to develop a national biodiversity strategy and action plan. The KPO partners strongly support the issue of biological diversity and are clearly aware of the effect that KPO's operations may have on biological resources. The effect may either be direct through habitat loss during construction activities or discharges, or indirect, by impacting air quality or contaminating the land through production processes. KPO is therefore focused on minimizing adverse environmental impacts. KPO actively uses Environmental Impact Assessments (EIA), a process that examines the biological environment in which any new project will be developed. The EIA process collects baseline data and identifies potential direct impact on habitat and species. Plans are then developed to institute mitigation measures. The EIA process also addresses indirect impacts such as emissions. Plans for these effects are included in KPO's Environmental Management System (KPO, 2007a).

The potential presence of contaminants in soil and groundwater arising from historical and current activities can affect the ecosystem. Reclaiming the land that was contaminated before KPO began development of the Karachaganak field, and ensuring that current operations do not result in further contamination, are major issues being addressed by KPO. As a result of a well blow-out in the Karachaganak field in 1987, 521 hectares of land were damaged. There was no vegetation except for some salt-tolerant plants growing where the contamination was not very deep. Having started development of the Karachaganak field in 1997, KPO invited experts from Kazakhstan and around the world to research methods to reclaim this land. A Soil Reclamation Plan was developed and research and field experiments were conducted. The experiments were conducted with various salt-tolerant plants. The experiments proved positive and full reclamation has started. In August 2001 barley was harvested in the former Gryphon area. What was barren is now able to support the growth of barley and wheat. KPO has completed the task of dismantling 28 abandoned rigs, the removal of 73,000 tons of waste to safe storage and the restoration of 52 hectares of land. The landscape in the Karachaganak field has changed thanks to an environment initiative that is seeing thousand of new trees planted throughout the area. Started in 1998, the programme has already seen more than 400,000 trees planted on nearly 160 hectares (KPO, 2007b).

The company noted that at present, due to the achievement of project capacity, the volume of wastes has been reduced considerably. Thus, in 2006, with a production increase of more than two times as compared to 2003, pollutant wastes dropped considerably. In 2004, the waste volume was 56.6 thousand tons, in 2005 – 45.5 thousand tons, in 2006 – 17.2 thousand tons. In accordance with the requirements of the law on oil, KPO developed the Gas Utilisation Program, which is a confirmation of KPO's ability to utilise 99.6 percents of the total gas produced. At present, the most modern technologies are being implemented at Karachaganak, which decreases the impact on the environment significantly.

The equipment at the Karachaganak Processing Complex (KPC) and Unit-2 (the gas and liquids separation and re-injection plant) is unique. The reduction of wastes and an absence of excess concentrations of pollutants in the atmosphere have been confirmed under the KPO Monitoring Programme, which provides for the two-fold collection of samples on the border of the sanitary protection zone in nearby villages. This programme is approved by national regulation agencies and is conducted by a contractor organisation with a state license. In addition, KPO installed eight automatic air monitoring stations on the border of the sanitary protection zone during 2006, which operate night and day. The data collected by these stations also confirm the absence of excess concentrations of pollutants in the atmosphere.

In addition, KPO pays special attention to waste utilisation. New methods are developed and introduced for environmental monitoring, waste recycling and utilisation. For example, during the operation of oilrigs industrial oil wastes are produced. KPO collects these, recycles them and produces a new product –various kinds' automobile oils, which are sold to west Kazakhstan consumers. During the past few years, KPO's total investments into the improvement of the ecological situation exceeded 117 million US dollars (KPO, 2007b).

4 PERSPECTIVES ON NATIONAL ENVIRONMENTAL MANAGEMENT

4.1 *National environmental management policy*

The state program "Kazakhstan Development Strategy 2030" identifies the following major public policy goals in the area of environmental protection and efficient natural resources use: "stabilizing the quality of the environment ensuring a favorable environment for human activity and preserving natural resources for future generation." Between 1994 and 2001 Kazakhstan ratified 19 environmental conventions embracing preservation of all natural components (UNDP, 2004). The following documents have been prepared with the assistance of the international donor community: National Environmental Action Plan for Sustainable Development, national strategies and action plans for biodiversity conservation, combating desertification, forest and mountain ecosystem protection. The main sustainable development strategy documents – Sustainable Development Framework and Agenda 21 – are currently being developed.

Environmental management in Kazakhstan is conducted by governmental bodies, such as the Ministry of Environmental Protection, the Ministry of Economy and Budgeting, the Ministry of Energy and Mineral Resources, the Ministry of Agriculture, and the Agency for Land Resource Management. National and local governments set out procedures for environmental protection and use, regulate activities of natural resource users in accordance with legislation and implement activities for the reproduction of biological resources. The Ministry of Environmental Protection (MEP) of Kazakhstan is the central executive body in the area of environmental protection. Its responsibilities include pursuing national environmental policy, enforcing laws, administering state supervision and state environmental impact assessment for projects in the area of environmental protection. The MEP oversees the country's compliance with ratified conventions and interstate agreements in the area of environmental quality and conservation of biodiversity. The MEP controls emissions and discharges of pollutants, issues permits to enterprises setting the volumes and composition of pollutants, and provides state environmental expertise for projects.

However, the country currently does not have an approved national environmental protection program. The National Environmental Action Plan for Sustainable Development was developed in 1998 with UNDP support and its major components were incorporated into the extended "Kazakhstan Development Strategy 2030."

4.2 *Environmental legislation*

The Constitution of the Republic of Kazakhstan, which was adopted in 1995, forms the basis of national environmental legislation. Article 31 of the Constitution states that "the state shall set an objective to protect the environment favorable for the life and health of the citizen." The access of the citizens of Kazakhstan to information on the environmental situation is viewed as a guarantee of the implementation of state policy on environmental protection. Another important document, essential for domestic and foreign policy in the area of environmental protection and nature use, is the Concept of Environmental Safety that approved by Decree Number 2967 of the President of the Republic of Kazakhstan on April 30, 1996. This document defines the fundamental principles and priorities of domestic and foreign policy, the priority directions of activity essential for ensuring and preserving sustainable environmental, economic and human development, and the prevention of disasters and industrial hazards in Kazakhstan.

Kazakhstan's basic law in the area of environmental protection is the "Law on Environmental Protection" of July 15, 1997. This law defines the basic terms in the area of environmental protection, determines the rights of civil society and the competence of state agencies and local governments in this area. This law also guarantees the right to environmental conditions conducive to life and health. Subsidiary laws regulate specific issues on environmental protection, such as permits and payments for environmental pollution (by the Government Resolution Number 1154 adopted on September 6, 2001), environmental monitoring, state registers and inventories of natural resources. A normative base for environmental standards is now being developed. For example, implementation of activities pertaining to "Establishment of Standards in the Republic of Kazakhstan" was continued by the MEP, while state agencies are preparing to use international environmental standards ISO 14000. On February 2, 2002, the technical committee for standardization, "Environmental Quality Management Number 39," was set up by the Resolution of the Committee on Environmental Protection.

A common problem in this area of environmental protection is the shortage of qualified experts in the field of environmental legislation, which affects the quality of normative drafts and has an adverse impact on their implementation and enforcement. As a result, many of normative environmental documents have legal defects. The situation is increasingly unstable when it comes to enforcement of legislation. In turn, Kazakhstan faces the problem of how to effectively implement the provisions of existing legislation. Kazakhstan needs to strengthen its environmental legislation, but more important is ensuring compliance with such legislation. Moreover, capacity has to be developed to implement Kazakhstan's international environment treaty obligations.

4.3 *Technology commercialization program*

With assistance from the World Bank, the government is preparing to implement a Technology Commercialization Project. The objectives of the project are to help Kazakhstan's national innovation system function more effectively and efficiently in a highly competitive global market environment so that Kazakhstan's legacy of scientific excellence, its still impressive stock of scientific knowledge, and its educated citizenry can be converted into a long term, sustainable resource for generating wealth, improving national competitiveness, diversifying the economy, and raising standards of living.

Innovation processes in the framework of the project supposes involvement technologies development in oil and gas industry as well. The essence of the project provides so-called environmental assessment of all the innovational technologies. That means that every new technology to be implemented in oil and gas industry will be checked for environmental protection compliance. This policy helps ensure that innovation projects are environmentally sound and sustainable. The environmental assessment is a process whose breadth, depth, and type of analysis depend on the nature, scale, and potential environmental impact of the research projects supported by the Kazakhstan Technology Commercialization Project. The environmental assessment process takes into account the natural environment (air, water, and land); human health and safety; social aspects (involuntary resettlement, indigenous peoples, and cultural property) and trans-boundary and global environmental aspects. The environmental and social impacts will come from the activities of the research projects that the Technology Commercialization Project will be financing.

Table 4 shows the implementation schedule of mitigation measures involving various parties proposed as a basis for initiating supervision:

Thus, Technology Commercialization Project constitutes integrated program of nationwide innovation development strategy and environmental protection management. Indisputable advantage of the project is ability to manage environmental protection on the stage of innovation development, which will help not only generate new competitive technologies within Kazakhstan's scientific base, but at the same time monitor the possible effect of the technologies on environment with opportunity of further improvement in order to prevent harmful technologies appearance.

4.4 *Public participation in environmental management*

Ratification of the Aarhus Convention, the Convention of UN European Economic Commission on Access to Information, Public Participation in Decision-Making and Access to Justice in Environmental matters, has made public participation in the area of environmental protection and has given additional impetus to cooperation between the state and the NGO sector. At present, there are over 170 environmental NGOs in Kazakhstan (UNDP, 2004), working in different environmental cooperation with Specially Protected Natural Territories (SPNT) aimed at preserving biodiversity, activities in the area of combating desertification and addressing climate change etc. The current

Table 4. Mitigation Measures Implementation Schedule to Kazakhstan.

Environmental issue	Mitigation measures	Monitoring strategy and contingency measures
1. Air emissions	• Lab staff will be provided with information and training on methods to minimize air emissions. • List of hazardous air pollutant sources and emissions and category will be provided to the laboratory. • A list of actual and potential emissions in the lab (fumes foods, stacks vents, etc.) will be prepared.	• Biannual exposure assessment of air pollutants will be developed. • Periodic verification of control systems will be undertaken. • Records of emissions will be kept and reviewed periodically by Bank supervision team and any other relevant authorities. It will be responsibility of EMS in charge for annual certification. • Regular inspection and maintenance of ventilation system.
2. Waste water discharges	• A comprehensive listing of sources and location of wastewater discharge will be prepared and maintained. • Appropriate operating procedure will be undertaken for minimization of wastewater such as neutralizing predisposal treatment. • On-site septic tank systems or appropriate waste water treatment system depending on the waste water characteristics will be encouraged for implementation. After treatment waste water will be discharged in to existing municipal sewer line. • Lab personnel will be trained in minimization and management of wastewater discharges.	• Periodic maintenance will be undertaken of the sewer system. • Periodic testing of lab procedures will be carried out to ensure compliance with regulatory measures. • Regular training will be provided to ensure waste minimization.
3. Storage of hazardous chemicals	• Procedure for segregation of chemicals will be developed and followed according to chemical classes and compatibility criteria. • Minimum inventory storage procedure of every hazardous chemical will be prepared. • Proper storage criteria for flammable, combustible and volatile chemicals will be identified. Filled and empty chemical containers will be segregated accordingly. • During reconstruction proper ventilation/exhaust system will be designed to avoid exposure to vapors and fumes of hazardous chemical. • Training program will be organized on proper storage and health effect for all employees.	• Periodic inspection criteria and regular visual inspection schedule to be developed and implemented. • Periodic review will be carried out to procure safer alternatives for highly toxic, carcinogenic, reactive or mutagenic material. If available. • Periodic checks will be done of the ventilation system by the lab.

Source: World Bank (2007), Environmental Management Plan to Kazakhstan. pp. 38–40.

stage of development of the environmental NGO sector is defined by the involvement of a greater number of organizations, which aspire to participate in environmental decision-making at different levels and who seek cooperation with state bodies and international organizations. For example, the Ecological Press-Center and the Greenwomen Agency of Environmental News are actively working to raise public awareness on existing environmental issues. The Environmental Society "Green Salvation" is taking part in legislative activities.

When it comes to public involvement in Environmental Impact Assessment (EIA), it should be noted that Kazakhstan does not have precedents in this area. Public involvement in EIA exists only in the course of major projects, for example, oil production projects in the Caspian Sea region. Local NGOs in the Caspian region are increasingly aware of the need to monitor the environmental aspects of oil industry activities and the companies in charge, such as TengizChevrOil, Agip KCO, and KazMunaiGas. Still, participation of the public in EIA remains insufficient in Kazakhstan due to lack of practical experience in this area. The few examples of public involvement in this sphere are rather acts of goodwill to proceed with initiators of economic activity rather than concentrating on environmental impacts. Current legislation does not indicate clearly the involvement of civil society initiatives in relation to norms of public participation.

5 FINDINGS

Oil extraction in the Caspian Sea presents a number of challenges in a complex environment:

- Very shallow waters, wide temperature variations and sea level change;
- Harsh environment (4–5 months of iced sea per year);
- Environmental sensitivity;
- High hydrogen sulphide content (16–20%).

As a result of investigations made on the three companies in the oil and gas industry of the case study country, the following findings can be summarized:

- Both Tengiz and Karachaganak are onshore oil fields, which extract oil and gas from sea and land respectively, whereas Kashagan is offshore one. Consequently, strategies implemented during the process of production significantly vary.
- The results of EIA showed the correspondence of methods used by TengizChevrOil and Karachaganak to accepted standards of performance, while Kashagan project does not meet requirements of EIA.
- Both TengizChevrOil and KPO focus on the natural gas separation and its re-injection, so that it would not be wasted. However, there are some technological differences due to the specifics of production.
- In the case of TengizChevrOil, two main challenges were emphasized because of specific features of Kazakhstan's oil. These challenges are necessity to flare natural gas extracted together with oil, and necessity to store sulfur got after oil and gas refining. Thus, the company's activities in sphere of environmental management were concentrated on solving these problems.
- In the case of Agip KCO, innovation in the environmental management occurs in chemistry field in production process. The company also deals with the problem of high sulfur content, in which new chemical agent allows making refining process more reliable and cost-effective.

Generally the process of government's environmental management and technology development in oil and gas industry can be divided into two main types of activities. The first one is monitoring of business operation's correspondence to environmental norms and standards. Another way of governmental involvement is organizing of special development programs within which innovation is stimulated. The following findings can be made:

- Kazakhstan has developed national strategies and action plans on protection of key elements of the environment and biological resources. A number of programs and projects are being implemented to assist in resolving specific problems and tasks in specific regions of the country.

- Environmental agencies make every effort to stabilize the quality of the environment,: sustainable use of resources and mitigation of adverse effects of man-caused pollution. State environmental agencies need adequate and stable measures for comprehensive long-term activities.
- In most cases, civil society does not have access to reliable and updated information on quality of the environment, social and economic consequences of environmental degradation. Participation in decision-making is not practiced and needs greater attention. Existing practice does not provide for participation of the public and NGOs in environment quality issues, in decision-making on use of natural resources.

There are several levels of environmental regulations that are found under a variety of national laws. At the same time government has an authority to provide rights for companies operations in the industry. The acquisition of these rights primarily provides the operator with the authority to explore and exploit a given area of land or seabed (United Nation Environmental Program, 1997). If hydrocarbons are discovered, the operating or contracted company will have to meet the requirements of various authorities and obtain, for example, a development consent approving the detailed development plans; a planning consent which usually incorporates the environmental assessment; and operational consent which provides detailed information on operational activities. Once operations start, monitoring regimes are required whether by legislation through authority inspection and enforcement, or through industry commitment to management system and self-regulation. However, policy and commitment alone cannot provide assurance that environmental performance will meet legislative and corporate requirements or best industry practice. To be effective, they need to be integrated with the formal management activity and address all aspects of desired environmental performance. Many companies operate in widely varying climatic, geographic, social and political circumstances. Companies need a consistent management approach but must be allowed sufficient flexibility to adapt to the sophistication of the existing infrastructure.

6 CONCLUDING REMARKS

Awareness of the importance of environmental issues has become more and more central to the oil and gas industry. The impact of oil and gas exploration on the environment depends not only on natural particularities of the region, local features of oil resources, but also on the effectiveness of planning, pollution prevention, and mitigation and control techniques.

Environmental program organized by the government jointly with the World Bank provides an opportunity for the oil operators to investigate and assess innovational technologies for the purpose of environmental sustainability. The important part of the sustainable growth is environmental management. Eco-efficiency states the necessity of success in both economic and environmental performance. It means turning over the capital stock and introducing new cleaner production processes and new products that would be environmentally friendly instead of the older polluting factories. It also means reducing waste, implementing recycling and reuse, using fewer and less toxic materials, and using energy and water more efficiently. A special program depending on formulated monitoring problems and tasks should be developed for each project in the oil and gas industry.

Overall, in order to achieve sustainable development, environmental sustainability should constitute an integral part of the growth process and cannot be considered in isolation from it. A good balance between economic development and environmental sustainability is one of the main tasks of contemporary management in the oil and gas industry. Much has already been achieved, but the industry recognizes that even more can be accomplished. A company which aspires to be sustainable in the market has to follow the principles of a learning business. We live in a world that is increasingly shaped by sustainable development issues, such as energy security, climate change, water availability, the degradation of ecosystems. These challenges are becoming central to business competitiveness and long-term success. Likewise, the need for cooperation between government, industry and civil society to foster a more sustainable future is becoming ever more urgent.

REFERENCES

Agency on Statistics of the Republic of Kazakhstan (2007). *Statistical Yearbook of Kazakhstan*. Astana, Kazakhstan

Agip KCO Oil and Gas Corporation (2006). Company's information booklet. presented at KIOGE 2006, Kazakhstan International Oil and Gas Industry Exhibition, Almaty, Kazakhstan, November 3–7, 2006

Akanayeva, S. (2004). Oil and gas industry in Kazakhstan, *Doing business in Kazakhstan*, vol. 3, no. 7, pp. 5–12

British Petroleum (2007). World Oil Production. (WWW document). accessed November 20, 2007, retrieved from http://www.bp.com/

CIA, World FactBook (2008). Kazakhstan. (WWW document). accessed April 5, 2008, retrieved from https://www.cia.gov/library/publications/the-world-factbook/geos/kz.html

Embassy of Kazakhstan to the USA & Canada (2008). Mineral Wealth. (WWW document). accessed March 15, 2008, retrieved from http://prosites-kazakhembus.homestead.com/mineralwealth.html

Energy Information Administration (2008). Country Analysis Brief: Kazakhstan. available at http://www.eia.doe.gov/cabs/Kazakhstan/pdf.pdf

International Crisis Group (2007). Central Asia's Energy Risks. Asia Report No. 133. (May, 2007). available at http://www.crisisgroup.org/

Kazakhstan Today (2007). Works on Kashagan may be stopped – Environment Protection Minister. (August 22, 2007). (WWW document), accessed November 23, 2007, retrieved from http://www.eng.gazeta.kz/art.asp?aid=94815

Kazakhstan Today Information Agency (2007). TCO developed three-year plan of environmental protection improvement. (WWW document). accessed November 21, 2007, retrieved from http://www.zakon.kz/our/news/news.asp?id=30094444

KPO (2007a). Biodiversity in the Karachaganak Field. (WWW document), accessed November 21, 2007, retrieved from http://www.kpo.kz/cgi-bin/index.cgi/139

KPO (2007b). Land Stewardship. (WWW document), accessed November 21, 2007, retrieved from http://www.kpo.kz/cgi-bin/index.cgi/141

LUKOIL (2008). Karachaganak Export Pipeline. (WWW image). accessed March 15, 2008, retrieved from http://www.lukoil.com/static.asp?id=76

Martynyuk, O. (2007). Sulfur challenge. (October 11, 2007). *Gazeta.kz,* (WWW document). accessed November 24, 2007, retrieved from http://www.gazeta.kz/art.asp?aid=97300

Oil and Gas Vertical (2006). Gas refining alternative. (November 15, 2006). (WWW document). accessed November 21, 2007, retrieved from http://www.ngv.ru/show_release.aspx?releaseID=3907

TengizChevroil (2006). For the benefit of Kazakhstan. TCO's official report. presented at KIOGE 2006, Kazakhstan International Oil and Gas Industry Exhibition, Almaty, Kazakhstan, November 3–7, 2006

TengizChevrOil (2007). TengizChevrOil (TCO) Sour Gas Injection and Second Generation Project, Tengiz Oil Field, Kazakhstan. (WWW document), accessed November 20, 2007, retrieved from http://www.hydrocarbonstechnology.com/projects/tengiz_chevr_oil

The Guardian (2002). By-product that blights Caspian life. (December 04, 2002). (WWW document). Accessed March 10, 2008, retrieved from http://business.guardian.co.uk/story/0,3604,853310,00.html

The National Bank of Kazakhstan (2007). Main Economic Indicators. available at http://www.nationalbank.kz/?docid=178&uid=74FDE468-802C-E8F0-E119F052EBEC6B88

United Nations Development Program (UNDP) in Kazakhstan (2004). Environment and Development Nexus in Kazakhstan. Publication series #UNDPKAZ 06. Almaty, Kazakhstan

United Nation Environmental Program (1997). Environmental management in oil and gas exploration and production: An overview of issues and management approaches. UNED IE/PAC technical Report 37, E&P Forum Report 2.72/254. available at http://www.ogp.org.uk/pubs/254.pdf

World Bank (2007). Kazakhstan – Innovation and Competitiveness Development Project: Environemental management plan. World Bank Report E1754. available at http://go.worldbank.org/ZXRO6MAEH0

World Resources Institute (2007). Poverty resources: Kazakhstan. Available at http://earthtrends.wri.org/povlinks/country/kazakhstan.php

Analyzing Critical Success Characteristics of South African Entrepreneurs

Louise van Scheers
School of Business Management, University of South Africa, South Africa

SUMMARY: This study was aimed at investigating the differences between successful and less successful entrepreneurs to identify trends between the prevelance, or lack thereof, of certain characteristics and the level of business success that had been achieved. Based on the answers to a questionnaire, the respondents were divided in four groups: very successful, successful, making a living and unsuccessful. The results were collated per group and then comparisons where sought. The research findings highlighted that there are notable differences between the qualities of the four groups of entrepreneurs. In particular very successful people took direct responsibility for their failures and never sought to blame those beyond their control. The majority of the unsuccessful and less successful entrepreneurs entered their current industries as a result of opportunities that had presented themselves. Very successful entrepreneurs, on the other hand, made conscious calculated decisions to enter particular industries.

1 INTRODUCTION

Contemporary interest in industrialists can be traced back as far as the nineteenth century. Empirical studies of capitalist behaviour, however, are relatively recent (Rabey, 2005). Most psychological research builds on McClelland's work on the theory of needs Rabey, 2005) attempting to predict entrepreneurship from individual characteristics such as locus of control, risk taking, personal values, job satisfaction, experience, role models, age, education and residency. Entrepreneurship can be defined as the initiation of change, an innovative art, creating a new company, where as management involves controlling and planning and running a business. Starting and running a small business involves both free enterprise and managerial skills. When entrepreneurs are developing a new product such as, for example, a solar powered cell phone; they are initiating an innovative change. When actually producing these phones, the role changes into a manager who performs planning, coordinating, leading and controlling functions. The person may return to the entrepreneurial role by altering the design of the cell phone.

The majority of previous examinations have been to determine aspects of background and personality characteristics that distinguish an entrepreneur from the general population. Far less research has been done on the association between these factors and the success of the former. Entrepreneurial research has been hampered by a failure to properly define this type of individual. The lack of agreement about the essential traits is equally problematic (Cromie, 2006). Entrepreneurs are thought to have numerous qualities that distinguish them from others. Hornaday and Knustzen (2000) mentioned the need for achievement, propensity for risk, tolerance of ambiguity, self-confidence, innovation, internal / external locus of control, ability to co-operate, always have a choice available, energy, focus, heightened awareness, imagination, ingenuity, initiative, knowledge of market, management skills, marketing skills, need for independence, raw intelligence, resolute sense of self determination, resourcefulness, vision, vivid imagination and willpower.

Success means different things to different people, but by and large, the success of a business is measured in terms of money. Measures of the amount of money a business made is measured in terms of turnover or net profit or cash flow or a combination of all of these.

Deamer & Earle (2006) noted that the small business owner was an individual who establishes and manages a company for the principal purpose of furthering personal goals. The business must be his primary source of income and consumes the majority of this time and resources. He sees it as an extension of his personality and intricately bound with family needs and choices. Dreamer and Earl also mentioned that the entrepreneur is an individual who establishes and manages a business for the principal purpose of profit and growth. He is characterized principally by innovative behaviour and will employ strategic management practice in business.

Owning or running a company does not make a person an entrepreneur. The key is creativity, innovation and being an agent of change. Many business owners are managers, running a business which they have copied from somebody. Business owners may be classified as entrepreneurs only when they comply with one or more of the following: introduces a new or improved product or service; opens up a new market; uses a new source of supply of raw materials; or creates a new industry.

Whatever specific business activity entrepreneurs are engaged in they are considered the heroes of free enterprise. However, they all need to obtain managerial skills. Capable entrepreneurs and managers achieve goals successfully, resourcefully and achieve targets with the minimum waste of resources.

Cunningham & Lischeron (2002) identified six major schools of thought on this subject. The "Great Person School" views an entrepreneur as a person with vigour, intuition, energy, persistence and self esteem, while the "Classical School" identifies such a person with innovation, creativity and discovery. The "Management school" refers to an entrepreneur as one who organizes, owns, manages and assumes risk. The "Leadership School" views an them as one who motivates, directs and leads whilst the "Entrepreneur School" focuses on skilful management with complex organizations. Lastly, the "Psychological Characteristics School" assumes that people behave in accordance with their values and that behaviour results from attempts to satisfy needs.

The aim of the current study was to identify the critical success factors or characteristics that successful entrepreneurs possess that differentiate them from less successful counterparts. The following hypotheses were drawn: There is a correlation between the characteristics of successful entrepreneurs and the level of business success that has been achieved. There is a correlation between the characteristics of less successful entrepreneurs and business failure.

2 METHODOLOGY

This study was conducted using a Trade and Industry list, consisting of entrepreneurs whose businesses ranged in size from achieving R3 million (i.e. 400 000 USD) turnover annually to in excess of R50 million (6 million USD) annually. The main secondary information sources used were journals, articles, press reports, and books. A qualitative research study was conducted. A sample frame of 100 was obtained. A structured questionnaire was e-mailed between 20 April and 30 May 2005. Of the 250 questionnaires e-mailed, 165 replies were received, which mean a response rate of 66 percent. This may be considered a good response rate considering the sensitivity of the topic, the nature of the problem under investigation and the inhibitions that management might have regarding the survey.

2.1 *Measures*

The main construct of this study (location) was measured through the use of Likert-type scales as opposed to demographic variables, like respondent income and age, whose questions delivered only nominal data. The basic design therefore consisted of a Likert-type with five scale points (with labels ranging from strongly agree to strongly disagree) and 11 degree items. This was found to be highly reliable with a Cronbach's Alpha of 0.07. The constructs measuring the characteristics of successful entrepreneurs and the level of business success that had been achieved were measured using five scale items. No items on any of these scales were reverse scored.

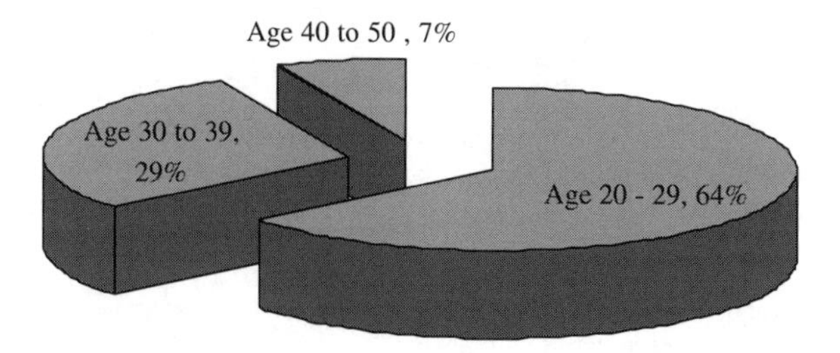

Figure 1. Age that respondents first became entrepreneurs.

2.2 *Inferential statistics*

Inferences were also made about the population based on what was observed in the sample. Inferential statistics allow researchers to make deductions concerning the true differences in the population. The following null and alternative hypotheses were formulated:

H1 = There is a positive correlation between the characteristics of successful entrepreneurs and the level of business success that has been achieved.
H2 = There is a correlation between the characteristics of less successful entrepreneurs and business failure.

3 RESULTS AND DISCUSSION

Respondents were asked at what age they first became entrepreneurs to establish whether this component has had any bearing on the success that they achieved (Figure 1). Results showed that 64 percent of the people first became entrepreneurs between the ages of 20 and 29 years. Twenty nine percent became entrepreneurs when they were 30 to 39 years old compared to 7 percent who started a business when they were older than 40 years. This suggests that to be an entrepreneur you need youthful energy.

The responses to the question, "What motivated the decision to become an entrepreneur", identified a variety of factors. The findings indicate that there was a common desire for independence and more money across all four categories of respondents. Although common among the four categories, desire for independence was most prevalent among the very successful respondents. There is however no apparent pattern or trend that could signify that the desire for independence has any effect on success (or the lack of) achieved. Money, although also a common factor across the categories, was prevalent in none. Therefore, although identified as a strong motivating factor for many people, it would appear that it has no effect on the probability of achieving success going forward.

When asked about their personality characteristics they felt contributed most to their current entrepreneurial success, 58 different qualities were listed with surprisingly few commonalities between them. The only characteristics universal to all were discipline, honesty, vision and self-confidence. A number of features were listed by only the very successful respondents in terms of business success measured by profit of the business, namely:

- Ability to sell
- Ambition
- Being risk averse
- Choosing the right people
- Competitive nature
- Flexibility
- Integrity

Table 1. Reasons behind previous failed business ventures.

Current Status of Entrepreneur				
	Very successful	Successful	Making a living	Unsuccessful
Reason behind previous business failure	Lack of financial acumen	No business background	External forces beyond control	Lack of capital
	Young, naïve and not street wise	No fully understanding necessary requirements	Lack of experience	Dishonest partner
	Did not research business	Market was not big enough for another entrant	Lack of necessary insight	External forces beyond control
	Laziness	Dishonest partner	Was not managing the business personally	
	Bad planning	Viewed as a sideline	Untrustworthy partner	

- Never being satisfied
- People skills
- Prepared to invest
- Problem solving skills
- Willingness to multitask

The only qualities listed by the very successful people that were not listed by the unsuccessful respondents were: leadership, creativity and understanding. Very successful respondents therefore identified more characteristics as being critical to their success than their less successful and unsuccessful social group.

In the case of previous business failures (Table 1), a key finding from the research was that the current very successful respondents always solely attributed previous failures to their own faults and weaknesses. In all the other categories at least one respondent in every group gave a reason for failure that was attributed to someone or something other than their own faults or weaknesses. In addition, it is worth noting that the reasons provided by the current unsuccessful respondents all attribute the failure to alternative factors other than themselves. It would be interesting to observe whether their current lack of success is attributed to similar reasons.

Entrepreneurs are born and not made was another statement put forward to gain insight into their thoughts on the topic. The findings indicated that a majority (67%) of the unsuccessful respondents agreed with the statement. In contrast, a majority (58%) of the very successful respondents disagreed with it. A small majority of 53% of successful respondents agreed with the statement while a small majority of 53% of making a living respondents disagreed with it. A point worth highlighting is that the vast majority of unsuccessful entrepreneur believe that entrepreneurs are born and not made.

A trend is evident on the perceptions of initiative. Figure 2 indicates that 80 percent of the very successful respondents, 50 percent of the successful respondent, and 40 percent of the making a living respondents and a mere 15 percent of the unsuccessful respondents believed initiative to be critical factor in success.

These locations allow us to compute the mean of successful entrepreneurs and initiative to assess the relationships between characteristics of successful entrepreneurs and the level of business success that has been achieved. Therefore the following null and alternative hypothesis can be formulated:

H0 = There is no correlation between the characteristics of successful entrepreneurs and the level of business success that has been achieved.

H1 = There is a positive correlation between the characteristics of successful entrepreneurs and the level of business success that has been achieved.

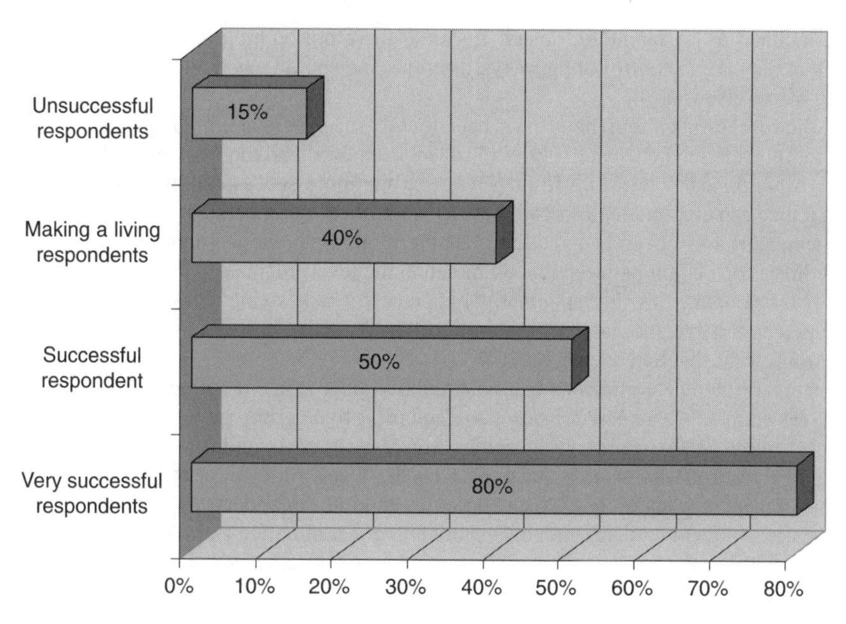

Figure 2. Importance of Initiative in Business Success.

Few differences were highlighted in the personality characteristics component of the research that had a strong correlation to the success of the entrepreneur. A number of components are however worth mentioning. The more successful entrepreneurs were found to seek advice from others most often. There is a direct correlation between the level of success achieved and the frequency of seeking advice. This component therefore has a direct correlation to entrepreneurial success. A further interesting finding is that the more successful entrepreneurs are of the belief that they can continually improve. Less successful people were, however, of the opinion that entrepreneurs are born and not made and are therefore incapable of improving. The research has shown a direct correlation between this opinion and the levels of success that have been achieved.

Of the 22 characteristics that were identified by a variety of authors, only two qualities, initiative and appetite for risk taking highlighted a direct correlation between perception and success. The opinion of how important respondents viewed initiative to be to their current success found that the more important initiative was perceived to be the greater the level of success that had been achieved. As an entrepreneur it should therefore be noted that there is a direct correlation between viewing initiative as critically significant and the levels of success achieved.

The findings also highlighted that the more successful the entrepreneur the greater the appetite for risk taking. It should be noted that there is an element of risk in all business activity and if one is not prepared to take calculated risks this can potentially limit success.

4 CONCLUDING REMARKS

The factors motivating people to become entrepreneurs showed a direct correlation to the levels of success that had been achieved. A conscious, calculated decision to enter a particular industry versus being a victim of circumstance has been shown to have a direct correlation to the level of success achieved. It should be noted however that although a definite correlation was shown to be evident, it is not an indication of definite entrepreneurial success. Nonetheless asking a person why they chose to enter a particular industry can give insight into the potential of an entrepreneur.

The study has highlighted that entrepreneurs who are prepared to take direct responsibility for their actions and shortcomings have proved to enjoy far greater success going forward than those who are inclined to lay blame on others. It should therefore be highlighted to people who seek success that they are in control of their own destinies and should always accept responsibility and never lay blame elsewhere.

The more successful entrepreneurs have been found to seek advice from others most often. There was a direct correlation between the level of success achieved and the frequency of seeking advice from others. A further interesting finding was that the more successful entrepreneurs are of the belief that they can continually improve. Less successful entrepreneurs are however of the opinion that entrepreneurs are born and not made and are therefore incapable of improving. The research has shown a direct correlation between this opinion and the levels of success that have been achieved.

Only two characteristics, initiative and an appetite for risk taking showed a direct correlation between perception and success. The more important an initiative was perceived to be the greater the level of success that had been achieved.

In closing, the study recommends that potential South African entrepreneurs should ensure that they possess many of the characteristics identified prior to entering the world of business. People should also make a calculated and conscious decision about the industry that they are entering as this is shown to have a significant impact on the levels of success achieved. Entrepreneurs should also accept responsibility for their successes and failures and try not to lay blame on others. Finally it is recommend that entrepreneurs should note that initiative and risk taking are important characteristics that can have a significant impact on their success.

REFERENCES

Cromie, S. (2006). 4th Edition. Assessing entrepreneurial inclinations: Some approaches and empirical evidence, European Journal of Work and Organizational Psychology. pp. 23–45.

Deamer, I. & Earle, L. (2006): "Searching for Entrepreneurship", Emerald Group Publishing Limited, 36, 3, pp. 99–103.

Drucker, P. (2005). Innovation and entrepreneurship, London: Heinemann.

Hornaday, J.A. and Knutzen, P. (2000): "Some Psychological Characteristics of Successful Norwegian Entrepreneurs in Ronstadt", Frontiers of Entrepreneurial research.

Rabey, G. 2005. MBA dissertation. Milpark business School & Brookes University. Johannesburg: Milpark.

Environmental Management and
Human Progress

Building a Sustainable Mindset

Siham El-Kafafi
Manukau Institute of Technology, Manukau Business School, Auckland, New Zealand

SUMMARY: This chapter discusses the rising movement of corporate responsibility reporting in the business environment due to the demand for sustainable management. It advocates the importance of building a sustainable mindset through considering how to develop people and organisations so that sustainability thinking and behaviour is a natural and ongoing part of the organisation's life. The chapter starts by explaining the evolving concept of the environment where organisations work. This sets the scene for the shift in ideology from traditional business strategies to contemporary strategies and its impact on business ethics, its economic and social role and how it could be used as a sustainability tool. Furthermore the research reports on techniques available for business to assist them in enhancing their honesty and integrity in their dealings. Finally the chapter provides suggestions for organisations to develop ethics at work as a means of building a sustainable business environment.

1 INTRODUCTION

As there is increasing demand for sustainable management, tools like Triple Bottom Line Reporting (TBL), reporting has become more important and widespread. Nevertheless, organisations have been evaluated not only for their performances on the bottom line, but also for their behaviour as worldwide corporate citizens. Global corporate scandals of high profile organisations like Enron Corp., WorldCom Inc. and Arthur Anderson LLP rocked stakeholder confidence and shamed business leaders and led to a higher scrutiny of organisations' integrity. Accordingly, a movement towards corporate responsibility reporting arose to drive transparency in the environment and social arena. In 2004, 1700 corporations filed responsibility reports, up from virtually none in the early 1990s. Some of those corporations are using these reports as a way to push their sustainability commitments further (e.g. Starbucks has set a goal of increasing the share of its Coffee and Farmer Equity Standards to 60% by 2007) (Assadourian, 2006). However, the success of these techniques depends on the commitment, skill and character of the people implementing them, and the culture of the organisations. This is why the consideration of sustainability in management and the professions has to address issues of business ethics thus including the concepts of corporate social responsibility, corporate sustainability, corporate citizenship and human resource development.

This research contributes to the field by explicitly considering how to develop people and organisations so that sustainability thinking and behaviour is a natural and ongoing part of the life of the organisation. How do we make sustainability a commitment rather than just a compliance cost? In order to answer such a question, the chapter starts by explaining the evolving concept of the environment where organisations work. This sets the scene for the shift in ideology from traditional business strategies to contemporary strategies and its impact on business ethics, its economic and social role and how it could be used as a sustainability tool. Furthermore the research sheds light on the latest reporting techniques available for business to help them enhance their honesty and integrity in their dealings. Finally the research provides suggestions for organisations to develop ethics at work as a means of building a sustainable business environment.

2 THE ENVIRONMENT IN WHICH WE WORK

Sustainability is a concept referring to a balance or wise use of how resources are exploited. The appropriateness of such an approach depends on the values and ideologies of various stakeholders. On the other hand, the history of resource management over the last century suggests that sustainable development is another term that has emerged in an attempt to reconcile conflicting value positions with regard to the environment and the perception that there is an environmental crisis that requires solution (Hall, 1998). Accordingly, sustainability could be viewed as a matter of applied ethics. It considers how we act to fulfil our obligations to the ecosystems, the earth, and its people in the long run (Costanza, 1992). This aspect of life has a historical context. Ethical thought has existed for millennia. However, recent decades has witnessed dramatic changes in the western world. Kluckhohn (1958) noted a change in society values. He did a professional survey of the available literature with the aim of finding out the shifts in the American values during the past generation. Through his empirical data, he detected a decline of the Protestant Work Ethic as the essence of the dominant middle-class value system.

Bell (1976), following his work in the U.K., stated that Protestant Ethic has been replaced by hedonism in contemporary society (i.e. the idea of pleasure as a way of life). He illustrated by saying that the culture was more interested in how to spend and enjoy and no longer concerned with how to work and achieve.

Lasch (1978) reinforced the previous thoughts by mentioning that the work ethic was gradually transformed into an ethic of personal survival. In other words, the emphasis on immediate gratification had created the narcissistic man of modern society.

Yankelovich (1981) asserted that people are committed to the search for self-fulfilment and have loosened their attachment to the ethic of self-denial and deferred gratification.

On the other hand, Buchholz (2000) gave the following justification for human change in beliefs and attitudes towards ethics. He commented that perhaps the development of the atomic bomb also had an impact on generations growing up after the Second World War, because the future has never been any certain since that time. Mankind all had to live with the knowledge that humans have the ability to destroy the planet. He further stipulated that one might as well live as well as one possibly can now rather than defer gratification for some future time that may not be there. Buchholz went on to say that changes in religious beliefs and the increasing secularization of society may also have weakened belief in an afterlife. More people came to hold the belief that you only go around once in life and might as well enjoy it to the fullest extent possible. This would confirm what Sargent (2002) stated that ethics as a field can be traced back to religious, philosophical, psychological and sociological sources.

Whatever we make of those interpretations, it is clear that widespread changes in social ethics and societal norms impact professional and management behaviour. Much has been made of the development of environmental ethics in recent years. In some sense green values have become widespread. However, it is important to remember this is only one strand of social life. It has to be considered how this fits with other characteristics of society in this century.

3 BUSINESS ETHICS

Company ethics is not a new concept since the Arthur Anderson and Enron scandal (Monks & Minow, 2004). The latter were involved in a vast cover up to hide the truth about Enron's finances from employees and investors. Since then more attention was paid to ethics or lack of it in business. It has been generally recognized that business ethics is more than compliance with law i.e. it requires a major shift in corporate culture by dealing with all stakeholders (i.e. employees, partners, clients, shareholders, communities and natural environments in which businesses operate). Steward (1996) noted that the Executive Vice President and Chief Executive Officer of Moody/Nolan LTD, Inc. described business ethics as follows. A company's business ethics are its self-established standards for its relationships with employees, customers, and the community as a whole. It was stated that it is the fabric, describing the integrity and values of a business, as exhibited by its leadership and its employees. Additionally, good business ethics lead to improved products and services, improved employees performance and greater customer loyalty. The continual well-being of business depends upon strong respect and support for its products and services by its employees and customers. He went on to state that loyalty is the glue which binds these separate constituents to a business. The source and catalyst of this loyalty is good ethics.

Traditional business strategy favoured the investors over the stakeholders; the main goal being to create organisations that maximized the value of the firm for its owners (Etzioni, 1988). This is consistent with the traditional neoclassical economics model. However, the relationship between investors and stakeholders is complex and it is this complexity which needs to be teased out in contemporary business ethics. As indicated by Steward (1976) above, there has been a shift in such an ideology in contemporary business ethics in which there is more emphases on relationships built on honesty and trust among all involved parties. This transition is further echoed by the Committee for Education in Business Ethics of the American Philosophical Association through the following key concepts in business ethics: obligation, autonomy-dependence-paternalism, freedom, justice, self-respect and dignity (Stewart, 1996). The following section discusses further the role of business ethics in society with its different roles.

3.1 *Role of business ethics in society*

Corporate or business responsibility is a vital issue in contemporary debate. Corporate influence can extend across the globe since business actions can determine the prosperity of communities and the health of environments. This is due to business involvement in the employment of people, the sales of products and a host of other interactions with communities.

The role of business corporations has been defined by a US Department of Commerce report (Task Force on Corporate Social Performance, 1980). A business corporation exist primarily to produce goods and services that society wants and needs. Achieving this objective is their first and foremost responsibility. If they are unsuccessful in this mission, they cannot reasonably be expected to assume others.

Society is expected to provide an environment in which businesses can develop and prosper. It is expected that society allows investors to earn returns and at the same time ensures that stakeholders can enjoy benefits of their involvement without fear of arbitrary or unjust actions. On the other hand, society is also expecting business to make its contribution through products and services produced by firms (Cannon, 1994).

However, business operates in a world of multiple challenges. Howitt (2001) summed it up when he said that sustainable improvements in human lives can be achieved through addressing the following four core values: social justice; ecological sustainability; economic equity; and cultural diversity. Presumably Howitt expected companies to do more than comply with the law in these areas. The challenge is to define what companies can and should do. The following section will discuss the economic and social role of business in society and how it could be used as a sustainability tool.

3.1.1 *Economic role*

The economy has a major impact on both the environment and society. Giddings et al (2002) stated that when governments, businesses and theoreticians talk about the economy, they usually are referring to the production and exchange of goods and services through the operation of the market (i.e. the capitalist economy).

Economic activities are both fundamental to and complementary to the activities within society. Cannon (1994) identified the following as some of the functions and tasks organisations perform in society.

- Economic and production tasks like land maintenance, food production, manufacturing and distribution of goods and services and all tasks associated with the creation and maintenance of wealth.

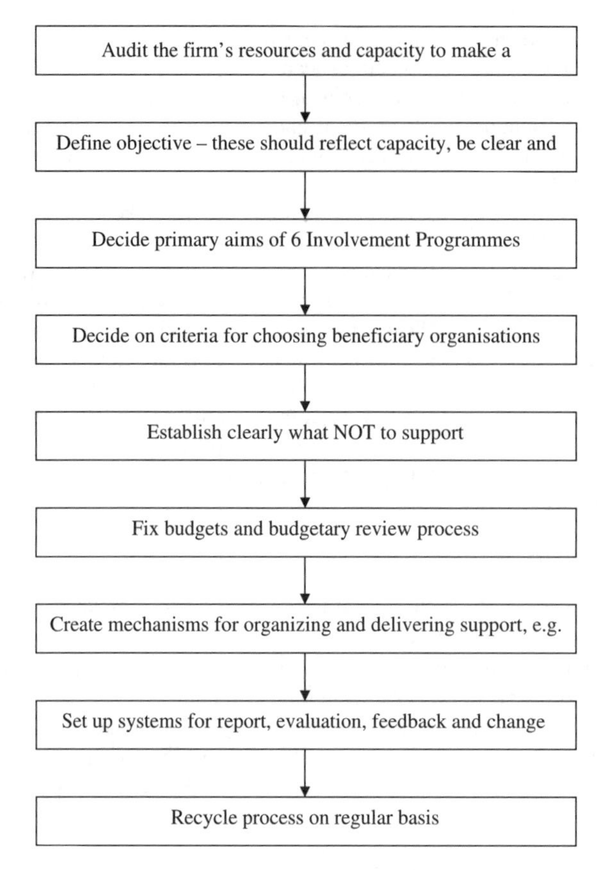

Figure 1. Community Involvement Choice Flow.
Source: Clutterbuck, 1992.

- Maintenance tasks to hold the society together through stability and ensuring continuation. Some examples of this task are education, religion, health and welfare service.
- Adaptive functions to help society respond to change. A good example of this category would be research organisations.
- Managerial or political tasks which are performed by government, political parties, the law and military. These organisations help implement policy for the group and related agencies to arbitrate and assess conflicting demands or expectations.

Clutterbuck (1992) proposed a systematic approach (Figure 1) showing that effective selection of projects, backed by sustained involvement and quality management is as important in managing corporate responsibility activities and individual programmes as it is in any other part of company activity. This takes organisations back to the question of what is their fundamental contribution to the community.

3.1.2 *Social role*

Society consists of the multitude of human behaviour and interactions that comprise human life. Some human needs are met through the construction of commodities, while others are met by other actions that take place partly or wholly outside what is described as the economy (Langley and Mellor, 2002). Society expects many things of its corporate sector. Cannon (1994) illustrates in Table 1 how stakeholders' analysis is used by some firms to identify and classify these expectations.

Table 1. Stakeholders and their Expectations.

Stakeholder	Expectations	
	Primary	Secondary
Owners	Financial return	Added Value
Employees	Pay and security	Work satisfaction, training
Customers	Supply of goods and services	Quality
Creditors	Credit worthiness	Security
Suppliers	Payment	Long-term relationships
Community	Safety and security	Contribution to community
Government	Compliance	Improved competitiveness

Source: Adapted from Cannon 1994.

It is rare to find a perfect fit between the aims of industry and the aspirations of the community. Accordingly, the community will discriminate in favour of certain enterprises and against others who are not supporting their goals and needs.

3.1.3 *Grassroots development as a sustainability tool*

One approach to integrating economic and social development is "Grassroots Development" (GRD). This term is used by Stevens and Morris (2001) to describe an approach used to offer populations means of empowerment which is not usually found in their current economic and social system. Adopting this strategy could lead towards a people-centred society motivated by meeting everyone's human needs.

This approach seeks to involve the uninvolved stakeholders. They also refer to it as the direct involvement of local people working to improve their often marginalized situations. Furthermore, for a project to be successful, the following three important criteria should be applied: there must exist a comprehensive understanding of the system; there must be an understanding of the values of those involved; and all participants must have a sense that working together as a team is imperative. The challenge is what does this mean in modern urbanised society. Not everyone wants to be involved in every decision. Many people feel consulted to death. Furthermore, the challenge is not only to involve the unmotivated but to engage them so they take responsibility.

3.2 *Integrating business ethics*

This section discusses the issue of integrating business ethics in order to create an organisation that can foster human growth and development.

3.2.1 *Interplay between business and ethics*

Castro (1995) stated that most of the researchers in business ethics usually either have a background in business or a background in ethics which makes those two groups look at each other with suspicion and frustration. Kurland (1995) mentioned that researchers in business ethics have concentrated a lot on applying theoretical doctrines to business decisions. Those doctrines are inaccessible to most business managers and are often hostile to capitalism. Hence, there is a great need for a pragmatic approach to help business managers resolve common ethical dilemmas they encounter.

Sargent (2002) reported that although experts from the two disciplines of business and ethics, may agree on the name 'business ethics', they cannot agree that both disciplines belong to the same entity. Actually each side feels that they need to defend themselves from perceived attacks. Nevertheless, there have been lots of efforts to create integration models. Jones (1995) sought to combine ethics and economics through considering the benefits of all stakeholders involved. Donaldson and Dunfee (1994) developed an integrated social contracts theory aimed at combining the field of ethics and the field of business. Kohls and Buller (1994) tried to bridge that gap between those who hold to cultural relativism and those who take a more absolute stance in dealing with cross-cultural ethical conflicts.

3.2.2 *Integrity*
Woolfolk (1995) explained that integrity is the final stage of individual development in accordance with Erikson's theory of psychosocial development; this ties in with Maslow's final stage of hierarchy of needs (Maslow (1970)). The author noted that none of his to be chronically unsure about the difference between right and wrong in their actual living. Whether or not they could verbalize the matter, they rarely showed in their day-to-day living chaos, the confusion, and the inconsistency, or the conflict that are common in the average person's ethical dealings. He went on to say that this may be phrased also on the following terms: these individuals are strongly ethical, they have definite moral standards, they do right and they do not do wrong.

Fromm (1947) reported that human beings are interested in developing their potentialities to the full. Hence, in order to serve their best interests, it is required for individuals to be fully mature and integrated with their surrounding environment. That is a good starting point for psychological maturity and health. Integrity is closely linked to integrating ethics and the work expectations. Accordingly integrity is number one ingredient in successfully running an organisation not the possession of a law degree, knowledge of the intricacies of Sarbanes-Oxley Act 2002 (e.g. it requires chief executive and chief financial officers of publicly held corporations to certify financial statements), or even an understanding of accounting standards and practices. For example, organisation executives should make integrity a prerequisite whether in building a senior management team or hiring a new college graduate by asking candidates to describe ethical dilemmas they have faced and how they handled them.

3.2.3 *Ethics in the organisation*
Sargent (2002) mentioned that individual ethical development does not grow in isolation from the environment. Hence, support and nurture is required for the ethical development of business people. There has been lots of research and studies conducted to investigate how individuals and organizations influence and affect each other and how an organization can have an influence on the ethical behaviour of their members. Nevertheless, influence does not only depend on the organizational structure, design and ethical climate, but also on the individuals' character and role played in the organization.

Sonnenberg (1994) raised the issue of trust and its relation to ethical behaviour in organisation. The author reported that trust is the fabric that binds us together, creating an orderly civilized society from chaos and anarchy. It was noted that trust is not an abstract, theoretical idealistic goal forever beyond our reach. Trust-or the lack of it-is inherent in every action we take and affects everything we do. It is, in fact, the underpinning for all ethical behaviour.

For ethics and business to become integrated, individuals and organisations must analytically and practically incorporate their comments into the daily activities. Research indicates that ethical business does not normally compromise an organisation's profit. On the contrary, organizations who emphasize values beyond the bottom line often have a more robustly sustainable future as they are more profitable in the long run than organizations who state their goals in purely financial terms (Boleman and Deal, 1997). Moreover, violating corporate integrity and business ethics could cost the company their reputation, their customers and accordingly great financial loses.

A recent example in 2005 of a company running afoul of rules and regulations is the drug company Merck & Co's allegedly nondisclosure of heart attack and stroke risks with the arthritis drug Vioxx. There were news reports that accused Merck of failing to pull the drug of the market when the company suspected something was wrong. The drug was not pulled until the company's findings were made public (Britt, 2005). Accordingly organisations need to be aware of the following seven signs of ethical collapse mentioned by Jennings (2006): pressure to maintain numbers; fear and silence; young 'uns and a larger-than-life CEO; a weak board; conflicts; innovation like no other; and belief that goodness in some areas atones for wrongdoing in others.

Jennings (2006) proposes the following micro companywide antidotes to prevent ethical collapse: voluntary changes in financial-reporting practices – clarity, honesty and full disclosure; value-based decision making; and restoration of virtue standards (e.g. fairness, honesty, humility, integrity, loyalty, responsibility, trustworthiness and zeal).

4 MORE THAN GOOD TECHNOLOGY

Various researchers (Casey, 1999; Jermier et. al, 1994; Barker, 1993; Kunda 1992; Hirschhorn, 1988; and LaBier, 1986) have indicated that employees are psychically affected by the organisation's cultural practices that categorize production and work. This raises queries on the effects of contemporary institutionalized organization practices and deliberately designed and programmatically implemented organizational cultural reform packages that shift the corporate organizational problem of control and compliance to the intra-psychic domain of the employee.

If people and organisations are not ethically robust, Triple Bottom Line Reporting TBL reporting and related activities will only be cosmetic. For a business to be sustainable, it needs to constantly review itself. It should be shedding its old skin (i.e. its old methods and paradigms) and replacing them with more sustainable methods and paradigms. For example the RepuTex Social Responsibility Rating System was introduced in 2003 to provide an independent measure of social responsibility performance (Laurel, 2004). RepuTex used four key indicators to measure an organisation's social performance: corporate governance: transparency, risk reporting and management and ethics; workplace practices: occupational health and safety, management systems, workplace culture and diversity; social impact: of products and services, policies and practices; and environmental impact: of operations, policies, procedures products and services.

The RepuTex rating system is different from other indexes or ethical investment appraisal systems because it can rate any type of organisation (e.g. government, private, NGOs, publicly listed or not). Furthermore, the system is community focused i.e. the assessment criteria in each category are made available for public comment. During the time of the company being rated by RepuTex, the community stakeholders, companies, expert bodies and interested parties may provide feedback and input (Laurel, 2004).

Another rising contribution in the field is the recent work conducted through the teaming up between IABC organisation and the International Organisation for Standardization (ISO) for the development of a future ISO 26000 Standard to provide organisations guidance for implementing social responsibility. The seven core social responsibility issues covered in the standard are environment, human rights, labour rights, organisational governance, fair operating practices, consumer issues, and community development. ISO 26000 is not intended to be a management system standard or to be used for certification. The final standard is expected to be published in 2009 (Anonymous, 2007).

Table 2 provides preliminary suggestions for building a sustainable business environment. Organisations can pursue several approaches to enhancing ethical depth. We identify five different approaches which have something to contribute. Normally people learn from problem centred cases where they clearly see there is an ethical problem/dilemma and get some understanding as to the causes. Furthermore, adverse events do not just occur i.e. they exist within a context. It is useful to explore the forces impacting on organisations, the environment and society. Analysis of people, problems and forces is unlikely to be in depth if it ignores the historical background and the evolution of the topic. If staff development is to be useful it needs to go on to facilitate the development of solutions and a culture that fosters transparency of communication and reporting. The challenge of this development is not only what should be covered but how to do it. Evidence suggests there is a strong case for case analysis of historic and emerging problems along with interaction with scholarly literature.

5 CONCLUDING REMARKS

Building sustainable firms and organisations requires a commitment to people's development. This development has a strong component. However, it has to be addressed gently as many citizens are

Table 2. Developing Ethics at Work.

Approaches	Methods
Problem centred ethics	Case studies of 'infamous' problems e.g. Bhopal gas tragedy
Analysis of 'forces' impacting on individuals, organisations and communities	Environmental, sociological and economic analysis of particular issues e.g. water quality studies
Philosophical and religious ethics	Find ways to familiarity with literature e.g. Bible, Kant, Singer
Solution centred ethics	Practical steps to achieve certain outcomes e.g. reducing hazardous waste pollution in New Zealand
Revitalising	Reading and renewal

anxious about even the use of the word 'ethics' unless some one has an obligation to them. Staff development programs can only be successful if organisations have a clear sense of their place in society to enable them from fostering a mindset for a sustainable ethical culture. Furthermore, this requires a strong self-identity and an understanding of stakeholder expectations. The development of such understanding with the organisation and community facilitates the building of trust and integrity which are fundamental to sustainability. In pursuing this agenda, there is a strong case for organisations to be proactive in developing their people by helping them analyse problem cases, analyse forces impacting on the organisation and society, become familiar with core ethical literature, develop solutions to ethical problems and revitalise the organisation. Only with those proposed ingredients would an organisation be on the road to building a sustainable mindset.

REFERENCES

Anonymous (July-August 2007). *Communication World.* http://commons.iabc.com/advocacy/ Retrieved on 13 December 2007.

Assadourian, E. (March-April 2006). Next Steps for the Business Community. *World Watch: Vision for a Sustainable World.* 19: 2, pp16–20.

Barker, J.R. (1993). Tightening the Iron Cage: concerti e control in self-managing items. *Administrative Science Quarterly*, 38: 408–437.

Bell, D. (1976). *The Culture Contradictions of Capitalism.* New York: Basic Books.

Bolman, L.G. and Deal, T.E. (1997). *Reframing organizations: Artistry, choice and leadership.* 2nd edition. San Francisco: Jossey-Bass Publishers.

Britt, P. (April 2005). Good Ethics Equals Good Business. *Customer Relationship Management.* p14. Retrieved on 13 December 2007 from (www.destination CRM.com).

Buchholz, R.A. (2000). *Toward a new ethic of production and consumption: A response to Sagoff.* In Reichart, J. and Werhane, P.H. (eds.). *Environmental challenges to business.* The Ruffin Series No. 2. Bowling Green: Society of Business Ethics.

Cannon, T. (1994). *Corporate responsibility: A textbook on business ethics, governance, environment: roles and responsibilities.* London: Pitman.

Casey, C. (1999). *New Organizational Cultures and Ethical Employment Practice.* In Wrhane, P.H. & Singer, A.E. (Eds.). *Business Ethics Theory and Practice: Contributions from Asia and New Zealand.* London: Kluwer Academic Publishers.

Castro, B. (1995). "Business ethics: Some observations on the relationship between training, affiliation and disciplinary drift." *Journal of Business Ethics.* 14(9), 781–786.

Clutterbuck, D. (1992). *Actions Speak Louder.* 2nd Edition. London: Kogan Page.

Costanza, R. (1992). *Ecological economics.* New York: Columbia University Press.

Donaldson, T. and Dunfee, T. (1994). "Toward a unified conception of business ethics: Integrative social contracts theory." *Academy of Management Review.* 19(2), 252–284.

Etzioni, A. (1988). *The moral dimension.* New York: The Free Press.

Fromm, E. (1947). *Man for himself: An inquiry into the psychology of ethics.* New York: Fawcett Premier.

Giddings, B., Hopwood, B. and O'Brien, G. (2002). "Environment, economy and society: Fitting them together into sustainable development". *Sustainable Development.* 10, 187–196. (www.interscience.wiley.com) Retrieved on 4 March 2007.

Hall, C.M. (1998). *Historical antecedents of sustainable development and ecotourism: new labels on old bottles?* In Hall, C.M. & Lew (editors), A.A. *Sustainable Tourism: A geographical perspective.* New York: Wesley Longman Limited.
Hirschhorn, L. (1988). *The Workplace Within: Psychodynamics of Organizational Life.* Massachusetts: MIT Press.
Howitt, R. (2001). *Rethinking Resource Management: Justice, sustainability and indigenous peoples.* London: Routledge.
Jennings, M.M. J.D. (2006). *The Seven Signs of Ethical Collapse: How to Spot Moral Meltdowns in Companies Before Its Too Late.* New York: St. Martin's Press.
Jermier, J., Knights, D. And Nord, W. (Eds.) (1994). *Resistance and Power in Organizations.* London: Routledge.
Jones, T. (1995). "Instrumental stakeholder theory: A synthesis of ethics and economics." *Academy of Management Review.* 20(2), 404–441.
Kluckhohn, C. (1958). "Have There Been Discernible Shifts in American Values During the Past Generation." *The American Style: Essays in Value and Performance.* E.E. Morrison (ed.). New York: Harper and Bros: 201–224.
Kohls, J. & Buller, P. (1994). Resolving cross-cultural conflict: Exploring alternative strategies. *Journal of Business Ethics.* 13, 31–38.
Kunda, G. (1992). *Engineering Culture: Control and Commitment in a High-Tech Corporation.* Philadelphia: Temple University Press.
Kurland, N. (1995). "Ethics, incentives, and conflicts of interest: A practical solution." *Journal of Business Ethics.* 14(6), 465–475.
LaBier, D. (1986). *Modern Madness: The Hidden Link Between Work and Emotional Life.* New York: Simon and Schuster.
Langley, P. and Mellor, M. (2002). "Economy, sustainability and sites of transformative space." *New Political Economy.* 7(1): 49–66.
Lasch, C. (1978). *The culture of narcissism American life in an age of diminishing expectations.* New York: Norton.
Laurel, G. (Aug. 2004). Rating corporate social responsibility. *Bussinessdate.* 12:4, pp 5–7.
Maslow, A.H. (1970). *Motivation and personality.* 2nd edition. New York: The Viking Press.
Monks, R.A. and Minow, N. (2004). *Corporate Governance.* 3rd Edition. USA: Blackwell Publishing.
Sonnenberg, F. (1994). "Trust me…trust me not." *Journal of Business Strategy.* 15(1), 4–16.
Sargent, T. (2002). "Towards integration in applied business ethics: The contribution of humanistic psychology." *Electronic Journal of Business Ethics and Organization Studies.* 7(1). (http://ejbo.jyu.fi/index.cgi?page=articles /0401_3) Retrieved on 4 March 2007.
Stewart, D. (1996). *"Business Ethics".* New York: The McGraw-Hill Companies, Inc.
Stevens, K. and Morris, J. (2001). "Struggling toward sustainability: Considering grassroots development." *Sustainable Development.* 9, 149–164. (www.interscience.wiley.com) Retrieved on 4 March 2007.
Task Force on Corporate Social Performance. (1980). *Report on Business and Society: Strategies for the 1980s.* Washington: US Department of Commerce.
Woolfolk, A.E. (1995). *Educational psychology.* 6th edition. Boston: Allyn and Bacon.
Yankelovich, D. (1981). *New rules: The search for self-fulfilment in a world turned upside down.* New York: Random House.

Role of the Arts and Environmental Management in the Secular West and in Islam

Jacqueline Taylor Basker
New York Institute of Technology, Amman, Jordan

SUMMARY: This chapter discusses the role the arts play in the sustainable management of the environment and its relationship to society. It examines the parallel attitudes in both the secular west and in Islam towards environmental stewardship and its influence on the work of artists. Artists are inevitably linked to the environment because their materials and tools come from nature. Their use of natural materials and the development of the technology of art has had a profound impact on human history and technological advancement. Human technologies paralleled the explorations and development of the artists and craftsmen, from the caves, through the iron and bronze ages, to present technologies. The natural world has been the fundamental source for art materials. The need to respect, nurture and preserve nature is inherent in the tradition of making art. Art materials themselves however, were often toxic for artists, and they were one of the earliest professions to examine the environmental and health hazards of their materials and working conditions. During the current ecological crisis, much art addresses the problem of the environment. New genres of art have emerged that confront the issues of environmental waste, hazards, and the destruction of nature. Artists have created forms of art using the natural environment itself as both subject and material for artworks. The secular west and Islam share a sense of ethical responsibility for the environment, and its impact on health and quality of life for the global human community. Their artists express these concerns in their art.

1 INTRODUCTION

Using natural materials to create tools by flaking and chipping flint pebbles into knives and pebbles was done over 2 million years ago in Africa by "homo habilis" (Handy Humans). About 100,000 years ago, Neanderthals made stone tools and carefully buried their dead, painted their bodies, adorning them with beads, and burying them with funerary objects. Stokstad (2005) noted that Cro-Magnons engraved, carved, drew and used ochers (iron ores ground into pigment) to paint. Prehistoric humans, hunter-gatherers, used all parts of slain animals, bone, fur, meat and sinew, for food, shelter, tools, weapons or art.

There has always been a close relationship between artists and the environment, as a source of raw materials to create art, or as inspiration for art. When nature and the environment were threatened by the excesses of the industrial revolution, it was artists whose work often challenged the environmental destruction, and created artwork to enshrine nature as sacred. It was in the artists' interest to protect nature and the environment, and to manage its resources wisely.

It is interesting to compare attitudes towards the environment, and the role of art in the secular west with Islamic tradition, especially in view of the recent controversy over Islamic law in Britain. Rowan Williams, (2008) the Archbishop of Canterbury, presented the foundation lecture at the Royal Courts of Justice, "Civil and Religious Law in England: A Religious Perspective," He noted that the series of lectures signalled the existence of what was very widely felt to be a growing challenge in society, the presence of communities which, while no less law-abiding than the rest of the population, related to something other than the British legal system alone. He suggested that these issues were not peculiar to Islam, and that there were many other religious authorities, particularly the Church of England, whose law was recognized with considerable independence.

He pointed out that there were large questions in the background about what we understood by and expected from law especially in a secular social environment. However, he concentrated on controversial issues relating to Islamic Law. Ironically, Rowan preached in Westminster Abbey whose recently restored 13th century Retable decorative style demonstrates a tradition of Islamic influence in English culture. The National Gallery of Art (2005) pointed out that this Retable, one of the most important Northern European panel paintings, carved in the French influenced Gothc style, is enhanced by a number of exotic motifs, Islamic in origin.

In the past, medieval art, architecture and design demonstrated the convergence of Islamic and western cultures. Today art reveals parallel attitudes that occur between the secular west's environmental consciousness and Islam's religious injunction to protect the environment. The ecologically conscious art produced recently in the United Kingdom and the United States, and in the Middle East visualizes these attitudes. It is art concerned with management of the resources of the environment and its impact on human communities. This art celebrates the land and confronts the threats to nature and to the environment.

2 THE ENVIRONMENT AND ART IN THE WEST: UNITED KINGDOM

David F. Peat (1997) asserted that the British traditionally have had an intense relationship with nature and their green island. He noted that the recently restored Wilton Diptych crafted in 1395, presented Richard the Lion Heart to the Virgin and Child below the silver knob on the top of the English flag showing England as a minute green island in a silver sea. It is this land where the Industrial Revolution was birthed, that perhaps has produced the strongest art that addresses nature and issues of the environment, strongly conscious of the impact on human health and well-being.

The drastic changes in rural England provoked a strong reaction from Romantic artists and writers. During the years Karl Marx was in the British Library writing his diatribes against rampant capitalist exploitation of the working classes, artists such as John Constable and Joseph Mallord Turner were producing canvases glorifying nature and the countryside which seemed under attack by capitalism and industrialization. Constable's canvases were tributes to the memories of rural England, whose lifestyle was endangered by the Industrial Revolution. These landscapes were not considered important in his own time, but later became the most famous of English paintings, as the land became more urbanized. Turner's work was in the mystical British tradition of finding the mystery and sacred in nature, seen in the work and writing of William Blake, and were also not appreciated, but ridiculed by the art academies. Yet nature was not pristine in Britain. Its countryside had long been manipulated, changed and denuded by the human presence since Neolithic times through deforestation, over-farming, grazing and enclosures.

Charoltte Bronte's windy heath in Wuthering Heights was a result of soil loss due to over-farming. Yet the British had an image of a former idyllic state of the British Isles when people lived together in small communities in harmony with nature. British literature is replete with the images of nature as a refuge for town-dwellers, from the Forest of Aden in Shakespeare's As You Like It to the adventures of Robin Hood to the science fiction of John Wyndam and John Christopher where civilization is destroyed by ecological disaster. This longing for the country and Camelot became more than just a subject for art and literature, but produced concrete community experiments in the 19th and 20th century especially in the Arts and Crafts movement of William Morris and Eric Gill.

Contemporary communities such as the Finhom Community (2008) in Scotland preserve that vision and define themselves as a spiritual society, ecovillage and an international centre for holistic education, helping to unfold a new human consciousness and create a positive and sustainable future. Peat (1997) claims that this shared Map in the Head of the British peoples is responsible for much of the contemporary Art and Ecology movement. Britain has long traditions that support environmental consciousness according to Peat.

Art within the environment was always an important part of English life, trimming hedge rows, decorating barges, whitewashing cottages, carving walking sticks, the annual cleaning of neolithic sites (such as the White Horse at Uffington), beating the bounds of the village, the celebration of seasonal rituals from Maypole Dancing in Spring to village bonfires at the onset of winter, cottage flower gardens, allotments for city dwellers, fruit, vegetables and baked goods exhibited at village fairs and, to top it all, the formal art of landscape painting. (Peat, 1997).

Thus the contemporary label of environmental artist does not emerge out of a vacuum, but from centuries of the British's deep sense of concern for the land. Artman.Net art directory (2008) defines an environmental artist as a person who uses images as a way to promote environmental awareness. This type of art is the creative face of the populist groups in the U.K. who group together to protest changes to the land by roads and modern construction to preserve the land, or even toads. BBC News (2006) reported that Toad Road Safety Patrols had been set up to help toads cross roads safely during their mating season, due to a drastic drop in the frog and toad population. Wildlife experts from Warwickshire Nature Trust called for volunteers for toad road patrols and urged people to install more ponds for them.

Peat (1997) reminds us that for over a century the English have been arrested or jailed to preserve their rights to walk on traditional roads and paths with access to common land, in the tradition of the movements of common people as the Luddites, the Peasants Revolt, the Levelers, the Diggers and Robin Hood and his Merry Men who took refuge walking and hiding in Sherwood Forest. Environmental artists such as Richard Long in this tradition of British walking created art that was a formal and holistic description of the real space and experience of landscape and its most elemental materials. Beginning his environmental art in 1967, five years after Rachel Carson's *Silent Spring* was published, Long stated that his intention was to make a new art which was also a new way of walking, walking as art. He explained that each walk followed his own unique route that realized a particular idea. He then arranged stones and brought back photographs and artifacts to the gallery.

Long's work, however, was criticized as being too connected to the art establishment, and that artists of his generation were still manipulating the environment in their art, as industrialization exploits natural materials. Artists began to examine art in context of community, as an expression of community cohesion in harmony with the environment. The radical German artist Joseph Beuys (1984) challenged the role of art in a consumer society, as something to trade, decorate walls or provide momentary uplift. He felt art was a genuinely human medium for revolutionary change in the sense of completing the transformation from a sick world to a healthy one. He developed the idea of Social Sculpture, influential in Britain.

In 2005 the Tate Museum sponsored a sold-out Social Sculpture Research Seminars, in collaboration with the Social Sculpture Research Unit at Oxford Brookes University, which investigated Beuys' work and ideas. The Tate Museum (2005) stated its purpose was to create new forms of socially engaged practice in the arts. Peat (1997) noted that using Beuys' definition of Social Sculpture, environmental art is everywhere, from demonstrations against the building of invasive new highways that damage the environment, to work of local craftspeople. The artists' work was to help bring about new communities, whose health and well-being would be protected, living in harmony with nature and the environment. He maintained that when artists become involved in a river cleaning project or working with living material they enter into a more direct communion with the spirit of the landscape.

3 AN UNUSUAL PARTNERSHIP: ART AND THE ENVIRONMENT IN ISRAEL AND PALESTINE

Beuys (1984) might see the unusual joint demonstrations by Israeli environmentalists and Palestinians against the Israeli separation barrier wall in Wadi Qelt as a form of art. The environmentalists oppose the barrier because they want the land unspoiled and uncut by a 7.5 mile of fence that will destroy the canyon and cut into the rocky Judean Desert near the Dead Sea. This land's destiny in a future Palestinian state is still not resolved, and the Palestinians do not want it divided. Besides the biblical history of this area where prophets like John the Baptist took refuge in its caves, and shared the springs with leopards, wolves and eagles, it is now one of the few open spaces left where wild animals still roam freely. Laurie Copens in The Boston Globe (Copens, 2008) quoted Imad Atrash, executive director of the Palestine Wildlife Society, who admitted that it is difficult for Palestinians to worry about wild animals and the environment when they are struggling for survival, but he appealled to their patriotism that they are going to protect nature for the Palestinians. Both the wild animals and the Arab farmers need the water.

Avraham Shaked, regional coordinator in Jerusalem for Israel's Society for Protection of Nature, lamented that this was the first time in history in which land that had been one land was taken and divided with a physical barrier that was meant to prevent the passage of anything larger than a rodent This damaged the connectivity that is one of the elements that many species rely on. The Israeli-Palestinian branch of Friends of the Earth brought a halt to work on the barrier in Wadi Fukin in the central West Bank, successful in the argument to the Israel Supreme Court, convincing them that the natural springs would be destroyed. Israeli settler and environmental activist Roee Simon, watching the gazelles graze from her Jewish settlement Kfar Adumim overlooking Wadi Qelt, agreed about the barrier destroying the existence of the wild animals, which need the water for survival.

Artists have actively united as well on this issue. The exhibit "Three Cities Against the Wall" (Laurson, 2005) brought together 56 painters, sculptors, filmmakers and graphic artists from Ramallah, Palestine, Tel Aviv, Israel and New York City. Artworks examining the human and environmental consequences of the separation barrier were exhibited in the three cities simultaneously during one month. The extraordinary event was organized through ABC No Rio, a community center for the arts on the Lower East Side, by a committee of artists and activists. In Ramallah, Tayseer Barakat, founder of the League of Palestinian Artists and curator of the Gallery Barakat, and Suliman Mansour, director of the Wasiti Art Center in Jerusalem, organized the exhibition. In Tel Aviv, the project was organized by a group of artists and activists associated with the Israeli Coalition Against the Wall, Ta'ayush, and Anarchists Against the Wall.

These artists have built relationships and networks that continue to work for peace and understanding, as well as for protection of the environment. However, environmental issues in the Middle East often receive less attention than in other areas of the world, in the middle of volatile political and economic issues. Royce J. Bitzer (2004) blames the problem on the lack of scientific information on the biodiversity of the region, and problems in sharing important research between countries in the area. This may also account for the small number of artists in the Middle East whose art reflects environmental issues; the environment surfaces usually in conjunction with politics as a subject for art. Surveying Jordanian art however the work of Jordanian artist Khaled Atieh, a deaf-mute carpenter, formerly from Palestine uses nature as the subject for much of his art.

When Princess Rania Al-Abdullah of Jordan formed a non-profit NGO to empower society, especially women and children, and in turn, to improve the quality of life to secure a better future for all Jordanians, she named it after the Jordan River (Al-Abdullah, 1995). The Jordan River Foundation has programs that promote local socio-economic productivity of communities across the Kingdom. Yet the water of the Jordan River and the aquifer of Jordan are in serious trouble, documented by a USAID Economic Development Program publication (USAID, 2008). This report concluded that the water crisis is imminent since current renewable water supply only meets about half of total water consumption in the Kingdom. Attendant with significant reductions in aquifer water levels is risk of permanent damage to the aquifer. The need for more public awareness and government intervention is urgent. The activity of artists in this area is critical as both human health and the environment is at great risk. USAID maintains that the first step in solving this critical problem is creating awareness among the Jordanian public and decision makers to help create behavior change. Artists can be vital in this need to disseminate the message at all levels of society. USAID warns that to continue growing its economy in a sustainable manner, Jordan needs to act now to optimize utilization of its water resources.

The Islamic concept of *taweed,* the unity of Faith, provides a theoretical framework for an ethical response to the problem since this assertion of the oneness of God implies that the ultimate goal of all human effort, taking care of the environment as well as art, becomes an issue of ethics. (Manzoor, 2003) There is a sound epistemological framework in Islam for a strong environmental conscience. Mustafa Abu-Sway (1998) maintains that within the Islamic world-view, a structure for a positive relationship with the environment emerges from an act of faith in harmony with our essential role as humans on earth. He believes this relationship belongs within the field of Islamic jurisprudence known as *Fiqh,* compatible with the aims (*maqasid*) of the *Shari'ah.* Abu-Sway prefers this approach, rather than a philosophical approach to environmental awareness, since some Islamic scholars regard philosophy borrowed from the Western world-view. This approach places the relationship between humans and environment in three categories: Vicegerentcy (*Khilafah*), Subjection (*Taskhir*) and Inhabitation (*T'mar*). He explains each category using sections of the *Qur'an* and the *Sunnah* to develop his case that taking care of the Environment is an act of faith and worship. While this protects it from the greed of humanity, its latitude permits sustainable development. Humans as vicegerents (*Qur'an, 6:165*) are trustees of the earth, paralleling the Biblical notion that humans are stewards of the earth. (*Genesis 1:26)* People are not permitted to cause corruption of any form in the environment, and they have great responsibility for its preservation. This responsibility is a type of test for humans to see how they would behave in God's creation (*Qur'an, 10:14*). Subjection (*Taskhir*) reminds humans that the earth is available for their use, but without abuse or misuse (*Qur'an, 16:12*). Inhabitation (*I'mar*) cautions against excesses or deficiencies in the use of the human habitat, and the need to establish a balanced way of life, while advancing activities for a prosperous life on earth. Islamic scholar Yusuf Al-Qaradawi even interprets this to be a prohibition *(al-tahrim)* against smoking (Abu-Sway, 1998). Nature (i.e. rivers, plants, birds) is extremely important because it contains the signs that point to the greatness of the Creator. (*Qur'an, 45:3–5*) For Islamic artists, the choice of nature as subject has more profound meaning than just aesthetic beauty, but is part of a world paradigm that views the universe as a book to be read, as the *Qur-an,* and reveals the beautiful forms of *dhikr,* signs open to the transparent heart. There are detailed sections in the *Qur'an* and the *Sunnah* regarding protection of human beings, animals, plants, land, water and air (Abu-Sway, 1998).

Iran has held the 15th Environmental Art Festival: Dream of Peace in the Persian Gulf in December, 2007, on Hormoz Island. This festival had no sponsor, and the young artists paid all their own expenses to create art works and events to respond to the environmental crisis in the Gulf. They wrote that ecological art is the art of the future. Their statement reviewed the serious damage to the environment of the Gulf by Saddam Hussein in 1981, then by the invasion of Kuwait in 1991, followed by the invasion of Iraq by the USA and its allies in 2003. Their art protested the militarization of the Gulf, and used collaborative environmental art to make their point. (Nadalian, 2008).

4 ART AND THE ENVIRONMENT IN THE UNITED STATES

The exhibit "Three Cities Against the Wall" (Laurson, 2005) is an example of what the American artist and art-critic Suzi Gablik describes in her essay "Conversations before the End of Time" (Gablik, 1999) where she noted that artists were expanding beyond the traditional gallery, and joining the community, addressing broad issues, and inviting dialogue between differing views. The political involvement of artists in this century, from the propaganda art of WWI and WWII, the anti-war movement of the Vietnam era, through the post 9/11 world and the Iraq war has continued this genre of art in the West from the Romantic artists glorifying war like Delacroix, or producing anti-war art like the etchings of Goya (Stokstad, 2005).

Modern and post-Modern American art reflects a broad diversity in relating the arts to the environment and to the community. The development of ecological sculpture that creates or changes the environment was a significant challenge to art that was limited to the Gallery or the Museum. Artists Jane Frank, Tony Smith and David Smith pioneered artwork that tried to break down the separation of art from life. Louise Nevelson developed large sculptures filled with objects, many she would find in New York and recycle in her assemblages. Another type of Environmental Sculpture emerged when sculpture was created for a specific site, designed to unify with its environment, and could not be moved to a Gallery or Museum. Beth Galston, (2008) contemporary sculptor, maintained that the site the acts as a catalyst to the creative process for this type of sculpture. Work by minimalist artist Richard Serra is included in this category. Another approach is practiced by

Andrew Rogers and Alan Sonfist, known as "Land Art" or "Earth Art." These pioneers worked in the 1960's and 70's, and the most famous of them, Robert Smithson, created his "Spiral Jetty" of rock, earth and algae protruding into the Great Salt Lake in Utah, in 1970. Their work was designed to change over time, part of the natural processes. However, he also produced land art that could exist in a gallery space.

Much of Land Art emerged from the attempt to separate from the art establishment but ended up being funded by wealthy patrons and foundations, and became part of establishment Public Sculptures. However, its concentration of natural forms and materials, in relation to its environment and community realized Joseph Beuys' concept of "Social Sculpture." Yet after the untimely death of Smithson in 1973, and the years of an anti-environmental administration and lack of funding for experimental art, American environmental art fell out of fashion. It now seems to have revived with the turn of the century. The work of James Turrell pushed Land Art to another level; his "Roden Crater" in Arizona, turned a volcanic crater into a naked-eye observatory to view the sky, enclosing the viewer to control their perception of light, to be ready in 2011. (Cook, 2001).

Art as a media for community and environment has expanded into every aspect of human life, and continually pushes the boundaries of its definitions. In 2007 three exhibitions were mounted simultaneously in the adjacent states of Pennsylvania and Ohio with unique approaches to ecological awareness: Global Anxieties in the College of Wooster Art Museum, Ohio, 6 Million Perps Held Hostage! Artists Addressing Global Warming at the Andy Warhol Museum, Pittsburgh, PA, and Beyond Green: Toward a Sustainable Art at the Cincinnati Contemporary Arts Center, Ohio. All three exhibitions were supplemented with educational events, and political activism. Projects were developed to explore solutions for sustainable living despite industrial and urban environmental damage. (Klein, 2007) Through post-industrial landscape photography Global Anxieties presented a toxic brew of images of scarred land, more like the nightmare landscapes of Hieronymus Bosch than Ansel Adams. Beyond Green presented solutions to the consumer society and hosted the German collective *WochenKlauser,* who formed Material Exchange with local students, an upcycling industry that rebuilt furniture for a homeless shelter and clothing pantry out of discarded materials, with an emphasis on design in the Bauhaus tradition. 6 Million Perps Held Hostage! was the most political of the exhibits including The Yes Men and their documentation of SurvivavaBall, 2006 where they impersonated Haliburton employees, Bush and Cheney in highly successful satirical performance art to the tune of Bobby Pickett's "The Monster Mash" (Klein, 2007). These diverse exhibits reflected the renewed focus of American artists to use their creativity to address the growing environmental crisis.

Artists have also lobbied to obtain government protection against hazardous art materials. Joy Turner Luke (2001) surveyd the history and current situation of health issues for artists, the labeling of materials and the struggle to obtain a national consensus on the labeling of art and craft materials for chronic health hazards. The American Society of Testing and Materials established a task group and eventually produced a standard called ASTM D4236, Practice for Labeling Art Materials for Chronic Health Hazards that finally became law in 1997.

Rensselaer Polytechnic Institute presented a series to produce collaborative work with interdisciplinary electronic artists addressing issues of environment and local communities and create connections between the communities of Troy, NY in the US, Iraq, Malaysia, Indonesia and England. Led by theorist Brian Holmes, (Holmes, 2008) it addressed questions as how the initial artistic act consists in establishing the environment and setting the parameters for a larger inquiry. Holmes stretched this to ask how this inquiry becomes expressive, multiple, and overflows the initial frame and opens up unexpected possibilities. Artists will ponder how these new kinds of practices produce a new definition of art as a mobile laboratory and experimental theatre for the investigation of social and cultural change.

Al Gore's film "An Inconvenient Truth" (2006) and Joon-ho Bong's horror film "The Host" (2006) challenged the irresponsibility and reluctance of power structures to recognize and confront these issues. The English artist Anish Kapoor used his art to create sacred spaces out of forms that interact with light and the environment. His large mirror sculpture "Sky Mirror" reflected the sky and surroundings, as an objection of contemplation and meditation on the immensity of creation.

Kapoor told a BBC reporter that the artist looks for a content that is on the face of it abstract, but at a deeper level symbolic, and that content is necessarily philosophical and religious (Tusa, 2008).

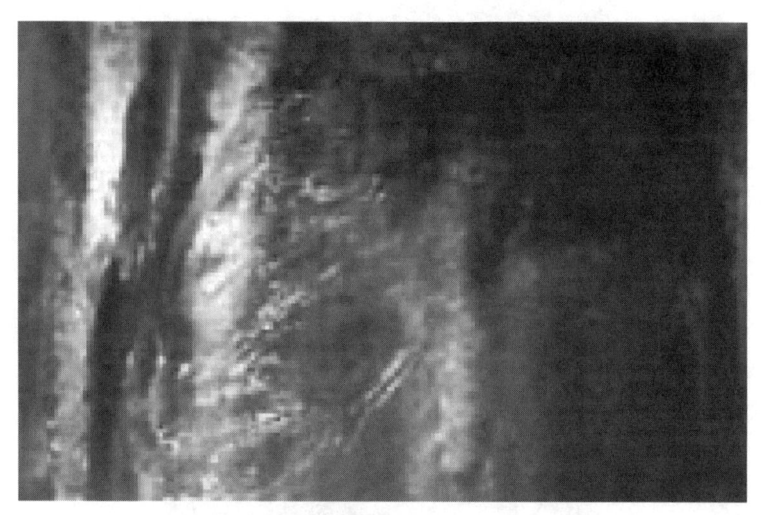

Figure 1. Mohammad Qaitouqa, "Jordan" pigments, paste, acrylic on wood, 208 × 130 cm. Born and working in Jordan, this artist's works are inspired from the colors and textures of the local environment in an attempt to capture and preserve the land and its spirit.
Source: Darat al Funum, Khalid Shoman Foundation, Jordan.

Figure 2. John Constable, "The Haywain," 1821, oil on canvas, 130.2 × 185.4.
Source: The National Gallery, London.

Figure 3. Richard Long, "A Line In Ireland", Walk Sculpture, 1974.
Origin: Richard Long Official Website.

Figure 4. Baptism Site, River Jordan, Left Israel, Right Jordan. Wildlife move freely in this rich nature preserve without the separation barrier wall.
Source: J. Taylor Basker, January, 2008.

Figure 5. Khaled Atieh, "Spirit of Nature" oil on canvas, 90 × 120, Jordan.
Source: Jordan Artists, http://www.jordan-art.nl/english/popup/khaledgallery.html.

Figure 6. Dream of Peace in Persian Gulf, Collaborative Artwork, December 2007.
Source: Google Earth.

5 CONCLUDING REMARKS

The looming ecological global tragedy requires radically new forms of art to confront the complacency and paralysis of much of humanity which remains oblivious, indifferent or helpless. Governments, business and individuals have not yet faced up to the enormous changes that must

Figure 7. "Three Cities Against the Wall," exhibit, New York, Ramallah, Tel Aviv, 2005. *Source*: http://www.commondreams.org/news2005/0718-03.htm

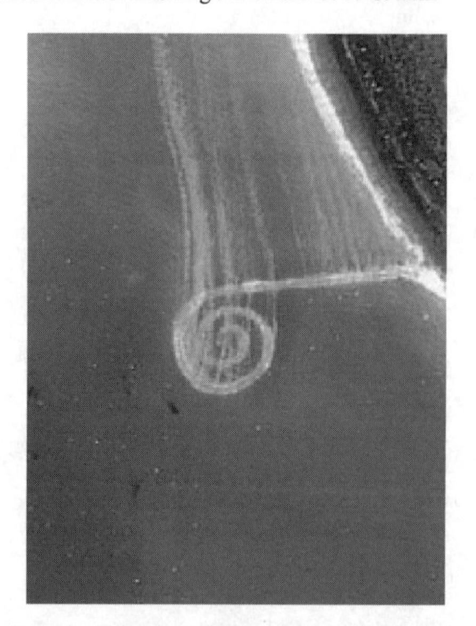

Figure 8. Robert Smithson, Spiral Jetty, Salt Lake, Rozel Point, Utah, 1970. *Source*: IKONOS satellite image, 11/11/02.

occur in society to prevent an appalling ecological calamity that will have an apocalyptic effect on human society. Artists around the world have addressed these problems with a broad variety of art forms. The sense that the earth is sacred occurs in both the secular west, and in Islam. This shared sensibility is reflected in the growth of environmental consciousness in both civil and religious law, as well as in art.

REFERENCES

Abu-Sway, Mustafa. 2008. Towards An Islamic Jurisprudence of the Environment, *Fiqh al-Bi'ah fil-Islam*. Lecture, Belfast Mosque, February.

Al-Abdulla, Rania. 1995. About Us, Vision, Mission Statement, The Jordan River Foundation, The Jordan River Foundation, http://www.jordanriver.jo/about.asp
Artmam.Net. 2008. Environmental Artist, http://artmam.net/Environmental_Artist.htm
BBC News, 2006. Call for Road Safety Patrols, http://news.bbc.co.uk/2/hi/uk_news/england/coventry_warwickshire/4881520.stm
Beuys, Joseph. 1984. *Quartetto,* exhibition catalog, Arnoldo Mondadori Editore,, Milano, p. 106.
Bible,. 1984. *Genesis 1:26*. . .New International Version, International Bible Society.
Cook, Earl. 2001. Roden Crater, The Art & Vision of James Turrell, http://www.lasersol.com/art/turrell/Roden Menuhtml
Copens, Laurie. 2008. Opposing Barrier They Find Common Ground. The Boston Globe,
Associated Press, March 6, http://www.boston. com/news/world/middleeast/articles/2008/03/06/opposing_barrier_the y_find_common_ground/
Darat al Funum Foundation. 2008. www.daratalfunun.org/main/resourc/exhibit/qaitouq/qaitouk.html
Findhorn Community. 2008. Vision, http://www.findhorn.org/whatwedo/vision/vision.php
Gablik, Suzi. 1999 Conversations before the End of Time: Dialogues on Art, Life, and Spiritual Renewal. *The Journal of Aesthetics and Art Criticism*, Vol. 57, No. 4, Autumn, 486–487.
Galston, Beth. Definition of Environmental Sculpture, http://www.bethgalston.com/env_sculpt.html.
Holmes, Brian. 2008. Art and Islam: Electronic Representations and Local Communities, lecture at iEAR Presents! Conference, Rensselaer Polytechnic Institute, Troy, N.Y., February, 2008.
Klwin, Jeannie. 2007. Curating Environmentalism for Post-Industrial America, *Art Papers*, 31 no.5, 19–21.
Laursen, Eric. 2005. ABCNio, Press Release. Three Cities Against the Wall, exhibition in New York, USA, Tel-Aviv, Israel and Ramalla, Palestine. In New York, http://www.commondreams.org/news2005/0718-03.htm.
Long, Richard. 2008. Official Web Site, http://www.richardlong.org/
Luke, Joy Turner, 2001. An Update on Art-material health and quality standards, *American Artist*, v. 65 no. 705, February, 18–22.
Manzoor, S, Parvez. 2003. Environment and Values: The Islamic Perspective, http://www.readingislam.com/servlet/ Satellite?c=Article_C&cid=1154235107723&pagename=Zone-English- Discover_Islam%2FDIEL ayout
Nadalian, Ahmad. 2008. Dream of Peace in Persian Gulf: 15th Environmental Art Festival in Iran at the Persian Gulf, January. Payvand's Iran News, http://www.payvand.com/news/08/feb/1111.html.
National Gallery of Art. 2008. Home Collection, "The Hay Wain, 1821, Constable, John, 1776–1837.
National Gallery of Art, Press Release. 2005. The Westminster Retable: England's Oldest Altarpiece, London.
Peat, David. F. 1997. Art and the Environment, http://www.fdavid**peat** .com/bibliography/essays/**art**brit.htm
Qur'an. Trans. By Dr. Muhammad Taqi-ud-Din al-Hilali and Dr. Muhammad Muhsin Khan, King Fahd Complex for the Printing of the Holy Qur-an, Madinah, K.S.A., 1427 A.H. Qur'an, 45:3–5
Tate Museum,.2005.Social Sculpture Research Seminars, 4 April. Tate Modern website: http://www.tate.org.uk/modern/eventseducation/coursesworkshops/socialsculpture2637.htm
Tusa, John. 2008. Transcript of the John Tusa Interview with the sculptor Anish Kapoor, BBC Radio, 8 March http://www.bbc.co.uk/radio3/johntusainterview/kapoor_transcript.shtml
Stokstad, Marilyn. 2005. *Art History*, Pearson Prentice Hall, Upper Saddle River, NJ, 2–3.
USAID. 2008.Economic Development Program, Responding To The Water Crisis In Jordan,Wash. D.C.
Williams, Rowan. 2008. Civil and Religious Law in England: A Religious Perspective, Foundation Lecture, Royal Courts of Justice, London, 7 February, http://www.archbishopofcanterbury.org/ 1575

Modeling of Artificial Emotions and its Application Towards a Healthy Environment

Khulood Abu Maria
College of Information Technology, Arab Academy of Business and Financial Sciences, Amman, Jordan

Raed Abu Zitar
Faculty of Engineering and Computer Science New York Institute of Technology (NYIT) Amman, Jordan

Issa Shehabat
Faculty of Information Technology, Philadelphia University, Amman, Jordan

Naim Ajlouni
Princess Ghazi College of Information Technology, VP, Al-Balqa Applied Science University, Jordan

SUMMARY: This chapter proposes modeling of artificial emotions through agents based on the symbolic approach. Two models were built for agent-based systems; one was supported with artificial emotions and the other one without. Both were used in solving a benchmark problem: orphanage care. The goal was to have a clean and healthy environment for an orphanage house. An application symbolic approach utilizes symbolic emotional rule-based systems (i.e. rule base that generates emotions) with continuous interactions with the environment and an internal "thinking" machinery that comes as a result of a series of inferences, evaluations, evolution processes, adaptations, learning and emotions. The two systems were simulated and results were compared. The study showed that systems with proper model of emotions could perform better than systems without emotions. The study sheds the light on how artificial emotions can be modeled in simple rule-based agent systems, and if emotions as they exist in real intelligence can be helpful for artificial intelligence.

1 INTRODUCTION

Emotions are an essential part of humans and other animals. The artificial agent in computer science sense is a piece of software that has its own inputs, outputs, and processes. It has a goal to achieve at a local level and an ultimate goal at global level. It may interact with the environment or with surrounding agents. However, why would we want or need to endow artificial agents with emotions? Two main answers are possible, depending on what our principal concern is when modeling emotions. Artificial agents can be used as a testing ground for theories about natural emotions in animals and humans. This provides a synthetic approach that is complementary to the analytic study of natural systems. Alternatively, we might want to exploit some of the roles that emotions play in biological systems in order to develop mechanisms and tools to ground and enhance autonomy, adaptation, and social interaction in artificial and agents-mixed (different types of agents) societies.

Emotions are an essential part of human life. They influence how we think, adapt, learn, behave, and communicate with each other. Without the preferences reflected by their positive and negative effects, our experiences would be a neutral gray. How unlikely is it that we are better without emotions than with? As El-Nasr et al noted (1998, 2000), the question also is not whether intelligent machines can have any emotions, but whether machines can be intelligent without emotions. In El-Nasr's work, (2000) neurological evidence was provided that proved that emotions do in fact play an important and active role in the human decision-making process. The interaction between the emotional process and the cognitive process may explain why humans excel at making decisions based on incomplete information; acting on our gut-feelings.

This chapter proposes modeling of artificial emotions through agents based on the symbolic approach. Two models were built for agent-based systems; one was supported with artificial emotions and the other one without. Both were used in solving a benchmark problem: orphanage care. The goal was to have a clean and healthy environment for an orphanage house.

2 CHARACTERIZATION OF EMOTIONS

Emotions are seen occurring when the cognitive, physiological and motor/expressive components are usually more or less dissociated in serving separate functions as a consequence of a situation-event appraised as highly relevant for an individual (Scheutz, 2004). For the more general definition of emotions, one crucial aspect is the distinctive features of emotions as compared with other psychological states that may have an affective element to them but that can hardly be considered full-fledged emotions. The different affective states can be categorized as follows (Scheutz, 2004): Emotions (e.g., angry, sad, joyful, fearful, ashamed, proud, elated, desperate); Moods (e.g., cheerful, gloomy, irritable, listless, depressed, buoyant); and Preferences/Attitudes (e.g., liking, loving, hating, valuing, desiring) The design features proposed for the differential definition of these states are partly based on: Response characteristics, such as intensity and duration or the degree of synchronization of different reaction modalities (e.g., physiological responses, motor expression, and action tendencies), antecedents (e.g., whether they are elicited by a particular event on the basis of cognitive appraisal) and consequences in terms of stability and impact on behaviour choices.

2.1 *Role of emotions in nature*

Emotional control is widespread in nature and seems to serve several crucial roles in animals and humans alike. In simple organisms with limited representational capacities, for example, emotions provide a basic evaluation in terms of hedonic values, often causing the organism to be attracted to what it likes and to avoid what it does not like. If another organism perceivably causes a threat, a fear-anger system may generate fight-or-flight behavior, depending on an estimate of the likelihood that a fight can be won. While emotional states such as fear and anger control immediate actions (Le Doux, 1996), other affective states operate on long-term behavioral dispositions. For example, anxiety leads to increased alertness without the presence of any immediate threat.

In humans, emotions, and more generally, affection, seem to be deeply intertwined with cognition in that they can influence, bias, and direct cognitive processes and, more generally, processing strategies. Negative affection, for example, can bias problem solving strategies in humans towards local, bottom-up processing, whereas positive affection can lead to global, top-down approaches (Nemani and Allan, 2001). Emotions also seem to play an important role in social contexts, ranging from signaling emotional states (e.g., pain) through facial expressions and gestures to perceptions of affective states that cause approval or disapproval of one's own or another agents' actions.

2.2 *Role of emotions in artificial agents*

Based on the functional roles of emotions proposed by researchers for natural systems, it is worth asking whether emotions could serve similar functional roles in artificial systems. Specifically, we can specify 12 potential roles for emotions in artificial agents:

- Action selection (e.g., what to do next based on current emotional state).
- Adaptation (e.g., short or long-term changes in behavior due to emotional states).
- Social Regulation (e.g., communicating or exchanging information with others via emotional expressions).
- Sensory Integration (e.g., emotional filtering of data or blocking of integration).
- Alarm mechanisms (e.g., fast reflex-like reactions in critical situations that interrupt other processes).
- Motivation (e.g., creating motives as part of an emotional coping mechanism).

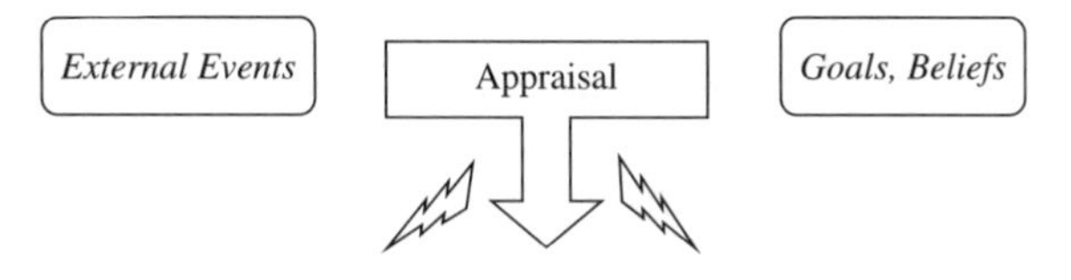

Figure 1. Emotions & Behavior Interaction.

- Goal Management (e.g., creation of new goals or reprioritization of existing ones).
- Learning (e.g., emotional evaluations as Q-values in reinforcement learning).
- Attention Focus (e.g., selection of data to be processed based on emotional evaluation).
- Memory Control (e.g., emotional bias on memory access and retrieval as well as decay rate of memory items).
- Strategic Processing (e.g., selection of different search strategies based on overall emotional state).
- Self-Model (e.g., emotions as representations of what a situation is like for agent).

2.3 *Emotions cognitive appraisal*

Humans are social and emotional beings by nature. Since our agents are fundamentally designed to mimic human behavior, modeling emotions is a core aspect of the current research. Emotions are interesting to model for a variety of reasons. Not only are they complex, but they are also dynamic, varying both episodically and longitudinally. A negative event can trigger an emotional response that may dissipate within a short time. Other emotions evolve slowly due to maturity, experience, or cumulative history. Some emotions are actually self-generating. For example, a person may become angry, with cause, but then remain so because anger feeds on itself. Some emotions fade quickly, while others persist long after the original cause has been pushed from the forefront. For example, surprise may be short-lived, while guilt may fester. In addition, once an emotion is activated, it becomes a driving force in subsequent actions.

Emotions arise only when an experience somehow matches with appraisal/evaluation structure as shown in Figure 1. This occurs when it has an impact on the elements. Not every experience makes direct contact with the appraisal/evaluation structure. People propagate implications of an experience, and sometimes these match with one's appraisal/evaluation structure. But these can just as well have no implications for one's goals, standards or attitudes. Only if a match is made, then the possibility of an emotional reaction arises.

2.4 *Categorization using the OCC model*

Since emotions are so vague and difficult to tell apart, and vary greatly in intensity, there is a lot of confusion and discussion about how we should order them. Approaches vary from pointing out some basic emotions and trying to reduce everything to these or trying to find underlying patterns so that they can be clustered into groups with similar features. An OCC (Orton, Core and Collins) (Ortony et al, 1988) model is considered the best categorization of emotions available. With this approach, the world is perceived as being divided in three different categories: events, agents, and objects. While there is no real justification on the choice of these categories, the idea itself is not so strange. The three are distinct factors with regard to behavior. There could be distinct handling (neurologically) of the three categories. This model is widely accepted and used. Let us go along with the proposed distinction. Emotions are considered valence reactions to one of these three perspectives on the world. Emotions are grouped in types corresponding to these categories, and are further divided by other criteria.

One can be pleased about the consequences of an event or not (*pleased/displeased*); one can endorse or reject the actions of an agent (*approve/disapprove*) or one can like or not like aspects of an object (*like/dislike*). A further differentiation consists of the fact that events can have consequences for others or for oneself and that an acting agent can be another or oneself. The consequences of

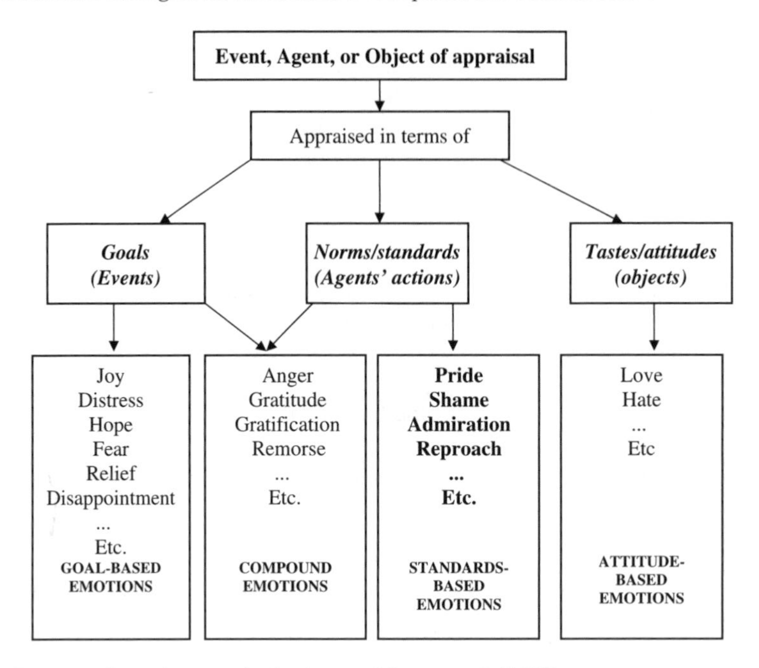

Figure 2. Structure of emotion types in the theory of Ortony et al. (1988).

an event for another can be divided into *desirable* and *undesirable*; the consequences for oneself as relevant or irrelevant expectations. Relevant expectations for oneself finally can be differentiated again according to whether they actually occur or not (*confirmed/disconfirmed*). This differentiation leads to the following structure of emotion types as shown in Figure 2.

The OCC model is simplified by collapsing the original categories down to five distinct positive and five negative specializations of two basic types of affective reactions–positive and negative ones. These categories have enough generative capacity to endow any affective agent with the potential for a rich and varied emotional life. Ortony et al (1988), too, points out the importance of personality which determines the consistency of emotional reactions over time and makes an agent believable.

2.5 *Related work*

As shown in previous sections, the topic of emotion is very challenging since it is hard to fully understand how we feel and why we feel that way. Part of the reason is that most of our emotions occur at the subconscious level. Moreover, it is still unclear how emotions shift from the subconscious to the conscious brain. Searching for a solution, Bates (1994) built a realizable agent (OZ project) using the model described by Ortony et al (1988) (Table 1). While the model only described basic emotions and innate reactions, it presented a good starting point for building computer simulations of emotion. The basic emotions that are simulated in the model were anger, fear, distress/sadness, enjoyment/happiness, disgust, and surprise.

El-Nasr et al (2000) proposed a model called FLAME (i.e. Fuzzy Logic Adaptive Model of Emotions). FLAME was modeled to produce emotions and to simulate the intelligence process. The model was built using fuzzy rules to explore the capability of fuzzy logic in modeling the emotional process. Fuzzy logic helped them in capturing the complex nature of emotions. They noted the advantages of using fuzzy modeling over conventional models to simulate a better illusion of reality. The authors concluded that the use of fuzzy logic did improve the believability of the model simulated.

What makes the human mind so complex is the interactions between its processes. The cognitive and emotional processes are not as separate as used in El-Nasr et al's (2000) model. In fact,

Table 1. Positive and negative reactions.

Positive Reactions:
. . . because something good happened (joy, happiness etc.)
. . . about the possibility of something good happening (hope)
. . . because a feared bad thing didn't happen (relief)
. . . about a self-initiated praiseworthy act (pride, gratification)
. . . about an other-initiated praiseworthy act (gratitude, admiration)
. . . because one finds someone/thing appealing or attractive (love, like, etc.)
Negative Reactions:
. . . because something bad happened (distress, sadness, etc.)
. . . about the possibility of something bad happening (fear, etc.)
. . . because a hoped-for good thing didn't happen (disappointment)
. . . about a self-initiated blameworthy act (remorse, self-anger, shame, etc.)
. . . about another-initiated blameworthy act (anger, reproach, etc.)
. . . because one finds someone/thing unappealing or unattractive (hate, dislike, etc.)

the complexity in the human mind lies in the complexity of the interaction between both the emotional and the cognitive processes. Therefore, there is a need to further study the possible ways of interactions between these two models.

Fellous (1999) reviewed experimental evidence showing the involvement of the hypothalamus, the amygdale and the prefrontal cortex in emotion. For each of these structures, he showed the important role of various neuro-modulatory systems in mediating emotional behavior. It was suggested that behavioral complexity is partly due to the diversity and intensity of neuro-modulation and hence depends on emotional contexts. Rooting the emotional state in neuro-modulatory phenomena allows for its quantitative and scientific study and possibly its characterization.

Poel et al (2002) introduced modular hybrid neural network architecture, called SHAME, for emotion learning. The system learned from annotated data how the emotional state was generated and changed due to internal and external stimuli. Part of the modular architecture was domain independent and part must be adapted to the domain under consideration. The generation and learning of emotions was based on the event appraisal model. The architecture was implemented in a prototype consisting of agents trying to survive in a virtual world. An evaluation of this prototype showed that the architecture was capable of generating natural emotions and furthermore that training of the neural network modules in the architecture was computationally feasible.

In their paper, Poel et al (2002) introduced a model that makes it possible to talk about an emotional state and changes because of appraisals of events that the agents perceive in their environment. It was based on the OCC model (Ortony et al, 1988), a cognitive theory for calculating cognitive aspects of emotions. Poel et al (2002) used the OCC model as the basis for the supervised learning approach of emotions.

Kort et al (2001) proposed a novel model by which they aimed to conceptualize the impact of emotions upon learning. They then built a working computer-based model that would recognize a learner's affective state and would respond appropriately to it so that learning would proceed at an optimal pace. Accurately identifying a learner's emotional/cognitive state is a critical indicator in how to assist in achieving an understanding of the learning process.

Ushida et al (1998) proposed a model for life-like agents with emotions and motivations. This model consisted of reactive and deliberative mechanisms. The basic idea of the model came from a psychological theory, called the cognitive appraisal theory. In the model, cognitive and emotional processes interact with each other. A multi-module architecture was employed in order to carry out the interactions. The model also had a learning mechanism to diversify behavioral patterns. Their expectation was to build a device that would be capable of seeing other facial features such as eyebrows, lips, and specific facial muscles; tracking them and reacting to them as they occurred. It was also expected that the device would be able to make immediate software-driven evaluations of the emotional state.

Doya (2002) presented a computational theory on the roles of the ascending neuro-modulatory systems from the viewpoint that they mediate the global signals that regulate the distributed learning mechanisms in the brain. Based on a review of experimental data and theoretical models, it was proposed that dopamine signals the error in reward prediction, serotonin controls the time scale of reward prediction, noradrenalin controls the randomness in action selection, and acetylcholine controls the speed of memory update. The possible interactions between those neuro-modulators and the environment were predicted on the basis of computational theory of meta-learning.

Clocksin (2004) explored issues in memory and their affect in connection with possible architectures for artificial cognition. The work represented a departure from the traditional ways in which memory and emotion have been considered in artificial intelligence research. It was formed by two strands of thought emerging from social and developmental psychology. First, there has been an increasing concern with persons, agencies and actions, rather than causes, behaviors and objects. Second, there was an emphasis on the self as a social construct, that persons are the result of interactions with significant others, and that the nature of these interactions is in turn shaped by the settings in which these interactions occur.

Gratch (2000) discussed an extension to command and control modeling architecture that addressed how behavioral moderators influence the command decision-making process. He described how behavior moderators such as stress and emotion can influence military command and control decision-making. The goal of his work was to extend this modeling architecture to support how a synthetic commander performs activities based on behavioral moderators such as stress and emotional state and how dispositional factors, such as level of training or personality differences, influence the level of stress or emotion. It differed from other computational models by emphasizing the role of plans in emotional reasoning.

Evans (2002) claimed that emotions enable human to solve problems by providing the right kind of search strategy for each kind of problem. The search hypothesis thus offers an account of the relationship between emotions and reason. Emotions played a positive role in aiding reason to make good decisions.

Nilsson (2001) described architecture for linking perception and action in a robot. It consisted of three towers of layered components. The perception tower contained rules that created increasingly abstract descriptions of a current environmental situation starting with the primitive predicates produced by the robot's sensory apparatus. These descriptions were deposited in a model tower which was continuously kept faithful to the current environmental situation by a truth maintenance system.

3 EMOTIONAL AGENT MODELING

Using an emotional rule base, the behavior of two agents and objects living in a simulated world were modeled. One represented a Regular Intelligent Agent (RIA) and the other one an Emotional Intelligent Agent (EIA), where the agent uses emotions as an important tool to guide its behavior. Both agents have the same objects in their world and main goals to achieve certain activities. Each agent gives priorities for its empirical goals so as to achieve the main goal. Agent's global variables and states can be monitored through simulation using graphs, plots and reports.

3.1 *Model hypothesis*

The hypothesis was based on the assumption that emotions in animals are mechanisms that enhance adaptation in dynamic, uncertain, and social environments, with limited resources and over which the individual has very limited control. When an artificial agent is confronted with an environment presenting similar features, it will need similar mechanisms in order to survive and enhance its behavior, survival, learning and adaptation.

Our aim, ultimately, is to build agents that behave, in a way, similar to intelligent creatures implementing several applications such as information, transaction, education, tutoring, business, entertainment and e-commerce. Therefore, we wanted to develop artificial mechanisms that could take part in the role that emotion plays in natural life. These mechanisms are called artificial emotions. We will investigate if emotional control can improve performance of the agent.

The tool employed was the NetLogo program that is an extension of the Logo program that allows users to give commands to a single agent. NetLogo allows the user to control many agents on the screen. The latest version is NetLogo 3.1. Due to the ongoing development nature of the program, there exists very little documentation and evaluation of the product. The main source of documentation on NetLogo is from the team at Northwestern University via the website: http://ccl.northwestern.edu/netlogo. NetLogo is an agent-modeling environment. It is particularly well suited for modeling complex systems developing over time. Modelers can give instructions to hundreds or thousands of independent "agents" all operating in parallel (agent and agent-set).

3.1.1 *Netlogo features*

The following features were taken from the NetLogo website at http://ccl.northwestern.edu/netlogo System:

- Cross platform: runs on MacOS, Windows, Linux, et al
- Models can be saved as applets to be embedded in web pages

Language:

- Fully programmable
- Simple language structure
- Language is Logo dialect extended to support agents and parallelism
- Unlimited numbers of agents and variables
- Many built-in primitives
- Integer and double precision floating-point math
- Runs are exactly reproducible cross platform

Environment:

- Graphics display supports turtle shapes and sizes, exact turtle positions, and turtle and patch labels
- Interface builder with buttons, sliders, switches, choices, monitors, text boxes

3.1.2 *Orphanage scenario*

The simulations required the development of a Netlogo environment providing a set of behavior rules that may be used for the simulation of agent behavior. This environment can be described as an emotional agent behavior toolkit. Two agents we employed in solving the orphanage care problem. Both agents lived in a simulation world with a set of objects, which behaved in similar way. The agents had to make sure that their main goal of keeping the orphanage status as healthy as possible was satisfied.

To achieve the main goal the agent should go to and take care of (work in) the Orphanage. Taking care of the Orphanage depends on the agent working capacity (agent attribute).The agent should spend money on the Orphanage. Taking care of the Orphanage depends on the agent earning level (agent attribute). It needs cash to support its work in the orphanage for any expenses of any type.

To preserve the earning level from declining to zero (simulation will stop) the agent should go to work to make some money. The working capacity can be improved at the Academy. The agent should go to the Academy (to improve its knowledge and, hence, improve its earning salary) and study their available courses. Assume that the working capacity does not decay over time.

The agent needs to raise (improve) its social capacity. The agent should go a social place such as club, restaurant, mall or parties. Clubs, restaurants, malls, parties cost money (expensive social activities).

The agent must pay fees when learning at the Academy, and by attending courses the agent will improve its working capacity and then its earning salary will be improved as a result. Social capacity does not decay over time and the agent has to keep socializing in the club (or any of social

Table 2. Regular Intelligence Agent Attributes.

Type	*Attributes*
• Working Memory Goals • Main Goal Priority • Empirical Goals • Agent Performance Measures	• Memory-Previous-Goal • Memory-Selected-Goal • Memory-Current-Goal • Orphanage Priority • Earning: Working Priority • Learning: Studying Priority • Socializing: Club Priority • Working: Current Working Capacity • Earning: Current Earning Level • Socializing: Current Social Capacity

Table 3. Emotional Intelligence Agent Attributes

Type	*Attributes*
• Working Memory Goals • Main Goal Priority • Empirical Goals • Agent Performance Measures • Emotion Aspects	• Memory-Previous-Goal • Memory-Selected-Goal • Memory-Current-Goal • Orphanage Priority • Earning: Working Priority • Learning: Studying Priority • Socializing: Club Priority • Working: Current Working Capacity • Earning: Current Earning Level • Socializing: Current Social Capacity • Causing Emotional Reactions

places) to keep it at an acceptable level. Improving social capacity will influence agent adaptation in work and with other agents, that would positively affect its career and hence salary.

4 SIMULATIONS SETTING

Emotions play important roles at the control-level of our agents' behavior. Emotion may lead a reflexive reaction, it may support the goals and motivation of an agent, it can create new motivations (e.g., new goals and then sub goals) and it can contribute in new criteria for the selection of the methods and the plans the agent uses to satisfy its main goals. Since emotions are processes that operate at the level of control of agent architecture, the behavior of the agent will improve if the emotional processes of the agent are included. Agent can generate emotion signals, evaluate and assesses events, take into account the integration of agent goals, personality and feeling, Then it can take the proper action (i.e. behavior). We will build two models for agent based systems; one is supported with artificial emotions and the other one without emotions.

The Netlogo models can be used to develop the setup procedures separately from the execution of the model. This allows the effects of varying initial conditions to be analyzed more easily. Agents attributes are outlined in Tables 2 and 3.

4.1 *Simulation of objects*

The current problem is about an agent whose main goal is to look after an orphanage. It is a voluntary job. To reach this main goal, the agent needs to arrange empirical goals

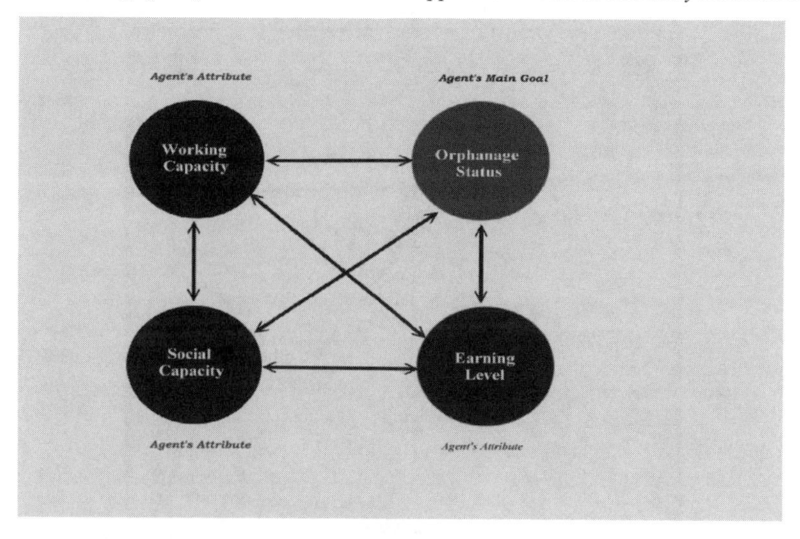

Figure 3. Relations between simulation objects.

that lead to main goal priorities. Simulation objects relations are shown in Figure 3. The objects are:

- The Orphanage: It is the agent focus point. It declines in status (i.e. decrease in healthy living conditions) over time. This can only be improved by the agent take care of the orphanage. The greater the working capacity of the agent, the healthier the orphanage gets.
- Job: It offers new jobs (opportunities) with a new randomly calculated salary. Agent earns money by working at job object. Higher social and working capacity will eventually result in raising agent's earning level.
- Club: For improving agent's social capacity, the agent must socialize in the club. In the club, the agent has to pay for drinks and food and learn to socialize. Costs and social capacity improvement are recalculated randomly every run.
- Academy: In the academy the agent takes the courses offered and receives a reward in terms of improvement in working capacity. First, it has to pay the fees (cores admission).

The performance measures for the simulations can be summarized as follows:

- Main Goal Status (Orphanage);
- Agent's Performance Measures:
  - Social Capacity (Gained)
  - Working Capacity (Gained)
  - Earning Level (Gained Cash)

Agents live in the simulation environment with their own set of objects, which behave similarly. The agents have to make sure that their orphanage status does not decay completely (i.e. become unhealthy for the parentless children) and that they do not run out of money. Please, see Tables 2 and 3 for the EIA and RIA.

5 SIMULATION RESULTS AND ANALYSIS

Simulations were run for the architectures (EIA's and RIA's). The goal was to find how artificial emotions in agents could improve performance behavior even with limited resources. The simulation world consisted of two agents and objects divided into two regions (Figure 4). Entities in both

Figure 4. Simulations Layout.

Table 4. *Regular Intelligence Agent (RIA)* Results.

Trial	Orphanage Status	Agent's Earning Level	Agent's Social Capacity	Agent's Working Capacity
1	74.769	63.808	44.820	84.240
2	64.100	100.000	61.940	80.265
3	80.700	84.081	55.740	83.228
4	92.400	96.000	36.120	84.254
5	70.700	100.000	62.720	85.292
6	64.600	64.642	74.240	80.339
7	65.400	61.194	71.120	86.376
8	85.900	100.000	4.700	81.169
9	48.400	98.000	59.080	83.268
10	81.300	100.000	65.620	86.340

regions were identical, and thus behaved in the same manner. Each agent had one orphanage, a job, an academy to go to and a club to visit. The agents were compared head to head by looking at their performance and attributes in real time, instead of comparing them after separate simulations. Do artificial emotions work for agent's benefit? Can they have artificial emotions? How does personality motivate agents to act?

As randomness plays a big role, the simulation was run ten times for every setting of the environment. A run consisted of 2000 iterations of the simulation. A test was considered a failure if the agent ran out of money (earning level <=0) or if the orphanage status dropped to zero.

Setting 1 maximized average values of the random numbers that were generated in the simulation for the salary, social and working capacities. The results are shown in Table 4 for RIA agent and Table 5 for EIA agent. The charts for a typical run of both agents are described in Figure 5.

As shown in the previous tables, the EIA agent performs better than the RIA agent under setting (1) does. Performance of the RIA agent is acceptable although it has difficulties keeping main goal status (Orphanage) between its desired values (90–100). RIA fails to maximize its social capacity. Earning level of the RIA agent was better because it performed according to a fixed threshold value. The EIA agent had its Main Goal Status stable (i.e. less fluctuation); whereas the RIA Main

Table 5. *Emotional Intelligence Agent (EIA)* Results.

Trial	Orphanage Status	Agent's Earning Level	Agent's Social Capacity	Agent's Working Capacity
1	83.800	87.476	82.760	98.733
2	88.100	89.627	82.860	100.000
3	94.700	91.185	75.900	9.708
4	90.200	88.017	86.600	94.513
5	85.300	88.712	85.880	93.750
6	90.200	92.275	72.275	100.000
7	99.500	89.520	82.660	95.521
8	97.300	84.840	83.760	93.450
9	88.000	77.139	93.340	95.491
10	84.400	70.789	87.940	91.871

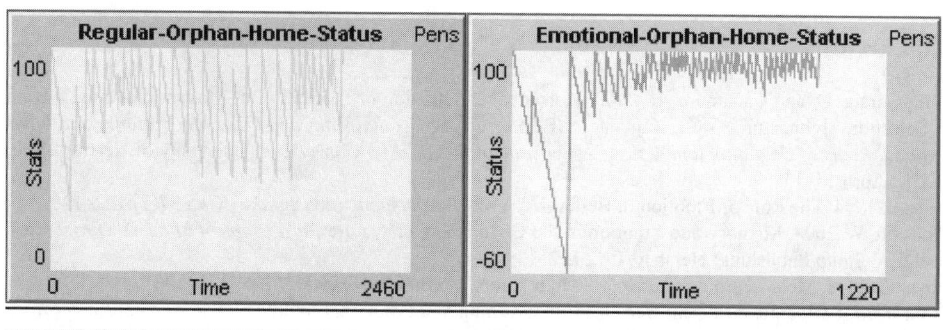

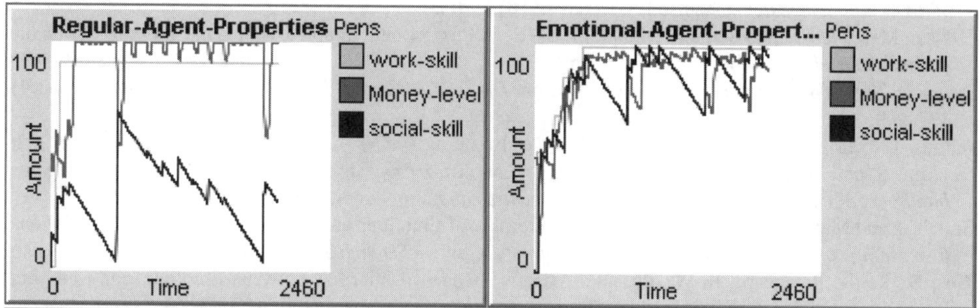

Figure 5. RIA EIA Agents status.

Goal Status had more fluctuations. This general trend was a positive one. EIA acted and guided by artificial emotion that helped the agent to reach its main goal in stable trend. The EIA thinking model that supported pleasant/unpleasant experience guided the agent to reach its major goals.

What do these charts tell us? First, it was evident that the EIA agent did a better job than the RIA. As for arterial emotions values, it seemed that the EIA agent was happier with the world in general; the results were more stable and smoother than for the RIA agent. In this respect, we have succeeded in finding settings that guaranteed the best results from both agents (Avila-Garcia and Canamero, 2005, and Franklin and Graesser, 1996).

6 CONCLUDING REMARKS

This work has shown that artificial emotions can be successfully modeled in agents. It was found that emotional agent could outperform its non-emotional counterpart. The simulations environment can be used as a framework for others since it allows for easy comparison of different agent architectures head to head in identical circumstances and in real-time. The ultimate goal of the agents, for the example used in this study, was to have a clean and healthy environment for the orphanage house.

It was observed that emotions could improve the performance of an agent. It was noticed that the emotional agent performed better than the non-emotional agent. While the regular intelligence agent took extremely good care of its main goal and financial situation, the emotional intelligence agents tried to maximize all their goals, and their level of capacities and cash. This must have been the result of a direct translation from perception into priorities, instead of relying on fixed priority values, as it is in the regular intelligence agent.

In closing, artificial emotions are applicable in modeling agents, but due to the very different mechanisms in implementing functions, the benefits of emotions existing in humans are not easily translated. A more advanced model is needed for artificial emotion mechanisms and new scenarios. Applications could be in information system, business, e-commerce, and games. Artificial emotions can be very beneficial to the development of agents and hence to many applications in information technology.

REFERENCES

Avila-Garcia, O. and Cañamero, L. 2005. Hormonal Modulation of Perception in Motivation-Based Action Selection Architectures. in: L. Cañamero (Ed.), *Proc. Agents that Want and Like: Motivational and Emotional Roots of Cognition and Action, Symposium of the AISB '05 Convention,* University of Hertfordshire, UK, April 14–15.

Bates, J. 1994. The Role of Emotion in Believable Agents. *Communications of the ACM*, 37(7):122–125.

Clocksin, W. 2004. Memory and Emotion in the Cognitive Architecture. *In: Visions of Mind* D. Davis, (ed.), IDEA Group Publishing: Hershey, PA., 122–139.

Doya, K. 2002. Meta-Learning and Neuro-Modulation. *Neural Networks 15*, (4–6) 495–506.

El-Nasr, M. S., Ioerger, T. R. and Yen, J. 1998. Learning and Emotional Intelligence in Agents. *Proceedings of AAAI Fall Symposium*. Madison, Wisconsin, 1017–1025.

El-Nasr, M. S., Yen, J., and Ioerger, T. 2000. FLAME – A Fuzzy Logic Adaptive Model of Emotions. *Automous Agents and Multi-agent Systems*, 3, 219–257.

Evans, D. 2002. The Search Hypothesis of Emotion. *The British Journal for the Philosophy of Science*, 53(4): 497–509.

Fellous, J-M. 1999. The Neuro-Modulatory Basis of Emotion. *The Neuroscientist.* 5(5):283–294.

Franklin, S. and Graesser, A. 1996. *Is it an Agent, or just a Program? Taxonomy for Autonomous Agents. Intelligent Agents {III}. Agent Theories, Architectures and Languages*, Springer-Verlag.

Gratch, J. and Marsella, S. 2003. Modeling the Interplay of Emotions and Plans in Multi-Agent Simulations. *Proceedings of the 23rd Annual Conference of the Cognitive Science Society,* Edinburgh, Scotland, 1–6.

Kort, B., Reilly, R., Picard, R. W. 2001. An Affective Model of Interplay between Emotions and Learning: Reengineering Educational Pedagogy – Building a Learning Companion, *ICALT*, 43–48.

LeDoux, J.E. 1996. *The Emotional Brain*. New York, Simon and Schuster.

Nemani, S. and Allan, V., 2003. *Agents and the Algebra of Emotion. Proceedings of the second international conference on AAMS.* July 14–18, Melbourne, Australia, 34–39.

Nilsson, N. 2001. Teleo-Reactive Programs and the Triple-Tower Architecture. *Electronic Transactions on Artificial Intelligence*, Vol. 5, Section B, 99–110.

Ortony, A., Clore, G. L., and Collins, A. 1988. *The Cognitive Structure of Emotions.* Cambridge, UK, Cambridge University Press.

Poel, M., op den Akker, R., Nijholt, A. and van Kesteren, A. J. 2002. Learning Emotions in Virtual Environments. in: R. Trappl, editor, *Proceedings of the Sixteenth European Meeting on Cybernetics and System Research*, Volume 2, Vienna, Austria, 751–756.

Scheutz, M. 2004. Useful Roles of Emotions in Artificial Agents: A Case Study from Artificial Life. in: *Proceedings of AAAI, AAAI Press*. San Jose, California, 42–48.

Ushida, H., Hirayama, Y. and Nakajima, H. 1998. Emotion Model for Life – Like Agent and Its Evaluation. *Proceedings* of *AAAI/IAAI Press*, Madison, Wisconsin, 62–69.

New Technologies for Human Development: A Case Study from Eurasia

Jung-Wan Lee
Kazakh-British Technical University, Republic of Kazakhstan

SUMMARY: This chapter presents a review of new technologies as a human development tool. The positive and negative effects of these on the growth of individual potential were assessed using the Eurasian country of Kazakhstan as a case study region. Technologies that are critical for human, economic and sustainable development were identified.

1 INTRODUCTION

Balanced and sustainable development is measured not only by income levels but also by factors such as the opportunity to live a long and healthy life, the ability to acquire knowledge, to be able to access resources required for a civilized life and to be able to participate in society (Delyagin, 2005). In a wider sense, the term human development embraces all aspects of personnel progress, ranging from health to economic and political liberties. Presently, individual growth is viewed as a progress of expanding human opportunities and capabilities. This is an emerging concept.

Society has come to understand that human development is an end, to which economic growth, education, medicine and a good environment are the means. New technologies, in turn, are a recognized tool for economic, educational, healthcare, and environmental improvement as well as overall personal maturity. In this regard, there is now a need for a comprehensive evaluation of new technologies from the perspective of their implications for individual growth. In addition, it is important to identify an environment where new technologies can become an effective tool for achieving this objective.

New technologies are the result of ground-breaking activity presented in the market as a new or improved product, so called "product innovation," or a new technological process used in practice, so called "process innovation." The value of new technologies lies in the mechanisms and intensity of human implications and resulting effects on social structure and social relations as a whole (Delyagin, 2005). Collective technologies are understood to be those used in social relations. Shared technologies cover several fields of human activity and can be used in for instance medicine, education, and governance. For example, in medicine, social technologies can improve the efficiency of doctor-patient communication; in education, administration-student/faculty and faculty-student communication; while in public administration, such technologies promote public participation in social life and enhance government-public communication.

Another technology classification criterion is relative effectiveness, depending on which technologies may be considered new or old. As noted, new technologies are more effective and efficient, can better meet customer needs and have more impact on human development. New technologies are gradually replacing old and inefficient ones. This results in growing productivity and more effective use of natural resources, reducing the energy intensiveness of products and environmental pollution, while exhaustible natural resources are replaced by others that facilitate comprehensive human development.

This chapter looks at the positive and negative effects of new technologies on the development of human potential using the Eurasian country of Kazakhstan as a case study. Examples of new technologies that were assessed include information and communication technologies and biotechnologies.

Table 1. Main indicators of ICTs in Kazakhstan (*unit: per 100 persons).

Year	Internet*	Fixed telephone line*	Mobile*	GDP growth rate (% of previous year)
1998	0.01	10.7	0.2	98.1
1999	0.14	10.5	0.5	102.7
2000	0.24	11.0	0.6	109.8
2001	0.63	11.7	5.6	113.5
2002	0.94	12.7	6.1	109.8
2003	1.23	13.6	8.8	109.3
2004	1.42	14.3	16.3	109.6
2005	2.04	15.5	36.0	109.7
2006	2.06	19.1	51.2	110.7

Sources: Agency on Statistics of the Republic of Kazakhstan (2007)

2 NEW TECHNOLOGIES FOR HUMAN DEVELOPMENT

2.1 *Information and Communication Technologies*

Information and Communication Technologies (ICT) are one of the fast developing areas. They are increasingly more important and have a significant impact on the economic and social life of numerous countries. An accurate definition of ICTs embraces all technologies that process information and enable communication. The scale of social development implications of ICTs becomes clearer when the description includes both old and new technologies, ranging from radio and television to computer and cell phones. ICTs are increasingly used to meet the basic needs of people around the world, such as access to education, employment opportunities, and public participation in social life. New technologies help to inform the public and facilitate immediate communication. If applicable, they enable access to better quality education. Many schools and vocational training institutions use them for students and teachers. New technologies change classroom methods through, for example, use of multi-media manuals, Internet-based materials and information, which makes the learning process more interactive.

Currently even at the global level, there are significant disparities in access to and use of ICTs. For example, according to 2006 statistics of the Agency on Statistics of the Republic of Kazakhstan (2007), the number of users with Internet access at home stands at only 32.5 percent. This breaks down to urban areas with 65 percent, while in rural areas it is only 5.3 percent. Of these, the number of mobile phone service subscribers per 100 persons is 51 percent of the population and the number of subscribers to the Internet per 100 persons was only 2 percent in 2006. Among them, more than 62 percent of the Internet access was made through the fixed-line telephone in households, which is very costly and slow. The level of computer literacy represents 50.5 percent of the population in Kazakhstan in 2006, of which the computer literacy in urban areas users represents 55.8 percent and in rural areas represents 37.4 percent in 2006 see Table 1 for details. To reduce the "digital divide" in Kazakhstan has adopted the "Program for Digital Divide Reduction in the Republic of Kazakhstan for 2007–2008" aiming to facilitate effective use of the Internet on a routine basis by at least 20 percent of the population and promote the social and economic importance of information resources in Kazakhstan (Kazakhstan Governmental Decree, 2006).

2.2 *Biotechnologies*

Biotechnology means a type of technology using biological systems, living organisms or their derivatives to produce or modify products or processes to then be applied in practice. Biotechnologies have developed into the most important technology for human development resulting from their use in sectors vital for human development. Biotechnological trends in science and production emerged as a result of the rapid progress of different elements of physical-chemical biology.

Biotechnologies have been establishing themselves over the last years and now are at a peak of the development. The increasing use of biotechnologies in different fields of medicine, veterinary science and agriculture is becoming an integral part of scientific and technological progress. An important factor distinguishing biotechnology from other fields of science and production is that it initially focuses on present-day concerns such as food production, primary proteins, and the maintenance of natural energy balance through a shift from non-renewable to renewable resources and environmental protection. Clearly, technologies significantly affect lifestyles, quality of life and human development. Recognizing this, many countries have come to pursue scientific and technological development policies in a more conducive context.

For example, the "Strategy for Industrial and Innovative Development of the Republic of Kazakhstan for 2003–2015" identifies biotechnology as a priority. Under the Strategy, the "Concept of Development of the National Biotechnology Center of the Republic of Kazakhstan for 2006–2008" and a science and technology program "Development of Modern Technologies to Shape a Biotechnology Cluster in the Republic of Kazakhstan for 2006–2008" were approved by the Governmental Decree. Below are some of examples of the application to biotechnology and agriculture.

Today over 40 percent of Kazakhstan's population live in rural areas (Agency on Statistics of the Republic of Kazakhstan, 2007). In 2004, one third of all employed people were employed in agriculture, which indicates that even a small-scale introduction of biotechnologies may result in large-scale implications for human life and development in Kazakhstan. Agricultural biotechnology includes plant protectors, veterinary biotechnology, transgenic plants and animals. Kazakhstan exports some crops such as wheat. At the same time, some regions are short of fresh vegetables. For example, in Spring 2006 allocation was made from the Karagandy Oblast budget to buy 100,000 tons of Uzbek vegetables (UNDP Kazakhstan, 2006). Agricultural revenue varies from year to year depending on yield, which is currently lower than in developed countries. For example, in 2004 wheat yield was only 8 hwt per hectare versus 29 hwt in the US. Losses of up to 30 percent in crop yield occur through pests, diseases and weeds (UNDP Kazakhstan, 2006). Agricultural biotechnology aims to increase crop productivity both in quantitative and qualitative terms, and to develop and use new sorts of crops. In 2004, areas under crops of Kazakhstan selection totaled 3.4 million hectares (27.8 percent) versus 4.6 million hectares (35 percent of the total territory) in 2005 (UNDP Kazakhstan, 2006). This indicates an upward trend in areas assigned to local crops.

Currently, however, the development and production of biotechnology products in Kazakhstan is inadequate. According to an analysis of UNDP Kazakhstan (2006), the range of biotechnology products is limited. Most of them cater for domestic needs because they cannot compete in the international market. Furthermore, over the last 15 years the range of research has narrowed, biotechnological research has shrunk and technologies have become somewhat obsolete. Kazakhstan inherited the Soviet model of scientific progress and production characterized by slow response to the fast changing needs of innovative sectors and poor links between the needs and priorities of science and technology policy. Of many new developments, only a few are finally introduced, with even fewer being successful. The wide audience of consumers and business remain unaware of products of the bioindustry development, which are still only discussed in scientific and government circles.

3 THE HUMAN DEVELOPMENT INDEX

Each year since 1990 UNDP's Human Development Report has published the human development index (HDI) that looks beyond GDP to a broader definition of well-being. The HDI provides a composite measure of three dimensions of human development: living a long and healthy life (measured by life expectancy), being educated (measured by adult literacy and enrolment at the primary, secondary and tertiary level) and having a decent standard of living (measured by purchasing power parity (PPP) income). The index is not in any sense a comprehensive measure of human development. It does not, for example, include important indicators such as gender or income inequality and more difficult to measure indicators like respect for human rights and political freedoms. What

Table 2. Main indicators of socio-economic development and human development.

Year	1997	1998	1999	2000	2001	2002	2003	2004	2005	2006
GDP[1]	101.7	98.1	102.7	109.8	113.5	109.8	109.3	109.6	109.7	110.7
Population[2]	15188	14955	14902	14866	14851	14867	14951	15075	15219	15397
GDP per capita, US$ at PPP	4626	4379	4293	4487	5219	5862	6532	7232	8068	9052
Life expectancy at birth[3]	64	64.5	65.7	65.5	65.8	66.3	65.8	66.2	65.9	66.2
Employment rate, as %	87	86.9	86.5	87.2	89.6	90.7	91.2	91.6	91.9	92.2
R&D works[4]	0.25%	0.25%	0.24%	0.23%	0.28%	0.37%	0.31%	0.32%	0.38%	0.35%
Not enrolled[5]	n.a.	10.8	11.6	9.8	10.6	4.5	3.0	1.1	1.1	0.05
Poverty index	n.a.	29.5	26.2	25.1	23.7	22.0	20.9	20.1	19.7	16.0
HDI[6]	n.a.	0.736	0.740	0.740	0.753	0.764	0.770	0.780	0.786	0.794

n.a.: data not available
[1] Gross domestic product, as percentage of previous year
[2] Total population, end of year, thousand persons
[3] Life expectancy at birth, years; total population
[4] Value of performed R&D works as percent of GDP
[5] Proportion of 16 year olds not enrolled in education
[6] Human development index
Source: Agency on Statistics of the Republic of Kazakhstan (2007)

it does provide is a broadened prism for viewing human progress and the complex relationship between income and well-being.

According to UNDP (2007)'s Global Human Development Reports, Kazakhstan has gone through two phases of human development. The first phase (1990–1995) was characterized by dramatic declines in the main human development indicators, sending Kazakhstan down from 54th to 93rd in the world HDI ranking. During the second phase (1996–2005), the human development indicators slowly recovered, raising Kazakhstan to 73rd out of 177 countries in 2006, with the HDI of 0.794 (see Table 2 for details).

The year 2005's HDI highlights the very large gaps in well-being and life chances that continue to divide regions of the country. Table 3 illustrates that the country have very different levels of income by regions. By looking at some of the most fundamental aspects of people's lives and opportunities it provides a much more complete picture of the country's development than other indicators, such as income per capita at PPP by regions. Of the components of the HDI, only income and gross enrolment are somewhat responsive to short term policy changes. For that reason, it is important to examine changes in the human development index over time by regions.

The HDI trends tell an important story in that aspect. Since 1999 Kazakhstan has been progressively increasing its HDI score, see Figure 1 for details. Development of human potential stands high on Kazakhstan's agenda. New ground was broken in 1997 in recognition of human development concept in Kazakhstan. The first two national Human Development Reports for 1995 and 1996 were a powerful spur to the use of the concept as a tool for social and economic analysis, resulting in the adoption of the concept as an underlying strategy for a number of national documents, of which the most important is the long-term national development strategy, "Kazakhstan-2030" (UNDP Kazakhstan, 2006). This shows that the Strategy's priorities centre on "health, education and well-being of Kazakhstan's people," which are the basic components of human development. It aims to ensure that the population live healthy lives in a healthy environment. To accomplish this, the government has been working to improve living standards and quality of life through modern high quality healthcare, environmental management, access to clean water, an improved epidemiological situation, improved quality of education, as well as employment schemes.

Table 3. Basic human development indicators and related indexes by region, year 2005.

Oblast	Life expectancy	Literacy rate	Enrolment rate	Income[1]	HDI
Akmola	63.5	99.4	79.3	4296	0.768
Aktobe	65.7	99.7	90.4	10292	0.801
Almaty	66.4	99.4	67.7	3228	0.766
Atyrau	66.7	99.7	87.4	32649	0.794
East Kazakhstan	64.8	99.2	79.6	5244	0.779
Zhambyl	67.0	99.7	78.8	2742	0.772
West Kazakhstan	66.5	99.4	85.8	13388	0.793
Karagandy	64.0	99.5	86.2	7729	0.784
Kostanai	65.3	99.4	78.0	5916	0.770
Kyzylorda	66.0	99.6	80.6	6900	0.764
Manghistau	65.0	99.5	95.3	20715	0.803
Pavlodar	65.6	99.4	86.0	8121	0.789
North Kazakhstan	64.9	99.2	71.5	4382	0.765
South Kazakhstan	67.4	99.9	83.3	2598	0.773
Astana city	70.7	99.7	89.1	19417	0.856
Almaty city	67.4	99.8	100.0	16629	0.834
Kazakhstan	65.9	99.5	85.0	8068	0.787

[1] Income per capita at PPP, US dollars
Source: United Nations Development Programme in Kazakhstan (2006), p.115

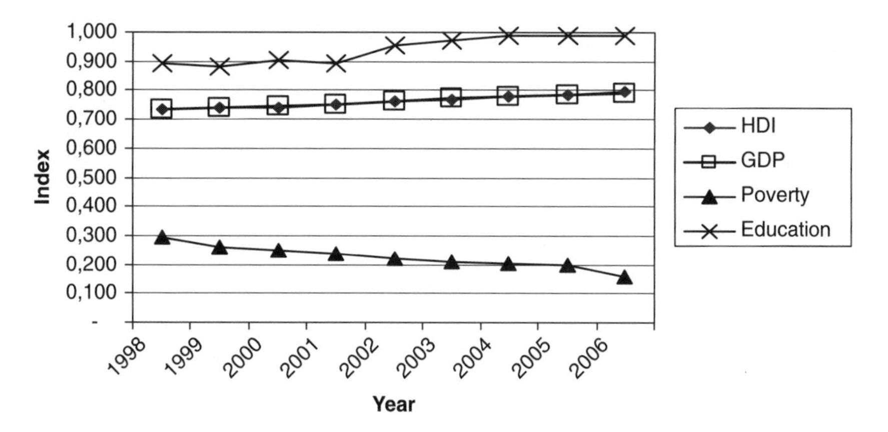

Figure 1. HDI Trends of Kazakhstan.
Source: Calculated by the author based on the data from the Agency on Statistics of the Republic of Kazakhstan (2007) and the UNDP Kazakhstan (2006)

However, it should be noted that to achieve this priority Kazakhstan should achieve four ambitious targets identified in the Millennium Declaration adopted by the UN General Assembly in 2000 in New York, which are:

1. Between 1990 and 2015 reduce by two thirds the under-five mortality rate;
2. Between 1990 and 2015 reduce maternal mortality by three quarters;
3. By 2015 halt and reverse the spread of HIV & AIDS;
4. By 2015 halt and reverse the spread of tuberculosis and other major diseases.

Table 4. Main indicators of science and innovations in Kazakhstan.

Year	2002	2003	2004	2005	2006
Number of organizations engaged in R&D	267	273	295	390	437
Number of specialists engaged in R&D	14109	14578	15127	17474	19127
Number of organizations in biology research	44	46	76	94	108
Number of organizations in agriculture research	51	35	87	84	92
Number of enterprises with technological innovations	n.a.	n.a.	184	352	420
The amount of investment for technological innovations (million US dollars, by official exchange rate)	n.a.	n.a.	252	516	661
Value of performed R&D works as % of GDP	0.37	0.31	0.32	0.38	0.35

n.a.: data not available
Source: Agency on Statistics of the Republic of Kazakhstan (2007)

The first and second targets do not seem feasible in Kazakhstan, since the infant mortality rate has fallen two-fold from 26.4 per 1,000 live births in 1990 to 15.15 in 2005 and the maternal mortality has fallen by only one quarter from 55 per 100,000 live births in 1990 to 40.5 in 2005 (UNDP Kazakhstan, 2006). The situation regarding the third target looks more disturbing. Registered HIV infection cases grew from 10 in 1998 to 964 in 2005. Achievement of the forth target is being hampered by growing TB rates, which increased nationally from 65.8 cases per 100,000 people in 1990 to 147.2 cases in 2005 (UNDP Kazakhstan, 2006). Other disease incidence rates such as STIs, alcohol and drug abuse have also been rising.

4 PROMOTING AN ENVIRONMENT OF NEW TECHNOLOGIES DEVELOPMENT

4.1 *Promoting technological development in Kazakhstan*

Building national capacity in science and technology is instrumental to achieving high economic productivity through new technologies to ensure human development combined with promoting economic and social welfare. To this end, it is important to review the current and future status of science and technology and look at the role of science and technology in the light of human development. Activities in science and technology mainly focus on the continuous upgrade of the technological base using the latest scientific findings in all spheres of life. Scientific and technological capacity has many components such as human resources, scientific infrastructure, funding, business participation in R&D, implementation of technological programs by the government.

To access Kazakhstan's current capacity in science and technology, the Table 4 shows its strengths and weaknesses. Weaknesses include poor technological capacity of scientific organizations and higher education establishments due to limited funding for equipment and facilities; limited inflow of young scientists as a result of low motivation; poor investment attractiveness of the science and technology sphere in nationwide. Strengths include Kazakhstan's high literacy rates and people recognizing the importance of education and technologies.

This indicates the need for action to improve scientific literacy, promote opportunities for developing research as a career for young talented scientists and researchers, as well as enhance the quality of higher education, particularly engineering education through the integration of science, higher education and business, and the promotion of market oriented research in higher education establishments (see Figure 2).

National science and technology policy created through active research and development and use of top-class new technologies should help address the issues raised in this chapter, in particular, the enhancement of human development. To this end, there is a question of how well the existing legal framework can facilitate the development and implementation of measures to improve living standards and quality of life through new technologies. Up to now, the national technological development has not been considered through the prism of human development, which explains why policy documents cannot fully address existing problems.

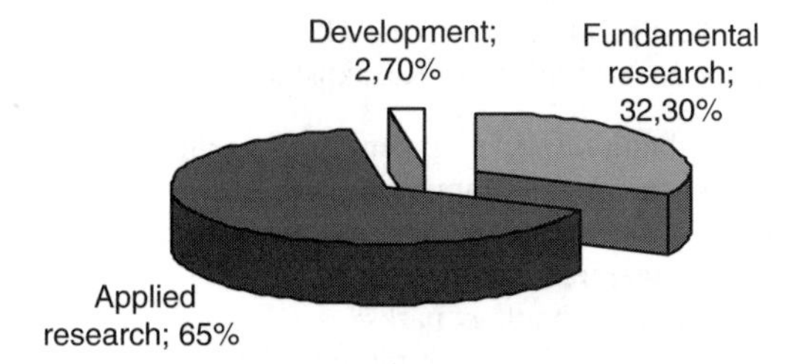

Figure 2. Breakdown of R&D by type of research, 1994–2006 in Kazakhstan.
Source: Raw data obtained from the National Centre for Scientific and Technical Information under the Ministry of Education and Science of the Republic of Kazakhstan (2007)

In terms of legislative measures, it would be advisable to develop legal control of the creation, transfer and protection of intellectual property and to create an enabling environment for those involved in the production and commercial use of scientific knowledge and technologies. Such measures should contribute to: beneficial credit of the most promising scientific and technological works; reducing taxation for individual science and technology projects; building national infrastructure for the transfer of intellectual products.

There are different mechanisms of the government contribution to an enabling innovation climate in the economy. For example, they can be funding through a system of government science and technology programs. In order to change and improve the situation, the following should be considered from the perspective of the government:

- continue active negotiation on WTO accession; harmonize national legislation with WTO rules and regulations and other related international structures and standards such as International Standard Organization (ISO) and World Intellectual Property Organization (WIPO);
- the government should pay more attention to internationalization of education, research and innovation system;
- stimulate investment in science and technology through an enabling environment both for national and international investors.

4.2 *Strategies to promote development of biotechnology in case study region*

Biotechnology allows transformation of new knowledge generated through fundamental research, into capital, triggering the development of different economic sectors. Biotechnology-based approaches to setting up new production processes allows significant increases in returns on financial and other investment. It is through the latest biotechnology achievements that economic performance can become outstanding. In many cases biotechnology is more efficient and effective than traditional technologies. Priorities in biotechnology development in Kazakhstan should take into account the specifics and prospects of economic development, such as the great share of the agroindustrial and mining complexes and rich natural resources. The reality is that the goal of becoming one of the world's 50 most competitive countries cannot be achieved without the engagement of leading international scientists and specialists, as well as biotechnology companies to transfer modern knowledge and technologies. The development and introduction of new scientific products requires a developed science and research base and local cadre to ensure the creation of competitive industrial products. The building of human capacity and technological progress are therefore mutually supportive and interdependent processes.

In addition to engaging internationally recognized biotechnology experts, Kazakhstan should provide opportunities for overseas study for its specialists. Today, biotechnology training is provided for Bolashak bursary recipients to study at international educational establishments. In addition, the National Biotechnology Center has a number of international scientific projects underway with leading universities of the US, Japan and France (UNDP Kazakhstan, 2006). This includes laboratories with Kazakhstani employees, who assist greatly with contact making, so that Kazakhstani biotechnologists can learn best practices overseas. Therefore, the following is a list of concrete recommendations to develop biotechnology in Kazakhstan:

- create an environment enabling the development of biotechnology companies and encourage investment through on improved legal framework and provision of benefits and guarantees on the part of the government;
- create a real links between research and development, and their practical application through biotechnology divisions within technology parks;
- introduce international standards of research through support to and building of technological capacity of organizations;
- create an enabling environment for international accreditation as an integral part of research projects undertaken by local scientific institutions and laboratories.

5 CONCLUDING REMARKS

Developing national capacity in science and technology is instrumental to achieving a high level of economic productivity through the use of new technologies to ensure the human development and economic and social well-being of the nation. The following challenges were identified: in the area of human resources, deteriorating prestige of the scientific profession, limited inflow of young talent in science and technology, and low qualifications of the cadre with technical education; in the area of funding of research and development, limited commitment of business and private sectors to participate in R&D programs and projects. An important factor in the introduction of new technologies is the infrastructure facilitating human development, that is, ensuring social receptiveness towards the introduction of technologies; equal access of a wide community and all groups of the population to social and technical resources; and extensive use in all spheres of life.

To ensure sustainable human development at the current stage, Kazakhstan needs to:

- Improve life expectancy. Not only should the country work to increase births, but also take care of those already born. In particular, the excessively large disparity between female and male life expectancies should be addressed;
- Address the health targets of the Millennium Declaration relating to reduced child and maternal mortality, halting the spread of HIV/AIDS and other diseases;
- Improve the quality of education based on the needs of the both economy and human development.

There should be criteria for the choice and use of new technologies in order to promote increased income levels, improved literacy and life expectancy, and other human development components. In addition, the national policy should aim to:

- Provide creative incentives and new forms of partnerships. Cooperation between academic establishments, research institutions and the private sector should be encouraged through benefit offerings such as tax exemptions, subsidies, and grants.
- Ensure commitment to laws related to the protection of intellectual property and the TRIPS Agreement, the Agreement on Trade-Related Aspects of Intellectual Property Rights, including trade of forged goods.

The ultimate goal is to use technologies to make human life more comfortable and improve human welfare, rather than simply promote overall economic development.

REFERENCES

Agency on Statistics of the Republic of Kazakhstan (2007). *Statistical Yearbook of Kazakhstan*. Astana, Kazakhstan.

Delyagin, M. (2005). *Globalization: Social crisis, social changes, social revolutions*. Moscow, Russia: Analytical Club of the Information Analysis and Management School.
Kazakhstan Governmental Decree (2006). On Approval of Programme for Digital Divide Reduction in the Republic of Kazakhstan for 2007–2009. Decree Number 995 on October 13, 2006.
Ministry of Education and Science of the Republic of Kazakhstan – National Centre for Scientific and Technical Information under the (2007). Data available at http://www.naukakaz.kz
UNDP – United Nations Development Programme (2007). *Human Development Report 2007/2008*. Available at http://hdrstats.undp.org/countries/country_fact_sheets/cty_fs_KAZ.html
UNDP Kazakstan – United Nations Development Programme in Kazakhstan (2006). *Human Development Report*. Almaty, Kazakhstan.

Integrating a Sustainable Environment into Human Development: Conceptual & Practical Approach

Fouad H. Beseiso
Former Governor of Palestine Monetary Authority

SUMMARY: This chapter aims at exploring why and how a sustainable environment can be integrated into human development. It focuses on the conceptual and practical lessons resulting from international and regional experiences pertaining to the environment and continual human development. The empirical works of the World Bank were presented which covered a wide range of environmental aspects in its relation to human growth. Practical and successful examples arising from these reviews focused on, the impact of economic, agricultural, food security, water, soil, and land systems' restructuring. Other areas covered included the impacts of institutional and legal reforms, privatization, designing systems for risk management and ecological fluctuations, disaster management, early warning system for environmental crisis and disaster mitigation.

1 INTRODUCTION

Sustainable development in Arab countries has had many achievements including economic, social, and environmental areas such as health, education and economic conditions, reduced illiteracy rates, increased life expectancy, issuing and upgrading legislation, improved capacity building, as well as the strengthening of regional cooperation in the Greater Arab Free Trade Zone, transportation, gas and electricity networks, and strengthening the Arab specialized councils and the role of civil society organizations (UN-ESCWA, 2002). However the Arab Ministers Declaration on Sustainable Development (2002) recognized that major constraints still exist. The declaration identified the following: Instability, escalating poverty and illiteracy; High population growth rates, unemployment and the dept burden; Continued population increase and the unbalanced distribution between rural and urban areas; Increased pressure on the natural resources base, as well as the public utilities and services; Air pollution, and solid waste accumulation; The severe arid nature of the region, limited areas available for agriculture, water scarcity; and shortage of non renewable sources of energy.

The Human Development Index measures the average achievements in a country in three basic dimensions (UNDP, 2008): a long and healthy life, as measured by life expectancy at birth; knowledge and education, as measured by the adult literacy rate (with two-thirds weighting) and the combined primary, secondary, and tertiary gross enrollment ratio (with one-third weighting); and a decent standard of living, as measured by the log of gross domestic product (GDP) per capita at purchasing power parity (PPP) in USD.

Serious challenges and gaps remain which handicap a successful strategy for integrating sustainable environment issues into human development. This chapter aims at exploring why and how a sustainable environment can be integrated into human development. It focuses on the conceptual and practical lessons resulting from international and regional experiences pertaining to the environment and continual human development. This study is also intended to improve public awareness of the potential benefits of adopting an integrated strategy. The analysis is based on micro and macro economic conceptual as well as a practical frame work. Methodologies and rating techniques adopted by the World Bank and intellectual achievements are carefully reviewed with respect to a strategic planning scheme for human development.

2 CONCEPTUAL FRAMEWORK FOR ENVIRONMENTAL ECONOMICS & MANAGEMENT

The history of economic thought as well as the developed concepts and instruments play a leading role in achieving a self sustained environment and paving the way to solid governance (Samuelson and Nordhaus, 2001). Economists have long studied the question of the relative importance of different factors in determining fiscal growth. Early economists like Adam Smith and T. R. Malthus stressed the critical role of land resources in growth (Samuelson and Nordhaus, 2001).

In the wealth of nations, Adam Smith (1776) (Samuelson and Nordhaus, 2001) provided a handbook of economic development. He began with a hypothetical idyllic age: that preceded both the appropriation of land and the accumulation of capital stock. This was the time when land was freely available to all, and before capital accumulation had begun to matter. What would be the dynamics of economic growth in such a golden-age? Since land is freely available, people would simply spread out onto more acres as the population increases, just as the settlers did in the American west. What about real earnings? The entire national income would go to wages because there is no subtraction for land rent or interest on capital. Output expands in step with population, so the real wage rate per worker would be constant over time. But this golden age cannot continue forever.

Eventually, as population growth continues, all the land will be occupied. Once the frontier disappears, balanced growth of land, labor, and output is no longer possible. New laborers begin to crowd into already worked soils. Land becomes scarce, and rents rise to ration it among different uses (Samuelson and Nordhaus, 2001).

Population still grows, and so does the national product. But output must develop more slowly than does population. Why? With new laborers added to fixed land, each worker now has less land to work with, and the law of diminishing returns comes into operation, the increasing labor- land ratio leads to a declining marginal product of labor and hence to declining real wage rates.

Malthus thought that population pressures would drive the economy to a point where workers were at the minimum level of subsistence (Samuelson and Nordhaus, 2001). It was reasoned that whenever wages were above the subsistence level, the population would expand; below subsistence wages it would lead to high mortality and a population decline. Only at subsistence wages could there be a stable equilibrium of population. Figure 1(a) shows the process of economic growth in Smith's golden age. Here, as population doubles, the production-possibility frontier (PPF) shifts out by a factor of 2 in each direction, showing that there are no constraints on growth from land or resources. Figure 1(b) shows the pessimistic Malthusian case, where a doubling of population leads to a less-than- doubling of food and clothing, lowering per capita output, as more people crowd onto limited land and diminishing returns drive down output per person (Samuelson and Nordhaus, 2001).

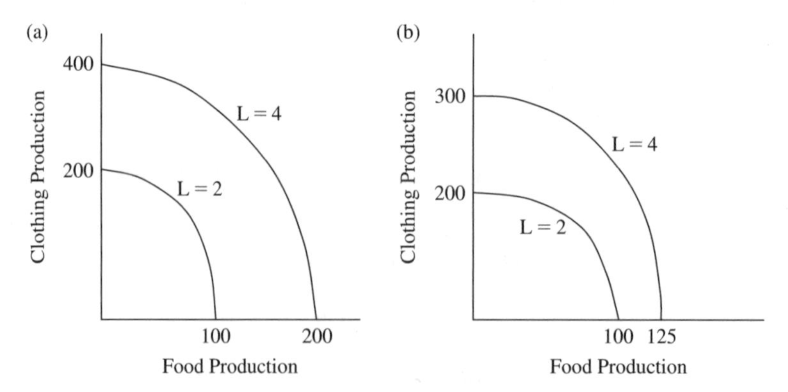

Figure 1. The Classical Dynamics of Smith and Malthus. (a) Smith's Golden Age (LHS) and (b) Malthus's Dismal Science (RHS) (Samuelson and Nordhaus, (2001), Economics. Seventeenth edition, U.S, Mc Graw Hill, p. 574).

The law of diminishing return states that the marginal product of a variable factor must eventually decline as a more of the variable factor is combined with other fixed resources. In Figure 1(a) unlimited land on the frontier means that when population doubles, labor can simply spread out and produce twice the quantity of any food and clothing combination. In Figure 1(b) limited land means that increasing population from 2 million to 4 million triggers diminishing returns.

In the last two decades, Malthusian ideas have surfaced as many antigrowth advocates and environmentalists have argued that economic growth is limited due to the finiteness of our natural resources and because of environmental constraints. Economic growth involves a rapid increase in the use of land and mineral resources and (if not controlled) in the emissions of air and water pollution. Economic experience shows why people are concerned that rapid economic growth may lead to resource exhaustion and environmental degradation.

Worries about the viability of growth surfaced prominently with a series of studies in the early 1970s by an ominous-sounding group called the club of Rome. Critics found a receptive audience because of mounting alarm about rapid population growth in developing countries and after 1973, the upward spiral in oil prices and the sharp decline in the expansion of productivity and living standards in the major industrial countries. This first wave of anxiety subsided with declines in natural resource prices after 1980 and slowing population growth in developing countries. (Samuelson and Nordhaus, 2001).

A second wave of growth pessimism emerged over the last decade. It involved not the depletion of mineral resources like oil and gas but the presence of environmental constraints on long-term economic growth. The possibility of global environmental problems arose because of mounting scientific evidence that industrial activity was significantly changing the earth's climate and ecosystems. Among today's concerns are global warming, in which use of fuels is warming the climate; wide spread evidence of acid rain; the appearance of the Antarctic ozone hole along with ozone depletion in temperate regions; deforestation, especially of the tropical rain forests, which may upset the global ecological balance; soil erosion, which threatens the long-term viability of agriculture; the species extinction, which threatens to limit potential future medical and other technologies.

Global environmental constraints are closely linked to the Malthusian constraints of an earlier age. Whereas Malthus held that production would be limited by finite land, today's growth pessimists argue that growth will be limited by the finite absorptive capacity of our environment. We can, some say, burn only a limited amount of fossil fuel before we face the threat of dangerous climate change. The need to reduce the use of fossil fuels might well slow our long-term economic growth.

The dilemma is illustrated in Figure 2. An economy begins in period 1 with the production-possibility frontier (PPF) between environmental quality and output labeled as AA. Economic growth without technological change moves the PPF to BB. In this new situation, society might experience higher output at the expenses of deteriorating environmental quality (Samuelson and Nordhaus, 2001). A more acceptable state occurs when technological change introduces, for

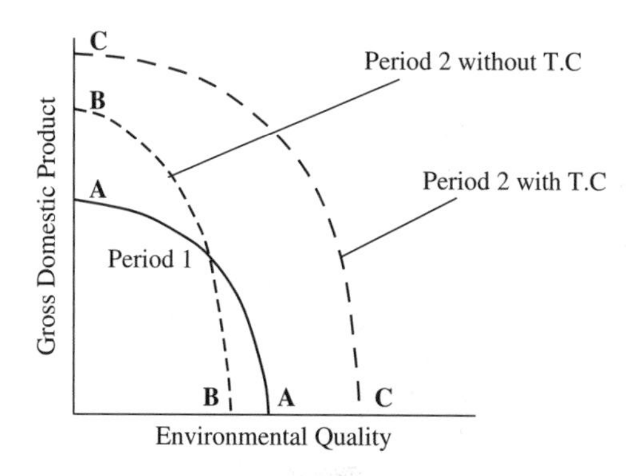

Figure 2. Environmental Constraints and Technology Role.

example, equipment to mine and burn low-sulfur coal, or requiring pollution-control devices on automobiles, or developing economical solar power. This pushes out the PPF to CC so that society can have both more output and cleaner environment (Figure 2).

What is the empirical evidence on the effect of resource exhaustion and environmental limits on economic growth? The quality of land and mineral resources has deteriorated over the last century such that we are required to drill deeper for oil, use more marginal lands, and mine lower-grade mineral ores. But until now, technological advances have largely offset these trends, so the prices of oil, gas, most minerals, and land have actually declined relative to the price of labor. Moreover, new environmentally friendly technologies have become increasingly important, and many of the worst environmental abuses have been alleviated. Nonetheless, ecological constraints have become more costly, and some economists believe that the United States, for example, has experienced a significant slowdown in measured productivity growth because of the costs of environmental regulations (Samuelson and Nordhaus, 2001).

2.1 *Environmental and welfare economics*

Environmental management needs economic theory and the economy needs the environment. This very simple principle underpins the realm of Environmental Economics. Long before the environmental revolution in the 1960s, the world of economics was generating analytical tools, which have become relevant for environmental management (World Bank, 2005a and 2005b). The most important contributions of economic theory can be synthesized into two elements: The concepts of externalities and public goods, and the economics of welfare.

The theory of externalities dates backs to the seminal work of Pigou in 1920 (Bolt et al., 2005). An externality occurs when a benefit or a cost incurred by a party is caused by somebody who does not take this effect into account in their decisions. Environmental science has shown that environmental externalities can be pervasive and effect individuals across space and time dimensions.

One of the objectives of economists has been to analyze solutions to externality problems such as through the use of taxes, named Pigouvian taxes after economist who proposed them, and regulations. Often governments need to intervene in the case of externalities. However, this perception was modified after the publication of the 1960 paper by Ronald Coase, who suggested that parties could negotiate a solution to externalities in the absence of government intervention (Bolt et al., 2005). However, such negotiation is unlikely when there are many individuals affected, as in the case of much pollution.

The concept of public good is intimately related to the notion of an externality. Public goods, such as clear air, coastal views and broadcast radio waves, have two main characteristics: Everybody can use them without depleting their availability for others. Economists call this "non-rivalry", and it is very difficult, technically, to prevent people from using them. In other words, public goods are non-excludable.

The problem with public goods is that everyone has a relatively small incentive to provide the good. Therefore, people will tend to free-ride on others providing it. As a consequence, public goods are under-provided or, reversing the argument, there will be an over-provision of public bads (such as air pollution or ozone layer depletion). As in the case of externalities, state action is usually required to solve the problem. The question is whether such action is justified or, in other words, whether the benefits balance the necessary costs of publicity providing the good.

Environment strategies have been a reflection of world realized vision corresponding to the required integration of plans, policies, programming between sustained environment activities and welfare economics. A welfare economy is the study of how the allocation of economic resources affects the material well-being of consumers and producers (Hirschey, 2006). It is society's best interest to have the largest benefits tied to consumption and production activities (Figure 3) (Bolt et al., 2005). Differences between private and social costs of benefits are called externalities; a negative externality is a cost of producing, marketing, or consuming a product that is not born by the product's producers or consumers. Environmental pollution is one well-known negative externality.

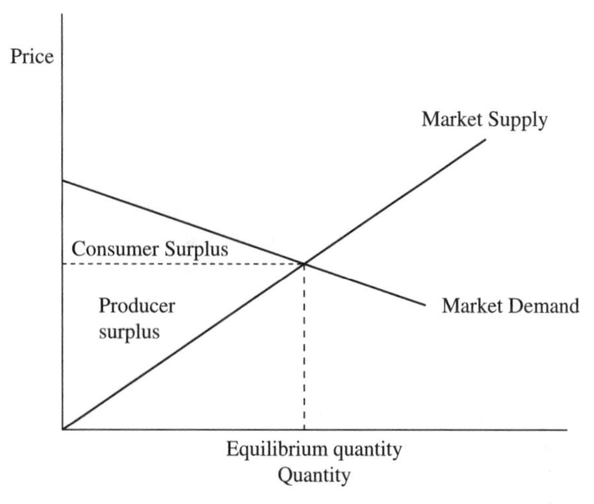

Figure 3. Consumer Surplus, Producer Surplus, and Social Welfare in Competitive Market Equilibrium. Social welfare is the sum of the consumer and producer surplus. (Hirschey, 2006).

The theory of welfare economics was developed by economists, such as Hicks and Kaldor, in the 1930s and 1940s (Hirschey, 2006). It provided for clear decision-making in virtually all cases where a public action would benefit somebody and cause costs to others. The compensation criterion established that an action was justified on efficiency grounds if the winners from the policy could potentially compensate the losers and still be better off compared to the initial situation. This was the case even if no real compensation took place.

Environmental economists are interested in the concept of value from a strictly anthropocentric point of view. What is being examined is the willingness of individuals to spend scarce resources on the environment that could very well be used for alternative purposes. This means that if individuals consider migratory birds as very important, their propensity to spend money for bird conservation will be high. But what is being valued is not the intrinsic worth of birds, which is totally independent of the existence of man, but rather, the importance that man attaches to such birds. Yet another form of positive economic value arises even when the good in question is consumed, exhausted or even seen. Individuals in fact may have a propensity to pay for a good they will never see, just to insure somebody in the future will.

Economists define economic value as the maximum willingness to pay for an environmental or natural resource. This is defined as the area below the demand curve for the resource. Again, assuming that the environmental good has some importance to individuals, a demand for the good must exist even if no explicit market transaction takes place. No markets for clean air can be observed in the real world, but if we look into individuals' behavior we may notice that they actually give up other resources to mitigate the effects of dirty air around them. For example individuals spend money on air filters to avoid exposure to air pollution or on asthma treatments to ameliorate its effects. This information is what allows economists to measure economic values. (Bolt et al., 2005).

2.2 *Role of government*

When considering the role of government in the market economy, it has been traditional to focus on how management influences economic activity through tax policies, trade policy, law enforcement, and infrastructure investments in highways, water treatment facilities, and the like (Hirschey, 2006). Therefore, the incidence, or placement, of taxes and the costs and benefits of regulatory decisions are important. If a given change in public policy provides significant benefits to the poor, society may willingly bear substantial costs in terms of lost efficiency. The government has a vital role in maintaining and establishing a healthy economic environment. It must ensure respect for the rule

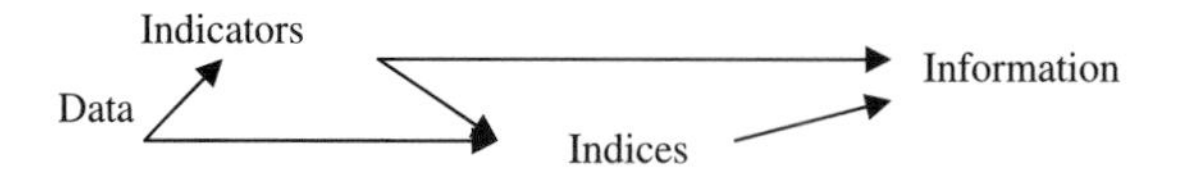

Figure 4. From data to information.

of law, enforces contracts and orient its regulations towards competition and innovation. A leading role is the investment in human capital through education, health and transportation.

Government should focus its efforts on areas where there are clear signs of market failures. It should dismantle regulatory impediments to the private sector in areas where government has comparative disadvantage (Samuelson and Nordhaus, 2001). Policy makers must decide when to intervene and when to adopt a hands-off strategy.

Free economy based on competitive markets may lead to distortions to cost benefits tied to production and consumption and disequilibrium that is reflected in market prices. Such situation necessitates public policy intervention (Hirschey, 2006). Consequently public policies and programs related to environmental and human development could achieve a self sustained quality well being by creating positive externalities to producers and consumers that are not reflected on prices. This concept necessitates the coordination between economic and environmental management.

2.3 *Sustainable development indicators*

The Rio Conference on Environment and Development in 1992 recognized the need for better knowledge and information about environmental conditions, trends, and impacts. New thinking and research were needed with regard to indicator frameworks, methodologies, and actual indicators (Segnestam, 2002).

In working with indicators, there are several terms that figure frequently. The most common ones are data, indicator, index and information (Figure 1). Indicators, which are derived from data, are commonly the first, most basic tools for analyzing changes in society. They can work as a basis for assessment by providing information on conditions and trends of sustainable development. Secondly, as a basis of such assessments, indicators can provide input to policy formulation processes. Thirdly, by presenting several data in one number that commonly is simpler to interpret than complex statistics, they can facilitate communication between different groups. If two or more indicators are combined an index is created. Indices are commonly used at more aggregated analytical levels such as at the national or regional levels (Figure 4) (Segnestam, 2002).

The World Bank and other regional and international institutions have been involved in various indicator projects (World Bank, March 2002). Wealth estimate and genuine saving, estimates, for example, have been combined into a synthetic indicator that measures both environmental and economic factors. The wealth measure is a stock measure and is a new way of estimating a country's total resources, including both produced assets, natural capital, and human resources. Genuine saving, a flow measure, adjusts gross savings numbers by the value of depletion of the underlying resource asset and pollution damages, and considers current educational spending as an increase in saving. For more information about these estimates, see http://www-esd.worldbank.org/eei/wealthgen.html.

As well as serving as an indicator of sustainability, genuine saving has other advantages as a policy indicator. It presents resource and environmental issues within a framework that finance and development planning ministries can understand. It reinforces the need to boost domestic saving, and hence the need for sound macroeconomic policies. It highlights the fiscal aspects of environment and resource management, since collecting resource royalties and charging pollution taxes both raise development finance and ensure efficient use of the environment. Measuring genuine saving also makes the growth-environment tradeoff more explicit because countries planning to grow and to protect the environment will have depressed rates of genuine saving (World Bank, 2002).

Hamilton (2000) explored the issue of measuring changes in per capital wealth. He developed a theoretical approach to estimating total wealth. Then cross-country estimates were presented of changes in per capita wealth. Based on preliminary estimates, it was concluded that in the majority of countries below the median per capita income, wealth was accumulating more slowly than the population was growing.

A project is under development within the World Bank whereby task managers will be able to get ideas for environmental performance indicators on the intranet. The indicators will be presented in lists conveying different sectors and cross-cutting issues (e.g., institutional issues) and will be made externally available as soon as possible. For information on these initiatives, as well as other parts of EEI's work program, refer to website at http://www-esd.worldbank.org/eei.

The Land Quality Indicator (LQI) program, and the global coalition being assembled to develop and disseminate the findings of the program, is an important contribution towards more sustainable management of the world's land, water and biological resources. Indicators developed by the program include nutrient balance, and yield gap (productivity) analyses. More information and publications can be found at http://www.ciesin.org/lw-kmn/.

Another project, which is a collaboration between CIAT, UNEP and the World Bank, has as its objective to develop, test and refine environmental, land quality indicators and information tools in a user-friendly geographic information system (GIS) interface. This will be employed for policy-making and planning in Central America countries (http://www.ciat.cgiar.org/indicators/index.htm).

The Compendium of Sustainable Development Indicator Initiatives and Publications site provides an overview of initiatives on sustainable state levels (http://iisdl.iisd.ca/measure/compindex.asp). It has been prepared by the International Institute for Sustainable Development, Environment Canada, Redefining Progress, and the World Bank.

2.4 *Community based development*

Enabling an institutional environment that is responsive to community needs is considered a pre-requisite for sustained and environmental development. In emergent countries, scaling-up focuses the agenda on more transparent and accountable governance at the national and local levels, as well as a legal and regulatory framework that recognizes community groups and protects them from arbitrary regulation, oversight and rent-seeking. Good information flows are also essential, as are continuity of efforts over time to bring about sustainable change in behavior and attitude (McNeil et al., 2004).

The World Bank Institute has initiated Capacity Enhancement Needs Assessments (CENAs) in Nigeria, Tajikistan, and Ghana in an attempt to develop viable assessment tools to better guide the institute in meeting the training and capacity building needs of a client country (McNeil et al., 2004). The CENA approach is action-oriented and relies on development actors to carry out their own needs assessments and propose remedial activities. In particular, the CENA focuses attention on the institutional environment in the thematic area identified. In addition, emphasis is placed on the relational aspects between institutions and communities, between local and central government, and between and among donors and NGOs involved in a country.

Institutional building for sustainable environment management has been internationally recognized as a strong pre-requisite for sustainable human development (UN-ESCWA, 2002). Biodiversity should be conserved by preserving habitats and creating information systems and gene banks. Integrated and intraregional approaches to coastal and marine resources management should be encouraged.

2.5 *Water management, population growth and education*

The supply of freshwater in the Economic and Social Commission for Western Asia (ESCWA) region is found to be highly variable depending on the season and the location of the source. Freshwater scarcity is in the region is clearly manifested by the low average per capita share of renewable water (slightly less than 1000 m^3/year), which is expected to decline further to an average of 570 m^3/year by the year 2025. Scarcity is also evident by the fact that six ESCWA member States (Bahrain, Jordan, Kuwait, Qatar, the United Arab Emirates and Yemen) are among the world's poorest countries in water, with per capita annual shares of less than 200 m^3 (UN-ESCWA, 2002).

Groundwater resources play a major role in satisfying water demand in the region. Fossil water sources meet about 15 per cent of water demand. However, easily accessible groundwater resources are increasingly over-exploited, which risks exacerbating damage to underground water reserves through saltwater intrusion and the seepage of pollutants. Alternatively, deep fossil groundwater basins offer the potential of additional groundwater reserves that are not fully explored or utilized because of the high investment cost required to reach them.

The main challenges to environmentally sound water resource management in the region are: rapidly increasing population pressures; accelerated development and competition for water between the urban, industrial and agricultural sectors; ineffective water management policies and practices; erratic precipitation; and the highly volatile regional peace and security situation. Growing water claims of upstream countries of shared water resources has resulted in further reducing the amount of water available for downstream ESCWA member countries, especially Syria, Iraq and Jordan.

The regional political and security situation also stymies efforts in water resource management. This is important since in 2000 it was estimated that nearly 80 per cent of the ESCWA region's annual renewable water resources (about 176 BCM), which provides water for 70 per cent of the region's population, flows from outside the region (UN-ESCWA, 2002).

Appropriate preparations should be made in infrastructure to support human resources in the region, including investment in education, and training. Fast population growth rates and imbalances in population distribution have resulted in increasing demands and pressures on finite natural resources and limited urban services. With the exception of the Arabian Gulf countries, the population growth rate is imposing a serious problem. Census and other social statistical data should be used in master planning, and linkages between populace expansion and available natural resources. Population issues need to be managed intelligently in order to achieve sustainable development.

The region needs to invest in capacity building and empowering its young population. Sound education, proper training, and proper management of human resources are required to produce the type and quality of human capital needed to fuel sustainable development. Scientific research and technology development should address the priorities of the region, and serve achieving sustainable development.

The development and transfer of appropriate technology will make a significant contribution to the achievement of sustainable development. The use of cleaner production strategies, investment in relevant indigenous technology, encouragement of research, the transfer of appropriate but not necessarily the most advanced technology, and linkage of research and development to market demand in order to provide the appropriate technology at reasonable price should all be encouraged. These activities will also assist in the reduction of brain drain. Information technology for sustainable development should be harnessed, with improved access and better training (FAO 2007).

It has become evident that the culture heritage of the region has been under-utilized, especially as a tool for development. Family values must be respected and protected against foreign norms. Media and civil society should be empowered to play a leading role in this area. The role of cultural heritage and Islamic values should be emphasized in pursuing sustainable development tracks. Countries of the region should intensify efforts to revive, preserve, maintain, and prudently manage natural and cultural heritage. Indigenous knowledge and cultural heritage should be used as a tool for economic growth, e.g., ecotourism. Cultural dimensions must be incorporated into the formulation of all development policies and project interventions.

The scarcity of financial resources to invest in preserving and developing this sector can be alleviated through policy and institutional reform for increasing the sector's self-financing, and by drawing the interest of the international community to invest in the region's culture heritage as a world heritage (UN-ESCWA, 2002).

2.6 *Globalization, trade and financing*

Globalization transcends geographical, financial and cultural barriers. Developing countries, indigenous peoples and cultures and Small and Medium Enterprises (SMEs) SMEs face the potential risk associated with globalization. In this connection, the international trade structure has witnessed major changes since the inception of the World Trade Organization. Trade liberalization should be pursued in the countries of the region according to a time plan associated with building the necessary infrastructure and the reform of institutions and policies necessary to achieve trade liberalization while minimizing possible negative impacts.

To improve competitiveness and increase market access, countries need to improve efficiency, product standards, and to align production technologies with international standards, including environmental standards (e.g., ISO 14000, eco-labeling). They must also diversify products and services, and export manufactured goods rather than raw materials. Furthermore, national economies should be integrated into regional and global economies to take advantage of international trade systems.

Countries need to create the right environment for full inclusion in globalization including essential infrastructure, services, and institutional set-up. Further, they should in cooperation with civil society, properly assess the impacts of international treaties on the region before signing on. To counter the undesirable values and consumption patterns that may result, it is important to prompt local cultural values (UN-ESCWA, 2002).

Achieving sustainable development requires not only legal and institutional changes but also financial arrangements to cover the initial cost of moving to new patterns and mechanisms of development. The following are some of the options that countries can undertake (UN-ESCWA, 2002):

- Encouraging the involvement of the private sector to invest in sound sustainable development projects;
- Intensifying efforts and developing programs aimed at mobilizing internal financial resources, such as mobilizing domestic savings and investment, enforcing the polluter pays principle, and adopting environmental fees and license schemes;
- Applying self-financing mechanisms to some sectors, such as the cultural heritage sector;
- Adopting policies, which increase exports, while rationalizing imports;
- Encouraging, with the help of civil society institutions, citizens to choose national and regional products and services.
- Undertaking feasibility studies by countries and the private sector prior to requesting loans and ensuring rational borrowing and efficiency in the management of debts;
- Ensuring that money from loans is used for sustainable development programs and projects, transparently and with full accountability;
- Improving the economic performance of governments through efficient

Financing for sustainable development in the region has remained limited, mostly because external indebtedness continues to sap financial resources. Falling tariff revenues caused by trade liberalization and economic restructuring has forced governments to finance national debt and public expenditures through alternative means, including taxation.

Foreign direct investment, privatization, environmental and social funds and micro-financing are means for increasing financial support for sustainable development. However, most firms still lack adequate access to resources from local financial markets to support environmental investments. Reducing the debt burden in view of tightening budgets would release funds for sustainable development programming. Furthermore, while regional and international financial institutions have provided important levels of financing for sustainable development aid levels have decreased in recent years.

It should be noted that one of the most important and neglected aspects of financing for sustainable development in the region concerns the lack of effective oversight and monitoring of allocated project funds. While individual donor institutions and agencies might require financial monitoring and auditing of grants and loans in a piecemeal fashion, none of the countries in the region have a comprehensive system for assessing the effectiveness of financial instruments for building national capacity or facilitating progress towards sustainable development. This is because beneficiary countries generally look at financial assistance in a piecemeal fashion and in terms of quantity, and not of quality. This belief, however, is beginning to change among donor institution in light of increasing fiscal constraints. Accordingly, beneficiary countries of monetary assistance should seek to improve both the quality and effectiveness of each donor dollar, as well as to increase the supply of aid contribution in order to more appropriately finance sustainable development (UN-ESCWA, 2002).

2.7 *Monitoring and evaluation systems*

The system of monitoring and performance evaluation (M&E) proved to contribute to the effective implementation of strategies and programs (UN, Secretary-General Bulletin, 1987). Developing countries face growing pressures to monitor changes in a broad range of environmental and natural resource conditions. Part of this demand stems from the need to comply with various international agreements on environmental issues (e.g., on biodiversity and green house gas emissions), and part from the need to improve the quality of the lives of their citizens.

Unfortunately, monitoring is not a costless activity, and public funds have to be used that have alternative uses (e.g., investments in schools, roads, and education). Moreover, many monitoring activities are passive and do not lead to the changes needed to rectify the problems they identify. Monitoring can all too easily become an end in itself, particularly once it has been institutionalized. If monitoring systems are to serve a viable social function, then priority should be given to environmental problems that offer a potentially large social payoff relative to the costs of monitoring.

Realizing favorable benefit/cost ratios is more likely if (Hazell, 2000) the environmental problems selected for monitoring have high social costs if left unchecked; the monitoring system is designed and used in a way that leads to rectification of the environmental problems that are being monitored, and the monitoring system is designed and operated so that it is cost effective.

3 PRACTICAL BASIS FOR A SUSTAINABLE ENVIRONMENT

The World Bank environment department reviewed forty interim and full Poverty Reduction Strategy Papers (PRSPs) on environment related issues from countries in Africa, Latin America and Eastern Europe, the Middle East, Central and East Asia. Four major questions: were posed: (i) What issues of environmental concerns and opportunities are identified in the PRSPs?; (ii) To what extent are poverty-environment causal links analyzed?; (iii) To what extent are environmental management responses and indicators put in place as part of the poverty reduction efforts?; and (iv) To what extent has the design and documentation of the process allowed for mainstreaming the environment? (World Bank, 2005a & 2005b). The environmental issues that were reviewed included many important activities such as: macroeconomic management, developing economic and financial instruments that promote sustainable management of natural resources participatory approach to implement the world agenda.

Strengthen the institutional mechanisms for participation and macroeconomic policies including subsidies, regulation, pricing, taxation, exchange rate, dept and other policies affect the environment in several ways (Tarrant, J. R. 1980):

- Management of natural resources, water resources, food security, land ownership and property rights, soil degradation, cost assessment of projects and programs, and limiting the impact of fluctuations .
- Promoting environmental early warning system.
- Environment legislations.
- Reducing ecological vulnerability.

The treatment and sensitivity to environmental issues were highlighted by the World Bank (World Bank, 2002, and Bojo, 2002). The Burkina Faso Poverty Reduction Strategies and Environment (PRSP) noted that climatic conditions, land locked status, low agricultural productivity, and degradation of soil and water resources, are major constraints to economic growth and contribute to massive poverty and severe food insecurity among rural inhabitants. Income from farming and livestock are highly dependent on rainfall which varies considerably from year to year.

Most poor households depend upon wells for drinking water and pit latrines for sanitary purposes. In 1998, for example, only 37 percent of the population had access to proper sanitary facilities in Gambia (World Bank, 2002, and Bojo, 2002).

Pollution resulting from lack of environmental regulation and the impacts on human health is well illustrated by the Honduras PRSP. Lack of land use and urban development planning, has contributed to increased problems of environmental deterioration and pollution in the main urban centers of the country, with impacts on human health... pollutants in soils and water cause high rates of diarrhea illnesses in Tegucigalpa (World Bank, 2002, and Bojo, 2002).

3.1 *Public policy and incentive structure*

Georgia's energy sector reform reflects the impacts of privatization on the poverty-environment relationship (World Bank, 2002, and Bojo, 2002). With the collapse of the Soviet Union, Georgia lost access to cheap energy resources. Privatization of the electricity market raised the energy tariff by 2.4 times up to about 20 percent of the average family income. Inability of the poor to pay for electricity resulted in increased demand for wood. A system of energy allowances to households adopted early in the reform was found to be inadequate and inefficient to meet the targeted budgetary support to poor.

The Ghana PRSP highlights the benefits of the Structural Adjustment program to natural resource management. Community water supply and sanitation benefited from the injection of capital and restructuring of the Ghana Water and Sewage Corporation. The water tariff reform was to take into account the ability to pay of poor households and the financial viability of the utility.

In Moldova, a series of natural disasters, terms-of-trade shocks from liberalized energy prices, and regional instability caused intermittent policy reversals in the implementation of the structural reform program and resulted in adverse impacts on the standard of living. A social protection plan was implemented to rationalize the energy prices, eliminate non-targeted energy subsidies and limit the energy subsidies to the most vulnerable groups. The plan is likely to benefit government finances and improve the targeting of social protection program.

Georgia's PRSP proposed privatization of land and water resources, promotion of a land market, creation of water user associations, and implementation of a rural credit policy establishing guarantee funds and insurance against climatic hazards. The PRSP was also aimed at improving soil and water resource management, restoration of agricultural infrastructure, ensuring property rights and income sources to vulnerable groups.

While process issues, participation and design cannot be expected specifically highlight the involvement and environmental constituencies, the more inclusive the design allows such voices to be heard. It is interesting to note that most of the countries that have been identified as providing good practices relating to process and participation also score high on mainstreaming: Bolivia, Kenya, and Nicaragua are such examples.

To strengthen the institutional mechanisms for participation, Armenia proposed to conduct focus group discussions, stakeholder analysis, social assessments, an information campaign on the poverty strategy, and to collect feedback from key stakeholders. In Bolivia, National Dialogue 2000 was initiated as participatory mechanism for implementing a social, economic and political agenda. The processes provided an opportunity for civil society consultation on a poverty reduction and development strategy. A series of workshops under the name of "Government Listens" was instrumental in providing society input on environment, capacity building, gender, participation, opportunities, vigilance, and monitoring.

3.2 *Environment management capacity and governance*

To reorganize development and management of national land resources and make property rights more secure, Burkina Faso adopted the environment code, forestry code, mining code, and water code. Plans are being adopted to implement the codes under the National Land Management Program (Bojo, 2002).

The Cambodia PRSP proposed to establish a hydrological information system of surface and ground water sources that supports strategic planning and environment-friendly development of water resources for irrigation, potable water, hydropower, fishery, and flood control. Honduras planned to reduce ecological vulnerability by improving risk management at both central and decentralized levels and develop economic and financial instruments that promote sustainable management of natural resources. Nicaragua PRSP stated that the government was strengthening its capabilities in risk reduction... geographic information system to map natural threats, developing early warning systems, producing geological and warning maps, and improving its monitoring of volcanoes and areas vulnerable to land slides.

In 1998, the Central American Environment and Development Commission estimated the annual economic losses of Honduras deforestation, in terms of damage to timber and non-timber products, biodiversity losses, and losses of affected water resources and ecotourism at about US$112 million (Bojo, 2002). In Yemen, per capita water supply was about 2 percent of the world average and 88 percent below the amount needed for domestic use. Ground water had reached a state of over exploitation as a result of over drilling simulated by diesel subsidies.

A range of vulnerabilities related to natural hazards presented by PRSPs are summarized as follow: An earthquake in 1988 affected 40 percent of Armenia's geographical area and one-third of country's population, with the poor being the most affected. In addition to loss of life, it disrupted critical services like housing, water supply, and sewage systems. The Honduras PRSP presented a detailed assessment of vulnerability due to hurricanes. The PRSP noted that Hurricane Mitch had a severe impact on living conditions in Honduras and this in turn affected poverty levels nationwide. It was estimated that the number of poor households rose from 63 percent in 1998 to 66 percent in the following year. The damage to total capital stock of the country was estimated at $3800 million, accounting for 7 percent of GDP. In case of the housing sector alone, the damages were estimated at US$344 million from the total damage of 35,000 houses and 10 to 50 percent damage to another 50,000 houses (Bojo, 2002).

Can differences in how political markets function and shape the incentives of political agents explain the large variation in the performance of developing country governments in delivering broad public services and reducing poverty? The main conclusion of a World Bank report (World Bank, 2005b) was that much of the variation in public service delivery across developing countries can be attributed to differences in the incentives of politicians. Even in emergent countries that are democracies, where politicians depend on the poor for support, public expenditures often fail to deliver basic services to the deprived because political agents have incentives to misallocate public resources to private rents and to inefficient transfers that benefit a few citizens at the expense of many (World Bank, 2005b). Such misallocation can be traced to imperfections in political markets that constrain the extent to which poor people can hold governments accountable for their actions, such as lack of information about service quality, lack of credibility of political promises, and polarization of voters on social and ideological grounds. Institutions such as decentralization, party systems, electoral regimes, and constitutional rules interact with these political market imperfections in determining outcomes.

3.3 *Corruption, public policies and enforcement*

Corruption is a substantial impediment to development in poor countries (World Bank, 2005a). Using individual-level data for 35 countries to investigate the microeconomic determinants of attitudes toward bribery, the World Bank report found that woman, the employed, the less wealthy, and older people were more averse to corruption. It also provided evidence that social effects play an important part in determining individual attitudes toward dishonesty, as these are robustly and significantly associated with the average level of tolerance. These findings lend empirical support to theoretical models in which corruption emerges in multiple equilibriums and suggests that big push policies might be particularly effective in combating dishonesty.

Another analysis looked at corruption and economic openness, investigating whether the presence of barriers to international trade and capital flows was associated with higher fraud. The evidence suggested that the main effect of trade barriers on bribery came through the incentive for collusive behavior between individuals and customs officials rather than from the reduced pressure from foreign competition. Interestingly, no clear association emerged between corruption and variables proxying for the presence and intensity of controls on capital flows.

A third analysis examined the role of trade tariffs, explicitly accounting for the interaction between importers and corrupt customs officials. It argued that setting tariffs at a uniform level not only limited the ability of public officials to misclassify imported goods and thereby extract bribes from importers, but also could deliver higher government revenue and welfare than a Ramsey tariff structure when corruption is pervasive. The empirical evidence suggested that a highly diversified menu of trade tariffs might fuel dishonesty, as a significant and robust association between an appropriately computed measure of tariff diversification and corruption in customs emerges across countries.

The results of the study should help to improve the conception, design, and support of policies and projects through greater participation of stakeholders and early planning for flexibility and effectiveness. The consequences may also prompt a shift in the focus of evaluation from increases in incomes to improvements in opportunities and capabilities. (World Bank, 2005a).

Finally, most local authorities in developing countries have not allocated sufficient resources to improving water quality, despite the known economic and financial benefits of such efforts. While many factors contribute to this situation, a critical one is lack of public pressure and lack of broad participation in the public decision-making process. Earlier experience with schemes to publicly disclose environmental performance showed that this lack of public pressure in turn results from insufficient access to information (World Bank, 2005a). This experience also suggests that a formal system for ranking the performance of local governments in protecting water resources can generate strong incentives for them to invest in maintaining or improving water quality.

3.4 *Capital markets and environmental performance*

Designing effective incentives for pollution control requires an understanding of what determines the environmental performance of industrial enterprises. Earlier research on agents that may exert pressure on enterprises to improve their environmental performance has shown that capital markets, including in such countries as Argentina, Chile, Mexico, and the Philippines, react to news relating to a firm's environmental performance (World Bank, 2005b). But do the reactions of capital markets, reflected in lower market values, and then induce enterprises to improve their environmental performance?

Since 1989 Korean environmental authorities have published a monthly list of enterprises that have violated the country's environmental rules and regulations. In 1993–2001 these monthly lists recorded more than 7,000 violations, involving more than 3,400 different companies. A descriptive analysis of this data set suggested that the news media have given important (through perhaps declining) coverage to the violations. When completed, the research will provide policy makers with clear guidelines on how public information can affect enterprises' environmental performance through the capital markets.

4 CONCLUDING REMARKS

Sustainable development in Arab countries in particular as well as in developing countries in general has had many achievements including economic, social, and environmental. However major constraints still exist: instability, escalating poverty and illiteracy; high population growth rates, unemployment and the debt burden; continued population increase and the unbalanced distribution between rural and urban areas; increased pressure on the natural resources base, as well as the public utilities and services; air pollution, and solid waste accumulation; the severe arid nature of the region, limited areas available for agriculture, water scarcity; and shortage of non renewable sources of energy.

Institution building for environmental management has been internationally recognized as a strong pre-requisite for sustainable human development. Achieving this requires not only legal and institutional changes but also financial arrangements to cover the initial cost of moving to new patterns and mechanisms of development.

Enabling an institutional environment that is responsive to community needs is considered a prerequisite for sustained and environmental development. Good information flows are essential, as are continuity of efforts over time to bring about sustainable change in behavior and attitude.

The system of monitoring and performance evaluation, for example, contributes to the effective implementation of strategies and programs.

Strengthen the institutional mechanisms for participation and macroeconomic policies including subsidies, regulation, pricing, taxation, exchange rate, and debt affect the environment in several ways: management of natural resources, water resources, food security, land ownership and property rights, soil degradation, cost assessment of projects and programs, and limiting the impact of fluctuations, promoting environmental early warning system, environment legislations, and reducing ecological vulnerability.

In the area of corruption and economic openness, the evidence suggested that the main effect of trade barriers on bribery came through the incentive for collusive behavior between individuals and customs officials rather than from the reduced pressure from foreign competition. Interestingly, no clear association emerged between corruption and variables proxying for the presence and intensity of controls on capital flows.

REFERENCES

Bojo, J and R.C. eddy (2002) World Bank, Poverty Reduction Strategies and Environment – A review of 40 Interim and Full Poverty Reduction Strategy Papers, Paper No. 86. June 2002.

Bolt, K., R. Giovanni and S. Maria (2005) Estimating the Cost of Degradation A training Manual, World Bank, September 2005. pp. E.1–E.14.

FAO (2007) High Level Conference on World Food Security and Global Challenges, Thirty Fourth Session, Rome, 17–24 November 2007.

Hamilton, Kirk (2000) Sustaining Economic Welfare: estimating changes in per capita wealth, policy Research working paper no. wps 2498.

Hazell, P. (2000) The Design of Policy Relevant Monitoring Systems for Natural Resources, World Bank Publications, Environmental Economics Indicators.

Hirschey, M. (2006) Economics for Managers, U.S, Thompson South Western, p. 243 & pp. 767–691.

McNeil, M., K. Kuehnast, A. O'Donnel and Kuehnast with O'Donnel, Anna (2004) Assessing Capacity for Community-Based Development, World Bank Institute, A pilot Study in Tajikistan, December 2004, Washington D.C., pp. 4–15.

Samuelson, P.A and W.D. Nordhaus (2001) Economics. Seventeenth edition, U.S, McGraw Hill, pp. 573–601.

Segnestam, Lisa (2002) Indicators of Environment and Sustainable Development – Theories and Practical Experience, Environmental Economic Series, paper no. 89, pp. 1–13.

Tarrant, J.R. (1980) Studies in Environmental Management and Resources Development, New York: John Wiley & Sons, pp. 296–317.

UNDP (2008) Human Development Report 2007, Human Development Index, from Wikipedia the free Encyclopedia, January 2008. pp. 1–5

UN-Economic and Social Commission for Western Asia (2002) Arab Ministerial Declaration on Sustainable Development, World Summit on Sustainable Development, Johansburg, South Africa, 26 August–4 September 2002.

UN-ESCWA (2002) The Effects of Socio Economic Inequity on Sustainable Development in the ESCWA Region, World Summit on Sustainable Development, Johannesburg, 26 August–4 September 2002, E/ESCWA/SDP/2002/17.

UN Secretary General Bulletin (March 1987) Regulations and Rules Governing Program Planning, The program Aspects of the Budget, The Monitoring of Implementation and the Methods of Evaluation, ST/ SGB/ PPBME Rules/1987.

World Bank (2002) Poverty Reduction Strategies and Environment – A review of 40 Interim and Full Poverty Reduction Strategy Papers, Washington D.C.

World Bank (2005a) The World Bank Research Program – Abstracts of Current Studies, Environment Department Papers, Washington D.C, pp. 65–144.

World Bank (2005b) Environmental Economic Series, pp. 9–23.

World Bank (March 2002) Environmental Indicators, An Overview of Selected Initiatives at the World Bank, Updates 3 March 2002, pp. 1–12.

Chemical Evolution of the Milky Way Galaxy and its Relevance to Life on Earth

Ahmad Abdelhadi
New York Institute of Technology, Amman, Jordan

SUMMARY: The environment in which mankind currently lives is the end result of a long process of chemical and biophysical development which is not, as yet, fully understood. The main aim of this chapter is to provide insight into star-molecular cloud encounters, and thus to improve our understanding of the evolution of the solar system. The scattering of stellar orbits by galactic molecular clouds was studied in hope of explaining isotopic peculiarities of presolar grains from Asymptotic Giant Branch (AGB) stars. Silicon isotopic anomalies found in the mainstream silicon carbide (SiC) grains, for example, were observed to have heavy isotopes enrichment. This work aims to seek a possible theory for such richness and peculiarity through scattering of AGB stars in molecular clouds. A successful explanation to this problem will influence how we think nuclides were formed and then distributed in the Galaxy. It will shed new light unto the chemical environment on Earth This type of study will allow us to better appreciate the earth's environment and how it has affected the evolution of life on the planet. The scientific community, not just astrophysicists, shares such interests.

1 INTRODUCTION

Life on earth has evolved from a complex chemical and biochemical process over millions of years (Binney and Merrifield, 1998; Pudritz, 2002; Solomon and Sanders, 1985; Ward-Thompson, 2002). The earth in turn has evolved from the Milky Way Galaxy over billions of years. The environment in which mankind currently lives is thus the end result of a long process of chemical and biophysical development which is not, as yet, fully understood.

The main aim of this chapter is to provide insight into star-molecular cloud encounters, and thus to improve our understanding of the evolution of the solar system. This will allow us to better appreciate the earth's environment and how it has affected the evolution of life on the planet.

2 EVOLUTION OF THE MILKY WAY GALAXY

Ancient stardust has made its way to Earth in meteorites. These silicon carbide grains are peculiar in their age and isotopic ratios; they formed before the Sun was born, and their isotopic signature indicates that they come from a different galactic region (Hoppe et al., 1994; Zinner et al., 2006). It is important that we know where do they come from and how do they end up in our backyard.

Many studies (Hoppe et al., 1994; Zinner et al., 2006) have shown that the mainstream presolar silicon carbide (SiC) grains extracted from meteorites formed in asymptotic giant branch stars. Ion probe studies show them to have $^{29}Si/^{28}Si$ and $^{30}Si/^{28}Si$ isotopic ratios larger than those found in solar material (Figure 1). The problem lies in the fact that the mainstream presolar SiC grains are richer in the heavier silicon isotopes than the solar composition even though the grains must have originated in stars that formed prior to the Sun. To deliver their grains to the solar birth cloud, the donor stars clearly must have been born and evolved prior to solar birth. Homogeneous chemical evolution of these Si isotope ratios increase monotonically with time and simply can not

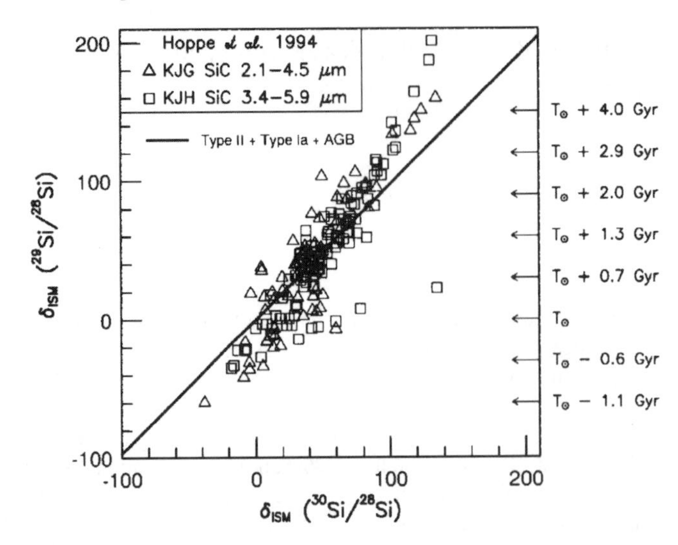

Figure 1. Silicon isotope deviations in a three-isotope plot. Murchison SiC samples measured by Hoppe et al., are shown and have a best-fit slope of 4/3. The renormalized mean Interstellar Medium (ISM) evolution is shown as the solid line, and by construction it passes through solar abundances at $t = t_{Solar}$. With such a construction, deviations with respect to solar abundances and with respect to interstellar abundances are the same $\delta_{Solar} = \delta_{ISM}$ (Clayton, 1997).

accommodate earlier stars having higher ratios. Therefore, one seeks reasons for the high isotopic ratios of the majority of donor stars. Clayton (1997) argued that this silicon isotopic richness is due in part to the diffusion of the parent AGB stars from more central birthplaces within the Galaxy. Due to a more evolved interior position, these central birthplaces are known to have a higher metallicity. The outward diffusion can be likened to the gravity assist used in solar system exploration.

Since this study proposes to deal with star-molecular-cloud encounters, one might ask if stars collide with one another? This question has been addressed by other researchers (Binney and Tremaine, 1987; Tayler, 1996). The interval between collisions was calculated to be 10^{19} years. This is considerably longer than the believed age of the Galaxy ($1 - 2x10^{10}$ years). Evidently, the stars in the Galaxy do not suffer significant individual deflections due to the presence of other stars and large-angle collisions between stars are very infrequent and do not contribute to migration of AGB stars.

To calculate the orbits of stars, a numerically convenient axisymmetric gravitational potential was used as proposed by Miyamoto and Nagai (1975).

$$\Phi(R, z) = -\sum_{i=1}^{3} \frac{GM_i}{\sqrt{R^2 + \left(a_i + \sqrt{z^2 + b_i^2}\right)^2}} \tag{1}$$

where R is the galactocentric radius and z is the height above and below the galactic mid-plane. Respectively, the three parameters a (in kpc), b (in kpc), and M (in 10^{10} M_{Solar}) are: 0.0, 0.3, and 1.8 for the bulge; 6.2, 0.4, and 17.4 for the disk; 0.0, 31.2, and 83.5 for the corona[4].

Depending on the choice of the two parameters a and b, Φ can represent the potential of anything from an infinitesimally thin disk to a spherical one. This expression is in good agreement with observational determinations of the Galactic rotational velocity curve (Zwitter et al., 2008), the vertical force K_z at the Sun's position, and the total local mass density[5] ($\rho_{Solar} \sim 0.15 M_{Solar}$ pc^{-3}). Molecular clouds individually produce fluctuations in the gravitational field seen by nearby stars, which in return produce changes in the magnitude and direction of each stellar velocity.

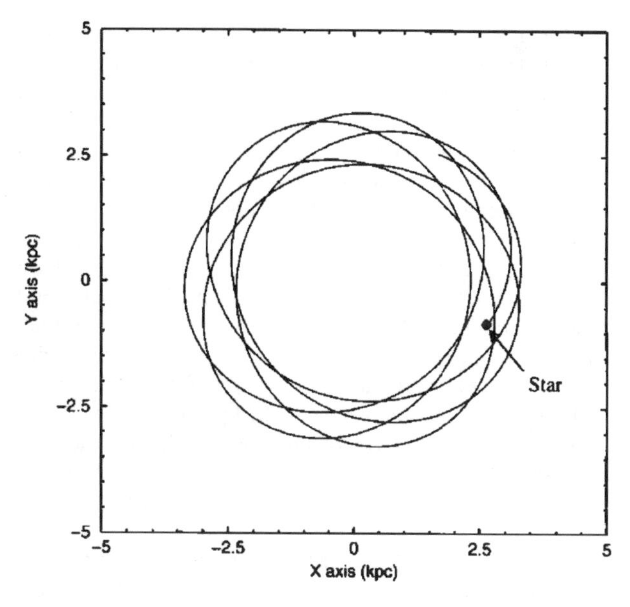

Figure 2. A face-on galactic plane orbit that forms a rosette. The star has an initial circular velocity 196.2 km/s and lived for 0.5 billion years.

A typical star's orbit is given in Figure 2. Molecular clouds are detected and mapped by observing millimetre emission from a trace molecule: carbon monoxide (CO) (Pudritz, 2002; Wielen et al., 1996). The results from the different galactic plane surveys were used to distribute the molecular clouds radially (Pudriz, 2002; Solomon and Sanders, 1985). The results showed that the molecular clouds are the dominant component of the interstellar medium in the inner half of the disk at $R < 0.8R_{solar}$ with a rise between $R = 5$ and $R = 7$ kpc and a sharp fall toward the larger radius where there is very little emission beyond $R = 15$ kpc. Most of the surveys suggest a median cloud mass of about $10^5 M_{Solar}$ (Binney and Merrifield, 1998; Pudritz, 2002; Solomon and Sanders, 1985; Ward-Thompson, 2002).

Since the problem concerns the effect of the molecular clouds on the motion of stars through an otherwise azimuthally symmetric potential, we can utilize a simple approximation of their masses and the radial distribution for that purpose. To a first approximation the molecular clouds can be taken to be spherical. It was not the intention to model the molecular clouds nor their radial distribution. For that purpose a Gaussian distribution was employed that randomly generates positions for the molecular clouds along the galactic mid plane. The transformation that was used for the Gaussian deviations is (Press et al., 1992):

$$R = \sqrt{-2 \ln a_1} \cos 2\pi a_2 \tag{2}$$

where R is the galactocentric radius, a_1 and a_2 are two uniform deviates on (0,1).

A typical calculation was set up to trace 10^4 molecular clouds and 10^3 stars for two billion years. The molecular clouds were placed randomly on circular orbits in the galactic mid plane. Since each molecular cloud is more massive than the star; a gravitational force from a nearby star on the molecular cloud is going to be small and thus neglected. To a first approximation the molecular clouds retain their circular orbits and constant angular momentum. The stars are initially distributed randomly in radial zones. A zone is a ring with 500 pc width; for example, the region between $R = 4.0$ kpc and $R = 4.5$ kpc is one zone. Each zone has 10^3 stars positioned randomly in R and in its azimuthal angle in the galactic mid plane. Each star is given a randomly generated peculiar velocity between 10 and 20 km/s. The direction of the peculiar velocity is also randomly chosen in θ and ϕ, where θ and ϕ are the known spherical coordinates. Added to the peculiar velocity is the proper circular velocity deduced from Miyamoto and Nagai (1975) potential. With these initial conditions the system evolves for two billion years. A numerical method was employed called the Bulirsch-Stoer algorithm (Press et al., 1992) with adaptive time step to trace the orbits of stars by integrating the equation of motion for the stars. The nearby molecular clouds' gravitational field mainly governs a star's trajectory. The total force $\vec{F}$ (per unit mass) on a star is the sum of the galactic gravitational force derived from Miyamoto and Nagai potential and the nearby molecular-cloud gravitational forces.

The results have shown that, for the most part, stars experience a small perturbation in their orbits and they are less likely to deviate largely from their initial circular orbits. Close to 40% of the stars suffered inward migration. Only a small fraction of AGB stars ~1% migrated to larger radii. Wielen et al. (1996) found that the Sun was born at galactocentric radius $R = 6.6$ kpc and has in 4.5 Gyr diffused outwards to its present location at $R = 8.5$ kpc. Calculations showed that in two billion years only 0.4% of stars born between 4.0–4.5 kpc migrate outward to 6.0–6.5 kpc zone where they give their AGB stardust to a presolar cloud. This result, in particular, is very interesting because Sellwood and Binney (2001) studying the scattering of stars by spiral waves, concluded that the grains must have migrated from $R \sim 4$ kpc to ~6 kpc where they were incorporated into a presolar cloud.

Migration of AGB (Asymptotic Giant Branch) stars from more central regions in the galaxy due to scattering by molecular clouds is deemed unlikely. The percentages the calculations have produced are small and cannot account for the observed richness in the heavier silicon isotopes when compared with solar values. This study has shown that only a few stars do this.

3 CONCLUDING REMARKS

The environment in which mankind presently lives is the end result of a long process of chemical and biophysical development of the galaxy which is not, as yet, fully understood. This study has, hopefully, provided some insight into star-molecular cloud encounters. It will help to improve our understanding of the evolution of the solar system. This will allow us to better appreciate the earth's environment and how it has affected the evolution of life on the planet.

ACKNOWLEDGMENTS

The author gratefully acknowledges the support of the NASA Origin of Solar Systems Program.

REFERENCES

Binney, J., & Merrifield, M. *Galactic Astronomy* (Princeton, New Jersey, 1998)

Binney, J., & Tremaine, S. *Galactic Dynamics* (Princeton, New Jersey, 1987)

Clayton, D. Placing the sun and mainstream SiC particles in galactic chemo-dynamics evolution. *The Astrophysical Journal* **484**, L67 (1997)

Hartmann, D., Epstein, R., & Woosley, S. Galactic neutron stars and gamma-ray bursts. *The Astrophysical Journal* **348**, 625 (1990)

Hoppe, P., Amari, S., Zinner, E., Ireland, T., & Lewis, R. S. Carbon, Nitrogen, Magnesium, Silicon, and Titanium isotopes compositions of single interstellar silicon carbide grains from the Murchison carbonaceous chondrite. *The Astrophysical Journal* **430**, 870 (1994)

Miyamoto, M., & Nagai, R. Three-dimensional models for the distribution of mass in our Galaxy. *Publ. Astron. Soc. Japan* **27**, 431 (1975)

Press, V., Teukolsky, S., Vetterling, W., & Flannery, B. *Numerical Recipes in Fortran* (Cambridge University Press, New York, 1992)

Pudritz, R. E. Clustered star formation and the origin of stellar masses. *Science* **295**, 68 (2002)

Sellwood, J. A., & Binney, J. Radial mixing in galactic discs. *Monthly Notices of the royal Astronomical Society* **336**, 785 (2001)
Solomon, P. M., & Sanders, D. B. Star formation in a galactic context: the location and properties of molecular clouds. *Protostars and Planets II* (University of Arizona Press, Tucson, 1985)
Tayler, R. *Galaxies: Structure and Evolution* (University Press, Cambridge, 1996)
Ward-Thompson, D. Isolated star formation: from cloud formation to core collapse. *Science* **295**, 76 (2002)
Wielen, R., Fuchs, B., & Dettbarn, C. On the birth-place of the sun and the places of formation of other nearby stars. *Astronomy and Astrophysics* **314**, 438 (1996)
Zinner, E., Nittler, L., Gallino, R., Karakas, A., Lugaro, M., Straniero, O., & Lattanzio, J. Silicon and Carbon Isotopic Ratios in AGB Stars: SiC Grain Data, Models, and the Galactic Evolution of the Si Isotopes. *The Astrophysical Journal* **650**, 350 (2006)
Zwitter, T., Siebert, A., Munari, U., Freeman, K. C., Siviero, A., Watson, F. G., Fulbright, J. P., Wyse, R. F. G., Campbell, R., Seabroke, G. M., Williams, M., Steinmetz, M., Bienaymé, O., Gilmore, G., Grebel, E. K., Helmi, A., Navarro, J. F., Anguiano, B., Boeche, C., Burton, D., Cass, P., Dawe, J., Fiegert, K., Hartley, M., Russell, K., Veltz, L., Bailin, J., Binney, J., Bland-Hawthorn, J., Brown, A., Dehnen, W., Evans, N. W., Re Fiorentin, P., Fiorucci, M., Gerhard, O., Gibson, B., Kelz, A., Kujken, K., Matijevič, G., Minchev, I., Parker, Q. A., Peñarrubia, J., Quillen, A., Read, M. A., Reid, W., Roeser, S., Ruchti, G., Scholz, R.-D., Smith, M. C., Sordo, R., Tolstoi, E., Tomasella, L., Vidrih, S., & Wylie-de Boer, E. The Radial Velocity Experiment (rave): Second Data Release. *The Astronomical Journal* **136,** 421 (2008)

Governance and Practical Approach for Integrating a Sustainable Environment into Human Development: A Regional Case Study

Fouad H. Beseiso
Former Governor of Palestine Monetary Authority

SUMMARY: The aim of this chapter was to explore the feasibility and practicality of how to integrate sustainable environment activities into human development on a national and regional basis for developing countries in general and the Western Asian (ESCWA) and Middle East North African (MENA) region in particular. The achievements in the region were reviewed with special focus on the role of environmental governance.

1 INTRODUCTION

Over the past decade, the Economic and Social Commission for Western Asia (ESCWA) member states have witnessed marked improvements in health, education, literacy, a decline in fertility rates, as well as a strengthening in the status of women, the traditional family and civil society (UN-Economic and Social Commission for Western Asia, 2002a). This progress has been achieved in face of increasing population pressures, fluctuating economic conditions, persistent unemployment, continued poverty and unresolved regional conflicts. The links between population, health and sustainable development are particularly complex in the ESCWA region where environmental resources are finite and population pressures are great.

Urbanization has increased, and is exacerbated by losses to agriculture, rural-urban migration and industrialization. The western Asian regional labor force is now largely skewed towards youth who require greater vocational and technical training to meet employer needs. Women have also increasingly entered the workforce, although expanding information technologies in the service sector may serve to set back gains in female employment. These trends have resulted in higher demand for food, increased consumption, greater production of industrial and municipal waste, strains on public services, increased unemployment, and expansion of the informal housing sector and the loss of green spaces to infrastructure development.

Economic growth in the ESCWA region regularly experiences cyclic fluctuations and volatility associated with regional instability and fluctuating oil prices. Economic growth in the region remains dependent upon natural resources. This has kept economic growth in the ESCWA region below the average recorded for developing countries. Furthermore, traditional approaches to economic growth have exacerbated poverty and debt. This is because while countries have grown marginally richer over the past decade, benefits have not been equally distributed. The unbalance provision and access to basic services includes a gender dimension, since women and children generally suffer the most under extreme poverty and in areas experiencing conflict.

It is thus fair to conclude that although the region has made considerable efforts to alleviate poverty and move towards sustainable development, challenges remain. Opportunities emerging from these challenges include the strengthening of non-governmental organizations (NGOs) and their transformation from being service providers to social advocates. Governments are forging partnerships with the private sector to improve public services.

There are specific driving forces that are directly impacting the sustainability of the regional environment. These factors are oil and gas production and mining of non-renewable natural resources, industrial and agricultural development, tourism and transport development, urbanization and urban stress, and production and consumption patterns (UN-Economic and Social Commission for Western Asia, 2002a). The ESCWA region has tremendous oil and natural gas reserves, which play a significant role in the region's economy and export portfolio, despite efforts at economic diversification. Mining and the processing of industrial minerals and metals have increased. However, the industrial sector remains heavily labor-intensive and energy-intensive, which illustrates the region's hesitance to update technologies or progress to more knowledge-based production of higher value added goods. Small and medium-sized enterprises (SMEs) are also unable to take advantage of trade liberalization opportunities or face increasing domestic competition.

Effective monitoring and management of industrial and hazardous waste remain problematic. These sectors impose direct impacts on air, water and marine quality. Secondary effects include increased migration to industrial centers and the expansion of transport infrastructure to support these industries.

Agricultural is another significant contributor to the regional economy. However, agricultural protectionism and subsidization of irrigation water, agrochemicals and land use reinforces unsustainable patterns of production. This has rendered the sector inefficient and vulnerable to trade liberalization. Furthermore, agriculture is the primary user of freshwater in the region, consuming more than 80 per cent of available resources despite growing municipal water demand (UN-Economic and Social Commission for Western Asia, 2002a). Regional fertilizer consumption has also risen, alongside pesticide and agrochemical use, despite global trends among developed countries to the contrary.

The ESCWA region's share of the world tourism industry is small However, tourist arrivals more than doubled during the 1990s. The challenge facing the region is how to support the tourism sector while preserving historic monuments, cultural heritage, natural resources and coastal areas. The number of motor vehicles in the ESCWA region has also increased, doubling and even tripling in some countries since the mid-1980s (UN-Economic and Social Commission for Western Asia, 2002a).

The aim of this chapter was to explore the feasibility and practicality of how to integrate sustainable environment activities into human development on a national and regional basis for developing countries in general and the Western Asian (ESCWA) and Middle East North African (MENA) region in particular. The achievements in the region were reviewed with special focus on the role of environmental governance.

2 ENVIRONMENTAL TRENDS, ACHIEVEMENTS AND CHALLENGES

The environmental pillar of sustainable development in the ESCWA region is primarily concerned with the management of natural resources and the preservation of cultural heritage. Water scarcity, for example, characterizes the region, with 80 per cent of the renewable water resources flowing from outside ESCWA member states. Over-exploitation, saltwater intrusion and wastewater discharges are impacting water quality and limiting the availability of already scare groundwater resources (UN-ESCWA, 2002a).

Water-borne diseases have become a threat to public health. As such, countries are developing non conventional water sources, such as desalination. Progress has been made in formulating water policies and using economic instruments such as water use restrictions, water services pricing, and subsidy reductions.

Regional conflicts, tourism, transportation infrastructure, industrial and agricultural activities, aquaculture, fishing, oil spills and urbanization have all had their tolls on marine and coastal areas. Some progress has been achieved in promoting integrated coastal zone management. However, these measures are hampered by inadequate awareness, ineffective enforcement and weak institutional arrangements.

Improvements in energy efficiency have been achieved and have helped to perk up air quality. Public and private partnerships have been forged and energy subsidies have been gradually removed. However, integration of sustainable energy policies into sectoral planning is lacking. Also, while most ESCWA member states have passed air quality laws, standards are often unreachable without implementation plans. A scientific basis for decision-making and air monitoring data and analysis are needed.

Land degradation and drought are the primary causes of desertification. Growing populations, urbanization, industrialization, and poor agricultural practices are also degrading land resources. The impacts are reflected in increased desertification, loss of soil fertility, deforestation, pollution of scarce land and water resources, increased rural-to-urban migration and poverty. The lack of comprehensive strategies to address land degradation, food security and water scarcity in an integrated manner is a major constraint to sustainable use of resources, and exacerbates the effects of desertification.

3 CHALLENGES FACING REGIONAL SUSTAINABLE DEVELOPMENT

Most ESCWA member states suffer from inadequate technical, human and financial resources as well as limited institutional capacity (ESCWA, 2002b). These obstacles, along with changing political and economic dynamics, prevent the region from actively engaging in sustainable development planning, implementation and follow-up. UN-ESCWA defined eight key challenges facing sustainable environmental development: governance, stakeholder participation and access to information, economic instruments and voluntary arrangements; environmental monitoring and information networks; environmental education; research and development; regional insecurity and conflicts; and trade liberalization and regionalization.

Regional countries over the past ten years have struggled to put into place effective institutions and instruments for managing sustainable development in an integrated manner (ESCWA, 2002b). However, as other regions have also discovered, the sustainable development process is neither smooth nor easy due to the sheer complexity of linking social, economic, and environmental development issues, and the difficulty of institutionalizing consultation between various public and private stakeholder groups. Nonetheless, the district has made significant progress in environmental management planning and implementation.

Since United Nations Conference on Environment and Development (UNCED) nearly all countries of the ESCWA region have adopted a national environmental strategy (NES) or policy for the purpose of moving towards sustainable development. At least eight of them have approved a national environmental action plan (NEAP) for the purpose of transforming strategies into actions. The majority of the region's environmentally related institutions also experienced changes in structure and mandate during the 1990s, indicating greater awareness and a rethinking of environmental concepts. States ratifying regional and multilateral environmental agreements also increased.

However, while ecological management has improved, efforts at integrated sustainable development have slowly stagnated. NES and NEAP reports prepared by ministries of environment or environmental authorities, assisted by international consultants and limited public consultation exercises, have become proxies for national sustainable development strategies (NSDSs) in nearly all countries of the region, with the exception of Jordan and the United Arab Emirates. This means that cultural, political, and socio-economic dimensions and interests are not always adequately represented in sustainable development strategy formulation. Criteria for policy instrument selection and evaluation are also not well-developed in the region, with little attention paid to policy prioritization, cost effectiveness, the technical practicality of the instruments proposed or their social, political or cultural acceptability (UN-General Assembly, 2006).

Limited foresight, coordination and complementarities of strategies mean that the level of NES and NEAP implementation in the region has been mostly disappointing. NSDS efforts generally have been piecemeal and receive little financial support, with *ad hoc* national sustainable development Commissions (NSDCs), if they exist, only being revived to respond to reporting demands for global conferences such as the WSSD. There are still several challenges to effective sustainable development governance (ESCWA, 2002a). These will be outlined next.

3.1 *Coordination & complementarity's of sustainable development institutions & instruments*

The main challenge facing most decision-makers in the ESCWA region is how to effectively formulate, integrate and implement multi-sectoral sustainable development policies (ESCWA, 2002b). This requires coordination and consultation between government institutions, as well as complementary and coherence between policy instruments being implemented by different ministries.

These difficulties are exacerbated by the centralized, yet compartmentalized nature of governance. For instance, national environmental agencies in the region are generally assigned responsibility for sustainable development policy formulation and implementation. This reinforces the sector-based bias regarding sustainable development, and makes social and economic ministries less engaged and committed to articulated sustainable development goals. This practice also reduces the importance of the issue since environmental institutions in the region are generally neither central to government decision-making nor able to exert influence over line ministries. Furthermore, there is limited communication between parliamentarians, the public and the bodies responsible for implementing and overseeing enacted legislation related to sustainable development, which also explains why policies are not effectively implemented.

While institutional coordination is an important dimension of the problem, the implementation of appropriate and effective policy instruments is another. For instance, most ESCWA member States are seeking to implement recently enacted water and land use legislation, but doing so with various degrees of success, mostly because of inadequate coordination between ministries and poor enforcement. Moreover, the region remains mired in command-and-control regulatory approaches.

Finally, sustainable development instruments that are being used are mostly applied piecemeal from a sector perspective, with little synergy, chronology or linkage sought between policies and programs implemented by different ministries so as to maximize their collective impact.

3.2 *Linking national and local sustainable development policies and programs*

While national environmental management and sustainable development processes have improved throughout the area, local initiatives have remained limited. The main reasons for this dichotomy are: the centralized nature of government decision-making and complexity of national bureaucracies; the limited or lack of financial autonomy of provinces (governorates) and municipalities from the national purse; the limited institutional capacity to serve rural and marginalized communities; the tendency for skilled human resources able to organize sustainable development initiatives to congregate in large urban centers, normally the capital; and the tendency for community groups and NGOs (non governmental organizations), which normally push for and participate in sustainable development efforts, to be more organized and vocal in capital cities and urban centers.

The disconnect between local and national sustainable development planning efforts is echoed in national to regional to global linkages, which should be better organized to reinforce one another. However, linkages between local-national-regional-global sustainable development initiatives need to take into consideration the social, economic, political and cultural sensitivities that are specific to each area. This is particularly important for ESCWA member countries given the need to encourage and adopt locally grown approaches to sustainable development that are innovative, appropriate, gradual and applicable to the region.

3.3 *Global governance*

There is a need to improve the governance system to be transparent and conducive to the needs of developing countries to be able to address the changes induced by globalization, the new world economic order. The role of the United Nations and other international organizations concerned with sustainable development needs to be elaborated to fit the requirements for achieving sustainable development. The WSSD (2002) provides the opportunity and a platform for such a global exercise.

3.4 *Follow-up and accountability of the sustainable development process*

Effective sustainable development governance also requires a system for monitoring progress in achieving stated targets and goals. While many countries of the district are progressing from the stage of planning to implementation, few mechanisms have been established to assess the quality or impact of policy and program outputs. Furthermore, oversight arrangements for ensuring the integration of national sustainable development goals into sector-based work programs are rare. Accordingly, public and/or private sector agencies should be assigned a watchdog role to monitor and report on the effectiveness of government institutions in supporting sustainable development (Samuelson and Nordhaus, 2001). Such a system of accountability could improve institutional performance, as well as inform and empower public stakeholders.

4 ENVIRONMENTAL MONITORING AND INFORMATION NETWORKS

Information is a prerequisite for reliable environmental monitoring, assessment and reporting. It is the key to identifying environmental concerns, root causes, impacts, trends, reactions, as well as emerging issues. Adequate information is also essential for formulating effective strategies, policy responses and action plans to priority environmental and development problems (Segnestam, 2002).

Environmental monitoring is not new to the ESCWA region. It can be found in the traditions of old civilizations as well as in recent history through practices developed to monitor water levels along riverbeds and record the intensity of seasonal floods. Most countries now have monitoring programs that operate with supporting laboratories to monitor coastlines, water resources and air quality. They also have established remote sensing and GIS organizations, centers or divisions that support environmental monitoring. However, the capacities of these facilities are still limited in terms of their technical expertise, geographic scope, utilization and financial support. Furthermore, most countries of the ESCWA region have not developed systematic approaches and systems for environmental assessment and reporting. While most endeavor to produce and update national state of environment reports, each is prepared using significantly different methodologies that are oftentimes influenced by the donor institution supporting the preparation of the report. The process is also hindered by many other factors, including the lack of institutional frameworks, inadequate capacity, shortage of experienced personnel, limited financially resources, lack of or inaccessibility to necessary data, absence of appropriate indicators, and weak ties with stakeholders in line ministries and community-based groups.

Positive efforts have been made by the League of Arab States assisted by regional United Nations organizations to identify harmonies, test and utilize sustainable development indicators and indices for monitoring and reporting on sustainable development. Several countries in the area have also established national environmental information systems and networks. At the national level, Lebanon for example, established a National Environment and Development Observatory.

At the regional level, eleven founding members from regional and international organizations established the Arab Region Environmental Information Network (AREIN). The intention is to expand the network to include national networks and information systems into one regional integrated network. There are also some efforts to establish sub-regional networks, such as one for the GCC countries.

5 FUNDRAISING, LOANS, INVESTMENT AND TAX REFORM

As financing for sustained development is well acknowledged to be main determinant for successful strategies. However, financing for sustainable development has remained limited, mostly because indebtedness continues to sap the energies of the region. Furthermore, while a variety of financial instruments and institutions for sustainable development have been strengthened over recent years, coordination and oversight of programs to achieve environmentally friendly development financing needs to be improved. The international conference on *Financing for Development*, which was held in Monterrey, Mexico, March 2002, marked an important step towards addressing this complex issue (ESCWA, 2002a).

Over the last decade, some countries in the region have established funds to support environmental protection. For instance, Egypt established the Environmental Protection Fund with Law No. 4 of 1994. Social and economic development funds also exist in several countries (e.g., Egypt, Saudi Arabia, and Yemen). However, more effort needs to be made in creating sustainable mechanisms for endowing these funds through pollution taxes, licensing fees and other mechanisms, as well as coordinating the objectives and activities of these funding programs (ESCW, 2002a).

While governments have recently sought to increase financing for technology based and export oriented countries (e.g., Egypt and Jordan), the vast majority of firms still lack sufficient access to investment financing from traditional banking institutions and local financial markets. Funding for small and medium-sized enterprises (SMEs) has remained particularly limited since commercial banks tend to maintain an elitist approach to their loan portfolios.

Government and private commercial banks in the region also do not require the completion of environmental impact assessments as part of their loan approval process, with environmental considerations generally only being taken into account when local funds are partnered with those from international financial institutions. Alternatively, micro-financing, the provision of financial services (credits, deposits, and savings) to the entrepreneurial poor, has achieved positive, but mixed success. By the end of 1998, six commercial banks maintained micro-financing facilities in the ESCWA region (in Egypt, Jordan, Lebanon, Palestine and Yemen), while over 50 other programs were being implemented by NGOs and government agencies through state-owned banks.

The ESCWA region had an estimated 120,000 active micro-finance borrowers at the end of 1997 maintaining about US$ 70 million in loans Egypt has the largest share of active borrowers in the micro-financing sector, most of whom reside in urban areas. Less than 10 per cent of existing micro-financing programs in the ESCWA region were in existence prior to 1985. Despite this growth, the financial sustainability of micro-finance programs remains tenuous, particularly in small markets and with programs that target specific beneficiaries. This is because challenges to micro-financing include social, political and religious attitudes, as well as negative perceptions by governments and financial institutions that the activity serves the informal economy. Studies also question whether micro-finance facilities adequately benefit the female poor.

Research has found that credit decisions tend not to focus on gender issues unless programs, usually maintained by NGOs, are specifically designed to provide micro-financing for women. Nevertheless, successes have been recorded. For instance, the United Nations Development Fund for Women and ESCWA joined forces in 1996 to strengthen institutions to support the development of women-owned enterprises through credit facilities and business support services and training. The program has served hundreds of women in Jordan, Lebanon, Palestine and the Syrian Arab Republic.

Structural economic reforms and tightening budgets have prompted governments to increasingly seek out privatization as a means to raise funds for sustainable economic development. Private sector participation in the provision of environmental services and the use of licensing fees has also become more common. While privatization programs have been implemented in Egypt, mixed support for these programs in Jordan, Lebanon and Saudi Arabia has slowed the process. A key issue of concern is how to ensure access to public and social services by poor and marginalized communities while allowing private sector service providers to secure reasonable profits. This has created social-political obstacles for regional governments seeking to privatize water, sanitation and transportation services. Another challenge is how to handle environmental liability tied to past pollution caused by enterprises identified for privatization. Governments of the region need to pay closer attention to this issue in order to avoid post-privatization problems.

The ESCWA region's share of global foreign direct investment (FDI) averaged one per cent during the 1990s, as compared to a share of two per cent of world GDP. Most FDI inflows were concentrated in four ESCWA member countries: Egypt, Jordan, Oman and Saudi Arabia; inflows into Egypt and Saudi Arabia alone represented 63 per cent of the total. Most FDI in the ESCWA region went towards the oil sector (Oman), petrochemicals (Saudi Arabia), tourism (especially in Egypt), and textiles, metals and minerals (Egypt and Jordan). FDI not only provides funds for economic and industrial development, but also generates technological spillover effects from investors acquainted with new technologies. However, although the potential exists, such spillovers have yet to be largely realized in ESCWA member States. For instance, Qatar, which is preparing to become world leader in clean fuels, particularly in gas-to-liquid technologies, and will likely become the recipient of new FDI inflows and matching technical know-how.

Falling tariff revenues caused by trade liberalization and economic restructuring requires that governments identify alternative means to finance the national debt and support government expenditures. While indirect taxation is already prevalent throughout the region, value added taxes and individual income taxes (which mostly comprise a small share of revenues) are now being viewed as alternative revenue streams. This might have adverse effects on individual savings, however, which is already very low in ESCWA member States as compared to other developing regions. Reduced savings could threaten liquidity for private investment, e.g., investment in technological upgrading or more environmentally friendly processes. Alternatively, tax reform could readjust market signals through graduated tariff systems that encourage sustainable behavior by encouraging conservation and reducing excessive consumption.

Debt-for-nature swaps have been another constructive way for ESCWA member States to secure financing for sustainable development while reducing their debt burden. Jordan has taken advantage of this mechanism several times, and is currently negotiating another swap with the assistance of the International Union for the Conservation of Nature and the Government of Germany. Debt forgiveness and limits on public expenditures for debt financing, particularly for highly indebted poor countries such as Yemen, are being touted as an alternative means for securing government allocations for sustainable development.

6 FINANCIAL INSTITUTIONS

Apart from national ministries and agencies designed to solicit financial support for sustainable development programs, regional and international institutions have also continued to provide important levels of funding for sustainable development initiatives in the ESCWA region. Aid provided by bilateral donors has also been significant, although its level has decreased in recent years.

Several regional development funds and institutions provide significant financial assistance to support the implementation of National Environmental Action Plans (NEAPs) in ESCWA member States. The region also continued to receive important financial and technical assistance for sustainable development from the bilateral aid institutions of industrialized countries, notably the European Union, the United States and Japan. International financial institutions (e.g., World Bank, European Investment Bank) and grant mechanisms also provide important technical and financial assistance to various countries in the region in support of sustainable development. Nevertheless, international donors have not lived up to the expectations raised at the Earth Summit.

The volume of development assistance has declined and several developed countries, for various reasons, have failed to live up to their commitments to assist less developed countries, including those in the case study region. Accordingly, despite the continued channeling of funds from international, regional and bilateral donors, levels remain well short of what is needed.

NGOs, community-based organizations, universities, think tanks and religious institutions can provide valuable technical and financial resources to complimenting government spending in support of sustainable development. If well-integrated into the National Sustainable Development Strategy (NSDS) planning process, local, regional and global NGOs can solicit funding from charitable groups and public agencies through avenues not accessible by governments. Coordination and complementarities of governmental and non-governmental activities in support of sustainable development is thus essential to reduce the financial gap.

The financial challenge limiting progress towards sustainable development not only involves increasing the net supply of available funds, but also improving the quality of assistance provided by donor agencies to recipient countries. In an effort to improve complementarities and co-ordination of programming in the constituency, several international, regional and national donor institutions joined forces to establish the Coordination Secretariat of Arab National and Regional Development Institutions.

Experience shows that the most successful initiatives at the regional level are those that are supported by international donors and/or regional institutions and that mobilize funds for specific component of a national action plan supported by national stakeholders. The key to the success of the activity is based on whether it is implemented through a consultative process that involves dialogue between relevant ministries and consultation with relevant local stakeholders and experts.

The region is plagued by frequent wars and conflicts of various scales. This has kept defense spending high and detracted funds from sustainable development initiatives. International politics have had regional repercussions on poverty, demography, trade and development, with populations plunged into poverty due to an internationally sanctioned embargo. Marginalized groups, namely migrants, women, children and elderly, tend to be the most affected during times of conflict. This complicates challenges related to social integration, the rebuilding of social and economic institutions, the reintegrating displaced populations and the physical reconstruction of human settlements.

Finally, one of the most important and neglected aspects of financing for sustainable development concerns the lack of effective oversight and monitoring of allocated project funds; While individual donor institutions and agencies might require financial monitoring and auditing of grants and loans in a piecemeal fashion, no countries in the region have a comprehensive system for assessing the effectiveness of financial instruments for building national capacity or facilitating progress towards sustainable development. This is because beneficiary countries generally look at financial assistance in a piecemeal fashion and in terms of quantity, and not of quality. This thinking, however, is beginning to change among donor institution in light of increasing fiscal constraints. Accordingly, beneficiary countries of financial assistance in the region should seek to improve the both quality and effectiveness of each donor dollar, as well as increase the supply of aid contribution in order to more appropriately finance sustainable development.

7 STRATEGIC PRIORITIES

Mainstreaming requires understanding the social and environmental aspects of policies, trends, and decisions in each sector of the economy, as well as identifying cost-effective solutions to key environmental issues (Bolt et al., 2005). The World Bank defined in its updated Environmental Strategy for MNA (Middle East North Africa) countries (i.e. 16 Arab countries and Iran) the strategic priorities and required actions which illustrate the required integratory approach for sustained development. In terms of strategic priorities, MNA should continue to shift away from the environment as a separate sector and toward considering the environment as a component of other sectors.

Seven strategic priorities were defined by the World Bank to be undertaken (World Bank, 2001):

- Improving water resource management.
- Controlling land, costal zones and natural habitats degradation.
- Reducing urban pollution.
- Strengthening the capacities of environmental institutions, local -communities and NGO's.
- Strengthening the capacity of the private sector.
- Continuing the support for regional initiatives.
- Integrating global environmental issues into the Bank's operations.

The implementation of the proposed strategic actions would require important leadership, collaboration, and coordination with different stakeholders, including NGOs, bi-lateral and multilateral donors and international financial institutions. Partnerships will become an important cornerstone for the Bank's environmental assistance in the MNA region. At the country level, the Bank should be prepared to participate or convene a donor- country coordination group on environment (and/or water issues) to achieve a greater integration of efforts and reduce overlaps between similar programs.

7.1 *Poverty alleviation and social integration*

Poverty alleviation remains a significant challenge for many countries of the region, and thus a major impediment to achieving sustainable development. Social integration must be part of the equation to resolve the poverty problem to ensure that social, economic and environmental benefits are equally shared. It is thus imperative to strengthen political commitments and efforts to implement sustainable development policies and enhance the quality of life for all sectors of the population, with special emphasis on vulnerable groups such as women, children and the disabled. The process of reaching this goal should seek to stabilize population growth, improve access and quality of education, restructure technical and vocational training to fill in gaps in the labor market, supply and train communities in information technology to facilitate knowledge transfer; and provide equitable access to public services (water, sanitation, etc.) among all social groups (UN-Escwa, 2002a).

Global efforts should also be made to ensure that less developed countries are able to provide basic social services to local communities, especially in rural areas, to improve educational and health services, and to empower women as resource managers at the community level (UN-Escwa, 2002a).

Poverty eradication necessitates the development of programs aimed at improving social welfare, eradicating illiteracy, promoting employment and equality of opportunity and protecting the environment. Specific programs are required for capacity building to promote income generation for the most vulnerable groups (e.g., women and children). The poor should be empowered to be active actors in developing opportunities to eradicate poverty. They should also be provided with free access to basic public services, such as safe drinking water, sanitation, healthcare, elementary education and technical training.

There is also a need to improve resource management to provide for the needs of the poor. Focus should be on the development of rural areas to improve shelter, infrastructure, services, investment and job creation. Private developers and investors should be encouraged to invest in low-income housing (World Bank, 2002a).

All United Nations organizations should adopt poverty as an issue of paramount concern, and should strengthen all agency programs that aim at addressing poverty. International aid programs should be developed to include poverty eradication and to involve civil society institutions. International soft loans should be encouraged, the debt burden should be reduced and economic embargoes should be removed since they lead to the further deterioration of living conditions of the poor (UN-Escwa, 2002a).

7.2 *Sustainable management of natural resources*

Water scarcity, land degradation and food security represent the major challenges in the ESCWA region. The three issues are interdependent and jointly could significantly influence biodiversity, population policies and security in the region (UN-ESCWA, 2002a)

Countries of the area are encouraged to adopt integrated water resources management, including demand management approaches. This requires coordination and cooperation between departments/agencies dealing with water issues. They are also encouraged to develop renewable and non-conventional resources, including harvesting of rain and fog water, exploring and developing deep groundwater, water recycling and water desalination.

States need to optimize and rationalize the use of water resources by reallocating water to higher value uses, growing water efficient crops, addressing the real value of water in all sectors by applying cost recovery of investment in water projects, and increasing the efficiency of irrigation through technical improvements. The use of expert farm management is also important to maximize land productivity and efficiency in the use of irrigation water. Stakeholders should be encouraged to participate in water management through public awareness campaigns, participatory programs for stakeholders and local communities and community based water associations. Rationalizing water consumption should also be encouraged. Concerned authorities, with the help of other stakeholders, are also encouraged to extend efforts to rehabilitate steppe areas, marginal lands and the areas of irrigated and rain fed agriculture, and expand the establishment of protected lands to allow the restoration of natural condition and converse biodiversity.

There is a need to intensify efforts to develop water and land related technologies, specifically for irrigation and water desalination using solar energy technologies. Pollution control measures, including integrated programs of pest management and control of chemical pollution, should to be established and enforced to protect water resources. Capacity building and improvement of institutional set-up are needed to effectively manage land and water resources, and protect biodiversity. Integrated social and economic land and water policies conducive to the rational use and development of land and water resources should be developed. Financing schemes should be developed to secure the funding necessary for the management of water resources and to implement water investment projects, considering cost recovery through services provided and public fees for wastewater treatment.

Water, land resources and food security should be addressed within a regional framework through the development of a unified regional strategy and alternative policies aimed at regional integration, especially agricultural production and trade policies. This should include the sustainable use of shared water resources (including aquifers), the activation of agreements between member States of the region concerning the distribution of agricultural products in terms of a "food integration" strategy and the unification of water legislation and standards.

In exploring the feasible pragmatic approach to Arab Gulf States Development cooperation, a piece meal partial and developmental approach has been proposed with top priority being given to regional cooperation on food security (Beseiso, 1981). Furthermore, in order to address the issue of food security, it is necessary to promote regional investment projects that take into consideration the comparative advantages of the countries (Beseiso, 1984a).

The establishment of the Greater Arab Free Trade Area and the removal of tariffs and non-tariff barriers would encourage trade in food and agricultural products in the region. Food security is connected to peace, and political stability. Countries of the region are urged to support regional action plans regarding selective agriculture in suitable zones, determined by climate and resources (i.e. land, water and labor), and to establish regional mechanisms to conserve and rationalize water consumption.

Academic research institutions and regional organizations should accord food and water a top priority and encourage, for example, cooperation to develop selected seeds that are resistant to drought, salinity and that have increased productivity. The district should support the role of specialized regional centers and organizations in the field of research and development and direct them to serve integrated development programs including the utilization of solar energy in the field of water desalinization.

Key proposals for actions emphasized at the international level are respect of the historical rights of countries to shared water resources (e.g., rivers basins, aquifers), cooperation among those countries in management, and protection from pollution, of shared resources including the development of regional strategies, master plans, and mechanisms for joint implementation.

The United Nations and other international organizations can play a catalytic and coordinating role to work out regional agreements and forums on shared water resources, to provide further technical assistance in capacity building, to assist in developing integrated water management policies and to strengthen water resources institutions (UNDP, 2008). Resolution of the conflicts on water rights in the Middle East on the basis of the United Nations' Resolutions and just and equitable sharing is essential for sustainable development, peace and security in the region.

7.3 *Land degradation and desertification*

Countries of the region should develop programs for the rehabilitation of degraded land including meadows and forests, and develop national desertification strategies and action plans in order to combat desertification. They also need to allocate more resources; initiate innovative solutions in support of land users in rural communities to deal with new global changes and overcome the constraints faced by the poor, marginalized and disadvantaged, in particular women, indigenous people and small farmers.

In order to understand and combat desertification, countries need to set up programs to monitor land resources using modern technologies, such as remote sensing and GIS. Countries of the district are urged to comply and implement acceded international conventions and Multilateral environmental agreements (MEAs) related to land resources, specially the Convention on Combating Desertification, as instruments of sustainable development by integrating them fully into national and regional socio-economic development planning, in coordination with the relevant regional and international agencies.

There should be regional cooperation in the implementation of the international Convention to Combat Desertification, and the harmonization and reconciliation of policies, strategies and programs for land use, combating desertification and integrated ecosystem management. It is also important to establish regional programs to monitor desertification based on scientific research and the use of modern technologies (Bolt et al., 2005).

7.4 *Marine and coastal zones*

Countries of the region should adopt an integrated approach to address coastal and marine resources issues, including the adoption of integrated coastal area management for the sustainable development of coastal and marine environment, increasing awareness, strengthening cooperation and integration between institutions and with stakeholders, and implementing the provision of the law of sea (UNCLOS).

Securing financial and technical resources is of paramount importance to implement the activities of integrated management. It is important to develop management plans and mechanisms for sustainable management of living marine resources, including fisheries and aquaculture at national and regional levels, and to take measure to mitigate pollution from land-based activities, e.g., development of waste treatment capabilities and rehabilitation of damaged habitats. There is a need to encourage research and development for sustainable development of coastal and marine areas and resources, and to expand monitoring, surveillance and assessment of coastal and marine resources. It is also important to support national and regional stock assessment studies.

Countries should promote interregional cooperation in the protection of the marine environment, including regional contingency planning and minimization of navigational and emerging marine pollution risks (Beseiso, 1984b). ESCWA member States are urged to cooperate with the regional seas action plans to implement the strategic action programs for the protection of the regional seas in the region. Furthermore, marine protected areas should be regionally considered and identified for protection. It is of paramount importance to implement the Global Plan of Action at the regional level with a view of eliminating sewage releases in the coastal and marine environment and control of other sources of land based pollution (UN-ESCWA, 2002a).

7.5 *Peace and security*

Peace and security has had and will continue to have significant impacts on the progress to achieve sustainable development in the region. The continued failure to resolve long-standing problems is a major limiting factor (ESCWA, 2002a).

It is essential for sustainable development to bring social and political stability, to address internal problems, to achieve equity in wealth and resource distribution and to respect the rights of citizens regardless of social or religious beliefs. People should be secure in their societies from prosecution based on their beliefs or their political stands. Peace and security based on the respect of human rights is a prerequisite for national security.

Achieving regional stability, peace, and security is required to advance sustainable development in the region and to reduce population displacement and migration. The peaceful settlement of the

Bahrain/Qatar dispute through the International Court of Justice provides a model for replication in settling other conflicts, disputes or tensions in the region.

A number of key areas regarding peace and security have been identified as requiring the support of the international communities and the United Nations system. The world community must be called upon to take a proactive role in the implementation of the United Nations resolutions on regional conflicts. Countries responsible for loss and damage of natural resources during wars should be accountable for cost of restoration (UN-ESCWA, 2002a).

8 INTEGRATING ENVIRONMENTAL ASPECTS INTO FOOD SECURITY

The one common agreement to come out of the food crisis was recognition of the need for international stocks of grain. Unfortunately, little progress has been made beyond this recognition and there are no agreements to who should be paid for the accumulation, the purposes to which they should be put and the trigger mechanisms which would result in their use (Tarrant, 1980). Many indicators reflected to the continued decline in cereal production per capita. This was marked in many developing countries, while at the same time per capita demand for meat and processed cereals rose fast (Tarrant, 1980).

The challenges arising from the close linkages among food security and environmental aspects including climate change and bio energy, has resulted in the demand for a more integrated and comprehensive response within broader strategies designed to address mitigation of and adaptation to the impacts of climate change and bio energy, especially for the most vulnerable populations.

In advocating for and supporting international and national efforts to achieve the food security objectives endorsed by the World Food Summit and reflected in the Millennium Development Goals, FAO and its partners are being called upon to assist the international community in facing new global challenges that relate to the close inter-linkages among food security, climate change and bio energy. Such challenges demand a more integrated and comprehensive response within broader strategies designed to address mitigation of and adaptation to the impacts of climate change and bio energy, especially for the most vulnerable populations (FAO, 2007).

Climate change affects agriculture performance by altering the availability of water, land, biodiversity and terrestrial ecosystem services, and heightens uncertainties throughout the food chain, from yields to trade dynamics among countries and ultimately the global economy. This will affect food security and the ability to feed 9 billion people by 2050.

Bio energy places further demands on agricultural products as well as on the natural resources necessary to meet food and rural employment requirements. Rapidly changing environmental conditions and increased competition over productive resources in turn affect human security in all its forms.

The FAO (Food and Agricultural Organization) will seek to support national and international responses to these challenges, including multilateral instruments, as well as national strategies and policies, ensuring that capacity building for mitigation and adaptation also adequately takes into account the nexus among food security, climate change and bio energy (FAO, 2007). An International High-Level conference on World Food Security and the Challenges of Climate Change and Bioenergy was held at FAO Headquarters in Rome in June 2008. All FAO members' countries, relevant inter-governmental and non-governmental organizations, as well as other relevant institutions participated 56. The overall purpose of the proposed Conference was to address food security and poverty reduction in the face of climate change and energy security. More specifically, the objective was to assess the challenges faced by the food and agriculture sectors from climate change and bioenergy in order to identify the steps required to safeguard food security within the broader context of action being recommended to address climate change and bioenergy at the global, regional and national levels.

The FAO has proposed to hold a High-Level Conference entitled Feeding the World in 2050. Such an event will bring together world leaders and international personalities to discuss options for a common vision and relevant forward-looking pathways for a broader focus that addresses, besides bioenergy and climate change, other drivers of global change. Dramatic changes such as uneven population growth (9 billion people in 2050, mainly in developing countries), migration and urbanization, new food market structures and consumption patterns require new strategies and actions as regards availability, access, stability and utilization of the food system. A High-Level Conference on Feeding the World in 2050 will strategically focus on future challenges and establish processes that will define FAO's long-term vision for food and agriculture, including forestry and fisheries.

9 GOVERNMENT INTEGRATOR ROLE FOR STRATEGIC DIRECTION

The challenge facing the developing countries is focused not only on issues related to develop and implement policies and investment programs that support economic growth but also on distributing the gains of development in a more equitable manner between the rich and the poor (Mc Neil, Mary. and Kuehnast with O'Donnel, Anna., 2004); on avoiding sacrificing the interest of future generations to meet the needs of the present; and on building on the emerging global consensus that natural resources and the environment are valuable assets that should be protected (World Bank, 2001).

The Middle East and North Africa Region (MNA) is disproportionately endowed with natural resources, being the world's richest in oil and gas reserves and one of the poorest in renewable water resources. It continues to rely excessively on natural resources for its sustainable development. The MNA countries share the following long-standing environmental issues, which only differ by magnitude and severity between the countries; water scarcity and quality; land and coastal degradation and desertification; urban and industrial pollution; and weak institutional and legal frameworks (World Bank, 2001).

The harmonization of Environment Impact Assessment (EIA) policies across the Western Asian region is emerging as the optimal modality for avoiding environmental dumping and threats to competitiveness. Furthermore, if ESCWA member countries fail to consider and mitigate potentially adverse trans-boundary environmental impacts triggered by large-scale development and infrastructure projects, serious disputes could arise between neighboring states. The countries involved might adopt unilateral conservation measures or impose bilateral sanctions: neither outcome could be described as a sustainable solution or comprehensive support for development goals. ESCWA member countries should thus work cooperatively to identify mutually shared environmental concerns and find appropriate ways to address them (UN-ESCWA, 2001).

An effective means of addressing transboundary concerns would be the application of "transboundary EIA" (TEIA) analysis as a regional policy for all proposed projects and activities with potentially adverse environmental impacts. Part of the problem lies with differences and inadequacies in environmental legislation, regulation and institutional capacity, lack of clarity, fear of additional cost, and poor enforcement and monitoring systems. This creates a rift in the region between those countries with strict EIA policies and those allowing lax implementation.

10 GOVERNANCE ROLE AND MONITORING

Good and effective governance is also challenging perquisites to achieve sustainable development. It includes strengthening the institutional and legal frameworks, nurturing equitable participation in decision making, approaching policy making from an inter-disciplinary approach, and promoting effective participation by civil society (World Bank, Bojo Reddy R. C. 2005a).

Population, health, education, employment, social integration, governance and empowerment are all socio economic issues of crucial importance to sustainable development. This is because each contributes to the ability of countries to effectively govern human settlements, manage natural resources, generate economic growth and respond to socio economic inequities and human development (Akrawi, 1969).

Generally, two parallel tracks have been pursued by governments to address these issues: direct actions to assist the most in need; and public reforms to improve the provision of health, education, infrastructure, information technology and employment (UN-ESCWA, 2001). While environmental management has made certain progress towards integrated sustainable development. This progress has been limited. The major conceptual as well as practical challenge is how to achieve policy integration. Coordination and consultation between institutions, as well as complementarily and coherence between policy instruments are a basic perquisite requirement. Decentralization in formulating and financing local programs is also a successful concept to have a successful agenda (ESCWA, 2002b).

A solid system for monitoring the applied activities of adopted strategic program planning; program budget is not only a perquisite for good governance but also contributes to building the required information networks and the application of economic instrument which has been severely handicapped by ineffective environmental monitoring systems. (Hazell, 2000).

Developing countries face growing pressures to monitor changes in a broad range of environmental and natural resource conditions. World Bank role and achievements pertaining to the required indicators for monitoring and assessing environmental developments on national, regional and international circles should be pursued (World Bank, March 2002b). Part of this demand stems from the need to comply with various international agreements on environmental issues. Unfortunately, monitoring is not a costless activity, and public funds have to be used that have alternative uses (e.g., investments in schools, roads, and education). Moreover, many monitoring activities are passive and do not lead to the changes needed to rectify the problems they identify. Monitoring can all too easily become an end in itself, particularly once it has been institutionalized.

If monitoring systems are to serve a viable social function, then priority should be given to monitoring environmental problems that offer a potentially large social payoff relative to the costs of monitoring. Realizing favorable benefit/cost ratios is more likely if:

- The environmental problems selected for monitoring have high environmental or social costs if left unchecked.
- The monitoring system is designed and used in a way that leads to rectification of the environmental problems that are being monitored.
- The monitoring system is designed and operated so that it is cost effective.

The policy relevant monitoring system proposed here has three primary objectives (Hazell, 2000). It is informative about changes in the condition of key natural resources. It should provide information about what is changing, how it is changing, and the timing of the change. Ideally, the monitoring system should give advance (or lead) warning about future changes. It is intelligent in that it identifies the causes of change and suggests appropriate responses by key stakeholders for fixing the problem. It is interactive and brings the key stakeholders together to obtain consensus on the problems to be addressed, their causes and solutions, and assign responsibilities for implementing the agreed solutions.

In order to achieve these objectives, the monitoring system requires three components (Hazell, 2000):

- A set of INDICATORS for monitoring resource condition. Two types of indicators are suggested to help contain costs. The first type comprises a low-cost set of indicators that are measured on a routine basis, and which sound a warning whenever a serious problem is emerging. Once a warning occurs, then a second set of diagnostic indicators is activated to provide more detailed information about the nature and probable causes of the emerging problem (Bolt, K. et al., 2005).
- An ANALYTICAL FRAMEWORK to provide a means of identifying the causes of an emerging problem and to evaluate alternative options for fixing it. Depending on the complexity of the

environmental system being monitored. The analytical framework can also provide a means to evaluate the potential environmental costs of different natural resource problems, and hence guide decisions about which problems are worth monitoring on a routine basis (Hirschey, M. 2006).

- An INSTITUTIONAL FRAMEWORK OR AGREEMENT to manage the collection and analysis of monitoring data, maintain and operate the analytical framework (or model), and provide a forum in which the different stakeholders can meet to resolve any disputes and to agree on the implementation of needed changes.

10.1 *Strategic program planning and budget, monitoring and performance evaluation*

The planning, programming, budgeting, monitoring and evaluation cycle constitute the main methodology to be adopted in conformity with the constitution and legal system, aiming at the following (UN Secretary General Bulletin, March 1987):

- To assess what is feasible and derive from the assessment objectives which are both feasible and politically acceptable.
- To translate those objectives into programs and work plans where the responsibilities and tasks of those who are to implement them are specified.
- To indicate the required resources needed to design and implement activities and to ensure that these resources are utilized according to legislative intent and in the most effective and economic manner.
- To provide a framework for setting priorities among activities.
- To establish effective system for monitoring implementation and verifying the effectiveness of the work actually done.
- To evaluate periodically the result achieved, with a view either to confirming the validity of the orientation chosen or to reshaping the programs towards different orientations.

In pursuance of the above aims, the following instruments are to be utilized: The introduction to the medium Term Plan whereby orientations are given to all environment stakeholders. The program budget and the program performance report, where the institution is committed to precise work plan involving delivery of output and where implementation there of is mentioned and reported. The evaluation system which allows for continuing critical review of achievements, collective thinking there on and formulation of subsequent plans. The medium term plan constitutes the legal mandate for the budget (UN Secretary General Bulletin, March 1987).

Instruments of Environment Integrated Management Activities undertaken by the Ministry of Environment, Related Institution should be submitted to an integrated management process reflected in the following instruments: Medium Term Plans, Program budget, Report program performance, and Evaluation reports.

Each of these instruments refers to one phase in a program planning cycle and, consequently shall serve as a framework for the subsequent phases. The planning, programming, budgeting and evaluation cycle shall form and integral part of the general policy making and management process of the government. No activity shall be included in the proposed program budget unless it is clearly in implementation of the plan strategy and likely to help to achieve the plan objectives or in implementation of legislation passed subsequent to the approval or revision of the medium term plan. Priority is a preferential rating for the allocation of limited resources.

The activities should be divided into highest priority and lowest priority. The highest priority activities are those that would be conducted even if total resources were significantly curtailed; activities with lowest priority are those that would be curtailed or terminated if all anticipated resources were not available or if activities with higher priority have to be commences or expanded.

Planning should be based on participatory contributions by all environmental public and private sectors. Target group members have to take an active part in the planning and decision-making, implantation and evaluation.

11 CONCLUDING REMARKS

There are eight key challenges facing sustainable environmental development in the Western Asian (ESCWA) and Middle East North African (MENA) region: governance, stakeholder participation and access to information, economic instruments and voluntary arrangements; environmental monitoring and information networks; environmental education; research and development; regional insecurity and conflicts; and trade liberalization and regionalization. Regional countries have struggled to put into place effective institutions and instruments for managing sustainable development in an integrated manner. However, as other regions have also discovered, the sustainable development process is neither smooth nor easy due to the sheer complexity of linking social, economic, and environmental development issues, and the difficulty of institutionalizing consultation between various public and private stakeholder groups. Nonetheless, the district has made significant progress in environmental management planning and implementation.

These difficulties are exacerbated by the centralized, yet compartmentalized nature of governance. However, linkages between local-national-regional-global sustainable development initiatives need to take into consideration the social, economic, political and cultural sensitivities that are specific to each area. This is particularly important for ESCWA member countries given the need to encourage and adopt locally grown approaches to sustainable development that are innovative, appropriate, gradual and applicable to the region.

The implementation of the proposed strategic actions would require important leadership, collaboration, and coordination with different stakeholders, including NGOs, bi-lateral and multilateral donors and international financial institutions. Partnerships will become an important cornerstone for the Bank's environmental assistance in the MENA region. At the country level, the Bank should be prepared to participate or convene a donor- country coordination group on environment (and/or water issues) to achieve a greater integration of efforts and reduce overlaps between similar programs.

Peace and security has had and will continue to have significant impacts on the progress to achieve sustainable development in the region. The continued failure to resolve long-standing problems is a major limiting factor. It is essential for sustainable development to bring social and political stability, to address internal problems, to achieve equity in wealth and resource distribution and to respect the rights of citizens regardless of social or religious beliefs. People should be secure in their societies from prosecution based on their beliefs or their political stands. Peace and security based on the respect of human rights is a prerequisite for national security.

Good and effective governance is also challenging perquisites to achieve sustainable development. It includes strengthening the institutional and legal frameworks, nurturing equitable participation in decision making, approaching policy making from an inter-disciplinary approach, and promoting effective participation by civil society.

REFERENCES

Akrawi, Matta. "The University and Government In the Middle East", In Science and Technology in Developing Countries, Editors Clair, Nader and Zahlan, A.B. Proceedings of International Conference held at the American University of Beirut in December 1967, Cambridge University Press, 1969, pp. 335–356.

Beseiso, F. (1981) Apragmatic Approach to Arab Gulf States Development Cooperation – The Coceptual and Pragmatic Basis, Ph.D Dissertation, University of Durham, U.K, December 1981, pp. 529–550.

Beseiso, F. (1984a) Development Cooperation Among Arab Gulf Cooperation Council Countries, The Proposed approach and conceptual and Pragmatic Basis, Beirut, Arab Unity Center Studies, Ph.D Dissertations Series, No. 6, May 1984 – (Arabic) pp. 370–423.

Beseiso, F. (1984b) Prospects of Agriculture and Fisheries Wealth Development in Arab Gulf States Council, In Journal of the Gulf and Arabian Peninsula Studies, Kuwait University, Vol. X, No. 38 – April 1984, (Arabic), pp. 125–168.

Bolt, K. et al., (2005) Estimating the Cost of Degradation A training Manual, World Bank, September 2005. pp. E.1–E.14.

FAO, (2007) High Level Conference on World Food Security and Global Challenges, Thirty Fourth Session, Rome, 17–24, November 2007.
Hamilton, Kirk. (2000) Sustaining Economic Welfare: estimating changes in per capita wealth, policy Research working paper; no. wps 2498.
Hazell, P. (2000) The Design of Policy Relevant Monitoring Systems for Natural Resources, World Bank Publications, Environmental Economics Indicators.
Hirschey, M. (2006) Economics for Managers, U.S, Thompson South Western, p. 243 & p. 696.
Interviews with the Ministry of Environment Jordan, The Minister & Director of Policies and Development, February 2008.
Mc Neil, Mary. and Kuehnast with O'Donnel, Anna., (2004) World Bank Institute, Assessing Capacity for Community-Based Development, A pilot Study in Tajikistan, December 2004, Washington D.C., pp. 4–15.
Samuelson, P. A. and Nordhaus, W. D. (2001) Economics. Seventeenth edition, U.S, Mc Graw Hill, pp. 573–601.
Segnestam, Lisa. (2002) Indicators of Environment and Sustainable Development – Theories and Practical Experience, Environmental Economic Series, paper no. 89, pp. 1–13.
Tarrant, J. R. (1980) Studies in Environmental Management and Resources Development, New York: John Wiley & Sons, pp. 296–317.
UNDP, (2008) Human Development Report 2007, Human Development Index, from Wikipedia the free Encyclopedia, January 2008. pp. 1–5.
UN-Economic and Social Commission for Western Asia, (2002a) World Summit on Sustainable Development, Assessment Report for the ESCWA Region, E/ESCWA/ENR/2002/19.
UN-Economic Commission for Western Asia, (2002b) The Effects of Poverty and Unemployment on Sustainable Development in the ESCWA Region-Briefing papers, E/ESCWA/SDP/2002/18.
UN-ESCWA, (2001) Development of Guidelines for Harmonized Environmental Impact Assessment suitable for ESCWA Region, E/ESCWA/ENR/2001/7.
UN-ESCWA, (2002c) The Effects of Socio Economic Inequity on Sustainable Development in the ESCWA Region, World Summit on Sustainable Development, Johannesburg, 26 August–4 September 2002, E/ESCWA/SDP/2002/17.
UN-General Assembly, (2006) follow up and implementation of the outcome of the international conference on financing for Development, A/C/61/L.34, 7 November 2006, New York, U.S.A.
UN Secretary General Bulletin, (March 1987) Regulations and Rules Governing Program Planning, The program Aspects of the Budget, The Monitoring of Implementation and the Methods of Evaluation, ST/SGB/ PPBME Rules/1987.
World Bank, (2001) The Middle East and North Africa Region Environment Strategy Update 2001–2005, December 2001.
World Bank, (2002a) Poverty Reduction Strategies and Environment – A review of 40 Interim and Full Poverty Reduction Strategy Papers, Washington D.C.
World Bank, (March 2002b), Environmental Indicators, An Overview of Selected Initiatives at the World Bank, Updates 3 March 2002, pp. 1–12.
World Bank, Bojo Reddy, R. C. (2005a) Environment Development Papers – Towards Environmentally and Socially Sustainable Development. Poverty Reduction Strategies and Environment. A review of 40 interim and full Poverty Reduction Strategy Papers, p. 86, June 2002, pp. 8–24.
World Bank, (2005b), The World Bank Research Program – Abstracts of Current Studies, Environment Department Papers, Washington D.C., pp. 65–144.

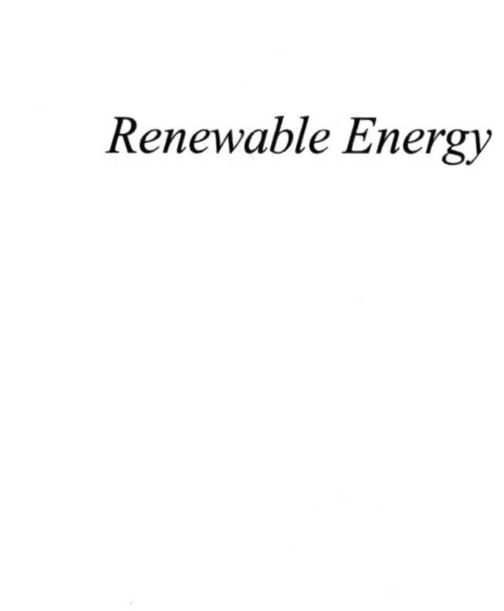

Renewable Energy

Renewable Energy in the Caribbean: A Case Study from Puerto Rico

José A. Colucci-Ríos, Efraín O'Neill-Carrillo & Agustín A. Irizarry-Rivera
University of Puerto Rico, Mayagüez

SUMMARY: An overview of the most mature renewable energy technologies will be given focusing on their potential implementation in the case study area, Puerto Rico, a US territory in the Caribbean. The discussion section will also include findings from an ongoing initiative at the municipality of Caguas which is becoming a sustainability model for the Island. The overall analysis includes some elements of social, technical, cultural, political and economic criteria. Also, sensitivity analyses will be performed regarding the energy generation potential of these processes. The technologies included are photovoltaic, wind energy, fuel cells, concentrated solar power and solar thermal water heating. These are referred to as near term implementation technologies. Other medium/long term ocean energy technologies will be discussed including tide, waves and ocean thermal. The last discussion subsection will briefly consider the area of transportation fuels (gasoline and diesel).

1 INTRODUCTION

Energy is the ability to do work. That basic tenet governs all activity in the world. Energy cannot be created nor destroyed, only transformed. That is another important tenet that determines many times how living beings behave and interact on Earth. The search for and control of energy sources, to guarantee food supply or to improve quality of life has always been a human struggle and source of numerous conflicts. The use of electricity as an energy carrier in the late 19th century, and the introduction of the combustion engine marked the dawn of a new energy struggle for humankind. The availability of cheap fossil fuel products in the 20th century motivated the development of world economies based on these non-renewable energy sources (Smil, 2008).

Electricity and transportation were the cornerstones of the economic development experienced in many countries in the 20th century. Electricity is commonly generated by using energy from a primary energy source such as fossil fuel products. On the other hand, most forms of transportation worldwide mostly depend on fossil fuel products. Many industries and national infrastructures worldwide depend and exist based on the extraction, processing or use of fossil fuels. Fossil fuels are non-renewable energy sources because they exist in nature in a limited supply, formed over millions of years from organic material that was trapped beneath the Earth's surface at specific pressure and temperature conditions (Smil, 2005). The worldwide dependence on fossil fuels presents an enormous challenge today and in the future as the price of oil reaches record levels (Appenzeller, 2004).

A renewed interest in renewable energy (RE) processes has emerged driven by the oil dependence/depletion mentioned earlier, strong environmental movements, and lately national security concerns (Tester et al., 2005). Renewable energy sources are those that can be used but are able to renew or recover themselves naturally. Several RE technologies such as wind, niche photovoltaic and biodiesel are presently very competitive in certain applications versus their oil counterparts especially in Europe and certain locations in the mainland United States. Others are slowly penetrating certain markets such as fuel cells.

In the case study area of Puerto Rico, a US territory in the Caribbean, renewable energy technologies have not penetrated to the same levels as the continental United States and Europe. This is puzzling given the ideal conditions present on parts of the island for these technologies especially solar thermal and photovoltaic. For example, the average yearly solar irradiation is approximately 5.52 kWh/m^2 (Jiménez-González and Irizarry-Rivera, 2005) There is also a continuous strip of more than 200 hundred miles along the north, east and south coast identified by an updated wind map from the National Renewable Energy Laboratory (NREL) with prevailing wind velocities more than sufficient for the implementation of wind turbines. Ocean energy also has potential especially wave and ocean thermal. In addition, Puerto Rico has strong environmental movements, oil dependence/depletion and national security concerns, which also drives development and commercialization of these technologies. However, RE technologies are almost nonexistent in the Island. There is some commercial activity in solar thermal for water heating ($<5\%$ of the potential market), energy efficiency (lighting), biodiesel, photovoltaic and even backup hydrogen fuel cell (5 kW) systems. Recent legislation for net metering implementation and the tripling of the record cost of a barrel of oil during 2008 is increasing the awareness for the need of these technologies. The government is also collaborating with the University of Puerto Rico Mayagüez in identifying achievable renewable energy targets that could lead to establishing Renewable Portfolio Standards.

In this chapter an overview of the most mature renewable energy technologies will be given focusing on their potential implementation in the case study area of Puerto Rico. The technologies included were photovoltaic (O'Neill et al., 2004), wind energy (Ramos-Robles and Irizarry-Rivera 2004; Ramos-Robles and Irizarry-Rivera 2005) fuel cells (Colucci et al., 2007), concentrated solar power and solar thermal water heating (Jiménez-González and Irizarry-Rivera, 2005). These are referred to as near term implementation technologies. Other medium/long term ocean energy technologies were also discussed including waves and ocean thermal. The last discussion subsection briefly considers the area of transportation fuels (gasoline and diesel). In the last sections, some barriers to RE implementation are presented and a holistic approach is proposed that integrates not only the technical and economic potential of alternatives, but also the social and environmental dimensions.

2 THE HIGH COST OF ELECTRICITY IN PUERTO RICO

For years the Industrial Association of Puerto Rico has identified the cost of electric energy as a major obstacle for doing business in the Island. Yet, long-term changes in the electricity cost have not occurred. The recent government effort towards supporting biosciences, and the investments of major biotechnology companies in Puerto Rico, stress the need to provide a business environment where the fixed operating costs are diminished as much as possible. Furthermore, other economies are also investing in attracting these same biosciences companies to their countries. Businesses will go wherever the investment environment is more opportune (Colucci et al., 2007).

However, the electricity needed by businesses, citizens, and visitors comes at a premium cost. Puerto Rico's average electricity cost per kWh is the highest in the United States. The average cost per kWh in the United States was \$0.0814 during the year 2005. In the case study site the average cost per kWh was \$0.1691 more than twice the U.S. average as shown in Table 1 (EIA, 2006). Notice that as of May 2008 the average price of residential electricity is approximately 0.24 \$/kWh. The increase in electricity price is directly related to the dependence on foreign oil to produce electricity. In contrast to the United States where only 3% of the electricity is generated from oil, in the case study area this dependence is approximately 80%. There are also issues of efficiency in the Island's utility-owned generation plants (many of them over 30 years old). Another reason is that some of the utility's oil-based plants are required to use fuel #6, which is more expensive, due to environmental restrictions in the area where the plants are located.

In the near future, especially in an island-environment, the traditional view of equating energy use to economic development is not sustainable. A new perspective on energy use at all levels, and its relationship to economic development must be established. On the other hand, the fuel diversification strategy for the electric industry of equal division among coal, gas and oil perpetuates our dependence on external sources of energy. Alternatives that create local jobs and keep the money in Puerto Rico need to be sought.

Table 1. 2005 average retail sales price by sector, in ¢/kWh, of electricity in the United States vs. Puerto Rico.

Sector	Puerto Rico	United States
Residential	16.57	9.45
Commercial	17.94	8.67
Industrial	14.64	5.75
Average	16.91	8.14

There are substantial benefits from increased use of renewable energy resources. Among the benefits those cited most frequently in the literature are: reduced costs of fuel for electricity; reduced reliance on imported oil supplies and exposure to the volatile prices of the world oil market; risk management by diversifying the portfolio of electricity generation options; job creation and economic benefits; and environmental benefits (Tester et al., 2005). Warnings about the danger of the Island's dependence on oil have been given since the 70's and 80's and resonate stronger today: "A general consensus is needed so that strategies and actions for oil substitution alternatives on a large scale may be implemented. Plans to implement alternative energy sources should be translated into action. Prudence and economics dictate that countries move toward energy self-sufficiency as rapidly as possible" (Bonnet, 1983). Furthermore, in 1980 a study from the National Academy of Sciences concluded that "Puerto Rico, in dealing with its own energy problems, should grasp its opportunity to become an international energy laboratory, seeking and testing solutions especially appropriate to the oil-dependent tropical and sub-tropical regions of the world. The Island's geographical position and its established energy research and development facilities can enhance this potential" (U.S. National Academy of Sciences, 1980). A similar statement was said about Hawaii in February 2008. Hawaii is a good benchmark for Puerto Rico because of the islanded nature of its electrical system, its high energy prices and its dependence on outside energy sources (DOE, 2008). The difference is that Hawaii's statement was backed by prompt state and federal action that allowed Hawaii to commit to a goal of 70% use of renewable energy by year 2030. Puerto Rico still have the potential mentioned in the National Academy of Sciences report, and can aim to a goal similar to Hawaii in terms of renewable energy use. The technologies discussed in the next section present near term opportunities for Puerto Rico given their advance state of development and commercialization.

3 NEAR TERM TECHNOLOGIES

Figure 1 shows an estimated foot print that would be required in the case study area to meet the total electrical power needs of the Island. The maximum total capacity considered was 5 GW. Presently the peak demand is approximately 3.5 GW with an average consumption of 2.8 GW. Notice in Figure 1 that in theory all the evaluated options require less than 1% of the total land to meet a capacity of 5 kW except wind turbine farms that would require close to 8%. This is a "macro" view that will be validated at a local level especially with photovoltaic and solar thermal. In the former residences and expressways are ideal sites for this technology. A sensitivity analysis was performed for these two cases. The results are shown below (Figure 2). In both cases the energy generation potential is very promising long term. For example, 30% of the residences (~300,000) with a 400 ft^2 system will generate approximately 1.2 GW during peak solar periods. This is slightly lower than the Puerto Rico residential sector peak consumption (45% of 3.5 GW). Also, it represents a 6 to 7 billion dollar business opportunity using \$6 to \$7/Watt investment unit per for the systems.

Wind energy is another area with immediate application potential. The 2007 version of the NREL map shows that the best wind for development is in the East Coast of the Island. However, there are other areas in Puerto Rico that also have wind energy resources that are technically and economically

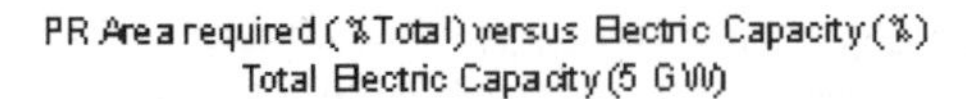

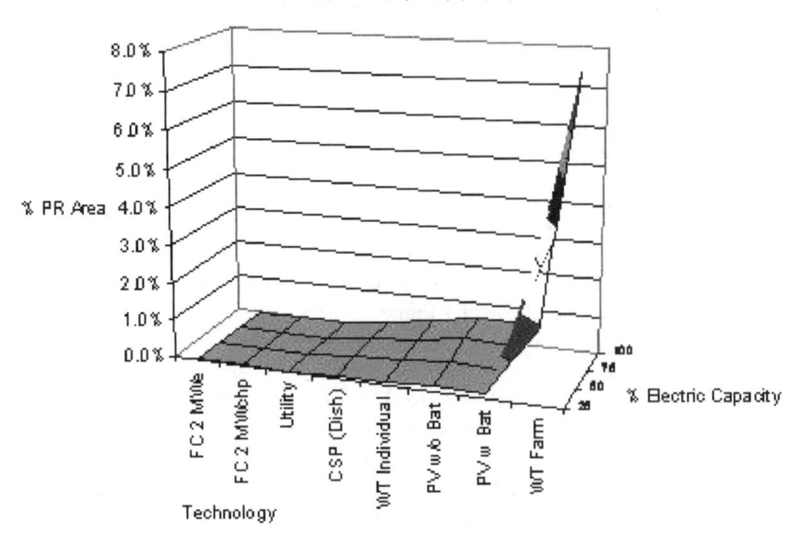

Figure 1. Estimated footprint that would be required in Puerto Rico to meet the total electrical power needs of the island.

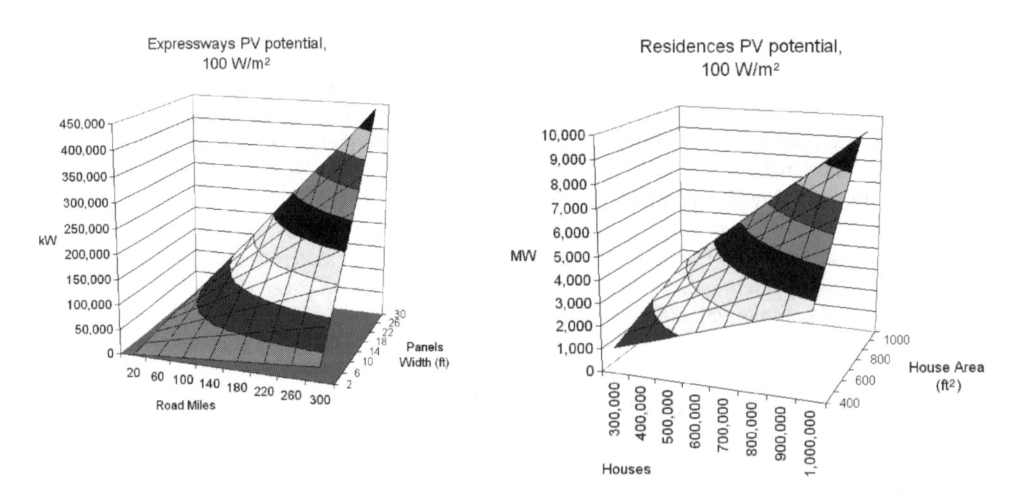

Figure 2. A sensitivity analysis was performed for photovoltaic (PV) renewable energy generation using expressways and residences as sites for this technology.

feasible for development. The Puerto Rico Energy Affairs Administration is supporting a privately financed wind farm development (~45 MW) in the southwest coast of the island. It is generally accepted that these developments are commercially feasible even at the Power Authority avoided cost of approximately \$0.10/kWh. In addition, in a recent study a continuous strip of more than 200 hundred miles along the north, east and south coast with prevailing wind velocities more than sufficient for the implementation of wind turbines was identified (Irizarry-Rivera et al., 2008). The strip follows the valley of the coasts, which are highly developed. There is potential for offshore wind farms in the southern portion of the Island. Besides technical considerations, site selection, environmental concerns and community acceptance are usually areas that need to be addressed in wind projects.

Table 2. Electric and emissions impact of the use of solar water heaters.

	Estimated reductions
Electricity Demand	10 MW
Electric System Losses	0.36 MW and 3.9 MVAR
Electric Generation	10.36 MW and 3.9 MVAR
Emissions, lbs	CO_2 – 8,600, NOX – 6.48 and SO_2 – 5.25

In a fuel cell analysis the most interesting result was that efficient heat recovery is critical (Colucci et al., 2007, 2004). Several companies (e.g., Abbot Pharmaceutical, Johnson & Johnson, Pfizer, Hewlett Packard) were considering 250 kW molten carbonate units for their facilities. Lack of a natural gas infrastructure and subsidies (federal and state) were hard to overcome. At best using propane resulted in a breakeven scenario for a ten year project. In addition, back up 5–10 KW hydrogen fueled Proton Exchange Membrane Fuel Cells (PEMFC) are been considered for remote applications by the communications industry.

Concentrated solar power is also an option with some potential based on the following: highly versatile 5 kW (residential) to 200 MW options; competitive foot print – cost index; effective in arid/desert regions with sustained high levels of direct normal insolation (i.e., Peñuelas/Guayanilla brownfields); over 20 years of operating experience at MW levels; and designed with energy storage systems to operate 24/7.

Another technology that was considered with near term implementation potential was solar thermal water heaters (STWH). The City of Mayagüez was the test case considered assuming the following (Jiménez-González and Irizarry-Rivera, 2005):

- Puerto Rico's average daily solar radiation is 5.52 kWh/m^2
- Puerto Rico's abundant solar resource could be exploited via Solar Thermal Water Heating to achieve generation displacement, reduced emissions and electric system losses.
- Total occupied housing units: 34,742
- Average family size: 3.41 persons
- Typical U.S. household hot water consumption for 4 persons is 240 liters/day (63 US gallons/day)
- Water temperature rise of 75°F (23.89°C), corresponding to an inlet water temperature of 60°F and water heater set point temperature of 135°F.
- 85% of occupied housing units can use STWH to replace its electric water heater.

The preliminary results are shown in Table 2. Based on these results the following benefits were identified for the implementation of solar water heaters: demand reduction (increased reliability, capital investment savings since electric grid utilization is increased); reduced power losses (savings for the customer thru the effect in fuel and energy adjustment factor, currently quantifying this); generation displacement (increased reliability, capital investment savings); reduced emissions into the atmosphere; and an alternative to fuel diversification.

4 MID AND LONG TERM TECHNOLOGIES

The technologies considered in this section were ocean waves and ocean thermal (Bonnet and Graves, 1980). The ocean wave's analysis sensitivity analysis preliminary results are shown in Figure 3. A design equation provided by Tester et al. was used to generate the graph. The critical variables for ocean waves are wave height and wave period. The graph used as the basis 1 km of coastline. Notice in the figure the excellent energy generation potential of this technology. On-going investigation indicates that the north-west of Puerto Rico has wave heights with considerable energy generation potential.

The last technology considered was ocean thermal. The south east corner of the island has one of the best locations for this technology in the world due to the proximity of deep coasts that would

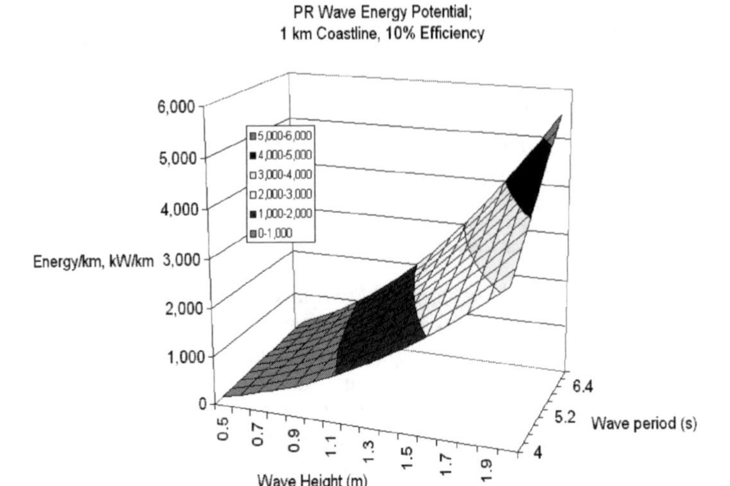

Figure 3. Renewable energy from ocean waves.

provide excellent temperature gradients required for thermal flows. This has also been mentioned by Tester et al and in past studies by the Center for Energy and Environmental Research (Bonnet and Graves, 1980; Colucci et al., 2007).

In summary, good wave power often correlates with offshore wind power sources, shoreline sites are expensive, cost structure is not well known (i.e. capital and operational), and aesthetic, fishing and navigation hazards concerns must be addressed.

5 LIQUID FUELS

This section considers alternatives to replace both diesel and gasoline. For the former, biodiesel was considered as a drop-in replacement (Colucci et al. 2003, 2005, 2007). An analysis was performed to estimate the amount of land required to produce significant amounts of biodiesel based on palm oil. This is a high productivity agriculture crop. However, the land requirements are not acceptable. At 1 million gallons per year per 1,000 hectares, approximately 300,000 to 500,000 hectares will be required to supply Puerto Rico's diesel equivalent needs. This corresponds to 30 to 50% of Puerto Rico's horizontal plane area. The same applies for other crops intended to produce fuels to replace gasoline. However, gasoline utilization in the case study area is approximately 1 billion gallons per year. Even at an optimistic 5 million gallons per 1,000 hectares per year, utilizing lignocellulose material will require 200,000 hectares, which again is unrealistic (Sáez et al., 2002).

One alternative that is being considered is microalgae. According to NREL these organisms can produce twenty times the amount of hydrocarbons than their land counterparts in open raceway ponds (Sheehan et al., 1998). Biophotoreactors are expected to double or even triple these productivities. Both of these technologies, however, are in the development and/or technology transfer stage. Nevertheless, at present they seem to be the only reasonable option for an agricultural based biofuel/biochemicals production economy.

Another alternative proposed a liquid fuel is the waste to fuel/electricity option. Concerning solid waste already deposited in Puerto Rico's landfills, their energy generation potential is limited to 50 to 100 MW_e. For example, in the San Juan landfill it is estimated that it can produce between 2–4 MW_e or compressed natural gas for approximately 100 trucks per day. The latter project is under the consideration by the municipality in collaboration with the University of Puerto Rico – Mayagüez campus and a landfill consultant company. The generation capacity for "new waste" is limited to 60–120 MW_e. The bases for this estimate are 2,000 to 4,000 tons of energy containing waste/day basis and 30 MW/1,000 tons/day. This corresponds to 40–50% of the total waste. It should be mentioned that the "Autoridad de Desperdicios Sólidos de Puerto Rico" or Puerto Rico Solid Waste Management Authority is recommending the establishment of two waste to energy plants in PR. This is described in detail in their publication "Itinerario Dinámico" (ADS, 2007). Several municipalities are considering Waste to Energy technological options such as Gasification and Plasma Arc. These include Caguas, San Juan, Carolina, Barceloneta, Manatí and Toa Baja.

It should be noted that biofuels utilization is an area of interest to several sectors. For example, during 2008 the Puerto Rico Electric Power Authority performed a two day biodiesel (B20) test at their 20 MW Jobos facility diesel fuels turbine. In addition, the government operated public transportation bus agency is considering converting to biodiesel (B20) in the same year. They utilize approximately 2.5 MGPY of diesel. As mentioned below, the municipality of Caguas is converting to biodiesel (B20) their diesel fleet (100,000 GPY). Also at the end of 2006 the first of Puerto Rico's biodiesel production plants started operating in Guaynabo. They had a 1 million GPY biodiesel production capacity utilizing used cooking oil as the raw material and are considering increasing their capacity utilizing other raw materials including microalgae based oils. Their main markets are Laundromats and transportation fleets.

Another important activity in this area was the establishment in 2008 of the Biorefinery Research Consortium by the University of Puerto Rico – Mayagüez campus at the old Aquaculture facilities in Lajas. The initial objective is the development of microalgae technology using both ponds and biophotoreactors. In addition, the facility will include a 2.5 million gallon per year lignocellulosic based ethanol semiwork process.

6 CAGUAS SUSTAINABLE ENERGY SHOWCASE

The Caguas Sustainable Energy Showcase initiative started in the fall of 2006 when the mayor of Caguas requested the authors of this chapter to identify the potential of his municipality regarding renewable energy implementation for 2010 and 2020. In addition, the initiative included identifying eighteen high visibility projects that would familiarize the residential, industrial, commercial, and government sectors with these technologies.

Four projects were started in the summer of 2007 including Villa Turabo, Caguas Fluoresce, Jardín Botánico Fuel Cell and Diesel Fleet Biodiesel Conversion at an approximate cost of $1 million. The former consists of implementing photovoltaic (~700 W) and solar heating systems in 100 residences. They will provide between 20–30% of the energy requirements of these residences. Caguas Fluoresce is very similar to the Environmental Protection Agency Change a Light program except that it also involves a monitoring phase before and after installation of the fluorescence light bulbs. Two thousand residences out of 50,000 total were targeted with ten light bulbs per household (eight 60 $W_{equivalent}$ and two 100 $W_{equivalent}$).

The fuel cell project consisted of installing a 5 KW hydrogen fueled proton exchange membrane back up fuel cell at the Jardín Botánico facilities at Caguas. The system will include six hydrogen cylinders with a total capacity of 48 kWh. It will provide back up to the computer servers and occasionally will run a conference room used for lecture series during special activities. The biodiesel conversion project is an extension of Caguas biodiesel demonstration project in 2003. They utilized biodiesel in their whole diesel fleet for three months.

The next phase will be migrating these developments to the municipality in general. The main hurdle is the high payback even at current electrical energy rates ($0.20 to $0.25/kWh) (Figure 4). Notice in the figure that compact fluorescent lamps equivalent to 100 and 60 W incandescent lamps (CF100W and CF60W in the figure) both have paybacks that are less than 0.1 years. These are the exception. However, solar water heaters have paybacks between 2 to 3 years (SH300W and SH200W in the figure) depending on the electricity reduction for an equivalent electric heater. These correspond to 300 and 200 W_e reduction, respectively. The same applies to refrigerators (Ref75W and Ref50W in the figure). Their payback is between 6 to 8 years again depending on the electricity savings, respectively. Notice that a residential non-subsidized 1 kW photovoltaic system

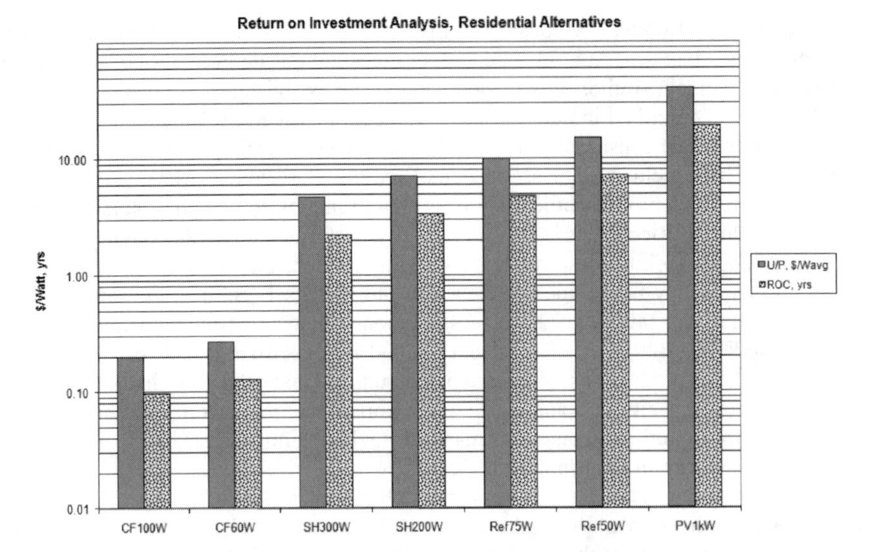

Figure 4. Capital Investment unit per and return on capital for residential renewable energy alternatives. CF – Compact Fluorescent, SH – Solar Heater, Ref – Refrigerator and PV – Photovoltaic.

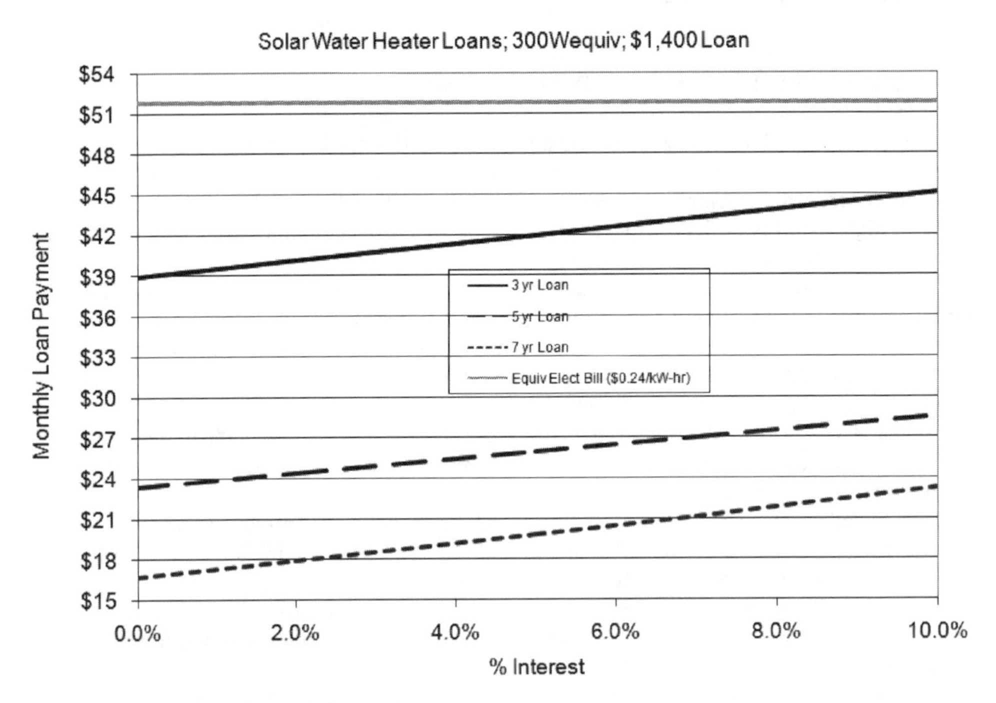

Figure 5. Residential solar water heaters loan payments.

has a 20 year payback. These are non-battery systems priced approximately at $10/W using Villa Turabo's quotations.

Based on the above scenarios, financing strategies were developed in order to identify the level of support the municipality must provide to the consumers to accelerate the commercialization of these technologies. The concept is to provide subsidized loans where the consumers will pay less for their loan than the equivalent energy expenses. This is shown in Figure 5 where for a

300 W_e electricity reduction even a 3 year loan results in lower monthly loan payments than the equivalent electrical bill even at high interest rates. Similar analyses were done for refrigerators and commercial/industrial photovoltaic systems.

Parallel to these efforts, the municipality is considering working very closely with UPRM's Institute for Tropical Energy, Environment and Society. Both will be organizing discussion forums with the communities and other potential stakeholders in order to define a consensus-based municipal energy public policy that not only will identify the best technologies but also include social and environmental considerations in their application and more important sitting. The last section of this chapter presents further discussion on this broader perspective for the energy dilemma.

7 CHALLENGES FOR RENEWABLE ENERGY

The dominant electric energy model of central generation has existed for over 100 years. The model has worked well. Any new energy alternative will face opposition from the industry establishment. The burden is on proposers of new practices and technologies to prove that their alternative is better than existing energy sources and practices (Tomain and Cudaht, 2004).

Second, the regulatory structure in the Island presents a huge challenge. On one hand, the 1941 law that established the Puerto Rico Electric Power Authority (PREPA) gives it ample powers over all things related to electric power, including being mostly self-regulated. There are historic reasons for that decision, and the strategy definitely was vital in the economic development of the Island. However, such powers and structures should be evaluated in light of the new global and local conditions in the energy industry.

The U.S. Public Utility Regulatory Policies Act (PURPA) of 1978 encouraged the use of renewable energy for electricity production, and diversified generation technologies. It fostered the growth of non-utility generators and independent power producers to reduce use of foreign fossil fuels in the U.S. Although its prime objective was energy efficiency and conservation, PURPA laid the groundwork for deregulation and competition by opening wholesale power markets to non-utility producers of electricity. PURPA and its aftermath also demonstrated that electricity generation is not a natural monopoly (Bosselman et al., 2006). It was through PURPA that some fuel diversity was achieved in Puerto Rico, with the establishment in recent years of the EcoElectrica (natural gas) and AES (coal) power plants.

The Energy Policy Acts (EPAct) of 1992 and 2005 also apply to the case study area. EPAct 92 encouraged states to open access of transmission lines for sales by private generators (known as wheeling) for inter-state power exchanges. It also encouraged the use of distributed generation (DG), generators operating at lower voltage levels closer to the points of use of electricity. EPAct 2005 reinforced federal programs on energy efficiency and renewable energy. It also stated that utilities and public service commission must consider important operating modes different from the dominant energy model: Interconnection of DG, and net metering (i.e., the sale of power by private producers at the same rate that utilities sale power to clients). It has been through EPAct 2005 that PREPA acted and approved interconnection of DG, and will also act on net metering. The challenge is to obtain just and reasonable regulations and rules for both interconnection of DG and net metering that effectively encourage the use of renewable energy. Another challenge is how Puerto Rico can be proactive in future energy alternatives instead of reactive to external energy markets and federal regulations.

On May 28, 2008 Law 73 for Incentives to Industry in Puerto Rico was signed. This law has created controversies and questions with regard to the impact on various sectors in the Island. A challenge and an opportunity this law presents is a mandate to PREPA to allow and enable the open access of their transmission lines (wheeling) mentioned in the previous paragraph. This is one of the most aggressive measures taken recently in terms of the electric power regulatory framework in Puerto Rico. It remains to be seen how wheeling will change the law that created PREPA's and how effectively wheeling can be used to reduce the cost of electricity for all Puerto Ricans. A challenge is how all sectors can participate in the decisions with regard to the implementation of wheeling in Puerto Rico and other electric energy reforms.

Third, there is also a need to be realistic in terms of using renewable energy sources: If there is no wind or sun, there is no energy. Renewable energy technologies have an environmental impact. For example, the manufacturing of PV panels is energy intensive and has some environmental impact since they involve semiconductors. Batteries used for storage in renewable systems present a challenge with regard to disposal after their lifespan is reached (typically 5–7 years). There are sitting issues, and space limitations with some renewable energy technologies. These remarks are not meant to discourage the use of renewable energy, but rather to engage in discussions of new energy alternatives that take into account these limitations. Conservation, efficiency & renewable energy sources could halt new construction of fossil fuel power plants in Puerto Rico, and must be included in the planning of our electric system. Incentives for residential and small commercial customers must be pursued, and PREPA could become an enabler of DG so that it complements central generation, and considered in the planning and operation of the Island's power system.

8 TOWARDS A SUSTAINABLE ENERGY FUTURE

Although the focus of this chapter has been on the technological potential of renewable energy, it is important to emphasize that technology alone is not enough to effectively deal with energy challenges. A well-known definition of sustainable development by the United Nations emphasizes on how we use resources today in a way that does not compromise the ability of future generations to meet their needs (United Nations, 1987).

Some of the issues that are related to sustainable energy are

- Technology literacy/acceptance
- Include environmental/health/quality of life costs (i.e., externalities) in the project estimates. Analytical techniques such as Life Cycle Analysis (LCA) and Internalization of Externalities are very powerful in this area.
- Consider all segments of the population economic resources
- Distributed versus centralized generation
- Population ageing/location
- Nature of new loads such as in transportation (electric vehicles) and entertainment (electronic).

For example population ageing is changing very quickly in Puerto Rico (Figure 7). For example, notice in the figure that the average age (vertical lines) is expected to increase to approximately 35 years old in 2010 versus 24 in 1990. This is a one year increase every two years. The graph also shows a 2% increase in the 65 to 74 age bracket from 1990 to 2010. Its implication regarding nature and load of their needs and their economic resources should be considered in future projections.

Regarding cost estimates, there are two powerful methodologies that are gaining acceptance especially in Europe; Life Cycle Analysis (LCA) and Internalization of Externalities. In the latter "external costs" are estimated. They are defined as follows: LCA is a process to evaluate the environmental burdens associated with a product, process or activity by identifying and quantifying energy and material usage and environmental releases, to assess the impact of those energy and material uses and releases on the environment, and to evaluate and implement opportunities to effect environmental improvements (Hohmeyer and Ottinger, 1991; Hohmeyer and Ottinger 1994; O'Neill-Carrillo, 2005). External costs on the other hand are as those actually incurred in relation to health and the environment and quantifiable but not built into the cost of the electricity to the consumer and therefore which are borne by society at large (Krewitt, 2002; UIC, 2004). Examples of these methodologies are shown in Tables 3 and 4. Notice that both LCA and external costs provide a better estimate of the impact of these technologies to society (Graedel and Allenby, 2003).

There are several other critical areas that must be addressed in order to begin a cultural change in the generation, distribution and utilization of energy. Puerto Rico's electric system is isolated therefore limitations imposed by the laws of physics, power flow, stability, power quality, power

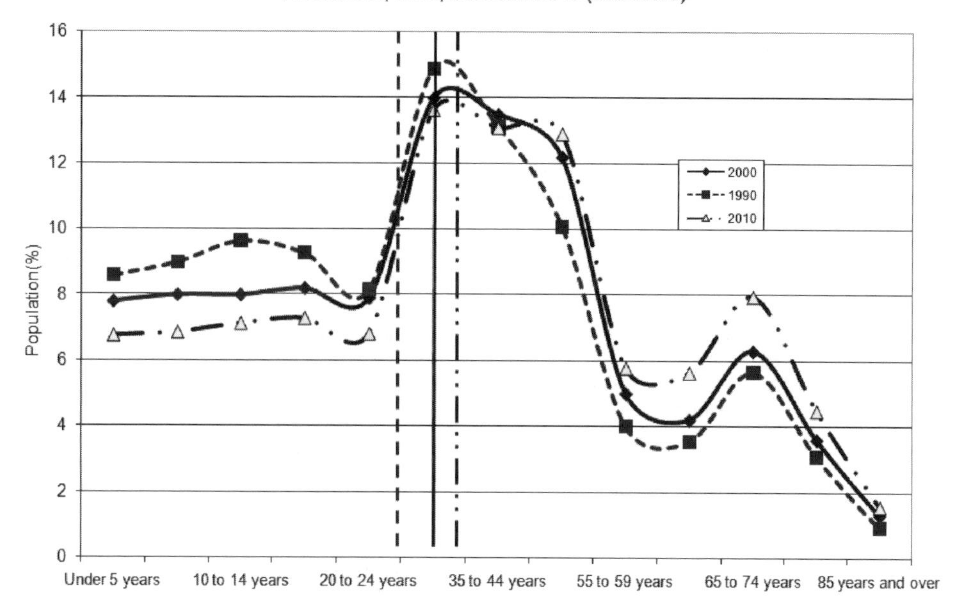

Figure 6. Puerto Rico census data.

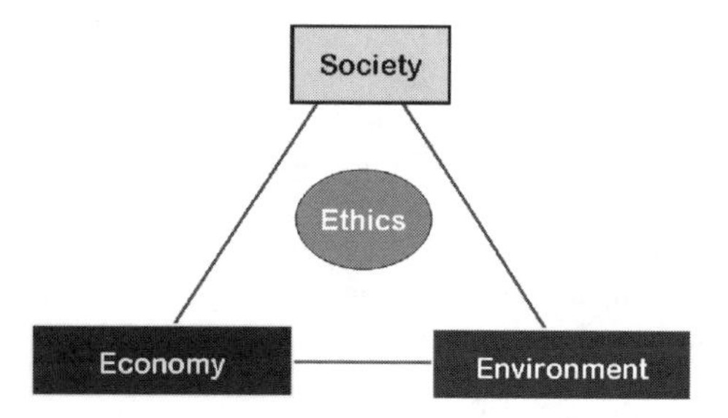

Figure 7. Energy ethics and sustainability.

electronics, among others must be seriously considered. Energy policy issues such as net metering, wheeling and distributed generation must be brought to the table for serious technical discussions.

It should be emphasized that these issues cannot be an excuse to do nothing. Again there is an urgent need for studies to understand what can be done, and how we can continue the diversification of energy sources and systems. It should be mentioned that there is little need to import talent to implement these changes. Puerto Rican engineers, scientists, professors and other local professionals have the expertise to study, lead and implement the changes needed.

There is another important dimension in this discussion that relates to the ethical dimension of the energy problem. The greatest impacts that humans have had on the world include the many consequences of the use of natural resources to ensure an energy supply to sustain the economic/technological development and the consumption patterns of the world's great powers (Humphrey et al., 2002). The search for energy supplies has historically being carried out and justified mostly from an economic perspective, obliterating in many instances human rights, the environment, communities, and paradoxically even long-range economic stability. The world's dependence on fossil fuels, and the transition to other energy alternatives represent a major challenge for humanity. How to face that challenge, avoiding individual, group and environmental injustices is perhaps one of the greatest challenges of humankind in the 21st century.

Table 3. LCA emission estimates for Germany (Krewit et al., 1998).

Generation type	SO_2 (g/MWh)	NOx (g/MWh)	Particulates (g/MWh)	CO_2 (g/MWh)
Nuclear	32	70	7	19,700
Coal	326	560	182	815,000
Gas	3	277	18	362,000
Oil	1,611	985	67	935,000
Wind	15	20	4.6	6,460
PV (Residential)	104	99	6.1	53,300

Table 4. External costs estimates in Minnesota (ME3, 2005).

	Urban	Metropolitan	Rural
SO_2 (\$/ton)	0	0	0
PM10 (\$/ton)	5,060–7,284	2,253–3,273	637–970
CO (\$/ton)	1.20–2.57	0.86–1.52	0.24–0.46
NOx (\$/ton)	421–1,109	159–302	20–116
Pb (\$/ton)	3,551–4,394	1,873–2,262	456–508
CO_2 (\$/ton)	0.34–3.52	0.34–3.52	0.34–3.52

The problems related to reaching a sustainable future are common to most countries, and a general consensus exist that energy, and water will cause instability in many parts of the world due to their direct relationship to health and the environment. With regard to energy, it is imperative that individuals, energy producers, businesses, governments, non-governmental organizations, and communities come together in a new scenario that allows ample and transparent participation in the tough decisions that need to be made. Consensus need to be reached, because there is no single or magic solution to the energy dilemma. The importance of consensus is even more important if a sustainable strategy is sought, where economic, environmental and social dimensions need to be conciliated and considered with equal weight (Norton, 2005; Zamot et al., 2005). Thus, participatory structures need to be created, that foster dialogue among all stakeholders.

If the goal is sustainable energy, and if consensus is sought among seemingly contradicting objectives, the search for sustainable technologies and practices must be embraced as a moral obligation with an integrative and global perspective in order to create the personal or institutional commitment necessary to withstand the hardships of consensus-building. Stakeholder engagement and involvement at all levels is essential in the successful application of this new energy ethic.

The main objective of the stakeholder engagement framework is to increase awareness so that a more responsible citizen may emerge, willing to contribute to the solution of the energy problems. The target audience includes individuals, energy producers, businesses, governments, non-governmental organizations, and communities. The activities and strategies present a multi- and trans-disciplinary perspective of energy as a global and moral problem which is usually approached from a narrow technical or economical perspective. Audiences are presented with the sustainable perspective as a way to conciliate the social, environmental and economical dimensions of the energy dilemma. Activities include discussions on energy policy, life cycle analysis, externalities, issues of justice, local considerations (island policies and consumption patterns), examples from the US and the rest of the world, energy practices and technologies. Whenever possible (especially with younger audiences), a demonstration of a photovoltaic system is performed. Strategies also include integration of the energy ethics to research and teaching activities at university settings, as well as outreach to communities, government and private entities.

An important target audience must be communities in Puerto Rico, especially vulnerable communities such as rural and low-income. A partner in this effort has been, for example, Casa Pueblo, a not-for-profit, community-based NGO in Adjuntas in the case study region, advocating natural resource conservation and the development of community-based projects. Through this collaboration, energy ethics has reached K-12 students and teachers, as well as members of the rural Adjuntas. If true community collaboration is sought, universities must involve themselves within the community.

A community-based sustainable energy lab has being established at Casa Pueblo to provide an environment for the study of the relationship of energy, communities and the environment. The building that houses the Community Institute provides a perfect setting for a photovoltaic installation that would also serve as a class laboratory. Not only UPRM (University of Puerto Rico) students would benefit from these projects, but also K-12 students and teachers, and also the general public. The aim of this collaboration is to educate the surrounding community on sustainable energy practices. There have also been outreach activities in other communities such as Caguas, San German, Añasco and Yauco.

Activities that integrate the new energy ethics at the undergraduate level include service learning projects in senior design courses, and ethics across the curriculum strategies in power engineering courses. At the College of Engineering at UPRM, there is a coordinator for Social, Ethical and Global Issues striving to change the engineering education paradigm from regulatory compliance to doing good (O'Neill-Carrillo et al., 2008). The new energy ethics and sustainability provide a global and practical context for the development of professional skills and a more service-committed student. There is also an emphasis in Global Challenges Education modules to empower individuals, cultivate social responsibility, and better prepare students for the interconnected, global workplace of the 21st century (Leskes, 2005). In terms of research and graduate education, the Institute for Tropical Energy, Environment and Society is committed to the new energy ethics. Engineering courses have integrated social and ethical dimensions of the energy problem, not only within UPRM, but also in graduate courses being taught through distance education to the Dominican Republic.

The new energy ethics is also disseminated through professional outreach activities. Articles in professional publications and participation in annual meetings of professional societies have allowed the new energy ethics to reach representatives from government and private organizations. This new energy ethics could inspire and commit the present generation with a new energy perspective that should be passed on to future generations as the base for a sustainable future. The thesis is that there is a circular relationship between stakeholder involvement and the successful implementation of an energy ethics. To go from engagement to involvement, personal or institutional commitment is needed. The commitment to be long-lasting and effective must be rooted in ethics fostering values that transcends time (future generations) and space (from local to global perspective). Each individual or institution must own the energy ethic, the universal principle that ties to the outside world being sustainability, while at the same time respecting space for diversity and particular perspectives within the search for consensus (Figure 7).

Some of the ethical issues that need to be resolved with regards to renewable energy and a sustainable future include social and environmental justice implications related to energy alternatives, ethics of energy expertise, the responsibility to the pursuit of truth as honest brokers, our social responsibility as policy entrepreneur that promote social change, and the intergenerational responsibility included in the sustainability definition of the United Nations (United Nations, 1987).

Focusing only on technological fixes for our energy problems has historically proved to be a wrong strategy. The authors firmly believe that the world's complex problems require a more holistic approach that integrates the expertise and will of many diverse fields and individuals. Efforts that describe one possible holistic approach are presented in O'Neill-Carrillo et al. (2008).

In order to move to a sustainable future, all energy stakeholders need to participate in the generation, evaluation and implementation of long-term strategies: Collaborations among government, industry, commerce and citizens need to be established, that allows us to go from an adversarial to a collaborative relationship, from mutual distrust, to a serious and lasting commitment for the public good, for social, environmental and economic welfare.

9 CONCLUSIONS

An integrated sustainable energy implementation plan should be the next step for Puerto Rico. In addition, ocean-based technologies should be studied given their high energy generation potential. All of the short, mid and long term technologies have gigawatt energy generation potential. Nevertheless, dealing effectively with the world's complex problems requires a more holistic approach that integrates not only the technical and economic potential of alternatives, but also the social and environmental dimensions. To expect that technology alone is the answer to our problems is to deny the painfully apparent consequences of this approach all over the world.

REFERENCES

ADS. 2007. Itinerario Dinámico para Proyectos de Infraestructura, Documento de Política Pública. Autoridad de Desperdicios Sólidos de Puerto Rico.

Appenzeller, T. 2004. The End of Cheap Oil. National Geographic 205: 80–106. (write the volume and number of the June 2004 issue)

Bonnet, J. A. 1983. The quest for energy self-sufficiency in Puerto Rico. Conference on Energy Planning for the U.S. Insular Areas. CEER-X-161.

Bonnet, J. A. and Graves, G. 1980. Planning and Initial Activities for the Utilization of Renewable Energy Sources in the Southern United States, Puerto Rico and the Virgin Islands. CEER Report.

Bosselman, F., Eisen, J., Rossi, D. Spence, J. Weaver. 2006. Energy, Economics and the Environment, 2nd Ed. Foundation Press.

Colucci, J. Alape, E., Borrero, E. 2003. Biodiesel for Puerto Rico. Green Chemistry & Engineering Proceedings. pp. 37–40.

Colucci, J. Alape, E., Borrero, E. 2005. Biodiesel from Alkaline Transesterification Reaction of Soybean Oil Using Ultrasound Mixing. JAOCS 82(7):465–542.

Colucci, J., Irizarry, A., O'Neill-Carrillo, E. 2007. Sustainable Energy in Puerto Rico. ASME Sustainability '07 Conference. Long Beach, CA.

Colucci, J., Pérez, R., López, Y., Ospinal, M. 2004. Fuel Cells Applications in Puerto Rico, an Environmentally Friendly Technology. Proceedings AIDIS 04.

DOE, 2008. U.S. Department of Energy and State of Hawaii Sign Agreement to Increase Clean Energy Technologies in Hawaii. Retrieved from http://www.doe.gov/5902.htm June 28, 2008.

EIA, 2006. Why are Electricity Prices Increasing? Electric Power Monthly. Energy Information Administration 4.

Graedel, T. and Allenby, B. 2003. Industrial Ecology. 2nd Ed., Prentice Hall.

Hohmeyer, O. and Ottinger, R. (eds.). 1991. External Environmental Costs of Electric Power. Springer-Verlag.

Hohmeyer, O. and Ottinger, R. (eds.). 1994. Social Costs of Energy. Springer-Verlag.

Humphrey, C., Lewis, T., Buttel, F. 2002. Environment, Energy and Society: A New Synthesis. Wadsworth.

Irizarry Rivera, A., O'Neill-Carrillo, E., Colucci-Ríos, J. 2008. Achievable Renewable Energy Targets in Puerto Rico. PREAA Project Final Report.

Jiménez-González, J. and Irizarry-Rivera, A. A. 2005. Generation Displacement, Power Losses and Emissions Reduction due to Solar Thermal Water Heaters. Proceedings of the Thirty-seventh Annual North American Power Symposium. Ames, Iowa.

Krewitt, W. 2002. External Costs of energy – do answers match the questions? Looking back at 10 years of ExternE. Energy Policy 30:839–848.

Krewitt, W., Mayerhofer, P., Friedrich, R., Trukenmüller, A., Heck, T., Greßmann, A., Raptis, F., Kaspar, F., Sachau, J., Rennings, K., Diekmann, J., Praetorius, B. 1998. ExternE – Externalities of Energy. National Implementation in Germany. IER, Stuttgart.

Leskes, A. 2005. The Art and Science of Assessing General Education Outcomes: A Practical Guide. Association of American Colleges and Universities.

ME3, "Minnesotans for an Energy-Efficient Economy," Retrieved from http://www.fresh-energy.org/about/focus/energy_efficiency.htm. June 28, 2008.

Norton, B. 2005. Sustainability. University of Chicago Press.

O'Neill-Carrillo, E. 2005. Externalidades en Energía y el Desarrollo Sostenible. Invited Paper, XXXIII Annual Meeting of the Puerto Rico Institute of Electrical Engineers, CIAPR. Carolina, PR.

O'Neill-Carrillo, E., Colucci-Ríos, J., Irizarry-Rivera, A. 2004. Opciones Energéticas de Puerto Rico Ante el Cambio Climático. Cambios Climáticos: Memorias del Segundo Simposio Rafael Echevarría. Universidad de Puerto Rico-Bayamón.

O'Neill-Carrillo, E., Frey, W., Jiménez, L., Rodríguez, M., Negrón, D. 2008. Social, Ethical and Global Issues in Engineering. Proceedings of the 38th Frontiers in Education Conference. Pittsburgh, PA.

O'Neill-Carrillo, E., Pérez-Lugo, M., Irizarry-Rivera, A., Ortiz-García, C. 2008. Sustainability, Energy Policy and Ethics in Puerto Rico. Proceedings of the Energy and Responsibility: A Conference on Ethics and the Environment. Knoxville, Tennessee.

Ramos-Robles C. A. and Irizarry-Rivera, A. A. 2004. Development of Eolic Generation Under Economic Uncertainty. Proceedings of the Eighth Probabilistic Methods Applied to Power Systems (PMAPS) International Conference. Ames, Iowa.

Ramos-Robles, C. A. and Irizarry-Rivera, A. A. 2005. Economical Effects of the Weibull Parameter Estimation on Wind Energy Projects. Proceedings of the Thirty-seventh Annual North American Power Symposium. Ames, Iowa.

Sáez, J. C., Schell, D. J., Tholudur, A., Farmer, J., Hamilton, J., Colucci, J. and McMillan, J. 2002. Carbon Mass Balance Evaluation of Cellulase Production on Soluble and Insoluble Substrates. Biotechnology Progress. 18:1400–1407.

Sheehan, J., Dunahay, T., Benemann, J., Roessler, P. 1998. A Look Back at the U.S. Department of Energy's Aquatic Species Program: Biodiesel from Algae. NREL/TP-580-24190.

Smil, V. 2005. Energy at the Crossroads. MIT Press.

Smil, V. 2008. Energy in Nature and Society. MIT Press.

Tester, J. E., Drake, E. M., Driscoll, M. J., Golay M. W., Peters, W. A. 2005. Sustainable Energy: Choosing Among Options. MIT Press.

Tomain, J. and Cudahy, R. 2004. Energy Law. Thomson Press.

United Nations. 1987. Our Common Future. The World Commission Report on Environment and Development. Oxford Press.

UIC. 2004. Energy Analysis of Power Systems. UIC Nuclear Issues Briefing Paper # 57.

U.S. National Academy of Sciences, 1980. Energy in Puerto Rico's Future. Report to the Center for Energy and Environmental Research.

Zamot, H. R., O'Neill-Carrillo, E., Irizarry-Rivera, A. 2005. Analysis of Wind Projects Considering Public Perception and Environmental Impact. Proceedings of the 37th North American Power Symposium, Ames, IA.

Contributions of the Energy and Environmental Sectors to Sustainable Economic Development: A Case Study from the Caribbean

Fred C. Schaffner
Office of Science and Technology, Puerto Rico Economic Development Company (PRIDCO)
Present address: Universidad del Turabo, School of Science and Technology, Gurabo, Puerto Rico

SUMMARY: This chapter examines the contribution of the energy and environmental sectors towards sustainable economic development using the Caribbean island of Puerto Rico, a US territory, as the case study site. The goal was to develop a strategic, long-term macroeconomic approach that reflects real-world requirements and conditions. Although the promotion of sustainable economic development is the official policy of the island's government, the real meaning of sustainability remains poorly understood. Linking this type of growth and business is essential for understanding that over the long term, economic security depends on environmental well-being. The territory must endeavour to achieve full participation in the global economic community. In addition, it must attain the development of a culture of quality, as well as a public policy agenda and a business agenda that will promote sustainable socioeconomic development, prosperity, better environmental compliance and an overall improved, and improving, quality of life. If the case study region can optimize the cultivation of its human capital, protect and conserve its water supply, and curb excessive energy costs, it will have a bright socioeconomic development and future.

1 INTRODUCTION

Faced with a rapidly changing economy, the world finds itself at a crucial moment in its socioeconomic development. Rapidly increasing monetary interconnectivity and international convergence of standards for environmental and quality compliance demand the development and implementation of long-term strategies for sustainable economic development and competitiveness that reflects real-world requirements and conditions. The energy and environmental sectors are essential, but as yet undervalued contributors to prosperity and sustainable financial progress. The purpose of the present analysis was to examine possible means of cultivating the contribution of these sectors to sustainable economic development in the context of the real-world conditions as they are likely to exist, using the Caribbean island of Puerto Rico, a US Territory, as a case study site.

Any meaningful approach to the analysis and design of effective strategies to promote sustainable economic development, and the potential contributions of the energy and environmental sectors must begin with a clear definition of terms. Here, we differentiate between the concepts of growth (an expansion or increase in size or number) and development (building capacity). As Drucker (1996) observed, even a refuse mound can grow, and an artist can develop without growth. We also differentiate between a view or focus on the short-term (tactical – a few years) versus the long-term (strategic – decades and generations) and between microeconomic versus macroeconomic considerations. Throughout, the goal is to develop a strategic, long-term macroeconomic approach to sustainable economic capacity building for the case study site of Puerto Rico. Any short-term, tactical, and microeconomic ideas that may be suggested are, in all cases, those believed to promote, or provide metrics (including incremental improvements) or objectives towards the long-term goal (Down, 1957; Hawken et al., 1999).

Long-term capacity building requires improvements in simple financial wealth, education, and the overall quality of life for all sectors of the population. This human development, and especially quality of life, is essential for attracting and retaining a pool of competitive talent in the workforce.

Everyone in the population must see themselves as potential beneficiaries of, and stakeholders in, the effort, if an economic development program is to be truly successful. It is imperative to avoid (and where necessary mitigate) socioeconomic polarization of the population – the juxtaposition of a layer of prosperity upon a layer of poverty (big winners and big losers). A stratum-diverse and interconnected socioeconomic structure promotes political stability, which in turn catalyzes further long-term economic development. This long-term, strategic focus, if implemented, will produce a positive feedback loop where the improved capacity of the population promotes further economic development that in turn builds greater human capabilities. Focus should be given both to job creation and job improvement. This underscores the importance of producing exports and acquiring external capital. Logically, activities that stimulate Puerto Rico's internal economy, foster internal efficiencies and economies, and provide support and enhancement to Puerto Rico's capacity to produce exports and acquire external capital must be encouraged. This requires the development of both a Policy Agenda and a Business Agenda.

2 CONFORMANCE WITH EXISTING OFFICIAL POLICY

The Puerto Rico Department of Economic Development and Commerce (DEDC in English, DDEC in Spanish) has established Sustainable Economic Development as an official goal and has articulated its twelve strategies for promoting economic development (DDEC, 2005–2007). The twelve strategies are as follows: Maximum use of our fiscal autonomy; Agility and flexibility in regulation, public permitting, and sustainable development; Lowering the cost of doing business in Puerto Rico; Decentralization of public investment through projects for the "Poles of Development"; Rehabilitation of the urban centres of the Municipalities of Puerto Rico; Establishment of high technology conglomerates ("clusters", or Alliances); Training and retraining of the labour force; Strengthening of the Puerto Rican entrepreneur and capital; Refocusing on the cooperative movement; Supporting strategic projects of economic impact; million-dollar investment in Special Communities; and a vision towards Puerto Rico in the year 2025.

The strategies of most immediate importance to the present effort are industry alliances (clusters), strengthening entrepreneurship, and the 2025 vision, a vision to promote sustainable economic and social development, and that emphasizes quality of life, competitiveness, and the integration of the island into the global economy.

The "Puerto Rico 2025" process that began in the year 2001 resulted in an official analytical and guidance document of the Commonwealth of Puerto Rico (Comisión Puerto Rico 2025, 2004), that establishes the specific objectives, strategies and new directions for public policy that are required for attaining the desired quality of life and sustainable socioeconomic development. This analysis and guidance places great emphasis on the same three elements of effort recognized by the European Union as essential for sustainable prosperity: Education, Science and Technology, and The Environment (U. Konig, pers. com.). This international convergence of vision forms the basis for the analyses and public policy recommendations embodied in this chapter.

2.1 *Linking sustainable development and business*

In November of 2003, World Bank President James Wolfensohn called on global business leaders to encourage environmentally and socially sustainable wealth creation as the world aims to fight poverty, in an effort to decisively contribute to peace and stability (Wolfensohn, 2004; WBCSD, 2005). Wolfensohn emphasized the need to move forward rapidly in establishing and implementing environmentally and socially sustainable economic development policy. The continuous focus on short-term goals, and the failure, or reluctance to develop and implement carefully designed long-term strategic policy serve only to trap an ever-increasing proportion of the population in inescapable poverty and threaten world peace and stability – all of which are bad for business. He also noted the importance of changing the ways we produce and consume energy, reducing subsidies, ensuring appropriate pricing, and adequately taxing environmentally damaging products.

2.2 *Sustainable economic development and capital*

Development (capacity building) can be thought of as an improvement in our ability or desire to satisfy our own legitimate desires, and those of others. A legitimate desire is any desire whose satisfaction does not reduce the ability or desire of others to satisfy their own necessities and legitimate desires.

The topic of sustainable development or sustainable economic growth has received great attention in recent years, but far too often its most enthusiastic proponents are unable to define it clearly. However, in general, the modern use of the word sustainability in the context used here was first articulated by Brundtland et al., (1987), as meeting the needs of the present without compromising the ability of future generations to meet their own needs.

Although a number of indicators of sustainability have been suggested (Neumayer, 2004), sustainable economic development can be described as progress without significant growth in resource throughput (the amount of resources a society uses *and discards*) and, therefore, without further depletion or degradation of natural resources. It does not mean, necessarily, that an economy must adjust itself to the local ecological carrying capacity, and may in fact imply quite the opposite – more interconnectivity with other regional economies. In a system focused towards sustainable development, emphasis must be placed on increasing the quality of goods and services, and obtaining greater operational efficiency and competitiveness. And again, for small island economies, attracting more external financial capital through exports must be a priority.

Intimately related to the sustainable development concept, especially for a small island economy, is the concept of developing a prosperous steady-state or near steady-state economic model characterized by balanced opposing forces that maintain a constant stock of physical wealth and people through a dynamic system of interactions and feedback loops (Daly, 1991, 1994, 1996, 1997, 1998). A low rate of flow *in situ* or throughput of matter and energy resources maintains this wealth and population size at some desirable and sustainable level, because high throughput usually incurs high costs and high waste. Again, efficiency is a key to sustainable prosperity – efficiency in use of biologic and mineral resources, efficiency in use of energy and material resources, and efficiency in the use of water. Sometimes referred to as "Ecological Economics", this approach recognizes that economic systems are not isolated from the natural world but are fully interdependent on ecosystems for the goods and services they supply, and subject to the negative consequences of not using these goods and services wisely (Jansson et al., 1994; Costanza, 2001; Daily and Ellison, 2002). Ignorance of this interconnectivity would be like a physician who monitored his patient's health by examining the circulatory system only, as if there was no digestive tract that connected it to the environment at both ends (Heal, 2000).

Capital can be defined as any form of wealth that contributes to the generation of more wealth, accepting implicitly the feasibility of positive feedback systems (Hawken et al., 1999; Heal, 2000; Figure 1). This capital includes both the natural resources provided by nature as well as the services that nature provides to humanity at no charge, although the costs to society can be very high if this capital is degraded or destroyed.

The concept of capital can be classified into four general categories: Financial Capital: cash, investments and monetary instruments; Manufactured Capital: buildings, tools, equipment, roads; Natural Capital: goods and services provided by nature, including "renewable" and non-renewable resources; and Human Capital: labour, intelligence, culture, expertise and organization.

These four categories can be further distilled into the latter two categories as the most essential, fundamental categories: Natural Capital and Human Capital, because all other forms of capital must begin with natural resources, and the human intellect and behaviour applied to them.

De Groot (1994) identified some of the principal ecosystem services, including a Regulated global energy balance and climate; Chemical composition of the atmosphere and oceans; Water catchment and groundwater recharge; Production, and recycling of organic and inorganic materials; maintenance of global biodiversity; Space and suitable substrates for human habitation, crop cultivation, energy conservation, recreation and nature protection; Oxygen, freshwater, food, medicine, fuel, fodder, fertilizer, building materials, and industrial inputs; and aesthetic, spiritual, historic,

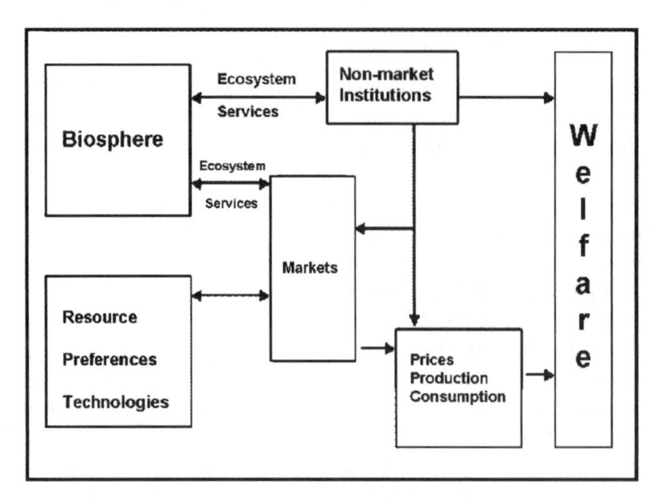

Figure 1. Adequate non-market institutions are needed to mediate between society, markets, and the environment (adapted from Heal, 2000). Non-market institutions such as government, international trade agreements and NGO's (e.g. GATT, WTO, etc.), the World Bank and other non-government institutions (public interest NGO's) can respond to environmental conditions and needs and influence them directly, or indirectly through adjustments to markets, prices, production and consumption to improve human welfare and economic capacity building, and a manner analogous to the Federal Reserve Banks influence on interest rate or NGO's organizing against the marketing of genetically modified crops in Europe (purposeful, and both direct and indirect), or by direct regulation. The governing authority (non-market institution) responds to the environment (biosphere) and influences markets, prices, production, consumption and welfare (= quality of life).

cultural, artistic, scientific, and educational opportunities and information (see also Jansson et al., 1994; and Foley et al., 2005).

Costanza et al., (1997) and Daily et al., (1997, 2000) estimated that the economic benefits enjoyed by the world's society from various ecosystem services totalled over $US 33.3 trillion (1997 dollars), of which some $US 17.1 were derived from soil formation alone. These services occur without human intervention, though human intervention can reduce them. Once one accepts the reality of this natural capital and that resources are finite, the meaning of sustainability becomes even clearer. Wealth and prosperity can be generated by the wise use and management of capital. As with financial capital, it is inescapable that consuming one's principal is unwise and is likely to lead to a downward trend in income, dividends and interest, and therefore is not sustainable.

The traditional focus on the generation of wealth has led to the misconceptualization of economic development as being synonymous with growth, which can have serious negative consequences when that growth is not sustainable. This traditional misconception, in turn, has lead to expansion and throughput (the amount of resources a society uses and discards) as the primary measures of economic development, expressed as Gross National Product (GNP) or Gross Domestic Product (GDP) (the sum of all products bought and sold in an economy).

The traditional neo-classical conception of the need for continuous growth is an artifact of past experience and economic practice that began during the period of European colonial expansion (populations grow exponentially while the resource base grows linearly), is inappropriate for the innovations that will be needed in the future (von Weiczacker et al., 1998; Tietenberg, 2006; Figure 2). These assumptions (growth in the neo-classical sense) are essentially artificial, and society's refusal to address these erroneous beliefs has led to the increasing externalization of producers' internal costs onto general society, avoidable waste, avoidable shortages and starvation, avoidable disease, suffering and morbidity, and continuing inequality, political instability and conflict – all of which have severely negative socioeconomic consequences. The classic example of social collapse at Easter Island, the Chatham Island massacres, and the widespread historical

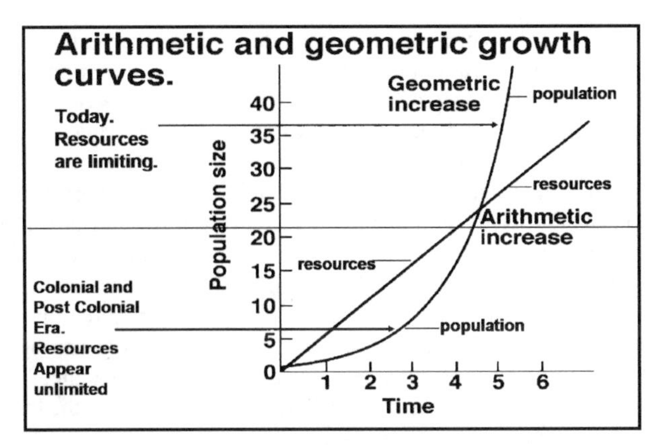

Figure 2. Populations grow geometrically, but resource availability, including renewable resources, grows arithmetically. Small populations see resources as unlimited and desire growth and greater exploitation. Although technological developments have steepened the resource line, the fundamental neo-classical presumption of infinite resources is known to be false, and current populations are already stressing the abundance and distribution of resources to unsustainable levels (adapted from Cunningham and Cunningham, 2004).

subjugation of technologically less advanced cultures by technologically advanced cultures are just a few examples (Diamond, 1999, 2005).

Serageldin (2002) revealed that the net upward flow of income and increased economic polarization that have accelerated with economic globalization have led to environmental degradation by increasing per capita consumption by the world's richest 20% of the population. (The world's richest fifth of the population receives more (about 85%) of the world's income than the remaining 80%, and the poorest fifth of the population receives only 1.3% of wealth. Consequently the world's poorest 80% increasingly use resources faster than they can be replenished, simply in order to subsist. This is causing the consumption of principal due to the over-consumption of resources for luxury, as well an inefficient, intensive, exploitation of natural resources for subsistence.

The perpetuation of poverty has additional negative consequences, including premature deaths and increasingly severe and preventable public health problems, the costs of which are eventually externalized from the producers to society in general (polluters seldom pay the medical costs of those affected by their pollution), creating a severe public burden and impediment to economic development. Clearly, innovative alternatives are needed. Alternatives such as industrial ecology (Graedel and Allenby, 1995, 2002; Castañeda-Muñoz, 2006) and ecological economics (Daly, 1991–1998), apply ideas of system functions and cycling to the concepts of resources and capital (Ashton and Chertow, 2004).

These views recognize the vital importance ecosystem functions, and recycling of materials and energy, as the residuals (waste or by-products) of one group become the raw materials of another. Ecological economics recognizes natural capital as part of the economy, and recognizes as "ecological services", those functions of particular use to man, such as purifying air and water, energy balance, the production of oxygen, water, soil and food. Traditionally, these resources have been taken for granted – excluded from conventional economic accounting, and not paid for directly – though we often pay for them when we suffer their absence (Hardin, 1968). Ecological management accounting, like environmental management systems within a company, strives to correct these deficiencies in order to establish the sustainable use of resources, reduction of throughput, and maximization of capital (lean operations and green operations) (Holly, 2001).

2.3 *Economic well-being depends on environmental well-being.*

The preceding analysis also reveals that there is no absolute categorical distinction between "renewable" and "nonrenewable" resources – only differences between their relative rates of exploitation and rates of replenishment. It also provides a framework in which prosperity and economic health can be enhanced and maintained without continuous growth in consumption and throughput, and provide stability and prosperity for societies with low or no annual population growth (substantially less than 1%), through efficiency and recycling of resources. This nearly steady-state, prosperous economy can be characterized by political and social stability, relatively greater reliance on replenishable resources, and a significantly enhanced quality of life and prosperity for an economically diverse and interconnected society with stable or decreasing birth and death rates and a stable or increasing life expectancy (examples – annual rates of population growth: Germany (0.02%), Finland (0.18%), Sweden (0.18%), Japan (0.08%) (CIA, 2006). The emerging small, prosperous European countries with declining, stable or very slowly growing populations are providing evidence of the modern steady-state concept (economies in post-industrial demographic transition). These economies also provide examples that the idea of achieving a "balance" between the economy and the environment is a fundamentally false dichotomy. In fact, over the long term, economic well-being is *dependent* on environmental well-being.

2.4 *Battling misconceptions – the environment and job loss*

Individual companies or industries suddenly faced with additional environmental regulations or restrictions have been very persuasive in creating the perception that protecting the environment is unacceptably costly, or results in job loss. This widely held misconception has sometimes led well-meaning policy makers to try to devise schemes or policies that seek to achieve a "balance" between the environment and the economy. Yet historical analyses show that the root causes of joblessness and industry failures are nearly always the result of market forces that already were in place and developing rapidly. For example, the demise of the American steel industry in the region sometimes called the "Rust Belt" was due primarily to the U.S.'s dependence on massive, expensive, integrated mills with antiquated blast furnace ironmaking technology, in the face of competition from companies using newer, more efficient blast furnaces or direct reduction ironmaking, with smaller and more efficient basic oxygen furnace (BOF) technology steelmaking, or EAF (electric arc furnace) mini-mill technology to re-mill scrap steel and iron. This lumbering, 19th Century, Edwardian megalith could not possibly survive the onslaught of 20th century industrial and technical advances (Drucker, 1996; Hawken et al., 1999; Myers and Kent, 2001; AISE, 1985; Mazurak, 2003).

Similarly, the demise of the timber industry, and especially the timber processing industry in the United States Pacific Northwest was the result of the expansion of the practice of direct export of raw logs to Japan, especially raw logs from Canada. This industry was well on its way to extinction long before the Spotted Owl controversy erupted (Myers, 1996; Hawken et al., 1999; Myers and Kent, 2001).

In their study of the effects of new and stringent air quality regulation during 1979–1991 by the South Coast Air Quality Management District in the Los Angeles Basin, California, Berman and Bui (2001) found that increases in air quality regulation that involved substantial increases in cost did not appreciably affect employment. In fact, they found small increases in employment that ruled out any suggestion of decreased employment due to regulation.

Morgenstern et al., (2002) examined the jobs-versus-the environment argument for four industries (paper pulp mills, plastics, oil refineries, and steel) and found that increased environmental stringency did not cause significant job loss, but, rather, job and industry transformation. An overall net increase in jobs mitigated the jobs lost. Also, of the total jobs lost in thee industries only about y 2% were due to environmental regulation.

In their study of the effects of new and stringent air quality regulation by the South Coast Air Quality Management District in California (Berman and Bui, 2001).

2.5 *Externalizing internal costs onto society, and the real costs of environmental degradation*

All economic goods and services have both internal and external costs. Externalities are any effects on commercial production or operation not included in the calculation of profit and loss. Air and water pollution are examples of socially negative externalities. They impose a cost on society that is not paid-for by the producers. For any good or service there may be both internal costs and benefits, as well as external costs and benefits, that sum to the total costs and benefits of the commercial activity. Most externalized costs are those that are negative or damaging to, public health or the environment and not included in the market price of product. They also can be viewed as private-social cost differences (Heal, 2000).

2.6 *The social costs of externalization*

2.6.1 *Rights and ownership*

A recent analysis by Heal (2000) has discussed the question of the establishment of the right of an operator to pollute versus the right the community has to enjoy clean air and water. If one assumed that (or assigned to) the community the right to clean air (or water) then the pollution emitted by an operator is a violation of the rights of the community that should be penalized and the community compensated:

$$\text{PRIVATE COST} + \text{TAX (OR SUBSIDY)} = \text{SOCIAL COST.}$$

Thus, the total cost to the polluting operator is equivalent to the social cost (Pigou, 1932). Pigou's approach has given rise to the dominant European policy approach of using corrective taxes and subsidies. A non-market institution or authority (government) must intervene because correction falls beyond the scope of market forces.

In contrast, Coase (1960) based his analysis on the philosophy that goods and services can be bought and sold only if they are owned – that they are someone's property, and that the problem of externalities arises from the absence of specific property rights, and therefore is immune to market influences. According to Coase, we must establish property rights to the environment so that anyone wanting to dump wastes into it must first purchase that right from the owner. This establishes a forum for bargaining between the polluter and the party affected by the externality. Coase's approach has inspired the American (USA) and Brazilian policy of tradable permits and quotas for emissions. By purchasing a tradable emission quota (TEQ), the polluter now owns the right to pollute and has established his own property rights to the environment. This approach is exactly opposite to the European approach, but theoretically has similar outcome – the total cost to the polluting operator approaches equivalency to the social cost:

$$\text{PRIVATE COST} + \text{QUOTA PRICE} = \text{SOCIAL COST.}$$

The tradable quota system requires the polluter to purchase a permit before discharging into the environment, thus raising the private cost of pollution in a way that appears to the polluter to be similar to a tax (i.e., quota pricing structures can be adjusted by the governing authority in a way that incentivizes clean operation.).

The two approaches, taxes versus tradable quotas, are somewhat equivalent in that they establish a common norm that "the polluter pays", a policy principle adopted by the Organization for Economic Cooperation and Development (OECD) for its member countries (OECD, 2007). However, the two approaches differ in some operational details, in their susceptibilities to different sorts of limitations and potential failures, and in the role of government as arbitrator for fairness in the management of pollution.

Under the tradable quota scheme it is presumed that in a competitive quota market, the price of a quota can be adjusted to exactly equate to the private-social cost differential. Prices of quotas can be raised or lowered by controlling the number of quotas available in the market, according to traditional supply and demand dynamics (and the health of the ecosystem). Thus, these quotas become commodities no different than stocks or bonds, and are identical to tradable commodities futures, except that these futures are freed from free-market volatility and are buffered and stabilized by the regulatory decision-making of a non-market institution – the government. In practice, government becomes an arbitrator in favour of buffering the negative economic impact to polluters (responding to the environment or biosphere). Polluters are the stakeholders with whom it has most frequent interaction. On the other hand, with the direct tax or subsidy approach, government plays a more active, central role, and may exhibit a natural tendency to raise revenues.

Clearly there exists the need to bring ecosystem services, and the natural infrastructure on which the population depends more within market, extending the influence of the scope of the market. This requires that policies and mechanisms be established by which both ecosystem services and socially negative externalities can be valued, internalized and monetized, such that markets can operate more effectively in the public good (welfare). This also requires incisive management and adjustments by non-market institutions such as governments (policy and regulation) and international agreements (Figure 1).

Indeed, Heilbroner (1994, 1999), Heilbroner and Millberg, 2001, and Heilbroner and Singer, (1998) have suggested that the public sector (government), far from being a drag on the marketplace, can be "an indispensable source for strength" (Heilbroner, 1994) that can set rules of fair play, stimulate and create new markets, protect public health, provide stability and direction and mitigate the deleterious effects of extreme socioeconomic polarization. Heilbroner's analysis points out the limitations of markets as self-ordering mechanisms, noting that in certain circumstances they can aggravate shortages. Nor does he put much stock in the capacity of profit-oriented activities to address (let alone solve) environmental problems, create infrastructure, ensure productive educational programs, harness potentially disruptive new technologies, and make other needed contributions to the common good. Indeed, the laissez-faire scenarios offered by economists like Hayek, Keynes, Marx, Schumpeter, and Smith have all proved either incomplete or in error. Again, management intervention and influence must be exerted by non-market institutions that can impart innovative, strategic policy (and eventually law).

2.6.2 *Cost-benefit analysis*

As noted previously, pollution of air and water sources can have very significant, negative health effects on the general population. This creates an increasing public health burden, the cost of which must be paid for through the public health and social welfare system. These externalized costs translate to increased expenditures by the state, higher tax burdens, and a generally increased drain on the economy (= retarded economic development). A classic analysis on the real costs and economic benefits of air pollution control (in Poland) demonstrated that the economic benefits to the general economy (relief of the state tax burden for public health) far outweighed the costs of pollution reduction, up to at least 85% clean smoke stack emissions (Cofala et al., 1991). Such results argue strongly for a shift in tax policy away from taxing wages and profits (= penalizing productivity), and towards taxing pollution and waste (EPA, 2001).

2.7 *Full cost pricing and internalization*

As with the Polish example the social costs of externalizing operators' internal costs can be severe. Thus, for most economists, the solution to the harmful costs of goods and services is to include, or internalize such costs in the market prices of goods or services. However, the internalization of external costs will not occur unless required by government regulation or policy, which are themselves influenced by public pressure and the marketplace. As long as operators receive subsidies and tax concessions for extracting resources and are not taxed on the contaminants they produce, few will volunteer to reduce short-term profits by becoming environmentally responsible.

Government, as the most significant non-market institution, must intervene. Its options for mitigating the harmful costs of externalization (public health, and markets) include levying taxes on pollution and waste, new laws and regulations (and/or adequate enforcement of those already in existence), properly adjudicated subsidies and incentives for activities with significant social value, and other strategies that that force or encourage operators to accept and include all costs in the market prices of goods and services, such that the market price would be the full cost of the goods or services – the internal costs plus the short- and long-term external costs. Adjustments and standards of practice also must be internationalized to prevent exportation of pollution problems (a severe problem currently being addressed by the World Bank – see WBCSD, 2004–2005).

Internalizing external costs of pollution, environmental degradation, over-consumption (luxury consumption far beyond need and waste (through taxes, incentives and quotas), would make pollution prevention and control more profitable than control and clean-up, and also would make waste reduction, recycling and reuse more profitable than the current disposal activities. But making this shift requires specific changes in existing policy. Government also may have to reduce taxes on wages and profits (taxes on productivity) to compensate for the costs of the new policy direction and withdraw current subsidies that mask externalized costs (past example: subsidies on fertilizer for sugar cane in Puerto Rico (Meyers and Kent, 2001). Long-term benefits of internalization include that operators would be encouraged to find ways to cut costs through innovative resource-efficient and less polluting methods, and the production or more environmentally beneficial products (so-called "green" products and "lean manufacturing") that help reduce the future costs of remediation or mitigation. Employment would be lost only in harmful industries that, could, but refuse, to innovate, but many more new jobs would be generated in the newer, innovative industries and businesses.

Government can address the problem of externalized costs directly by levying taxes, passing laws and developing appropriate regulations, providing subsidies and through other strategies that encourage or force producers to include currently externalized costs in their internal costs (e.g. sheep methane head taxes in New Zealand, for example). Internalizing the externalized costs of environmental pollution and degradation would make preventing pollution more profitable than cleaning it up and would make waste reduction, recycling, and reuse more profitable than burying or burning most of the waste we produce. This shift requires action of the principle non-market institution – the government – because few companies will intentionally increase their cost of doing business unless their competitors must do so as well. Public Education is important. People who want to run air conditioners, TV's and numerous appliances, etc., usually do not make the mental connection that their activities contribute to worsening air pollution by power generating plants, and many are willing to pay high prices for electricity rather than reducing their consumption. On a global scale, high-consumer countries that do not internalize costs of consumption can skew competition, resulting in poorer countries paying a significant portion of the externalized costs. International agreements such as the Kyoto (greenhouse gasses) and Montreal (CFC) Protocols represent attempts to remediate some of the environmentally damaging effects of externalization in the high-consumption developed nations. Additional examples of the transition to full cost pricing and improved economic performance are available. These contrast the choice of taxing waste, pollution and over-consumption (essential for future development) versus the predominant current policy of taxing productivity. This highlights accelerated economic activity through efficiency and consumer choice versus the repressive drag on the general economy caused by the current taxation policy that is used to finance the internalizable costs that are currently externalized to society in general.

2.7.1 *Single-minded policy and the need for strategic policy planning*

As noted by Hawken et al., (1999) sometimes single-problem, single-solution approaches do work, but often, optimizing one element in isolation pessimizes the entire system. Hidden connections that have not been recognized and turned to advantage will eventually tend to create disadvantage. Cities may find that the cause of their problems are prior solutions that have either missed their mark or boomeranged, like the bigger road that invites more traffic, the river channelization that worsens floods, the homeless shelter that spreads tuberculosis, and the prison that trains criminals in more sophisticated techniques. Rather, our goal should be to solve each problem in a way that also addresses many more simultaneously – without creating new ones. This "systems approach" not only recognizes underlying causal linkages but sees places to turn challenges into opportunities, and should be incentivized (Ackoff, 1999). Communities and whole societies need to be managed with the same appreciation for integrative design as buildings, the same frugally simple engineering as lean factories, and the same entrepreneurial drive as great companies.

A wider focus can help people protect not only the natural capital they depend upon but also their social fabric – their own human capital. Just as ecosystems produce monetized "natural resources" as well as the far more valuable but thus far unmonetized "ecosystem services", so too social systems (behaviour and group interactions) have a dual role. They provide not only the monetized "human resources" of educated minds and skilled hands, but also the far more valuable but usually unmonetized "social system services" – culture, wisdom, honour, love, curiosity and a whole range of values, attributes, and behaviours that define our humanity and make our lives worth living. Just as unsound ways of extracting wood fiber can destroy the ecological integrity of a forest until it can no longer regulate watersheds, atmosphere, climate, nutrient flows, and habitats, unsound methods of exploiting human resources can destroy the social integrity of a culture so it can no longer support the happiness and improvement of its members (this is manifested in decreased productivity, high turn-over of employees and high costs of training). Industrial capitalism can be said to be liquidating, without valuing, both natural and human capital – capturing short-term economic gains in ways that destroy long-term human prospect and purpose. An overworked but undervalued workforce, outsourced parenting, the unremitting insecurity that threatens even the most valued knowledge-workers with fear of layoffs – these all corrode and undermine civil society. Clearly, the former short-term, short-sighted policy approach is insufficient. Tactics cannot be substituted for strategy, and real, long-sighted strategic policy planning must be developed (Hawken et al., 1999).

2.7.2 *Jobs*

A recent (2000) analysis of job loss in the US, conducted by the Economic Policy Institute showed that less than about 0.1% of job loss in the US was the result of environmental or safety regulations (EPA, 2001).

Environmental requirements may be inconvenient for some individual operators, but they have not caused significant net job loss in any documented market or industry sector, and, in fact have catalyzed new businesses, more jobs, and overall enhanced economic performance.

Political and business leaders see sales of environmental protection goods and services (currently worth about $ US 600,000,000 per year (2002)) as a major source of new markets and future income because environmental standards and concerns are expected to rise everywhere.

Largely due to stricter air pollution regulations, German companies have developed some of the world's cleanest and most efficient gas turbines and invested in the world's first steel mill that used no coal to make steel. Germany marketed these and other environmental technologies globally (von Weiczacker et al., 1998).

Worldwide, environmental protection is a major growth industry that creates new jobs. The World Watch Institute (WWI, 2004) estimated that annual sales of global ecotechnology industries were about $600 billion (equal to the auto industry) and employed about 11 million people. During the year 2000, this industry in the US employed about 1.4 million people (nearly equal to the total Puerto Rico workforce) and generated annual revenues of more than $185 billion in services rendered to a variety companies and government and non-government institutions.

Analyses by the EPA (2001) indicate that environmental laws create far more jobs than have been lost. The Clean Air and Clean Water Acts had created about 300,000 new jobs in pollution control by 2000.

A recent analysis by the US Congress concluded that investing about $US 115 billion annually in solar energy and improving energy efficiency in the United States would eliminate about 1 million jobs in oil, gas, coal, and electricity production, but *create* over 2 million new jobs. Investing the money saved by reducing energy waste would create another 2 million jobs.

2.8 *The meaning of innovation*

The key concept embodied in the successful examples above is "innovation", what Peter Drucker has defined as "the purposeful response to change." As Drucker (1996) has pointed out, it is change that always provides the opportunity for the new and different. "Systematic innovation therefore consists in the purposeful and organized search for changes, and in the systematic analysis of the opportunities such changes might offer for economic or social innovation."

Drucker identified seven sources of innovative opportunity that must be monitored in order to achieve systematic innovation. The first four sources of innovative opportunity lie within enterprise, whether business or a public sector institution, or within an industry or service sector. They are therefore visible primarily to people within that industry or sector. They are essentially symptoms, but they are also highly reliable indicators of changes that already have happened or can be made to happen, with little effort. These four sources are: (1) the unexpected – the unexpected success, the unexpected failure, the unexpected outside event, (2) the incongruity – between reality as it actually is and reality as it is assumed to be or as it "ought to be", (3) innovation based on process need; and (4) changes in industry structure or market structure that catch everyone unaware.

The other three sources of innovative opportunity involve changes outside the enterprise or industry: (1) demographics (population changes), (2) changes in perception, mood, and meaning; and (3) new knowledge, both scientific and nonscientific.

The lines between these seven sources of innovation are often blurred, and overlapping, and none is inherently more important than any other. Drucker also points out that, contrary to almost universal belief, new knowledge, and especially new scientific knowledge, is *not* the most reliable or most predictable source of successful innovations. Despite all its visibility, glamour and apparent importance, scientific-based innovation, is actually the least reliable and least predictable source of innovation. Conversely, the mundane and unglamorous analysis of such symptoms of underlying changes as an unexpected or unexpected failure carry fairly low risk and uncertainty, and the innovations that arise from them typically have the shortest lead time between the start of a venture and its measurable results, whether success or failure (Drucker, 1996). Major innovations are likely to emerge from the analysis of symptoms of change as they are from the massive application of new knowledge resulting from a great scientific breakthrough. The post-WWI scenario for the re-emergence of the Japanese and European steelmaking mentioned previously by Drucker (1996) is a case in point. Rather than re-build a ferrous metals industry from the ground up (beginning with primary process ironmaking) these countries started with existing EAF (electric arc furnace) technology, taking advantage of an abundance of steel war debris, and only later expanded to direct reduction ironmaking coupled with BOF (basic oxygen furnace) steelmaking technology. The US steel industry was highly critical of this approach, noting that the steel this technology would produce was of significantly poorer quality (true enough), but the quality was adequate for the immediate market's needs at the time, and cheap, and eventually, these industries developed into quality producers.

Despite the widespread belief that new jobs come from "high tech", things are not that simple. Of the 40 million-plus jobs created in the US economy between 1965 and 1985, high tech contributed only 5–6 million. High tech thus contributed no more than the traditional "smokestack" industries lost. All the additional jobs in the economy were generated elsewhere, and only one to two of every 100 new businesses – a total of 10,000 per year during that period – were even remotely "high tech" (Drucker, 1996).

It is interesting that in the environmental field in particular, opportunities will arise not solely from the development of new technologies, but much more likely from the application of existing technologies, even mundane, unglamorous or artesanal technologies, in ways that have thus far not been considered or exploited. For example, currently both federal and local environmental regulations require car wash operators to recycle their wash water, yet few seem to be able to do this efficiently, as evidenced by several streams and ditches loaded with soapy drain water. At the same time numerous trailered "portable" car wash operations have sprung up in the San Juan metropolitan area to wash cars in parking lots or along the roadside. This business is destined for more intense regulatory scrutiny. At the same time, transporting water, and refilling the trailer water tanks can be expensive. A simple and efficient innovation to capture and recycle this wash water would be economically viable, both for the user as well as for the idea developer and should be encouraged. If the innovator also seeks and receives ISO certification (see below), the system also would be exportable and would generate external capital for Puerto Rico. This system then could be scaled-up to stationary operations.

Many of the new opportunities also will come from the "Fourth Sector" of public-private partnerships in which government units, either the state or municipalities, determine performance standards and provide funds and then contract-out a service – fire protection, garbage collection, transportation or recycling, to a private business on the basis of competitive bids, thus ensuring both better service and substantially lower costs (Drucker, 1996; Ackoff, 1999, see also Collins and Porras, 2002), ideally, with careful monitoring and regulation.

2.8.1 *Regulation creates jobs*

During the 1980's, one prominent Puerto Rican construction company, when faced with severe penalties for violation of Human Resources and Labour Law 1980's chose to make the required corrections, and subsequently marketed it's solution, as a consultant, to other companies.

Similarly, FDA and EMEA (European Medical Examining Authority) regulations for bio-pharmaceutical products and manufacturing process have catalyzed a lucrative business for consultants in the Validations and Quality Control/Quality Assurance fields, etc. – providing enhanced public health benefits and at no detriment to industry profitability.

The desirability of promoting a transition from Forced Acceptance (1980's–1990's), to Pollution Control and Confrontation, then Acceptance without Innovation, should be obvious and leads to Innovation-Directed Management. Innovation Directed Management consists of the following elements:

- Total Quality Management – pollution prevention and resource productivity.
- Life Cycle Management – Environmental product stewardship ("take back" rules where producers are required to take back and recycle products that finish their useful life spans) throughout a product's life cycle. Marketing services instead of things. In other words, leveraging a movement towards service flow economies instead of material flow economies. This is already happening with the luxury car market, where products (cars) increasingly are leased and returned reduced to component parts and materials and recycled, rather than purchased and discarded.
- Process Design Management – Clean technology. Totally redesigning existing products and manufacturing processes, or developing new ones in order to eliminate or reduce pollution and resource waste, and decrease production, waste management, product liability and pollution compliance costs.
- Total Life Quality Management (industrial ecology) – companies become involved in "eco-industrial networks" where they exchange resources and wastes in industrial webs, in physical proximity, whenever possible (clusters) (Graedel and Allenby, 1995, 2002; Ashton and Chertow, 2004; Castañeda-Muñoz, 2006).

The above requires an innovation in policy and practice – away from attempting to achieve economic development via producing more goods to utilizing information and knowledge to improve resource productivity by reducing or avoiding waste, through miniaturization, replacing chemical plants with "green" plants to produce the needed chemicals and learning how to do more with less (Repetto and Austin, 2000).

According to Monsanto CEO Robert B. Shapiro, understanding the concept of sustainability changed the way he thought (Magretta, 1997; Holliday et al., 2002). Furthermore, the economist Robert Heilbroner (1994, 1999) observed that there is a limit beyond which acquisitiveness no longer serves, and may well disserve, the adaptability of order.

Clearly, Puerto Rico is a very small "planet", and the development of economic and environmental sustainability is imperative. Yet as can be appreciated from the numerous prosperous small economies of countries like Finland, Ireland and Singapore, Puerto Rico's small size is no impediment to economic development and prosperity.

2.9 *Unlimited growth is not a sustainable activity*

Growth, and in particular, population growth, are not requirements for sustainable economic development. As recent experience points out, the popularity of the neo-classical single-minded focus on continuous growth is nothing more than an artifact of the European colonial expansion (Brown et al., 1998; Tietenberg, 2006), and where one's economic state lies on the Arithmetic – Geometric growth curve relationship (Figure 2).

Additional examples (mostly European) are available of how small, modern near steady-state economies and ZPG populations (zero population growth, countries in stage 4 (post-industrial)) demographic transition can generate and maintain prosperity, high and improving qualities of life, and improving environmental conditions.

2.10 *Quality of life*

Esty (2004) points out that many national policymakers worldwide are searching for more meaningful metrics of quality of life than the traditional metrics of Gross National Product (GNP) or Gross Domestic Product (GDP), recognizing that these indicators do not adequately capture the essence of the gamut of factors that are important in keeping and retaining productive talent, nor sustaining a national economy and improving the quality of life.

2.10.1 *Looking for the missing index*

Over three decades ago the king of Butan proposed a metric he called "Gross National Happiness (GNH)", which consisted of four key elements: (1) promotion of equitable and sustainable socio-economic development, (2) preservation and promotion of cultural values, (3) conservation of the natural environment, and (4) establishment of good governance. All are admirable goals but difficult to measure objectively. Traditional measures, such as GNP, and the more widely used GDP, however, also have many blind spots, including difficulty in measuring the value of volunteer work, the value of vacation and leisure, and the cost of lost environmental resources due to environmental degradation. Other measures that have been discussed (Veenhoven, 2004) include "Happy Life Years", "GPP – Gross Personal Product", "Average Life Satisfaction" and the United Nations Human Development Index, an approach that addresses life expectancy, adult literacy, education and various other economic indicators (United Nations Population Reference Bureau, 2003).

Diener and Seligman (2004) also emphasize the need to develop a more accurate measure of national progress and suggest the combination of three categories of indicators: (1) GDP, (2) social indicators such as education and health care, and (3) more subjective measures that reflect life satisfaction. They note that the per capita GDP of the Unites States tripled since World War II, yet life satisfaction rates have remained constant. These psychologists note that while numbers for the economy are available, data on life satisfaction are limited or not available and the authors suggest that we should be conducting surveys of "experience sampling", and they note that happy employees are generally healthier and more productive than those with lower levels of well-being. Currently, GDP in Puerto Rico is growing at a rate of nearly 1–3% per year while population is growing at just 0.5% per year, yet few among us would say that they feel their quality of life is improving, much less at the *ca.* 2.48% per year that the simple arithmetic suggests it should (BGF – 2007). The concept of quality of life also may be distinct from the idea of "happiness" and recent reports in the popular media that Puerto Ricans are among the "happiest" people in the world.

The United Nations Department of Economic and Social Affairs (DESA), Division for Sustainable Economic Development also has produced a framework of indicators for sustainable development. The DESA matrix includes social, environmental, economic and institutional indicators. The technical cooperation provided by DESA to developing countries places significant emphasis on the development of water policy, legislative frameworks, and technical assistance, in particular to Island economies, in the areas of (1) Energy, Transport and Atmosphere, and (2) Water and Natural Resources, thus underscoring the imperative of attending to environmental and energy issues as requisites for economic development (DESA, 2004–2005).

2.11 *The world bank classification of economies*

Though perhaps not completely fulfilling the lofty goals of the GNI (Gross National Income), the World Bank has developed an indicator that seems more informative than GNP or GDP, and approaches some of the considerations suggested by Diener and Seligman (2004), and Ditella and MacCulloch 2006). For its operational and analytical purposes, the World Bank's main criterion for classifying economies is gross national income (GNI) per capita (gross national product, or

GNP, in previous editions). Every economy is classified as either low income, middle income (subdivided into lower middle and upper middle), or high income. (Other analytical groups, based on geographic regions and levels of external debt, also are used).

Low-income and middle-income economies are sometimes referred to as developing economies (or even "Third World" countries). Though the use of the term "developing economy" is convenient; the World Bank says it does not intend it to imply that all economies in the group are experiencing similar development or that other economies have reached a preferred or final stage of development. Classification by income does not necessarily reflect development status (World Bank, 2004).

In the most recent World Bank analysis, economies are classified according to 2003 GNI per capita, calculated using the World Bank Atlas method. The groups are: low income, $735 or less; lower middle income, $US 736–2,935; upper middle income, $US 2,936–9,075; and high income, $ US 9,076 or more. By this classification, Puerto Rico Ranked 53rd of the 56 high income nations with a per capita GNI of $US 10,950, less than half that of Singapore ($US 1,230), Canada ($US 23,930), France ($US 24,770), Germany ($US 25,250), Hong Kong ($US 25,430), the Netherlands ($US 26,310), Ireland ($US 26,960), Finland ($US 27,020), the UK ($US 28,350), Denmark ($US 33,750), Japan ($US 34,510), the USA ($US 37,610), Switzerland ($US 39,880), Norway ($US 43,350), and Luxembourg ($US 43,940). [Once again, size doesn't matter]. The prominence of European Union countries, and other nations with stable, declining or slight population growth (regardless of current population density), *and* strict environmental standards and compliance is extremely noteworthy. Puerto Rico shares all the dynamic demographic characteristics of the most prosperous post industrial economies, but differs from them primarily in (1) its poor compliance and application of environmental standards and technologies, and (2) in its relatively low level of engagement with the global economic community. Thus, participation in the global economy appears essential for Puerto Rico's economic development and Puerto Rico's lack of environmental compliance and standards are obstacles to its full participation in the global economic community. Once again, leaving the environmental situation unattended is a significant threat to Puerto Rico's economic development.

It is worth noting that while some EU countries, like Ireland, are lagging behind in environmental compliance, they have been put on notice that becoming compliant in the very near future, and implementing a national plan to accomplish this, are required for full enjoyment of EU benefits. In addition, one also must distinguish between the "openness" Puerto Rico's economy versus its level of participation in the world economy. By most estimates (BGF, 2007) Puerto Rico's economy is the world's 8th most open economy, and thus has few obstacles to its potential participation. This means that it has great potential to develop practices that will allow its fuller participation and socioeconomic development. Essentially, the door is wide open, but we must begin moving more products (out) and more capital (in) through it.

2.12 *Towards full participation in the global economic community*

Accomplishing the above goals of sustainable economic development requires, among other things, the ability to produce exportable products and services and the generation of improved internal efficiencies within Puerto Rico's own domestic economy. This necessitates improvements in quality assurance in order to be able to develop export markets and bring in much-needed capital (ISO, 2006) and improvements in corporate Environmental Management Systems practices and policies. This requires enlightenment and leadership from corporate boards down through all levels of management; giving clear statements of the company's environmental principles and objectives. Companies also must involve employees, environmental groups and the local community in developing and evaluating a company's environmental policies, strategies for improvement and progress, and making the improvement of environmental quality a priority for all employees. Policies can be developed in-house or with the assistance of consultant services (a new market), much in the way that FDA (Food and Drug Administration, US) and EMEA (European Medicines Agency) regulation of the pharmaceutical industry catalyzed a diverse and lucrative service industry in validations and other vital professional services.

2.13 *Developing a culture of quality – why standards matter*

A useful metric for understanding the concept of quality is an economy's application of standards, both voluntary and regulatory. Many standards exist, including the developing European Union standards like the new EMEA standards, US Federal government regulatory standards of the FDA, FCC, EPA, OSHA, etc., and the voluntary standards of the International Organization for Standardization (ISO) and the International Electrotechnical Commission (IEC). ISO standards, in particular are useful metrics of participation in the global economy because they are voluntary, yet in practical terms, very necessary in order to effectively market products overseas and attract external capital. They can be interpreted as measures of the cultural acceptance of the concept of quality that is essential for entry into global markets.

2.13.1 *ISO 9000/9001–2000 quality management and conformity assessment*

ISO 9000 is concerned with "quality management" – what the organization does to enhance customer satisfaction by meeting customer and applicable regulatory requirements, and to continually improve its "performance management". It also means what the organization does to minimize harmful effects on the environment caused by its activities, and continually to improve its environmental performance.

What makes conformity assessment so important is that it's simplest, "conformity assessment" means checking that products, materials, services, systems or people measure up to the specifications of a relevant standard. Today, many products require testing for conformance with specifications (standards) or compliance with safety, or other regulations before they can be put on many markets. Even simpler products may require supporting technical documentation that includes test data. With so much trade taking place across borders, conformity assessment has become an important component of the world economy. Over the years, ISO has developed many of the standards against which products are assessed for conformity, as well as the standardized test methods that allow the meaningful comparison of test results so necessary for international trade. ISO itself does not carry out conformity assessment. However, in partnership with IEC (International Electrotechnical Commission), ISO develops ISO/IEC guides and standards to be used by organizations that carry out conformity assessment activities (Repetto and Austin, 2000; ISO, 2003, 2005, 2006).

The voluntary criteria contained in the ISO guides and standards represent an international consensus on what constitutes best practice. Their use contributes to the consistency and coherence of conformity assessment worldwide and therefore facilitates trade across borders.

2.13.2 *ISO 14000/14001*

ISO 14000/14001 provides a systematic approach to meeting a company's environmental and business goals through the development of an in-house Environmental Management System (EMS) (ISO, 2003). Key EMS benefits include improved environmental performance, reduced liability, competitive advantage, improved compliance, reduced compliance, fewer accidents, employee involvement, improved involvement, improved public image, enhanced customer trust, and better access to capital

According to ISO, an effective EMS makes good business sense. By helping a company identify the causes of environmental problems (and then eliminate them), an EMS can help it save money. ISO suggests the following questions: Is it better to make a product right the first time or to perform a lot of re-work later (Collins and Porras, 2002)? Is it more economical for a society to prevent a spill in the first place or clean it up afterwards? Is it more efficient for a society to prevent pollution or to manage it after it has been generated?

An EMS also can be an investment in long term viability of a commercial organization. An EMS will help a company be more effective in achieving environmental goals. And, by helping businesses to keep existing customers and attract new ones, an EMS adds value (ISO, 2003).

Much of what a company needs for an EMS-type may already be in place. The management system framework described in the ISO 14000 Guide contains many elements that are common to managing other business processes, such as quality, health, safety, finance, or human resources.

Many companies already have many EMS processes in place, but for other purposes (such as quality). Integrating environmental management with other key business processes can improve the organization's financial and environmental performance.

The key to effective environmental management is the use of a systematic approach to planning, controlling, measuring and improving an organization's environmental efforts. Potentially significant environmental improvements (and cost savings) can be achieved by reviewing and improving an organization's management processes because not all environmental problems need to be solved by installing expensive pollution control equipment.

Of course, there is some work involved in planning and implementing an EMS. But many organizations have found that the development of an EMS can be a vehicle for positive change. These organizations believe that the benefits of an EMS can be a vehicle for positive change. As they say in the Total Quality Management (TQM) world, "quality is free" – as long as you are willing to make the investments that will let you reap the rewards. The same holds true for environmental management (ISO, 2003, 2005, 2006). Companies that develop an effective in-house EMS also can then provide consulting services to other businesses wishing to develop their own EMS, as with the Puerto Rico human resources example mentioned previously).

An important element for implementing these newly developed policies is to conduct annual "cradle to grave" environmental audits of all operations and products, including a detailed strategy for making improvements, disseminating results to employees, stockholders and the public.

Additional services can be provided in helping customers safely distribute, store, use, and dispose of or recycle company products.

All of the above can provide new business opportunities and employment, whether conducted by contracting consulting services or in-house within a company.

Such improvements will encourage innovation; expand markets; improve profit margins; develop satisfied and loyal customers; attract and retain the best-qualified employees; and help sustain companies and the economy.

As of 31 December 2002 Puerto Rico held just 55 ISO 9000/9001 certificates, compared to Costa Rica with 60, and Finland, Ireland and Singapore with 1,870, 3,700, and 3,513 respectively. The number of 9000/9001 certificates fell to just 29 by the end of 2005, with the implementation of updated 9001 standards (ISO, 2005; ISO, 2006).

By 2002 Puerto Rico held only 3 ISO 14000/14001 certificates, compared to 38 for Costa Rico, and 750, 280, and 441 for Finland, Ireland and Singapore, respectively (ISO, 2003), and rose to just 5 by 2005 (ISO, 2006). Clearly, unless Puerto Rico rapidly and forcefully places more emphasis on quality and international standards it will not be able to diversify and improve its economy, nor take control of its own economic destiny. While ISO currently is a prominent and important standards metric, it is not the only one. Others exist and may be gaining in importance. However, it is important that Puerto Rico begin to develop a culture of quality; to acknowledge the importance of quality compliance with international standards as means of participating in the global export market.

2.14 *Moving forward*

Changes in traditional practice and moving towards an economy that emphasizes exports and global competitiveness are essential for Puerto Rico. A recent history of Puerto Rico's economic evolution (Puerto Rico Industrial Development Company, Office of Science and Technology, internal analysis, 2005, unpublished data) has classified Puerto Rico's Economic evolution into the following five stages:

- 1940's, 1950's → mid 1960's. Low-cost labour-driven economy ("Operation Bootstrap", etc.).
- Late 1960's to mid 1970's. US Petroleum import quotas drive economic growth in PR (chemical, petrochemical industries).
- 1976–1998. Federal tax incentive-driven economy.
- 1998 →? Knowledge-based and Innovation-driven economy.
- 2005 →? Next steps? Towards economic and environmental sustainability, diversification and global competitiveness.

Manufacturing has always been the largest sector of the economy and exports to generate foreign capital focused on shoes and apparel during the 1950's and 1960's, petrochemicals during the 1970's, pharmaceuticals during the 1980's and 1990's, with the emergence of biotechnology manufacturing during the late 1990's and early 2000's.

The island's population achieved an increase in the average length of schooling from 3.7 years in 1950, to 12 years in 2002, yet maintains a standard of living and per capita income lower than any US state (Dietz, 2003). Most surprisingly, tourism and recreation employ a lower proportion of the workforce than in any of the 50 states (Collins et al., 2006, The Economist, 2006) and represents only about 4% of Gross Domestic Product (Puerto Rico Government Development Bank data (BGF, 2007) – $US 3.2 billion of a total GDP of $US 82 billion).

3 NEW BUSINESS OPPORTUNITIES AND ECONOMIC DEVELOPMENT: NATURAL AND HUMAN CAPITAL

Ideas for the development of new business opportunities and economic development have emerged from meetings and discussions among participants in the Energy and Environment Alliance (or "cluster") of the Puerto Rico Industrial Development Company. The Alliance was formed in 2005 and consisted of a "think tank" of leaders from the business, academic, community and government sectors. The results are organized into the areas of Natural Capital – Water, Space and Waste, Air and Energy, and Human Capital. These ideas are discussed with the aim of developing more specific approaches to developing both a public policy agenda and a business agenda that will promote sustainable socioeconomic development, prosperity, better environmental compliance and an overall improved, and improving, quality of life.

3.1 *Natural capital: water, space and waste, air and energy*

All other forms of capital are based on human capital and natural capital. Puerto Rico lacks significant natural resources such as minerals, petroleum or timber, and due to a scarcity of arable agricultural land, long ago lost any hope of agricultural self-sufficiency. The island's principle industrially relevant natural capital is its water and scarce land resources. The condition of Puerto Rico's remaining land resources has an unappreciated, but vitally profound effect on the quality and availability of usable industrial and potable water. Therfore, policies related to protecting the land, the condition of the land, and regulation construction project siting will be increasingly and critically important for Puerto Rico's industrial development. Forcing the protection of greenspace obliges *redevelopment* of existing the urbanized platform as the means of constructing new "built" capacity. This will create an aesthetically improved urban environment and enhanced quality of life. This is advantageous for attracting and retaining a capable workforce (see below).

Data from the Puerto Rico Planning Board published in the "Puerto Rico 2025" report (Comisión Puerto Rico, 2025, 2004) show that during the past five decades, the rate of horizontal expansion of the concrete and asphalt platform of The San Juan Metropolitan area and other urban centres in Puerto Rico has proceeded at a rate three to four times more rapidly than its demographic growth. While Puerto Rico's population density has always has been high, the current rate of population growth is in fact modest and linear. Annual population growth in Puerto Rico is currently about 0.5% – a rate considerably lower than that of the USA and approaching some northern European countries (UNESCO, 2003, BGF, 2006a–d, 2007; Table 1; Figure 3).

3.1.1 *Water and waste water – making a business case for water resources*

Puerto Rico, with its small land area and high population density possesses few renewable natural resources for economic development other than land and water, and these are very intimately interconnected. Puerto Rico long ago lost any possibility of agricultural self-sufficiency, and the remaining land and water resources must be meticulously managed and protected in order and develop its industrial capabilities. The business community, and in particular the manufacturing

Table 1. Comparison of Demographic parameters in seven countries of importance in Puerto Rico's economic development (CIA, 2006). Puerto Rico's population growth is slower than that of the USA

Country	Persons/ km^3	Total Population	Annual Growth %/year	% Urban Population
USA	30	290,342,55	0.92	77.2%
Mexico	50	104,907,991	1.43	74.4%
Ireland	54	3,924,140	1.03	59.0%
Japan	335	127,214,499	0.14	78.8%
Puerto Rico	**435**	**3,897,960**	**0.49**	**75.7%**
Singapore	6,503	3,885,877	3.42	100.0%
Hong Kong	17,001	7,394,170	1.22	100.0%

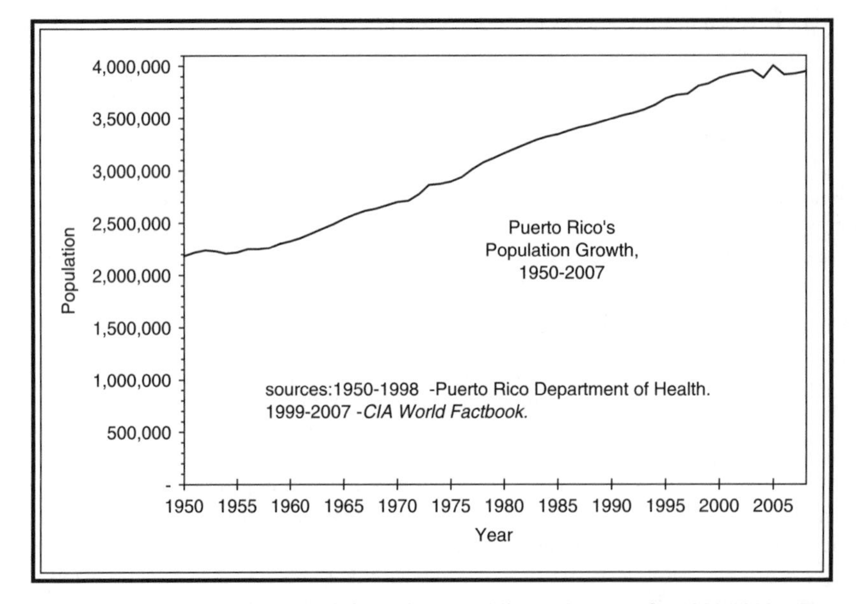

Figure 3. Puerto Rico's population growth is moderate and linear. Sources: for 1950-1998 – Puerto Rico Department of Health data (DSPR, 2000); for 1999–2007 – CIA World Factbook, 1999–2007. (Tourism is included within the "Services" sector)

sector that produces over 42% of Puerto Rico's GDP, has identified the lack of industrial water resources of sufficiently reliable quantity and quality as a factor that discourages new capital investment in Puerto Rico.

The disposition of solid and hazardous wastes are areas that also impact the availability of the remaining usable land and water resources of Puerto Rico. A result of rapid post-war (WWI) urban expansion in the San Juan metropolitan area now includes an enormous number of under-utilized, empty and abandoned buildings, in both residential and commercial areas. The metropolitan areas need revitalization and redevelopment of their deteriorated urbanized land and the full protection of the remaining remnants of greenspace. This approach will provide socioeconomic benefits, including additional employment, better health and an improved quality of life for a large portion of the population, and it will not deny the construction sector the opportunity to generate profits. However, this requires that the construction sector operate in a socially and environmentally responsible manner (Comisión Puerto Rico, 2025, 2004; Table 1).

The improvement of quality of life requires that people enjoy adequate living space, meaningful employment, health, clean air and sufficient useable potable and industrial water. On a small island like Puerto Rico, the quality and abundance of water is essential for maintaining an adequate quality of life and the economy, and thus, the preservation of vegetative cover is essential for assuring an adequate quality and quantity of potable, industrial, and environmental water. When vegetative cover is removed from an area, this and adjacent areas of land become dangerously susceptible to suffer "natural" disasters, significant deterioration, and the loss of economically useable water.

3.2 *The economic impact of vegetative cover on the water cycle*

Vegetative cover, even so-called "weed patches" of exotic plants, provide highly significant economic services of vital importance, among these can be mentioned:

3.2.1 *Reducing soil erosion and preventing and ameliorating floods*

Plant roots help to retain soil particles and augment soil permeability. In this way surface runoff is reduced, as is soil erosion and sedimentation of bodies of water (lakes, streams, and ocean). Without this vegetative cover society would have to build additional rain gutters and drainage systems and increase the capacity of sanitary (sewage) treatment systems. All types of vegetative cover provide these services, but in general, forests are the most effective systems. This is because tree roots hold the land and tree leaves reduce the impact of rain as it falls to the ground. Obviously, in order to reduce or buffer ground erosion and flooding, vegetative cover must be must be protected.

3.2.2 *Reducing water pollution*

Even in heavy rains, vegetative cover creates friction that serves to reduce the velocity of runoff, increase residence time and filter particles, preventing landslides, sedimentation, and water flows that might otherwise be strong enough to move homes and other buildings. Conversely, when the ground surface is covered-over with concrete and asphalt, rain water flows rapidly and directly to the various drainages, meanwhile acquiring contaminants such as oil, volatile toxins from paving, and other wastes from point and non-point sources, thus polluting the waters of local streams, rivers, and eventually ocean bays and estuaries such as San Juan Bay. Preserving the vegetative cover necessary to avoid this pollution will thereby protect human health and avoid additional costs to the public health system (Perkins, 2004). High population density does not have to be accompanied by the complete destruction of greenspace and preserving this greenspace is essential for preserving industrial water supplies, as Japan's example demonstrates (Diamond, 2005).

3.2.3 *Aquifer recharge*

The land, together with its vegetative cover acts like a great sponge that absorbs, filters, and stores water. Vegetative cover reduces the velocity of surface runoff and maintains water longer in the local area longer (increases residence time), where it can percolate or filter into the soil, thereby recharging the local aquifer, and helping keep wells productive. Contrarily, impermeabilization with concrete and asphalt prevents percolation of water to the aquifer and prevents local recharging of groundwater and potable water wells.

3.2.4 *Positive modification of local climate*

A significant effect of vegetative cover is the buffering or ameliorating of climatological extremes. Maintaining vegetative cover can reduce temperatures by 6°C lower than in adjacent urban areas without vegetation, and help maintain normal rainfall patterns.

3.2.5 *Promoting biodiversity*

Conserving vegetative cover provides habitat for wildlife and provides the opportunity to conserve additional forest habitat and in turn increase populations of native flora and fauna. Many studies have demonstrated that the number of species present in a given ecological reserve is proportional to the geographic size of the reserve, the diversity and structure of the reserve's topography, and the level of interconnectivity of the reserve's various units. The connectors permit the dispersal of wildlife and thus also the seeds of various species of trees, and other plants among various areas, thereby increasing even more the biodiversity of the reserve. In this way the reserve is continually improved, increasing its value for conservation and recovering native threatened and endangered species. If its size is sufficient, it also can supply species to other reserves.

In spite of the above considerations, in recent decades Puerto Rico has allowed run-away, chaotic patterns of urban development that have negatively impacted water resources continuously and accumulatively (Comisión Puerto Rico, 2025, 2004). The excuse offered to justify these damaging development patterns has been "the economy" pointing out the number of jobs and personal profits supposedly generated by construction. But as the data clearly show (Puerto Rico Government Development Bank), the perception of short-term economic benefits of construction activity is highly misleading, while the damage caused is long-term and affects the economic well being of the entire population. Abusive construction practices also represent the transfer of public goods to private control and individual profit, at the net expense of the general public – something prohibited by the Constitution of Puerto Rico.

This situation has been identified in various economic analyses (the Puerto Rico, 2025 report, Comisión Puerto Rico, 2025, 2004) and in the recent "Sustainable Industrial Development Model for Puerto Rico" Ashton and Chertow 2004), sponsored by the Federal Economic Development Administration, the Puerto Rico Office of Management and Budget, (Oficina de Gerencia y Presupuesto – OGP) the Puerto Rico Industrial Development Company (PRIDCO) and the Luis Muñoz Marin Foundation. Water availability and water pollution are major concerns. Despite the heavy rainfall the island receives in mountainous regions and its abundant ground water aquifers and surface water reservoirs, high population density makes the island's per capita water availability low when compared to that of other countries. A 2003 study by UNESCO ranks Puerto Rico 135th out of 180 countries and territories surveyed in this regard (UNESCO 2003). Beyond water availability, delivery holds additional problems at an estimated loss of 50% due to leakage in pipe-borne distribution. Puerto Rico has previously imposed water-rationing measures, as was the case during the 1990s drought periods. In that same decade annual rainfall reached the lowest levels of the 20th century, according to a study by the US Geological Survey (Carter et al., 2003).

Water pollution constitutes one of the most significant environmental problems on the island. A significant portion of surface waters in Puerto Rico is heavily contaminated – 19% of them do not support aquatic life and 21% are impaired for swimming. Due to the large aquifer system formed by porous karst rock on the northern part of the island, Puerto Rico has significant groundwater supply, accounting for 30% of its water consumption (USGCRP 2001). Nonetheless, groundwater systems have also been polluted, especially by septic tanks, livestock operations, agriculture, storage tanks, and landfills.

The pharmaceutical industry is highly water intensive, with utilization rates on the order of hundreds of thousands cubic metres per day for each facility

Most notable is the fact that since 1950, the urbanized horizontal "paved platform" of Puerto Rico has been growing three to four times faster than population growth and at the same time, population density within this platform has decreased by 72% (to only 28% of the original population density, Figure 3, Table 1). This has created significant environmental and economic problems, including problems with the quality and quantity of water available to industry and the population, and problems with the treatment of wastes and waste water. Despite its moderate population growth Puerto Rico's aquifers continue to deteriorate (Comisión Puerto Rico, 2025, 2004).

The available data indicate quite clearly that Puerto Rico can meet all of its construction needs quite easily by building in, or re-developing, existing urban underutilized or deteriorated urban areas. There is, quite literally, no need or justification to continue negatively impacting the remaining greenspace (the areas that effectively generate the potable and industrial water that is essential to the economy).

According to the analyses conducted by the Puerto Rico Government Development Bank (BGF, "Puerto Rico en Cifras", 2007 and Figure 4), the construction industry represents only about 2.4%

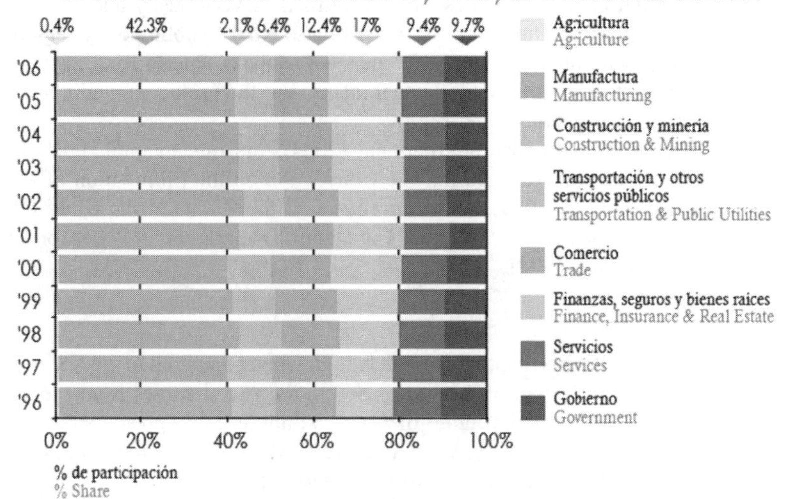

Figure 4. Analysis of the relative contributions of Puerto Rico's major industrial sectors to the island's economy. Note that construction and mining (third block from the left and listed third in the vertical key) accounts for about 2.1% of GDP, and that tourism, construction and mining and other services, together, account for less than one-third to one-half the contribution of manufacturing (BGF, 2007, page 7).

(and nearly always no more than 3%) of Puerto Rico's economic base (Gross Domestic Product – GDP), while manufacturing represents 42.1% of our economic base. Moreover, manufacturing figures do not include the very substantial indirect effects and induced effects on other sectors such as services, commerce, finance and transportation, etc., that depend heavily on manufacturing. Within the Manufacturing sector, the biopharmaceutical sector has particular importance (about 25% of GDP).

In terms of job creation, manufacturing (including biopharmaceuticals) provides substantially more jobs than both construction and tourism, (and as previously indicated, manufacturing is the base that generates a very substantial portion of the non-tourism service and commerce activities). Clearly the favoured, privileged, and priority status traditionally afforded to construction by the Puerto Rico Planning Board and much of the rest of Puerto Rico's government agencies over several decades has been done at the expense of the rest of the economy. Nor can Puerto Rico count on tourism for its future economic development, given that this sector is very small and is projected to continue to shrink in the future, due to external, international market forces beyond Puerto Rico's control (BGF- 2007).

The continuation of preferential treatment for the construction sector will, over the long term and incrementally and cumulatively, endanger the country's economic and social development and will be responsible for transforming Puerto Rico into much poorer country than it would be if it were to enjoy wiser and more strategic planning policy. Puerto Rico's true economic development, prosperity and well-being depend on reforming the current erroneous policy and orientation. Puerto Rico cannot afford to continue to jeopardize the economic future of the great majority of the population in order to satisfy the desires of a privileged minority sector. This should not be a cause of concern to the construction sector. If the majority of the economy is prosperous, then the construction sector also will prosper, because construction is correctly, and fundamentally a service industry to other economic sectors, and not an end in itself. This looming problem has not escaped the notice of both conservationists and industrialists (McPhaul, 2004), both of whom have indicated that protecting and conserving Puerto Rico's Water Sources is necessary for the island's economic health.

It is especially worth noting that unlike Singapore, Puerto Rico cannot import freshwater through a pipeline from a foreign country, and desalination remains a very costly and energy intensive proposition. Thus, the transformation and reform of Puerto Rico's public policy regarding water security, and its implementation are essential. In the final analysis, Puerto Rico's economic prosperity depends on the cultivation of its intellectual talent, and the protection and conservation of it's vegetative cover and endogenous water supply.

As observed in recent Green Cross International (2004), the fundamental right of access to water and sanitation has been recognized in: The Convention on the Elimination of all forms of Discrimination against Women (1979), The Convention on the Rights of the Child (1989), The Mar del Plata Action Plan for Water (1977), and The Dublin Declaration on Water for Sustainable Development Implementation (1992).

3.2.6 *Alliance input*

The cost of water in Puerto Rico is disproportionately high when compared to the US or competing countries such as Ireland. Nearly all of the water used in the island comes from precipitation in the form of rain. Although the total volume of water falling as rain appears sufficient, only a small portion is captured in our aquifers, lakes and rivers. Most of the water flows overland and through the rivers to the ocean. There is a pressing need to improve our water sourcing and water conservation and reuse. This situation presents us with opportunities to develop new and innovative solutions, for example, water treatment and reuse technologies and policies. Perhaps the waste heat from the power generation plants can be used to produce fresh water by using the heat energy currently vented to the atmosphere and ocean.

There is also the need for legislation to allow for the privatization and regulation of wastewater treatment plants for industries and residential areas; establish a database for water related issues (storage, distribution, treatment requirements, regulations, etc.); promote research on methods to recharge aquifers in coastal areas subject to saltwater intrusion, and articulate better guidelines on construction siting.

There must be further developments and activity by industry, public opinion and other driving forces such as product selection and awareness of economic benefits in order to leverage changes in law and policy to facilitate greater and more diverse wastewater treatment options. The availability of sufficient industrial water, both in terms of reliability of quality and quantity must be improved in order to be able to attract sufficient capital investment in manufacturing in Puerto Rico.

3.2.7 *Mitigating the water monopoly*

The Puerto Rico Aqueducts and Sewers Authority (PRASA) currently faces serious operational and financial difficulties, due primarily to a lack of long-term planning on the use of water resources, combined with the lack of an agile, efficient administration and a failure to incorporate modern technologies (McPhaul, 2004, 2005). Despite these severe problems PRASA still enjoys a near monopoly on providing water and waste water treatment, while at the same time trying to maintain the integrity of its infrastructure and financial solvency, yet it was widely recognized by nearly all Alliance participants that the regulated diversification and privatization of the system, especially of wastewater treatment, can provide greater security and financial solvency for PRASA (by relieving some of the pressure on it), and provide greater reliability and economic efficiency, and greater potential for environmental compliance (below).

3.2.8 *The convergence of entrepreneurship, security, environmental compliance and economic development*

Diversification of the potable water systems and wastewater treatment systems (including constructed wetlands), by private, municipal or cooperative groups will provide opportunities for local entrepreneurship and job creation. Establishment of a diversity of non-PRASA water providers and treatment services, also would provide the leverage needed to bring about greater environmental compliance and operational efficiency, create greater overall security for the system, and greater buffering against unexpected interruptions (any individual interruption would be limited to a smaller group of clients). Similarly, the system, industry, commerce, and the general population would be significantly better-protected against vandalism and terrorist attacks and sabotages.

3.3 *Solid waste management (space and waste)*

The case study sites high population density increases its impact on the natural environment (Comisión Puerto Rico 2025, 2004). It generates more waste per person than the USA and about twice as much waste per person as most western European nations. Yet despite an increasing environmental awareness in Puerto Rico, there is still little cultural consciousness regarding high consumer consumption and high waste generation.

Most of the currently operating landfills are not in compliance with federal and local environmental regulations. Before 1993, the number of known landfills operating on the island was about 60, but in that year about half of them were closed when the revision of the subtitle D of the Resource Conservation and Recovery Act (RCRA) was promulgated. Currently, there are no plans for landfill construction in Puerto Rico. Officials from the Puerto Rico Solid Waste Management Authority have said that only 4 of the remaining 29 landfills on the island fully comply with all environmental regulations. These landfills are located in Humacao, Ponce, Fajardo, and Carolina. Of these, only the Carolina operation has landfill gas (LFG) collection facilities. LFG collection and utilization for small-scale energy generation are expected to begin in 2006 (Carter et al., 2003). Capturing the methane produced at landfills has additional importance when one realizes that methane's per-molecule greenhouse effect is many times greater than CO_2 and therefore, keeping it out of the air is extremely important.

The major types of solid residues are yard residues, organic residues (e.g., food scraps), cardboard, paper, and grit; while metal, aluminium, plastic, and glass combined comprise approximately 25% of the solid residue stream. In total the waste stream consists of approximately 53% biomass, 28% recyclable material, and 19% material that can only be deposited in landfills.

In 1998, the island generated approximately 110,000 tonnes of solid residues per day, most of which was sent to landfills. This equates, according to the Caribbean Recycling Foundation, to more than 17 million m^3 of solid waste land filled annually in Puerto Rico. Three quarters of an acre are lost to landfills daily on the island, and landfill fees. At the current pace, without the addition of new landfill sites, best estimates anticipate that all current landfills will likely be filled within twenty-five years (Carter et al., 2003; Ashton and Chertow, 2004; Castaneda-Muñoz, 2006). Currently, tipping fees range from $US 5.25 per yard to $US 90 per ton with an average of $US 37 per ton (Carter et al., 2003). With no current plans for new landfills to be constructed on the island in the near future, alternative approaches to solid waste disposal are needed. Puerto Rico must develop options for waste reutilization and create incentives for recycling. Reducing Puerto Rico's waste stream is essential.

3.3.1 *Alliance input*

The solid waste situation in Puerto Rico is urgently in need of help. The hazardous waste must be sent to the US at great cost and the landfills are rapidly filling. Although a number of efforts have been made to deal with this situation, no successful solutions have been instituted. This need represents an opportunity to develop solutions that can lead to the formation of new businesses. The following should be considered:

- Develop a hazardous waste handling facility in Puerto Rico. Although difficult, this could be a viable business.
- The amount of garbage generated is over 5 lbs. per capita. Reuse and recycling strategies and infrastructure are needed and this also represents an opportunity.
- The volume of waste, if properly segregated and managed, could be renewable fuel for the generation of power.

Landfills produce methane (LFG) that is being vented. Methane is a greenhouse gas and could be captured to generate power and possibly qualify for carbon credits.

The Alliance also recommended the preparation of a new recycling initiative and the formation of a task force to articulate effective policies for the conversion of solid waste to energy. Economic incentives to promote recycling, such as charging for garbage collection and passing a comprehensive law by placing deposits on all recyclable containers also should be developed.

Real-cost pricing is essential for developing incentives to recycle. Diversification of the solid waste disposal system also will be essential for success. Garbage disposal is an externalized cost that is not properly adjudicated (internalized), and leads to a general lowering of quality of life.

3.4 *Air quality*

Issues of air quality in Puerto Rico, and relative to its economic development overlap broadly with the solid waste and energy generation areas. In addition, several public health issues, including cancer and the possibility of deleterious heritable mutations resulting from exposure to airborne contaminants must be addressed in greater detail (Quinn and Somers, 2004). This is an area in which seldom-appreciated externalized costs, as well as internalized costs to can have a significant influence on economic development and productivity.

3.4.1 *Alliance input*

There is an opportunity to develop strategies such as development of clean and renewable fuels; emission control technologies that can be added on to existing emission sources; propose new regulations that will benefit co-generation and facilitate the taking of emission credits for these projects; and identify renewable clean fuel that also will provide carbon credits to the generator. Subsequently it was decided to incorporate these ideas into the solid waste and energy generation areas.

3.5 *Energy conservation and generation*

Energy concerns have been addressed by several recent authors (Carter et al., 2003; Rodriguez, 2005; and DOE, 2004). All point out the island's electric generating capacity is already rapidly approaching its limit, is very costly, and very unreliable and unstable (power outages and fluctuations in voltage, etc.). Puerto Rico's per kilowatt-hour electricity cost is among the highest in the world, and substantially higher than in nearly all 50 US states (including Hawai'i) and most of Europe. Electricity supply has been identified as a critical area for improvement by members of Puerto Rico's Techno-Economic Corridor. Another simultaneous study revealed the high cost of energy is a significant potential barrier to new industries. Power shortages or inconsistencies have forced many private companies to acquire and operate backup power generators at significant expense.

The PRIOS (Puerto Rico Island of Sustainability) project final report (Ashton and Chertow, 2004) stated that the reliability of power generation and distribution systems is a decisive factor for businesses that are either energy intensive or that employ precision machinery, the latter being vulnerable to power fluctuations and also that studies have repeatedly shown that the reliability of the power generation sector, as well as the comparatively high prices paid for electricity on the island are crucial issues for businesses. The Report indicates that electricity costs are 180% more in the island than in the US mainland.

These observations are very much in line with the conclusions of Casten and Downs (2005) for US national and global energy production reform, and the specific recommendations provided by the Puerto Rico Chamber of Commerce to the incoming government in 2005, that focus on diversification, a movement towards proportionally greater use of replenish able fuels, increasing capacity, reducing emissions and/or improving the quality of combustion, reliability, reducing interruption frequency and duration, and reform of PREPA (Puerto Rico Electric Power Authority) policy away from current monopolistic practices and towards the regulated diversification and privatization of power producers (multiple small suppliers) and greater cogeneration in shared facilities, in addition to making other adjustments required for compliance with the Public Utility Regulatory Policies Act (PURPA). The disproportionately high cost of electricity is Puerto Rico's greatest disincentive for attracting new capital investment. Those industries that choose to remain in Puerto Rico will mitigate these costs and other unnecessarily high operating costs from their employees' pay checks.

3.5.1 *Alliance input*

The cost of electricity is a significant consideration in determining the viability of locating a manufacturing operation in Puerto Rico. This high cost can be viewed as an opportunity to develop alternate sustainable sources. There are currently numerous such technologies already in operation around the world that may be applicable to Puerto Rico and represent an opportunity to develop a new business/industry. The following are just a few ideas that can be explored: wind power, and cogeneration. Colucci-Ríos et al. *(this volume)* point out that the current efficacy of photovoltaic solar power generation is competitive with wind generation and is likely to become significantly more competitive in the near future. The hurdles to overcome for implementation of this alternative are neither economic not technical, but rather managerial, and of public policy. As of this writing, Puerto Rico still lacks a public policy (common in other jurisdictions) to incentivize the use of solar power for residences and businesses. Interestingly, an internal proposal from the PRIDCO (Puerto Rico Industrial Development Company) Office of Science and Technology to establish a solar energy demonstration project using photovoltaic panels on the roof and parking structure of the PRIDCO building was rejected, not for economic nor technical reasons, but for political reasons (Schaffner, unpublished data). PRIDCO management feared the inevitable negative reaction and militancy of the PREPA labour union, which had the power to paralyse the entire industrial capacity of the island.

The Alliance also recommended the formation of a task force to develop policies and guidelines for Puerto Rico Electric Power Authority (PREPA) to work with private energy generators, and, especially generators of energy from renewable sources, to conduct a review of policy, regulations and responsibilities of the Puerto Rico Office of Energy, to work with PREPA to determine the effect of the signing of the Kyoto Protocol and global warming on its future development plans, and to greater use of Performance Contracting by energy services companies.

Establishing financial solvency for the Puerto Rico Electric Power Authority (PREPA) is essential. We must develop policies and recommendations that assure that PREPA is paid for the electricity it provides to all government agencies so that (1) PREPA receives full compensation for the energy it produces and maintain a balanced book, and (2) so the private sector does not have to continue subsidizing the cost of energy to public users, and pays only for what it uses.

3.5.2 *Mitigating the PREPA (Puerto Rico Electric Power Authority) monopoly*

The diversification of the electric power system and the development of real cost pricing will provide incentives for conservation, wise use and new business generation. Reducing the PREPA monopoly will greatly enhance the availability of reliable energy and energy from more renewable sources, lower and more realistic costs to most consumers, greater compliance with environmental standards and law, financial solvency for PREPA, and promoting new technologies such as waste to energy technologies and plasma torch detoxification.

3.6 *The convergence of entrepreneurship, security, environmental compliance and economic development*

Diversification of electric power supply by private, municipal or cooperative groups employing a variety of technologies and fuels would provide opportunities for local entrepreneurship and job creation. Establishment of a diversity of non-PREPA energy providers, also would provide the leverage needed to bring about greater environmental compliance and operational efficiency, create greater overall security for the system, and greater buffering against unexpected interruptions (any individual interruption would be limited to a smaller group of clients). Similarly, the system, industry, commerce, and the general population would be significantly better protected to vandalism and terrorist attacks and sabotage.

Other key areas of policy development that should be developed soon (these also are DESA Technical Cooperation Areas (The United Nations Department of Economic and Social Affairs (DESA, 2004–2005).

3.6.1 *Energy efficiency*

integrated resource planning, demand-side management, energy standards and labelling, promotion of energy service companies, and access to financing.

3.6.2 *Renewable energy*

commercialization of renewable energy, entrepreneur development, innovative financing schemes, standards and best practices, and community participation.

3.6.3 *Clean cities*

environmental monitoring, waste to energy, transportation energy planning, and clean industrial production processes.

3.6.4 *Public transport*

integrated transport system design, public awareness, modal shifts, emission reduction.

3.6.5 *Cleaner fuels and vehicles*

lead phase-out, sulphur reduction, inspection and maintenance policy.

3.6.6 *Cleaner fossil fuel technologies*

clean coal technologies including coal bed methane recovery, clean production technologies for industry, advanced hydrocarbon recovery techniques, and transboundary air pollution monitoring.

All of the factors mentioned above will have very significant influences on Puerto Rico's economic development, but in the final analysis, water, are the most important, and perhaps the only native natural resource that Puerto Rico possesses in economically relevant quantities. Protecting the quality and availability of water is essential to the island's socioeconomic future.

4 HUMAN CAPITAL: EDUCATION AND PROFESSIONAL DEVELOPMENT

4.1 *Developing, attracting and retaining talent*

Developing, attracting and retaining a competitive pool of adequately trained and trainable employees is essential for Puerto Rico's sustainable economic development. Programs in human resource development and job training such as those funded through the federal Workforce Investment Act (WIA) should be pursued more aggressively.

The development of the new "knowledge-based economy" has received a great amount of attention recently, yet for all the attention it has received, a surprisingly few media observers actually understand its meaning. It does not mean that we are developing an economy based on Research and Development and that everyone will need doctorates and high-end research skills in order to find employment. Simply put, this knowledge-based economy is an economy in which workers will be required to have greater skill sets and higher levels of education than in the past. Low-skill, labour-intensive manufacturing has left Puerto Rico for areas with lower labour and operating costs, and Puerto Rico's economic base is formed on industries where the vast majority of the workforce has a university degree, including 3–5% with doctorates. In regards to attracting new industries to Puerto Rico, the slogan "build it and they will come", ***does not apply*** any longer. Rather, "provide an adequately trained and trainable workforce, at a reasonable price, and they will come" is much more appropriate. Further, even Puerto Rico's deep tax concessions are becoming less influential in site location for new industries. Where major capital investments are required for start-up, companies will accept high rates of taxation in exchange for up-front capital assistance, all other factors being more or less equivalent. This also is an example of a knowledge-based economic determinant, because it requires that that government planners themselves possess sufficient knowledge of the industry to be innovative, agile and adaptable to the current rapidly-changing conditions.

Government Development Bank data (BGF, 2007) indicate that the per capita productivity of the "typical" Puerto Rican worker in the private sector is about 85% that of his US counterpart, yet this worker receives only about 50–65% of his counterpart's salary, and with similar costs of living. Industry pays these relative low salaries, in part, to offset other disproportionately high operating costs, especially electricity and water. But these low wages create a circular problem of attracting and retaining sufficient talent to improve and maintain efficiency and productivity. The nefarious role of the Puerto Rico government's utilities monopolies here is obvious, as it, (the government) is the single most influential agent in unnecessarily driving up operating costs through inefficiency and waste. This drives down the private sector. Currently only 28% of Puerto Ricans are employed in the private sector, versus 58% in the US (Collins et al., 2006). This has occurred despite a period of 25 years of high growth, productivity and educational attainment from 1950–1975.

The large gap between GNP in Puerto Rico (gross national product – the value generated in Puerto Rico) and GDP (gross domestic product – the value retained in Puerto Rico and accrued locally) is very substantial, as a great deal of the value generated is transferred out of Puerto Rico.

In addition, the Puerto Rico government has a nearly monopolistic control of the primary and secondary education systems that are rapidly and increasingly falling behind those of their US counterparts. The gap in educational achievement between Puerto Rico and the US has been growing alarmingly in recent decades and the gap widens progressively with each year of primary and secondary education (i.e., the gap worsens with each grade of school) (Ladd and Rivera-Batiz, 2006).

Given all of the above considerations, support of economically relevant educational programs is, and will be increasingly important to Puerto Rico's economic future. This includes environmental fields, and all other fields that can help maintain and improve our quality of life – the key factor in attracting and retaining crucial talent. Standards matter here as well, as can be appreciated by the promotional information offered by a major human resources development company (i.e., Brainbench, 2008).

4.1.1 *Alliance input*

Prepare a guide for developers and industry on working with communities during site selection, permitting, construction, and operation of projects, as well as the develop an educational program or initiative for the media (newspapers, radio, television), including capsules using local artists. This also could be piggy-backed on existing programs.

Closer linkage and greater implementation of the Human Resource development and Job Training programs at the local and federal levels should be developed. This may include programs such as those established under the Work Incentives Act (WIA). Greater focus should be directed to enhancement of societal quality of life and it's influence on the attraction and retention of talent.

Puerto Rico's most important resource, and capital, is its people, the actual and potential intellectual and productivity capabilities of its local talent. Economic development *requires* that this capital be cultivated and developed to its fullest potential. Puerto Rico's second most important resource is water. Over the long term this natural capital is our only economically significant natural resource, and its supply is increasingly limited. It must be protected, and properly managed. If Puerto Rico can optimize the cultivation of its Human Capital (education, quality of life and the environment), and protect and conserve it's water supply, its socioeconomic development and future will be positive. Contrarily, neglect or mismanagement of these two key factors is a guaranteed formula for disaster.

5 CONCLUDING REMARKS

Although the promotion of sustainable economic development is official policy with many governments, the real meaning of sustainability remains poorly understood. Linking this type of growth and business is essential for comprehending that over the long term, economic security depends on environmental well-being. Countries and regions must endeavour to achieve full participation in the global economic community. In addition, they must reach the development of a culture of quality, as well as a public policy agenda and a business agenda that will promote sustainable socioeconomic growth, prosperity, better environmental compliance and an overall improved quality of life. If the cultivation of human capital, and the protection and conservation of water supplies, for example, can be optimized then mankind will have a brighter socioeconomic future.

ACKNOWLEDGEMENTS

Special thanks to Dr. Gary Gervais, whose inspiration lead to the birth and development PRIDCO Energy and Environment Alliance, and to Dr. L. Michael Szendrey, for his meticulous revision and many excellent ideas for improving the manuscript. I also would like to thank Victor Rivera, José Muratti, Richel Garcia, Marcos Polanco, Raymond Laureano, Janis González, Iván Lugo, Soraya Portillo, and Dr. Paul LaTortue, Ivette Laborde, Pedro Gelabert, Mabel Rodríguez, Hector Arana, Rosa Hilda Ramos, G. I. Ramírez, H. A. Minnigh, J. M. El Koury, P. Panzardi, F. Pérez, F. Mateo, G. Rodrguez, T. Rosario, R. De Jesus, W. Palermo, B. Pérez de Gracia, C. Amador, R. Figueroa, Sheila Ward, Eddie Laboy and Mattheus Goosen their ideas, input and contributions and/or in manuscript review.

REFERENCES

Ackoff, R. 1999. Ackoff's Best. Norton.

AISE – Association of Iron and Steel Engineers. 1985. The Making, Shaping, and Treating of Steel, 10th Edition, Association of Iron and Steel Engineers, Pittsburg.

Ashton, W., and M. Chertow. 2004. Sustainable Industrial Development Model for Puerto Rico. EDA Project No.: 01-79-07795. Final Report, July 31, 2004. Yale School of Forestry and Environmental Studies.

Berman, E., and L. T. M. Bui. 2001. Environmental Regulation and Labor Demand: Evidence from the South Coast Air Basin. *Journal of Public Economics 79: 265–295.*

BGF – Banco Gubernamental de Fomento. 2006a. Puerto Rico en Cifras. Disponible en línea: http://www.gdb-pur.com/economia/prcifras/prcifras.htm.

BGF – Banco Gubernamental de Fomento. 2006b. Pulso Turístico junio 2004. http://www.gdb-pur.com/economia/Pulso/documents/Pulsojunio2004.pdf.

BGF – Banco Gubernamental de Fomento. 2006c. Indicadores Económicos. http://www.gdb-pur.com/economia/indicadores/documents/prie110304.pdf.

BGF – Banco Gubernamental de Fomento. 2006d. Compendio de Datos. http://www.gdb-pur.com/ economia/datoseconomicos/documents/PR.Compendio.Datos.pdf.

BGF – Banco Gubernamental de Fomento. 2007. Puerto Rico en Cifras. Available online: http://www.gdb-pur.com/economia/prcifras/prcifras.htm.

Brainbench. 2008. Why Brainbench? – ISO Certified. available online at: http://www.brainbench.com/xml/monster/common/iso.xml.

Brown L.R., G. Gardner G., and B. Hailwell. 1998. Beyond Malthus. World Watch Institute. W. W. Norton & Company. New York.

Brundtland, G. H., M. Khalid, S. Agnelli, S. A. Al-Athel, Chidzero, Fakida, L. M., V. Mauff, I. Lang, M. Shijun, M. Marino de Botero, N. Singh, P. Nogueira-Neto, S. Okita, S. S. Okita, S. S. Ramphal, W. D. Ruchelshaus, M. Sahoun, E. Salim, B. Shaib, V. Sokolov, J. Stanounik, M. Strong, and J. MacNeill. 1987. "Our Common Future" – Report to the World Commission on Environment and Development. UN General Assembly, Forty-second session, Item 83 (e) of the provisional agenda (Development and International Economic Co-Operation: Environment. 318 pp. Available online at: http://www.are.admin.ch/are/en/nachhaltig/international_uno/unterseite02330.

Carter, B., S. A. Carty, A. Krishnan, A. Lubow and M. Zannis. 2003. The Rum Industry in Puerto Rico: Use of Industrial Symbiosis to Improve the Environmental and Economic Performance of the Industry. Research Paper for the Industrial Ecology Spring 2003 Course, Yale School of Forestry and Environmental Studies. New Haven, USA.

Castañeda-Muñoz, M. 2006. Industrial Ecology Approach to Management of Fly Ash from Fluidized Bed Combustion: Production of Slow-Release Fertilizer and Soil Conditioner. Doctoral Dissertation, Department of Civil Engineering, University of Puerto Rico, Mayagúez.

Casten, T. R., and B. Downs. 2005. Critical Thinking About Energy. The Skeptical Inquirer, Jan. 2005. Available online at: http://www.csicop.org/si/2005-01/energy.html.

CIA – Central Intelligence Agency (US). 2006. World Factbook. Available online at the CIA website: https://www.cia.gov/library/publications/the-world-factbook/.

Coase, R. 1960. The Problem of Social Cost. *Journal of Law and Economics 3: 1–44.*

Cofala, J., T. Lis, and H. Balandynowicz. 1991. Cost-benefit Analysis of Regional Air Pollution Control: Case Study for Tarnobrzeg. Polish Academy of Sciences, Warsaw.

Colucci-Ríos, J.A., E. O'Neill-Carillo, and A. Irizarry-Rivera. *(forthcomming 2009?)*. Renewable Energy in Puerto Rico: Caribbean Potential Showcase. *Taylor & Francis, this volume.*

Comisión Puerto Rico 2025. 2004. Puerto Rico 2025: Una nueva visión para el futuro de Puerto Rico. Oficina de la Gobernadora, San Juan, Puerto Rico.

Collins, J., and J. Porras. 2002. Build to Last: Successful Habits of Visionary Companies. W. W. Norton and Company.

Collins, S., B. Bosworth, M. Soto-Class (eds.). 2006. The Economy of Puerto Rico: Restoring Growth. Center for the New Economy, San Juan, PR and Brookings Institution Press, Washington. D.C.

Costanza, R, R. d'Arge, R. de Groot, S. Farber, M. Grasso, B. Hannon, K. Limburg, S. Naeem, R. V. O'Neill, J. Paruelo, R. G. Raskin, P. Suttonkk and M. van den Belt. 1997. The Value of the World's Ecosystem Services and Natural Capital. *Nature 387: 253–60.*

Costanza, R. 2001. Visions, values, valuation, and the need for ecological economics. *BioScience 51 (6): 459.*

Cunningham, W.P., and M. A. Cunningham. 2004. Principles of Environmental Science. McGraw Hill. Boston.

Daily, G. C., S. Alexander, P. Ehrlich, D. Tilman, and G. M. Woodwell. 1997. "Ecosystem Services: Benefits Supplied to Human Societies by Natural Ecosystems". Issues in Ecology Núm., 2. Ecological Society of America, Washington, D.C.

Daily, G. C., T. Soderquist, S. Aniyar, K. Arrow, P. Dasgupta, P. R. Ehrlich, C. Folke, A. Jansson, B. Jansson, N. Katusky, D. Tilman, and B. Walker. 2000. The Value of Nature and Natural Value. *Science 289: 395–396.*

Daily, G. C., and K. Ellison. 2002. The New Economy of Nature – The Quest to Make Conservation Profitable. Island Press, Washington, D.C.

Daly, H. E. 1991. Steady- State Economics. 2nd. Island Press. Washington. D.C.

Daly, H. E. 1994. Operationalizing Sustainable Development by Investing in Natural Capital. p. 22 in Jansson, A., M. Hammer, C. Folke, and R. Costanza, (eds.), "Investing in Natural Capital". Island Press. Washington, D.C.

Daly, H. E. 1996. Beyond Growth: The Economics of Sustainable Development. Island Press. Washington, D.C.

Daly, H. E. 1997. "Uneconomic Growth" From Empty World to Full World Economics. Rice University. De Lange-Woodlands Conference, "Sustainable Development: Managing the Transition". Houston TX, Mar 3, Columbia University Press.

Daly, H. E. 1998. Beyond Growth: Avoiding Uneconomical Growth. International Society fro Ecological Economics. 5th biennial conference, Santiago, Chile.

DESA – The United Nations Department of Economic and Social Affairs (DESA), Division for Sustainable Economic Development. 2004-2005. Indicators of Sustainable Development. Table 4. The CSD Theme Indicator Framework. Available online.
See: http://www.un.org/esa/sustdev/natlinfo/indicators/isdms2001/table_4.htm

DDEC – Departamento de Desarrollo Económico y Comercio de Puerto Rico. 2005–2007. Estrategias – Implementando doce estrategias formuladas para promover un desarrollo económico sostenible. Available online at: http://www.ddecpr.com/4.0_estrategias/4.12_pr/

De Groot. 1994. Investing in Natural Capital. "Environmental Functions and the Economic Value of Natural Ecosystems". pp 151-168, in, Jansson, A., M. Hammer, C, Folke, and R. Costanza (eds.), "Investing in Natural Capital". Island Press. Washington. D.C.

Diener, E. and M. E. P. Seligman. 2004. "Beyond Money: Toward an Economy of Well Being." *Psychological Science in the Public Interest 5 (1): 1–31.*

Diamond, J. 1999. Guns, Germs and Steel: The Fates of Human Societies. W.W. Norton & Company. New York.

Diamond, J. 2005. Collapse: How Societies Choose to Fail or Succeed. Viking. New York.

Dietz, J. 2003. Puerto Rico: Negotiating Development and Change. Lynne Rienner Publisher, Boulder, CO.

DiTella, R. and R. MacCulloch. 2004. Some Uses of Happiness Data in Economics. *Journal of Economic Perspectives 20 (1): 25–46.*

DOE – Department of Energy (federal), Energy Information Association (EIA) Country Analysis, Puerto Rico, 2004. Available online at: http// www.eia.doe.gov/emeu/cabs/prico.html.
Down, A. 1957. An Economic Theory of Democracy. Harper Collins Publishers. NY.
DSPR – Departamento de Salud de Puerto Rico. 2000. Informe Anual de Estadísticas Vitales de Puerto Rico, 2000. Departamento de Salud (Puerto Rico Department of Health), SAPEESI, San Juan. Puerto Rico.
Drucker, P. F. 1996. Innovation and Entrepreneurship. (2nd.). Harper Business. NY.
EPA – Environmental Protection Agency (US), National Center for Environmental Economics. 2001. The United States Experience with Economic Incentives for Environmental Pollution Control Policy. available online at: http://www.yosemite.epa.gov/ee/epa/eed.nsf/webpages/homepage.
Esty, A. 2004. Quality of Life. *American Scientist. 92 (6): 513.*
Foley, A.F., R. DeFries, G. P. Asner, C. Barford, Go. Bonan, S. R. Carpenter, F. S. Chapin, M. T. Coe, G. C. Daily, H. K. Gibbs, J. H. Helkowski, T. Holloway, E. A. Howard, C. J. Kucharik, C. Monfreda, J. A. Patz, I. C. Prentice, Na. Ramankutty, P. K. Snyder. 2005. Global Consequences of Land Use. *Science 309: 570–574.*
Graedel, T. and G. Allenby. 1995. Industrial Ecology. Prentice Hall. Englewood Cliffs, NJ.
Graedel, T. and B. Allenby. 2002. Industrial Ecology. 2nd edition. Pearson Education, 2002.
Green Cross International. 2004. Fundamental Principles For a Global Convention on the Fundamental Right to Water. Available at: *http://www.greencrossinternational.net/Tools/petition/principes_eng.pdf*.
Hardin, G. 1968. Tragedy of the Commons. *Science 162: 1243–1248.*
Hawken, P., A. Lovins, and L. H. Lovins. 1999. Natural Capitalism – Creating the Next Industrial Revolution. Little, Brown and Company, Boston.
Heal, G. 2000. Nature and the Marketplace. Capturing the Value of Ecosystem Services. Island Press. Washington, D.C.
Heal. G. 2000. Valuing Ecosystem Services. *Ecosystems 3: 24–30.*
Heilbroner, R. 1994. 21st Century Capitalism. W.W. Norton & Company.
Heilbroner, R. 1999. The Worldly Philosophers: The Lives, Times, and Ideas of the Great Economic Thinkers.
Heilbroner, R., and W. Millberg. 2001. The Making of Economic Society (11th). Prentice Hall, Upper Saddle River, NJ.
Heilbroner, R. and A. Singer. 1998. The Transformation of America: 1600 to Present.
Holliday, C. O., S. Schmidheiny, and P. Watts. 2002. Walking the Talk: The Business Case for Sustainable Development. Greenleaf Publishing.
Holly, C. 2001. Green Taxes. *Energy Daily 29: 4.*
ISO – International Organization for Standardization. 2003. The ISO Survey of ISO 9000 and ISO 14001 Certificates – Twelfth cycle: up to and including 31 December 2002.
ISO – International Organization for Standardization. 2005. The ISO Survey – 2005. Available online at: http://www.iso.org/survy2005.pdef.
ISO – International Organization for Standardization. 2006. The ISO Survey – 2006. Available online at: http://www.iso.org/survy2006.pdf.
Ladd, H. F., and F.L. Rivera-Batiz. 2006. Education and Economic Development, pp 189–238 in The Economy of Puerto Rico: Restoring Growth, S. Collins, B. Bosworth and M. Soto-Class. (eds.). Center for the New Economy, San Juan, PR and Brookings Institution Press, Washington. D.C.
Jansson, A., M. Hammer, C, Folke, and R. Costanza. 1994. Investing in Natural Capital: The Ecological Economics Approach to Sustainability. Washington, D.C. Island Press.
Magretta, J. 1997. Growth through Global Sustainability: An interview with Monsanto's CEO Robert B. Shapiro. Originally published in the Jan-Feb, 1997 of the Harvard Business Review, reprint #97110. Republished in: *Anonymous.* 2000. The Harvard Business Review on Business and the Environment. Harvard University Press, Cambridge, MA.
Mazurak, R. E. 2003. Ironmaking industry trends and directions. *Mining Engineering; 55 (4): 12–16.*
McPhaul, J. 2004. "Local Manufacturers and Conservationists show concern over water". *Caribbean Business, 17 March, 2004.*
McPhaul, J. 2005. "Puerto Rico Aqueduct and Sewer Authority in Dire Straits". *Caribbean Business, 13 May, 2005.*
Morgenstern, R.D., W. Pizer, and J. Shih. 2002. Jobs Versus the Environment: An Industry-Level Perspective. *Journal of Environmental Economics and Management* 43(1): 412–436.
Myers, N. 1996. Ultimate Security: The Environmental Basis of Political Security.
Myers, N., and J. Kent. 2001. Perverse Subsidies: How Misused Tax Dollars Harm the Environment and the Economy. Island Press. Washington, D.C.

Neumayer, E. 2004. Indicators of Sustainability. pp 139–188 in TheInternational Yearbook of Environmental and Resource Economics: 2004/2005. Tietenberg, T., and H. Folmer (eds.), Cheltenham, UK. Edward Elgar.
OECD – Organization for Economic Cooperation and Development. 2007. The Environmental Effects of Trade. http://www.oecd.org/document/8/0,3343,en_2649_34183_36629256_1_1_1_1,00.html.
Perkins, S. 2004. Paved Paradise? *Science News 166: 152–143.*
Pigou, A. C. 1932. The Economics of Welfare. London: Macmillan.
Prugh, T., R. Costanza, and J. H. Cumberland. 1999. National Capital and Human Economic Survival. CRC Press (ISEE / Lewis, ISBN:1566703980). Also available online via Google Books at: http://books.google.com/books?id=Uy0cazmTTiAC
Quinn, J. S., and C. Somers. 2004. Particulate Air Pollution and Inheritable Mutations in Mice: Possible Health Effects? *Discovery Medicine l 4 (22), 1390143.*
Repetto, R., and D. Austin. 2000. Pure Profit: The Financial Implications of Environmental Performance. World Resources Institute. Washington, D.C.
Rodriguez, M. 2005. Adding Variety to the Grid: Industrial Solutions for the Coming Years. *Industriales Magazine, March 2005.*
Serageldin, I. 2002. World Poverty and Hunger – The Challenge for Science. *Science 296:54–58.* data from the UN World Development Programme.
The Economist. 2006. Trouble on Welfare Island – Puerto Rico. *The Economist, May 25, 2006.*
Tietenberg, T. 2006. Environmental and Natural Resource Economics. 7th. Addison Wesley, Boston.
United Nations Population Reference Bureau. 2003. Report on population growth of s4electes countries. Available online.
UNESCO – United Nations Educational, Scientific and Cultural Organization. 2003. The United Nations World Water Development Report. Paris, France. http://www.unesco.org/water/wwap/wwdr/index.shtml.
USGCRP – United States Global Climate Change Research Program (2001, 2004), Regional Paper: US-Affiliated Islands of the Pacific and Caribbean. US National Assessment of the Potential Consequences of Climate Variability and Change. Washington, D.C. http://www.usgcrp.gov/usgcrp/nacc/education/islands/default.htm.
Veenhoven, R. 2004. A good time for Gross National Happiness. Proceedings of the Thimphu Conference.
von Weiczacker, E, A. B. Lovins, and L. H. Lovins. 1998. Factor Four: Doubling Wealth – Halving Resource Use: A Report to the Club of Rome. Publisher: Kogan Page (July 15, 1998). ISBN: 1853834068.
WBCSD – World Business Council for Sustainable Development. 2004-2005. Linking Development and Business. Available online at: (http://www.wbcsd.ch/plugins/DocSearch/details.asp?type=DocDet&DocId=3083). Acessed 25 March, 2005.
Wolfensohn, J. D. 2004. Making Growth Green. Environment Matters magazine. Available online at http://www.greenbiz.com. Accessed 25 March, 2005.
World Bank. 2004. Classification of Economies. Global Economic Prospects. Available online at: http://siteresources.worldbank.org/INTRGEP2004/Resources/classification.pdf.
WWI – World Watch Institute. 2002–2007. State of the World. (Annual series) World Watch Institute.

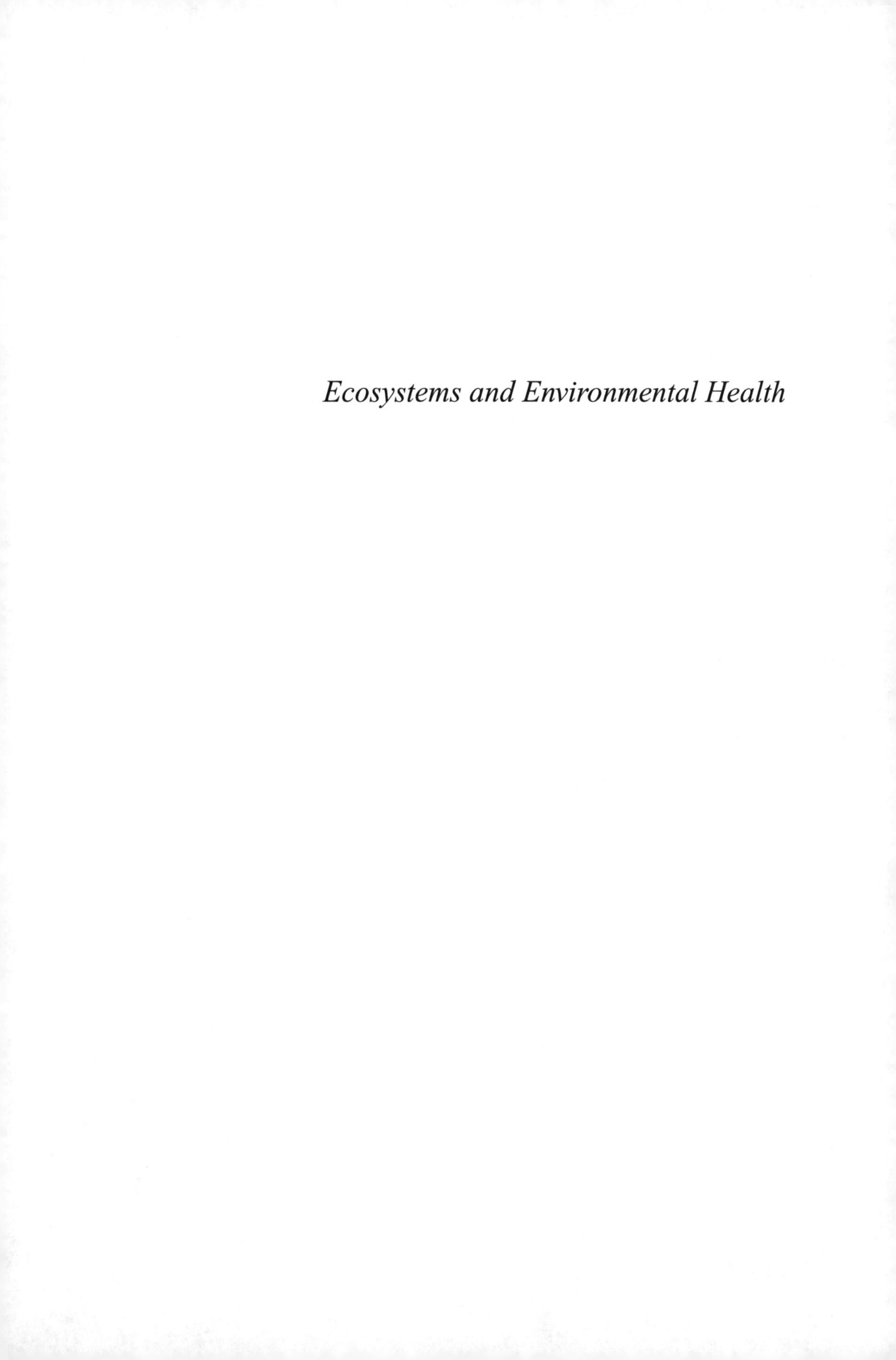

Ecosystems and Environmental Health

Human Activities and Ecosystem Health

Fuad A. Abdulla
New York Institute of Technology, Amman, Jordan

Mohammed H. Abu-Dieyeh & Esam Qnais
The Hashemite University, Zarqa, Jordan

SUMMARY: Maximizing the world's energy and agricultural capacities in the face of growing human population is the most critical task facing mankind and the environment. The aim of this chapter is to present information on how exponential growth impact global ecosystems and how human activities increase the presence of harmful pests and pathogens, magnify the effect of toxic chemicals, change the biotic and abiotic components of ecosystem cycling as well as change the worldwide climate. Although green solutions are available for a deteriorated environment, an overall healthy and sustainable solution is difficult to achieve.

1 INTRODUCTION

Mankind is living in ecosystems that could be recognized at many different spatial scales, ranging from local units such as a small forest to the entire globe (Townsend *et al.*, 2003). Natural ecological systems are dynamically stabilized based on balanced inputs and outputs. All ecological systems are controlled by the same processes including natural and anthropogenic (human caused) disturbances, however with increasing population, human activities create more disturbances and add more unbalanced situations for ecosystems. Humans are among the most successful living things on the earth. The present rate of growth in size of the global human population is unsustainably high. This stresses our future by further energy demands; a greater drain on non-renewable resources and more pressure on renewable resources. Therefore, worldwide national strategic plans contain goals that directly or indirectly alter ecological systems and end with unexpected surprises even if the plan includes very well organized preliminary investigations. This is due to the behavior of ecosystems which is a result of interactions among many process, most of which are incompletely understood. The response of disturbed ecosystems, in consequence, includes local, regional and/or global problems to human health and environment.

The aim of this chapter is to discuss and assess atmospheric pollution, its consequences on humans and the environment, and perspectives toward its control; pesticide problems, their persistence in the environment, the direct effects on individual health and proposed alternatives; the link between environmental pollution and diseases; toxological aspects of water pollution; and food borne diseases and protective measures.

2 ATMOSPHERIC POLLUTION

Through human history, the principal sources of energy have evolved from using wood to coal to oil to nuclear fuel. Both coal and oil are exhaustible resources as well as pollutants. From the mid 19th to the mid 20th century, the burning of fossil fuels together with deforestation supplied about 9×10^{10} metric tons of carbon dioxide to the atmosphere. A further 9×10^{10} metric tons has been added since 1950 (Townsend *et al.*, 2003).

The atmosphere contains the earth's biggest pool of nitrogen (78.08%), large supplies of oxygen (20.95%) and a very small fraction of carbon dioxide (0.03%). In addition to that argon (0.93%) and traces of hydrogen, methane and certain inert gases are also present in the atmosphere. The 0.03% of carbon dioxide represents the source of carbon used by all photosynthetic organisms.

The concentration of carbon dioxide has increased from about 280 PPM (parts per million) to about 370 PPM and it is projected to continue to rise to 700 PPM by the year 2100 (Townsend *et al.*, 2003). In consequence, it is estimated that the present air temperature at the land surface is about 0.6°C warmer than in pre-industrial times. This is predicted to continue to rise by a further 1.4–5.8°C by 2100 (IPCC, 2001).

Rain water has a pH of about 5.6 but pollutants like sulphur dioxide (SO_2) and oxides of nitrogen (NO_x) interact with water and oxygen to form acid rain by lowering the pH to less than 5.

2.1 *Greenhouse effect and global warming*

The major causes of CO_2 increase have been fossil fuels and the clearing and burning of tropical forest to make way for agriculture or timber production. The replaced communities like crops, grasslands and secondary succession organisms are all with lower biomass than forests and so recapturing of CO_2 by photosynthesis is not effectual to surmount the increase in CO_2. While the increase of CO_2 in tropics and subtropics was mainly due to land use changes, in the northern hemisphere it was due to combustion of fossil fuel. In 1980, it was estimated that the atmosphere of North America contained 1481×10^6 tons carbon per year due to fuel combustion but only 112 t c per year was due to deforestation (UNEP, 1993). The United States, Australia and Canada emit the most carbon which is close to more than 3 tons of carbon per head of population per year (Schipper *et al.*, 1993).

Recent analyses indicated that terrestrial vegetation has been fertilized by the increased atmospheric CO_2 so that a substantial amount of additional carbon has become fixed in vegetation biomass (Kicklighter *et al.*, 1999). However, because C_3 plants are more limited than C_4 plants by CO_2 availability, one consequence of increasing global CO_2 may be the spread of C_3 species into terrestrial habitats (Campbell and Reece, 2005).

During the day solar radiation warms up the earth's surface, which then radiates energy to the atmosphere mainly as infrared radiation. Carbon dioxide and water vapour absorb much of the reflected radiation, reflecting some of it back toward earth surface. This is a natural phenomenon responsible for warming the atmosphere. If this phenomenon did not exist, the temperature on earth's surface would be −18°C.

The industrial revolution elevated CO_2 concentrations and other pollutant gases like nitrous oxide, methane, ozone (O_3), and chlorofluorocarbons (CFCs) all of which absorb infrared radiation. Like the glass on a greenhouse, these gases prevent the radiation from escaping and so keep the temperature high. Several studies proposed that by the end of the 21st century, the CO_2 concentration will double in the atmosphere and the average global temperature will increase by about 2°C. However, an increase of only 1.3°C would make the world warmer than at any time in the past 100,000 years. Such changes will probably lead to melting of polar ice and consequently raising the sea level by an estimated 100 m, gradually flooding inland areas about 150 km or more from the coastline. Global warming as a problem of uncertain consequences is now, apparently, under way due to the non-stop addition of CO_2 to the atmosphere and unfortunately no simple solution.

2.2 *Acid rain and depletion of atmospheric ozone*

Sulfur oxides and oxides of nitrogen are atmospheric pollutants released from burning of wood, coal and other fossil fuels mainly in electrical generating plants. These gases react with water in the atmosphere forming sulphuric and nitric acid, respectively. The acid falls to the earth's surface as precipitation of rain, snow, sleet, or fog that has a pH less than the normal value of 5.6 for rain water. Acid rain acidifies the water in aquatic ecosystems especially when the underlying soil or rock does not contain buffering compounds such as bicarbonate to neutralize the acidity.

Low pH is critical to life of many delicate organisms but the indirect effect could be more harmful. For example at a pH below 4–4.5 the concentration of aluminium (Al^{+3}), iron (Fe^{+3}) and manganese (Mn^{+2}) becomes toxic to delicate tissues such as gills of fishes. Fish populations have declined in thousands in lakes of Norway and Sweden where the pH has dropped below 5 (Campbell and Reece 2005). More acidity in an already acidic lake could be lethal for all living organisms. In terrestrial ecosystems, such as the deciduous forests of New England, the change in soil pH due to acid rain causes calcium and other nutrients to leach from the soil and so limit the growth of trees and other plants. Acid rain is a regional and even global problem rather than a local one. The pollutants may drift hundreds of kilometers before falling as acid rain. Lake dwelling organisms in eastern Canada were dying because of air pollution from factories in the Midwestern United States (Townsend *et al.*, 2003).

In another problem area, ozone, the lower part of stratosphere, protects life on earth by absorbing most of the ultraviolet (UV) radiation. Evidence from satellite studies suggest that the ozone layer has been depleted with time since 1975. Ozone thinning is most apparent over Antarctica where cold winter facilitates the reaction of ozone (O_3) degradation to oxygen (O_2). An ozone hole was first described in 1985 over Antarctica and its size has generally increased in recent years and extends as far as southern most portions of Australia, New Zealand, and South America (Campbell and Reece 2005).

Scientists refer the degradation of ozone layer to accumulation of chlorofluorocarbons (CFCs) chemicals developed as propellants in aerosols and refrigerants which have been used on a very large international scale. In the stratosphere these chemicals react with ozone, reducing it to molecular O_2. Subsequent series of reactions end with chlorine which has the ability to react with other ozone molecules in a catalytic chain reaction. There is also evidence that nitric oxide produced by supersonic aircraft might contribute massively to ozone depletion (Townsend *et al.*, 2003).

The consequences due to ozone depletion may be unpredictable but scientists expect increases in skin cancer for humans, negative impacts on phytoplankton of aquatic ecosystems and certain delicate terrestrial organisms. Accumulation of ozone in the lower atmosphere due to oxidation of pollutants like carbon monoxide, industrial hydrocarbons, NO_x gases and methane, can be toxic to plants, contribute to smog and work as a greenhouse gas (Townsend *et al.*, 2003).

2.3 *Air pollution and health problems*

Concerns are growing over the health threat posed by extremely harmful airborne toxic chemicals produced by automobiles and various industrial processes. The USA Environmental Protection Agency (EPA) announced that 1.2 billion kilograms of hazardous pollutants were released into the air by industries in 1987, including 107 million kilograms of carcinogens which can cause cancers and birth and genetic defects. However according to later reports this number is underestimate, as it omits sources such as waste dumps, dry cleaners, cars and chemicals releases into soil or water and ends up in the air by evaporation and the true number may be exceeding 2.2 billion kilograms annually (French, 1990).

In the above mentioned EPA report the agency estimates that emissions of toxic materials into the air causes around 2,000 cancer deaths a year with higher rates in communities near factories and an incidence of respiratory cancer of about 20% above the USA national average (French, 1990). Unfortunately specific data is not available in developing countries but the airborne toxic substances are likely to rise rapidly as industrialization continues without restrict measures of control. However airborne toxic chemicals can be carried great distances before falling to the ground and the etiology for the effect could be a distant country.

Health effects caused by high levels of carbon monoxides include impaired perception and thinking, slow reflexes, drowsiness, unconsciousness, death and impaired growth and mental development of fetus if inhaled by pregnant women. Nitrogen dioxides irritate the lungs, cause bronchitis and pneumonia and increase susceptibility to viral and microbial respiratory diseases (Worldwatch Institute, 1985).

Suspended particulate matter in atmosphere due to natural dust and the black, particulate-laden smoke spewed out by diesel-fuelled vehicles can cause serious respiratory problems especially in the developing world. Thirty seven out of the 41 cities monitored for particulates averaged either borderline or excessive levels than the WHO standard. In Kuwait, New Delhi and Bejing, the level exceeds the WHO standard by as much as five times and this excessive level could extend for 294 days of the year in New Delhi (French, 1990). It was estimated that air pollution costs the United States as much as $40 billion annually in health care and lost productivity (Cannon, 1985).

Air pollution can cause serious health problems. Ninety four percent (94%) of all respiratory ailments are caused by polluted air according to the American Medical Association, which also reported that one-third of American people national health bill is for causes directly attributable to indoor air pollution. The term "sick building syndrome" (SBS) is used to describe situations in which building occupants experience acute health and comfort effects that appear to be linked to time spent in a building, but no specific illness or cause can be identified. The complaints may be localized in a particular room or zone, or may be widespread throughout the building. In contrast, the term "building related illness" (BRI) is used when symptoms of diagnosable illness are identified and can be attributed directly to airborne building contaminants.

Most people are aware that outdoor air pollution can damage their health but may not know that indoor air pollution can also have significant effects. In general, studies of human exposure to air pollutants indicate that indoor air levels of many pollutants may be 2–5 times higher than outdoor levels. The cause behind this is that most people spend as much as 90% of their time indoors. In recent years, scientific studies have consistently ranked indoor air pollution among the top five environmental risks to public health (U.S. Environmental Protection Agency, 1991). Indicators of SBS include:

- Building occupants complain of symptoms associated with acute discomfort, e.g., headache;
- Eye, nose, or throat irritation;
- Dry cough; dry or itchy skin;
- Dizziness and nausea;
- Difficulty in concentrating;
- Fatigue; and
- Sensitivity to odors.

Most of the complainants report relief soon after leaving the building.

The causes of SBS could be lack of adequate ventilation or chemical contaminants from indoor or outdoor sources, or biological contaminants. The latter includes bacteria, molds, pollen and viruses. This kind of science is known as "Aerobiology". These contaminants may be enriched in stagnant water that has accumulated in ducts, humidifiers and drain pans, or where water has collected on ceiling tiles, carpeting, or insulation. Sometimes insects or bird droppings can be a source of biological contaminants.

Physical symptoms related to biological contamination include cough, chest tightness, fever, chills, muscle aches, and allergic responses such as mucous membrane irritation and upper respiratory congestion. Toxic mold inhalation has been found to result in allergenic reactions, serious breathing difficulties, memory and hearing loss, dizziness, flu-like symptoms, and bleeding in the lungs (Burge *et al.*, 1991). One of these devastating effects had already happened in the Cleveland Outbreak wherein many infants suffeedr from Pulmonary Hemosiderosis-a severe bleeding of the lungs caused by inhalation of toxic fungus *Stachybotrys* (Etzel *et al.*, 1998).

2.4 *Air pollution control*

Stabilizing emissions of air pollutants will require demanding international effort and the acceptance of changes in both personal lifestyles and industrial processes. Concerns are increasing about climate change and human health risks from smog and other pollutants. These concerns are driving the demand for emission control technologies and alternative fuels to reduce air pollution. The present approaches of air pollution control have tended to be technological to reduce the effect rather than efforts to address the causes.

Technologies have been developed and marketed to address three main issues: industrial stack emissions, transportation emissions and indoor air quality. The global market for air pollution prevention and control exceeds US$50 billion annually. Demand is growing for emissions control, fuel switching and process engineering technologies. Canada exports over $300 million annually in air pollution control services and products, which represent almost one third of all Canadian environmental international sales (Environment Canada, 2008). The industrial technologies include: electrostatic precipitators and baghouse filters to control particulate emissions from power plants; efficient scrubbers and flue gas desulphurization technology mainly in coal-burning power plants which can remove as much as 95% of SO_2 emissions; Control systems for nitrogen oxides and volatile organic compounds which modify the combustion process and reduce emissions of nitrogen gases by 30–50%.

The above technologies provide necessary reductions but they are not the ultimate solutions since they can create problems of their own, such as the need to dispose of scrubber ash as a hazardous waste and they do little to reduce carbon gases.

Importantly, maximizing energy efficiency and seeking for less polluted or pollution free energy alternatives are healthier solutions for dropping emissions. Reducing energy consumption through more electrically efficient house wares and lightening; efficient car engines, intelligent transport system and reducing reliance on using automobiles in urban areas are all important to reduce energy consumption and cut gas emission.

Many options were opened recently to introduce vehicles powered by alternative fuels like methanol, ethanol, natural gas, hydrogen and electricity. However not one of these alternatives is still without concerns and/or problematic wastes. Methanol would help in reducing emissions related to ozone problem but its a highly poisonous gas and could be leaked from the storage tanks and contaminate ground water. When used to power vehicle it produces two to five times more formaldehyde, a carcinogen to human beings. Natural gas reduces hydrocarbon emissions 40–60% and carbon monoxides 50–95% but increases nitrogen oxides by 25% (French, 990).

Production of ethanol from living stocks (biodiesel) reduces carbon dioxide emissions by 63% compared with gasoline production. Ethanol is produced from grains like corn, soybean, sunflower, rapeseeds and canola. Its widespread adoption could result in a competition between food and fuel. As emphasis switched to production of natural oils for biodiesel, microalgae became the exclusive focus of the research, this is due to; much greater productivity than terrestrial crops, non-food resource is needed, non-productive land is used, can utilize saline water and can be in conjunction with waste water treatment, and can utilize waste CO2 streams (Tickell, 2006).

To reduce transportation emissions, automobiles should be provided with catalytic converters and cleaner, smaller but more efficient engines in accompany with intelligent transportation systems to enhance traffic flow. Alternative fuel vehicles that use propane, natural gas or Photovoltaic automobiles that use light-based energy should be seriously considered.

To enhance indoor air quality, highly-efficient dust collection and suppression system, air quality monitoring, analytical and management systems, chemical recovery and ventilation systems and alternative insulation technologies and products should be considered.

3 PESTICIDES AND PERSISTENT ORGANIC POLLUTANTS

The success of modern agriculture in feeding people all over the world would not have been possible without pesticides. Pesticides, by nature, are risky, but their benefits are real so if people want a total ban of pesticides, they must be ready to accept pests like weeds, insects and pathogens in their crops, gardens, parks and other amenities. Pesticides are poisons and can be hazardous not only because of accidental events or misuse. Some pesticides can harm the environment and non-targeted living things even though they are properly used (Delaplane, 2000).

Pesticide categorization is based on the targeted pest so they could be classified as: fungicides, algicides, herbicides, acaricides, insecticides, nematicides, avicides, rodenticides, to control targeted pests of fungi, algae, weeds, arachnids, insects, nematodes, birds, and rodents, respectively. The most widely pollutant pesticides are those used to control pests and weeds that damage agricultural and horticultural crops, forestry and human livestock with herbicides being the most. In the United States, the estimated average annual loss caused by weeds in 46 crops grown in 1991 was \$4.1 billion and this value would have been \$19.6 billion if herbicides had not been used (Bridges, 1992). In Canada the estimated annual loss caused by weeds was \$984 million (Swanton *et al.*, 1993) and so the herbicides comprised 85% of total pesticides sale in 1998 (Floate *et al.*, 2002).

A second classification is by the type of chemical compound (Blanpied, 1984), which includes a complicated list of categories such as:

- Chlorinated hydrocarbons or organochlorines. These pesticides break down chemically very slowly and can remain in the environment for long periods of time. Dieldrin, chlordane, aldrin and helptachlor are pesticides of this type.
- Organic phosphates or organophosphates. These pesticides are highly toxic to humans but do not remain in the environment for long periods of time. Parathion, malathion, thimet and trichlorphone are pesticides of this type.
- Carbamates compounds. These pesticides are considered highly toxic to humans.

A third classification, which is recommended by the WHO, is based on acute risk to human health (that is the risk of single or multiple exposures over a relatively short period of time). The toxicity of the technical compound and its common formulation were also including in the risk. The classification is based primarily on the acute oral and dermal toxicity to the rat with a consideration to the physical state of the active ingredient being classified. Based on hazard, pesticides could be extremely, highly, moderately or slightly hazardous (World Health Organization, 2005).

A good pesticide classification should give information needed to evaluate harmful consequences of using certain pesticide and hence the following criteria should be considered: how poisonous they are to human health and the environment (toxicity), how long they last in the environment (persistence), their concentrations and number of application times, how they are used, and their container size.

3.1 *Pesticides and their implications for the environment*

While recognizing the tremendous benefits of pesticides in modern agriculture, public and environmental concerns have been raised to ban or restrict their use. In the early stages of pesticide evolution, manufacturers were not worried about pesticide specificity. A good example was the insecticide Dieldrin, applied in massive doses to control the Japanese beetle in Illinois farmlands from 1954 to 1958. The result was a disaster on domestic animals and wild life, 90 percents of the cats and a number of dogs were killed, cattle and sheep were poisoned and 19 species of birds suffered losses (Luckman & Decker, 1960).

Problems of non-target effects arise when pesticides are toxic to living organisms other than the target pest and particularly when they drift beyond the application areas and persist in the environment beyond the application time (Townsend *et al.*, 2003). Sometimes the pest population may explode after a pesticide application. This may occur due to the ability of a pesticide to kill not only the target pest but also large numbers of its natural enemies. The surviving pest populations then flourish due to a plentiful food resource and suppressed or eradicated natural enemies. Moreover a secondary pest may evolve in certain areas due to the eradication of a primary pest.

In 1950 cotton crops of Central America were suffering from insect pests, the Alabama leafworm and the boll weevil (Smith, 1998). Organochlorine and organophosphate were applied up to five times per year. Consequently, by 1955, three secondary pests evolved and forced an increase in application times to 8–10 per year. Moreover the secondary pests had increased to eight and required about 28 applications per year. In another example, applying a herbicide mix of 2,4-D, Dicamba and Mecoprop in turfgrass to kill the dominant weed (common dandelion) for two years as two applications per year led to a weed shift hence other species, common mallow and yellow woodsorrel started to fill the studied plots instead of dandelion (Abu-Dieyeh & Watson, 2007b).

In a natural selection process, a pest may develop a degree of resistance to one or more pesticides. If there is a repeated application of a pesticide, a large proportion of the resistant pest population will evolve, taking into consideration the high intrinsic rate of reproduction of pests. Therefore the resistance will spread very rapidly in a population.

The number of pesticide resistance species have increased from very few cases in the period of 1940–1950 to 500 species of insects and mites and about 100 species of weeds in 1990 (Gould, 1991). Four populations of wild oats *Avena fauta* in intensive cropping systems of wheat, barley and canola in Alberta (Canada) were found to be completely resistant to all herbicides registered for use in wheat (Beckie *et al.*, 1999). Indeed the large scale and repeated application of non-specific pesticides raises other concerns like the transfer of resistance genes to wild relatives and pest shift toward more tolerant species.

Applying long persistence pesticides will introduce chemical residues into the environment with undetermined consequences (Alikhanidi and Takahashi, 2004). Chlorinated hydrocarbon pesticides such as DDT (used to kill insect pests after World War II) persist in the environment and are transported by water far from the areas of application. Consequently it becomes a global problem. DDT is responsible for the declining population of birds that feed at the top of the food web like pelicans and eagles. Scientists found that DDT accumulation in tissues of these birds interfered with the deposition of calcium, producing weak egg shells. Therefore the reproduction rate of these birds declined due to the inability of the egg shells to support incubation by parents. DDT was banned in 1972 in the United States and many other parts of the world.

Chlorinated hydrocarbon pesticides accumulate in specific animal tissues, particularly, fat and become biomagnified in successive trophic levels of food chains. Top predators in aquatic and terrestrial food chains can then accumulate extraordinarily high doses of pesticides. Polychlorinated biphenyls (PCBs) can accumulate in herring gull eggs, at the top of the food web, concentrations up to 5,000 times greater than that found in phytoplankton at the base of the food web (Campbell & Reece, 2005). PCBs were biomagnified in the livers of whales and dolphins up to 1000 times greater than the concentrations found in the primary consumer invertebrates.

Pesticide residues can be translocated to surface and ground water via rainfall, agricultural canals, streams and rivers thus affecting phytoplanktons and delicate living organisms and also to human health. Herbicides such as 2,4-D, bromoxynil and dicamba were frequently present in rainfall at concentrations that may have adverse effects on sensitive plant species and on the quality of surface water in Alberta, Canada (Hill *et al.*, 1999). According to the Netherlands National Institute of Public Health and Environmental Protection (RIVM, 1992) groundwater is threatened by pesticides in all European states. Loading of a significant amount of agricultural pesticides ($\sim$190000 tons) plus additional loadings of non-agricultural pesticides that are released by riparian countries bordering the North Sea, eventually are transported into the North Sea by a combination of river, groundwater, and atmospheric processes.

Since the 1970s the increased rate of disease, deformities and tumours in commercial fish species were reported in highly polluted areas of the North Sea and coastal waters of the United Kingdom. These symptoms are consistent with effects known to be caused by exposure to pesticides (WWF, 1993).

3.2 *Pesticides and health*

Effects on organisms are usually considered to be an early warning indicator of potential human health impacts. Different pesticides have markedly different effects on aquatic life which makes generalization very difficult. Many of these effects are gradual and chronic and include the following: death of the organism; cancers, tumours and lesions on fish and animals; reproductive inhibition or failure; suppression of immune system; disruption of endocrine (hormonal) system; cellular and DNA damage; teratogenic effects (physical deformities such as hooked beaks on birds); poor fish health marked by low red to white blood cell ratio, excessive slime on fish scales and gills, etc; and egg shell thinning (Ongley, 1996). These effects are not necessarily caused solely by exposure to pesticides or other organic contaminants, but may be associated with a combination of environmental stresses such as eutrophication and pathogens.

UNEP (1993) linked the effects of pesticides to the level of oncological (cancer), pulmonary and haematological morbidity, as well as on inborn deformities and immune system deficiencies. Human health effects are caused by skin contact through handling of pesticide products, inhalation through breathing of dust or spray, ingestion through consumption as a contaminant on and/or in food or in water. Farm workers have higher risks due to inhalation and skin contact during preparation and application of pesticides to crops. However, ingestion of food that is contaminated by pesticides is the principal route for the majority of the population.

Degradation of water quality by pesticide runoff has two principal human health impacts. The first is the consumption of contaminated fish and shellfish; especially those that are collected from areas around agricultural systems. The second is the direct consumption of pesticide-contaminated water. Many health and environmental protection agencies have established acceptable daily intake (ADI) values which indicate the maximum allowable daily ingestion over a person's lifetime without appreciable risk to the individual.

3.3 *Pesticide alternatives*

The USA, Canada and European countries have adopted a variety of measures to reduce the abuse of pesticides (FAO/ECE, 1991; Floate *et al.*, 2002): reduction in use of pesticides (by up to 50% in some countries); reviewing already registered pesticides to match the new standard that reasonable certainty that no harm will result from aggregate exposure to each pesticide from dietary and other sources; banning on certain active ingredients; revising pesticide registration criteria; training and licensing of individuals that apply pesticides; reduction of dose and improved scheduling of pesticide application; testing and approval of spraying apparatus; limitations on aerial spraying; and promoting the use of mechanical and biological alternatives to pesticides.

Progress in cultural methods of pest control, mainly weeds, has included the use of suppressing cover crops and the identification of specific crop traits for weed inhibition, sophisticated machine guidance and weed detection technology, and steaming and thermal techniques have all helped to maintain weed populations at manageable levels (Bond and Grundy, 2001).

Environmental and public concerns about the use of chemical herbicides and the increasing number of weeds that have developed resistance to one or more herbicides have led to development of new herbicides with relatively short half-lives, low toxicity to non-target organisms and with different mechanism of actions. The use of biological control agents (natural enemies and pathogens) or biologically-based products (allelochemicals) have also been implemented.

Allelochemicals are usually considered to be secondary metabolites or waste products of the main metabolic pathway in plants, released into the environment by various mechanisms: exudation from roots, volatilization from leaves, and leaching from leaves and plant litter on the ground by precipitation (Wu *et al.*, 1999). The first approach in studying allelopathy is to screen accessions of allelopathic crops for their ability to reduce weeds (Kebede, 1994). Fifty two Chilean cultivars of *Triticum aestivum* and *T. durum* were screened for the production of the allelochemical DIMBOA (2, 4- dihydroxy-7-methoxy-1, 4-benzoxazin-3-one) which was found to vary among cultivars from 1.4–10.9 mmol kg^{-1} fresh weight (Copaja *et al.*, 1991).

Significant progress in isolating rice allelochemicals has been achieved; 12.000 rice accessions against common rice weeds have been evaluated (Dilday *et al.*, 2001). Progress has been achieved in isolation and evaluation of rice allelochemicals under field conditions. The ability of these chemicals to suppress both monocot and dicot weeds was observed (Olofsdotter, 2001).

The second approach of allelopathy is to discover herbicides based on new natural products. Some of these are exploited commercially (triketone, cinmethylin, bialaphos, glufosinate and dicamba) (Bhowmik and Inderjit, 2003). Most of these products are at least partially water-soluble, active at low concentrations (as a result of natural selection) and mostly produced from microbial sources rather than higher plants (Vyvyan, 2002).

The third approach is the use of allelopathic activity in various cropping systems such as: mixed cropping, cover cropping, multiple cropping, crop rotations and minimum and no-tillage systems as part of an integrated pest management.

Biocontrol uses natural enemies to control insect, pathogen and weed pests. Recently, the definition of biocontrol has been broadened to the use of natural or modified organisms, gene or gene products to reduce the effect of undesirables organisms (pests) and to favour desirable organisms such as crops, trees, animals and beneficial insects and microorganisms (Cook 1987). Biocontrol involves one or more natural processes (e.g. antibiosis, parasitism, competition, predation and induced host resistance) that are influenced by abiotic and biotic factors from the surrounding environment. Biocontrol is successful when the biotic components and the environment interact in such a manner that pest control or suppression occurs (Kennedy & Kremer 1996). Given the high costs of such programs, the success rate should be maximized. This task cannot be achieved without understanding the ecology of the components of the host: pathogen system (Cousens and Croft 2000).

The term bioherbicide refers to the inundative application of microbial (pathogens) formulations to pests (mainly weeds). Currently, five fungi and one bacterium are registered and formulated as bioherbicides for weed control (Charudattan & Dinoor 2000). DeVine® (*Phytophthora palmivora*) is used to control *Morrenia odorata* (milkweed vine) in citrus fields in Florida. Collego® (*Colletotrichum gloeosporioides* f.sp. *aeschynomene*) is used to control *Aeschynomene virginica* (northern jointvetch) in Arkansas, Mississippi and Louisiana. BioMal® (*Colletotrichum gloeosporioides* f.sp. *malvae*) was registered in Canada for control of *Malva pusilla* (round-leaved mallow), but has not been commercialized due to production problems. Dr. BioSedge® was formulated based on the rust fungus *Puccinia canliculata* and registered in the United States to control *Cyperus esculentus* (yellow nutsedge). Stumpout® is a stump-treatment product based on the wood-infecting basidiomycetes, *Cylindrobasidium laeve* is registered in South Africa to control resprouting of cut trees in natural and trees plantation areas. CAMPERICO® an isolate of a wilt-inducing bacterium, *Xanthomonas campestris* pv. *poae*, is registered in Japan for the control of annual bluegrass in golf courses (Charudattan and Dinoor 2000). *Sclerotinia minor* (asporogenic ascomycete) has been registered to control broadleaf weeds in turfgrass systems after being extensively evaluated under different biotic and abiotic interactions (Abu-Dieyeh and Watson, 2005, 2007a and b). Although research is continuous on many potential bioherbicides, problems with mass-production, formulation and commercialization continue to prevent their implementation (Cook 1996; Charudattan and Dinoor 2000).

Integrated pest management (IPM) is an ecologically based management that involves combinations of several control methods including physical, cultural, chemical (in lower rates), biological and crop resistant varieties.

4 ENVIRONMENTAL POLLUTION AND DISEASE

It is often difficult to establish a direct relationship between environmental pollution and disease. The relationship is fairly clear for certain pollutants, such as the link between radon and lung cancer, or between lead and disorders of the nervous system. However, the evidence is less definite for many pollutants, and scientists can only suggest there is an association between the pollutant and a specific illness. One reason it is so difficult to establish a direct cause and effect is that other factors—such as a person's genetic makeup, diet, level of exercise, and whether he smokes—complicate the picture. These factors are often difficult to quantify. Other complications include the fact that certain segments of society (such as children, elderly people, people with chronic diseases and people living in poor neighbourhoods) may be more susceptible to bad health effects from an environmental pollutant (Khopkar, 2005; Belanger *et al.*, 2002; Cox, 1999).

Environmental pollutants that are dangerous to both the environment and human health can be classified into two categories; First toxic chemicals that persist and accumulate in the environment and magnify their concentration in the food web. These substances include certain pesticides (such as DDT, or dichlorodiphenyltrichloroethane), radioactive isotopes, heavy metals (such as lead and mercury), flame retardants (for example, PBDEs, or polybrominated diphenyl ethers), and industrial chemicals (such as dioxins and PCBs, or polychlorinated biphenyls) (Harrison, 2001; Ware, 2004). The effects of the pesticide DDT on many bird species first demonstrated the problems with these chemicals. Falcons, pelicans, bald eagles, ospreys, and many other birds are very sensitive to traces of DDT in their tissues. A substantial body of scientific evidence indicates that one of the effects of DDT on these birds is that they lay eggs with extremely thin, fragile shells that usually break during incubation, causing the chicks' deaths.

The impact of DDT on birds is the result of three characteristics of DDT: its persistence, bioaccumulation, and biological magnification. Some pesticides, particularly chlorinated hydrocarbons such as DDT take many years to be broken down into less toxic forms. When a pesticide is not metabolized (broken down) or excreted by an organism, it is simply stored, usually in fatty tissues. Over time, the organism may bioaccumulate,or bioconcentrate, high concentrations of the pesticide. Organisms at higher levels on food webs tend to have greater concentrations of bioaccumulated pesticide stored in their bodies than those lower on food webs. This increase as the pesticide passes through successive levels of the food web is known as biological magnification or biological amplification (Ware, 2004).

Second, pollutants may affect the body's endocrine system (which produces hormones to regulate many aspects of body function). Hormones are chemical messengers produced by organisms in minute quantities to regulate their growth, reproduction, and other important biological functions. Many endocrine disrupters (such as DDT, phthalates, heavy metals) appear to alter reproductive development in malesand females of various animal species by mimic the estrogens, a class of female sex hormones. Accumulating evidence indicates that fishes, frogs, birds, reptiles such as turtles and alligators, mammals such as polar bears and otters, and other animals exposed to these environmental pollutants exhibit reproductive disorders and are often left sterile (Ware, 2004; Guillette Jr and Crain, 2003).

The effects of toxicants or toxic chemicals following exposure can be immediate (acute toxicity) or prolonged (chronic toxicity). Acute toxicity, which ranges from dizziness and nausea to death, occurs immediately to within several days following a single exposure. In comparison, chronic toxicity generally produces damage to vital organs, such as the kidneys or liver, following a long-term, low-level exposure to chemicals. Toxicity can be measured by the dose at which adverse effects are produced. A dose of a toxicant is the amount that enters the body of an exposed organism. The response is the type and amount of damage that exposure to a particular dose causes. A dose may cause death (lethal dose) or may cause harm but not death (sublethal dose). Lethal doses, usually expressed in milligrams of toxicant per kilogram of body weight, vary depending on the organism's age, sex, health, metabolism, genetic makeup, and how the dose was administered (all at once or over a period of time). Lethal doses in humans are known for many toxicants because of records of homicides and accidental poisonings.

One way to determine acute toxicity is to administer various doses to populations of laboratory animals, measure the responses, and use the data to predict the chemical effects on humans. The dose lethal to 50% of a population of test animals is the lethal dose-50%, or LD50. It is usually reported in milligrams of chemical toxicant per kilogram of body weight (Hayes, 2007). There is an inverse relationship between the LD50 and the acute toxicity of a chemical: The smaller the LD50, the more toxic the chemical and, conversely, the greater the LD50, the less toxic the chemical. The LD50 is determined for all new synthetic chemicals as a way of estimating their toxic potential. It is generally assumed that a chemical with a low LD50 for several species of test animals is toxic in humans (Hayes, 2007).

5 WATER POLLUTION

Over two thirds of Earth's surface is covered by water; less than a third is taken up by land. As Earth's population continues to grow, people are putting ever-increasing pressure on the planet's water resources. In a sense, our oceans, rivers, and other inland waters are being squeezed by human activities—not so they take up less room, but so their quality is reduced. Poorer water quality means water pollution (Eckenfelder, 1989; Vigil, 2003; Ongley, 1996).

Water pollution can be defined in many ways. Usually, it means one or more substances have built up in water to such an extent that they cause problems for animals or people (Viessman, 2005). Oceans, lakes, rivers, and other inland waters can naturally clean up a certain amount of pollution by dispersing it harmlessly. If you poured a cup of black ink into a river, the ink would quickly disappear into the river's much larger volume of clean water. The ink would still be there in the river, but in such a low concentration that you would not be able to see it. At such low levels, the chemicals in the ink probably would not present any real problem. However, if you poured gallons of ink into a river every few seconds through a pipe, the river would quickly turn black. The chemicals in the ink could very quickly have an effect on the quality of the water. This, in turn, could affect the health of all the plants, animals, and humans whose lives depend on the river. Thus, water pollution is all about quantities: how much of a polluting substance is released and how big a volume of water it is released into (Viessman, 2005). A small quantity of a toxic chemical may have little impact if it is spilled into the ocean from a ship. But the same amount of the same chemical can have a much bigger impact pumped into a lake or river, where there is less clean water to disperse it.

Water resources like oceans, lakes, and rivers are called surface waters. The most obvious type of water pollution affects surface waters. For example, a spill from an oil tanker creates an oil slick that can affect a vast area of the ocean. Not all of earth's water sits on its surface, however. A great deal of water is held in underground rock structures known as aquifers, which we cannot see and seldom think about. Water stored underground in aquifers is known as groundwater. Aquifers feed our rivers and supply much of our drinking water. They too can become polluted, for example, when weed killers used in people's gardens drain into the ground. Groundwater pollution is much less obvious than surface-water pollution, but is no less of a problem (Vigil, 2003).

If pollution comes from a single location, such as a discharge pipe attached to a factory, it is known as point-source pollution. Other examples of point source pollution include an oil spill from a tanker, a discharge from a smoke stack (factory chimney), or someone pouring oil from their car down a drain.

A great deal of water pollution happens not from one single source but from many different scattered sources. This is called nonpoint-source pollution (Lull *et al.*, 1995). Sometimes pollution that enters the environment in one place has an effect hundreds or even thousands of miles away. This is known as transboundary pollution (Vigil, 2003, Rana, 2006). One example is the way radioactive waste travels through the oceans from nuclear reprocessing plants in England and France to nearby countries such as Ireland and Norway.

There are two main ways of measuring the quality of water. One is to take samples of the water and measure the concentrations of different chemicals that it contains. If the chemicals are dangerous or the concentrations are too great, we can regard the water as polluted. Measurements like this are known as chemical indicators of water quality (Harrison, 2001; Eckenfelder, 1989). Another way to measure water quality involves examining the fish, insects, and other invertebrates that the water will support. If many different types of creatures can live in a river, the quality is likely to be very good; if the river supports no fish life at all, the quality is obviously much poorer. Measurements like this are called biological indicators of water quality (Richardson, 1987, Rana, 2006).

Most water pollution does not begin in the water itself. Take the oceans: around 80 percent of ocean pollution enters our seas from the land. Virtually any human activity can have an effect on the quality of our water environment. When farmers fertilize the fields, the chemicals they use are gradually washed by rain into the groundwater or surface waters nearby. Sometimes the causes of water pollution are quite surprising. Chemicals released by smokestacks (chimneys) can enter the atmosphere and then fall back to earth as rain, entering seas, rivers, and lakes and causing water pollution. Water pollution has many different causes such as sewage, nutrients, waste water, chemical waste, radioactive waste, oil pollution, plastics and invasive species (are animals or plants from one region that have been introduced into a different ecosystem where they do not belong) (Harrison, 2001, Vigil, 2003).

There is no easy way to solve water pollution; if there were, it wouldn't be so much of a problem. Broadly speaking, there are three different things that can help to tackle the problem, education, laws, and economics, and they work together as a team.

One of the biggest problems with water pollution is its transboundary nature. Many rivers cross countries, while seas span whole continents. Pollution discharged by factories in one country with poor environmental standards can cause problems in neighbouring nations, even when they have tougher laws and higher standards. Environmental laws can make it tougher for people to pollute, but to be really effective they have to operate across national and international borders. This is why we have international laws governing the oceans, such as the 1982 UN Convention on the Law of the Sea (signed by over 120 nations), the 1972 London Dumping Convention, the 1978 MARPOL International Convention for the Prevention of Pollution from Ships, and the 1998 OSPAR Convention for the Protection of the Marine Environment of the North East Atlantic. The European Union has water-protection laws (known as directives) that apply to all of its member states. They include the 1976 Bathing Water Directive, which seeks to ensure the quality of the waters that people use for recreation. Most countries also have their own water pollution laws. In the United States, for example, there is the 1972 Water Pollution Control Act and the 1974 Safe Drinking Water Act. (Chaterjee, 2002).

Most environmental experts agree that the best way to tackle pollution is through something called the polluter pays principle. This means that whoever causes pollution should have to pay to clean it up, one way or another. Polluter pays can operate in all kinds of ways. It could mean that tanker owners should have to take out insurance that covers the cost of oil spill cleanups, for example. It could also mean that shoppers should have to pay for their plastic grocery bags, as is now common in Ireland, to encourage recycling and minimize waste. Or it could mean that factories that use rivers must have their water inlet pipes downstream of their effluent outflow pipes, so if they cause pollution they themselves are the first people to suffer. Ultimately, the polluter pays principle is designed to deter people from polluting by making it less expensive for them to behave in an environmentally responsible way (Eckenfelde, 1989).

6 FOOD BORNE DISEASES

The World Health Organization regards illness due to contaminated food as one of the most widespread health problems in the contemporary world. Food poisoning or food borne diseases may occur after the consumption of food containing toxins or organisms that multiply to cause disease (Mims, 2004). Several hundreds of food borne diseases have been described. Most of these diseases are infections, caused by a variety of bacteria, viruses, and parasites that can be food borne. Others are poisonings, caused by harmful toxins or chemicals that have contaminated the food, these different diseases have many different symptoms. Nausea, vomiting, abdominal cramps and diarrhea are common symptoms in many food borne diseases (Chief Medical Officer, 1992).

6.1 *Common food borne pathogenic microorganisms and natural toxins*

6.1.1 *Campylobacter*

Campylobacter is Gram-negative bacteria. Campylobacter is the most commonly identified food-borne bacterial infection encountered in the world (Adak *et al.*, 2005). There are several species of campylobacter (11 in total); however, most reported Campylobacter-related human illnesses are caused by the C. jejuni strain. It is most commonly transmitted by raw poultry, raw milk and water contaminated by animal feces.

Campylobacteriosis is an acute bacterial enteric disease ranging from asymptomatic to severe, with diarrhea, nausea, vomiting, fever and abdominal pain. It is the most commonly identified bacterial cause of diarrheal illness in the world. These bacteria live in the intestines of healthy birds, and most raw poultry meat has Campylobacter on it. Eating undercooked chicken or other food that has been contaminated with juices dripping from raw chicken is the most frequent source of this infection.

Symptoms usually last 2–5 days. Campylobacter is in itself a relatively harmless pathogen; however, it can cause post-infectious complications which could be serious. For example, about

0.1% of those infected with Campylobacter will develop Guillian Barrie Syndrome (an autoimmune disease which usually leads to complete paralysis of all body muscles) within three weeks (Allos, 1997). 5–10% of people affected with GBS die mainly due to respiratory failure (Allos, 1997). Reitter's syndrome, a form of reactive arthopathy, can also occur in up to 1% of campylobacteriosis patients (Murray and Baron, 2003.).

In general, campylobacteriosis is a self-limiting disease; i.e. no therapy is required (Mandal *et al.*, 1984). In elderly, infants and immunocompromised individuals more severe infections may develop, therefore antimicrobial agents may be needed.

6.1.2 *Salmonella*

Salmonella have been some of the most frequently reported etiological agents in fresh produce associated outbreaks of human infections in recent years (Adak *et al.*, 2005). It is more common in the very young (under 1 year) and the very old (older than 70 years).

Most infections are acquired by eating contaminated poultry, eggs or dairy products. It has been estimated that about 75% of all broiler chickens are contaminated with Salmonella during defeathering, slaughtering and evisceration, when feces splatter the skin (World Health Organization, 2002.). The Salmonella can also spread via contaminated hands, utensils or cutting boards. One important aspect of Salmonella is its ability to multiply in a wide variety of foods; therefore, it is important to be able to isolate the organisms even when present in very small numbers in the feces.

Salmonella infection usually leads to abdominal pain, diarrhea, mild fever, chills, headache, nausea and vomiting. The symptoms usually develop within 72 hours (but occasionally as long as 7 days) after infection. Salmonella infection may be serious for the elderly, infants and the immunocompromised, who may become extremely ill. Salmonella has been linked to reactive arthritis, a painful, chronic and potentially debilitating condition that causes joint inflammation and degeneration (Hill *et al.*, 2003).

In general, Salmonella infection is a self-limiting disease (presenting as acute gastroenteritis), and therapy is mainly replacing the lost fluid. No evidence to suggest that antibiotics can provide any clinical benefits in the case of Salmonella infection. However, antibiotic therapy is crucially needed for infants less than 3 months old with Salmonella gastroenteritis, and also in immunocompromised patients and patients with septicemia. Therapy should start as early as possible (Murray and Baron, 2003).

6.1.3 *Escherichia coli O157*

The genus Escherichia consists of five species, of which E. coli is the most common and clinically most important. It is one of the most dangerous pathogens. The incidence of E. coli O157 tends to fluctuate, reflecting the outbreak-specific nature of disease.

The main sources of E. coli O157 are ground beef, unpasteurised milk and juice, sprouts, lettuce and salami. Contact with cattle could also transmit the pathogen. Additionally, the organism is easily transmitted from person to person this fact explain it spread with ease in schools and highlight the importance of early detection and management (Pulz *et al.*, 2003).

E. coli O157 can cause a severe and bloody diarrhea and painful abdominal cramps; the fever is usually lacking or very mild. The symptoms may last up to 10 days.

In general, most people recover without the need for antibiotics. Fluid replacement therapy may be needed. In cases of the elderly, infants and the immunocompromised, complications may develop. E. coli infection has been associated with hemorrhagic colitis and hemolytic uraemic syndrome (Boyce *et al.*, 1995) which is a major reason for an acute loss of kidney function in childhood. Such complications may also include temporary anemia, profuse bleeding, and kidney failure.

6.1.4 *Clostridium perfringens*

Clostridium perfringens is often under-reported as a cause of bacterial food poisoning because it can exist as part of the normal gut flora in humans. Clostridium perfringens are an important cause of food poisoning and non-food-borne human gastrointestinal diseases, e.g., sporadic diarrhea and antibiotic-associated diarrhea (Miyamoto *et al.*, 2006).

Clostridium perfringens as food borne diseases pathogen is usually associated with food held at improper temperatures. Infection is most likely to occur when large quantities of food are prepared several hours before serving, such as in school cafeterias and nursing homes. Good food handling practices reduce the risk of the disease.

Acute gastroenteritis, watery diarrhea are common symptoms experienced in the first 24 hours after eating contaminated food.

The illness is usually self limiting, with the symptoms lasting less than 24 hours. However, there are individuals at higher risks like pregnant women, the very young, the elderly and the immuno-compromised (Meer *et al.*, 1997). Such people may experience prolonged or severe symptoms and may need hospitalization.

6.1.5 *Staphylococcus aureus*

The most common toxins implicated in staphylococcal food poisoning are sea to sej, which cause 95% of all outbreaks (Letertre *et al.*, 2003). The incidence of disease is underreported because most incidences are not reported.

Nausea, vomiting, diarrhea and abdominal cramping are the most common symptoms. The symptoms usually appear within hours after the consumption of the contaminated food. The severity of the symptoms depends on: amount of contaminated food eaten, amount of toxin ingested, individual's susceptibility to it. The symptoms are usually resolved with 48 hours (Hawker, 2005).

Food is usually contaminated during the handling stage after cooking. Many food types have been associated with staphylococcal food poisoning, including meat, eggs, bakery and dairy products. The symptoms are usually self limiting and does not require and specific antibiotics.

6.1.6 *Bacillus cereus*

Two types of illness have been attributed to the consumption of food contaminated with Bacillus; Namely cereus emetic and cereus diarrheal food poisoning syndromes.

Crude cereals, starchy food, dairy products, meat, dehydrated foods and spices are among the food types that are preferentially contaminated with Bacillus cereus. Bacillus cereus grows well after cooking and cooling (<48°C).

The common symptoms of the emetic strains are: Nausea, vomiting and malaise (sometimes followed by diarrhea, due to additional enterotoxin diarrhea). The emetic activity is extremely stable, being unaffected by heating or by extremes of pH. The common symptoms of the diarrheal strains are: Abdominal pain, watery diarrhea and occasional vomiting.

6.1.7 *Calicivirus (or Norwalk-like virus)*

Very common cause of food borne illness, however, it is under diagnosed due to difficulty in detecting. Usually spread from one infected person to another. Infected people can also spread it to the food they are preparing. Gastrointestinal illness, vomiting and diarrhea are the common symptoms. Symptoms usually resolve within two days.

6.2 *Food borne disease outbreaks*

A food borne disease outbreak can occur when a group of people consume the same contaminated food and two or more become sick with the same illness. A combination of events contributes to the outbreak. A contaminated food may be left out at room temperature for many hours, allowing the bacteria to multiply to high numbers, and then be insufficiently cooked to kill the bacteria.

Outbreaks are usually local in nature; A group of people becoming sick after attending an event or eating in a common place. However, an outbreak can be also national or even international. In this case the type of infecting microorganism could be the main evidence linking such cases together. The vast majority of reported cases of foodborne illness is not part of recognized outbreaks, but occurs as sporadic cases. It may be that many of these cases are actually part of unrecognized widespread or diffuse outbreaks.

Food-sharing practices link particular households in rural villages and have implications for the spread of food-borne pathogens. The food-sharing networks in remote rural villages are heterogeneous and clustered, consistent with contemporary theories about disease transmitters. Network-based measures may offer tools for predicting patterns of disease outbreaks, as well as guidance for interventions (Trostle *et al.*, 2008; Ho *et al.*, 2002).

A food borne outbreak is an indication that something wrong in the food safety and it needs to be improved therefore all outbreaks should be reported even if was a small one. There always a chance that the next outbreak might be bigger or more dangerous. Prevention of such chance is only possible by investigating the previous ones and correcting the problem in the safety procedures.

Once an outbreak is suspected, an investigation begins. A search is made for more cases among persons who may have been exposed. The symptoms and time of onset and location of possible cases are determined, and a description of the typical case is developed. A graph is drawn of the number of people who fell ill on each successive day to show pictorially when it occurred. A map of where the ill people live, work, or eat may be helpful to show where it occurred; calculating the distribution of cases by age and sex of the affected people. If the causative microbe is not known, samples of stool or blood are collected from ill people and sent to the public health laboratory to make the diagnosis.

There is now a very strong scientific consensus that global warming is occurring (Trenberth, 2001). The spectrum of food borne diseases is constantly changing. A century ago, typhoid fever, tuberculosis and cholera were common food borne diseases. Improvements in food safety, such as pasteurization of milk, safe canning, and disinfection of water supplies have conquered those diseases. Recently other food borne infections have taken their place, including some that have only recently been discovered. For example, in 1996, the parasite Cyclospora suddenly appeared as a cause of diarrheal illness related to Guatemalan raspberries. These berries had just started to be grown commercially in Guatemala, and somehow became contaminated in the field there with this unusual parasite (Lee, 2000).

Climate change and rising average global temperatures threaten to disrupt the physical, biological and ecological life support systems on which human health depends. The global warming has affected the number of heat stroke in Europe (Ishigami *et al.*, 2008). Around the world, the relationship between microbes and human beings is changing in response to many factors: social, demographic, environmental and economic. Some of the change reflects rising average temperatures and is manifested in altered incidence rates, seasonality and geographic range of a number of infectious diseases. For example, malaria, a disease that makes an enormous contribution to the world's burden of disease, may be changing its transmission pattern partly in response to global warming including its extension to higher altitudes of Eastern and Southern Africa (Pascual *et al.*, 2006). Other examples of change in infectious disease distribution include tick borne encephalitis in Sweden (Lindgren and Gustafson, 2001), and gastroenteritis due to Campylobacter pylori (Kovats *et al.*, 2005).

In the last two decades, several important diseases of unknown cause have turned out to be complications of food borne infections. For example, we now know that the Guillain-Barre syndrome can be caused by *Campylobacter* infection, and that the most common cause of acute kidney failure in children, hemolytic uremic syndrome, is caused by infection with *E. coli* O157:H7 and related bacteria. In the future, other diseases whose origins are currently unknown may turn out be related to food borne infections.

6.3 *Protection against food borne diseases*

Protection against food borne diseases is simple and can be done by taking a few simple precautions. The American Department of Health and Human Services has the following advice:

- Make sure that meat, poultry and eggs are cooked thoroughly. Well cooking meat and eggs will kill bacteria. For example, ground beef should be cooked to an internal temperature of 71°C and eggs should be cooked until the yolk is firm.

- Do not cross-contaminate one food with another. Wash hands, utensils, and cutting boards after they have been in contact with raw meat or poultry and before they touch another food. Cooked meat should be placed on a clean platter, rather back on one that held the raw meat.
- Leftovers should be refrigerated promptly. Bacteria can grow quickly at room temperature. Remember, large volumes of food will cool more quickly if they are divided into several small containers for refrigeration.
- Wash fresh fruits and vegetables in running tap water to remove visible dirt and grime. Remove and discard the outermost leaves of a head of lettuce or cabbage. Remove the cut surface of fruit or vegetable and be careful not to contaminate these foods while slicing them up on the cutting board. (Remember the cut surface can be contaminated easily with bacteria).
- Wash your hands with soap and water before preparing food. Avoid preparing food for others if you yourself have a diarrheal illness.
- All suspected food borne illnesses should be promptly reported.

Vaccination is an effective way to destroy the pathogen before transmission to human. For example vaccination of cattle with Type III-secreted proteins resulted in decreased shedding of E coli O157:H7 following both experimental infection as well as under conditions of natural exposure (Asper *et al.*, 2007).

Diarrhea is a symptom of many diseases. Defining etiology of acute diarrhea is critical to disease therapy and prevention (Marcos and DuPont, 2007). A health care provider should be consulted if diarrhea is accompanied by:

- High fever (temperature over 38.6°C, measured orally).
- Blood in the stools.
- Prolonged vomiting that prevents keeping liquids down (which can lead to dehydration).
- Signs of dehydration, including a decrease in urination, a dry mouth and throat, and feeling dizzy when standing up.
- Diarrheal illness that lasts more than 3 days.

6.4 *Food associated with food borne diseases*

Raw foods of animal origin are the most likely to be contaminated. Raw meat and poultry, raw eggs, unpasteurized milk, and raw shellfish are few examples. Foods that mix the products of many individual animals (e.g., bulk raw milk, pooled raw eggs, and ground beef) are especially high in the list because a pathogen from one animal may contaminate the whole batch. A single hamburger may contain meat from several animals. A single restaurant omelet may contain eggs from several chickens. A cup of milk may contain milk from several cows. Raw fruits and vegetables are of particular concern. Unfortunately, washing fruits and vegetables can only decrease but not completely eliminate contamination. Few pathogens getting to the human body may lead to food borne diseases. Using unclean water can contaminate many boxes of produce (Banda *et al.*, 2007). Alfalfa sprouts and other raw sprouts pose a particular challenge, as the conditions under which they are sprouted are ideal for growing microbes as well as sprouts, and because they are eaten without further cooking. That means that a few bacteria present on the seeds can grow to high numbers of pathogens on the sprouts. Unpasteurized fruit juice can also be contaminated if there are pathogens in or on the fruit that is used to make it.

7 CONCLUDING REMARKS

Earth as a balanced system continues to be deteriorated mainly by man made (i.e. anthropogenic) factors. Overpopulation, for example, has had a negative impact on the environment and consequently on human well being. The spread of pollution and diseases; depletion of resources; destruction of habitats and biodiversity crises in addition to climate change have all added to the ecosystem predicament. This chapter has presented an overview of the causes as well as suggesting solutions and green alternatives for environmental problems. Understanding of these issues may help scientists and decision makers to make positive modifications in their behavior. Cooperative efforts are required at the local, regional and international levels, through legislation and treaties, to help solve these ecosystem troubles and to contribute to their solutions.

REFERENCES

Abu-Dieyeh, M.H. and Watson, A.K. 2005. Impact of mowing and weed control on broadleaf weed population dynamics in turf. *J. Plant Interactions* **1(4):** 239–252.

Abu-Dieyeh, M.H. and Watson, A.K. 2007a. Efficacy of *Sclerotinia minor* for dandelion control: Effect of dandelion accession, age, and grass competition. *Weed Res.* **4(7):** 63–72.

Abu-Dieyeh, M.H. and Watson, A.K. 2007b. Population dynamics of broadleaf weeds in turfgrass as influenced by chemical and biological control methods. *Weed Sci.* **55:** 371–380.

Adak, G.K., Long, S.M. and O'Brien, S.J. 2005. Trends in indigenous food borne disease and deaths, England and Wales: 1992 to 2000. *Gut* **51:**832–841.

Alikhanidi, S. and Takahash, Y. 2004. Pesticide persistence in the environment – collected data and structure-based analysis. *J. Comput. Chem. Jpn.* **3:** 59–70.

Allos, B.M. 1997. Association between Campylobacter infection and Guillain–Barré syndrome. *J. Infect. Dis.* 176 Suppl. 2:S125–S128.

Asper, D.J., Sekirov, I., Finlay, B.B., Rogan, D. and Potter, A.A. 2007. Cross reactivity of enterohemorrhagic Escherichia coli O157:H7-specific sera with non-O157 serotypes. *Vaccine* 25(49): 8262–8269.

Banda, K., Sarkar, R., Gopal, S., Govindarajan, J., Harijan, B.B., Jeyakumar, M.B., Mitta, P., Sadanala, M.E., Selwyn, T., Suresh, C.R., Thomas, V.A., Devadason, P., Kumar, R., Selvapandian, D., Kang, G., and Balraj, V. 2007. Water handling, sanitation and defecation practices in rural southern India: a knowledge, attitudes and practices study. *Trans R Soc Trop Med Hyg.* **101(11):** 1124–30.

Beckie, H., Thomas, A., Legere, A., Kelner, D., Van Acker, R. and Meers, S. 1999. Nature occurrence and cost of herbicide resistant wild oat in small grain production areas. *Weed Tech.* **13:** 612–625.

Belanger, R., Jarvis, W. and Traquair, J. 2002. Sphaerotheca and Erysiphe spp., Powdery Mildews (Erysiphaceae) In: Mason PG, Huber JT, editors. Biological control programs in Canada, 1981–2000.NY: CABI Publishing. pp. 501–505.

Bhowmik, P. and Inderjit, A. 2003. Challenges and opportunities in implementing allelopathy for natural weed management. *Crop Prot.* **22:** 661–671.

Blanpied, N. 1984 Farm Policy, the Politics of Soil, Surpluses, and Subsidies (Washington, DC: Congressional Quarterly Inc.).

Bond, W. and Groundy A.C. 2001 Non-chemical weed management in organic farming systems. *Weed Res.* **41:** 383–405.

Boyce, T.G., Swerdlow, D.L., and Griffin, P.M. 1995. Escherichia coli O157:H7 and the hemolytic-uremic syndrome. *N. Engl. J. Med.* **333:** 364–368.

Bridges DC. 1992. Crop losses due to weeds in the United States. Champaign, III: World Science Society of America.

Burge, D., Harriet, A. and Feely, J.C. 1991. "Indoor Air Pollution and Infectious Diseases." In: Samet, J.M. and Spengler, J.D. eds., Indoor Air Pollution, A Health Perspective (Baltimore MD: Johns Hopkins University Press), pp. 273–84.

Campbell, N.A. and Reece, J.B. 2005. Biology 7th ed. Benjamin Cummings, San Francisco.

Cannon, J.S. 1985. The Health Costs of Air Pollution. The American Lung Association, New York.

Charudattan, R. and Dinoor, A. 2000. Biological control of weeds using plant pathogens: accomplishments and limitations. *Crop Prot.* **19:** 691–695.

Chaterjee, B. 2002. Environmental laws: implemetation problems and perspectives. Deep and Deep Publications (p) Ltd.

Chief Medical Officer. 1992. Definition of food poisoning. London: Department of Health.

Cook, R.J. 1987. Research Briefing from the Panel on Biological Control in Managed Ecosystems. Washington DC: Natl. Acad. Press. 12pp.

Cook, R.J. 1996. Assuring the use of microbial biocontrol agents: a need for policy based on real rather than perceived risks. *Can. J. Plant Pathol* **18:** 439–445.

Copaja, S., Barria, B. and Niemeyer, H. 1991. Hydroximic acid content of perennial Triticeae. *Phytochem.* **30:** 1531–1534.

Cousens, R. and Croft, A.M. 2000. Weed populations and pathogens. *Weed Res.* **40:** 63–82.

Cox, C. 1999. Herbicide Fact sheet, 2,4-D: Exposure. *J. Pesticide Ref.* **19:** 14–19.

Delaplane, K. S. (2000). Pesticide Usage in the United States: History, Benefits, Risks, and Trends. Bulletin 1121. Cooperative Extension Service, The University of Georgia College of Agricultural and Environmental Sciences. http://pubs.caes.uga.edu/caespubs/pubs/PDF/B1121.pdf

Dilday, R., Mattice, J., Moldenhauer, K. and Yan W. (2001) Allelopathic potential in rice germplasm against Ducksalad, Redstem and Barnyard grass. In: Allelopathy in Agroecosystems (Kohli, *et al.*, Eds). Food Products Press. New York. Pp. 287–301.

Eckenfelder, W. 1989. Industrial water pollution control. McGraw – Hill, New York.

Environment Canada: Air pollution control Technologies (date visisted : 14 Feb. 2008. http://canadainternational.gc.ca/dbc/documents/SF_Envir_AirPollution_Nov03_e.pdf

Etzel, R.A., Montaña, E., Sorenson, W.G., Kullman, G.J., Allan, T.M. and Dearborn, D.G. 1998. Acute pulmonary hemorrhage in infants associated with exposure to Stachybotrys atra and other fungi. *Arch. Pediatr. Adolesc. Med.* **152:** 757–761

FAO/ECE. 1991. Legislation and Measures for the Solving of Environmental Problems Resulting from Agricultural Practices (With Particular Reference to Soil, Air and Water), Their Economic Consequences and Impact on Agrarian Structures and Farm Rationalization. United Nations Economic Commission for Europe (UNECE) and FAO, Agri/Agrarian Structures and Farm Rationalization Report No. 7. United Nations, Geneva.

Floate, K., Berube, J., Boiteau, G., Dosdall, L., van Frankenhuyzen, D., Gillespie, D., Moyer, J., Philip, H. and Shamoun S. 2002. In: Masom PG, Huber JT, editors. Biological Control Programs in Canada, 1981–2000. NY: CABI Publishing.

French, H.F. 1990. World Watch Paper 94, Clearing the Air: A global agenda. World Watch Institute, Washington, USA.

Guillette Jr, L. J. and Crain, A. 2000. Environmental Endocrine Disruptors. CRC.

Gould, F. (1991) The evolutionary potential of crop pests. *Am. Scientist* **79:** 496–507.

Harrison, R. 2001. Pollution: causes, effects and control. 4th ed. Royal Society of Chemistry.

Hawker, J. 2005. Communicable disease control handbook. 2nd ed. Oxford: Blackwell Science.

Hayes, A. W. 2007. Principles and Methods of Toxicology. 5th ed. Informa Healthcare.

Hill, B.D., Inaba, D.J., Harker, K.N., Moyer, J.R. and Hasselback, P. 1999. Phenoxy herbicides in Alberta rainfall: cause for concern [internet]. [cited 2005 May 20]. Available from: http://www.traceorganic.com/2000/abst/hill1.pdf

Hill Gaston, J.S. and Lillicrap, M.S. 2003. Arthritis associated with enteric infection. *Best Pract. Res. Clin. Rheumatol.* **17:** 219–239.

Ho, A.Y., Lopez, A.S., Eberhart, M.G., Levenson, R., Finkel, B.S., da Silva, A.J., Roberts, J.M., Orlandi, P.A., Johnson, C.C., and Herwaldt, B.L. 2002. Outbreak of cyclosporiasis associated with imported raspberries, Philadelphia, Pennsylvania, 2000. *Emerg. Infect. Dis.* **8(8):** 783–788.

Ishigami, A., Hajat, S., Kovats, R.S., Bisanti, L., Rognoni, M., Russo, A., and Paldy, A. 2008. An ecological time-series study of heat-related mortality in three European cities. *Environ Health* **7:** 5.

Kebede, Z. 1994. Allelopathic chemicals: Their potential uses for weed control in Agroecosystem. http://www.colostate.edu/Depts/Entomoloogy/courses/en570/papers_1994/Kebede.html.

Kennedy, A.C. and Kremer, R.J. 1996. Microorganisms in weed control strategies. *J. Prod. Agr.* **9:** 480–485.

Kicklighter D.W., Bruno M. and Donges S. 1999. A first order analysis of the potential role of CO2 fertilization to affect the global carbon budget: a comparison of four terrestrial biosphere models. *Tellus* **51B:** 343–366.

Khopkar, S. M. 2005. Environmental pollution: monitoring and control. New Age International pvt Ltd.

Kovats, R.S., Edwards, S.J., Charron ,D., Cowden, J., D'Souza, R.M., Ebi, K.L., Gauci, C., Gerner-Smidt, P., Hajat, S., Hales, S., Hernández Pezzi, G., Kriz, B., Kutsar, K., McKeown, P., Mellou, K., Menne, B., O'Brien, S., van Pelt, W., and Schmid, H. 2005. Climate variability and campylobacter infection: an international study. *Int. J. Biometeorol.* **49:** 207–14.

Lee, M.B. 2000. Everyday and exotic foodborne parasites. *Can. J. Infect. Dis.* **11:** 155–158.

Letertre, C., Perelle, S., Dilasser, F., and Fach, P. 2003. Detection and genotyping by real-time PCR of the staphylococcal enterotoxin genes sea to sej. *Mol. Cell Probes* **17:** 139–147.

Lindgren, E. and Gustafson, R. 2001. Tick-borne encephalitis in Sweden and climate change. *Lancet* **358:** 16–18.

Luckman W.H. and Decker G.C. 1960 A 5-year report on observations in the Japanese beetle control area of Sheldon, Illinois. *J. Eco. Entomol.* **53:** 821–827.

Lull, K., Tindall, J. and Potts, D. 1995. Assessing nonpoint-source pollution risk. *J. Forestry* **93:** 35–40.

Mandal, B.K., Ellis, M.E., Dunbar, E.M. and Whale, K. 1984. Double-blind placebo-controlled trial of erythromycin in the treatment of clinical Campylobacter infection. *J. Antimicrob. Chemother.* **13:** 619–623.

Marcos, L.A. and DuPont, H.L. 2007. Advances in defining etiology and new therapeutic approaches in acute diarrhea. *J. Infect.* **55(5):** 385–93.

Meer, R.R., Songer, J.G. and Park, D.L. 1997. Human disease associated with Clostridium perfringens enterotoxin. *Rev. Environ. Contam. Toxicol.* **150:** 75–94.

Mims C.A. 2004. Medical microbiology. 3rd ed. Elsevier Science Publisher, Edinburgh.

Miyamoto, K., Fisher, D.J., Li, J., Sayeed, S., Akimoto, S. and McClane, B.A. 2006. Complete sequencing and diversity analysis of the enterotoxin-encoding plasmids in Clostridium perfringens type A non-food-borne human gastrointestinal disease isolates. *J. Bacteriol.* **188(4):** 1585–98.

Murray, P.R. and Baron, E.J. 2003. American Society for Microbiology. Manual of clinical microbiology. 8th ed. ASM Press Washington, DC.

Olofsdotter, M. 2001. Rice-A step toward use of allelopathy. *Agron. J.* **93:** 3–8.

Ongley, E.D. 1996. Control of Water Pollution from Agriculture. FAO Irrigation and Drainage, Paper 55, FAO, Rome.

Pascual, M., Ahumada, J.A., Chaves, L.F., Rodo, X., and Bouma, M. 2006. Malaria resurgence in the East African highlands: temperature trends revisited. *Proc. Natl. Acad. Sci.* **103:** 5829–5834.

Pulz, M., Matussek, A., Monazahian, M., Tittel, A., Nikolic, E., Hartmann, M. Bellin, T., Buer, J., and Gunzer, F. Comparison of a shiga toxin enzyme-linked immunosorbent assay and two types of PCR for detection of shiga toxin producing Escherichia coli in human stool specimens. *J Clin Microbiol* 2003; **41:** 4671–4675.

Rana, S. 2006. Environmental pollution: health and toxicology. Alpha Science International, Ltd.

Richardson, D. H. 1987. Biological Indicators of Pollution. Royal Irish Academy, Dublin.

RIVM. 1992. The Environment in Europe: A Global Perspective. National Institute of Public Health and Environmental Protection (RIVM), Netherlands.

Schipper, L., Martishaw S. and Unander, F. 1993. International comparisons of sectoral carbon dioxide emissions using a cross-country decomposition technique. *Energy J.* **22:** 35–75.

Smith J.W. 1998. Boll Weevil eradication: area-wide pest management. *Ann. Entomol. Soc. Am.* **91:** 239–247.

Swanton, C.J., Harker, K.N. and Anderson, R.L. 1993. Crop losses due to weeds in Canada. *Weed Tech.* **7:** 537–542.

Tickell, J. 2006. Biodiesel America. The Biodiesel America Organization and the National Biodiesel Bord. USA. 48pp.

Townsend, C.R., Begon, M. and Harper, J.L. 2003. Essentials of Ecology. Blackwell Publishing. MA, USA. 530pp.

Trenberth, K.E. 2001. Climate variability and global warming. *Science* **293:** 48–49.

Trostle, J.A., Hubbard, A., Scott, J., Cevallos, W., Bates, S.J., and Eisenberg, J.N. 2008. Raising the Level of Analysis of Food-Borne Outbreaks: Food-Sharing Networks in Rural Coastal Ecuador. *Epidemiology, in press.*

UNEP. 1993. The Aral Sea: Diagnostic study for the development of an Action Plan for the conservation of the Aral Sea. Nairobi.

U.S. Environmental Protection Agency, Office of Air and Radiation. Indoor Air Facts No. 4: Sick Building Syndrome, revised, 1991.

Viessman, W. 2005. Water supply and pollution control. 7th ed. Pearson Education, Upper Saddle River:

Vigil, D. and Kenneth, M. 2003. An Introduction to Water Quality and Pollution Control. 2nd ed. Oregon State University Press.

Vyvyan, J. 2002 Allelochemicals as leads for new herbicides and agrochemicals. *Tetrahedron* **58:** 1631–1646.

Ware, G. 2004. Pesticide Book. Meister Pub Co.

World Health Organization. 2002. Food and Agriculture Organization of the United Nations. Risk assessments of Salmonella in eggs and broiler chickens. Geneva: World Health Organization.

World Health Organization. 2005. The WHO recommended classification of pesticides by hazard and guidelines to classification: 2004. WHO Library Cataloguing-in-Publication Data. http://www.inchem.org/documents/pds/pdsother/class.pdf

Worldwatch Institute.1985. The Clean Air Act: A briefing Book for Members of Congress. Washington D.C.

Wu, H., Partley, J., Lemerle, D. and Haig, T. 1999. Crop cultivars with allelopathic capability. *Weed Res.* **39:** 171–180.

World Wide Fund (WWF). 1993. Marine Update 13: Marine pollution and pesticide reduction policies. World Wide Fund for Nature, Panda House, Godalming, Surrey, UK.

Environmental Profile and Management Issues in an Estuarine Ecosystem: A Case Study from Jobos Bay, Puerto Rico

E.N. Laboy-Nieves
Universidad del Turabo, School of Science and Technology, Gurabo, Puerto Rico

SUMMARY: Elements of a physical and biological nature exert influences to create and maintain the dynamics of estuarine ecosystems such as in the case study area of Jobos Bay, an ecological unit sculpted by natural and anthropic factors, located in the southeastern shore of Puerto Rico, an island in the Caribbean. This chapter will provide a compilation of the processes and elements that define the structure and dynamics of the submerged and terrestrial components of this unique ecosystem and the struggles that Jobos Bay has faced for the management of its resources. The first part describes abiotic environmental elements, like geology, hydrology, and climate. The second portrays the five major ecological communities: mangrove forests, littoral woodlands, mud flats, seagrass beds, and coral reefs. Aspects of the energetics that characterize these vulnerable and complex communities are explained. Finally, environmental management issues are discussed to depict the influence of human activities in the peripheral land cover. Baseline information is provided for future scientific, education, conservation, and management projects.

1 INTRODUCTION

For centuries, peripheral urban developments have affected the evolution of estuaries at a point that today some ecosystems exhibit a mosaic of areas that remains almost pristine while others are degraded (Cruz-Báez and Boswell, 1997). These estuarine ecosystems are natural laboratories for examining and understanding the communities defined by the littoral, upland, intertidal and submerged habitats and their respective biota, and anthropogenic influences. The aquatic and terrestrial zones and the populations that inhabit them are tightly linked, showing complex interactions between abiotic and biotic factors. Like many countries, in Puerto Rico those zones and interactions are been jeopardized by a plethora of factors, from which Laboy-Nieves (2008) emphasizes high demographic density, urban sprawling, poor waste management and social indolence with respect to the environment. This chapter provides a compilation of some of the processes and elements that define the structure and dynamics of the submerged and terrestrial components of the Jobos Bay Estuary, Puerto Rico, and the struggles this ecosystem has faced for the management of its resources.

1.1 *Case study area*

Puerto Rico is the eastern-most and smallest (~8,900 km^2) of the Greater Antilles. The Island has a rectangular shape and lies between the Atlantic Ocean (north) and the Caribbean Sea (south). Its center is located 18°15′N and 66°30′W and its longest extension is about 190 km. Numerous smaller islets, keys and reefs surround the main island. Forty-five percent of the land surface is above the 150 m contour and half of that area has slopes greater than 45 percent (Birdsey & Weaver, 1982). Cerro Punta is the maximum elevation (1338 m). Humid rich easterly trade winds cause orographic and convective rainfall, mostly in the northern and central regions, which are the cradle for more than a thousand small rivers and streams. These freshwater bodies are more abundant and longer, and discharge a greater volume of water in the north coast than in the drier south coast (Cruz-Báez and Boswell, 1997). They form estuaries in the serrated coastline. These estuaries define most of the physical conditions of the main embayments of the Island (Figure 1).

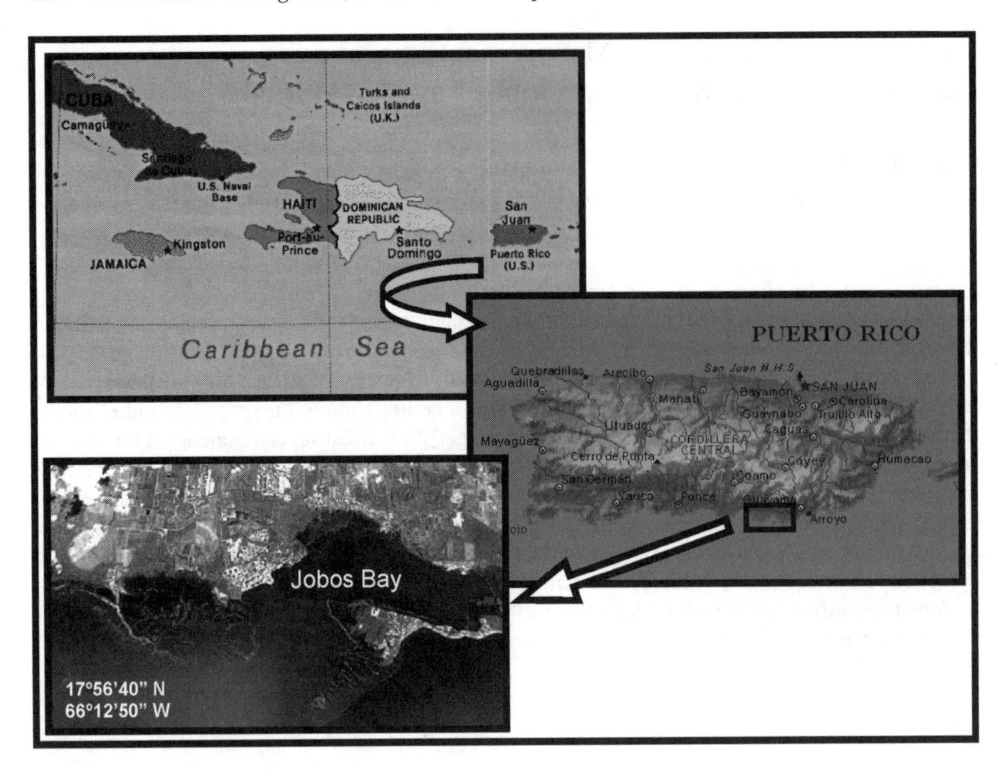

Figure 1. General location of Jobos Bay.

Jobos Bay is one of the few natural harbors in the southern coast and the second largest estuarine bay of Puerto Rico. It is located at 17°56′40″N, 66°12′50″W. It borders the municipalities of Salinas and Guayama (Figure 2). The bay covers an area nearly 31 km^2, encompassed mainly within the natural unit formed by the biological communities associated to the mangrove forest of the Aguirre State Forest and the Jobos Bay National Estuarine Research Reserve. Bordering the western limits of the Bay is the Mar Negro wetland (Figure 2). The 287 ha reef barrier that extends from Cayos Caribes to Cayo Morrillo presents the southern limits, while mangrove forests confine the northern and eastern sections of the Bay. Most of the peripheral upland north of Jobos Bay is cultivated for agriculture. Aguirre, Puerto de Jobos and Pozuelo are the main settlements. The main power plant and garbage disposal facility of the southeast lie north of the Bay.

The sigmoid-shaped Bay extends approximately 15 kilometers from the mangroves of Guayabal in the east to Cayo Morrillo in the west. The Bay can be distributed in three different zones: Inner Bay, Mid Bay and the Navigational Channel (PRWA, 1972). Average depth is around four meters (Figure 3). The Inner Bay is the eastern-most end of Jobos, separated from the Mid Bay by an imaginary line between the Aguirre Power Plant Dock and Punta Rodeo. This portion is relatively shallow ($\leq$3 m). The Mid Bay is border by Pozuelo on the east, Cayos Caribes on the south and Punta Colchones on the west. A seven hectares dead coral reef forms a wall about seven meters deep in front of Punta Colchones. The Navigational Channel is open to the Caribbean Sea on the west between Cayo Morrillo and Cayo de Pájaros, which channels are about 15 meter. On the east there is a four meters deep channel known as Boca del Infierno.

As interpreted from Figure 2, the Bay watershed is surrounded by a mosaic of different land covers. Agriculture and urbanization characterize most of the northern boundaries, while mangrove wetlands and their associated communities represent the rest of the land peripheral to Jobos Bay. The climatic, hydrologic and edaphic conditions are the key elements for the wide variety of flora on the upland marginal to the Bay. The terrestrial vegetation indicates an ecotone between the

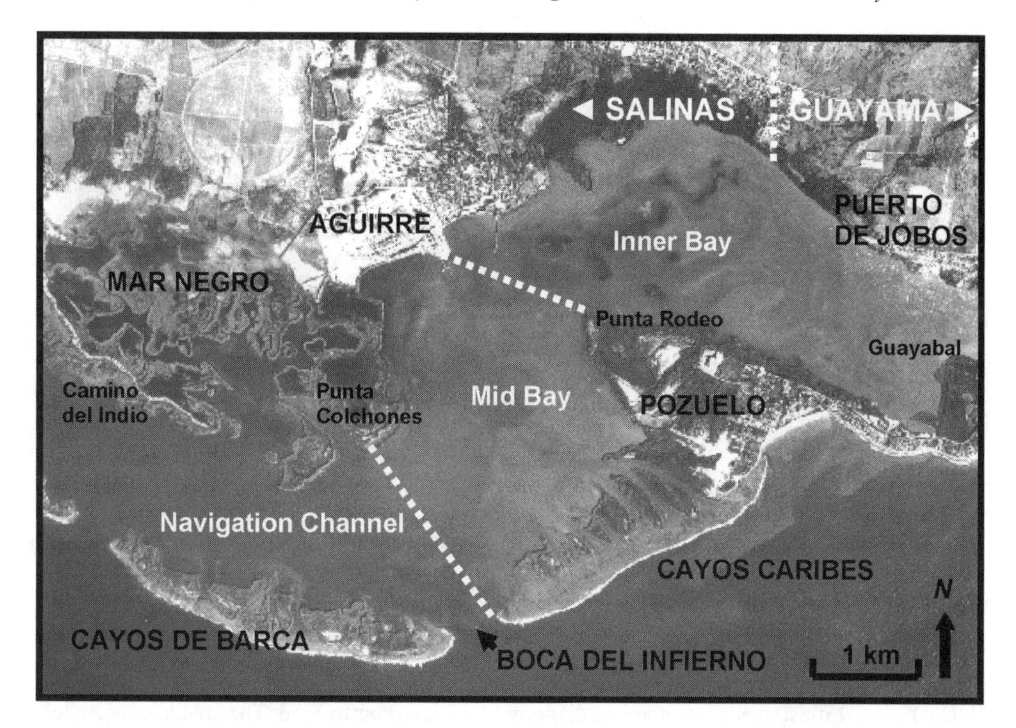

Figure 2. Geographical references of the Jobos Bay estuary.

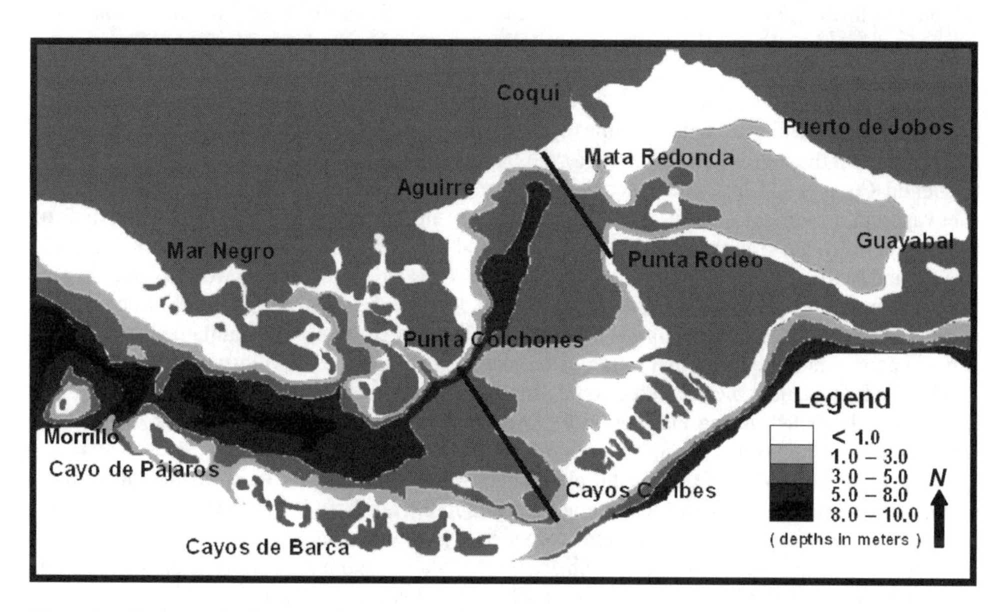

Figure 3. Bathymetric diagram of Jobos Bay (Laboy-Nieves, 2001).

subtropical dry forest and the subtropical moist forest as described by Ewel and Whitmore (1973). Mud flats, microbial mats, seagrass beds, fringing mangrove forests, and coral reefs comprise the submerged communities. These areas provide habitat, and feeding and foraging ground for many species. The interaction among the upland forests and the submerged communities promotes a very vulnerable and complex ecological dynamic in a very small area (Laboy-Nieves, 1983).

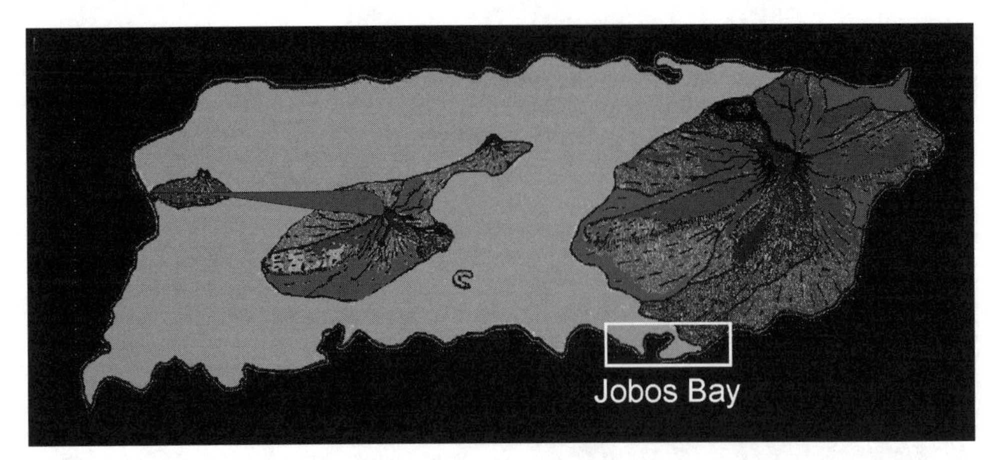

Figure 4. Artistic conception of the volcanic origin of Puerto Rico (adapted from Meyerhoff, 1933).

2 GEOLOGY, OCEANOGRAPHY AND HYDROLOGY OF THE CASE STUDY AREA

The geologic history of Puerto Rico dates from the Cretaceous period, nearly 135 million years ago, as the product from the collision between the Caribbean the North American plate (Joyce, 1992). Figure 4 illustrates a hypothetical diagram of ancient Puerto Rico. The area now occupied by the Island was a shallow sea or basin of deposition surrounded by islands built up by the lava and ash poured forth from a series of active volcanoes (Berryhill, 1960). Young volcanic mountains were weathered by rainfall, heat, and winds. Clays were washed while spreading waters sorted the soils and rock fragments, concentrating pebbles and boulders at the shoreline and carrying the finer material into the central part of the basin. Meyerhoff (1933) theorized that subsidence was prolonged but intermittent. When it came to a temporary halt, the depression filled with sediments (especially near the shore). Marshes were formed, and as they extended inland, their brackish waters freshened. Coral reefs fringed the volcanic islands and during periods of quiescence, they extended over vast shallow areas (Berryhill, 1960). A tropical swamp flora grew luxuriantly, forming mats of partly decayed vegetation that buried and transformed into peat. Clays and marls concealed basal gravels and sand containing petrified wood, which now extends into the hills west of Jobos Bay.

Limestone along the southern coast began their formation underneath a warm tropical sea, approximately 60 million years ago. These coastal deposits continued accumulating through the Tertiary period until the Quaternary period, and thereafter to the present (Cruz-Báez and Boswell, 1997). The fine-grained rocks and reef-type limestone found in the uplands adjacent to Jobos Bay, were either deposited during that period or formed at some distance from a center of volcanic activity (Berryhill, 1960).

The geological history of the coastal plains region is tied to the beginning of the Quaternary period, after the accumulation of ice in huge glaciers in the northern hemisphere. This caused the sea level to fall, which in turn resulted in steeper river gradients that permitted the limnetic flows the carving of deep valleys, some of which are found today as submarine canyons on the northern and southern marine flanks of the Island (Cruz-Báez and Boswell, 1977). These valleys were filled with sediments when the sea level eventually rose after the partial melting of the northern ice caps. This process complemented the formation of the lowlands of the southern coast. At times when the sea level arose, water covered the lowest plains, as evidenced by the large amount of shell deposits found in the upland adjacent to Jobos Bay and the coral reef fossils found in the hills of Cerro Aguirre (54 m) to the northwest of the Bay (Laboy-Nieves, 2001).

The predominant soil from Jobos consists of lagoon and swamp deposits from unconsolidated clay, silt and humus (Boccheciamp, 1977). These deposits are almost entirely covered by mangroves

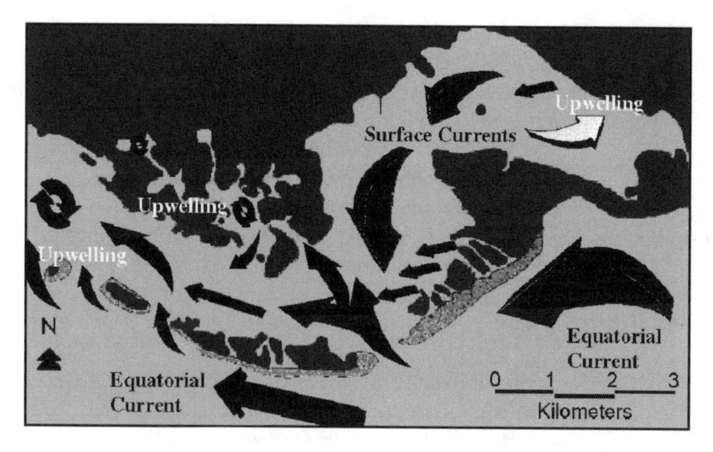

Figure 5. Schematic diagram of marine currents in Jobos Bay (Laboy-Nieves, 2001).

and seagrass beds. Beach deposits of calcareous sand, volcanic cobbles and coral reef debris distinguish most of the upland.

An inactive fault, known as the Esmeralda Fault, runs under the Bay in a northwest-southeast direction north of Jobos Bay (PRWA, 1972). There are no metallic mineral deposits of commercial value in the lands adjacent to the Bay. However, small calcite, hematite, and gypsum veins can be found in adjoining hills (Laboy-Nieves, 2001).

Jobos Bay is the second largest estuary in Puerto Rico, only preceded by the San Juan Bay. Tides at Jobos are primarily diurnal with a complex pattern composed of two tidal waves, one with a daily cycle and another with a cycle of 13.3 days (PRWA, 1972). Average tide height is 13.7 cm, ranging from −17.0 cm to 36.0 cm (Lugo et al., 2007). Lowest tides occur early in the year, while the highest are around October. A deep current brings water from the west into the Inner Bay during the flood tide. During the ebb tide, water moves out of the Bay in the entire water column (PRWA, 1972). The tidal excursion (the average distance traveled by a water particle during a half-tide cycle) is around 600 meters, decreasing eastward. Considering tidal and wind effects, the mean residence time of a water mass in Jobos Bay is approximately 5.5 days. The volume of water displaced daily from the Bay averages 30.5 million cubic meters. The peak tidal surge elevation that can be expected to occur at Jobos during a probable maximum hurricane would be about three meters (PRWA, 1972).

Geostrophic influences, tides and wind effects are factors that generate and control currents and the exchange and renewal of water in Jobos Bay, either separately or by interaction. The North Equatorial Current, flowing in a west-northwesterly direction, dominates the entire south coast of Puerto Rico. However, on a localized basis, special hydrographic conditions can modify the near shore coastal currents in Jobos Bay. Trade winds generate westward surface currents that are deflected by the labyrinth of reef channels, and prop and aerial roots from mangrove trees. The strong Equatorial Current from dissipates northward in the reef channels while eroding and excavating the islets, forming relatively deep (~3 m) depressions in the shoreline (Figure 5). The deep current enters the Inner Bay from the Navigational Channel and the Mid Bay (PRWA, 1972). The Inner Bay and part of the Mid Bay have a silt bottom, which gets stirred up during the normal 10-knot trade winds. These winds act with the Equatorial Current to produce upwelling spots (Suarez-Caabro and Shearls, 1973).

The reef crest and shallow substrata are exposed daily as a result of low tides and or diminishing winds. The relative height of the water surface is highest around October, a period which coincided with higher rainfall water storage in the mangrove forest. Early in the year, this stored brackish and detritus-rich water flushes back to the Bay. Geologically, hydrologically and biologically, Jobos Bay is being closed by the advancement of mangrove forests, as it happened in the former Las Mareas Bay, presently limited to a series of interconnected lagoons and known today as Mar Negro.

The South Coastal Plain consists of a series of successive fan deltas about 5 km wide, bordered to the north by intensely faulted low hills (Quiñonez-Aponte et al., 1997). These deltas are formed by steep gradients and intermittent streams that have small drainage areas. The different hydrogeologic characteristics of this plain have lead to its classification into three prototypical groups: the Salinas fan delta, the Jobos area, and the Arroyo fan delta. Jobos Bay is within the first group, characterized by mangrove swamps and tidal flats (Quiñones-Aponte, 1991).

Several intermittent creeks (Lapa, Majada, Río Seco, Quebrada Amoros, Quebrada Coquí, Quebrada Mosquito, and Quebrada Aguas Verdes) embody Jobos Bay. Almost all of them loose their flow completely between the mid- and upper valley reaches as they enter the fan deltas. According to McClymonds and Díaz (1962), fresh water available for Jobos Bay comes from three major sources: rainfall (15.2×10^7 m^3), limnetic bodies (5.8×10^7 m^3) and irrigation channels (4.93×10^7 m^3).

The existence of a sand-gravel dominated crust in the tilted terrain that defines the upland north of Jobos Bay has permitted the percolation of rain water and stream flows, forming the Great Southern Aquifer, which extends over 60 kilometers between the municipalities of Ponce and Arroyo (Figure 1). It is presumed that water quality is low, because most aquifers in Puerto Rico are polluted with toxic chemicals (Skanavis, 1999).

Three main geohydrologic units constitute the aquifer of the South Coastal Plain: (1) the shallow coastal water-table aquifer composed of sand, gravel, and clay; (2) the principal flow zone composed of fan delta and alluvial deposits and; (3) the regolith composed of weathered bedrock of varied types (Quiñonez-Aponte et al., 1997). In each of these areas, the aquifer extends inland about 3 km, and it is confined near the coast by a fine-grained, poorly permeable clay or silt bed. This bed is less than 18 m thick and lies at a depth of 14 m below the land surface. Freshwater inflow to the mangrove wetlands occur mainly from the last two units (Quiñonez-Aponte et al., 1997). The input of water from springs, intermittent creeks and irrigation channels define the peculiar estuarine nature of the Jobos Bay.

Due to high evapotranspiration (around 2.0×10^7 m^3/year) and the permeability of the alluvial soil of the area, water from creeks typically ceases its surficial flow before reaching the Bay but, seaward, that water replenishes the aquifer at a mean rate of 6.7×10^7 m^3/year (Quiñonez-Aponte et al., 1997). Stream flow infiltration to the aquifer system represents 19 percent of the water budget, supplied from irrigation applications and seepage from irrigation canals. Río Seco is the only freshwater input that discharges directly to the Bay. However, although data are not available, springs from the aquifer flourish within the Jobos Bay mangrove forests and benthic substrata, and seem to be the main source of limnetic inflow for the Bay (Laboy-Nieves, 2001).

The mangrove forest on Mar Negro receives only seepage from irrigation. Morris (2000) concluded that this forest is not flushed, and not affected by river flood events. However, extreme rain pulses associated to meteorological events, like hurricanes, produce abnormal runoffs that subsequently promote the flushing of almost all mangroves and upland areas in Jobos Bay (Laboy-Nieves, 2001). Access to freshwater in the form of groundwater discharges and about 1 m of annual rainfall, play a role in the high productivity of the Jobos mangroves (Lugo et al., 2007).

3 CLIMATE AND ECOLOGICAL COMMUNITIES OF JOBOS BAY

Jobos Bay is located within the subtropical dry forest zone, as described by Ewel and Whitmore (1973). The mountains of the Cordillera Central entrap the moisture carried by easterly trade winds, creating a low precipitation regime along most of the south coast. East of Jobos, the coast is moist, thus the Bay is in a climatic transition zone. Mean annual rainfall is 113 cm, September–October being the wettest months and February-March their dry counterpart (Figure 6).

Mean relative humidity is about 65% early in the year and 80% during the rainy season (July and October). Mean air temperature at Jobos is around 27°C and shows little annual fluctuation (Lugo et al., 2007). July and August are the hottest months while the coolest are January and February (Figure 7). Extreme temperatures range around 16°C and 38°C.

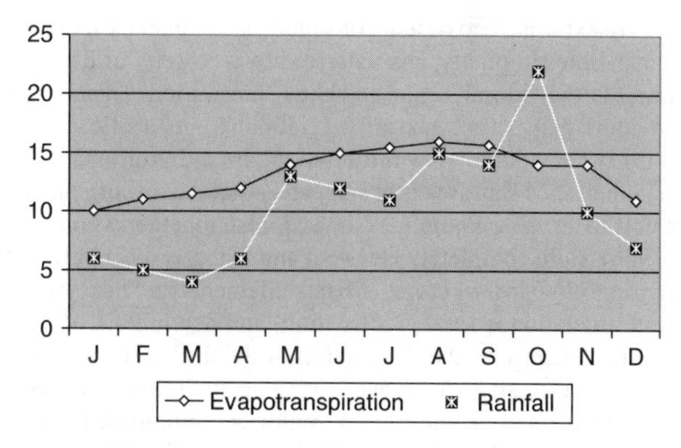

Figure 6. Annual mean evapotranspiration and rainfall in Jobos Bay.

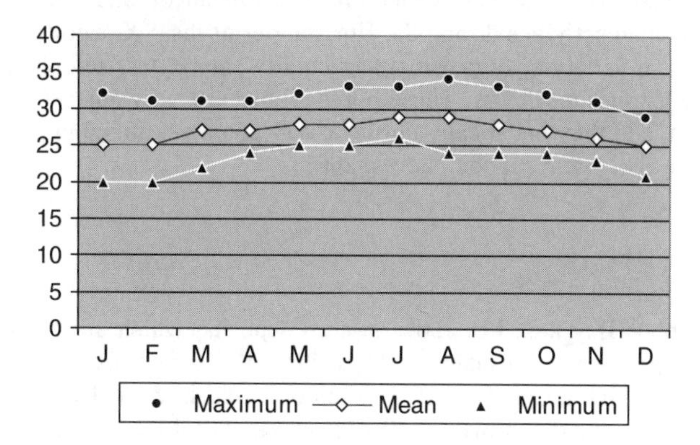

Figure 7. Annual mean temperature (°C) for Jobos Bay.

Trade winds blow mostly (46.8%) from easterly directions, averaging nearly seven knots. Tropical storms and hurricanes are one of the most important climatic factors on the Island of Puerto Rico. They produce extreme rainfall and geomorphological changes that inflict coral reefs, mangroves and seagrass beds. For instance, tropical storm Hortence in 1995 and hurricane Georges in 1998 produced rainfall over 45 cm in a 24 hour period over the Bay watershed. The latter impacted the land and marine vegetation with its gusts, causing mass mortality due to mechanical disturbances (Laboy-Nieves, 2001).

The south coast of Puerto Rico west of Jobos Bay is a xeric region (Murphy, 1916; Gleason and Cook, 1926; Dansereau, 1966), classified within the subtropical dry forest life zone (Ewel and Whitmore, 1973). It shows steep environmental gradients and a sharp community zonation (Lugo, 1983). Small and/or extreme variations in weather, hydrology, oceanography, edaphic condition, elevation, and salinity seem to be the main natural abiotic factors responsible for the existence of various terrestrial and aquatic communities within and peripheral to Jobos Bay. However, Jobos Bay has been substantially altered by decades of natural and anthropic catastrophic events (Laboy-Nieves, 2001).

Both the plant cover and the marine environment of Puerto Rico are almost entirely controlled by human activities. Virgin vegetation, whether forest, savanna, scrub or grassland, are scarcely existent. Nevertheless, some areas contain stands that actually have never been clear-cut or have been free of serious interference for a long period, to convey to the contemporary observer some idea of the original composition and quite possibly of their primeval structure as well (Dansereau, 1966). Although fuel wood was extracted in Jobos until early 1980, aerial photography records reveal that mangrove forests on Jobos Bay apparently are one of those barely altered rare stands (Laboy-Nieves, 2001).

When disturbances are extreme, ecosystems of immense complexity undergo rapid transformation to systems of remarkable simplicity, characterized by a scarcity of life forms. Environmental stressors increase towards the littoral zone, and thus, the littoral vegetation exhibits the lower species diversity and complexity of all coastal vegetation (Lugo, 1983). Stressed ecosystems do not recover; rather, further degradation may follow (Rapport and Whitford 1999). For instance, the existing degradation in the Jobos Bay area is at such level that the ox horn bucida (*Bucida buceras*) and the West Indian elm (*Guazuma ulmifolia*) climax forest, together with the marsh described by Gleason and Cook (1926), were completely clear-cut and filled, respectively, for the establishment of sugarcane (*Saccharum officinarum*) crops, urban settlements and heavy industries. As a result, almost all present land vegetation is seral (25% exotic) and nearly 90% of the coral reefs is dead and eroding, not growing. Therefore, the description of the different communities in and around Jobos Bay represents a degraded stage in an ever changing environment (Laboy-Nieves, 2001).

The Jobos Bay area presents a mosaic of five major communities: the mangrove forest, the evergreen littoral woodland, mudflats, seagrass beds, and coral reefs. A sixth unit, the marsh, is limited to very small patchy stands, which are the remnant of what they used to be before being inflicted by human activities during the first quarter of the XX century. Each one of these communities shelters a variety of organisms, particularly halophytes, xerophytic shrubs, marine and terrestrial invertebrates and birds. These populations exist very adjacent to each other in a relatively small area, and they holistically define a very complex, although paradoxically highly disturbed, cybernetic ecosystem (Laboy-Nieves, 2001).

4 THE MANGROVE FOREST

Over 90% of the Jobos Bay area lies above swamp deposits, which are deep and very poorly drained soils. This type of soil forms narrow strips adjacent to the ocean, which are slightly above sea level, but are wet at high tide and affected by salt or brackish water (USFWS, 1972). The high concentration of salt inhibits the growth of all vegetation, except the halophytes, the salt tolerant plants, of which mangroves are the dominant plant association.

Mangroves are circum-tropical intertidal forest ecosystems. They have a net import of nutrients, freshwater, and sediments from terrestrial environments and a net outflow of organic matter and water to marine or estuarine waters. Mangroves act as sediment traps that reduce water movement and retain suspended materials, gradually raising the land level by producing a rich organic soil. Being swampy and salty, mangrove ecosystems do not support a diverse flora, but contribute to estuarine and terrestrial food chains, to water quality, and to the maintenance of coastal geomorphology (Cintrón et al., 1985). The rich protected substrate provides habitat for a large variety of organisms, which in turn serve as the food base for marine and estuarine heterotrophic species.

The distribution, structure and physical conditions of the 400 hectares mangrove forest of the Jobos Bay estuary were extensively studied by Martínez et al. (1979). These authors concluded that the mangrove forest of Jobos represents 42.6% of this habitat type on the southern coast of Puerto Rico, thus it is a major ecological site.

Of the six physiographic types of mangrove forests (Lugo and Snedaker, 1974), three are found in Jobos Bay: basin, fringe and overwash forests (Figure 8). Basin forests develop inland and are characterized by slow sheet flows over wide areas of low topographic relief. This forest is normally separated from direct contact with the ocean, except during high tides or storm surges. It receives substantial amounts of freshwater during the rainy season, and in the dry season, salinity increases as the salt water intrudes. Fringe forests occur along the seaward edge and along the coastal lagoons and channels that connect the entire system to the open sea. The fringe forest at Jobos Bay shows a salinity gradient from the ocean (35‰) to the salt flat (100‰) and the basin (Lugo et al., 2007). Overwash mangrove forests develop offshore over shallow calcareous deposit platforms. Both fringe and overwash mangrove forests are flooded daily by high tides and by periodic storm surges; thus, they present a relatively constant salinity regime, oligotrophic conditions, and exposure to winds (Laboy Nieves, 2001). The degree and structural dynamic of these two forests are controlled

Figure 8. Aerial view depicting the three physiographic types *of mangrove forests in Camino del Indio, Jobos Bay.*

primarily by the quality of the soil and wave intensity. The water temperature at the fringe and basin forests in Jobos is directly proportional and higher than air temperature (Lugo et al., 2007). The geometry, shallowness, and high surface area of the mangrove fringe foment a high degree of physical exchange with the surrounding water, thus affecting the water temperature and oxygen diffusion (Laboy-Nieves, 1997). It has been demonstrated that the concentration of some leaf elements increased with salinity (N, P, Mg, Na) while Ca decreased (Lugo et al., 2007).

Tidal energy dissipates rapidly with distance from shore. The inland reduction of flushing results in a salinity gradient, with higher levels prevailing inland. Tidal amplitude determines the degree of flow and renewal of surficial and interstitial water. This water dynamic promotes the aeration of the substrate and the outflow of salts and noxious gases. Because of the low elevation leveled terrain, tidal changes and dissipation, flood frequency and depth gradients are generated. These conditions further affect the zonation patterns of the four mangrove species that inhabit the mangrove forest (Cintrón and Schaeffer-Novelli, 1983) at Jobos Bay: red mangrove (*Rhizophora mangle*), black mangrove (*Avicennia germinans*) white mangrove (*Laguncularia racemosa*) and buttonwood (*Conocarpus erectus*).

4.1 *Red mangrove* (Rhizophora mangle)

The red mangrove is the most conspicuous tree in the Jobos Bay wetland. It frequently pioneers the formation of the mangrove swamp, growing in shallow, silt soils under the influence of salty and brackish tidal water. This species is viviparous. Fully developed seedlings are rod-shaped, elongated and composed of two parts: a short plumule that consists of a pair of stipules protecting the first pair of leaves, and a long, heavy hypocotyl make up mainly of endospermous aerenchyma tissue. A small root forms before the seed drops from the parent tree (Figure 9). When released, the seedling floats on the currents for long distances and periods of time. After it reaches shallow waters or runs aground in mud, and it begins rapid growth (Juncosa, 1982).

According to Krohne (1998), red mangrove produces many prop roots that spread out to form a mass of interconnecting roots and to provide above-water growth. During the juvenile stages, red mangroves develop a short-lived subsystem of primary terrestrial roots, while in mature stages the tree generates an arching aerial root system that extends around the trunk or branches and emerges perpendicular to the ground. A lenticels system is responsible for aerating the roots while flooded. The roots provide anchorage for the tree while producing an extensive capillary root system that traps intertidal debris, which further produces a thick fibrous soil (Jiménez, 1985a).

Small islets and fringing stands along the shoreline are formed. The root labyrinth slows the flow of water, allowing silt and humic material to precipitate, gradually building up the substrate.

Figure 9. Three faces of the red mangrove (from left to right): seedlings, the rhizophere of the submerged root, and prop roots.

Behind the fringe, trees lower their size inland as interstitial salinity increases, establishing a forest with low (~2 m high) canopies and dispersed individuals (Laboy-Nieves, 2001).

This conspicuous tree forms a crested fringe along the shoreline of Jobos Bay (Figure 2). In the borders of Cayo Caribe, the species attains a height and basal area of 4.9 m and 11.5 m^2/ha, respectively, while in Mar Negro the respective values are 7.6 m and 18.5 m^2/ha (Villamil and Canals, 1981). After many years, a stable mudflat develops inland from the fringe and on the basin of the mangrove islets. Hypersaline conditions are established; the red mangrove no longer tolerates this environment and it is substituted by black and white mangroves.

4.2 *Black mangrove* (Avicennia germinans)

Black mangrove grows in tidal areas with salty or brackish waters in basin forests, over a broad range of soil salinity (0–100‰). It germinates on sandy, silty or clay soils, but in low areas inland from the fringe where tidal flooding is less frequent. Under high salinity conditions, structural development is suppressed and leaves excrete salt through specialized glands, contributing to salty through fall. Flowers are found in axillaries and terminal inflorescences, with one to 15 pairs of flowers per spike. They are small and with imbricate bracts. The corolla has four lobes with fragrant yellow petals. Flowering is sporadic throughout the year. The species is considered viviparous, because germination occurs while the embryo is still enclosed in the oblong fruit. Fallen seedlings float and are transported by tidal currents. The propagule sheds its pericarp and produces roots within three weeks of dispersal. Seedlings can become waterlogged and their establishment is limited to areas above water level at low tide and temperatures below 39°C (Jiménez and Lugo, 1985).

Black mangrove is characterized by a shallow underground root system with thin sinker roots and negatively geotropic pneumatophores that develop from lateral horizontal roots (Figure 10). As in the white mangrove, pneumatophores are responsible for gas exchange processes, but unlike that species, the number of pneumatophores in black mangroves reaches over 600/m^2. The surficial root systems make this species easily windthrown; therefore most of the trees at Jobos are short (about five meters tall), with exceptional individuals ranging up to 12 meters (Laboy-Nieves, 2001). Black mangrove is highly susceptible to changes in hydrological regimes; drought or flooding can cause extensive mortality. Despite the vulnerability of this species, in the middle of the Mar Negro basin lives a black mangrove individual that has a dbh of 76 cm, representing the largest tree of this species on the south coast of Puerto Rico.

The pneumatophore mat within the basin permits the accumulation of sand and humic deposits, which elevates the terrain, providing habitat for the least salt-tolerant mangrove species, the buttonwood, which lives as a shrub or tree peripheral to landward areas of the basin. This natural succession

Figure 10. Pneumatophores matt from the roots of the black mangrove.

pattern has been altered by anthropic factors, especially the construction of permanent structures and irrigation channels. For instance, Villamil and Canals (1981) reported that the construction of the Aguirre Railroad in the 1940's caused a detour in the flow of ground water to such a degree that interstitial salinity south of the railroad was 36‰ while north of it the values were around 2‰.

4.3 *White mangrove* (Laguncularia racemosa)

White mangrove grows under a wide variety of conditions. It is generally found on the inner fringe of the mangrove forests, on elevated soils where tidal inundations are less frequent and intense, and in basin mangrove forests where tidal flushing is limited. The fragrant flowers are pollinated by mostly by insects and bats; fruits production peaks in September and October. Normally, the fruit drops from the parent tree and the ridicule protrudes after a few days. Seedlings float and are dispersed by water. Floating is aided by a thick pericarp. Fruits sink after a floating period of about four weeks, and growth begins while the seed is still submerged (Jiménez, 1985b).

White mangroves possess peg-roots (pneumatophores) that radiates from the trunk. These seem to be associated with tidal fluctuations. They are club-shaped and their terminal heads contain ventilating tissues. In individual white mangroves growing in basin forests, aerial adventitious roots sometimes emerge from the lower section of the trunk (Cintrón and Shaeffer-Novelli, 1983). In the basin of Cayo Caribe, this species attains a height and basal area around 3.0 m and 1.5 m^2/ha, respectively, while in Mar Negro the respective values are 9.1 m and 20.5 m^2/ha (Villamil and Canals, 1981). However, there are two stands (northeast of Mar Negro and Pozuelo) inhabited by individuals about 15 meters high (Laboy-Nieves, 2001).

4.4 *Buttonwood* (Conocarpus erectus)

Buttonwood is not considered a real mangrove species, but a peripheral tree that constitutes the ecotone between the mangrove wetland and the dry upland. It occurs in elevated and low salinity

(<5‰) soils. It frequently grows as a shrub, but trees in Cayo Caribe and Camino del Indio reach heights around 8 m. The species has terminal globular inflorescences with bisexual greenish white fragrant flowers (Little et al., 1977). Each globule becomes a rounded brown aggregate fruit containing many seeds.

5 THE EVERGREEN LITTORAL WOODLAND

Most of the littoral upland vegetation (other than mangroves) around Jobos Bay is shrubby and secondary, and has been altered almost entirely by agriculture, industrialization, urban development, and recreation (Laboy-Nieves, 2008). Gleason and Cook (1926) described the original vegetation surrounding the Bay as a semi-evergreen seasonal forest dominated by *Bucida buceras* and *Guazuma ulmifolia*. García-Molinari (1952) pointed out that the coastal plain between Juana Díaz and Guayama is characterized by the almost complete absence of trees and the presence of only a few scattered shrubs and forbs. Villamil and Canals (1981) concluded that the vegetation of the scarce upland in Jobos Bay characterizes the seral evergreen littoral woodland.

In the temporal and spatial scale, the mosaic of populations in the upland observed at Jobos, defines transitional phases which are the by-product of the dynamic of natural and anthropic factors. Small variations in elevation, hydrology and climate, and pulses from catastrophic events, such as hurricanes, seem to be responsible for the physiographic diversity of the Jobos Bay upland. For instance, salinity or factors related to salinity play a major role in the fate of nitrogen compounds thus affecting the nitrifiers and denitrifiers (Mondrup, 1999). Migrating birds may contribute with their droppings to the phosphate regime in mud flats. Therefore, trees, shrubs, herbs, and vines develop structural and physiological adaptations to deal with salinity, winds, shifting nutrients, and water logging, aside the impact of human activities upon their survival.

The evergreen littoral woodland on Jobos is represented by more than 220 species (Table I), of which 25% are trees and the majority are native. As stated before, anthropic and natural phenomena have continuously disturbed this community. For example, only a single individual of the native milktree (*Plumeria alba*) is found on Cayo Caribe, while the exotics aloe plant (*Aloe vera*) and the wild tamarind (*Leucaena leucocephala*) are gaining terrain on that islet. The cactus *Cephalocereus royenii* is scarcely found on Cayos Caribes or Cayos de Barca, but it is relatively abundant on Pozuelo, less than two kilometers apart and having similar upland substrata (Laboy-Nieves, 2001).

Grasslands are the most representative community on the upland of Jobos Bay. They are short-grass type and occur in sparsely forested savanna of the southern coastal plain. Extensive saline areas along the coast line are clothed mainly with beachgrass (*Sporobolus indicus* and *S. virginicus*) while the slender grama grass (*Bouteloua heterostega*) and the Guinea grass (*Panicum maximum*) dominate the dry savanna. García-Molinari (1952) estimated that *Sporobolus* and other associated halophytic grasses and forbs, compress 15.9% of the vegetation of the southern coastal plain, the remaining was (*Bouteloua heterostega*), which occupies up to 130 tufts per square meter.

Almost nothing is known about the complex forests that grow behind the littoral zone of Puerto Rico, and there is confusion about the classification of the remaining stands. This vegetation complex is considered an ecosystem where plant associations in the seaward environments provide protection to the plants in the landward zones, by buffering the full impact of the marine climate. In turn, those associations provide stable and protective conditions that subsidize the seaward plants, by acting as seed sources, exporting organic matter, and acting as refuges during extreme environmental conditions (Lugo, 1983).

Jobos Bay is surrounded by a splendid mosaic of over 200 plant species. Table 1 shows the littoral woodland classified by Laboy-Nieves (2001). Herbs are the dominant vegetation (49%), followed by trees (27%) and shrubs (24%). The 57 species of trees identified by Laboy-Nieves (2001) inhabiting the Jobos littoral woodland represent less than eight percent of the native and introduced trees identified in Puerto Rico by Little et al. (1977). It is inferred that this relatively low taxonomic diversity is due to edaphic, climatic and anthropic stress experienced in the last century. European settlers gradually cut the coastal and lowland forest for pasture or cropland. Removals

Table 1. Phytobiota on the littoral woodland of Jobos Bay (T = Tree, S = Shrub, H = Herb, E = Endemic to Puerto Rico, N = Native to the Caribbean Region, P = Pantropical, E = Exotic).

Family and scientific name	Common english name	Type	Abundance	Origin
Acanthaceae				
Ruellia tuberosa	Many roots herb	H		N
Ruellia tweediana	Wild petunia	H	A	X
Aizoaceae				
Sesuvium portulacastrum	Sea purslane	H	A	N
Amaranthaceae				
Achyranthes indica	Amaranth	H	A	P
Amaranthus dubius	Amaranth	S	A	P
Amaranthus spinosus	Amaranth	S	A	P
Amaryllidacea				
Hymenocallis caribaea	Beach lily	H	I	N
Anacardiaceae				
Comocladia dodonaea	Poisson ash	S	I	N
Comocladia glabra	Carrasco	S	R	N
Magnifera indica	Mango	T	I	X
Spondias purpurea	Purple mombin	T	I	X
Annonaceae				
Annona muricata	Soursop	T	R	N
Annona reticulata	Custard apple	T	R	N
Apocynaceae				
Catharanthus roseous	Periwinkle	H	A	X
Plumeria alba	Nosegay tree	S	VR	N
Rauvolfia nitida	Milk bush	A	A	N
Urechites lutea	Wild allamanda	H	A	N
Araceae				
Colocasia esculenta	Coco yam	H	A	X
Aristolochiaceae				
Aristolochia trilobata	Santiago's vine	H	A	N
Asclepiadaceae				
Asclepias curassavica	Butterfly milkweed	H	A	N
Calotropis procera	Giant milkweed	H	A	X
Criptostegia grandiflora	Purple allamanda	H	A	X
Bataceae				
Batis maritima	Saltwort	H	V	P
Bignoniaceae				
Crescentia cujete	Calabash tree	S	R	N
Crescentia linearifolia	Calabash tree	S	R	N
Macfadyena ungis-cati	Cat-claw	H	A	P
Spathodea campanulata	African tuliptree	T	VR	X
Tabebuia heterophylla	White-cedar	T	A	N
Bombaceae				
Ceiba pentandra	Silk-cottonwood tree	T	VR	N
Boraginaceae				
Cordia colloccoca	Red manjack	T	VR	X
Cordia sebestena	Geiger-tree	T	VR	N
Heliotropium curassavicum	Heliothrope	H	A	P

(*Continued*)

Table 1. (Continued)

Family and scientific name	Common english name	Type	Abundance	Origin
Bromeliacea				
Bromelia pinguin	Pinguin	H	I	N
Tillandsia recurvata	Bunch moss	H	A	N
Burseraceae				
Bursera simaruba	West Indian birch	T	A	N
Cactaceae				
Hylocereus trigonus	Strawbery-pear	H	I	N
Opuntia dillenii	Erect pricklypear	S	A	N
Cephalocereus royenii	Dildo	S	A	N
Capparaceae				
Capparis flexuosa	Limber caper	S	A	N
Cleome aculeata	Jasmine	H	I	P
Cleome gynandra	Spider flower	H	I	X
Casuarinacea				
Casuarina equisetifolia	Australian pine	T	R	X
Celastraceae				
Cassine xylocarpa	Marble tree	S	A	N
Crossopetalum rhacoma	Coral	S	A	N
Chrysobalanaceae				
Crysobalanus icaco	Coco plum	S	R	N
Combretaceae				
Bucida buseras	Oxhorn bucida	T	A	N
Conocarpus erectus	Buttonwood	T	A	P
Laguncularia racemosa	White mangrove	T	A	P
Terminalia catappa	West Indian almond	T	A	X
Commelinaceae				
Commelina diffusa	Climbing dayflower	H	A	P
Commelina elegans	French-weed	H	R	P
Compositae (Asteraceae)				
Bidens alba	Romerillo	H	A	N
Elephantopus mollis	Elephant's foot	H	A	P
Emilia sonchifolia	Lilac tassel flower	H	I	P
Mikania cordifolia	Florida Keys hempvine	H	A	P
Pluchea odorata	Sweet scent	S	A	N
Symphyotrichum expansum	Saltmarsh aster	H	A	N
Tridax procumbens	Coatbuttons	H	A	P
Vernonia cinerea	Bitter bush	H	A	P
Wedelia trilobata	Bay Biscayne creeping-oxeye	H	A	N
Convolvulaceae				
Ipomea batatas	Sweet potato	H	A	X
Ipomoea carnea	Tree morning glory	H	A	N
Ipomoea indica	Oceanblue morning-glory	H	A	X
Ipomoea macrantha	Coast moon vine	H	A	P
Ipomoea pes-caprae	Beach morning glory	H	A	
Merremia aegyptia	Beach vine	H	A	N
				N
Crassulaceae				
Bryophyllum pinnatum	Life plant	H	I	P

(Continued)

Table 1. (Continued)

Family and scientific name	Common english name	Type	Abundance	Origin
Cucurbitaceae				
Cucumis anguria	Wild cucumber	H	A	X
Curcubita moschata	Squash	H	R	X
Mormodica charantia	Wild balsam apple	H	A	X
Cyperaceae				
Cyperus rotundus	Nut-grass	H	A	P
Cyperus planifolius	Beach sedge	H	A	P
Eleocharis mutata	Scallion grass	H	A	N
Fimbristylis cymosa	Spikerush	H	A	N
Fimbristylis dichotoma	Forked fimbry	H	A	P
Fimbristylis ferruginea	West Indian fimbry	H	A	P
Mariscus ligularis	Alabama swamp flatsedge	H	A	P
Erythrocylaceae				
Erythroxylum rotundifolium	Brisselet	T	A	N
Euphorbiaceae				
Chamaesyce articulata	Seaside spurge	S	A	N
Chamaesyce glomerifera	Spurge	H	A	N
Chamaesyce prostrata	Snake-weed	H	R	P
Cnidocalus acotinifoliums	Spurge nettle	S	A	X
Croton lobatus	Croton	S	R	N
Croton lucidus	Fire bush	S	R	N
Croton rigidus	Croton	S	A	N
Euphorbia heterophylla	Wild poinsettia	H	A	N
Hippomane mancinella	Manchineel	T	A	N
Hura crepitans	Sandbox	T	A	N
Jatropa gossypifolia	Tautuba	S	A	N
Phyllanthus acidus	Gooseberry tree	T	R	X
Ricinus comunis	Castor-oil plant	S	R	X
Fabaceae				
Abrus precatorius	Crab eye vine	H	A	X
Mucuna urens	Oxeye bean	H	R	N
Mucuna pruriens	Cow itch	A	A	P
Goodeniaceae				
Scaevola plumieri	Inkberry	H	A	N
Gramineae/Poaceae				
Andropogon bicornis	Bear grass	H	A	N
Arundo donax	Giant reed	H	A	X
Axonopus compressus	Tropical carpet grass	H	A	P
Bouteloua heterostega	Grama grass	H	A	N
Cenchrus brownii	Slimbristle sandbur	H	A	X
Cenchrus echinatus	Bur grass	H	A	N
Chloris inflata	Swollen fingergrass	H	A	P
Cynodon dactylon	Bermuda grass	H	A	P
Dactylotenium aegyptium	Crowfoot grass	H	A	X
Dichantium annulatum	Shedagrass	H	A	X
Digitaria decumbens	Fingergrass	H	A	X
Echinohcloa colonum	Hungle rice	H	A	P
Eleusine indica	Piepul	H	A	P
Leptochloa filiformis	Red sprangletop	H	A	X

(*Continued*)

Table 1. (Continued)

Family and scientific name	Common english name	Type	Abundance	Origin
Panicum adspersum	Redtop millet	H	A	X
Panicum fasciculatum	Browntop panicum	H	A	N
Panicum maximum	Guinea grass	H	A	X
Panicum purpurascens	Para grass	H	A	N
Panicum virgatum	Switchgrass	H	I	N
Paspalidium geminatum	Egyptian panicgrass	H	A	N
Paspalum conjugatum	Creeping wheatgrass	H	A	P
Paspalum distichum	Knotgrass	H	A	P
Paspalum laxum	Coconut paspalum	H	A	N
Pennisetum ciliare	Buffle grass	H	A	N
Saccharum officinarum	Sugar cane	H	A	X
Sporobolus indicus	Beachgrass	H	A	N
Sporobolus virginicus	Seashore dropseed	H	A	N
Tricholaena repens	Natal grass	H	A	X
Zea mays	Corn	H	I	N
Guttiferae				
Calophylum brasiliense	Alexandrian laurel	T	R	N
Labiatae				
Leonotis nepetifolia	Lion's ear	H	A	P
Lauraceae				
Cassytha filiformis	Devil's gut	H	A	P
Leguminosae				
Acacia farneciana	Sweet acasia	T	A	P
Albizia lebbeck	Lebbek	T	R	X
Bauhinia monandra	Purple bauhinia	T	VR	X
Caesalpinia bonduc	Gray nickers	S	VR	P
Caesalpinia pulcherrima	Flowerfence	S	R	X
Cassia alata	Ringworm shrub	S	R	N
Cassia obtusifolia	Java-bean	S	R	N
Cassia occidentalis	Stinking weed	S	A	P
Cassia siamea	Senna tree	T	R	X
Delonix regia	Flamboyant-tree	T	VR	X
Enterolobium cyclocarpum	Earpod-tree	T	VR	X
Erythrina glauca	Swamp immortelle	T	I	N
Haematoxylum campechianun	Logwood	T	A	X
Leucaeba glauca	Leadtree	T	A	X
Machaerium lunatum	Palo de hoz	T	A	N
Mimosa pudica	Sensitive plant	H	A	N
Neptunia plena	Water dead and awake	T	A	P
Parkisonia aculeata	Horsebean	S	A	P
Pithecellobium dulce	Guamuchil	T	R	X
Pithecellobium saman	Raintree	T	A	N
Pithecellobium ungis-cati	Cat's claw	T	A	N
Prosopis juriflora	Mesquite	T	A	X
Tamarindus indica	Tamarind	T	A	N
Liliaceae				
Aloe vera	Common aloe	H	I	N
Sansevieria hyacinthoides	Bowstring hemp	H	A	X
Malpighiaceae				
Stigmaphyllon periplocifolium	Monarch Amazonvine	H	A	N

(*Continued*)

Table 1. (Continued)

Family and scientific name	Common english name	Type	Abundance	Origin
Malvaceae				
Gossipium barbadense	Cotton tree	S	A	X
Hibiscus tiliaceus	Sea hibiscus	T	A	P
Malachra alceifolia	Mallow	S	A	N
Sida acuta	Southern sida	H	A	P
Thespesia populnea	Cork tree (emajaguilla)	T	A	P
Urena lobata	Caesarweed	S	A	P
Meliaceae				
Swietenia mahogani	Mahogany	T	R	X
Moraceae				
Artocarpus altilis	Breadfruit tree	T	VR	X
Ficus laevigata	Shortleaf fig	T	R	N
Myoporaceae				
Bontia daphnoides	White-alling	S	A	N
Myriaceae				
Myrica cerifera	Southern bayberry	S	I	N
Myrtaceae				
Psidium guajaba	Guava	S	R	N
Nyctaginaceae				
Boerhavia diffusa	Spreading hog-weed	H	A	P
Pisonia albida	Corcho bobo	T	I	E
Oleaceae				
Jasminum fluminense	Jasmine	S	R	X
Onagradaceae				
Ludwigia octavalvis	Primrose Bellow	S	A	N
Orchidaceae				
Vanilla planiflora	Vanilla	H	A	N
Palmaceae				
Cocus nucifera	Coconut palm	T	I	P
Phoenix dactylifera	Date palm	T	I	P
Roystonea borinquena	Royal palm	T	I	E
Papaveraceae				
Argemone mexicana	Prickly poppy	H	A	N
Papilionaceae				
Andira inermis	Cabbage angelin	T	I	N
Cajanus cajan	Pigeon pea	S	I	X
Canavalia maritima	Baybean	H	A	P
Centrosema pubescens	Butterfly pea	H	A	N
Centrosema virginianum	Butterfly pea	H	A	P
Crotalaria palida	Rattleweed	S	A	P
Gliricidia sepium	Mother-of-cocoa	T	A	X
Lonchocarpus domingensis	Geno-geno	T	R	N
Macroptilium lathyroides	Wild bushbean	S	A	X
Pictetia aculeata	Fustic	S	A	N
Sesbania sericea	Papagayo	S	A	X
Tephrosia cinerea	Tephrosia	H	A	N
Vigna luteola	Hairypod cowpea	H	A	N

(*Continued*)

Table 1. (Continued)

Family and scientific name	Common english name	Type	Abundance	Origin
Passifloraceae				
Passiflora edulis	Passion fruit	H	R	X
Passiflora foetida	Fetid passionflower	H	A	N
Passiflora suberosa	Passiflora	H	A	N
Polygonaceae				
Antigonum leptopus	Coral	H	A	X
Cocoloba diversifolia	Doveplum	T	A	N
Cocoloba uvifera	Sea grape	T	A	N
Polypodiaceae				
Acrostichum aureum	Mangrove fern	H	A	P
Acrostichum danaefolium	Mangrove fern	H	A	P
Portulacaceae				
Portulaca oleracea	Common purselane	H	A	P
Pteridaceae				
Acrostichum danaeifolium	Giant leather fern	S	A	P
Rhamnaceae				
Colubrina arborescens	Greenheart	S	A	N
Colubrina reclinata	Soldierwood	T	A	N
Rhyzophoraceae				
Rhizophora mangle	Red mangrove	T	A	P
Rubiaceae				
Chiococa alba	Alba	H	A	N
Randia aculeata	Ink berry plant	S	A	N
Rutaceae				
Amyris elemifera	Tea	S	A	N
Sapindaceae				
Melicoccus bijugatus	Kinep	T	A	X
Serjania polyphylla	Basketwood	H	A	N
Sapotaceae				
Manilkara zapota	Sapodilla	T	R	X
Schrophulariaceae				
Capraria biflora	Tea	H	R	N
Simaroubaceae				
Suriana maritima	Baycedar	S	A	P
Solanaceae				
Datura stramonium	Jimsonweed	S	A	P
Solanum erianthum	Dove plant	S	A	N
Solanum torvum	Turkeyberry	S	A	P
Sterculiaceae				
Guazuma ulmifolia	West Indian elm	T	A	N
Helicteres jamaicensis	Screwtree	S	A	N
Melochia pyramidata	Pyramid flower	S	A	P
Melochia tomentosa	Teabush	S	A	N
Sterculia apetala	Panama tree	T	VR	X
Waltheria indica	Malva branca	S	A	P
Theophrastaceae				
Jacquinia arborea	Braceletwood	S	A	N

(*Continued*)

Table 1. (Continued)

Family and scientific name	Common english name	Type	Abundance	Origin
Thyphaceae				
Typha dominguensis	Cattail	H	A	N
Tiliaceae				
Corchorus hirsutus	Jackswitch	S	A	P
Trimfetta semitriloba	Bur bark (cadillo)	S	A	P
Turneraceae				
Turnera diffusa	Damiana	S	A	N
Vervenaceae				
Avicennia germinans	Black mangrove	T	A	P
Citheraxylun fruticosum	(Péndula)	T	A	N
Lantana involucrata	Buttonsage	S	A	N
Lippia nodiflora	Matt limpia	H	A	P
Lippia stoechadifolia	Southern fogfruit	H	A	P
Stachytarpheta jamaicensis	Vervain	H	A	N
Vitaceae				
Cissus trifolliata	Sorrelvine	H	A	N

for timber, charcoal, and fuel wood significantly modified the forest (Wadsworth, 1950). The abandonment of sugarcane crops and the increase of urbanization, paradoxically have promoted the maturation of remnant stands of native flora, and the proliferation of exotic species (40% of all trees). Hence, tree species diversity apparently has increased in the last decades, when compared to the description provided by Gleason and Cook (1926).

6 MUD FLATS, SEAGRASS BEDS AND CORAL REEFS

Mudflats are important soft-bottom littoral systems. They are formed inland from the mangrove forest, as a result of reduced water runoff, higher evaporation rates and drought. Their existence is also a byproduct of two other factors: the prevalence of offshore barriers (typically mangrove fringe forests and coral reefs) to moderate the waves, and the upward sloping of the bottom so that tidal waters can spread and deposit the silt, mud and organic debris imported from open waters. Salt and mud flats are at least superficially depauperated in species; i.e., the diversity of emergent plant life is not conspicuous (Krohne, 1998). In Jobos, where vegetation exists, it consists mainly of the salt and heat tolerant grasses like *Sporobolus virginicus, Fimbristylis cymosa, Batis maritima* and *Sesuvium portulacastrum*.

Mud flats are common to all intertidal marshes. The flats exposed at low tide or drought, contain considerable quantities of detritic mud, salts, and plant and animals remains resulting from the action of the water. As the tide ebbs, it exposes broad areas of seemingly barren mud or sand; but the appearance is deceiving. From a few millimeters below the surface of the mud flat, down to more than a meter, the moist humic bottom supports bacteria, fungi, diatoms and a veritable universe of marine animals, including crabs, clams, worms and nematodes.

Mud flats are the least-studied community in Jobos Bay. A detailed analysis of aerial photographs from 1937 and 2004, revealed that the surface area of mud flats is increasing behind the mangrove fringes of Jobos Bay, particularly in Mar Negro and Punta Pozuelo. Natural (hurricanes and drought) and anthropic (deforestation, urbanization, and pollution) factors may be contributing to this phenomenon (Laboy-Nieves, 2001). Field observations of old structures built near shore early in the 20th century, also indicates that the sea level have risen. Ellison and Stoddart (1990) argued that low island mangrove ecosystems in the past have been able to keep up a sea level rise of up to 9 cm per century, but that at rates over 12 cm per century, they have not been able to keep up with sea level rises. On this basis, they concluded that mangroves will eventually collapse as viable coastal ecosystems, as a result of sea level rise. Goudie (2000) postulated that rises in sea level will increase near shore water depths and thereby modify wave refraction patterns. It is deduced that this sea level rise eventually may force the inland longshore migration of salt marsh and mud flat systems. However, mangroves may respond rather differently from other marshes in that they are composed of relatively long-lived trees and shrubs, which means that speed of zonation change will be less (Woodroffe, 1990).

Sea grasses are aquatic angiosperms that are generally restricted to soft sediment habitats. Their leaves and stems provide the primary substratum for the attachment of epiphytes within seagrass meadows, and they are the basic structural component in which the grazer-epiphyte interaction occurs (Jernakoff et al., 1996). The seagrass beds cover around 70% of the shallow (<3 m) substrata in Jobos Bay, and about 30% in deeper areas down to 10 m (PWRS, 1972). Turtle grass (*Thalassia testudinun*), manatee grass (*Syringodium filiforme*) and the Caribbean seagrass *Halophila decipiens* are the three main angiosperms. These plants share their habitat with nearly 60 algae species (Almodovar, 1964), mostly from the genus *Caulerpa, Diplanthera*, and *Halodule*. Seagrass beds have long been recognized as one of the most productive biological communities in the world (Zieman et al., 1989). They provide nursing ground, food and shelter for most fish and invertebrate species, and large animals like the West Indian manattee (*Trichechus manatus*) and the hawksbill sea turtle (*Eretmochelys imbricata*).

Seagrass beds develop mainly in low energy zones protected by fringing reefs and mangroves. Light penetration, salinity, temperature and turbidity of the water column (Zieman, 1976) and the natural and anthropic mechanical disturbance of the substratum are the major factors that impoverish their distribution (Zieman, 1976; Duarte et al., 1997; Laboy-Nieves, 1997). Although seagrass beds form blowouts (bare muddy or sandy benthic depressions), they are prone for the accumulation of sediments by binding the substrate and protecting it from erosion. This creates a baffle on which fine sediments can settle and provides suitable habitat for many benthic species (Patriquin, 1975).

Migration of the blowouts must result in periodic disruption of the major part of the seagrass community and turnover of associated sediments. The most important effect of this instability on the ecology of the seagrass beds is to limit seral development. The absence in the blowouts of a well-developed epifauna and flora, a characteristic of advanced stages of seral development in the beds, is evidence of this phenomenon. *Syringodium* frequently precedes *Thalassia* in the colonization of blowouts, apparently because of its greater tolerance to environmental stress and higher rhizome growth rates, but *Thalassia* is invariably the terminal dominant of the Caribbean seagrass community in shallow waters (Patriquin, 1975).

Sea grasses are the primary producers in the submerged habitat of Jobos Bay. It is infer that the productivity of the beds may be due to the shallowness of the Bay, which permits an abundance of light penetration to provide energy to the phytoplankton, algae and fanerogams. In this habitat, phytoplankton is the foundation of all food production, and the zooplankton, particularly copepods, make up the broad pasturing that further turns into plankton explosions that attract other primary and secondary consumers. Herbivores consume the bulk aboveground production, but herbivory can stimulate new shoot growth at intermediate grazing levels. Its also appears that destructive overgrazing of sea grasses is prevented by the grazing-induced loss of seagrass shelter for juvenile herbivores, therefore plant-animal interactions in seagrass-dominated ecosystems require substantial revision (Heck and Valentine, 1999).

The coral reef is a near shore community confined to shallow circumtropical waters. Together with mangroves and seagrass beds, coral reefs form one of the must complexes, diverse and productive coastal associations in the world. Their superstructure is the product of tiny colonial animals (polyps from Phylum Cnidaria) that attach to the hard surfaces of the sea floor. By releasing calcium carbonate from seawater, they build skeletal structures in an infinite variety of shapes and sizes. For coral reefs to develop, the delicate polyps must flourish. This requires several critical environmental factors: constant movements of warm water, salinity, high transparency, and a firm base for attachment (Humann, 1996). The coral reef is one of the best examples of mutuality symbioses in tropical marine waters, performed by the polyps and the alga *Zooxanthellae* spp. Finger corals, gorgonians, zoanthids, and seagrass beds characterize the scoured channels between the Cayos Caribes and Cayos La Barca islets, where most of the coral reefs in Jobos Bay are located (Figure 2).

Corals at Jobos Bay present the typical zonation of Caribbean reefs, as described by Colin (1978). The reef flat, usually less than 0.5 m deep and exposed during low tides, is dominated by an intermixed association composed of isolated coral heads of finger coral (*Porites porites*), fire corals (*Millepora complanata*) and zoanthids (*Zoanthus sociatus*). The reef surge zone is a high wave energy area that ranges in depth from one to six meters. The dominant coral species in this area are the fire coral and the elkhorn coral (*Acropora palmata*). The reef slope drops off to about 18 m, and contains massive heads of the brain coral (*Diploria clivosa*), boulder star coral (*Montastrea annularis*), great star coral (*Montastrea cavernosa*) and sea fans (Laboy-Nieves, 2001).

Global climate changes seem to be jeopardizing more than 60% of the world coral reefs. Rising sea temperatures fueled by climate change, is causing coral to lose its symbiotic relationship with the zooxanthelae, or the photosynthetic pigmentation of the algae. The resulting bleaching phenomenon indicates that corals are starving, beyond a point of recovering (Van Putten, 2000). The reefs at Jobos are not an exception. Sedimentation, thermal and chemical pollution, and mechanical disturbances are the factors that have almost exterminated the coral reefs in the Mid Bay zone. Recent hurricanes have broken massive pieces of the typical surf zone corals. Extreme flooding discharges from Río Seco, Río Malanía and Río Guamaní, associated to abnormal rainfall, seems to have lowered the physiological limits for transparency and salinity that corals can tolerate, producing an extended mass mortality, similar to that reported by Laboy-Nieves and Conde (2001). All of these factors have lead coral reefs to less than 30% of live cover (Morelock, 1988).

7 FAUNA ASSOCIATED WITH INTERTIDAL UPLAND HABITATS AND MUDFLATS

The emergent segments of the prop roots of the red mangrove, as well as the stems of the white and black mangroves, the littoral woodland and the salt and mud flats are used as substrate, feeding, perching, and nesting sites by a wide variety of invertebrates especially insects, crustaceans and mollusks, as well as birds, reptiles and bats. These animals represent endemic, native, migrant and exotic species that enrich the trophic scenery of the Jobos Bay.

Associated with the submerged prop roots of the red mangrove is a rich epibiota (Figure 9). Competition for space on these roots is high (Kolehmainen, 1972). Among the most abundant groups are oysters, tunicates, sponges, crustaceans, cnidarians, and algae. Yoshioka (1975) identified common species inhabiting the submerged rhyzosphere of the red mangrove in the channels of Mar Negro. The stripped and the Poli's stellate barnacles (*Chthamalus stellatus* and *Balanus amphitrite*, respectively) are the dominant species. According to Díaz et al. (1985) and Laboy-Nieves (2001) it is inferred that the abundance and distribution of these species in the rhyzosphere community is influenced by currents, transparency, dissolved oxygen, salinity, nutrients and other physical, chemical and biological conditions. The existence of an epibenthic community on the red mangrove rhyzosphere is more imminent in clearer and less disturbed waters, which in Jobos are predominantly found in the inner channels of Mar Negro and the reef lagoons.

Terrestrial invertebrates from Jobos Bay have barely been studied. Laboy-Nieves (2001) presented a preliminary list of land macroinvertebrates. Mangroves are commonly inhabited by snails (*Melampus coffeus*), termites (*Coptotermes brevis*) and bees (*Apis melliphera*). The littoral woodland shows more biodiversity. Among the most conspicuous invertebrates are black tarantula (*Avicularia laeta*), hairy spider (*Cyrtopholis portoricae*), giant millipede (*Orthocricus arboreus*), centipede (*Scolopendra alternans*), giant ant (*Odontomachus raematoda*), black butterfly (*Calisto nubila*) and ground wasp (*Stictia signata*).

Because of the predominantly saline and xeric environmental conditions, few amphibians inhabit Jobos Bay, compared to the rest of the Island. The dominant species identified by Laboy-Nieves

(2001) are two tree frogs (*Eleutherodactylus antillensis* and *E. coqui*), a toad (*Bufo marinus*) and the white-lipped frog (*Leptodactylus albilabris*). The endemic garden snake (*Alsophis portoricensis*) and the snapping turtle (*Pseudemys terrapen*) share the land habitat in Jobos Bay with 11 lizards. These lacertid fauna is integrated by the teiid (*Ameiva exsul*), anoles (*Anolis cristatelus, A. poncencis, A. pulchellus*, and *A. stratulus*), two geckos (*Hemidactylus brooki* and *Phyllodactylus wirshingi*), worm lizard (*Amphisbaena caeca)*, the exotic green iguana (*Iguana iguana*), and two dwarf geckos (*Sphaerodactylus macrolepis* and *S. nicholsi*). Only six species of wild mammals inhabit the Jobos Bay upland. These are the fruit bat (Artibeus jamaicencis), fishing bat (*Noctilio leporinus*), mongoose (*Herpestes javanicus*), house bat (*Molosus molosus*), mice (*Mus musculus*), and rat (*Rattus norvegicus*). Feral cats, dogs, pigs and goats also roam in the upland of Jobos. Birds are the most prominent vertebrate fauna in Jobos Bay. Mangroves, xeric vegetation, marshes, beaches and intertidal zones provide habitat to more than 100 native, migrant and exotic bird species (Table 2).

For many years, it was generalized that deposit-feeders are more abundant in muddy habitats than filter-feeders, because the latter are excluded by the former through amensalistic interactions. Snelgrove and Butman (1994) conducted a critical re-examination of the data on animal-sediment relationships. They suggested that many species are not always associated with a single sediment type, and that suspension- and deposit-feeders often coexist in large numbers. Many species alter their trophic mode in response to water flow and food flux. Species distribution and abundance at the mud flat, particularly the infauna (benthic invertebrates that live largely within the mud bed) seem to be directly related to sediment pore, organic matter content, pore-water chemistry, microbial abundance, larval supply, and climatic and hydrologic conditions.

The muddy flats exposed at low tides contain considerable quantities of detritus, resulting from plant and animal remains carried by currents and filtered in the mangrove fringe. Around the mudflat edges and above the reach of the highest tides, crabs forage almost continuously for detritic material, digesting nutrients from the organic portion and excreting the sand, mud and undigested organic remains. The surface of the flat is inhabited by bacteria, diatoms, fungi, nematodes, copepods, ostracods, and turbellarians, while underneath naupli, sipunculids, clams, and worms are prevalent. These organisms serve as food sources for deposit-and filter-feeders, carnivorous birds, such as herons and plovers and omnivorous decapods, like the land and the blue crabs, *Cardisoma guanhumi* and *Callinectes sapidus*, respectively.

One of the relevant trophic events of the mud flats occurs when diatoms and cyanobacteria form microbial mats in the inundated pannes. The carpet of these microorganisms covers the mud at different periods of the day. Diatoms (*Nitzchia* spp.) remain planktonic during the daytime (Talbot et al., 1990), thus they are available to filter-feeders like clams. Cyanobacteria from the genus *Lyngbya* and *Schizotrix* form a carpet over the substrate, which is further broken by the ascension of air bubbles or currents. These microorganisms exhibit microbial loops and form mucilages that are the main food source for copepods, isopods, polychaetes and mollusks inhabiting the panne (Hart et al., 2000). The enrichment of the panne attracts macroinvertebrates and birds to feed in the flat, and a complex food web is established. Another relevant trophic episode in Jobos is the parasitism of fish by the isopods *Alcirona krebsii*, *Excorallana tricornis*, *E. quadricornis*, and *E. sexticornis*, as reported by Anonymous (1975).

8 FAUNA ASSOCIATED WITH SEAGRASS BEDS AND CORAL REEFS

The presence of mangrove fringes, seagrass beds and coral reefs adjacent to each other in Jobos Bay allow for the establishment of a rich and complex faunal structure. Animals from these communities usually move in and out of them, thus it could be erroneous to assign planktonic, nektonic, and benthonic species as isolated residents of one these communities. For instance, fishes use the red mangrove submerged roots for nursing; as the individuals grow, they move to the coral reefs and usually forage in the seagrass beds. Most of the organisms living in seagrass beds and their associated blowouts, receive nutrients directly or indirectly from the coral reefs and mangrove forests, and vice versa, hence defining a cybernetic biotope.

Table 2. List of birds inhabiting Jobos Bay.

Order	Scientific name	Common english name
Pelecaniformes	*Fregata magnificens*	Magnificent frigatebird
	Sula leucogaster	Brown booby
	Pelecanus occidentalis	Brown pelican
Ciconiiformes	*Ardea herodias*	Great blue heron
	Bubulcus ibis	Cattle egret
	Butoroides striatus	Green heron
	Casmerodius albus	Great egret
	Egretta caerulea	Little blue heron
	Egretta thula	Snowy egret
	Egretta tricolor	Louisiana heron
	Nycticorax nycticorax	Black-crowned night heron
	Nycticorax violaceus	Yellow-crowned night heron
	Plegadis falcinellus	Glossy ibis
	Phoenicopterus rubber	Flamingo
Anseriformes	*Anas bahamensis*	White-cheeked pintail
	Anas discors	Blue-winged teal
Falconiformes	*Buteo jamaicensis*	Red-tailed hawk
	Cathartes aura	Turkey vulture
	Falco peregrinus	Peregrine falcon
	Falco sparverious	Kestrel
	Pandion haliaetus	Osprey
Gruiformes	*Gallinula chloropus*	Common moorhen
	Porzana carolina	Sora rail
	Rallus longirostris	Clapper rail
Charadriiformes	*Actitis macularia*	Common sandpiper
	Anous stolidus	Brown noddy
	Arenaria interpres	Ruddy turnstone
	Calidris canutus	Red knot
	Calidris himantopus	Stilt sandpiper
	Calidris mauri	Western sandpiper
	Calidris melanotus	Pectoral sandpiper
	Calidris minutilla	Least sandpiper
	Calidris pusilla	Semipalmated sandpiper
	Charadrius alexandrinus	Snowy plover
	Charadrius semipalmatus	Semipalmated plover
	Charadrius vociferus	Killdeer
	Charadrius wilsonia	Wilson's plover
	Gallinago gallinago	Common snipe
	Haematopus palliatus	American oyster catcher
	Himantopus mexicanus	Black-necked stilt
	Larus atricilla	Laughing gull
	Limnodromus griseus	Short-billed dowicher
	Numenius phaeopus	Whimbrel
	Pluvialis squatarola	Black-bellied plover
	Rynchops niger	Black skimmer
	Sterna antillarum	Least tern
	Sterna hirundo	Common tern
	Sterna maxima	Royal tern
	Sterna sandvicensis	Sandwich tern
	Tringa flavipes	Lesser yellowlegs
	Tringa melanoluca	Greater yellowlegs

(Continued)

Table 2. (Continued)

Order	Scientific name	Common english name
Columbiformes	*Columba leucocephala*	White-crowned pigeon
	Columba livia	Rock dove
	Columbina passerina	Ground dove
	Patagioenas inornata	Puerto Rican plain pigeon
	Zenaida asiatica	White-winged dove
	Zenaida aurita	Zenaida dove
	Zenaida macroura	Morning dove
Psittaciformes	*Amazona ventralis*	Hispaniolan parrot
	Amazona vidirigenalis	Red-crowned parrot
	Melopsittacus undulatus	Parakeet
	Myopsitta monachus	Monk parakeet
Cuculiformes	*Coccyzus americanus*	Yellow-billed cuckoo
	Coccyzus minor	Mangrove cuckoo
	Crotophaga ani	Smooth-billed ani
Strigiformes	*Asio flammeus*	Short-eared owl
Caprimulgiformes	*Chordeiles gundlachi*	Antillean nighthawk
Micropodiformes	*Anthracothorax dominicus*	Green mango
	Anthracothorax viridis	Antillian mango
	Chlorostilbon maugaeus	Puerto Rican Emerald
	Cypseloides niger	Antillean black swift
Coraciiformes	*Megaceryle alcyon*	Belted kingfisher
Piciformes	*Melanerpes portoricensis*	Puerto Rican woodpecker
Passeriformes	*Agelaius xanthomus*	Yellow-shouldered blackbird
	Coereba flaveola	Bananaquit
	Dendroica atriata	Blackpoll warbler
	Dendroica discolor	Prairie warbler
	Dendroica petechia	Yellow warbler
	Elaeina martinica	Caribbean elaeina
	Estrilda melpoda	Scarlet-cheeked weaver finch
	Eupletes orix	Red bishop
	Geothlypis trichas	Common yellowthroat
	Hirundo rustica	Barn swallow
	Hirundo fulva	Cave swallow
	Lonchura cuculata	Bronze manikin
	Margarops fuscatus	Pearly-eye thrasher
	Mimus polyglotus	Northern mockingbird
	Mniotilta varia	Black and white warbler
	Molothrus bonariensis	Glossy cowbird
	Myarchus antillarum	Puerto Rican flycatcher
	Quiscalus niger	Greater Antillean grackle
	Parula americana	Northern parula
	Progne dominicencis	Caribbean martin
	Seiurus aurocapillus	Ovenbird
	Seirus noveboracensis	Northern watertrush
	Setophaga rutinilla	American redstart
	Spindalis zena	Stripped-headed manager
	Tiaris bicolor	Black-faced grassquit
	Tiaris olivacea	Yellow-faced grassquit
	Tyrannus dominicencis	Gray kingbird
	Vireo altiloquus	Black-whiskered vireo

The benthos (animals that live on or in the sea floor) is dominated by foraminifers *Quinqueloculina lamarckiana* and *Archais angulatus* on the soft bottoms, and mollusks, crustaceans and echinoderms on the hard substratum of the seagrass beds. Kolehmainen (1972) found 21 species of ophiurids, representing the largest diversity of echinoderms in the *Thalassia* beds. Laboy- Nieves (2001) reported 16 species of holothurians. Crustaceans (brachyuran larvae, amphipods and copepods), tunicates and gastropod larvae compose most of the zooplankton in Jobos Bay (PRWRA, 1972). In the coral reefs, 20 species of anthozoans and 10 species of soft corals have been identified, and associated with them, over 250 fish species (Villamil and Canals, 1981). The largest animals that frequently forage in the seagrass beds of Jobos Bay are the West Indies manatee (*Trichechus manatus*), bottlenose dolphin (*Tursiops* truncatus), hawksbill sea turtle (*Erectmochelys imbricata*), and the green sea turtle (*Chelonia mydas*). The most spectacular animal phenomenon in Jobos Bay is bioluminescence, due to the presence of the dinoflagellates *Pyrodynium bahamensis* and *Ceratium furca*.

9 ENERGY AND NUTRIENT DYNAMICS

All life is energy limited. For instance, the *Thalassia* seagrass and the coral reef community develop ultimately as a result of changes in the ability of plants to obtain solar energy. The reproductive strategy of fiddler crabs, mangroves, herons, fishes and many representative estuarine species is also constrained by energy. Species interactions, a major organizing force in a community is often a contest of energy, too. Thus at the ecosystem level, the process of accumulating and dissipating energy defines the structure and dynamic of the communities that represent the estuary. The rate of accumulation of energy in organic molecules by photosynthesis is known as primary production. Productivity is a difficult quantity to measure in the field, because is limited by six major factors: light, temperature, water, nutrients, stress, and disturbance. These factors affect ecosystems at different spatial and temporal scales. Nevertheless, among the different units to express productivity, biomass represents the most practical approach, because of the fact that the total dry weight of organic matter is directly correlated with energy content. It has been generalized that ecosystems with high biomass also have high productivity. However, it is difficult to interpret this result, because it may be that a large biomass is needed for production to occur at a high rate, or perhaps high production leads to high biomass. After determining the ratio of productivity and biomass (P:B) for various terrestrial and aquatic systems, Whitakker (1975) noted that the ratios fall into three general groups: low P:B (forests and shrubland habitats), average P:B (grassland-type habitats), and high P:B (freshwater and marine habitats); this last group due primarily by single-celled phytoplankton. Krohne (1998) stated that estuaries and coral reefs have about the same productivity as most terrestrial ecosystems. The production in estuaries is more likely to be by multicellular plants, such as mangroves and sea grasses. The primary producers in reefs are algae in symbiotic association with corals.

The lack of substantial data from Jobos Bay limits the holistic interpretation of the energy flow within this ecosystem. In order to understand the energy scenario, the scarce information on energy dynamics in Jobos will be complemented with data bank from homologous tropical estuaries.

10 PRIMARY PRODUCTIVITY

The microbial mats, together with the 90 species of algae (Almodóvar 1964) and sea grasses represent the primary producers within the submerged environment of Jobos Bay. Diatoms and dinoflagellates dominate the phytoplanktonic community. Preliminary data suggest that phytoplankton exhibits a spatial pattern in it's abundance, with densities ranging from 200 cell/ml in the oceanic habitat near Cayos Caribes to 1150 cells/ml in the mangrove channels within Mar Negro (DNER, 1998).

According to Talbot et al. (1990) diatoms show a daily abundance rhythm. They form dense mucilagous patches over sandy and muddy substrates. In the morning, they release mucous and migrate to the aerobic zone of the reef, becoming the prey for suspension feeders. At dusk and while migrating to the sea floor, they synthesizes the polysaccharide rich mucilagous. After adhering to the sediment, they become the prime food source for many deposit feeders. But at time passes the mat turns anoxic and sulfidic by the interaction of cyanobacteria (Stal, 1995), and this implies inhospitable conditions for suspension and deposits feeders, like bivalves, holothurians, polychaetes and sipunculids. It is suggested that these organisms may migrate or cease metabolic activities to avoid anoxic stress and that the rhythmic appearance of diatoms and cyanobacteria seems to trigger the distribution and abundance of predators in a daily basis (Laboy-Nieves, 1997).

The abundance and distribution of most aquatic primary producers is closely related to a combination of salinity, temperature and light conditions (Dawes et al., 1985). Epibenthic algal mats are characteristics from mangrove lagoons and closed embayment at Jobos. Algae species like *Enteromorpha clathrata, E. flexuosa* and *Ulva fasciata* are common inhabitants in front or under mangroves, rocks, buoys, and mangrove channels (Almodóvar, 1964; Santana Ferrer et al., 1996). Under very calm conditions and humic inputs, phytoplanckton and periphyton blooms. In warm temperatures, the dinoflagellate *Pyrodinium bahamense* flourish, accounting for the spectacular bioluminescence within Mar Negro waters during the summer months (Laboy-Nieves, 2001). During the hurricane season (June-November), sluggish, high currents and windy conditions dominate the water of the Bay, and the macrophytes drift westward (Laboy-Nieves, 2001). Therefore, most of the among-site variation in planktonic and substrate primary producers seems to be related to the local energy regime, as suggested by Bell and Hall (1997). Also, the bloom of phytoplanktonic species and the macroalgae observed in the Bay could be an indicator of eutrophication, a condition that could act as a pathway by which stressed ecosystems give off excess nutrients and biomass (Pizzolon, 1996), or a way by which bacteria and protozoa supply carbon to metazoans (Hart et al., 2000).

Seagrass beds are among the most productive ecosystems in the oligotrophic coastal waters of the Caribbean (Zieman et al., 1989). They support complex food webs by virtue of both their physical structure and primary production (Short and Willie-Echeverria, 1996). Seagrass density, growth, biomass and primary production vary, in response to local fluctuations in environmental conditions such as salinity, exposure to air, water clarity, sediment depth, nutrients (van Tussenbroek, 1995; Lee and Dunton, 1996) and natural and human-induced disturbance (Short and Willie-Echeverria, 1996). Turtle grass (*Thalassia testudinum*) dominates these beds and is the main contributor to primary production. Total production of this species is difficult to assess, since most of its biomass is below the substratum (van Tussenbroek, 1995). Lee and Dunton (1996) reported that biomass of *T. testudinum* in the coast of Texas changed significantly with season: values ranged from 454 to 885 g dry wt/m^2 from March to September, and suggested that temperature and underwater irradiance have been considered as a major factor controlling seasonal leaf growth at an exponential trend. A substantial fraction of seagrass carbon enters coastal and estuarine food webs through microbial transformation and particulate detritus.

Natural disturbances most commonly responsible for seagrass loss include hurricanes, earthquakes, diseases and herbivory. Human activities mostly affecting sea grasses are those which alter water quality or clarity, nutrients, sediment loading from runoff, sewage disposal, dredging and filling, pollution, upland development and certain fishing and recreational practices (Short and Willie-Echeverria, 1996). Jobos Bay experiences almost each year the impact of tropical storm winds and siltation by heavy rains runoff. These two factors, together with the agricultural, thermoelectric production, urbanization and filling activities immediately adjacent to the Bay, and the weekly traffic of boats, barges and jet skies, are the main threats of this ecosystem (Laboy-Nieves, 2001).

The mechanical perturbation of seagrass beds promotes the exodus of invertebrates (Roenn et al., 1988), accelerates sediment resuspension and its transport to coral reefs (Wolanski and Gibbs, 1992), diminished primary productivity (Short and Willie-Echeverria, 1996), inflicts species abundance (Laboy-Nieves, 1997) and buries macrophytes with sediments (Duarte et al., 1997). The decline of these beds will carry the wanton destruction of submerged habitats and will jeopardize the fragile throphic dynamic within the Jobos Bay.

Seagrass beds growth and production are also limited by nutrients and disturbances (Zieman et al., 1989; Short and Willy-Echeverria, 1996; Alcoverro et al., 1997). Similar dynamics occurs in phytoplankton from the Atlantic Ocean, were productivity seems to be limited by iron inflows (Behrenfeld and Kolber, 1999). Nutrient limitations affect primary production, thus the availability of organic compounds to marine heterotrophs seems to be directly or inversely related to nutrients inflows. Coral reefs also respond to pulsating nutrient inflows, to a point that hurricanes, upwelling of cold waters and freshwater inflows have been accelerating the degradation of this community in the whole Caribbean Region (Hughes, 1994; Laboy-Nieves et al., 2000). The reefs in Jobos Bay are not the exception. Sediment influx has been the major factor in altering the physical environment, causing changes in the coral reef cover and the rest of the benthic community. A combination of natural and anthropic disturbances, have destroyed almost 70% of the coral reefs in Jobos (Morelock and Williams, 1987). The chain events tighten to primary productivity, respiration and degradation of nutrients are as complex as the Jobos Bay ecosystem itself, and ought to be long-term monitored to start interpreting its crucial dynamics.

Cycling of organic matter in mangrove forests depends on the magnitude of abiotic factors that regulate mineral inputs and flows into, through, and out of the system, together with abiotic factors that incorporate those elements essential for the functioning of the forest (Lugo and Snedaker, 1974). Hydrological flows transport and distribute nutrients throughout the mangrove forest were they are sequestered in plant tissues or the underlying sediments. These nutrients are fixed into organic compounds by plants and are returned to the forest via retranslocation, litterfall and grazing. Under favorable conditions, mangroves at Jobos Bay are capable of high rates of production and fast circulation of nutrients (Lugo et al., 2007). The same abiotic and biotic factors act in the recycling of nutrients. Animals move and redistribute nutrient and organic compounds within and outside the systems. The flushing if forest floor litter maintains a low litter stock, which coupled to the extraordinary rate of litterfall (10.2 Mg ha^{-1} yr^{-1}) results in high productivity and turnover of nutrients and organic matter (Lugo et al., 2007). Consequently, the net results of this cycling process in mangroves, is the import of inorganic components from adjacent systems and the export of organic products to the sea (Lugo and Snedaker, 1974). Mangroves also import some inorganic components from the sea due to tidal flooding and from the air by wet and dry deposition (López et al., 1988).

Autotrophic production by phytoplankton and phytobenthos in the adjacent aquatic system is partly the result of mangrove leaf and litter decomposition (López et al., 1988). Trees are not large and there is an abundance of stunted individuals growing around hypersaline lagoons. However red mangrove leaves are relatively large, and seedling density and growth in the forest floor are high and luxuriant, suggesting favorable growth conditions (Lugo et al., 2007).

The only nutrient dynamic study conducted in Jobos Bay by Lugo et al. (2007), demonstrated that peak litter fall followed climax rainfall events in May and September. Late in the year, water level decreases and flushes the forest floor. About 60% of nitrogen and 80% of phosphorous is retranslocated prior to leaf fall. Litter productivity is high and its flushing maintains a low storage which results in high rapid turnover. Nutrients are not loss, because mangroves show high rate of nutrient use efficiency. Lugo et al. (2007) reported that annual litterfall was around 19 t/ha, and a high portion of this litterfall occurs during the hurricane season, consistent with the findings reported by Sharpe (1999), after the forest experience heavy rains and high winds. The flushing of forest floor litter maintains a low biomass storage, which coupled to the high rate of litterfall, results in high litter turnover. This could cause problems to nutrient recycling, because of leaching and loss to the system. These problems are mitigated by the high rates of retranslocation of nutrients before leaf falling; black mangroves have one of the highest reported phosphorous retranslocation rate (~85%).

Tidal regime and upland runoff regulate the residence time of litter on the forest floor, and partially control litter decomposition rates and export. Fringe forests are subjected to continuous flooding and litter is exported with decomposition occurring elsewhere. Basin forests develop in more protected inland areas and litter accumulates, decomposes *in situ*, and less particulate debris is exported from the system. Decomposition starts with leaching of soluble inorganic and organic compounds. Soon after, bacteria and fungi colonize the litter surface reducing the C:N ratio (López et al., 1988). Litter material is gradually reduced in size by microbial and grazing activities. These particles are then consumed by detritivores that derive their nutrition from the microbial and meiofaunal assemblage associated to the debris.

11 TROPHIC INTERACTIONS

Any diagram of a food web pretends to illustrate the interrelationships of species within the same community, from adjacent communities or from non-related communities. Species lists for macrofaunal invertebrates are available and from them, it is known that the taxonomic (genera) composition of West Indian mangrove communities is similar to other tropical and subtropical areas (López et al., 1988). Describing a food web for Jobos Bay ought a complex inventory of the species present in its five main communities: littoral woodlands, mangrove forests, seagrass beds, coral reefs and salt/mud flats, which coexists in among sharp ecotones in areas less than a half square kilometer. At present, this inventory from Jobos Bay is very incomplete, because benthic and demersal invertebrates associated with this ecosystem, particularly protozoans and meiofauna, have barely been examined and trophic interactions have not been studied. Invertebrates certainly represent a critical link between detrital and/or dissolved organic carbon sources and the macrofauna.

According to López et al. (1988) information on the pelagic community of the Caribbean estuarine/wetland system is limited to independent characterizations of phytoplankton, macrozooplankton, and fish populations. The taxonomic structure of the bacterioplankton, ciliate protozoans, larval stages of planktonic copepods or jellyplankton, and interactions among them are not fully known. The lack of a systematic approach has precluded the understanding of their processes in the marine food webs. Although these limitations, hypothetical models for trophic interaction, can be adapted to Jobos Bay. The most suitable models are those explained by López et al. (1988) and Joglar (2005). A focus on strong interactions can simplify food web structure and identify those species responsible for most of the energy flow in communities, but many factors contribute to define such interactions. It is well known that pollution promotes the loss of sessile and mobile species in mangroves (Ellison and Farnsworth, 1996), that environmental variability inflicts the distribution and abundance of benthic species in seagrass beds and coral reefs (Laboy-Nieves et al., 2000), and that the feeding activities of a few keystone species may control the whole structure of a community (Molles, 1999). Modeling the food web or trophic pyramids for Jobos Bay or any other system requires holistic approaches to understand the n-dimensional biological and physical factors that regulate the complex dynamics of intra and interspecific interactions.

12 REFLECTIONS ON MANAGEMENT ISSUES

Marine and coastal environments are among the most ecologically and socio-economically important habitats on Earth. However, climate change associated with a variety of anthropogenic stressors may interact to produce combined impacts on biodiversity and ecosystem functioning, which in turn will have profound implications for marine communities and the economic and social systems that depend upon them (Cardoso et al., 2008). The Jobos Bay estuary is not the exception.

In 1890's, the Aguirre Corporation Sugar Mill was established in a 1,205 acres land west side of Jobos Bay, known as the Lugo Viñas farm. The plant triggered local population growth that led to the establishment of an independent town (Aguirre), which flourished between 1925 and 1970. In 1918, the Aguirre State Forest, which main property surrounds the eastern portion of Jobos Bay was designated under the jurisdiction of the Public Works Department. For nearly half century, the Forest protection was under the responsibility of a park ranger.

During the early 1970's, Jobos Bay was considered as a potential site for an oil transshipment port, a nuclear power plant and an additional fossil fuel power plant. Scientific surveys were conducted in Jobos, but interests declined as the oil embargo was lifted, and as the Puerto Rico Environmental Policy Act took place. However, the non-peer-reviewed publications produced by that research provided the prime scientific data of Jobos Bay, further compiled by Williams and Bunkley-Williams (2003).

In 1974, the DNR proposed the U.S. National Oceanic and Atmospheric Administration (NOAA) to participate in the National Estuarine Sanctuary Program. Considerations were not taken seriously until 1981, when the Aguirre Corporation ceased operations and its executives proposed to sell the Lugo Viña Farm to the Sanctuary Programs Division of OCRM/NOAA. The Corporation was advised to offer the farm to the Department of Natural Resources. The DNR submitted NOAA a proposal for the establishment of the Aguirre Estuarine Sanctuary. The Jobos Bay location was finally chosen because of its ecologic and socio-economic factors. The eleventh National Estuarine Sanctuary, with an area of 2,800 acres, was designated in September 1981. To meet NOAA's requirements of land area, the state government planned to incorporate the Cayos Caribe into the Sanctuary. Fifteen of the seventeen mangrove islets were transferred to the Sanctuary and two remained part of the Aguirre State Forest. In the 90's, the Sanctuary was renamed to its present name: Jobos Bay National Estuarine Research Reserve (JBNERR). In 2006, 72 acres from four of the La Barca Islets where incorporated to the Reserve.

For many decades, the surrounding region of JBNERR has attracted developers, particularly because of it ground water, leveled topography, road access and natural landscape. Private and governmental development plans have led to the establishments of ports, two power generating plants, one of the largest pharmaceutical alleys in Puerto Rico, two gulf resorts, agricultural farms, a waste-disposal site, commercial centers, and residential complexes. Consequently, the environment has been altered and degraded mainly by pollution and habitat destruction. These structures have also caused the interruption of the natural water flow, most directly by the removal of wetlands. Even though the land surrounding Jobos Bay experiences a constant population growth and structural developments, a sewage treatment plant has not been build, and all the raw sewage is discharged directly into the Bay.

Historically, Jobos has shown a high degree of conflicts for the use of aquatic and terrestrial resources. Conscious of this fact, the Puerto Rico Coastal Zone Management Program designated Jobos Bay as a Special Planning Area (SPA). This designation points out the situation of "areas subject to serious present or potential use conflicts, and therefore, requiring detailed planning".

The implementation of management actions for specific resources within the Bay is the main responsibility of the Department of Natural and Environmental Resources (DNER), the government agency that has two natural areas under its jurisdiction: the Aguirre State Forest (established in 1918) and the Jobos Bay National Estuarine Research Reserve (established in 1981). The main goal of the Forest is the conservation of the mangrove for recreational and educational purposes, while for the Reserve, it is the protection of its terrestrial and submerged ecosystems for their natural conservation and to promote research and education activities.

Jobos Bay is protected by several Federal (USA) and Commonwealth (Puerto Rico) statutes and regulations. This legal framework aims the protection of the natural integrity of the ecosystems, from unduly disruptive or unlawful activities occurring inside its boundaries.

The establishment of the Reserve has been a great asset for the study and protection of the Jobos Bay natural resources. A visitor center with a public library, a museum, a research and monitoring laboratory and quarters for investigators have given a very good standing to the DNER in terms of increasing public awareness and scientific activities, which consequently will protect the Bay environment. DNER and NOAA have institutionalized the administration of the Reserve through a management plan, which is been reviewed and updated every five years since the publication of its first edition (Laboy-Nieves, 1983). Although the Aguirre State Forest was established nearly fifty years before the Reserve, the former has never had a management plan.

As inferred from above, the DNER manages Jobos Bay under a multiple use strategy, with more operational emphasis and budget allocation to the Reserve than to the Forest. The Reserve Management Plan has divided the Bay in three sectors. These are known as the Preservation, Conservation, and the Limited Use Sector (Figure 11). Each sector has recognized compatible uses allowed.

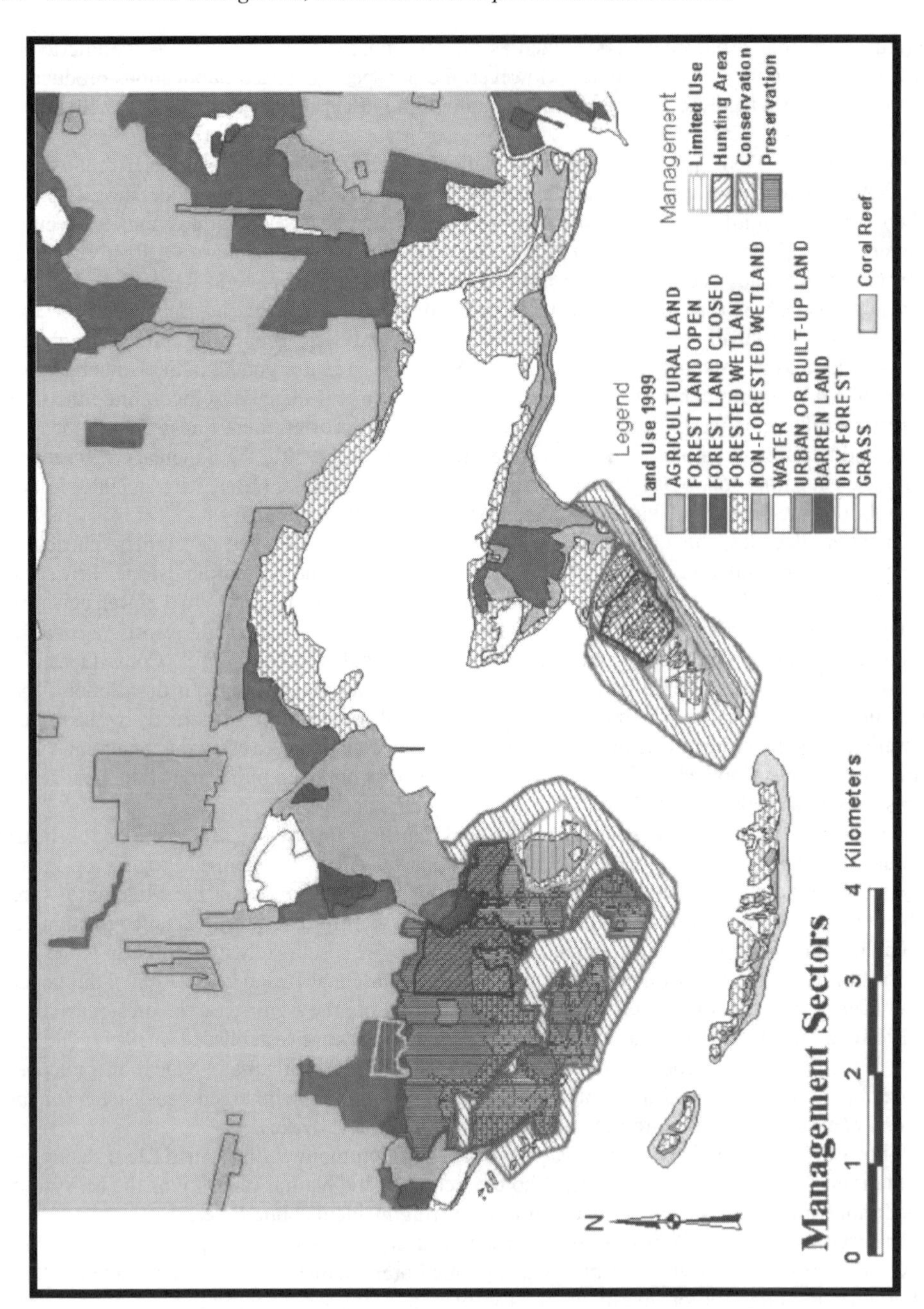

Figure 11. Management sectors and land use in Jobos Bay (DNER 2000).

However, there have been discrepancies in the range of activities permitted, particularly because Jobos Bay has two different management approaches from the DNER to a single ecological unit.

The DNER has shown discrepancies and inconsistencies in the administration of the Bay. While the Forest has had five on-site managers in nearly 90 years, the Reserve has had seven managers in 25 years. In 1982 the present author was assigned a double task as the first and fourth Manager of the Reserve and the Aguirre State Forest, respectively. Aware that the Forest and the Reserve are integral ecological units, Laboy Nieves unsuccessfully tried to convince the DNRA administration to fuse their management strategies. Presently, there is almost no communication between the Forest and the Reserve managers. This conflict is extrapolated to ecosystem management. For instance, hunting and the use of jet skis, are not allowed in any region of the Reserve, but in some areas of the Forest, while ironically huge barges weekly transport tons of oil within the Reserve waterways. DNER has very little coordination with other public agencies (Marine Unit of the Police Department, Aguirre Power Plant, the Ports Authority) or private corporations (Guayama Yatch Club, Applied Energy System Coal Power Plant, Salinas Garbage Disposal Site, and several pharmaceutical industries) that constantly uses the Bay or their peripheral resources for their operational activities.

According to the Reserve's Management Plan (DNER, 2000), the Preservation Sector requires a high degree of protection due to its natural vulnerability, ecological importance, presence of flora and fauna, and/or historic or archeological values. Physical and biological assessments have shown the need to designate preservation sectors in the Mar Negro component that meet these criteria. Many of the sites classified for preservation include coves, shallow semi- enclosed areas, and fringing mangroves within the lagoon system. They have been identified as spawning areas and nursery grounds for valuable commercial fish species and habitat for endangered species. The preservation of these sites is vital to maintain the population dynamics of the estuary. Activities in this Sector are limited to authorized research and monitoring. Fishing, hunting and the use of motor vehicles are not allowed in these areas. However, traditional shellfish harvesting from mangrove roots are allowed and shellfish fishermen are encouraged to provide harvesting reports at the Reserve.

The Conservation Sector encompasses the vast majority of the Reserve and the Forest property. It is composed of environmentally sensitive areas, and includes wetlands, mangrove areas, and scenic outlooks that require protection against inappropriate or excessive use. For instance, fishermen have been granted with a right-of-way for boat traffic through the central corridor of Mar Negro. The Reserve Administration has proposed to limit boat size to a maximum of 22 feet, and speed to five knots. Hook and line fishing will be allowed, but with emphasis on the importance of releasing small or immature specimens in order to ensure sustainability. Pot and net fishing will not be allowed inside the Mar Negro area. Hunting of aquatic birds will be permitted in designated areas, complying with the measures established for their management. Passive recreation activities such as bird watch, snorkeling, kayaking and diving will be promoted as part of Reserve's education outreach plan.

The Limited Use Sector represents areas designated for education, outreach and passive recreation activities. Interpretative trails, boardwalks, limited docking piers for public use, and minimum facilities are provided in both in the State Forest and the Reserve. There is no picnic or camping areas designated in the Reserve, but in the Forest. Anchoring in designated areas is limited to a maximum of three hours. Docking and anchoring capacities have been established and appropriate signage has been posted. The Reserve's Management Plan clearly states that no activities or uses will be permitted in the Reserve or adjacent to it that will be detrimental or adverse to the maintenance, improvement, or conservation of the Bay existing water bodies, air quality, and land. Excavation, mining, or removal of sand, gravel, clay, or other materials shall be prohibited except when closely related to a research project. A closer look to Figures 12, 13, and 14 unmask the contradiction between theory and reality.

Despite the statements of the Management Plan, the Environmental Quality Board (EQB) of Puerto Rico approved in 2007 the Environmental Impact Statement (EIS) for the Southern Gas Duct Project of the Puerto Rico Energy and Electric Authority (PREEA). This project implies digging land adjacent to the northern boundaries of Jobos Bay, to place and bury a natural gas pipeline that will feed the Aguirre Thermoelectric Power Plant for its conversion from oil to gas. PREEA made very clear that the pipeline construction will have a high impact on the natural environment. Even though that statement, added to community opposition, the EQB approved the EIS.

The natural history of Jobos Bay and that of many other valuable ecosystems of Puerto Rico, is taking a wrong heading, mostly because of the ignorance, indolence and conflictive interests on the environment. Abiotic and biotic resources continue to be the silent victims of a holistic

Figure 12. Air and water pollution in Jobos Bay as a by-product of the Aguirre Power Plant operations.

Figure 13. Solid wastes disposal facility adjacent (0.5 km) to the northern boundary of the Jobos Bay National Estuarine Research Reserve.

management approach for Jobos Bay. Deterioration, habitat destruction, and the depletion of species are some of the indicators of the above asseveration. Nazario (2002) and Massol Deya (2003) reported the presence of heavy metals, trace elements and carcinogens in Puerto Rico. Landfills, toxic dumps and petrochemical facilities, have endangered public and environmental health. Anthropogenic radionuclide activities done by the US Department of Energy (US DOE) and the Department of Defense (US DOD) have also been found to contaminate and deteriorate natural resources in Puerto Rico (US DOE, 2003; US EPA, 2001; PRNC, 1970). Water quality has decreased (Hunter & Arbona, 1995), soil has been eroded through deforestation and habitat lost by development projects (Ithier-Guzman, 2004). Accidental releases to the environment, such as petroleum-related products and chemicals including benzene and xylene, also impact negatively natural resources on the island (US EPA, 2001). The Bay serves as an unquestionable nursery and feeding ground for many marine, estuarine and terrestrial species. But this estuarine community has strong anthropogenic influences, characterized by alterations of its physical attributes, disturbances in the biota and diminishing environmental quality. Although it is widely interpreted that the general

Figure 14. Land crabs over-harvested from the mudflats of Jobos Bay.

health of the Jobos Bay environment is declining, it is not clearly understood the extent of that degradation. The lack of baseline data, which provides knowledge of conditions in the past, has not allowed specifying restoration targets. These degrading conditions have been reported by several authors elsewhere (Lafite et al., 2007).

Integrated resource management is of particular concern with environmentally sensitive areas of coastal ecosystems such as mangrove forests, seagrass beds and coral reefs. Management practices have been implemented and enforced in severely impacted coastal areas in order to restore the ecosystem's integrity and balance (Islam et al., 2004). This is the case of Jobos Bay, where local and federal agencies have begun to manage and conserve its most threatened and environmentally critical resources, but with a wrong approach in two management units.

Amid signs that estuarine ecosystems are increasingly degraded and may reach new thresholds of irreversible decline, restoration ecologists and coastal managers world-wide have joined the debate on how best to reverse the trends of the recent past (Weinstein, 2008). Any meaningful effort at reversal must recognize that humans are an integral part of the landscape, particularly in urban estuarine settings, and that natural resource baselines have permanently shifted. Consequently, the human dimensions component of sustainability has become an integral part of ecological restoration/rehabilitation planning. New schema evolving out of coastal governance and management are not only increasingly underpinned by transdisciplinary science, but are beginning to address the sacrifices and compromises that will be necessary to achieve a balance between human uses of estuarine resources and ecological integrity. The challenge will be to preserve ecosystem functions and use natural capital at a variety of scales while simultaneously sustaining local communities, social and formal institutions, economies and markets at the highest levels of system organization. Thus, it is of utmost importance that residents of Puerto Rico refocus their individual and collective way of thinking about the unique natural heritage of this Island, if the paradigm of sustainability is to be achieved (Laboy-Nieves, 2008).

13 CONCLUDING REMARKS

The case study area of Jobos Bay is a very dynamic ecosystem. The natural history of the Bay has been sculpted by physical, biological and anthropological factors. The aquatic and terrestrial zones and the biota that inhabit them are tightly linked, showing complex interactions between abiotic and biotic elements. For centuries, peripheral urban developments have affected the evolution of the estuary to a point that today this ecosystem exhibits a mosaic of areas that remains almost pristine while others are degraded. Nevertheless, Jobos Bay is a natural laboratory for examining mangroves and upland forests, submerged communities and anthropogenic influences. For that reason, the Bay has two management units, but with different and sometimes conflicting management approaches. The byproduct of such conflicts is poor institutionalization of programs to manage the estuarine natural resources as a single ecological unit. It is hoped that this chapter will help to increase the environmental literacy and awareness of Jobos Bay, so that individual and collective efforts be focused to conserve the unique natural heritage of this estuarine bay. This ecosystem should be treated and conserved as a single ecological unit.

REFERENCES

Alcoverro, T., Romero, J., Duarte, C. M., and López, N. I. 1997. Spatial and temporal variation in nutrient limitation of seagrass *Posidonia oceanica* growth in the NW Mediterranean. Marine Ecology Progress Series 146:155–161.

Almodóvar, L. R. 1964. The marine algae of Bahía de Jobos, Puerto Rico. Nova Hedwigia VII(1/2):33–52.

Anonymous. 1975. Aguirre Environmental Studies, Jobos Bay, Puerto Rico, Final Report. Puerto Rico Nuclear Center Volume I:1–95.

Behrenfeld, M. J., and Kolber, Z. S. 1999. Widespread Iron Limitation of Phytoplanckton in the South Pacific Ocean. Science 283:840–843.

Bell, S. S., and Hall, M. O. 1997. Drift macroalgal abundance in seagrass beds: investigating large-scale associations with physical and biotic attributes. Marine Ecology Progress Series 147:277–283.

Berryhill, H. L. 1960. Geology of the Central Aguirre Quadrangle. U.S. Geological Survey. Misc. Geol. Inv. Map I-319.

Birdsey, R. A., and Weaver, P. R. 1982. The forest resources of Puerto Rico. U.S.D.A. Forest Service. Resource Bulletin SO-85. Southern Forest Experiment Station. New Orleans, Louisiana.

Boccheciamp, R. A. 1977. Soil Survey of Humacao Area of Eastern Puerto Rico. United States Department of Agriculture, Soil Conservation Service and the University of Puerto Rico, College of Agricultural Sciences. Mayaguez, Puerto Rico.

Cardoso, P. G., Raffaelli, D., Lillebø, A. I., Verdelhos, T., and Pardal, M. A. 2008. The impact of extreme flooding events and anthropogenic stressors on the macrobenthic communities' dynamics. Estuarine, Coastal and Shelf Science 76(3):553–565.

Cintrón, B. 1983. Coastal freshwater swamp forests: Puerto Rico's most endangered ecosystems? In A.E. Lugo (ed.) Los Bosques de Puerto Rico. Instituto de Dasonomía Tropical. Río Piedras, Puerto Rico.

Cintrón, G., and Schaeffer-Novelli, Y. 1983. Introducción a la Ecología del Manglar. UNESCO- ROSTLAC. Montevideo, Uruguay.

Cintrón, G., Lugo, A. E., and Martínez, R. 1985. Structural and Functional Properties of Mangrove Forests. In W. G. D'Arcy and Mireya D. Correa, eds. The Botany and Natural History of Panama: IV Series: Monographs in Systematic Botany, Vol. 10. Missouri Botanical Garden, Saint Louis, Missouri.

Colin, P. I. 1978. Caribbean Reef Invertebrates and Plants. T.F.H. Publications. Neptune City, New Jersey.

Cruz-Báez, A. D., and Boswell, T. D. 1997. Atlas Puerto Rico. The Cuban American National Council. Miami, Florida.

Dansereau, P. 1966. Studies on the Vegetation of Puerto Rico: Description and Integration of the Plant Communities. University of Puerto Rico. Institute of Caribbean Science. Special Publication No.1. Mayaguez, Puerto Rico.

Dawes, C. J., Hall, M. O., and Riechert, R. K. 1985. Seasonal biomass and energy content in seagrass communities on the West Coast of Florida. Journal of Coastal Research 1:255–262.

Díaz, H., Bevilacqua, M., and Bone, D. 1985. Esponjas en Manglares del Parque Nacional Morrocoy. Acta Científica Venezolana. Caracas, Venezuela.

DNER. 1998. Management Plan for the Jobos Bay National Estuarine Research Reserve, Guayama/Salinas, Puerto Rico. Department of Natural and Environmental Resources. San Juan, Puerto Rico.

Duarte, C. M., Terrados, J., Agawin, N. S. R., Fortes, M. D., Bach, S., and Kenworthy, W. J. 1997. Response of mixed Philippine seagrass meadows to experimental burial. Marine Ecology Progress Series 147:285–294.

Ellison, A. M., and Farnsworth, E. J. 1996. Anthropogenic Disturbance of Caribbean Mangrove Ecosystems: Past Impacts, Present Trends, and Future Predictions. Biotropica 28(4a):549–565.

Ellison, J. C., and Stoddart, D. R. 1990. Mangrove ecosystem collapse during predicted sea level rise: Holocene analogues and implications. Journal of Coastal Research 7:151–165.

Ewel, J. J., and Whitmore, J. L. 1973. The Ecological Life Zones of Puerto Rico and the U.S. Virgin Islands. U.S. Forest Service Research Publication USDA-ITF-18.

Galiñanes, M.T. 1977. Geovisión de Puerto Rico: Aportaciones Recientes al Estudio de la Geografía. Editorial Universitaria. Río Piedras, Puerto Rico.

García-Molinari, O. 1952. Grassland and grasses of Puerto Rico. University of Puerto Rico. Agricultural Experiment Station. Bulletin 102:5–167.

Gleason, H. A., and Cook, M. T. 1926. Plant ecology of Porto Rico. Scientific Survey of Porto Rico and the Virgin Islands. Volume VII (1,2):1–173. New York Academy of Sciences. New York.

Goudie, A. 2000. The Human Impact on the Natural Environment. 5th Edition. MIT Press. Cambridge, Massachusetts.

Hart, D. R., Stone, L., and Berman, T. 2000. Seasonal dynamics of the Lake Kinneret food web: the importance of the microbial loop. Limnology and Oceanography 45(2):350–361.

Heck, K. L., and Valentine, J. F. 1999. Plant-animal interactions in seagrass-dominated ecosystems. American Zoologist 39(5):90A.

Hughes, R. P. 1994. Catastrophes, Phase Shifts, and Large-Scale Degradation of a Caribbean Coral Reef. Science 265:1547–1551.

Humann, P. 1996. Reef Coral Identification: Florida, Caribbean & Bahamas. New World Publications. Jacksonville, Florida.

Jernakoff, P., Brearley, A., and Nielsen, J. 1996. Factors affecting grazer-epiphyte interactions in temperate seagrass meadows. Oceanography and Marine Biology: an Annual Review 34:109–162.

Jiménez, J. A. 1985a. *Rhizophora mangle* L. SO-ITF-SM-2. USDA Forest Service. Río Piedras, Puerto Rico.

Jiménez, J. A. 1985b. *Laguncularia racemosa (L.)*. SO-ITF-SM-3. USDA Forest Service. Río Piedras, Puerto Rico.

Jiménez, J. A., and Lugo A. E. 1985. *Avicennia germinans (L.)L*. SO-ITF-SM-4. USDA Forest Service. Río Piedras, Puerto Rico.

Joglar, R. 2005. Biodiversidad de Puerto Rico: Vertebrados Terrestres y Ecosistemas. Editorial Instituto de Cultura Puertorriqueña. San Juan, Puerto Rico.

Joyce, J. 1992. Geology of the East Coast of Puerto Rico. Unpublished Manuscript. University of Puerto Rico. Mayaguez, Puerto Rico.

Juncosa, A. M. 1982. Developmental morphology of the embryo and seedling of the *Rhizophora mangle* L. (Rhizophoraceae). American Journal of Botany 69(10):1599–1611.

Kolehmainen, S. 1972. Ecology of Sessile and Free-living Organisms on Mangrove Roots in Jobos Bay. In Aguirre Power Project, Environmental Studies. Puerto Rico Nuclear Center Annual Report 162:141–173.

Krohne, D. T. 1998. General Ecology. Wadsworth Publishing. New York.

Laboy-Nieves, E. N. 1983. Jobos Bay National Estuarine Sanctuary Management Plan. Department of Natural Resources. San Juan, Puerto Rico.

Laboy-Nieves, E. N. 1994. Guía de Educación Ambiental para Maestros/as de Escuela Elemental e Intermedia del Sureste de Puerto Rico. CECIA. Universidad Interamericana de Puerto Rico. Guayama, Puerto Rico.

Laboy-Nieves, E. N. 1997. Factores Ambientales que Limitan la Distribución y Abundancia de *Isostichopus badionotus* y *Holothuria mexicana* en el Parque Nacional Morrocoy. Unpublished Doctoral Dissertation. Venezuelan Institute for Scientific Research. Caracas, Venezuela.

Laboy-Nieves, E. N. 2008. Ética y Sustentabilidad Ambiental en Puerto Rico. Memorias del Primer Foro Nacional del Agua. INAPA. Dominican Republic. ISBN 978-9945-406-80-1:63–76.

Laboy-Nieves, E. N. 2001. Historia Natural de la Bahía de Jobos, Puerto Rico. Universidad Interamericana de Puerto Rico. Guayama, Puerto Rico.

Laboy-Nieves, E. N., Conde, J. E., Klein, E., Losada, F., and Bone, D. 2000. Mass mortality of tropical marine communities in Morrocoy, Venezuela. Bulletin of Marine Science 68(2):163–179.

Lafite, R., Garnier, J., and de Jonge, V. N. (eds). 2007. Consequences of estuarine management on hydrodynamics and ecological functioning: ECSA 38th Symposium – Rouen 2004 Co-organisation Seine-Aval Programme and ECSA. *Hydrobiologia*, 588. Springer: The Netherlands.

Lee, K. S., and Dunton, K. H. 1996. Production and carbon reserve dynamics of the seagrass *Thalassia testudinum* in Corpus Christi Bay, Texas, USA. Marine Ecology Progress Series 143:201–210.

Little, E. L., Wadsworth, F. H., and Marrero, J. 1977. Árboles Comunes de Puerto Rico y las Islas Vírgenes. Editorial Universitaria. Río Piedras, Puerto Rico.

López, J. M., Stoner, A.W., García, J. R., and García-Muñiz, I. 1988. Marine food webs associated with Caribbean Islands Mangrove Wetlands. Acta Científica 2(2–3):94–123.

Lugo, A. E. 1983. Coastal Forests of Puerto Rico. In A.E. Lugo (ed.) Los Bosques de Puerto Rico. Instituto de Dasonomía Tropical. Río Piedras, Puerto Rico.

Lugo, A. E., and Snedaker, S. C. 1974. The Ecology of Mangroves. Annual Reviews of Ecology and Systematics 5:39–64.

Lugo, A., Medina, E., Cuevas, E., Cintrón, G., Laboy-Nieves, E. N., and Schaffer-Novelli, Y. 2007. Ecophysiology of a fringe mangrove forest in Jobos Bay, Puerto Rico. Caribbean Journal of Sciences 43(2):200–219.

Martínez, R., Cintrón, G., and Encarnación, L. 1979. Mangroves in Puerto Rico: A Structural Inventory. Department of Natural Resources. San Juan, Puerto Rico.

McClymonds, N. E., and Díaz, J. R. 1972. Water Resources of the Jobos Area, Puerto Rico: A Preliminary Appraisal. Water Resources Bulletin 13.

Meyerhoff, H. A. 1933. The Geology of Puerto Rico. The University of Puerto Rico. San Juan, Puerto Rico.

Molles, M. C. 1999. Ecology. WCB-McGraw-Hill. Boston, Massachusetts.

Mondrup, T. 1999. Salinity effects on nitrogen in estuarine sediment investigated by a plug-flux method. The Biological Bulletin 197:287–288.

Morelock, J., and Williams, L. B. 1987. Sediments and physical oceanography at Jobos Bay. Unpublished paper. University of Puerto Rico. Mayaguez, Puerto Rico.

Morris, G. L. 2000. Hydrologic-Hydraulic and Biological Analysis of Jobos Estuarine Mangrove Mortality Jobos, Puerto Rico. Unpublished Report. Greg L. Morris & Associated. San Juan, Puerto Rico.

Murawski, S. 2005. Strategies for Incorporating Ecosystem Considerations in Fisheries Management. In D. Witherell, (ed.), Managing Our Nation's Fisheries II. Proceedings of the Conference on Fishery Management. Washington, DC.

Murphy, L. S. 1916. Forests of Puerto Rico: past, present and future, and their physical and economic environment. U.S. Department of Agriculture Bulletin 354:1–99. Washington, D.C.

Patriquin, D. G. 1975. "Migration" of blowouts in seagrass beds at Barbados and Carriacou, West Indies, and its ecological and geological implications. Aquatic Botany 1:163–189.

Pizzolon, L. 1996. Importancia de las cianobacterias como factor de toxicidad en las aguas continentales. Interciencia 21(6):239–245.

Puerto Rico Water Resources Authority (PWRA). 1972. Aguirre Power Plant Complex Environmental Report. San Juan, Puerto Rico.

Quiñones-Aponte, V. 1991. Water Resources Development and its Influence on the Water Budget for the Aquifer System in the Salinas to Patillas Area, Puerto Rico. In: F. Gómez-Gómez, V. Quiñonez-Aponte and I. Johnson (ed.), Aquifers of the Caribbean Islands. American Water Resources Association Monograph Series 15. Bethesda, Maryland.

Quiñones-Aponte, V., Gómez-Gómez, F., and Renken, R. A. 1997. Geohydrology and Simulation of Ground Water Flow in the Salinas to Patillas Area, Puerto Rico. Water Resources Investigations Report 95-4063.USGS, San Juan, Puerto Rico.

Roenn, C., Bonsdorff, E., and Nelson, W. G. 1988. Predation as a mechanism of interference within an infauna predator *Nereis diversicolor*. Journal of Experimental Marine Biology and Ecology 116(2):143–157.

Santana-Ferrer, L. F., Montero Acevedo, L. L., and Dieppa Ayala, A. R. 1996. Macroalgas de la Reserva Estuarina de Bahía de Jobos. Unpublished Paper, University of Puerto Rico-Humacao. Biology Department.

Sharpe, J. M. 1999. Impact of Hurricane Georges on the Mangrove Fern *Acrostichum danaeifolium*. Jobos Bay National Estuarine Research Reserve Newsletter 5(1).

Short, F. T., and Willie-Echeverria, S. 1996. Natural and human-induced disturbance of seagrasses. Environmental Conservation 23(1):17–27.

Skanavis, C. 1999. Ground Disaster in Puerto Rico – The Need for Environmental Education. Journal of Environmental Health 62(2):29–37.

Snelgrove, P. V. R., and Butman, C. A. 1994. Animal-Sediment Relationships Revisited: Cause Versus Effect. Oceanography and Marine Biology: an Annual Review 32:111–177.

Stal, L. J. 1995. Physiological ecology of cyanobacteria in microbial mats and other communities. New Phytology 131:1–32.

Suarez-Caabro, J. A., and Shearls, E. A. 1973. Macrozooplankton of Jobos and Guayanilla Bays, southern Puerto Rico, and its fluctuations under special conditions. Puerto Rico Nuclear Center. University of Puerto Rico Report PRNC – 162.

Sumich, J. L. 1999. Marine Life. WCB/McGraw-Hill. Boston, Massachussetts.

Talbot, M. B., Bate, G. C., and Campbel, E. E. 1990. A review of the ecology of surf-zone diatoms, with special reference to *Anaulus australis*. Oceanograpghy and Marine Biology: an Annual Review 28:155–175.

USFWS. 1972. Classification of Wetlands and Deepwater Habitats of the United States. FWS/OBS-79-31.

Van Putten, M. 2000. Warming Beneath the Sea. International Wildlife (Jan.–Feb.):7–11.
van Tussenbroek, B. I. 1995. *Thalassia testudinum* leaf dynamics in a Mexican Caribbean coral reef lagoon. Marine Biology 122:22–40.
Villamil, J., and Canals, M. (eds). 1981. Suplemento Técnico para el Plan de Manejo de Bahía de Jobos. Department of Natural Resources. San Juan, Puerto Rico.
Wadsworth, F. H. 1950. Notes on the climax forests of Puerto Rico and their destruction and conservation prior to 1900. Caribbean Forester 12:93–114.
Weinstein, M. P. 2008. Ecological restoration and estuarine management: placing people in the coastal landscape. Journal of Applied Ecology 45(1):296–304.
Whitakker, R. H. 1975. Communities and ecosystems. Macmillan Publishers. London, England.
Williams, E. H., Jr., and Bunkley-Williams, L. 2003. Biobliography of research publications concerning Jobos Bay. Retrieved from http://www.caribjsci.org/epub6/
Wolanski, E., and Gibbs, R. 1992. Resuspension and clearing of dredge spoils after dreadging, Cleveland Bay, Australia. Water Environment Research 64(7):910–914.
Woodroffe, C. D. 1990. The impact of sea-level rise on mangrove shorelines. Progress in Physical Geography 14:483–520.
Yoshioka, P. M. 1975. Mangrove root communities in Jobos Bay. Aguirre Environmental Studies, Jobos Bay, Puerto Rico. Final Report. Puerto Rico Nuclear Center Volume 1:50–65.
Zieman, J. C. 1976. The ecological effects of physical damage from motor boats on turtle grass beds in southern Florida. Aquatic Botany 2:127–139.
Zieman, J. C., Fourqurean, J. W., and Iverson, R. L. 1989. Distribution, abundance and primary production of seagrass and macroalgae in Florida Bay. Bulletin of Marine Sciences 44:292–311.

Impact of Urban Wastewater on Biodiversity of Aquatic Ecosystems

Evens Emmanuel, Ketty Balthazard-Accou & Osnick Joseph
Laboratoire de Qualité de l'Eau et de l'Environnement, Université Quisqueya,
BP 796 Port-au-Prince, Haïti

SUMMARY: Discharge of chemical substances in aquatic ecosystems (rivers, lakes, oceans) may cause changes in biotic community structure and function; otherwise know as biotic integrity. Among the main effects of pollutants on aquatic organisms, the literature reports severe pathologies, behavioural problems, and species migration and disappearance. These effects not only alter the functioning of communities and ecosystems, but also mean the loss of irreplaceable heritage potentially useful to health and sustainable development. The conservation of biological diversity essentially demands the conservation *in situ* of natural ecosystems and habitats. The aim of this work was to carry out a synthesis of the literature on urban liquid effluents in view to highlighting their potential impact on aquatic ecosystems, and thus obtain better understanding of their consequences on the quantitative and qualitative degradation of water resources, especially regarding the loss of biodiversity. This work reported in this chapter has four main sections. In the first section is presented a physicochemical characterisation of urban waters. The second section describes the different mechanisms implemented in the treatment of urban waters. The third section presents the regulatory aspects intrinsically related to urban effluents. The fourth section is devoted to studying the impact of urban effluents on aquatic biodiversity. The "complete ecotoxicological approach" is used to present the acute and chronic effects of certain pollutants present in urban waters on aquatic organisms and food chains.

1 INTRODUCTION

Anthropic activities that utilise natural resources generate solid wastes, liquid discharges (wastewater) and gaseous effluents. Consequently they lead to transfers of pollutants to natural environments and in turn are liable to jeopardise the biological equilibrium of ecosystems (Emmanuel, 2004). The physicochemical aggressions sustained by surface waters due to contact with pollutants contained in liquid discharges lead to a loss of aquatic biodiversity in many cases.

Indeed, the continual discharge of chemical substances in aquatic ecosystems can bring about changes in the structure and functioning of the biotic community, i.e. on biotic integrity (Karr, 1991). As a function of their bioavailability, the pollutants present in effluents cause a large number of harmful effects on the biodiversity of aquatic environments (Forbes and Forbes, 1994).

The main substances involved in chemical pollution phenomena are heavy metals, organic compounds, especially organohalogenated substances (Emmanuel *et al.*, 2004), detergents-"*surfactants*" (Lewis, 1992), pesticides, Polycyclic Aromatic Hydrocarbons (PAHs) and Polychlorobiphenyls (PCBs) (Förstnet and Wittmann 1979; Boisson, 1998; Garnaud, 1999; Février, 2001; Lassabatère, 2002), nitrates and phosphates (Zimmo *et al.*, 2004, Lacour *et al.*, 2006), and drug residues (Halling-Sorensen *et al.*, 1998; Kümmerer, 2001). The impacts of these pollutants on aquatic ecosystems are reported in the literature (Scott *et al.*, 2003). They can be at the origin of long term disturbances of host aquatic ecosystems, especially for primary producers sensitive to both organic and metallic pollutions (Feray, 2000).

Among the main effects of pollutants on aquatic organisms, the literature reports severe pathologies (Forbes and Cold, 2005), behavioural problems (Heckmann and Friberg, 2005), and species migration and disappearance (McIntyre *et al.*, 2005). These effects not only alter the functioning of communities and ecosystems, but also mean the loss of irreplaceable heritage potentially useful to health and sustainable development (United Nations, 1992).

The conservation of biological diversity essentially demands the conservation *in situ* of natural ecosystems and habitats (United Nations, 1992). The scientific lag in the development of technologies capable of efficiently controlling the pollutants contained in urban wastewater simply exacerbates concern. The aim of this work was to carry out a synthesis of the literature on urban liquid effluents in view to highlighting their potential impact on aquatic ecosystems, and thus obtain better understanding of their consequences on the quantitative and qualitative degradation of water resources, especially regarding the loss of biodiversity.

This work reported in this chapter has four main sections. Initially, we present a physicochemical characterisation of urban waters. This first section first proposes a synthesis of information on the two main components of urban water (wastewater and rainwater), followed by a review of the main pollutants detected in these aqueous matrices. The second section describes the different mechanisms implemented in the treatment of urban waters. It also describes the diffusion in the natural environment of certain substances that escape the control of wastewater treatment plants, while taking care to underline the risk for the species living in natural ecosystems. The third section presents the regulatory aspects intrinsically related to urban effluents, in order to permit better understanding of the stakes of this study. The fourth section is devoted to studying the impact of urban effluents on aquatic biodiversity. The "complete ecotoxicological approach" is used to present the acute and chronic effects of certain pollutants present in urban waters on aquatic organisms and food chains.

2 PHYSICOCHEMICAL CHARACTERISATION OF URBAN WATERS

Generally, urban waters group rainwater and residual urban wastewater. The latter, as defined by directive 91/271/CEE, is urban household wastewater (black or grey) that comes from establishments and residences and mainly produced by human metabolism and household activities), or the mixture of household wastewater with industrial wastewater and/or runoff water (European Commission, 1991).

2.1 *Grey water and black water*

Grey water does not contain any substance originating from lavatories. Thus the term covers wastewater produced in bathrooms, baths, sinks, washing machines and kitchens, in houses, offices, schools, etc. Generally, grey water accounts for an estimated 75% of residential drainage and contains low levels of organic matter in comparison to black water (ordinary wastewater), which comprises urine, faecal matter, toilet paper, etc. Eriksson *et al.*, 2002).

Urban wastewater contains mineral and organic materials. The content of these pollutants in the matrices can be evaluated by characterising heavy metals (copper, zinc, lead, cadmium), total suspended matter (TSM), total dissolved solids (TDS), nitrous and ammoniacal compounds (total N, N-NH_4) and phosphate compounds (total P) (Tardat-Henry, 1984; Gray and Becker, 2002).

Metals are present in numerous products used in the home and liable to be discharged down drains such as cosmetics, cleaning products, drugs, and paints (Lester, 1987). Water used for cleaning, especially for clothes, is the main source of metals in domestic wastewater (Grommaire-Mertz, 1998). Table 1 sums up the concentrations of certain pollutants (conventional and non conventional) measured in urban waters.

2.2 *Rainwater*

By definition, rainwater is the precipitation of liquid atmospheric water. It groups meteorological water and water running off urban surfaces (roads, roofs). Effective rainfall can be divided into two

Table 1. Concentrations of pollutants in urban wastewater (Grommaire-Mertz, 1998).

Parameters	Symbol	Concentrations
Total suspended matter	TSM	100 to 500 mg/L
Chemical oxygen demand	COD	250 to 1000 mg/L
Biochemical oxygen demand	BOD_5	100 to 400 mg/L
Cadmium	Cd	1 to 10 μg/L
Copper	Cu	83 to 100 μg/L
Lead	Pb	5 to 78 μg/L
Zinc	Zn	100 to 570 μg/L

Table 2. Pollution of runoff water – origin and content of heavy metals (Valiron and Tabuchi, 1992).

Elements	Mean content (mg/L)	Origin	Phase
Pb	0.1 to 0.8	Gasoline Industry: 35% Rain: 50% Suspended Solids	Suspended Solids
Cd	–	Industry: 35% (combustion) Rain: 20% Tyre wear	Dissolved
Zn	0.3 to 0.8	Industry: 35% (waste incineration) Rain: 30% Tyre wear Corrosion of metal objects	Dissolved

types of water: runoff water and infiltration water (Valiron, 1990). Rainwater contains atmospheric gases in dissolved state (N_2, O_2 and above all CO_2) and also small quantities of the different chemical combinations found in the atmosphere (H_2SO_4, NaCl close to coastal areas, Ca salts and Mg, PO_4, etc.) and a multitude of organic dusts and micro-organisms. Furthermore, they are loaded by different contaminants (Valiron and Tabuchi, 1992).

The presence of high concentrations of certain heavy metals, such as cadmium, lead and zinc in rainwater is reported in the literature (Valiron and Tabuchi, 1992; Plassard *et al.*, 2000; Lassabatère, 2002). Table 2 provides a summary of the values measured for heavy metals detected in runoff water.

The pollution of urban runoff water originates on the one hand from atmospheric leaching and the leaching and erosion of urban surfaces. Indeed, the quantification and characterisation of the pollution of different types of runoff water (roofs, roads, etc.) is necessary given that certain data show that runoff can be a not inconsiderable source of micropollutants. Table 3 shows the physicochemical modifications sustained by these waters when running over urban surfaces.

Road runoff water carries mineral and organic materials into hydrosystems in chronic fashion (Boisson, 1998). These additions modify the physical, chemical and biological characteristics of the host environment, and thus cause eutrophication phenomena or have toxic effects on organisms (Figure 1). These phenomena are mainly due to nitrate and phosphate salts (Menoret, 1984).

In addition, certain pollutants found in runoff water can stem from the erosion or corrosion of urban surfaces by rain. The following are examples of this: the inflow of earth, sand and gravel

Table 3. Mean quality of rainwater (Colandini, 1997).

Parameters	Symbol & unit	Rainfall	Roof runoff	Road runoff
Hydrogen potential	pH	4.9	6.2	6.4–7.5
Electric Conductivity	CE (μS/cm)	32	80	108
Suspended Matter	SM (mg/L)	17.5	22–40	64–140
Chlorines	Cl^- (mg/L)	0.9–1.6	0.8	6–125
Iron	Fe (μg/L)	3–4.8	5.6	16–62.2
Sulphates	$SO4^{2-}$ (mg/L)	160–223	1200	4200–10400
Lead	Pb (μg/L)	5–76	23–104	128–311
Cadmium	Cd (μg/L)	0.6–3	0.7	1.9–6.4
Copper	Cu (μg/L)	1.5–12	27–235	62–108
Zinc	Zn (μg/L)	5–80	24–290	220–603
Polycyclic Aromatic Hydrocarbons	PAH (ng/L)	86–145	500	240–3100

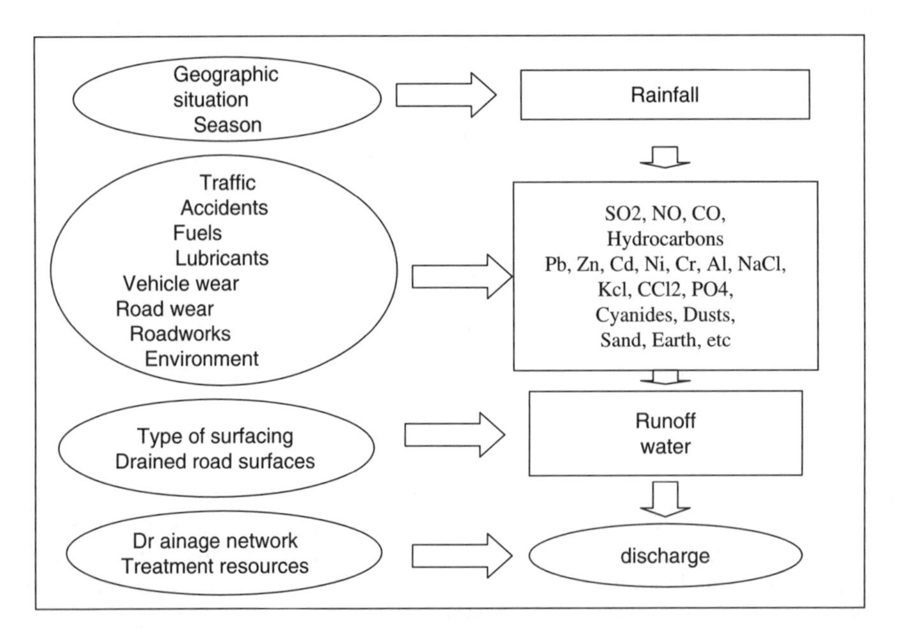

Figure 1. Pollution cycle of runoff water (Boisson, 1998).

from non-sealed surfaces, the inflow of hydrocarbons from asphalt wear, and the inflow of metals from metal surfaces (Garnaud, 1999).

In conclusion, rainwater is a potential source of contamination for host environments (Mikkelsen *et al.*, 1996). During dry periods, deposits of different substances accumulate on roads, in gutters, pavements and squares. All these substances are carried massively by the first rainfall flush which washes them from the surface. The pollutant elements are dispersed in the mass of water and accumulate with all the erosion materials that they pollute at the same time. Studies have shown that pollution due to the first flush of water can be very high, since analysis has shown it to be of the same magnitude as that of urban effluent and often even higher (Chocat *et al.*, 1993). The first flush is defined as being the quantity of rainfall required to ensure that the first 30% of the volume flowing carries 80% of the mass of pollution (Saget *et al.*, 1996).

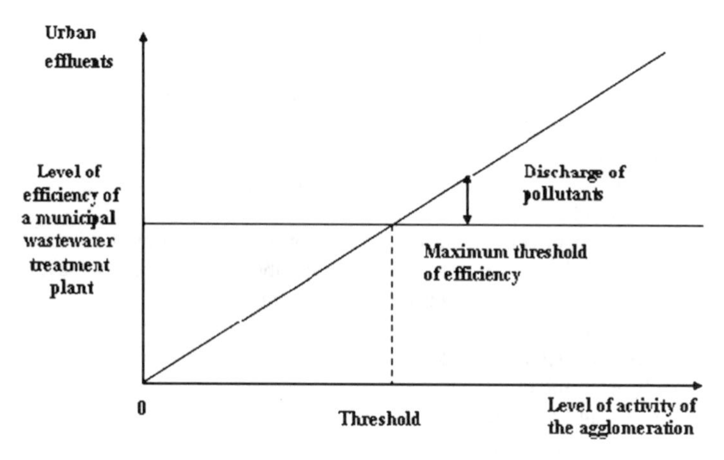

Figure 2. Impacts of human activities on ecosystems (Emmanuel *et al.* 2002).

3 WASTEWATER TREATMENT

In many countries, urban effluents are subjected to physiochemical and biological treatments in municipal sewage stations. The treatment capacity of such wastewater treatment plants ensures the degradation of organic substances and triggers a change in the chemical state of the mineral substances contained in urban wastewater (Emmanuel, 2004). Wastewater treatment in agglomerations can be performed by using various processes all of which rely on physical, chemical and biological phenomena. Generally, four types of treatment grouping processes are applied to wastewater: pre-treatment (e.g. screening, degritting); primary treatment (e.g. settling, sedimentation); secondary treatment (e.g. biological treatment); and tertiary treatment or physicochemical treatment (e.g. coagulation, flocculation, filtration, disinfection).

The level of treatment required for different types of wastewater can be determined by comparing the load of pollution tolerated in natural waters and the pollution produced by municipal wastewater and different industrial activities. This approach makes it possible to identify the treatment process best adapted to the analytical characterisation of the municipal liquid effluents of a particular locality and its industry. Treatment mechanisms are thus adjusted to stabilise the total pollutant load of the wastewater (industrial wastewater + municipal liquid wastes) of the locality considered. Treated water, also called treated wastewater, is obtained according to the treatment method used, and is either reused or discharged into a host environment, while the sludges are recycled, stored, converted or incinerated (Fresenius *et al.*, 1990).

However, the results of studies performed on municipal sewage plants have shown that certain substances fall outside the treatment capacities of the latter and are discharged into the natural environment (Kosmala, 1998; Kümmerer, 2001). Once the maximum threshold of a plant's efficiency has been exceeded (Figure 2), it becomes saturated by the flow rate and pollutant load of the urban effluents entering it, resulting in the release of these pollutants into the natural environment (Emmanuel, 2002).

The chemical compounds released by municipal sewage plants can therefore pollute the natural environment and lead to biological imbalance. If the environmental conditions permitting the degradation of these substances do not exist, the pollutants are liable to remain in the natural environment for long periods of time and lead to risks in the short, medium and long terms for the species living in these ecosystems (Emmanuel, 2004).

4 LEGAL AND REGULATORY FRAMEWORK

International conventions, regional directives and even national laws relating to environmental issues are generally difficult to apply in certain countries as they are poorly adapted to the socio-economic, political and eco-climatic realities of the contracting parties. Nonetheless, they remain the reference frameworks in the combat against pollution and the degradation of water, which in turn lead to the loss of biodiversity.

The stipulations of the Convention on the prevention of marine pollution resulting from the immersion of wastes signed at Washington on 29 December 1972, touch on several generalities concerning water pollution. Article 1 of this convention stipulates that the signatory countries must seek to promote alone or jointly the control of all sources of pollution of the marine environment and commit themselves in particular to taking every possible measure to prevent pollution of the sea by the immersion of wastes and other materials liable to endanger human health, damage biological resources, marine fauna and flora, damage the amenities, and to manage all other legitimate uses pertaining to the sea. Article 2 also states that these countries should combine their efforts and policies as a function of their scientific, technical and economic capacities in view to adopting appropriate measures to prevent marine pollution due to immersion (COHPEDA, 1995).

Nearly twenty years later, in June 1992 the Convention on Biological Diversity held in Rio de Janeiro encouraged each contracting party, insofar as possible and according to their possibilities, to adopt procedures capable of evaluating the impacts on the environment of the projects they propose and that are liable to significantly harm biological diversity, in view to avoiding and reducing such effects to a minimum, and, if necessary, to permit the public to participate in these procedures.

In its preamble, European Commission Directive 91/271/CEE, since amended by Directive 98/15/CEE on 27 February 1998, relating to the treatment of urban wastewater, mentions industrial wastewater that enters drainage systems and insufficiently treated urban wastewater, as being among the causes of damage to the environment. Articles 3 and 4 of this directive fix in terms of Equivalent Inhabitant (EI) the thresholds for member States of wastes of different origins and for different ecoclimatic conditions. Article 5 deals with the specific case of "sensitive areas" for which stricter standards should be applied. A sensitive area means a mass of water known to be eutrophic or which could become eutrophic in the short term if protective measures are not taken. This is the case of estuaries, bays and other coastal waters where the exchange of water is known to be poor, or which receive large quantities of nutritive elements (European Commission, 1998).

With reference to the European Commission's directives, legislators in France (an industrialised country) drew up restrictive quantitative standards in view to protecting its environment and combating the pollution of its water resources, whereas those of Haiti (a developing country), for example, approached the issue from a general and qualitative angle. The following considerations on these two very different legislations illustrate this observation.

Article L.35–8 of the French Health Code defines the authorisation for connection to drainage networks as follows: "Any discharge of wastewater other than domestic into public drains must be authorised beforehand by the local authority owning the infrastructures used for this wastewater before it reaches the natural environment".

Article 34 of the Ruling of 2 February 1998 of the French Ministry of Territorial Development and the Environment stipulates: "The connection to a municipal sewage plant, whether urban or industrial, can only be considered in the case where the municipal sewage infrastructure (drainage network and sewage plant) is capable of conveying and treating the industrial effluent under good conditions. The impact study includes a section relating to connection that sets out the abovementioned aptitude, determines the characteristics of the effluents allowed in the network and stipulates the type and dimensioning of the pretreatment structures planned, if any, to reduce the pollution at its source and minimise pollution flows and the flow rates connected to the network. The impact of connection on the operation of the sewage plant, sludge quality, etc., and their subsequent utilisation, if any, are given particular attention with respect to the possible presence of mineral and organic micropollutants in the effluents". In its section 3: the pollution of surface water, and in articles 31 and 32, the ruling sets the threshold values to be conformed with for discharging liquid effluents into the urban network.

As for Haitian regulations, the Ruling of 8 April 1977 sets the limit of the territorial waters of the Republic of Haiti. Article 7 of this ruling empowers the Haitian government to exercise any control judged necessary to "prevent pollution, contamination and other risks liable to jeopardise the biological equilibrium of the marine environment".

Article 140 of the Rural Code of François Duvalier (1962), relating to surface waters, stipulates that the discharge of wastewater from industrial installations and dwellings into natural water courses and irrigation canals is formally forbidden. Nonetheless, a request for authorisation to carry out such discharges can be made (COHPEDA, 1995).

Furthermore, the discharge standards stipulated by the different regulatory texts set the values that should not be exceeded for parameters based on estimations of the efficiency of treatment facilities. The latter are evaluated as a function of the rate of treatment of the load of organic pollutants, content in metal (heavy metals, etc.) and non metallic mineral species and certain physicochemical parameters (pH, conductivity, redox potential, etc.).

From the strict viewpoint of the real impact of discharge into the aquatic environment, the limitations of this practice quickly became apparent insofar as it is seldom possible on this basis to identify and thus take into account all the chemical species liable to be found in an effluent. These parameters do not constitute in themselves a sufficiently reliable approach to toxicity, given the phenomena of synergy and antagonism and the difficulty of evaluating global toxicity on the basis of each component taken individually (Perrodin, 1988; Emmanuel, 2004).

5 IMPACT OF THE MAIN POLLUTANTS CONTAINED IN URBAN WASTEWATER ON AQUATIC BIODIVERSITY

5.1 *Ecosystem and Biodiversity*

The term "Ecosystem" is defined as the basic ecological unit to which more complex ecological systems can be reduced. Structurally, an ecosystem is composed by the association of two components in constant interaction: a specific physicochemical, abiotic environment clearly defined in space and time, named *biotope*, associated with a living community characteristic of this biotope, named the *biocenosis*, hence the relationship: "Ecosystem = biotope + biocenosis" (Ramade, 1998). This functional unit evolves constantly and autonomously, fuelled by flows of energy.

In addition, every entity of the living community, i.e. all the genes, species and the ecosystem, is a component of biodiversity or biological diversity (MDR, 2005). Indeed, the term biodiversity designates a number of entities of variable and increasing complexity. Scientific literature reports three levels (Figure 3) of increasing complexity or characterisation of biodiversity (Ramade, 2002; MDR, 2005).

Genetic diversity: The diversity of genes expresses the diversity of characters of a population. This level of complexity comprises the characteristics of genes and their distribution within a species but also the comparison of genes of different species (MDR, 2005).

Specific diversity: This level serves as the basic reference in all actions aimed at conserving species biodiversity (Ramade, 2002). It takes into account (MDR, 2005): (i) the number of living species; (ii) the position of species in the classification of living organisms; (iii) the distribution in number of species by surface units and the numbers of each species.

Ecosystemic biodiversity: This level has specific characteristics. It is not only characterised by the number of species that it encompasses but above all the properties stemming from the assembly of the species composing it as such. The assembly of species interconnected with each other and specific to a given ecosystem gives rise to the particularities that differentiate it from other analogous ecosystems (Ramade, 2002).

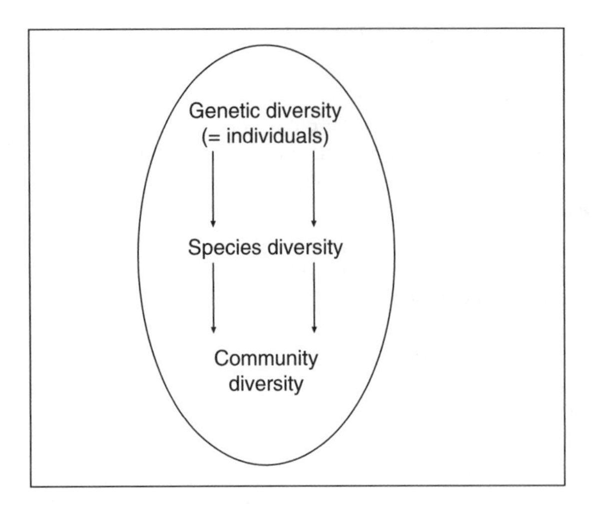

Figure 3. Levels of biodiversity (Ramade, 2002).

5.2 *Pollution and ecotoxicity*

The word *"pollutant"* designates any substance of strictly anthropogenic nature introduced into a given biotope from which it was absent or whose content it modifies (in water, air or soils according to the biotope) when present spontaneously (Ramade, 2000). Generally, the effect of a pollutant on the biodiversity of an ecosystem depends on its content and bioavailability.

Pollutant toxicity and concentration: The toxicity of a product is not intrinsic. It depends on its concentration and the type of organism that absorbs it. In the 16th century, Paracelsus announced the principle "*Sola dosis fecit venenum*", much used ever since as the basis of modern toxicology, now expressed as "All things are poison and nothing is without poison; only the dose makes a thing a poison". The notion of dose-response or effect due to an exposure-dose illustrates this principle very well. According to this principle, the same toxic substance can lead to different effects, hence the terms "acute toxicity" (single adsorption of a high dose of a chemical substance) with a lethal or sub-lethal effect, and chronic toxicity" (exposure to low but repeated doses for a period of time) causing progressively severe problems (Keck and Vernus, 2000). Finally, the label "special toxicity" is given to substances to which long-term exposure can lead to effects on reproduction or cancers.

For a toxic sample, the relation between concentrations and response at the end of an ecotoxicological test takes the shape of a sigmoid curve. Usually, the results are summarised and expressed by a concentration inducing a toxic effect for 50% of persons (immobilisation, mortality) or inhibition of 50% of the activity of organisms (luminescence, growth) in comparison to a control. This concentration, known as Efficient Concentration 50 (EC_{50}), is the median calculated as a function of the dose-response relation. The higher the toxicity of the sample is, the lower its efficient concentration.

The disadvantage of the correspondence between ecotoxicity and EC_{50} is that they are inversely proportional. To express toxicity in a form that both facilitates handling and is directly proportional to effect, the United States Environmental Protection Agency (USEPA) also proposes the use of *Toxic Units (TU)*. The latter are calculated by the formula:

$$\text{Toxic Units } TU_{50} = 100/EC_{50} \text{ } (EC_{50} \text{ in \% volume}) \quad (1)$$

To characterise the ecotoxicity of chemical substances, Directive 93/21/CEE proposes a classification based on the results of bioassays on fish, daphnia (crustaceans) and green algae (European Commission, 1992):

- Very toxic : $EC_{50} < 1$ mg/L;
- Toxic : 1 mg/L $< EC_{50} <$ 10 mg/L

- Noxious : $10\,mg/L < EC_{50} < 100\,mg/L$
- Non toxic : $EC_{50} > 100$ mg/L

Pollutant toxicity and bioavailability: Bioavailability is defined as the degree to which a contaminant can be assimilated by an organism (Newman and Jagoe, 1994). This notion is governed by three processes (Campbell, 1995; Hudson, 1998): dispersion from the solution to the surface of the membrane, fixing to transmission vectors and assimilation in the organism through the lipid membrane.

5.3 *Evaluation of impact*

It is not always possible to perform chemical analyses of the pollutants found in the different compartments of aquatic ecosystems because of the myriad different components in them, often at concentrations lower than the capacities of analytical detection (Narbonne, 1988; Flammarion *et al.*, 2000). What is more, such an approach does not provide information on the risks to animal and plant populations exposed to pollutants, and when used alone, it can neither predict the biological effects of mixtures of contaminants (synergies, antagonismes, etc.), nor simply quantify the bioavailability of pollutants for living organisms (Dutka, 1998). Consequently, decision-makers lack information on the urgency of the measures to be taken to improve the health of these ecosystems (Lascombe, 1997), or protect their biodiversity and integrity (Flammarion *et al.*, 2000). Physicochemical characterisation of the pollutant matrix and biological evaluation methods (*complete ecotoxicological approach – implementation of bioassays and biomarkers*) used to assess the toxic potential of certain pollutants detected, permit responding to these problems.

The ecotoxicity of urban effluents: Bioassays on raw effluent are used to measure the effective toxicity of these effluents and to estimate the potential impacts of complex effluents on aquatic ecosystems (Birge *et al.*, 1989). The advantage of tests relies on the possibility of controlling certain biotic (species, age, predation) and abiotic conditions (photoperiod, temperature, physicochemical composition of environments) (Kosmala, 1998). Some tests are aimed at determining the short-term effects of an effluent, while others focus on longer term sub-lethal or specific effects. Table 4 provides a list of the main and most often used standardised monospecific tests and bioassays.

Each level of biological organisation can be subjected to a range of methods intended to highlight the nature and intensity of effects. The choice of a method depends on the aim desired. Nearly all ecotoxicity tests have been designed from the outset to determine the fate and behaviour of products, the biological effects of pollutants or both. Ecotoxicological tests can be classified according to their protocols, level of biological organisation, exposure time and target (Forbes and Forbes, 1994).

Among the different pollutants present in urban effluents and which are capable of disturbing the functioning of host aquatic ecosystems, it was decided in the framework of this study to focus on the undesirable effects of the following substances on aquatic organisms: heavy metals, pesticides, nutriments (nitrates and phosphates), drug residues, radioelements and detergents (surfactants). The absence of other pollutants detected in urban effluents does not mean that they are less important in terms of ecological impact, it is simply because they are present in such great numbers. The following paragraphs summarise the physicochemical properties of the pollutants chosen and their fate in aquatic ecosystems. Given that urban effluents are a pollutant matrix in themselves, a summary of the theory of the combined effects of pollutants is presented.

Heavy metals: Modifications of the structure of a plant and animal population are expressed in terms of "biodiversity loss". Among the substances responsible for this phenomenon, metals and non metals play a decisive role. In particular they are held to account in the case of the death, growth inhibition and reproduction of aquatic and terrestrial animals (Académie des Sciences, 1998). As a function of their nature, the reactivity of the biotope and the metabolic capacities of the biocenosis, they can evolve through time according to different mechanisms when combined with other constituents of the environment:

- transformation of chemical forms resulting in the modification of mobility and bioavailability;
- biotransformation by micro-organisms, with the possibility of forming organometallic derivatives;

Table 4. Main standardised mono-specific tests.

Reference	Title
NF EN ISO 7346-1 March 1998	Water quality – Determination of the lethal acute toxicity of substances on a freshwater fish (*Brachydanio rerio* Hamilton-Buchanan (Teleostei, Cyprinidae)) – Part 1: static method
NF EN ISO 7346-2 March1998	Water quality – Determination of the lethal acute toxicity of substances on a freshwater fish (*Brachydanio rerio* Hamilton-Buchanan (Teleostei, Cyprinidae)) – Part 1: semi-static method
NF EN ISO 7346-3 March 1998	Water quality – Determination of the lethal acute toxicity of substances on a freshwater fish (*Brachydanio rerio* Hamilton-Buchanan (Teleostei, Cyprinidae)) – Part 1: method with continual renewal
NF T90-377 December 2000	Water quality – Determination of chronic toxicity on *Brachionus calyciflorus* in 48 h – Inhibition test of population growth
NF EN ISO 6341 May 1996	Water quality – Determination of the mobility of *Daphnia magna Strauss* (*Cladocera, crustacea*) – Acute toxicity test
NT T90-378 December 2000	Water quality – Determination of chronic toxicity on *Daphnia magna Strauss* in 7 days – Simplified inhibition test of population growth
ISO 10706:2000 April 2000	Water quality – Determination of long term toxicity on *Daphnia magna Strauss* (*Cladocera, Crustacea*)
NF T90-376 December 2000	Water quality – Determination of chronic toxicity on *Ceriodaphnia dubia* in 7 days – Simplified inhibition test of population growth
NF T90-375 December 1998	Water quality – Determination of chronic toxicity on of water by inhibiting the growth of freshwater algae *Pseudokirchneriella subcapitata* (*Selenastrum capricornutum*)
NF EN 28692 May 1993	Water quality – Inhibition test of the growth of freshwater algae with *Scenedesmus subspicatus* and *Selenastrum capricornutum*
NF EN ISO 10253 April 1998	Water quality – Inhibition test of marine algae growth with *Skeletonema costatum* and *Phaeodactylum tricornutum*
NF EN ISO 11348-3 February 1999	Water quality – Determination of the inhibitive effect of water samples on the luminescence de *Vibrio fischeri* (luminescent bacteria test. Part 3: Method using lyophilised bacteria)

– biodegradation, in the case of organometallic derivatives, with progressive mineralisation of the organic structure.

The presence of metals in the environment is a natural phenomenon, though amplified by certain anthropic activities the chief of which is industrialisation (Sigg *et al.*, 2000). In water, metals present in hydrated form (M^{2+}), chelating with organic ligands (fulvic and humic acids), and inorganic form are adsorbed onto particles (Tessier and Turner, 1995). The physicochemical characteristics of water (pH, redox potential, light, temperature, hardness, ionic strength, concentration in organic and inorganic ligands, etc.) act on the degree of dissociation between chelated and ionic forms. Chelating with organic ligands and minerals as well as competition with other divalent cations (Ca^{2+}, Mg^{2+}, Fe^{2+}) influence the fate of metals in particular (Tessier and Turner, 1995; Gilbin, 2002, Devez, 2004). Figure 4 summarises the different interactions affecting metals during their transfer in aquatic environments. It should be emphasised that one of the main environmental problems raised by metals is that they do not degrade biologically like certain organic pollutants, and that they bioaccumulate in food chains.

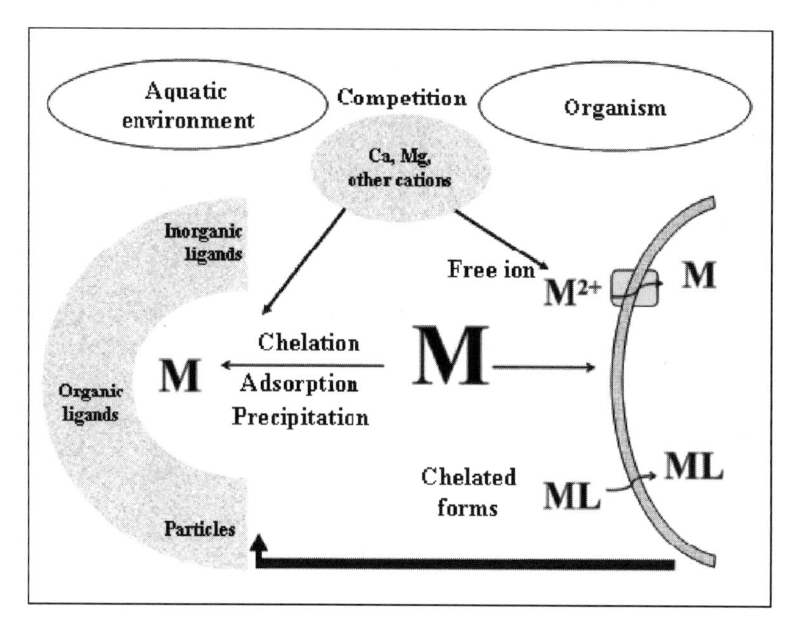

Figure 4. Physicochemistry of metals in an aqueous environment (Devez, 2004).

Food chains are constituted by networks formed between primary organisms (that draw their energy from solar radiation and their mineral environment), secondary organisms that feed on the former, the consumers of these secondary organisms, and so forth. Bioamplification phenomena have been highlighted for certain bioaccumulative pollutants, as living organisms form a food chain whose pollutant content increases according to their position in the trophic hierarchy (Zakrzewski, 1997).

This phenomenon results from the direct bioaccumulation of the pollutant from the environment to the organism (property associated with the accumulative nature of the pollutant) and the concentration of the pollutant in the organism at each step of the food chain. A large number of organochlorine compounds, such as dioxins, and mercury found in methylmercury form, are typical bioaccumulative toxins, due to their remanent and liposoluble natures. Since their solubility is generally relatively low, heavy metals discharged into water are more frequently found after adsorption on particles and deposited with sediments. This is the case with mercury part of which is transformed into methylmercury by bacterial activity on suspended matter and in sediments. The liposolubility of this compound is greater, thus it easily passes through biological membranes, accumulates in aquatic organisms, and reaches increasing concentrations as one follows the food chain (Keck and Vernus, 2000).

Table 5 provides the results of ecotoxicity tests implemented on the three levels of trophic organisation and on the decomposers of aquatic environments.

Pesticides: Agricultural activities contribute towards the entry and accumulation of trace elements and pesticides that are not without consequence for the surrounding environment (Alloway, 1995). One of the main characteristics that determines risks of contamination and the impact of pesticides is their varying persistence through time in a given environment. Table 6 shows the acute and chronic toxicity data of three herbicides and two insecticides on certain aquatic organisms.

Nitrate and phosphate substances: Dystrophisation (or the excessive proliferation of algae and aquatic macrophytes) is one of the major disturbances caused by nutrient pollution (nitrate and phosphate substances). It is often linked to the excessive addition of nitrates and phosphates.

Table 5. Toxicity of certain metals on aquatic organisms (AdS, 1998).

Species Elements	Chemical forms	Micro-organisms	Daphnia	Fish	Algae
Mercury	Inorganic derivatives	EC50 0.01–0.05 mg/L	EC50 48 h: 0.02–0.6 mg/L	LC50 96 h: 0.001–0.1 mg/L	EC50: 0,007–0.05 mg/L
	Organic derivatives	EC50 0.005 mg/L	EC50 48 h: 0.02 mg/L		EC50 0.001 mg/L
Lead	Inorganic derivatives	Inhibition of respiration: 0.08 mg/L	EC50 48 h: 0.45–65 mg/L	LC50 96 h: 0.6–242 mg/L	Inhibition of growth: 0.5–10 mg/L
	Tetraethyl lead		Inhibition of reproduction: 0.012 mg/L	LC50 96h: 0.05–300 mg/L	
Cadmium			EC50 48 h: 0.006–20 mg/L Inhibition of reproduction: 0.002–0.01 mg/L	LC50 96 h: 0.002–16 mg/L	EC50: 0.0015–0.25 mg/L
Copper			LC50: 0.01–0.07 mg/L Inhibition of reproduction: 0.03 mg/	LC50 96 h: 0,02–10 mg/L LC50: 1.85 mg/L	Inhibition of growth: 0.0004 mg/L
Zinc			EC50 48 h: 0,15–30 mg/L Inhibition of reproduction: 0.002–0.01 mg/L	LC50 96 h: 9.9–52 mg/L	EC50 72 h: 0.6–25 mg/L CE50 72 h: 0,05–0.1 mg/L
Arsenic	Pentavalent	EC50: 0.04–9.7 mg/L	EC50 48 h: 7.4 mg/L EC50 21 days: 0.5–3.2 mg/L	LC50 96 h: 10,8–150 mg/L	EC50 72 h: 0,26–4.7 mg/L
	Trivalent		EC50 48 h: 1.4–6.2 mg/L EC50 21 days: 0.96 mg/L	LC50 96 h: 11–50 mg/L	EC50 72 h: 1.7–4 mg/L
Nickel		EC50: 0.7 mg/L	EC50 48 h: 0.7–86 mg/L	LC50 96 h: 2–40 mg/L	EC50: 0.3–0.8 mg/L
Chrome	Potassium dichromate	Inhibition of heterotrophic potential: 12 mg/L	EC50 48 h: 0.1–2.5 mg/L	EC50 96 h: 20–300 mg/L	EC50 72 h: 0.16–1.35 mg/L
	Chrome derivatives		EC50 48 h: 30–500 mg/L	EC50: 63–375 mg/L	EC50: 1 mg/L

The latter often defines the limiting factor of this process in freshwater as well as in upstream watersheds (Manahan, 2000). Indeed, algae can develop with phosphate concentrations as low as 0.05 mg/L (Rodier, 1996) and/or with nitrate concentrations of 1 mg/L (Miquel, 2003). However, generally speaking, municipal wastewater contains about 25 mg/L phosphate (orthophosphates, polyphosphates and insoluble phosphates) (Manahan, 2000).

Table 6. Toxicity of certain pesticides on aquatic organisms (Devez, 2004).

Species Substances		Daphnia magma	Rainbow Trout	Selenastrum capricornutum	Minnows
Azimsulfuron	Effect	Immobilisation	LC50	LC50	
	Duration	21 days	96 h	120 h	
	Concentration	0.013 μM	0.36–2.26 mM	0.03–0.05 μM	
Oxadiazon	Effect	Immobilisation	Inhibition of growth	IC50	Inhibition growth
	Duration	48 h	43–63 days	120 h	43–63 days
	Concentration	> 7 mM	0.0025 μM	0.024 μM	0.096 μM
Pretilachlor	Effect		CL50		
	Duration		96 h		
	Concentration		2.9–9.0 μM		
Fipronil	Effect	IC50	LC50		
	Duration	48 h	96 h		
	Concentration	> 0.43	0.57 μM		
	Effect	Inhibition growth and reproduction			
	Duration	21 days			
	Concentration	0.02; 0.046 μM resp.			
Alphacypermethrine	Effect	IC50	LC50	IC50	LC50
	Duration	24 and 48 h	96 h		96 h
	Concentration	0.0026 and 0.0007 μM	0.0022 μM	0.24 μM	0.0067 μM

In coastal waters, increasingly high additions of nitrates stimulate anthropic eutrophisation in estuaries (Ryther and Dunstan, 1971; Howarth, 1998; Nixon; 1995). This phenomenon leads to ecological disasters: the destruction of large phanerogame communities and coral reefs (Bell, 1992; Short and Burdick, 1996; Hauxwell *et al.*, 2001), and the eventual disappearance of aquatic fauna due to drastic reductions of crustaceans, molluscs and finfish (Nixon *et al.*, 1986).

The case of pollution by nutrients, especially by nitrates and phosphates due to natural and/or anthropic causes, generates disturbances in biochemical cycles resulting in an undesirable accumulation of sometimes toxic intermediaries in these cycles capable of generating ecological imbalances (Kuenen and Robertson, 1988; Heathwaite, 1993) that depend on their chemical form and concentration (Féray, 2000).

Excessive levels of ammonia can be harmful to aquatic life. Fish can suffer loss of equilibrium, hyper-excitability, increased respiratory activity and oxygen consumption, and faster cardiac rhythm. Different sub-lethal effects may occur: reduced hatching, reduced growth rates and morphological development, injuries to gills, liver and kidneys, and so forth. At extreme ammonia levels, they can suffer from convulsions followed by coma and death. The lethal concentration (LC_{50} 96 h) for a certain number of fish species varies from 0.2 and 1.1 mg NH_3/L for Salmonidae and from 0.7 to 3.4 mg NH_3/L for Cyprinidae (Garric, 1987).

Regarding phosphorous, it is not intrinsically toxic for terrestrial and aquatic fauna and flora. On the contrary, "eutrophisation", which is the direct consequence of excess phosphorous in the environment, has a very wide range of effects leading to serious concern. For example, the effects of the phosphates contained in household detergents are dangerous for the environment.

Drug residues: The presence of traces of drug residues in effluents of municipal sewage plants has been observed since the 1980s. Figure 5 shows the circuit followed by the drugs used in

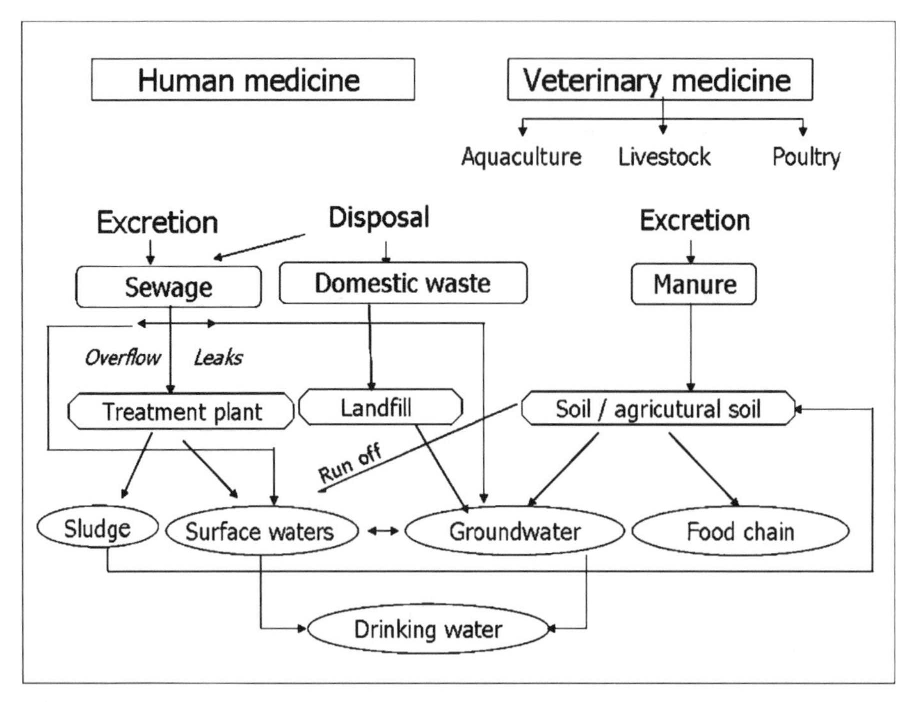

Figure 5. Contamination circuit of aquatic ecosystems by drugs used in human and veterinary medicine (Diaz-Cruz *et al.*, 2003).

human and veterinary medicine, which finally end up in water washed down the household drain (Diaz-Cruz *et al.*, 2003).

These compounds are considered as micropollutants for the environment because they have been developed with the intention of producing a biological effect on the organism (Halling-Sorensen *et al.*, 1998). The fate of drugs in the natural environment has led to considerable interest from the scientific community. Concentrations of drugs higher than 1 μg/L have been detected in aquatic ecosystems (Richardson and Bowron, 1985). Their very presence in aquatic ecosystems represents a danger for organisms.

For this study, we have chosen three major groups of drugs from the different drugs subject to impact studies on aquatic biodiversity: sex hormones (endocrine blockers), antibiotics (multi-resistance to bacteria), and anti-cancer and antineoplastic agents (genotoxicity).

Study of the fate of drugs in water has shown that certain sex hormones (endocrine blockers) have effects on aquatic organisms at concentrations lower than 1 μg/L. For example, Estradiol, the female sex hormone (and a hormonal marker of aquatic pollution), can modify the sexual characteristics of certain fish at concentrations of 20 ng/L (Raloff, 1998).

Antibiotic (multi-resistance to bacteria) residues in the environment are suspected of being the causal agent of the development of types of resistance in bacteria chains. Therefore these substances represent a serious threat to pubic health regarding the treatment and control of certain infectious diseases.

Anti-cancer and antineoplastic agents (genotoxicity) are by far the most often used drugs in medical departments. Given their potential impact on the environment, antineoplastic agents form a major group of drugs in terms of health and environmental risks (Kümmerer, 2001). Their mutagenic, carcinogenic and teratogenic characteristics have already been demonstrated (Skov *et al.*, 1990). Indeed, anticancer agents are acknowledged as being the most toxic.

Table 7. Ecotoxicity of the 6 main groups of drugs on aquatic organisms (CSTEE, 2001).

Groups	Extremely toxic $EC_{50} < 0.1$ mg/L	Very toxic $EC_{50} < 0.1$–1 mg/L	Toxic EC_{50} 1–10 mg/L	Slightly toxic EC_{50} 10–100 mg/L	Non toxic $EC_{50} > 100$ mg/L
Analgesics			D	D,E	
Antibiotics	A	B			
Antidepressants		D			
Anti-epileptics			C		D,E
Antineoplastic agents		A		D,E	
X-ray contrast agents					A,B,D,E

The most sensitive taxonomic groups: A- Bacteria; B- Algae, C- Cnidaria, D- Crustaceans, E- Fish.

The fate of drugs in aquatic ecosystems can take one of three possible main states; A completely oxidised substance that results in carbon dioxide and water (Halling-Sorensen *et al.*, 1998). This state can be assimilated with the ecocompatibility of the substance (Navarro *et al.*, 1994). The substance is lipophilic, hardly degradable, though some of the substance is adsorbed by sewage plant sludge (Halling-Sorensen *et al.*, 1998). Under the action of metabolisation, these substances produce metabolites whose hydrophilic structure is different from that of the mother compounds (lipophilic). However, the two mother and daughter compounds are remanent and pass through the sewage treatment processes of the plant to be discharged into host environments, giving rise to a risk for aquatic organisms in the case where the metabolites are active (Halling-Sorensen *et al.*, 1998). The well-known example of this state is Clofibrate and its metabolite clofibric acid (Hignite et Azarnoff, 1977; Richardson and Bowron, 1985). Table 7 shows the list of the six main groups of drugs used in human medicine most toxic for aquatic environments (CSTEE, 2001).

Radioelements: These substances are most usually used in nuclear medicine and in the energy industry. In medicine, radioelements ^{90}Y and ^{198}Au are injected in the form of colloidal solutions in body cavities, most often in quantities in the region of 100 to 200 mCi. Most of this concentration remains in the organism and is not excreted. However, from 60 to 70% of ^{131}I, administered orally, is excreted in the urine; the doses used vary from 100 μCi for diagnostics to more than 100 mCi for treating thyroid cancers (Rodier, 1971; Erlandsson and Matsson, 1978). Regarding radioelements used for nuclear diagnoses $^{99}Tc^{m}$ and ^{201}Tl, they can be easily found at different points of the drainage network (Erlandsson and Matsson, 1978).

Studies carried out on the radioactive pollution of aquatic ecosystems show indications of the bioamplification phenomenon of certain radioelements on the aquatic biocenosis. Indeed, it has been proven in the United States that the salmon in the Columbia river, exposed to discharges of ^{32}P had an average contamination of 1.5 $Bq.g^{-1}$ liable to result in "isolated" individuals who may have consumed 40 kg of salmon a year, "critical" irradiation of bones – organs – of 0.3 mSv per year, i.e. 20% of the admissible dose (Ramade, 1998).

Detergents-surfactants: A detergent is a product used to *"remove dirt"* by physical and chemical action. The ruling of 28 December 1977 published in the Official Gazette of 18 January 1978, relating to the biodegradability of surfactants, stipulated that the biodegradability of detergents must be equal to or higher than 90%. Surfactants are the surface or tensioactive agents of detergents. They constitute the greater fraction of the organic part composing detergents. In water their critical micellar concentration (CMC) forms a discontinuous and fragile film on the surface that more often than not prevents sunlight and oxygen from penetrating into the water, leading to a biological imbalance and, in the medium and long terms, in the case of an aerobic biological treatment reactor, the occurrence of anaerobic areas. It can also facilitate the onset of eutrophisation in natural aquatic environments.

Table 8. Acute toxicity of surfactants on *Vibrio fischeri* and *Daphnia* (Emmanuel *et al* 2005).

Bioassays	Units	CTAB	TX 100	SDS
EC_{50} 30-mn *Vibrio fischeri*	mg/L	1.02	63.6	0.276
EC_{50} 24-h *Daphnia*	mg/L	0.024	38.05	29.21

Table 9. Classification of the interactions of two substances (Plackett and Hewlett, 1948).

	Similar joint action	*Dissimilar joint action*
Non-interactive	*Simple similar*	*Independent*
Interactive	*Complex similar*	*Dependent*

Surfactants are composed of an easily hydrated hydrophilic polar part (negatively charged ions), and a hydrophobic or lipophilic non polar part (insoluble in water – hydrocarbon chain). They are toxic substances for aquatic organisms (Talmage, 1994) and can be synthesised chemically (synthetic surfactants): cationic, anionic, non ionic, amphoteric; or produced by micro-organisms such as bacteria and fungi (bio-surfactants) (Edwards *et al.*, 2003).

The toxicity of surfactants in invertebrates and crustaceans has been reported in the literature (Cserháti *et al.*, 2002) and the undesirable effects of surfactants on the three levels of the aquatic food chain (algae, crustaceans, fish) have been studied (Schwarz and Vaeth, 1987; Talmage, 1994).

Generally, cationic and anionic surfactants are more toxic than non ionic surfactants (Rouse *et al.*, 1994; Park and Bielefeldt, 2003). This fact has been demonstrated (Table 8) by studies carried out by Emmanuel *et al.* (2005) on three types of surfactant: Sodium Dodecyl Sulfate (SDS – anionic), Cethyltrimeyhylamonium Bromide (CTAB – cationic) and Triton X-100 (TX-100 non ionic).

Combined effects of pollutants on aquatic organisms: Bliss (1939) identified three modes of action of the constituents of a mixture on living organisms: "*Independent joint action*" – in this type of action the constituents act on different action sites and the biological response of a constituent is not influenced by another; "*Similar joint action*" – the constituents act on the same action sites and the biological response is not influenced by another. This is the approach most often used for studying mixtures; and finally "*Synergistic action*" in which the response of a mixture cannot be known from the isolated responses of its constituents. The response of a mixture depends on the combined effects of its constituents. Groten *et al*, (2001) showed that these interactions are above all apparent when the substances composing the mixture play a role in enzymatic transformation or act on metabolism.

Basing themselves on the results of the initial work performed by Bliss (1939), Plackett and Hewlett (1952) formulated the following classification (Table 9) for the mixture of two substances. In this approach "*Similar joint action*" means the same action sites, "*Dissimilar*" means different sites, "*Non-interactive*" means that the biological response of a product is not influenced by that of the other (in other words interaction is additive), while "*interactive*" means that the effects are more than additive (synergistic) or less than additive (antagonistic).

Indeed, the combined toxic effects of mixtures of pollutants on a target organism can be the simple addition of action, but also antagonistic or potential – "*synergy*" (Hermens *et al.*, 1984). Antagonism means that the ecotoxicity of a mixture leads to a toxic response less than the total of the toxicities observed when the substances are absorbed individually. The additivity of the responses occurs when the total of the responses is equal to the toxic response of the mixture. As for synergetic responses, they are observed when the combination of responses of a mixture is higher than the total of the responses of each component (Mercier, 2002; Warne 2003).

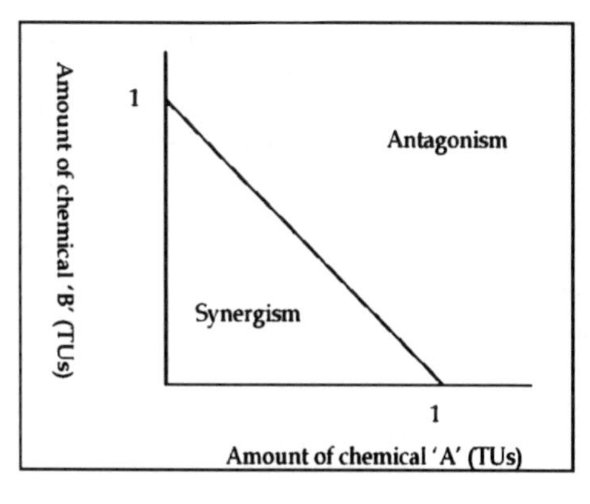

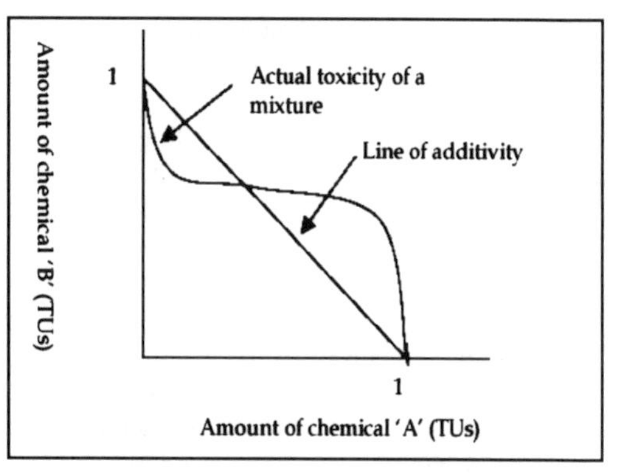

Figure 6. Example of isobolograms (Warne, 2003).

Several models have been formulated to perform the quantitative analysis of mixtures. Brown (1968) and Sprague (1970) introduced the concept of Toxic Unit (TU). The TU of a substance is the dose used divided by its EC_{50} on the organism considered. The toxicity of a mixture of two constituents can then be expressed as the total of each of their TUs, since these are dimensionless relations. If the sum of the TU is equal to 1, it is assumed that the toxicity is additive and it will be more than additive if the sum of the TUs is > 1 and less than additive if this sum is < 1.

By extending the TU model, Marking (1977) developed that of the Additivity Index. This model is a modification intended to solve the problem of non-linearity of TU around 1. Belkhadir (1979) developed the toxicity index model by using the following formula:

$$TI = \sum_{i=1}^{n} (Cmél/Cseul)_i \quad (2)$$

where TI: Toxicity index, Cmél: concentration of the substance in the mixture and Cseul: Concentration of the substance in a pure solution

For:

TI = 1: the interaction of the effects is additive

TI > 1: the interaction of the effects is antagonistic

TI < 1: the interaction of the effects is synergetic

The Toxicity Index method is also used to identify the general characteristics of interactions in a mixture of more than two compounds. On the other hand, the hypothesis TI = 1, implying the additivity of effects is often criticised and widened. Thus, according to Deneer (2000), the hypothesis can be widened to 0.5 < TI < 2. EIFAC (1980) proposes an additivity of effects for 0.5 < TI < 1.5.

The graphic tool most often used to study the combined effects of two substances on an organism is the isobologram, which is a two dimensional chart used to highlight the effects of interactions between two compounds on an organism. The ordinates and abscissa represent the UT relative to each substance. For a mixture of two compounds A and B, there is an ordinate UT_B and an abscissa UT_A. Although the effects are additive, the curve is a straight line (solid line); if the effects are synergetic, a curve at the bottom is obtained, to the left of the additivity isobole, and if they are antagonistic, there is a curve at the top, on the right (Calamri and Alabaster, 1980). This representation is very often used for studies of the toxic behaviour of binary mixtures (Warne, 2003) It is a compromise between TU and TI that permits viewing the interaction of the effects of a binary mixture with the different ratios of its two components, since, as ca be seen on the right hand isobol (figure 6), the interaction of the effects in the binary mixture can be quite different as a function of the ratio of each of the components.

Most models are based on the additivity hypothesis. They generally permit concluding that the ecotoxicological effects of the components of a mixture are additive or else there is a deviation of additivity, i.e. antagonism or synergy of interactions. Many studies recommend the additivity of effects as being sufficient for modelling the ecotoxicity of mixtures. The additivity of effects can be verified for 70 to 80% of mixtures, whereas 10 to 15% of mixtures have antagonistic accumulative effects and the same proportions hold for accumulative synergetic effects (Ross et Warne, 1997).

Experimental approaches in situ: Monospecific tests performed in the laboratory (bioassays) with pure products administered to laboratory species are the most common form of experimental test, but it is possible to create larger and more complex models, ranging from multispecific tests to microcosms and mesocosms (integrated tests). The results obtained with these models are more difficult to use to evaluate the ecological impact of pollutants (Rivière, 1998).

Bioassays are limited since they do not take into account the interactions between species and are performed under physicochemical conditions that differ from those of the natural habitat (Pontasch *et al.,* 1989). Experimental laboratory approaches are also criticised for their lack of ecological realism (Forbes and Forbes, 1994) and their inability to estimate the direct and indirect effects of contaminants on higher levels of biological organisation. Given the uncertainties linked to chemical and biological measurements, researchers have recommended integrated approaches that call on chemical analyses, toxicity tests and field measurements of biokinetic behaviour (Kosmala, 1998).

In situ experimental approaches are performed in view to establishing a direct link between contaminants and biological effects and they enhance impact assessment by adding the realism of field work. This approach includes eco-epidemiological observations aimed at highlighting the toxic effects of pollutants and performed on sentinel species. This approach also includes the observation of toxic effects on plants and caged animals on the study site. The United States Academy of Science defines sentinel species "As a system set up in which data on the exposure of animals to pollutants in the environment are regularly and systematically collected and analysed to identify potential dangers for human beings and other animals (NRC, 1991)".

In situ studies of the effects of certain pollutants on aquatic organisms, biomarkers are used to observe the behaviour of fish exposed to organic and mineral substances. A biomarker is a parameter measured at molecular, cellular and functional levels on individuals taken from a population and which indicates either that the individuals have been exposed to pollutants or toxic substances, or that the individual develops pathological effects within varying periods of time (NRC, 1989).

Indeed, biological markers provide information on the nature and level of chemical contamination. They also permit diagnosing the health of organisms and populations living in ecosystems (Flammarion *et al.*, 2000). Biological markers are considered as indicators that signal events in biological systems and samples and can be of different types. Exposure biomarkers indicate the contamination of biological systems by one or more xenobiotics. The monoexygenase activities of fish, for example, are biological markers of exposure to major environmental pollutants and indicators sensitive to water quality (induction). Certain P450 are induced by major environmental pollutants such as PCBs, PCDDs, PAHs and pesticides. Toxicity biomarkers indicate more or less long term biological effects on biological systems, for example, the P450 of family 1A. These are most probably precursor indicators of later toxic effects (induction leading to carcinogenesis). Finally, individual sensitivity biomarkers indicate different sensitivities for certain individuals of a population. They permit studying the effects of a product on a percentage of the population on the basis of metabolic rate (fast or slow). Variations of metabolic activity are the main cause of intraspecific variability.

On the one hand, the specificity of a biomarker for certain families of chemical compounds (polycyclic aromatic hydrocarbons, polychlorobiphenyls, heavy metals, phytosanitary products, etc.) reveals their presence, while on the other it provides information on the bioavailability of these pollutants and early biological effects on organisms (Kramer and Botterweg, 1991). Flammarion *et al.* (2000) note that the main biomarkers studied in fish respectively express the exposure of organisms to certain families of compounds: changes in enzymatic activity, breaking of single and double DNA strands (genotoxic effect), disturbance of vitellogenin synthesis (damage to reproductive capacity).

A large number of compounds, organic chlorines (DDT, PCBs, etc.), and substances present to a great extent in sewage plant effluents, such as the degradation products of detergents containing alkylphenol polyethoxylates or phtalates, are acknowledged as causing disturbances to the endocrine system by setting off varying levels of oestrogen-mimetic activity in male fish (Jobling *et al.*, 1996; Tyler *et al.*, 1996; Flammarion *et al.*, 2000). In situ studies performed downstream of sewage plant discharge outlets have highlighted such oestrogenetic activity in male trout exposed to sewage plant effluents, in relation with environmental contamination by alkylphenols and/or oestrogens (Harries *et al.*, 1997).

6 CONCLUDING REMARKS

In the framework of this study, the analysis of the toxic potential of certain pollutants has been performed on the basis of a non exhaustive synthesis of information available on the physicochemical characteristics of urban effluents and their impacts on aquatic ecosystems. The presence of pollutants in urban effluents can in several unlikely cases have a low impact on the environment. In contrast, in some situations it can significantly disturb terrestrial and aquatic ecosystems, and significantly participate in the quantitative and qualitative degradation of water resources, especially in terms of loss of biodiversity.

The physicochemical characterisation of urban effluents has led to the observation that most of the pollutants chosen are found at concentrations higher than the threshold values dictated by the regulations governing the discharge of wastewater into natural environments. Given that the toxicity of a substance depends on its available concentration, we postulate that the discharge of urban wastewater containing pollutants at detected concentrations into natural ecosystems causes short-term toxic effects (mortality) and long term effects (occurrence of cancers, reproductive disorders, etc.) on living organisms. In order to validate this hypothesis, we have decided to present the results obtained from toxicity tests and field measurements of the biokinetic behaviour of effluents. The information reported on the biological effects of pollutants confirms the presence of dangerous substances in urban effluents.

Monospecific tests on urban effluents show that the toxicity of these matrices is high for the first three levels of the trophic organisation of aquatic ecosystems. The results provided by biological markers have also permitted measuring the state of health of living organisms, particularly fish, exposed to pollutants contained in urban effluents. These results not only confirm the existence of dangerous substances in urban effluents, but also explain the contribution of these pollutant matrices to the loss of genetic, specific and ecosystemic biodiversity of aquatic organisms.

The results of the study on the degradation by human activities of the three main ecosystems of the Bay of Port-au-Prince, i.e. coral reefs, communities of phanerogames and mangrove swamps, clearly illustrate the problem of aquatic biodiversity loss linked to anthropic pollution. The resulting major disturbance of the equilibrium of the marine environment and the dynamics of animal and plant populations is manifested by the proliferation of brown algae. The presence of these algae highlights a major biological imbalance of the marine environment and their growth reaches about 1 cm a day, allowing them to submerge the coral which only grows by 1 cm per year (Vermande et Raccurt, 2001). These brown algae are of no interest to the fish, as they are inedible and they stifle the coral.

Existing regulations stipulate the conditions for treating urban effluents and their discharge into natural environments. However, the results of the works performed over the last ten years on methods to eliminate liquid wastes have demonstrated the inefficiency of sewage plants in degrading certain pollutants. On the other hand, toxicity tests and field biokinetics measurements carried out on these effluents provide interesting responses leading to better understanding of the disturbances affecting aquatic ecosystems. It appears that terrestrial and marine water resources no longer tend to facilitate the development and continuation of life.

The involvement of politicians, decision-makers, industry and scientists in actions aimed at achieving the integrated management of urban liquid waste and aquatic biodiversity would be a positive step. Furthermore, in the context of tropical countries, methodologies for implementing toxicity tests and field measurements of biokinetic behaviour should be adapted to indigenous species (adapted to existing temperature ranges). This scientific contribution to the development of tropical ecotoxicity will permit better evaluation of the impact of the pollutants contained in the urban effluents generated in tropical countries on the biocenosis of these areas.

ACKNOWLEDGEMENTS

The authors would like to acknowledge the office of the Honorable Jacques Edouard Alexis, Prime Minister of Haiti for supporting the research program on urban wastewater of Laboratoire de Qualité de l'Eau et de l'Environnement (LAQUE), Université Quisqueya (Haiti).

REFERENCES

Académie des Sciences. Contamination des sols par les éléments traces: les risques et leur gestion. Raport no. 42, Paris: Lavoisier Tec&Doc, 1998, 440 p.

Alloway B.J. Soil processes and the behaviour of metals. In: Heavy metals in soils. Edited by B.J. Alloway, 2nd ed., London: Blackie Academic and Professional, 1995, pp 11–37.

Belkhadir E.M. Etude sur l'écotoxicologie des hydrocarbures aromatiques legers en milieu dulçaquicole. *Thèse*. Université de Metz, Metz, 1979.

Bell P.R.F. Eutrophication and coral reefs—some examples in the Great Barrier Reef Lagoon. Water Research, 1992, 5:553–568.

Birge W.J., Black J.A. et Short T.M. A comparative ecological and toxicological investigation of a secondary wastewater treatment plant effluent and its receiving stream. Environmental toxicology and chemistry,1989, 8: 437–450.

Bliss C. I. The toxicity of poisons applied jointly *Ann. Appl. Biol.,* 1939, 26, 585–615.

Boisson J.-C. Impacts des eaux de ruissellement de chaussées sur les milieux aquatiques, Etat des connaissances. Bulletin des laboratoires des Ponts et chaussées, 1998, 214: 81–89.

Brown V. M. The calculation of the acute toxicity of mixtures of poisons to rainbow trout. *Water Research,* 1968, 2, 10, 723–733.

Calamari D. and Alabaster J.S. An approach to theoretical models in evaluating the effects of mixtures of toxicants in the aquatic environment. Chemosphere, 1980, 9:533–538.

Campbell P.G.C. Interactions between trace metals and aquatic organisms: a critique of the free-ion activity model. In: Metal Speciation and Bioavailability in Aquatic Systems. Edited by A. Tessier and D.R. Turner. New York: John Wiley and Sons, 1995, pp 45–102.

Chocat B., Thibault S., Seguin D. Hydraulique urbaine et assainissement. Lyon:Institut National des Sciences Appliquées, 1993.

COHPEDA (Collectif Haïtien pour la Protection de l'Environnement et un Développement Alternatif). Haïti: législation environnementale – Compilation de testes légaux haïtiens sur l'environnement. Port-au-Prince: COHPEDA, 1995, 274 p.

Colandini V. Effet des structures réservoirs à revêtement poreux sur les eaux pluviales : qualité des eaux et devenir des métaux lourds. Thèse de doctorat, Université de Pau et des pays de l'Adour, LCPC Nantes. 1997, 171p. + annexes.

Commission Européenne. Directive 91/271/CEE du Conseil, du 21 mai 1991, relative au traitement des eaux urbaines résiduaires Journal officiel n° L 135 du 30/05/1991 p. 0040–0052.

Commission Europénne. Directive 93/21/CEE de la Commission du 27 avril 1993 – Rapprochement des dispositions législatives relatives à la classification, à l'étiquetage et à l'emballage des substances dangereuses. Journal officiel L 110, 04.05.1993.

Cserháti T., Forgács E., Gyula O. Biological activity and environmental impact of anionic surfactants. Environment International, 2002, 28:337–348.

CSTEE Opinion on "Draft discussion paper on environmental risk assessment of non-genetically modified organism non-gmp; containing medicinal products for human use". Brussels:European Commission, Directorate General, 2001.

Deneer J. W. Toxicity of mixtures of pesticides in aquatic systems. *Pest Management Science,* 2000, 56, 6, 516–520.

Devez A. Caractérisation des risques induits par les activités agricoles sur les écosystèmes aquatiques. *Thèse*, Ecole Nationale du Génie Rural, des Eaux et des Forêts, 2004, 269p.
Diaz-Cruz M.S., Lopez de Alda M.J., Barcelo D. Environmental behavior and analysis of veterinary and human drugs in soils, sediments and sludge. Trends in Analytical Chemistry, 2003, Vol. 22, 6:340–351.
Dutka B.J. Foreword. In. Wells P.G., Lee K., Blaise C. (ed.), Boca Raton: CRC Press, 1998.
Edwards K.R., Lepo J.E., Lewis M.A. Toxicity comparison of biosurfactants and synthetic surfactants used in oil spill remediation to two estuarine species. Marine Pollution Bulletin, 2003, 46:1309–1316.
Emmanuel E., Perrodin Y., Keck G., Blanchard J.-M., Vermande P. Effects of hospital wastewater on aquatic ecosystem. Proceedings of the XXVIII Congreso Interamericano de Ingenieria Sanitaria y Ambiental. Cancun, México, 27–31 de octubre, 2002. CDROM.
Emmanuel E. Evaluation des risques sanitaires et écotoxicologiques liés aux effluents hospitaliers. Thèse. Institut National des Sciences Appliquées de Lyon, 2004, 260 p.
Emmanuel E., Blanchard J-M., Keck G., Vermande P., Perrodin Y. Toxicological effects of sodium hypochlorite disinfections on aquatic organisms and its contribution to AOX formation in hospital wastewater. Environment International, 2004, 30:891–900.
Emmanuel E., Hanna K., Bazin C., Keck G., Clément B., Perrodin Y. Fate of glutaraldehyde in hospital wastewater and combined effects of glutaraldehyde and surfactants on aquatic organisms. Environment International, 2005, 31:399–406.
Ericksson E., Auffarth K., Henze M., Ledin A. Characteristics of grey wastewater. Urban Water. Urban Water, 2002, 4:85–104.
Erlandsson B., Matsson S. Medically used radionucides in sewage sludge. Water, Air, and Soil Pollution, 1978, 9:199–206.
European Commission. Directive 98/15/EEC amending Council Directive 91/271/EEC: Urban wastewater treatment. Brussels; 1998. Off. J. of European Communities N° L 67/29–30 (7 March 1998).
Féray C. Nitrification en sédiment d'eau douce: incidence de rejets de station d'épuration sur la dynamique de communautés nitrifiantes. Thèse. Université Claude Bernard-Lyon I, 2000, 204 p.
Février L. Transfert d'un mélange Zn-Cd-Pb dans un dépôt fluvio-glaciaire carbonaté, Approche en colonnes de laboratoire. Thèse. Institut National des Sciences Appliquées de Lyon, 2001.
Flammarion P., Devaux A., Garric J. Marqueurs biochimiques de pollution dans les écosystèmes continentaux. Exemples d'utilisation et perspectives pour le gestionnaire. Bull. Fr. Pêche Piscic, 2000, 357/358:209–226.
Forbes V.E. and Forbes T.L. Ecotoxicology in theory and practice. New York: Chapman et Hall,1994, 220 p.
Forbes V.E. and Cold A. Effects of the pyrethroid esfenvalerate on life-cycle traits and population dynamics of chironomus riparius – importance of exposure scenario. Environmental toxicology and chemistry, 2005, Vol. 24, 1:78–86.
Förstnet U., Wittmann G.T.W. Metal pollution in the aquatic environment. Berlin: Springer-Verlag, 1979, 486 p.
Fresenius W., Schneider W., Böhnke B., Pöppinghaus K. Technologie des eaux résiduaires – Production, collecte, traitement et analyse des eaux résiduaires. Berlin:Springer-Verlag, 1990, 1137 p.
Garnaud S. Transfert et évolution géochimique de la pollution métallique en bassin versant urbain. Thèse. Ecole Nationale de Ponts et Chaussées, 1999.
Garric J. Toxicité de l'azote ammoniacal pour la faune aquatique dans les eaux continentales. Rapport CEMAGREF Lyon, Génie de l'Environnement–Ecodéveloppement, Lyon: CEMAGREF, 1987, 75 p.
Gilbin R. Caractérisation de l'exposition des écosystèmes aquatiques à des produits phytosanitaires – Spéciation, Biodisponibilité et Toxicité. Thèse. des Universités de Montpellier I et de Genève (Suisse), 2002, 192 p.
Gray S.R., Becker N.S.C. Contaminant flows in urban residential water systems. Urban water, 2002, 4:331–346.
Grommaire-Mertz M. C. La pollution des eaux pluviales urbaines en réseau d'assainissement unitaire, Caractéristiques et origines. Thèse. Ecole Nationale de Ponts et Chaussées, 1998.
Groten J. P., Feron V. J. et Sühnel J. Toxicology of simple and complex mixtures. *Trends in pharmacological sciences,* 2001, 22, 6, 316–322.
Halling-Sørensen B., Nielsen N., Lanzky P.F., Ingerslev F., Holten-Lützhøft H.C., Jørgensen S.E. Occurrence, fate and effects of pharmaceutical substances in the environment – A review . Chemosphere, 1998, 36: 357–393.
Harries J.E., Sheahan D., Jobling S., Mathiessen P., Neall P., Sumpter J.P., Tylor T., Zaman N. Oestrogenic activity in five United Kingdom rivers detected by measurement of vitellogenesis in caged male trout. Environ. Toxicol. Chem., 1997, 16, pp. 534–542.
Hauxwell, J., Cebrian, J., Furlong, C., Valiela, I. Macroalgal canopies contribute to eelgrass (Zostera marina) decline in temperate estuaries. Ecology, 2001, 82:1007–1022.

Heathwaite A. L. Nitrogen cycling in surface waters and lakes. In: Burt T. P. *et al.* (Eds) Nitrate: Processes, patterns and management. Chichester: Wiley J. & sons, 1993, p. 99–140.

Heckman L.-H. and Friberg N. Macroinvertebrate community response to pulse exposure with the insecticide lambda-cyhalothrin using in-stream mesocoms. Environmental toxicology and chemistry, 2005, Vol. 24, 3:582–590.

Hermens J., Canton H., Jansen P., De Jong R. Quantitative structure-Activity relationships and toxicity studies of mixtures of chemicals with anaesthetic potency: acute lethal and sublethal toxicity to *Daphnia magna*. Aquatic Toxicology, 1984, 5:143–154.

Hignite C., Azarnoff D. L. Drugs and drugs metabolites as environmental contaminants: chlorophenoxyisobutyrate and salicyclic acid in sewage water effluent. Life Sci., 1977, 20:337–342.

Howarth, R.W. Nutrient limitation and net primary production in marine ecosystems. Annual Review of Ecology and Systematics, 1998, 19:89–110.

Hudson R.J.M. Which aqueous species control the rates of trace metal uptake by aquatic biota ? Observations and predictions of non-equilibrium effects. Sci. Total Environ, 1998, 219:95–115.

Jobling S., Nolan M., Tyler C.R., Brighty G., Sumpte J.P. Inhibition of testicular growth in rainbow trout exposed to oestrogenic alkylphenol chemicals. Environ. Toxicol. Chem., 1996, 15:194–202.

Karr J.R. Biological integrity: a long-neglected aspect of water resource management. Ecol. Appl., 1991, 1:66–84.

Keck G., Vernus E . Déchets et risques pour la santé . Techniques de l'Ingénieur, traité Environnement G 2450, 2000, 17 p.

Kosmala A. Evaluation écotoxicologique de l'impact des effluents de stations d'épuration sur les cours d'eau: intérêt d'une approche intégrée. Thèse. Université de Metz, 1998.

Kramer K.J.M., Botterweg J. Aquatic biological early warning systems: an overview. In. Jeffrey D.W., Madden B. (ed.) Bioindicators and environmental management. London: Academic Press, 1991, p. 95–126.

Kuenen J. G., Robertson L. A. Ecology of nitrification and denitrification. In: Cole J. A. and Fergusson S. J. (eds) The nitrogen and sulfur cycles. Cambridge: Cambridge University Press, 1988, pp 161–218.

Kümmerer K. Drugs in the environment: emission of drugs, diagnostic aids and disinfectants into wastewater by hospitals in relation to other sources – a review. Chemosphere, 2001, 45:957–969.

Lacour J., Joseph O., Plancher M. J., Marseille J. A., Balthazard Accou K., Pierre A., Emmanuel E (2006). Evaluation des dangers environnementaux liés aux substances azotées et phosphatées contenues dans les effluents urbains. RED, 1(3): 6–13.

Lascombe C. Les variables biologiques au service de la gestion des ecosystems aquatiques. In. CEMAGREF (Eds) Séminaire national ≪ Les variables biologiques: des indicateurs de l'état de santé des écosystèmes aquatiques ≫ Ministère de l'environnement, 2–3 Novembre 1994, Paris, 1997, p. 27–37.

Lassabatere L. Modification du transfert de trois métaux lourds (Zn, Pb et Cd) dans un sol issu d'un dépôt fluvio-glaciaire carbonaté par l'introduction de géotextiles. Thèse. Institut National des Sciences Appliquées de Lyon, 2002.

Lester J.N. Heavy metals in wastewater and sludge treatment processes. Boca Raton: CRC Press, 1987, 183 p.

Lewis M.A. The effects of mixtures and other environmental modifying factors on the toxicities of surfactants to freshwater and marine life. Review paper. Wat. Res., 1992, 26:1013–1023.

Manahan S. E. Environmental Chemistry, 7^{e} edition. Boca Raton: Lewis Publisher, 2000, 1492 p.

Marking L. L. Method for assessing additive toxicity of chemical mixtures. *Aquatic Toxicity and Hazard Evaluation,* 1977, 634, 99–108.

McIntyre P.B., Michel E., France K., Rivers A., Hakizimana P. & Cohen A.S. Individual – and assemblage – level effects of anthropogenic sedimentation on snails in Lake Tanganyika. Conservation Biology, 2005, Vol. 19, 1:171–181.

MDR (Ministère Délégué à la Recherche). Conférence internationale, Biodiversité: Science et Gouvernance. Paris: MDR, 2005.

Menoret M.C. Elimination biologique du phosphore. Thèse. Institut des Sciences Appliquées de Lyon, 1984.

Mercier T. *Avis de la Commission d'Étude de la Toxicité concernant les mélanges de produits phytopharmaceutiques. Réponses aux questions faisant l'objet d'une saisine de la Commission par la Direction Générale de l'Alimentation.* projet v 10. Versailles: INRA, 2002, 29 p.

Mikkelsen P.S., Häfliger M., Ochs M., Tjell J.C., Jacoben P., Boleer M. Experimental assessment of soil and groundwater contamination from two old infiltration systems for road run-off in Switzerland. Sci. Total Environ, 1996, 189/190:341–347.

Miquel G. Rapport sur la qualité de l'eau et de l'assainissement en France. Tome II – Annexes. Paris: Office Parlementaire d'évaluation des Choix Scientifiques et Technologiques, 2003, 293 p.

Narbonne J.F. Historique – fondements biologiques de l'utilisation de biomarqueurs en écotoxicologie. In: Lagadic L., Caquet T., Amiard J.C., Ramade F. (ed.) Utlisation de biomarqueurs pour la surveillance de la qualité de l'environnement. Tec et Doc Lavoisier, Paris, 1998, pp. 1–7.

Navarro A., Blanchard J.-M., Bouster C., Gourdon R., Manfe C., Maraval S., Mathurin D., Mehu J., Murat M., Naquin P., Perrodin Y., Revin P., Rousseaux P., Veron J. Gestion des déchets. Techniques de l'Ingénieur, traités Généralités et Construction, 1994, A8660 à C4260, 32 p.

NCR (National Research Council) Animals as Sentinels of Environmental Health Hazards. Washington DC:National Academy Press,1991, 176 p.

Newman M. and Jagoe C. Ligands and the bioavailability of metals in aquaticenvironments. In: Hamelin J.L.L, Bergman P.F. and Benson H.L. (Eds.) Bioavailability – Physical, Chemical and Biological Interactions. Boca Raton: Lewis Publishers, 1994, pp 39–41.

Nixon, S.W. Coastal marine eutrophication: a definition, social causes, and future concerns. Ophelia, 1995, 41:199–219.

Nixon, S.W., Oviatt, C.A., Frithsen, J., Sullivan, B. Nutrients and the productivity of estuarine and coastal marine ecosystems. Journal of the Limnological Society of South Africa, 1986, 12:43–71.

Park S.-K. and Bielefeldt A.R. Equilibrium partioning of a non-ionic surfactant and pentachlorophenol between water and a non-aqueous phase liquid. Water Research, 2003, 37:3412–3420.

Perrodin Y. Proposition méthodologique pour l'évaluation de l'écotoxicité des effluents aqueux: Mise au point d'un Multi-Test Macroinvertébrés (M.T.M.) – Application aux lixiviats de décharges et à leurs composants caractéristiques. Thèse. Institut National des Sciences Appliquées de Lyon, 1988, 135 p.

Plackett R.L., Hewlett.P.S. 1952. Quantal responses to mixtures of poisons. J. R. Stat. Soc. (Lond.) B14: 141–163.

Plassard F., Winiarski T., Petit-Ramel M. Retention and distribution of three heavy metals in a carbonated soil: comparison between batch and unsaturated column studies. Journal of Contaminant Hydrology, 2000, 42:99–111.

Pontasch K.W., Niederlehner B.R., Cairns J. Comparisons of single species, microcosm and field responses to a complex effluent. Environmental Toxicology and Chemistry, 1989, 8:521–532.

Raloff J. Drugged Waters. Science News, 1998, Vol. 153:187–189.

Ramade F. Dictionnaire encyclopédique des sciences de l'eau – Biochimie et écologie des eaux continentales et littorales. Paris: Ediscience international, 1998, 800 p.

Ramade F. Dictionnaire encyclopédique des pollutions. Paris: Ediscience international, 2000, 690 p.

Ramade F. Dictionnaire encyclopédique de l'écologie et des sciences de l'environnement. 2ème édition, Paris: DUNOD, 2002, ISBN 2 10 006670 6.

Richardson M. L. and Bowron J. M. The fate of pharmaceutical chemicals in the aquatic environment. J. Pharm. Pharmacol., 1985, 37:1–12.

Rivière J-L. Évaluation du risque écologique des sols pollués. Paris: Association RE.C.O.R.D. – Lavoisier Tec&Doc, 1998, 230 p.

Rodier J. La protection des eaux contre la radioactivité. Bulletin de l'Association pharmaceutique française pour l'hydrologie. Paris, 4 mars 1971, 40 p.

Rodier J. L'analyse de l'eau. Paris: DUNOD, 8e édition, 1996.

Rouse J.D., Sabatini D.A., Suflita J.M., Harwell J.H. Influence of surfactants on microbial degradation of organic compounds. Crit. Rev. Environ. Sci. Technol. 1994, 24:325–370.

Ryther, J.H., Dunstan, W.M. Nitrogen, phosphorus, and eutrophication in the coastal marine environment. Science, 1971, 171:1008–1013.

Saget A., Chebbo G., Bertrand-Krajewski J.L. The first flush in sewer systems. Water Science and Technology, 1996, 33, 9:101–108.

Schwarz G. and Vaeth E. Analysis of surfactants and surfactant formulations. In: Falbe J. (Ed.) Surfactants in consumer products: theory, technology and application. Heidelberg: Springer Verlag, 1987.Scott D. Dyer, Peng C., McAvoy D.C., Fendinger N.J., Masscheleyn P., Castillo L.V., Lim J.M.U. The influence of untreated wastewater to aquatic communities in the Balatium River, the Philippines. Chemosphere, 2003, 52: 43–53.

Short, F.T. and Burdick, D.M. Quantifying eelgrass habitat loss in relation to housing development and nitrogen loading in Waquoit Bay, Massachusetts. Estuaries 1996, 7:358–380.

Sigg L, Behra P, Stumm W. Chimie des milieux aquatiques – Chimie des eaux naturelles et des interfaces dans l'environnement, 3e édition. Paris:DUNOD, 2000: 567 p.

Skov T., Lynge E., Maarup B., Olsen J., Roth M., Withereik H. Risks for physicians handling antineo-plastic drugs. Lancet, 1990, 336:14–46.

Sprague J. B. Measurement of pollutant toxicity to fish. II. Utilizing and applying bioassay results. *Water Research,* 1970, 4, 1: 3–32.

Talmage S. Environmental and human safety of major surfactants, alcohol ethoxylates and alkylphenol ethoxylates. Boca Raton: Lewis Publishers, 1994.

Tardat-Henry M. Chimie des eaux. Québec: Les éditions le Griffon d'Argile Inc., 1984.

Tessier A. and Turner D.R. (Eds.) Metal Speciation and Bioavailability in Aquatic Systems. New York: John Wiley and Sons, 1995, 679 p.

Tyler C.R., Van Der Eerden B., Jobling S., Panter G., Sumpter J.P. Measurement of vitellogenin, a biomarker for exposure to oestrogenic chemicals, in a wide variety of cyprinid fish. J. Comp. Physiol. B., 1996, 166:418–426.

United Nations. Convention sur la diversité biologique. New York: United Nations, 1992.

Valiron F. Gestion des eaux, principes – moyens – structures. Paris: Presses de l'Ecole Nationale des Ponts et Chaussées, 1990.

Valiron F., Tabuchi J.P. Maîtrise de la pollution urbaine par temps de pluie: Etat de l'Art. Paris: Tec et Doc Lavoisier, 1992, 564 p.

Vermande P., Raccurt C. Haïti: Ecosystème et ressources marines. Audience Magazine, Haïti, 2001: pp. 12–14.

Warne M. S. J. *A review of the ecotoxicity of mixtures, approaches to, and recommendation for, their management.* In: A. Langley, M. Gilbey and B. Kennedy. Fifth national workshop on the assessment of site contamination, Adelaide, Australia. 2003: pp. 253–276.

Zakrzewski S. F. Principles of environmental toxicology. 2nd ed., Washington DC: American Chemical Society, 1997, 352 p.

Zimmo O. R., Van der Steen N. P., Gijzen H. J. Nitrogen mass balance across pilot-scale algae and duckweed-based wastewater stabilization ponds. Water Research, 2004, Volume 38, 4:913–920.

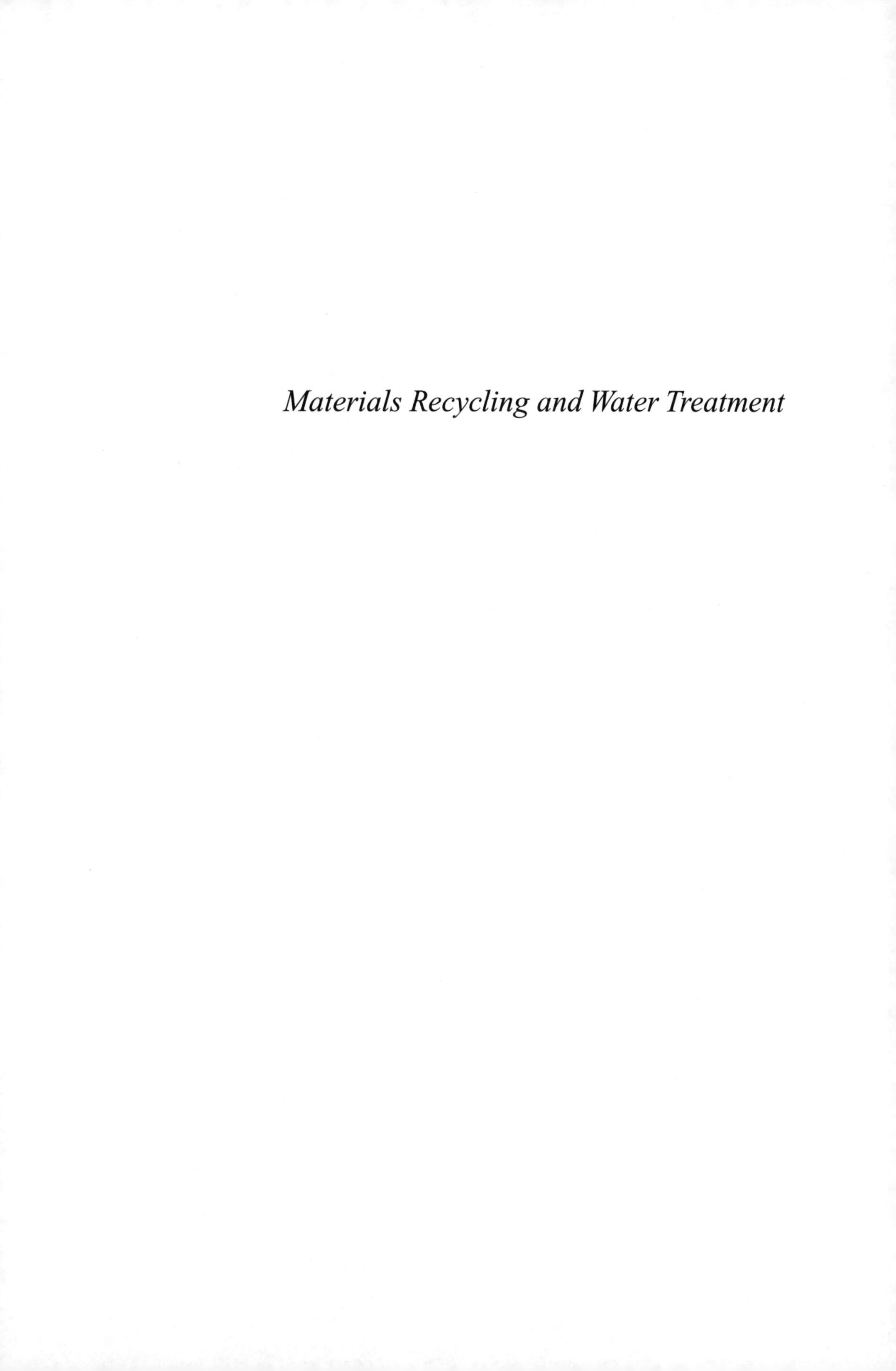

Materials Recycling and Water Treatment

Utilization of Solid Waste Materials in Highway Construction

Serji Amirkhanian, Carl Thodesen & Khaldoun Shatanawi
Clemson University, SC, USA

SUMMARY: As natural resources start to dwindle and landfill space gets filled up, the importance of proper waste reduction and management systems increases in importance. One solution to this problem has been studied and implemented by the highway construction industry. The use of waste byproducts as a replacement for virgin materials could provide relief for some of the burden associated with disposal and may provide a cost effective construction product exhibiting all the properties of virgin products. This chapter presents the current situation of waste generation in the United States; also discussed are the current applications of some of the materials in question. The materials covered are: Recycled Concrete Aggregate (RCA), Fly Ash, Glass, Plastics, Reclaimed Asphalt Pavement (RAP), and Slag. While further research is still necessary in the field of utilizing waste materials for construction purposes, there can be no doubt that the utilization of waste materials in the construction industry offers a solution to sustainable waste management practices. The concept of utilizing reclaimed materials as a construction material is particularly relevant in developing countries. Economical and ecological solutions are particularly important in situations where growing populations (and thus also infrastructure) are coupled with finite economic and natural resources.

1 INTRODUCTION

In the United States, the field of waste disposal is relatively new. Approximately 100 years ago, the main waste disposal systems were dumping or open burning. It was only in 1965 that the first federal legislation (i.e., the Solid Waste Disposal Act) was enacted to directly approach the waste problem (Ruiz, 1993). As such, the country is only now starting to come to terms with the amount of waste it produces.

In their feasibility study of the use of waste materials in highway production, Amirkhanian and Manugian identified the following thirteen waste materials with specific applications for the highway construction industry: bottom ash, compost, construction debris, fly ash, plastics, reclaimed asphalt pavement, shingle scarps, slag, sludge, and tires (Amirkhanian and Manugian, 1994).

There are a variety of current practices of using waste materials in the highway construction sector. Recycled products have emerged as a viable alternative to virgin materials in many areas, including the highway construction field. Examples of waste materials put to use in highway applications include, the use of furnace bottom ash as a fill material in embankments, the use of construction debris as a mineral filler in portland cement stabilized roads, the use of waste roofing shingles to produce asphalt concrete, and the recycling of plastics to make fence posts, to name but a few (Schroeder, 1994).

This chapter presents the current situation of waste generation in the United States; also discussed are the current applications of some of the materials in question. The materials covered are: Recycled Concrete Aggregate (RCA), Fly Ash, Glass, Plastics, Reclaimed Asphalt Pavement (RAP), and Slag.

2 BACKGROUND

Approximately 229 million metric tons of municipal solid wastes (MSW) were generated in 2006 in the United States; this is approximately equal to 2.1 kilograms per person per day (Environmental Protection Agency, 2006). A portion of this waste is currently reused, recycled, or reprocessed; however, a majority of it is disposed in landfills. The practice of using landfills for waste disposal purposes can no longer be considered sustainable, as 75% of all landfills in the United States have closed in the past 18 years and further closures are anticipated (Figure 1).

This problem is also magnified by the fact that the United States currently produces more solid waste than ever before in its history. As seen in Figure 2, the generation and disposal of MSW over the last 45 years has steadily increased. Although, the amount of MSW recycled in the last 15 years has increased, the amount generated has outpaced the recycling efforts.

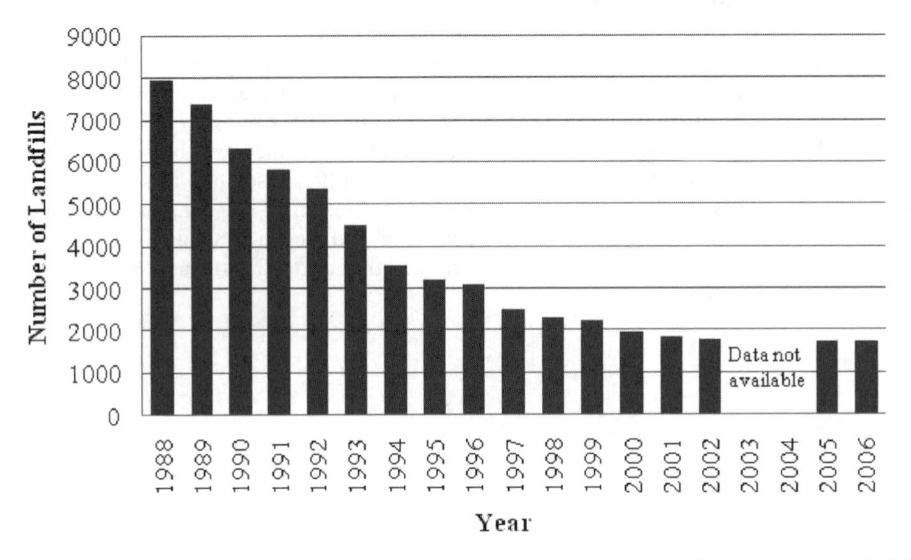

Figure 1. Number of Landfills in the United States, 1988–2006 (Environmental Protection Agency, 2006).

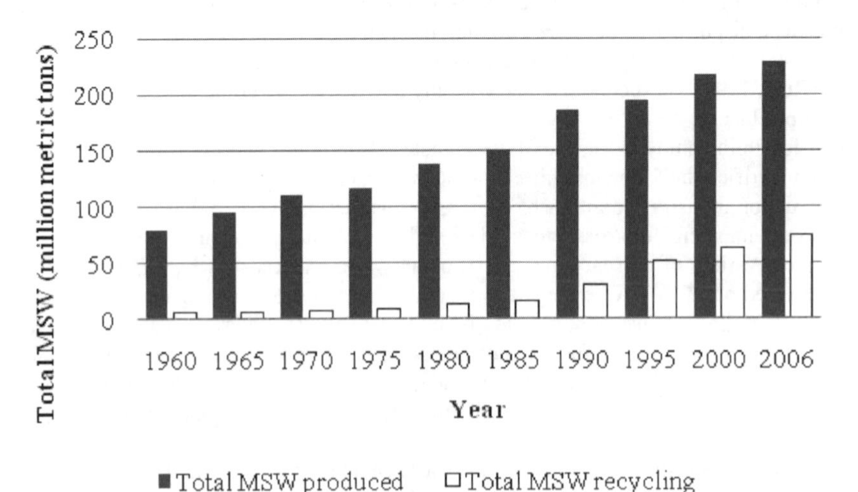

Figure 2. Annual Production of MSW in the United States, 1960–2006 (Environmental Protection Agency, 2006).

2.1 *Green House Gas (GHG) emissions*

The transportation industry and Portland cement industry are the two largest producers of carbon dioxide in the world, where the latter is responsible for approximately 7% of the world's carbon dioxide emissions (Mehta, 1999). As seen in Figure 3, significant amounts of energy savings could be made if recycled materials were utilized. The concrete industry is also the world's largest consumer of virgin materials such as sand, gravel, crushed rock, and fresh water. It consumes an annual rate of 1.6 billion metric tons of Portland and modified Portland cements, and thus also a significant portion of limestone, clay, and energy required to produce cement (Mehta, 2004).

The effect of waste management on the environment affects not only the soil, water, and air quality; it also has a significant impact on energy consumption. Studies have shown that if the waste was managed with energy implications in mind, significant energy savings could be made. Furthermore, it needs to be considered that products entering the waste stream also have associated energy requirements (and thus also GHG emissions) at each stage of their life cycle. These stages include the acquisition of raw materials, the manufacture of raw materials into products, product use by consumers, and finally product disposal.

2.2 *Hierarchy of recycling options*

Given the recognized environmental issues facing society, green alternatives to waste management are actively pursued in many parts of the world. One widely accepted model to waste minimization is the recycle disposal hierarchy as seen in Figure 4. This model advocates the preservation of resources at all stages of the products' design life, as well as encouraging recycling instead of landfilling.

2.3 *Legislation*

There has also been a response to the amount of waste produced and waste management at the federal, state, and local levels. For instance, in October 1991, President Bush signed Executive Order 12780, which requires all federal agencies to establish recycling programs and increase

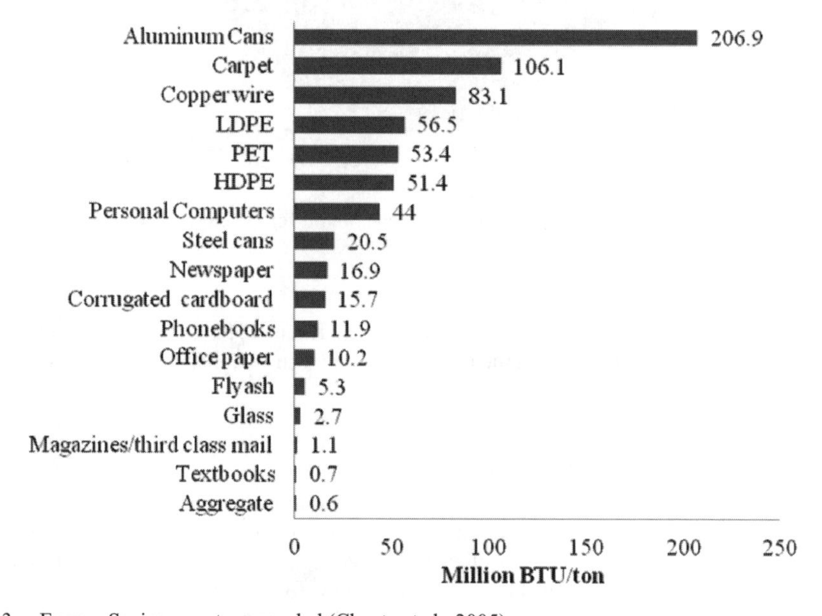

Figure 3. Energy Savings per ton recycled (Choate et al., 2005).

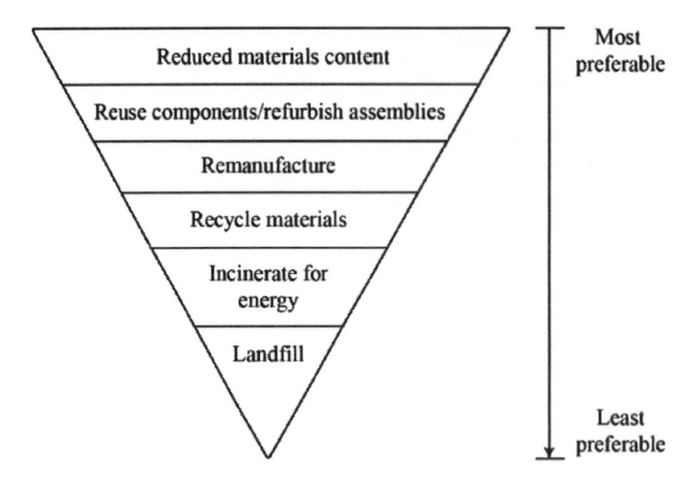

Figure 4. Hierarchy of Recycling Options (Bishop, 2000).

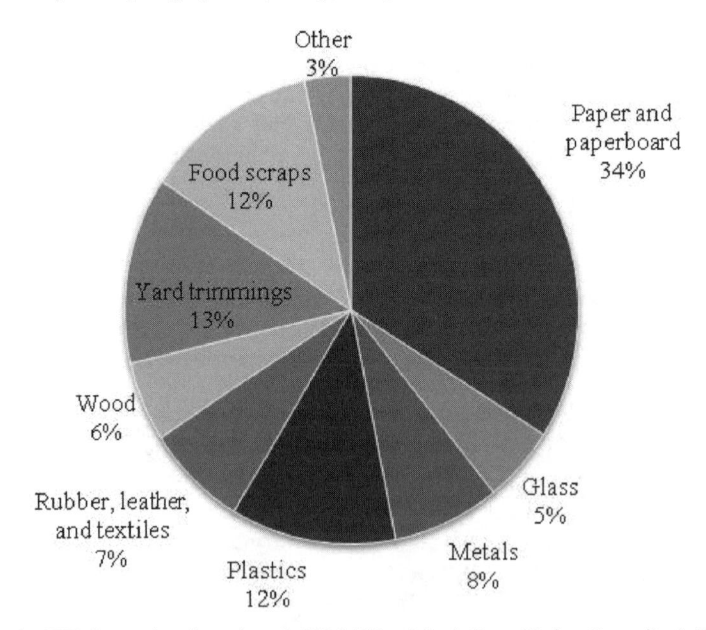

Figure 5. Total MSW Generation (by volume), 229 Million Metric Tons (Before Recycling) (Environmental Protection Agency, 2006).

the use of products containing recovered materials (Scharman, 1992). President Clinton issued Executive Order 13101, which built on the prior orders and also established the White House Task Force on Waste Prevention and Recycling.

2.4 *Recycling in the highway construction industry*

In order to protect the environment, new areas of recycling must be investigated. During the last 25 years, one industry which has been very active in the use of SWM is the highway industry. For example, some highway agencies have been using SWM (e.g., slags, reclaimed asphalt pavement, construction debris, etc.) for use in highway construction for many years. Figure 5 shows the materials (by volume) in the MSW stream before recycling. Of these materials, glass, plastics, wood, and rubber, which take up over 30% of SWM landfill space, are used in a limited capacity by many state highway agencies in their pavements.

3 CHARACTERIZATION OF SELECTED WASTE MATERIALS

Six waste materials were selected for discussion based on their applicability to the highway construction industry. The aim of this section is to provide a general description of the waste material as well as to identify current and potential uses, as well as the benefits and limitations of applying certain waste materials in highway construction.

3.1 *Recycled Concrete Aggregate (RCA) from construction debris*

The FHWA reports that the construction waste generated from building demolition alone is estimated to be approximately 112 million metric tons per year. Due to the shear quantity of material produced, a growing interest in environmentally friendly building practices has arisen. These include the use and recycling of RCA (recycled concrete aggregate) generated from the construction and demolition (C&D) market.

RCA makes up approximately 67% of the material recovered from C&D material. Typically, concrete is composed of cement, water, coarse and fine aggregates, and admixtures. RCA is a high-volume, low-cost material that is used in extremely large quantities. In 2003, it was estimated that 50 to 60% of concrete waste was recycled, while the remainder was landfilled (Environmental Protection Agency, 2003).

Construction debris has been divided into nine types: steel and iron, copper, lead, aluminum, concrete brick, wood, glass, and plastic. Typically, the debris content is composed of approximately 67% concrete, 17% wood, and 14% brick, with small percentages of steel and iron. Figure 6 shows the percentages, by weight, of the eight most widely used materials from a variety of buildings (Wilson et al., 1976).

A number of factors control the amount of usable construction debris generated. From construction debris, "clean fill" is extracted and can be used in highway applications. Clean fill does not include any wood, plastics, or metals. The amount of clean fill will depend on both the economic climate and whether or not clean fill materials were primary construction materials at the time of construction (Dougan, 1988).

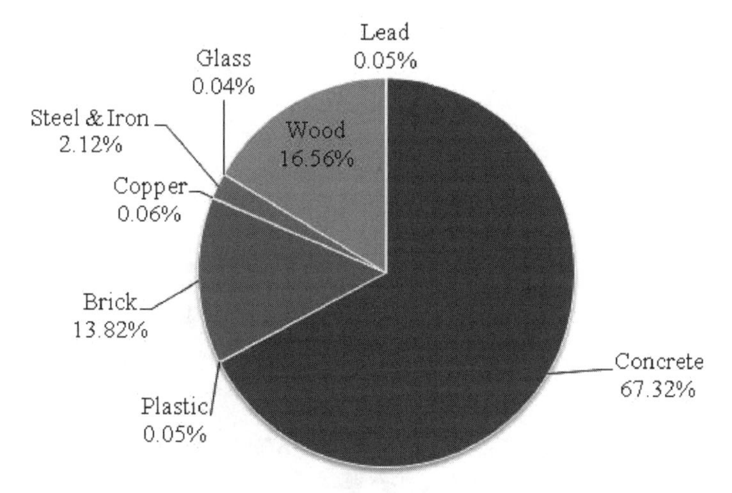

Figure 6. Percentages (by weight) of the eight most widely used materials in building construction (Wilson et al., 1976).

3.1.1 *Concrete*

According to the FHWA, various transportation agencies experiences indicate that under specific conditions RCA has the potential to produce strong durable materials suitable for use in highway infrastructure (Federal Highway Administration, 2004). Some studies have suggested that using recycled fines from RCA in a Portland cement concrete (PCC) mix may result in lower compressive strengths; this is because natural fine aggregates tend to have higher strength than recycled fines. One reason for this might be that a significant portion of the fines in a recycled aggregate is derived from mortar in the original concrete mix (Federal Highway Administration, 2004).

3.1.2 *Asphalt*

According to the FHWA, a number of states allow the use of recycled concrete aggregate in hot mix asphalt. However, due to the high absorption of asphalt by recycled concrete aggregate (and thus increased asphalt demand and price), there does not seem to be much promise for this practice (Federal Highway Administration, 2004).

3.1.3 *Fill material*

RCA has been used in highway construction as a fill material in road bases, while crushed concrete has been used as filler in portland cement-stabilized road bases (Collins & Ciesielski, 1992). Other uses for RCA include their use as an aggregate for a flowable fill, shore line protection material (rip rap), a gabion basket fill, or a granular aggregate for base and trench backfill.

3.2 *Fly ash*

The American Coal Ash Association reported that in 2006 approximately 72.4 million tons of fly ash were produced in the United States. It is further estimated that approximately 46% (over 15 million tons) were consumed for the production of concrete, concrete products, and grout. It is also reported that an additional 4.1 million tons were used to produce cement (American Coal Ash Association, 2007). The main uses of fly ash in the United States are shown in Figure 7.

Fly ash is the inorganic, noncombustible residue of powdered coal after burning in power plants. Today, it is well accepted as a pozzolanic material that may be used as a partial replacement for Portland cement or as a mineral admixture in concrete. ASTM C618 specifies that fly ash is divided into two categories, Class C and Class F. Class C ash is produced from the combustion of bituminous coal; it contains more lime and has both pozzolanic and cementitious properties. Class F ash is produced from the combustion of lignite or sub-bituminous coal; it has less lime and pozzolanic

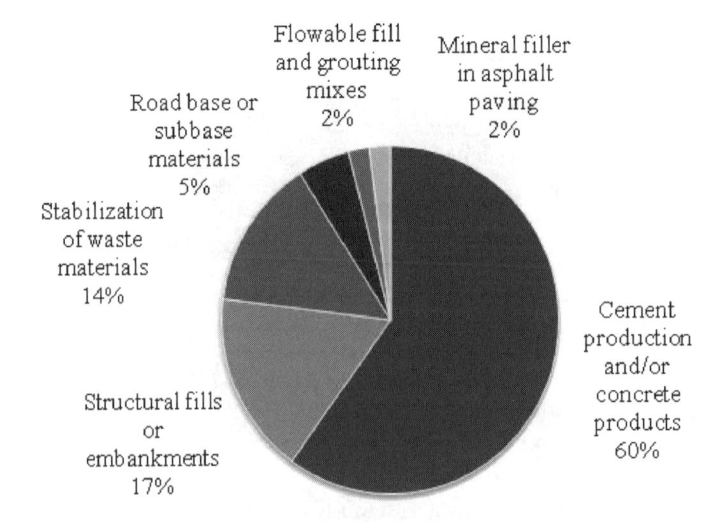

Figure 7. Percentage of reclaimed fly ash used in different highway construction applications.

properties but little or no cementitious properties. The differences between Class C and Class F fly ash do not produce significant effects on properties of pavements.

3.2.1 *Concrete*

As seen in Figure 7 the largest use of fly ash is as a mineral admixture in PCC. While the main benefit of using fly ash in concrete is economic (fly ash costs less than one half the price of cement), there are also many technical benefits. Initially, concrete containing fly ash will gain strength at a slower rate than concrete using only Portland cement; however, numerous studies have shown that in the long term concretes using fly ash replacement exhibit significantly higher compressive strengths than conventional concretes. These studies also indicate that Class C ash usually produces concretes with a higher compressive strength (Mindess et al., 2003). Other advantages usually associated to using fly ash replacement in concrete include reduced heat of hydration, improved workability, decreased permeability, improved sulfate resistance, control of alkali-silica reaction, decreased chloride diffusion, and reduced leaching and efflorescence (Mindess et al., 2003).

In his paper, Mehta identified High-Volume Fly Ash (HVFA) concrete, which had 50% or more cement replacement by fly ash, as a particularly attractive material (Mehta, 2004). He reported that, field experience and laboratory testing have shown HVFA exhibits the following properties:

- Improved flowability, pumpability, and compactibility.
- Very high durability with respect to the reinforcement corrosion, alkali-silica expansion, and sulfate attack.
- Better cost economy due to lower material cost and highly favorable life cycle cost.
- After three to six months of curing, much higher electrical resistivity, and resistance to chloride ion penetration, according to ASTM Method C1202.
- Superior dimensional stability and resistance to cracking from thermal shrinkage, autogenous shrinkage, and drying shrinkage. In unprotected concrete a higher tendency for plastic shrinkage cracking.
- Superior environmental friendliness due to ecological disposal of large quantities of fly ash, reduced carbon-dioxide emissions, and enhancement of resource productivity of the concrete construction industry.

3.2.2 *Asphalt*

When used in asphalt concrete, fly ash is used as a mineral filler to create a denser mix. Fly ash consists of very fine particles which can fill the voids in an asphaltic surface. This creates a pavement which is less permeable to water and, hence, less susceptible to moisture damage. If it has sufficient qualities, Class C fly ash (containing Calcium Oxide) may also be used as a partial anti-strip agent in sufficient quantities, replacing part of the anti-strip additive needed to control moisture susceptibility of the pavement (Rosner et al., 1982).

Researchers at the University of Michigan used fly ash as a partial replacement of asphalt cement in asphalt mixtures. They found that its use reduced the optimum amount of asphalt cement required in the mix. By replacing up to 30% by volume of asphalt cement with fly ash, the researchers concluded that the mix characteristics were improved considerably over the control mixes (with no fly ash). They also observed that when asphalt cement was mixed with fly ash, the mix became less viscous, thus allowing better bonding between the aggregate and the binder. Another finding of this study was that the addition of fly ash created a more economical mix by decreasing the amount of binder, as well as, acting as a partial antistrip agent (Suheibeni, 1986).

3.2.3 *Pavement base courses*

Highway contractors have used fly ash in Lime-Fly Ash-Aggregate (LFA) mixes as pavement base courses for many years. Advantages include increased subgrade stabilization and shear strength, reduced plasticity, and improved workability.

Base mixes are produced in conventional mix stabilization plants where quality control is easily monitored. Conventional paving equipment places the mix and adds a pre-determined amount of water. The mix is then compacted to its optimum density; this is the most important factor in producing a strong base. The mix must be cured from 7–14 days to ensure that the lime reacts to form a cementitious gel that imparts a steady, long term strength development to the base (Brackett, 1978).

3.2.4 *Fill material*

Fly ash can be used to correct landslide problems and to backfill trenches, retaining walls, embankments, and dams. This recycling method has the potential to use more ash than any other method. Fly ash density is approximately 80% of most soils, which makes it easier to handle, and it is more workable than soil. Additional advantages of using ash over borrow materials include greater availability, higher shear strength, moisture insensitivity, and lower costs (Electric Power Research Institute, 1992).

3.3 *Glass*

Glass can be defined as the fusion product of inorganic materials which have been cooled to a rigid condition without crystallizing. Types of glass can range from molten magma cooled as it is hurled out of a volcano to cane sugar cooled into a lollipop. The most common is the type used to make what is generically known as glass. This type of glass is formed by fusing together oxides at high temperatures (approximately 1480°C). Different oxides impart different qualities to the glass.

In 2006, approximately 11.5 million metric tons of glass was disposed of in the municipal solid waste stream in the United States. In recent decades, the utilization of recycled glass in highway applications has gained popularity. Crushed recycled glass (cullet) has been used independently, and has also been blended with natural stone construction aggregate at different replacement rates.

The equipment used to crush the waste glass is generally similar in nature to the equipment used for rock crushing. Magnetic separation and air classification may be necessary to remove any residual metal or paper retained in the cullet. Once crushed to a fine particle size, the properties of the waste glass are similar to those of natural sand.

3.3.1 *Concrete*

Generally, it is recommended that glass not be placed in PCC due to alkali reactions. Due to the chemical make up of the glass, it readily reacts with alkaline materials around it. The lime from the cement has a tendency to react with the glass, thus weakening the concrete. Also, the factors of thermal expansion in glass are much higher than in conventional aggregates. Therefore, during the curing of the PCC, the high temperature coupled with the reactivity of the CaO tends to expand the glass. This expansion can be around 20 times the expansion of normal aggregate (Larsen, 1991).

While glass is not recommended for use in structural concrete, the following modifications to the standard method of producing PCC have been studied (Clean Washington Center, 1996): Cleaning of glass aggregate; Use of low alkali cement; Use of low alkali pozzolan or set retarder; Use of an air entrainment system; Reduction of moisture content; and Increasing surface area to volume ratio of glass aggregate.

3.3.2 *Asphalt*

Glasphalt is the term popularly used to describe hot mix asphalt incorporating 10–15% crushed glass. Studies suggest that when waste glass is used as a fine aggregate substitute material, glass performance in hot mix asphalt should be comparable to conventional mixes. Where larger, gravel-sized glass particles are used, raveling and stripping in particular could be a problem. The stripping issue is seen as one of the main concerns regarding the use of glass in asphalt. As glass does not absorb any asphalt binder and is also hydrophilic, mixes with more than 15% waste glass are prone to excessive stripping of the asphalt binder from the aggregate. The addition of lime as an antistripping additive has been recommended when high concentrations (greater than 25% weight of the mix) of glass are used, but results are still inconclusive.

3.3.3 *Fill material*

When used as a fill material, properly sized and processed cullet exhibits properties similar to gravel or sand; it is therefore a suitable material for use as a road base or fill. When used as a filler material, the recycled glass must be crushed and screened to produce a uniform product with an adequate gradation (Finkle et al., 2007).

3.4 *Plastics*

Plastics are polymers (long chains of atoms bonded to one another) composed of carbon and hydrogen alone or may contain oxygen, nitrogen, chlorine or sulfur in the backbone. Plastics are divided into three categories: durable goods, non-durable goods and containers and packaging (Environmental Protection Agency, 2006). Figure 8 shows the amounts of plastic produced in these categories. The plastics (which are called thermoplastics) usually come from durable goods and packaging containers. Thermoplastics polymers are defined as polymers that can be deformed and reformed by heating and cooling (Pearson, 1993).

The EPA reports that in 2006, the United States generated more than 12.5 million metric tons of plastics in the MSW stream as containers and packaging, more than 5 million metric tons as nondurable goods, and in excess of 8 million metric tons as durable goods. Plastics contributed approximately 12% by volume of all MSW generated in 2006, Table 1 provides an indication of the quantities of various thermoplastics produced in the United States in 2006.

3.4.1 *Concrete*

Studies have shown that recycled plastics can successfully be used in concrete as a replacement for conventional aggregates. The use of recycled plastic in concrete tends to reduce the bulk density; however, decreases in compressive strength are also common as the recycled plastic concentration increases. In one study, reductions in the compressive strength were seen to vary between 34 and 67% for concrete containing 10–50% recycled plastic. Concrete containing plastic aggregates exhibited more ductile behavior than concrete made with traditional aggregates. Such ductile behavior may contribute to the reduction of crack formation and propagation (Al-Manaseer & Dalal, 1997).

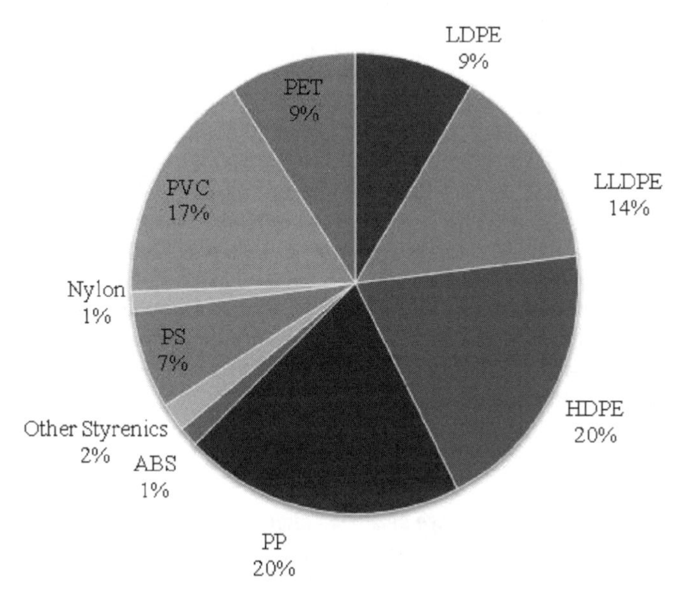

Figure 8. Production of various thermoplastics for 2006, total 41.2 million metric tons (Plastic Industry Producers' Statistics Group, 2006).

Table 1. Plastic Acronyms, recycling codes, and common uses.

Plastic	Acronym	Recycling number	Common uses
Polyethylene terephthalate	PET	1	Bottles, films.
High – density polyethylene	HDPE	2	Plastic bags, fuel tanks for vehicles, milk jugs.
Polyvinyl chloride	PVC	3	Plumbing material, magnetic stripe cards, vinyl siding.
Linear Low – density polyethylene	LLDPE	4	Plastic bags, stretch wrap, geomembranes.
Low – density polyethylene	LDPE	4	Computer components (hard drives, screen cards and disk-drives), six-pack soda can rings, plastic bags.
Polypropylene	PP	5	Food packaging, ropes, textiles.
Polystyrene	PS	6	Pharmaceutical bottles, foam caps, cafeteria trays.

Precast concrete performance has also been seen to improve with the addition of an unsaturated polyester resin based on recycled PET. This material was shown to achieve more than 80% of its ultimate strength in one day. Similarly, unsaturated polyester resin based on PET produced good quality mortar (Rebeiz, 1995).

The addition of polypropylene fibers was seen to have an adverse effect on the air content of the concrete. Specifically, the addition of 0.5% polypropylene fibers resulted in an increased air content, but decreases workability. However, the recycled polypropylene fibers were seen to increase the impact resistance of the concrete (Bayasi & Zeng, 1993).

Studies suggest that reinforcing concrete with shredded mixed plastic, milled mixed plastic, and melt processed plastic fibers resulted in significant increases in the resistance of concrete to impact and shrinkage cracking. Improvements were also noted for recycled plastic reinforced concrete's impermeability and deicer salt scaling resistance. However, the recycled plastic reinforced concrete also yielded a lower abrasion resistance (Soroushian et al., 2003).

3.4.2 *Asphalt*

Research has been done on the use of incorporating recycled plastics as an aggregate replacement (Plastiphalt) in continuously graded asphaltic concrete. In this study, recycled waste plastics, predominantly composed of LDPE in pellet form, were used in dense graded bituminous mixes to replace (by volume) a portion of the mineral aggregates of an equal size (5.00–2.36 mm). The mechanical properties of the recycled mix were found to be equal to that of the conventional plastiphalt and better than the control mixes (Zoorob & Suparma, 2000).

Studies have also shown that waste plastics may be used as an additive for modifying asphalt cement. Such techniques are not new; however, a number of new modifiers are being produced using LDPE resin. The resin is derived from waste plastic (plastic trash, sandwich bags, etc.) and once introduced to the virgin asphalt produces asphalt binder with improved properties such as increased resistance to rutting.

3.4.3 *Other uses*

A number of state highway agencies have also been attempting to use recycled plastic to produce items such as guard rail posts, delineator posts, fence posts, noise barriers, sign posts, and snow poles. Highway rest areas have also used recycled plastics in picnic tables and trash receptacles, while other construction uses of waste plastics have been identified as geotextiles, plastic sheeting, and safety fences (Smith & Ramer, 1992).

3.5 *Reclaimed asphalt pavement*

Generally, reclaimed asphalt pavement (RAP) is not considered a waste material due to its widespread use in the highway construction industry. However, due to the volume produced and the importance of the material, a portion of the chapter is dedicated to the material.

In 2001 approximately 623,000 tons of RAP were used, the USDOT spent approximately $19,940,000 on RAP. RAP is bituminous concrete material removed and/or reprocessed from pavements undergoing reconstruction or resurfacing. Reclaiming the bituminous concrete may involve either cold milling a portion of the existing bituminous concrete pavement or full depth removal and crushing (Chesner et al., 1998; Illinois Department of Transportation, 1994).

Pavement recycling is divided into three areas: surface recycling, in-place surface and base recycling, and central plant recycling. Each type of recycling has a hot and a cold process (National Cooperative Highway Research Program, 1978).

3.5.1 *Surface recycling*

Surface recycling is the reworking of the surface of a pavement to a depth of less than one inch using a heater-planer, heater-scarifier, hot-milling, cold-planing, or cold milling devices. This operation is continuous (using one single pass and a number of steps) and may involve using new materials such as aggregate, modifiers, or mixtures (National Cooperative Highway Research Program, 1978).

Studies have concluded that this type of recycling is not suitable for pavements with severe distress extending more than 19 mm below the original pavement surface. The ride quality is somewhat inferior to that obtained with a one inch asphalt concrete overlay. However, hot surface recycling did provide a good bond between the recycled mix and the old asphalt concrete surface (Collins & Ciesielski, 1992).

3.5.2 *Cold in-place recycling*

In-place surface and base recycling involves the pulverization of the pavement to a depth greater than one inch, followed by reshaping and compaction (National Cooperative Highway Research Program, 1978). This operation may be performed with or without the addition of a stabilizer. This method has also been called cold mix recycling (Ahmed, 1991). Studies suggest that pavements utilizing cold in-place recycling tend to experience some wheel path rutting.

3.5.3 *Hot central plant recycling*

Central plant recycling is the scarification of the pavement material, removal of the pavement from the roadway prior to or after pulverization, processing of the material with or without the addition of a stabilizer or modifier, and laydown and compaction to a desired grade. This operation may involve the addition of heat, depending on the type of material and the stabilizer used (National Cooperative Highway Research Program, 1978). This method has also been called hot mix recycling (Ahmed, 1991).

For central plant recycling, properties of the RAP are heavily dependent on the pavement from which the material is milled. RAP is generally finer than the aggregate from which it was derived; this is due to traffic loading and the method of processing when obtaining the RAP. As aggregate quality, size, asphalt content, and mix consistency vary from pavement to pavement there is significant variability within the RAP (Chesner et al.,1998).

During the production of RAP, the reclaimed pavement is crushed and screened to a 1/4- to 1/2-inch size. Separate stockpiles of the RAP are maintained to guarantee that a homogenous blend of RAP is used in the mix. QC/QA is performed on the RAP to ensure the gradation and quality of the RAP. As asphalt content varies from pavement to pavement, it is also necessary for the asphalt producer to determine the asphalt content of the RAP. Upon determining the asphalt content and passing inspection, the RAP is blended with virgin aggregate and asphalt to produce Hot Mix Asphalt (HMA) (Saylak et al., 1996; Chesner et al., 1998).

Table 2. Iron and steel slag sold or used in the United States (million metric tons and million dollars) (van Oss, 2003).

	Air-cooled Blast Furnace Slag[1]	Granulated Blast Furnace Slag[1]	Total[2]	Steel furnace slag	Total iron and steel slag
Quantity[3]	7.3	3.6	10.9	8.8	19.7
Value[e,4]	49	212	261	35	296

[e] Estimated.
[1] Excludes expanded (pelletized) slag to protect company proprietary data. The quantity is very small (less than 0.1 units).
[2] Data may not add to totals shown because of independent rounding.
[3] Quantities are rounded to reflect inclusion of some estimated data and to reflect inherent accuracy limitations of reported data.
[4] Values are rounded because of the inclusion of a large estimated component.

The amount of RAP allowed in a mix depends on the end use of the mix. Some state DOTs do not allow the use of RAP in the highest class bituminous concrete or in polymer-modified mixes; however, the do allow the incorporation of RAP into some SUPERPAVE mixes. The amount of RAP allowed in a road is: 30% for low volume roads, up to 50% for non-critical mixes (shoulder, base, and subbase, up to 25% for high-type binder courses, between 10 and 15% for all but the highest volume highway surface courses (Illinois Department of Transportation, 1994; Saylak et al., 1996).

3.6 *Slag*

Almost 20 million metric tons, totaling approximately $300 million, of iron and steel slag were consumed in the United States in 2006 (Table 2). However, the use of slag as a construction material is not a current event. Slags have been used as construction material since Roman times; however, since the advent of the industrial revolution and the subsequent increase in steel production, the volume of slag produced has greatly exceeded the amount consumed (van Oss, 2003).

In a blast furnace, crude or pig iron is made by stripping the oxygen and other impurities from iron ore by means of high temperature reactions with reducing agents (mainly) carbon and fluxes (van Oss, 2003). Slag is a byproduct of these metallurgical processes; these occur during the production of metals from ore or the refinement of impure metals. As seen in Table 3, slag has numerous applications as a highway construction material.

There are three main types of blast furnace slags: Air cooled blast furnace slag, granulated slag, and pelletized or expanded slag.

3.6.1 *Air cooled slag*

Air cooled slag is derived by allowing molten slag to cool relatively slowly under ambient conditions. The cooled material is hard and dense, but might exhibit vesicular texture with closed pores. Upon completion of crushing and screening the air cooled slag is typically used as an aggregate for asphaltic paving, concrete, and road bases and fill. In asphalt pavements, it is primarily and successfully used as an aggregate in open-graded friction courses. (Collins & Ciesielski, 1992; Ahmed, 1991; Federal Highway Administration, 1977).

3.6.2 *Granulated slag*

Granulated slag is produced by quenching molten slag in water, the very rapid cooling of causes the slag as sand sized particles of glass. Ground granulated blast furnace slag (GGBFS) is produced by grinding the granulated slag very finely (finer than most grades of Portland cement). As GGBFS is known to develop strong cementitious bonds in the presence of free lime, there is a ready market for the material as a partial substitute for Portland cement in ready mixed concrete. Alternatively, it may

Table 3. Sales of ferrous slag in the United States in 2003, by use[1] (van Oss, 2003).

Use	Blast Furnace Slag[2]	Steel Slag[3]
Ready Mixed Concrete	9.3	0
Concrete Products	6.4	0
Asphaltic Concrete	20.6	17
Road Bases and Surfaces	34.7	46.4
Fill	3.1	11.1
Clinker raw material	5.7	5.4
Miscellaneous[4]	4	2.5
Other or unspecified	16.2	17.6

[1] Data contain a large component of estimates and are reliable to no more than two significant digits.
[2] Air-cooled slag only. Expanded or pelletized slag would have been sold mostly for lightweight aggregate or (when ground) as a cementitious additive for cement or concrete. Almost all granulated slag is sold as a cementitious additive for cement or concrete.
[3] Steel slag use is based on the 89% of total tonnage sold for which usage data were provided.
[4] Reported as used for railroad ballast, roofing, mineral wool, or soil conditioner.

be mixed with Portland cement to make finished blended cement. Concrete containing a portion of GGBFS generally takes longer to develop strength than conventional concrete; however, in the long term GGBFS concrete tends to exhibit higher strengths (Sakai, 1992). Concretes containing GGBFS are also known to demonstrate a low heat of hydration, reduced permeability, and increased resistance to chemical attacks (Nagataki, 1992).

3.6.3 *Expanded slag*

Expanded slag (also known as pelletized slag) is produced by cooling the molten slag through a water jet. This cooling method leads to rapid steam generation and the consequent development of vesicles within the slag. Expanded slags exhibit good mechanical binding with hydraulic cement; this is due to the vesicular texture and lower overall density of the slag. While generally used as lightweight aggregate, when finely ground expanded slag may also be used as a supplementary cementitious material (SCM) similar to GGBFS (van Oss, 2003).

The reactivity of the slag is known to vary from slag to slag, as it is dependent on the method of production as well as the metal source. Generally, blast furnace slags have significant amounts of lime, silica, and alumina. The mass percentages at which these compounds are found are typically $CaO = 35–45\%$; $SiO_2 = 32–38\%$; $Al_2O_3 = 8–16\%$; $MgO = 5–15\%$: Fe2O3 < 2%, and sulfur 1–2% (Mindess et al., 2003).

4 CONCLUDING REMARKS

As natural resources start to dwindle and landfill space gets filled up, the importance of proper waste reduction and management systems increases in importance. One solution to this problem has been studied and implemented by the highway construction industry. The use of waste byproducts as a replacement for virgin materials could provide relief for some of the burden associated with disposal and may provide a cost effective construction product exhibiting all the properties of virgin products.

As seen in this chapter, a significant amount of research has already been done on reusing reclaimed materials for highway construction. While further research is still necessary in the field of utilizing waste materials for construction purposes, there can be no doubt that the utilization of waste materials in the construction industry offers a solution to sustainable waste management practices. The concept of utilizing reclaimed materials as a construction material is particularly relevant in developing countries. Economical and ecological solutions are particularly important in situations where growing populations (and thus also infrastructure) are coupled with finite economic and natural resources.

REFERENCES

Ahmed, I. (1991). *Use of Waste Materials in Highway Construction, Report No. FHWA/IN/JHRP-91/3.* Washington, DC: Federal Highway Administration.

Al-Manaseer, A. A., & Dalal, T. R. (1997). Concrete containing plastic aggregates. *Concrete International 19(8)*, 47–52.

American Coal Ash Association. (2007). *ACAA Releases 2006 CCP Production and Use Survey.* Aurora, CO: American Coal Ash Association.

Amirkhanian, S. N., & Manugian, D. M. (1994). *A Feasibility Study of the Use of Waste Materials in Highway Conctruction.* Columbia, SC: South Carolina Depatment of Transportation/Federal Highway Administration.

Bayasi, Z., & Zeng, J. (1993). Properties of polypropylene fiber reinforced concrete. *ACI Materials Journal 90 (6)*, 605–610.

Bishop, P. L. (2000). *Pollution Prevention: Fundamentals and Practice*. New York, NY: McGraw-Hill.

Brackett, C. (1978). Production and Utilization of Ash in United States. *Coal Ash Utilization-Fly Ash, Bottom Ash and Slag Pollution Review.* Pittsburgh, PA: Material Research Society.

Chesner, W. H., Collins, R. J., & MacKay, M. H. (1998). *Users Guidelines for Waste and By-Product Materials in Pavement Construction. Report No. FHWA-RD-97-148.* Washington, DC: FHWA.

Choate, A., Pederson, L., Scharfenburg, J., & Ferland, H. (2005). *Waste Management and Energy Savings: Benefits by the Numbers.* Washington, DC: Environmental Protection Agency.

Clean Washington Center. (1996). *Best Practices in Glass Recycling: Recycled Glass in Portland Cement Concrete.* Seattle, WA: Clean Washington Center.

Collins, R., & Ciesielski, S. (1992). Utilization of Waste Materials in Civil Engineering Construction. *American Society of Civil Engineers* .

Dougan, C. (1988). *Report to the General Assembly on the Feasibility of Expanding the Use of Demolition Materials in Projects Undertaken by the Department of Transportation, Report No. 343-20-88-13.* Wethersfield, CT: Connecticut Department of Transportation.

Electric Power Research Institute. (1992). *Fly Ash Design for Roads and Site Applications.* Palo Alto, CA: Electric Power Research Institute.

Environmental Protection Agency. (2003). *Background Document for Life-Cycle Greenhouse Gas Emission Factors for Clay Brick Reuse and Concrete Recycling, EPA530-R-03-017.* Washington, DC: Environmenal Protection Agency.

Environmental Protection Agency. (2006). *Municipal Solid Waste Generation, Recycling, and Disposal in the United States: Facts and Figures for 2006.* Washington, DC: Environmental Protection Agency.

Federal Highway Administration. (1977). *Availability of Mining Wastes and Their Potential for Use as a Highway Material–Executive Summary, Publication No. FHWA-RD-78-28.* Washington, DC: Federal Highway Administration.

Federal Highway Administration. (2004). *Transportation Applications Of Recycled Concrete Aggregate: FHWA State of the Practice National Review.* Washington, DC: Federal Highway Administration.

Finkle, I., Ksaibati, K., & Robinson, T. (2007). Recycled Glass Utilization in Highway Construction. *Transportation Research Board Annual Meeting 2007 Paper #07-0929*.

Illinois Department of Transportation. (1994). *Special Provision for RAP Mixtures for Class I, Type 1 and 2 Bituminous Concrete Binder, Leveling Binder and Surface Course (Mixture C and D).* Springfield, IL: Illinois Department of Transportation.

Larsen, D. (1991). *Feasibility of Utilizing Waste Glass in Pavements, Report Number 343-21-89-6.* Wethersfield, CT: Connecticut Department of Transportation.

Mehta, P. K. (1999). Concrete Technology for Sustainable Development. *Concrete International 21(11)*, 47–52.

Mehta, P. K. (2004). High-Performance, High-Volume Fly Ash Concrete for Sustainable Development. *International Workshop on Sustainable Development and Concrete Technology.* Beijing, China.

Mindess, S., Young, J. F., & Darwin, D. (2003). *Concrete Second Edition.* Upper Saddle River, NJ: Pearson Education, Inc.

Nagataki, S. (1992). Properties of Concrete Using Newly Developed Low-Heat Cements and Experiments with Mass Concrete Model. *Proceedings of the Fourth International Conference on Fly Ash, Silica Fume, Slag and Natural Pozzolans in Concrete.* Istanbul, Turkey.

National Cooperative Highway Research Program. (1978). *Recycling Materials for Highways.* Washington, DC: Transportation Research Board.

Pearson, W. (1993). *The McGraw - Hill Recycling Handbook.* New York, NY: McGraw – Hill.

Plastic Industry Producers' Statistics Group. (2006). *PIPS Year-End Resin Statistics for 2006: Production Sales, and Captive Use 2006 vs 2005.* Arlington, VA: American Chemistry Council.

Rebeiz, K. S. (1995). Time-temperature properties of polymer concrete using recycled PET. *Cement and Concrete Composites 17*, 119–124.

Rosner, J., Chehovits, J., & Morris, G. (1982). Fly Ash as a Mineral Filler and Anti Strip Agent for Asphalt Concrete. *The Challenge of Change – Sixth International Ash Utilization Symposium Proceedings.* Washington, DC: United States Department of Energy and National Ash Association.

Ruiz, J. (1993). Recycling Overview and Growth. In H. Lund, *The McGraw-Hill Recycling Handbook.* New York, NY: McGraw-Hill Inc.

Sakai, K. (1992). Properties of Granulated Blast-Furnace Slag Cement Concrete. *Proceedings of the Fourth International Conference on Fly Ash, Silica Fume, Slag and Natural Pozzolans in Concret.* Istanbul, Turkey.

Saylak, D., Estakhri, C., Viswanathan, R., Tauferner, D., & Chimakurthy, H. (1996). *Evaluation of the Use of Coal Combustion By-Products in Highway and Airfield Pavement Construction. Report No. TX-97/2969-1F.* Austin, TX: Texas Dept. of Transportation.

Scharman, B. (1992). EPA Federal Procurement Guidelines and the Impact of RCRA Reauthorization. *Proceedings of the Second Interagency Symposium on Stabilization of Soils and Other Materials.* Washington, DC: Federal Highway Administration.

Schroeder, R. L. (1994). *United States Department of Transportation – Federal Highway Administration.* Retrieved January 4, 2008, from The Use of Recycled Materials in Highway Construction: http://www.tfhrc.gov/pubrds/fall94/p94au32.htm

Smith, L., & Ramer, R. (1992). Recycled Plastics for Highway Agencies. *Transportation Research Board*, Washington, DC.

Soroushian, P., Plasencia, J., & Ravanbakhsh, S. (2003). Assessment of reinforcing effects of recycled plastic and paper in concrete 100 (3). *ACI Materials Journal*, 203–207.

Suheibeni, A. (1986). The Use of Fly Ash as an Asphalt Extender in Asphalt Concrete. *PhD Dissertation.* Department of Civil Engineering, University of Michigan.

van Oss, H. G. (2003). *Slag —Iron and Steel.* Reston, VA: U.S. Geological Survey Minerals Yearbook.

Wilson, D., Foley, P., Weisman, R., & Frondistou - Yannas, S. (1976). Demolition Debris: Quantities, Composition, and Possibilities for Recycling. *Proceedings of the Fifth Mineral Waste Utilization Symposium.* Washington, DC: Bureau of Mines.

Zoorob, S. E., & Suparma, L. B. (2000). Laboratory design and investigation of the properties of continuously graded Asphaltic concrete containing recycled plastics aggregate replacement. *Cement and Concrete Composites*, 233–242.

Activated Carbon in Waste Recycling, Air and Water Treatment, and Energy Storage

Sarra Gaspard & Axelle Durimel
Université des Antilles et de la Guyane, Guadeloupe

Mohamed Chaker Ncibi
High Institute of Agronomy, Chott Meriem

Sandro Altenor
Université des Antilles et de la Guyane Guadeloupe, Université Quisqueya, Haïti

SUMMARY: The beneficial use of activated carbons for environmental purposes is reviewed. Activated carbons can be the product of waste recycling from agricultural by-products, industrial or municipal wastes. They are commonly used as an efficient and versatile adsorbent for air purification, water and wastewater treatment, and hydrogen storage. The main discussion is divided into six major parts: (i) the activated carbons production from diverse precursors and then their diverse applications, with (ii) water purification and (iii) wastewater treatment, (iv) gas purification, (v) activated carbon regeneration and (vi) hydrogen storage.

1 INTRODUCTION

Activated carbons (ACs) are carbonaceous materials characterized by a large specific surface area, typically in the range of 500–2500 m^2/g. Commercial ACs are prepared from various precursors, including lignite, coals, asphalt, petroleum residues, wood residues and coconut shell. Activation can be accomplished by one of two distinct processes: (1) chemical activation or (2) physical activation. Thus, due to their high production costs, wastes of diverse origin have been investigated as AC precursors (Ioannidou, 2007). Wastes conversion into AC, adds economic value, helps to reduce the cost of waste disposal and provides a potentially inexpensive alternative to the existing commercial activated carbons.

The manufacturing process and the intended application of the carbon will also be important considerations. Their highly developed inner porosity, with a large specific surface area coupled with the presence of active functional groups such as acido-basic sites (Debyshire et al., 2001), make AC very versatile sorbent-materials. They are commonly used as an efficient and versatile adsorbent for air purification, water and wastewater treatment and hydrogen storage. As presented in Table 1, activated carbons have industrial uses in the purification of liquids and gases by the adsorption of gaseous substances from vapour phases and, of dissolved or dispersed substances from liquids such as for water purification, wastewater cleaning operations (Gelderland, 2001; Robinson, 1990). Their applications involve flowing liquid and vapor-phase streams purification systems in activated carbon-filled canister, column, or filtration apparatus. ACs can be used as membranes, filters, and catalyst support in both gas and liquid phases. Besides, to enhance the overall economical benefit in using activated carbons, other researchers developed some innovative techniques to regenerate those costly materials. Additionally, activated carbon has extensively been used as electrode material for energy storage devices.

Activated carbons (ACs) are important materials known for their great ability to adsorb various molecules on their surface. Despite its frequent use AC remains an expensive material. Petroleum residues, natural coal and woods were for a long time the main AC precursors (Guo and Lua,

Table 1. Samples of organic and inorganic compounds amenable to absorption by activated carbon.

Categories	Examples	References
Aromatic solvents	Benzene, Toluene	(Choi et al., 2007; Wibowo et al., 2007; Lillo-Rodenas et al., 2006)
Polynuclear aromatics	Naphthalene, polychlorinated biphenyls	(Ania et al., 2007a; Valderrama et al., 2007; Ania et al., 2007b)
Chlorinated aromatics	Chlorobenzene, endrin, DDT, toxaphene, pentachlorophenol	(Ornad et al., 2007; Shin et al., 2002; Slaney and Bhamidimarri, 1998)
Phenolics	Phenol, cresol, resorcinol, nitrophenols, chlorophenols	(Dabrowski et al., 2005; Mohanty et al., 2008; Qu et al., 2007; Ayranci and Duman, 2005; Brasquet et al., 1999; Slaney and Bhamidimarri, 1998)
Soluble organic dyes	Acid and basic dyes	(Vazquez et al., 2007; Emad et al., 2008; Thinakaran et al., 2008 ; Chan et al., 2008)
Petroleum products	Gasoline, kerosene, oil	(Mohammadi and Esmaeelifar, 2005; Ayotamuno et al., 2006; Sheldon et al., 2002)
Pesticides	2,4-D, atrazine, simazine, DDT, aldicarb, alachlor, Carbofuran, lindane, HCH	(Ornad et al., 2007; Shaalan et al., 2007; Ayranci and Hoda, 2005; Kouras et al., 1998)
Inorganonic substances	Cu^{2+}, Cd^{2+}, Pb^{2+}, Ni^{2+}, Cr^{6+}, Hg^{2+}, Zn^{2+}, etc.	(Kang et al., 2008; Sirianuntapiboon and Ungkaprasatcha, 2007; Park and Kim, 2005; Starvin and Rao, 2004; Amuda et al., 2007)

1999). But, since a few years, other precursors at low or null-cost materials have been investigated as AC precursors including biological resources (Nishimo et al., 1983) and agricultural by-products, industrial or municipal wastes (Okada et al., 2003; Rozada et al., 2005; Nakagawa et al., 2004).

2 WASTE RECYCLING

The preparation of activated carbons from wastes exemplifies waste recycling because high value products are obtained from low cost materials, and simultaneously, brings solutions to waste management. This is an alternative to expensive precursors, and a way to protect the environment by reducing wastes accumulation. Agricultural and wood by-products and wastes offer an inexpensive and renewable source of AC. Wastes from municipal and industrial activities have also been used for AC preparation. Since few decades many research works has been performed on the preparation of activated carbons using various precursors, and different experimental procedures. Samples with different textural, chemical, and adsorptive properties can be elaborated.

Granular and powder activated carbons can be prepared from many agricultural wastes (Ioannidou et al., 2007) such as bagasse (Valix et al., 2004; Juang et al., 2001; Juang et al., 2002; Tsai et al., 2001; Ahmedna et al., 2000), coir pith (Namasivayam & Kavitha, 2002; Namasivayam et al., 2001), banana pith (Kadirvelu et al., 2003), sago waste (Kadirvelu et al., 2003), silk cotton hull (Kadirvelu et al., 2003), corn cob (Juang et al., 2002), maize cob (Kadirvelu et al., 2003), straw (Kannan & Sundaram, 2001), rice husk (Mohamed, 2004; Malik, 2003; Kannan & Sundaram, 2001), rice hulls (Ahmedna et al., 2000), fruit stones (Aygü n et al., 2003), nutshells (Aygü n et al., 2003; Ahmedna et al., 2000), pinewood (Tseng et al., 2003), sawdust (Malik, 2003), coconut tree sawdust (Kadirvelu et al., 2000; Kadirvelu et al., 2003), bamboo (Wu et al., 1999) and cassava peel (Rajeshwarisivaraj et al., 2001), Vetiver roots (Gaspard et al., 2007; Passé-Coutrin et al., 2008).

Among municipal wastes and industrial by products, PET bottle (Nakagawa et al., 2004; Lopez et al., 2007), waste tires (Nakagawa et al., 2004), refuse derived fuel (Nakagawa et al., 2004), waste carbon slurries and blas furnace slag (Jain et al., 2003; Gupta et al., 2003), waste newspaper (Okada et al., 2003) or biological wastes generated during lactic acid fermentation from garbage (Nakagawa et al., 2004), and sewage sludge (Rozada et al., 2005; Otero et al, 2003; Rio et al., 2006) has also been used for granular activated carbon (GAC) and powder activated carbon (PAC) production.

Carbon activation can be accomplished by chemical and physical processes. The porosity of chemically activated products is generally created by dehydration reactions occurring at significantly lower temperatures with chemical agents such as phosphoric acid, potassium hydroxide or zinc chloride. In the physical activation process the precursor is carbonized under an inert atmosphere leading to char. The second step in the preparation of activated carbon by physical activation involves a controlled gasification of the char at high temperature with steam, carbon dioxide, air or a mixture of these. This gasification selectively eliminates most reactive carbon atoms from the sample, generating the porosity and the final carbon with the pore structure sought. Activated carbons produced by physical activation are typically more microporous (<2 nm); while carbons produced by chemical activation are typically more mesoporous (pore of diameter between 2 and 50 nm). Generally, the choice of a precursor will largely depend upon its availability, cost and purity (Dias et al., 2007). The impact of the selected activating process on the environment is crucial for the feasibility of industrial production of ACs.

3 WATER PURIFICATION

Application of activated charcoal for the removal of undesirable compounds, odours and taste in drinking water has been recognized at the dawn of civilization. Using bone char and charred vegetation, gravel, and sand for the filtration of water for domestic application has been practised for thousands of years. However, over the last century research on the production and use of activated carbon for water treatment has grown and it becomes one of the most adsorbents used. Activated carbon has the advantage of removing pesticides including by-products, chemicals, chlorine, tastes, odours and colours, without affecting naturally occurring minerals in water. GAC for drinking water treatment needs a pore structure to allow the adsorption of a wide range of organic compounds including specific micro-pollutants and natural organic matter. The GAC must also possess a suitable amount of transport pores which allow the molecules to be transported to the adsorption site. The adsorption capacity for drinking water applications is very difficult to quantify without laboratory evaluation. Some physical parameters such as: specific surface area, total pore volume, iodine index and, experiments in liquid phase in batch or column with phenol (Aygun et al., 2003; Zhang et al., 2002) or methylene blue (Aygun et al., 2003; Giles et al., 1969; Wang et al., 2005; Hameed et al., 2007; Kavitha and Namasivayam, 2007; Zabaniotou et al., 2008) considerated like standard molecules, indicate the overall adsorption capacity of the activated carbon.

In recent decades, the extensive use of chemical fertilizers and improper treatment of wastewater from the industrial sites has led in many countries, to several environmental problems, such as an increase in the concentration of nitrate in underground and surface waters. A high concentration of nitrate in drinking water leads to the production of nitrosamine, which is related to cancer occurrence and the risk of numerous of diseases such as methemoglobinemia in new-born infants. Several methods that serve to reduce nitrate in drinking water have been presented (Dhab, 1987; Feleke and Sakakibara, 2002; Leakovic et al., 2000; Wasik et al., 2001). The use of biological reactor seems to be the most promising technique in the treatment of high nitrate concentration. However, maintaining biological processes at their optimum conditions is difficult, and the problems of contamination by dead-bacteria have to be solved to make such processes safety enough to be utilized in drinking water treatment. Adsorption is a very feasible process for in situ treatment of underground and surface water, primarily due to its convenient of application. Mizuta et al., (2004), show the adsorption effectiveness of nitrate-nitrogen by powder activated carbon from bamboo waste. The effects of the temperature on nitrate removal by activated carbon, was also investigated. The results suggest that powder activated carbon from waste bamboo is effective in removing nitrate from underground and surface water.

Humic substances are a major fraction of natural organic matter (NOM), which is very common in surface waters. They account for up to 90% of NOM. They are formed during the degradation of plant and animal materials and are yellow to black in colour and refractory. Humic substances are a heterogeneous group of high molecular weight organic species that adversely affects drinking water treatments and water quality of produced water. Humic substances are thus known to increase disinfectant and coagulant demands that generate potentially harmful disinfection by-products, foul membrane and favour biological growth in the distribution network. Several works describe the use of activated carbon for NOM removal (Lee et al., 1981; Weber et al., 1983; Kilduff et al., 1996).

For dissolved organic matter removal, an invention of Karanfil and Dastgheib (2006) was set up for improving the NOM uptake of granular activated carbons. The methods include treating starting materials so as to provide a combination of physical and chemical characteristics favourable for NOM uptake. The utilized method depends upon the characteristics of the starting materials, but it also considers: increase in surface area in pores greater than 1 nm; increase in overall alkalinity; and impregnation with an iron species. The processed materials exhibit improved uptake of NOM from natural waters as compared to previously known GAC materials (Karanfil and Dastgheib, 2006).

Humic substances can react with chlorine added as a disinfectant, producing harmful halogenated organic compounds such as trihalomethanes and holoacetic (Reckhow et al., 1990). Another application discloses a method for enhanced removal of chloramines from a chloramines-containing fluid media by contacting said media with a catalytic activated carbon characterized by having present in the graphene structure of the carbon from 0.01 to 10 wt % of aromatic nitrogen species (Baker, 2005). The catalytic activated carbons used in the invention may be prepared from carbon materials that have been contacted or otherwise exposed to ammonia, with or without simultaneous exposure to an oxygen-containing vapor or gas at temperatures above 700°C and preferably, are in the form of a solid carbon block.

Ion removal from water sources is generally achieved using ion exchange membranes (Namasivayam et al., 2007; Daifullah et al. 2007). Indeed, an electrode for deionization of water is made of a continuous activated carbon structure (Gadkaree et al., 2001). It is important to note that, in addition to municipal and/or industrial treatment systems, water is also treated at domestic scale. Activated carbon is used at home as filter for removal of taste and odour in drinking water. In recent years, activated carbon is been employed for removal of some contaminants such as 2-methylisoborneol (MIB) and geosmin (Mackenzie et al., 2005; Rangel-Mendez and Cannon, 2005) discovered in water supplies.

Home activated carbon treatment systems are quite simple. The material is normally packaged in filter cartridges which are inserted into the purification device. Water needing treatment passes through the cartridge, contacting the activated carbon on its way to the faucet. Activated carbon filters eventually become fouled with contaminants and loose their ability to adsorb pollutants. At this time, they need to be replaced. Activated carbon filters used for home water treatment contain either granular activated carbon or powdered block carbon. Although both are effective, the block activated carbon filters were more effective in removing chlorine, taste and halogenated organic compounds (Environmental Protection Agency, 2000).

It is agreed that the activated carbon filtration, used alone or in hybrid system remains one of the most effective ways for most pollutants removal in liquid phase. Effectiveness is linked to its physical and chemical properties. In addition to the traditional activated carbon from wood, coal or oil residue, new activated carbon from plant parts (roots, stems, nuts, bark, fruit clusters) are available. Hence, there is a need to know the characteristics of activated carbon to be used, depending of the pollutants nature to be treated before any water treatment (drinking water or wastewater).

4 WASTEWATER TREATMENT

The use of activated carbon in wastewater treatment facilities can either involve granular or powder carbon. The granular form is placed inside a cylindrical steel vessel with screens in the bottom and top to confine the carbon like a packed bed. More than one vessel may be used, with the vessels being connected either in series or in parallel. The water usually flows downward in each column, either pressure driven (with a pump) or gravity driven. Most real feed streams contain suspended solids, which are largely filtered out as the feed passes through the carbon bed. Therefore the bed must be periodically backwashed, to wash out these solids. Alternatively a pressure-driven upflow mode may be used with the flow rate being such as to expand the bed slightly so as to prevent or minimize plugging due to suspended matter (Cooney, 1999).

Granular activated carbon (GAC) adsorption has been often used successfully for the treatment of municipal and industrial wastewater (Environmental Protection Agency, 2000). GAC is used to adsorb soluble organics and inorganic compounds such as nitrogen, sulfides, and heavy metals remaining in the wastewater following biological or physico-chemical treatment. Typically, GAC adsorption is utilized in wastewater treatment facilities as a tertiary process following conventional secondary treatment or as one of several unit processes composing physical-chemical treatment. In wastewater treatment plants utilizing biological secondary treatment, GAC adsorption is generally located after filtration and prior to disinfection. Using of GAC in a physical-chemical treatment process is generally located following chemical clarification and filtration and prior to disinfection. GAC adsorption systems have a relatively small dimension making them suitable for facilities with limited land availability. The successful application of carbon adsorption for municipal wastewater treatment depends on the quality and quantity of the wastewater delivered to the adsorption system. Wastewater constituents that may adversely affect carbon adsorption include suspended solids, BOD5, and organics such as dye active substances, phenol for example and dissolved oxygen (Environmental Protection Agency, 2000). Environmental factors that must be considered include pH and temperature because they may impact solubility, which affects the adsorption properties of the wastewater components onto carbon (Cooney, 1999; El Qada et al., 2008; Environmental Protection Agency, 2000).

The use of GAC presents some advantages. For wastewater flows which contain a significant quantity of industrial flow, it is a proven to be a reliable technology for dissolved organics removal. Space requirements are low, it can be easily incorporated into an existing wastewater treatment facility, and GAC could be regenerated. However, GAC adsorption systems could present some inconvenient (Environmental Protection Agency, 2000) such as:

- Generation of hydrogen sulfide from bacterial growth, creating odours and corrosion problems.
- Land disposal problem of surplus carbon, if not regenerated,
- Wet GAC is highly corrosive and abrasive.
- Requirement wastewater pre-treatment to obtain a low suspended solids concentration.
- variations in pH, temperature, and flow rate may also adversely affect GAC adsorption.

Digested molasses spentwash which represent a large source of aqueous pollution in sugar cane industries can be treated using activated carbons (Ahmedna et al., 2007; Figaro et al, 2006). Figaro et al. (2006), show that an activated carbon, with an acidic character containing an adequate repartition of both micropores and mesopores and a significant amount of macropores to their access, have a good adsorption efficiency for compounds such as tannic acid melanoidins (Figaro et al., 2006). Although this wastewater contains a majority of low-size compounds (<1 nm) which may be preferably adsorbed in micropore, microporous AC may not be suitable for such an application. The selection of mesoporous activated carbon can be of great interest for adsorption wastewaters. The authors concluded that, according to the application for which they are destined, meso- and macro-porosity has to be considered, in order to evaluate the potential adsorption uses of the sorbent for wastewater treatment (Figaro et al., 2006).

Devi et al., (2008) show that activated carbon from agricultural by-products is an effective adsorbent for reduction of chemical oxygen demand (COD) and biological oxygen demand (BOD) concentration from effluent of coffee processing plant and domestic wastewater. In batch experiments, adsorption of COD and BOD was found to be dependent on treatment time, adsorbent dose, pH, initial COD and BOD concentration, agitation speed and adsorbent particle size. Under optimum operating conditions using activated carbon, the maximum percentage reduction of COD and BOD concentration was up to 98–99%.

The use of powder activated carbon (PAC), as opposed to GAC, is advantageous when required periodically or seasonably. The PAC is ordinarily prepared as a slurry, and added to the water in a mixing tank. After a suitable period of contact, with continuous mixing, the carbon is allowed to settle. Sedimentation may ocurr in separate unstirred tank, or may be done in the mixing tank by simply turning off the mixing device (Cooney, 1999). When activated carbon is used for urban waste water treatment, powder activated carbon is generally applied before an active sludge step. The powdered activated carbon wastewater treatment system can combine the biological treatment, as the activated sludge process with adsorption on powdered activated carbon. Typically, the contact between living microorganisms, PAC contact and wastewaters is done in an aeration basin or anaerobic tank. Biomass removes biodegradable organic contaminants through biological assimilation, while the carbon physically adsorbs conventional and toxic organics. The benefits of this process include:

- Improved removal of Biochemical Oxygen Demand (BOD), Chemical Oxygen Demand (COD), non-biodegradable organic compounds and toxicity
- Increased stability to shock loads
- Removal of inhibitory compounds including for nitrifying bacteria
- Improved sludge dewatering
- Reduced aerator foaming.

In activated sludge system, powdered activated carbon is added to aid the removal of substances which are either toxic or inhibitory to the biological process (Ilett, 1995). The adsorbent is introduced after the main oxygenation stage. Since peak concentrations of the target substances in the wastewater are reduced by virtue of the mixing and residence time within the oxygenation stage, the quantity of adsorbent needed is advantageously reduced. The technology removes organic contaminants from wastewater and minimizes the inhibitory effects of process wastewater containing toxic organic compounds. Mobile powdered activated carbon wastewater treatment systems can be furnished as continuous flow or batch treatment systems.

Hybrids methods coupling activated carbon with others conventional methods have been investigated (Mohammadi and Esmaeelifar, 2005; Lesage et al., 2008; Vazquez et al., 2007). Mohammadi and Esmaeelifar (2005) reported a hybrid ultrafiltration-activated carbon process for the treatment of wastewater of vegetable oil factory. In their work, they compared three processes: the conventional biological method, the conventional ultrafiltration (UF) and the hybrid ultrafiltration-powdered activated carbon (UF-PAC). Their experiences show that a pressure difference more than 3×10^5 Pa, a high cross flow velocity (depending on economic considerations), a temperature of 30°C and an optimum pH of 9 are the best operating conditions. Table 2 shows that UF is better than conventional biological method, and UF-PAC is better than UF. The reduction of the phosphate concentration by UF-PAC is considerable.

Lesage et al. (2008) compared a membrane bioreactor (MBR) and a hybrid membrane bioreactor (HMBR), coupling membrane separation, biological activity and adsorption on powdered activated carbon (PAC) in order to remove a toxic compound. The two processes were compared in terms of water treatment efficiency, membrane fouling and biological sensitivity to the toxic compound. Experiments were performed with a synthetic wastewater and 2,4-dimethylphenol (DMP) was chosen as a molecule representative of toxic compounds present in effluents from the oil industry. This study showed the advantages of the hybrid process in comparison with a conventional MBR, due to the positive effects of PAC adjunction. Results show that PAC adjunction slightly reduces

Table 2. Analysis of wastewater and treated wastewaters by different methods (Mohammadi and Esmaeelifar, 2005).

Feed Permeate (ppm)	COD 550	TOC 300	TSS 120	PO_4^{3-} 20	Cl^- 350
Biological	90	130	50	25	400
Reduction (%)	84	57	58	–	–
UF (ppm)	50	40	~0	3	210
Redcuction (%)	91	87	~100	85	40
UF-PAC (ppm)	35	20	~0	0.2	200
Reduction (%)	94	93	~100	99	43

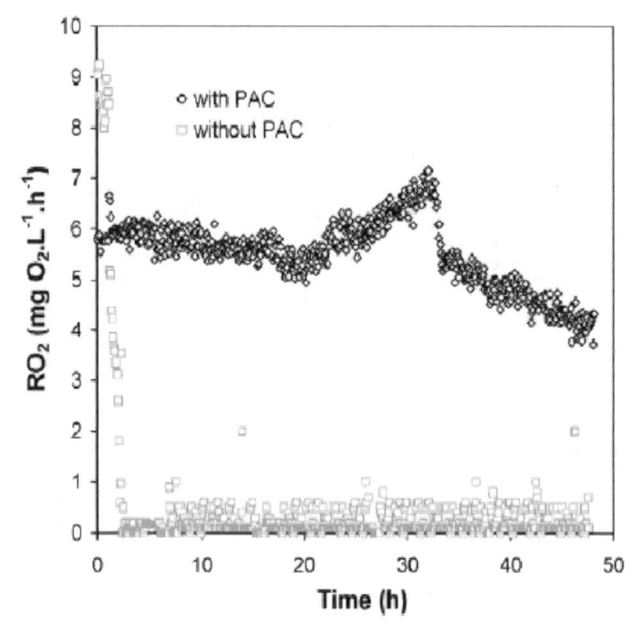

Figure 1. Variation of microorganisms respiration rate after toxic injection for the MBR and HMBR (Lesage et al., 2008).

sludge production in the HMBR. Toxic injection inhibits the biological activity in the MBR whereas the biological activity is maintained in the HMBR (Figure 1), with a biodegradation of the toxic compound after an acclimation period.

Various physico-chemical methods have been proposed for the treatment of wastewaters containing phenolic compounds. The choice of treatment is often governed upon effluent characteristics such as phenolic compounds concentration, pH, temperature, flow volume, biological oxygen demand, the economics involved and the social factor like standard set by government agencies. However, adsorption onto the surface of activated carbons is the most widely used method (Vazquez et al., 2007; Mukherjee et al., 2007; Phan et al., 2006; Singh et al., 2008).

Mohanty et al. (2008) have investigated a method using a multi-stage external loop airlift reactor (Figure 2) for removal phenol in wastewater. This method was used successfully to remove phenol from wastewater. Nearly 95% removal of phenol was achieved with a lower removal time as well as the activated carbon loading for this system as compared to simple batch adsorption systems. The chemical oxygen demand (COD) removal reported for this system is nearly 95%. The optimum

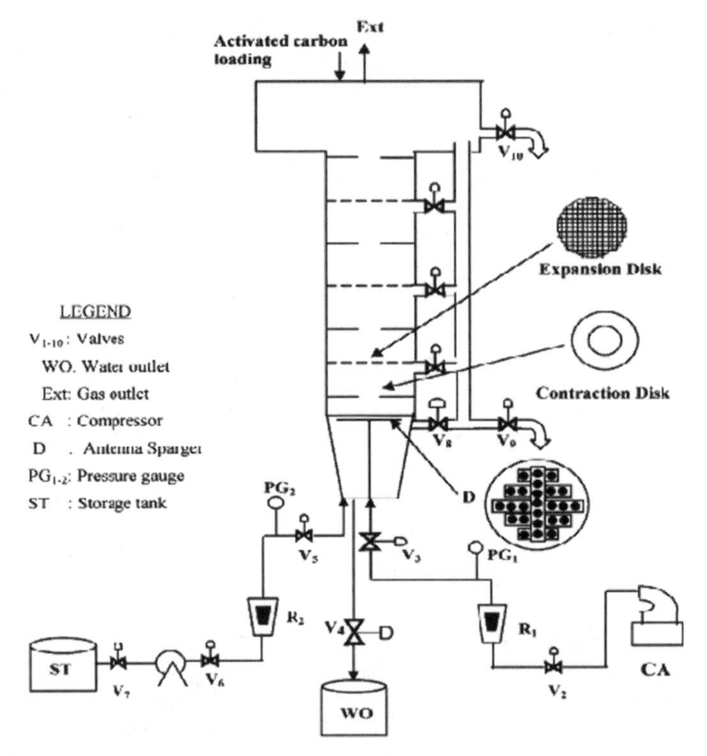

Figure 2. Schematic diagram of experimental setup for removal of phenol from wastewater (Vazquez et al., 2007).

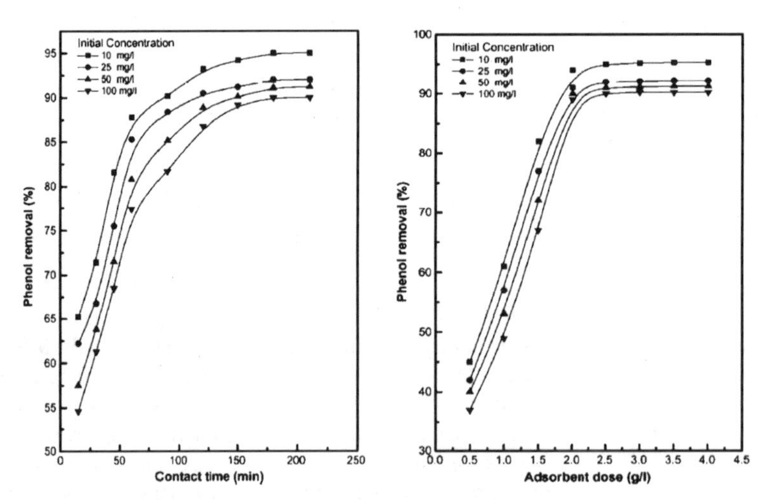

Figure 3. Effect of contact time and dose on removal of phenol in a MS-ELALR, $UG = 0.0219$ m/s, adsorbent dose = 2 g/l, pH 3.5. (Vazquez et al., 2007).

operating parameters for the removal of phenol were found to be: pH of 3.5, adsorbent dose of 2 g/l and superficial gas velocity of 0.0219 m/s. The results obtained confirmed that phenol can be effectively removed using the method of adsorption in a multi-stage external loop airlift reactor.

Qu et al. (2007) demonstrated that ozonation combined with activated carbon fiber (ACF) adsorption (Figure 4) could significantly improve rate removal of phenol in water if compared

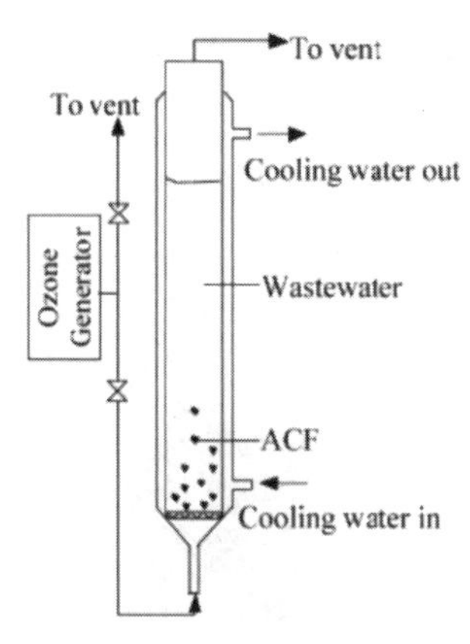

Figure 4. Schematic of the experimental setup (Qu et al., 2007).

to the adsorption of ACF or ozonation alone, and led to mineralization of organic matter. With the use of 2 g of ACF and ozonation, phenol removal could reach 99% for an initial phenol concentration of 100 mg/L in less than 10 min, and in the same time, 95% COD removal could be achieved under the same operating conditions.

5 GAS PURIFICATION

Fossil fuel fired power plants, auto exhausts and other industrial chemical emissions are the major sources of air pollutants. Furthermore, incinerators, dealing with municipal solid waste, hazardous industrial waste, medical waste and sewage sludge results in the formation of a flue gas containing a range of gaseous pollutants.

Sulfur dioxide (SO_2) and nitric oxide (NO) are two major air pollutants whose emissions from fossil fuel fired power plants have been linked with the formation of acid rain, urban smog and many other undesirable environmental and health hazards such as hair loss, throat inflammation, impaired vision and respiration, and even critical illnesses.

Activated carbon can simultaneously remove of NO and SO_2. In such a process, SO_2 is removed in the first charcoal bed of a two-stage adsorption and NO is catalytically reduced to nitrogen using ammonia as the reductant in the second bed. The active charcoal, as used in this process, is prepared from hard coal.

The activated carbon process has the added advantage that it operates at low temperatures, and is also cheaper compared to the combined cost of selective catalytic reduction process, using V_2O_5 based catalysts for NO removal, and limestone scrubbing process for SO_2 removal. Costs of carbon based processes can be reduced by using less expensive sources of activated carbon (Finqueneisel, et al., 1998; Dalai et al., 1996; Sakintuna et al., 2004). Nitrogen monoxide can be adsorbed but as well reduced by AC with formation of N_2 and O_2. The use of transition metals such as Cu, Fe, Ni and Co supported by AC significantly increase NO removal (de la Puente et al., 1998; Goncalves et al., 2004; Illan Gomez et al.,1995). NO removal by the aforementioned impregnated AC is enhanced in the presence of volatile organic compounds (Lu and Wey, 2007). Activated carbon is able to adsorb the sulphur dioxide that is then oxidized by the oxygen contained in the flue gas, with the carbon as catalyst. The resultant oxides are then absorbed in water, to thereby form sulphuric acid, which is removed from the carbon material. Oxygen containing surface functional groups at the carbon surface enhance the carbon adsorption capacity (Chattopadhyaya et al., 2006).

Volatile organic compounds (VOCs) are emitted by chemical industries dealing with paints, lubricants and liquid fuels processing. With continued global industrialization, the atmospheric concentrations of VOCs due to primary as well as fugitive emissions have constantly been on rise.

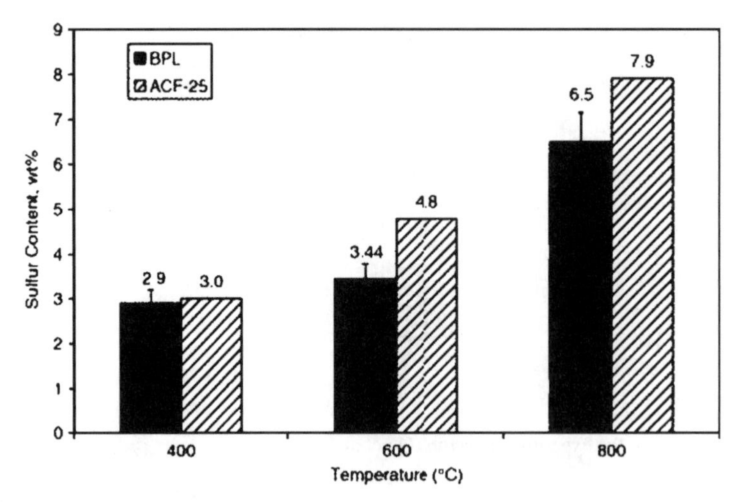

Figure 5. Effect of temperature on sulfurization of ACF25 and BPL in the presence of H_2S during stable temperatures only (S only), (Feng et al., 2006).

Those compounds have harmful impacts, such as eye and throat irritation, damage to liver, central nervous system, and carcinogenic effects. It is thus necessary to achieve effective abatement technologies such as adsorption by activated carbon (Cal et al., 1997). Activated carbon fibbers exhibit larger adsorption for VOC than granular activated carbon under identical operating conditions. Furthermore, regeneration of ACF equilibrated with VOC can be carried out by electrical heating (Dwivedi et al., 2004).

Carbon-based adsorbent have been evaluated for the removal of hydrogen sulphide due to the need to control odorous gases generated in sewer systems and wastewater plants. Feng et al., (2005) reported that ACFs with higher surface area showed greater adsorption and retention of sulphur, and heat treatment further enhanced adsorption and retention of sulphur. The retained amount of hydrogen sulfide was correlated well with the amount of basic functional groups on the carbon surface, while the desorbed amount reflected the effect of pore structure. Feng et al., (2006) showed that sulphur uptake and stability generally increased with the increase in temperature from 200 to 800°C (Figure 5) due to a shift in the reaction mechanism. The sulfurization process is associated with the decomposition of surface functionalities, which creates active sites for sulphur bonding. The presence of H_2S during the cooling process increased the sulphur content by increasing the presence of less stable sulphur forms.

Gasoline typically includes a mixture of hydrocarbons ranging from higher volatility butanes (C_4) to lower volatility C_8 to C_{10} hydrocarbons. When vapour pressure increases in the fuel tank due to conditions such as higher ambient temperature or displacement of vapour during filling of the tank, fuel vapour flows through openings in the fuel tank. The fuel vapour is a mixture of the gasoline vapour and air. To prevent fuel vapour loss into the atmosphere, the fuel tank is vented into a canister that contains an adsorbent material such as activated carbon granules. Mesopores are necessary for the adsorption of large molecules. ACs prepared from carbonized PET by steam activation via pretreatment by mixing PET with a metal salt [$Ca(NO_3)_2{\cdot}4H_2O$, $Ca(OH)_2$, $CaCO_3$, ZnO, and $AlNH_4(SO_4)_2{\cdot}12H_2O$], and with acid treatment after carbonization show large adsorption capacity for nC_4H_{10} and iC_4H_{10}. These carbons are suitable as adsorbents for canisters (Nakagawa et al., 2003).

6 ACTIVATED CARBON REGENERATION

Removal of pollutants by adsorption on activated carbons (ACs) is definitely widely used for water or gas purification. However, this process enables to eliminate the compound from water or gas phases but do not degrade it. After saturation of adsorption capacity of AC, the adsorbents must be treated or stored in order to avoid an additional environmental pollution. Therefore, they can be stored at a place envisaged for this purpose, burnt in a specific installation followed of purification of gases obtained, or regenerated for reuse as adsorbents. Due to the high cost of AC, it is always preferable from an economical point of view, to regenerate it. Regeneration can be defined as a process allowing the desorption of molecules from the AC porous structure. The aim of this method is to remove maximum of molecules adsorbed onto AC without influencing its physical properties (weight loss, chemical groups on the surface, etc.). A maximum number of adsorption-regeneration cycles should be performed while preserving a greatest of adsorption capacity. The spent AC can be regenerated by thermal (Maroto-Valer et al., 2006; Moreno-Castilla et al., 1996; Pelech et al., 2005), chemical (Zhang, 2002; Zhou, 2006), and biological (Aktas and Ceçen, 2007) methods. The more often applied process is thermal regeneration but this method is expensive and carbon consuming (Lim and Okada, 2004). Moreover, during regeneration, air emissions from the furnace contain volatiles stripped from the carbon. Carbon monoxide is formed as a result of incomplete combustion. Therefore, afterburners and scrubbers are usually needed to treat exhaust gases.

Chemical regeneration is widely used too but it requires much more extraction steps. It is as well generally very difficult to extract the desorbed molecules for a further degradation step and regeneration yields does not exceed 70% (Hamdaoui et al., 2005). Recently, some other methods were investigated in order to decrease the disadvantages of the common ACs regeneration such as regeneration by ultrasound (Lim and Okada, 2004), microwaves (Ania et al., 2004, C.O. Ania et al., 2005), catalytic regeneration (Matatov-Meytal et al., 1997; Matatov-Meytal and Sheintuch, 2000). The limit of all the physico-chemical regeneration techniques remains in the fate of the desorbed pollutant. Bioregeneration is the most promising process in an environmental point of view, regarding to the use of microorganisms that may achieve, in some cases complete degradation of the pollutant, but it remains a much longer treatment. Several factors, that are not fully understood, may influence the bioregeneration process such as reversibility of adsorption, biodegradability of organic compound, properties of substrate (molecular weight, polarity, pKa etc), activation process, AC porous structure, AC surface properties, solutions conditions, pH, and adsorbate concentration (Aktas and Cecen, 2007). Research efforts should be performed for a better knowledge of the bioregeneration mechanism and for setting up its large scale application.

7 HYDROGEN STORAGE

Due to the worldwide petroleum and energy crisis occurring since the last decades, hydrogen storage on activated carbon materials has been attracting attention because of the importance of hydrogen as an ideal eco-friendly substitute for fossil fuels (Bénard and Chahine, 2007). Besides, from economic point of view hydrogen technology will be able to revolutionize the transport and energy market (Dillion et al., 1997) if its storage, the main technical problem, will be mastered adequately.

Hydrogen storage problem heeds to be conquered for the successful implementation of fuel cell technology in transport applications (Takagi et al., 2004). Several methods and techniques of storage are under investigation including high pressure gas, liquid hydrogen, adsorption on porous materials, complex hydrides and hydrogen intercalation in metals (Ross, 2006). However, none of these methods completely satisfy all the criteria for the amount of hydrogen that can be supplied from a given weight or volume of tank for transport purposes.

The use of hydrogen sorption on porous materials is one of the main methods being considered for vehicle applications, since it allows the storage of large amounts of hydrogen at near-ambient temperatures and pressures.

Several researches have been centered on storing hydrogen into solid porous media (Hynek et al., 1997). The use of activated carbons was deeply investigated for both its efficiency and also for economical considerations. Maeland and Skjeltorp (2001) proposed a method for storing hydrogen in a carbon material containing microstructures in the form of cones with cone angles being multiples of 60°, the carbon material is introduced in a reaction vessel which is evacuated while the carbon material is kept at a temperature of 295–800 K, after which pure hydrogen gas is introduced in the reaction vessel, the carbon material being exposed to a hydrogen gas pressure in the range of 300–7600 torr (in vacuum applications: 760 mmHg = 760 torr) such that the hydrogen gas is absorbed in the carbon material, and after which the reaction vessel is left at the ambient temperature with the carbon material under a fixed hydrogen gas pressure. (split this 120 words statement into two or three simple ones) When used, the hydrogen is released in the form of a gas from the carbon material either at ambient temperature or by heating the carbon material in the reaction vessel.

In the sorption process, many parameters might influence the hydrogen adsorption amount and its stability in the porous structure. Among those key parameters, Xu et al. (2007) studied the effect of temperature on the hydrogen storage capacity of various carbon materials, including regular activated carbon, single-walled carbon nanohorn, single-walled carbon nanotubes, and graphitic carbon nanofibers (Figure 5), under 77 and 303°K. The results showed that hydrogen storage capacity of carbon materials was less than 1 wt% at 303°K, and by lowering adsorption temperature to 77°K, hydrogen storage capacity of carbon materials increased significantly up to 5.7 wt% at a relatively low pressure of 3 MPa. Besides, it was proven that hydrogen storage capacity of carbon materials was proportional to their specific surface area and the volume of micropores, and the narrow micropores was preferred to adsorption of hydrogen, indicating that all carbon materials adsorbed hydrogen gas through physical adsorption on the surface.

Furthermore, always within the thermal effect, Hermosilla-Lara et al. (2007) investigated this parameter during high-pressure charging of a packed bed hydrogen storage tank. The proposed experimental setup is illustrated in Figure 6. The studied column is packed with a coconut derived activated carbon, which has an average surface area of 2600 m^2/g. The temperature at six locations in the storage tank and the pressure value at the bottom of the tank are recorded during the charging stage. Several experiments were carried out to investigate the effect of the initial flow rate on the temperature field in the reservoir and on the duration of the charging process.

As well, other parameters were studied to monitor the hydrogen storage into activated carbon. In that issue, Takagi et al. (2007) tested the influence of both the porosity and the activating agent on optimizing the hydrogen uptake. The experimental results showed that the amount of adsorbed hydrogen by weight both at 77 and 303°K seemed to be linearly related to the micropore volume of the sample. Besides, the amount of hydrogen adsorbed on the carbonaceous sample increased remarkably upon treatment with nitric acid. Such acid treatment resulted in an increase in the number of sites with high interaction potentials for hydrogen adsorption, and these sites can be considered to be the inside of tubes or the interstitial space between the tubes.

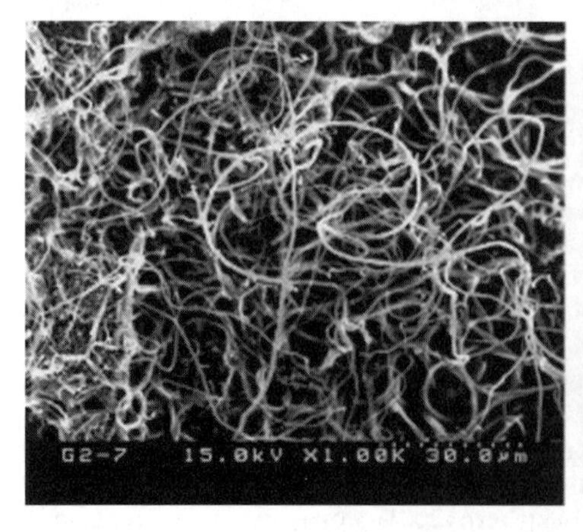

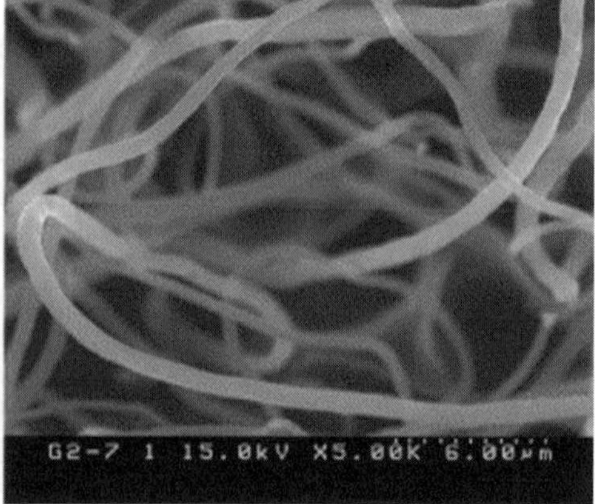

Figure 6. SEM photographs of graphitic nanofibers (Source: Xu et al., 2007).

The major limitations for H_2 adsorption was reported by Thomas (2007), who revealed that the maximum amount adsorbed is limited by the hydrogen adsorbate density, adsorbent pore structure and pore volume available in the narrowest pores. Adsorption of hydrogen at ambient temperatures and high pressures has produced inconsistent reports of uptakes. This is due to major experimental difficulties with the adsorption of impurities, isotherm corrections at high pressure, etc. However, recent results (Erdogan and Kopac, 2007) suggest that the amounts of hydrogen adsorbed are very small under high pressure and ambient temperature conditions. Fast adsorption/desorption kinetics and relatively small (<10 kJ/mole) adsorption enthalpies are observed for hydrogen adsorption on many porous materials which indicates that physisorption (??) on porous materials is suitable for fast recharging with hydrogen. The narrowest pores make the biggest contribution to hydrogen adsorption capacity whereas mesopores contribute to total pore volume, but little to hydrogen capacity, and are detrimental for volumetric capacity. Thus, porous materials with very narrow pores or pore size distributions are required for enhanced hydrogen capacity at low pressures.

Jin et al. (2007) proved that the hydrogen adsorption capacities depend almost linearly on porous structure parameters such as specific surface area and total pore volume (Figure 7). Due to the fast adsorption and desorption kinetics and almost complete reversibility, physical adsorption by activated carbon is a promising method for hydrogen storage at moderate temperature, however, the highest measured hydrogen adsorption capacity is less than 1 wt.% at 100 bars, 298°K, in spite of their highly developed porosities reaching up to 2800 m^2/g in specific surface area. New materials with ultra-high porosity, even higher than existing activated carbons, should be investigated to achieve sufficient hydrogen storage capacities at room temperature.

8 CONCLUDING REMARKS

Adsorption on activated carbons is widespread due to their versatile applications. It is a challenge to develop waste materials as precursors and to expand activated carbon fabrication processes which have limited environmental impact and are still affordable. Other factors that have to be taken

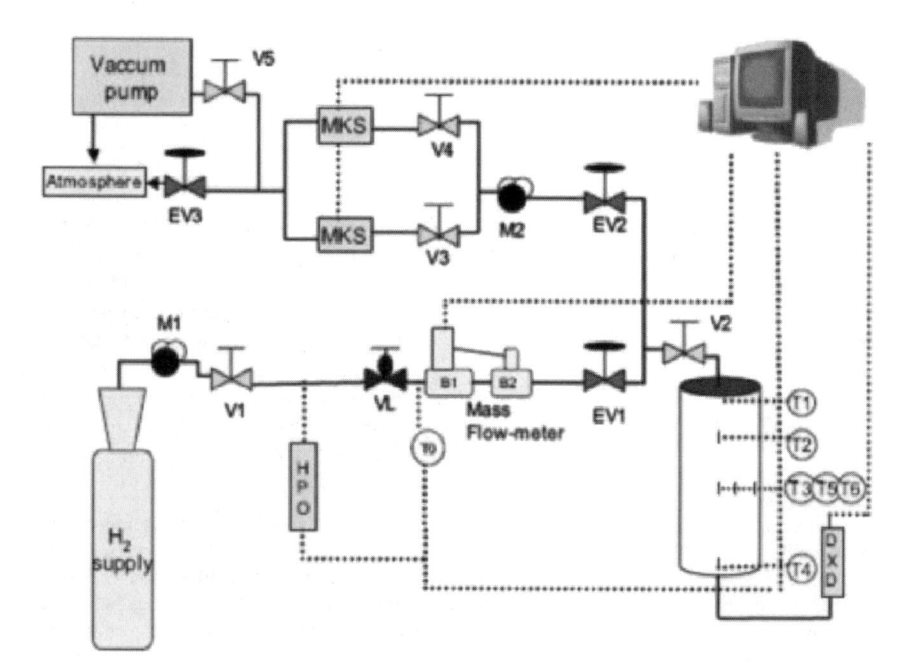

Figure 7. Experimental setup for hydrogen storage (Source: Hermosilla-Lara et al., 2007).

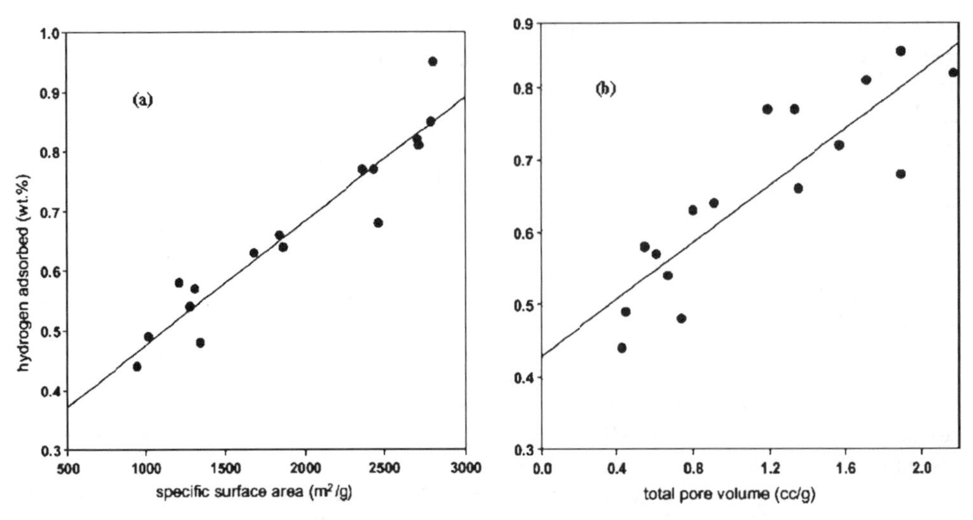

Figure 8. Effect of specific surface area (a) and total pore volume (b) on the hydrogen adsorption capacity of activated carbon at 100 bars, 298 K (Source: Jin et al., 2007).

in account include: the abundance and local availability of the precursor, and the impact of the selected activating process on the environment and its production cost at industrial scale. From a fundamental point of view, the factors (i.e. raw material composition, nature of oxidizing gas or chemical agent, temperature) governing the textural and chemical characteristics of the activated carbon obtained for a given application should be studied more rigorously. Bioregeneration is the most promising regeneration process from an environmental point of view. Research efforts should be directed at obtaining a better knowledge of the bioregeneration mechanism and for scaling up applications.

REFERENCES

Ahmedna M, Marshall WE, Rao RM. (2000). Production of granular activated carbons from select agricultural byproducts and evaluation of their physical, chemical and adsorption properties. Bioresource Technology 71: 113–123.

Amuda O.S., Giwa A.A., Bello I.A., (2007), Removal of heavy metal from industrial wastewater using modified activated coconut shell carbon, Biochemical Engineering Journal 36 (2): 2174–181.

Ania, C.O. Menéndez J.A., Parra J.B., Pis J.J. (2004), Microwave-induced regeneration of activated carbon polluted with phenol: A comparison with conventional thermal regeneration. Carbon 42: 1383–1387.

Ania C.O., Parra J.B. Menéndez J.A., Pis J.J. (2005). Effect of microwave and conventional regeneration on the microporous and mesoporous network and on the adsorptive capacity of activated carbons. Microporous and Mesoporous Materials 85: 7–15.

Ania C.O., Cabal B., Pevida C., Arenillas A., Parra J.B., Rubiera F., Pis J.J. (2007a), Removal of naphthalene from aqueous solution on chemically modified activated carbons, Water Research 41 (2): 333–340.

Ania C.O., Cabal B., Pevida C., Arenillas A., Parra J.B., Rubiera F. and Pis J.J., (2007b), Effects of activated carbon properties on the adsorption of naphthalene from aqueous solutions, Applied Surface Science 253 (13): 5741–5746.

Aktas Ö. and Ceçen F. (2007). Bioregeneration of activated carbon: A review. International Biodeterioration & Biodegradation, 59: 257–272.

Aygun A, Yenisoy-Karakas S, Duman I. (2003), Production of granular activated carbon from fruit stones and nutshells and evaluation of their physical, chemical and adsorption properties. Microporous and Mesoporous Materials 66: 189–195.

Ayotamuno M.J., Kogbara R.B., Ogaji S.O.T., Probert S.D., (2006), Petroleum contaminated ground-water: Remediation using activated carbon, Applied Energy, 83: 1258–1264.

Ayranci E., Duman O., (2005), Adsorption behaviors of some phenolic compounds onto high specific area activated carbon cloth, Journal of Hazardous Materials 124 (1–3): 125–132.
Baker, F. S. (2005) Patent US 20050167367.
Bénard, P. and Chahine, R. (2007). Storage of hydrogen by physisorption on carbon and nanostructured materials. Scripta Materialia 56: 803–808.
Brasquet C., Subrenat E., Le Cloirec P., (1999), Removal of phenolic compounds from aqueous solution by activated carbon cloths, Water Science and Technology, 39(10–11): 201–205.
Cal, M.P., Rood, M.J., Larson, S.M. (1997) Gas phase adsorption of volatile organic compounds and water vapor on activated carbon cloth Energy and Fuels 11 (2): 311–315.
Chan L.S., Cheung W.H., McKay G., (2008), Adsorption of acid dyes by bamboo derived activated carbon, Desalination, 218 (1–3): 304–312.
Chattopadhyaya, G., Macdonald, D.G., Bakhshi, N.N., Mohammadzadeh, J.S.S., Dalai, A.K. (2006) Adsorptive removal of sulfur dioxide by Saskatchewan lignite and its derivatives. Fuel 85 (12–13): 1803–1810.
Choi J.W., Choi N.C., Lee S. J., Kim D.J., (2007), Novel three-stage kinetic model for aqueous benzene adsorption on activated carbon, Journal of Colloid and Interface Science 314 (2): 367–372.
Cooney, D. O. (1999), Adsorption design for wastewater treatment, Lewis Publisher: Boca Raton, pp. 190
Dąbrowski, A., Podkościelny, P., Hubicki, Z., Barczak, M. (2005) Adsorption of phenolic compounds by activated carbon – A critical review. Chemosphere 58 (8): 1049–1070.
Daifullah, A.A.M., Yakout, S.M., Elreefy, S.A., (2007), Adsorption of fluoride in aqueous solutions using $KMnO_4$-modified activated carbon derived from steam pyrolysis of rice straw Journal of Hazardous Materials 147 (1–2): 633–643.
Dalai, A.K., Zaman, J., Hall, E.S., Tollefson, E.L., (1996) Preparation of activated carbon from Canadian coals using a fixed-bed reactor and a spouted bed-kiln system Fuel **25** (2): 227–237.
Derbyshire F, Jagtoyen M, Andrews R, Rao A, Martin-Gullon I, Grulke E, (2001). Carbon Materials in Environmental Applications. Marcel Decker, New York.
Devi R., Singh V., Kumar A., (2008), COD and BOD reduction from coffee processing wastewater using Avacado peel carbon, Bioresource Technology 99 (6): 1853–1860.
de la Puente G., J.A. Menéndez, (1998), On the distribution of oxygen-containing surface groups in carbons and their influence on the preparation of carbon-supported molybdenum catalysts, Solid State Ion. **112**: 103–111.
Dhab, F., (1987) Treatment alternatives for nitrate contaminated groundwater supplies. Environ. Syst. 17: 65–75.
Dias JM, Alvim-Ferraz MCM, Almeida MF, Rivera-Utrilla J, Sanchez-Polo M. (2007) Waste materials for activated carbon preparation and its use in aqueous-phase treatment: A review. Journal of Environmental Management 85: 833–846.
Dillion, A.C., Jones, K.M., Bekkedahl, T.A., Kiang, C.H., Bethune, D.S. and Heben, M.J. (1997). Storage of hydrogen in single-walled carbon nanotubes. Nature 386: 377–383.
Dwivedi, P., Gaur, V., Sharma, A., Verma, N. (2004) Comparative study of removal of volatile organic compounds by cryogenic condensation and adsorption by activated carbon fiber. Separation and Purification Technology 39 (1–2): 23–37.
El Qada E., Allen S. J., Walker G. M., (2008), Adsorption of basic dyes from aqueous solution onto activated carbons, Chemical Engineering Journal, 135 (3): 174–184.
Environmental Protection Agency, (2000), Wastewater Technology Fact Sheet, USA, EPA 832-F-00-017
Erdogan, F.O., Kopac, T (2007) Dynamic analysis of sorption of hydrogen in activated carbon, International Journal of Hydrogen Energy 32 (15): 3448–3456.
Feleke Z., Sakakibara Y., (2002) A bio-electrochemical reactor coupled with adsorber for the removal of nitrate and inhibitory pesticide, Water Research, 36: 3092–3102.
Feng, W., Kwon, S., Borguet, E., Vidic, R. (2005) Adsorption of hydrogen sulfide onto activated carbon fibers: Effect of pore structure and surface chemistry. Environmental Science and Technology 39 (24): 9744–9749.
Feng W., Eric Borguet and Radisav D. Vidic . (2006) Sulfurization of carbon surface for vapor phase mercury removal – I: Effect of temperature and sulfurization protocol Carbon, 44 (14): 2990–2997.
Figaro S., Louisy-Louis S., Lambert J., Ehrhardt J.-J., Ouensanga A., Gaspard S., (2006), Adsorption studies of recalcitrant compounds of molasses spentwash on activated carbons Water Research 40 (18): 3456–3466.
Finqueneisel G., Zimny T., Vogt D., Weber J.V., (1998), Feasibility of the preparation of effective cheap adsorbents from lignites in rotary kiln Fuel Process Technol **57**: 195–208.
Gadkaree K.P., Mach J.F., Stempin J.L. (2001) US Patent 6214204.
Gaspard S, Altenor S, Dawson EA, Barnes P, Ouensanga A. (2007). Activated carbon from vetiver roots: gas and liquid adsorption studies. Journal of Hazardous Materials 144: 73–80.

Gelderland SMR, Marra(2001), J: US Patent 2001/0052224A1.

Giles CH, da Silva AP, Trivedi AS., (1969), In: Proceeding of the International Symposium on Surface Area Determination. Bristol. pp. 317–323.

Goncalves F., J.L. Figueiredo, D (2004), Development of carbon supported metal catalysts for the simultaneous reduction of NO and N_2O, Appl. Catal., B Environ. **50**: 271–278.

Guo J, Lua AC. (1999), Textural and chemical characterisations of activated carbon prepared from oil-palm stone with H_2SO_4 and KOH impregnation. Microporous and Mesoporous Materials; 32: 111–117.

Gupta VK, Ali I, Mohan D. (2003). Equilibrium uptake and sorption dynamics for the removal of a basic dye (basic red) using low-cost adsorbents. Journal of Colloid and Interface Science; 265: 257–264.

Hamdaoui O., Naffrechoux E., Suptil J., Fachinger C. (2005). Ultrasonic desorption of p-chlorophenol from granular activated carbon. Chemical Engineering Journal, 106: 153–161.

Hameed, B.H., Din, A.T.M., Ahmad, A.L. (2007) Adsorption of methylene blue onto bamboo-based activated carbon: Kinetics and equilibrium studies Journal of Hazardous Materials 141 (3): 819–825.

Hermosilla-Lara, G, Momen, G., Marty, P.H., Le Neindre, B. and Hassouni, K. (2007). Hydrogen storage by adsorption on activated carbon: Investigation of the thermal effects during the charging process. International Journal of Hydrogen Energy, 32: 1542–1553.

Hynek, S., Fuller, W. and Bentley, J. (1997). Hydrogen storage by carbon sorption. International Journal of Hydrogen Energy, 22: 601–610.

Ilett, K.J., UK patent GB 2286 824 A.

Illán-Gómez M.J., Linares-Solano A., Radovic L.R. and Salinas-Martínez de Lecea C., (1995), NO reduction by activated carbons. 3. Some influence of catalyst loading on the catalytic effect of potassium, Energy Fuels **9**: 104–111.

Ioannidou O, Zabaniotou A. (2007) Agricultural residues as precursors for activated carbon production – A review. Renewable and Sustainable Energy Reviews 11: 1966–2005.

Jain AK, Gupta VK, Bhatnagar A. (2003), Utilization of industrial waste products as adsorbents for the removal of dyes. Journal of Hazardous Materials;101: 31–42.

Jin, H., Lee, Y.S. and Hong, I. (2007). Hydrogen adsorption characteristics of activated carbon. Catalysis Today, 120: 399–406.

Juang R, Tseng R, Wu F. (2001). Role of microporosity of activated carbons on their adsorption abilities for phenols and dyes. Adsorption; 7: 65–72.

Juang R, Wu F, Tseng R. (2002), Characterization and use of activated carbons prepared from bagasses for liquid-phase adsorption. Colloids and Surfaces A: Physicochemical and Engineering Aspects; 20: 191–199.

Kadirvelu K, Palanival M, Kalpana R, Rajeswari S. (2000). Activated carbon from an agricultural by-product, for the treatment of dyeing industry wastewater Bioresource Technology; 74: 263–265.

Kadirvelu K, Kavipriya M, Karthika C, Radhika M, Vennilamani N, Pattabhi S. (2003). Utilization of various agricultural wastes for activated carbon preparation and application for the removal of dyes and metal ions from aqueous solutions. Bioresource Technology; 87: 129–132.

Kannan N, Sundaram MM. (2001).Kinetics and mechanism of removal of methylene blue by adsorption on various carbons – a comparative study. Dyes and Pigments; 51: 25–40.

Kang K.C., Kim S. S., Choi J.W., Kwon S.H., (2008), Sorption of Cu^{2+} and Cd^{2+} onto acid- and base-pretreated granular activated carbon and activated carbon fiber samples, Journal of Industrial and Engineering Chemistry, 14 (1) pp. 131–135.

Karanfil T., Dastgheib S. A., (2006) US Patent 20060157419.

Kavitha D., Namasivayam C., (2007), Experimental and kinetic studies on methylene blue adsorption by coir pith carbon, Bioresource Technology, 98 (1): 14–21.

Kilduff JE, Karanfil T, Weber WJ, (1996).Competitive Interactions among Components of Humic Acids in Granular Activated Carbon Adsorption Systems: Effects of Solution Chemistry. Environ. Sci. Technol.; 30: 1344–1351.

Kouras A., Zouboulis A., Samara C., Kouimtzis Th., (1998), Removal of pesticides from aqueous solutions by combined physicochemical processes the behaviour of lindane, Environmental Pollution, 103 (2–3): 193–202.

Leakovic S., Mijatovic, I., Cerjan-Stefanovic, S., Hodzic, E., (2000) Nitrogen removal from fertilizer wastewater by ion exchange. Water Research, 34 (1): 185–190.

Lee S. H.D., Kumar R., Krumpelt M., (2002), Sulfur removal from diesel fuel-contaminated methanol, Separation and Purification Technology, 26 (2–3): 247–258.

Lesage N., Sperandio M., Cabassud C., (2008), Study of a hybrid process: Adsorption on activated carbon/membrane bioreactor for the treatment of an industrial wastewater, Chemical Engineering and Processing, 47 (3): 303–307.

Lillo-Ródenas M.A., Fletcher A.J., Thomas K.M., Cazorla-Amorós D., Linares-Solano A., (2006), Competitive adsorption of a benzene–toluene mixture on activated carbons at low concentration, Carbon, 44 (8): 1455–1463.

Lim, J.-L. Okada M. (2004). Regeneration of granular activated carbon using ultrasound, Ultrasonics Sonochemistry 12: 277–282.

Lopez Garzon FJ, Fernandez MI, Domingo Garcia M, Perez Mendoza MJ, Almaza Almazan M (2007): Patent ES2277565.

Lu, C.-Y., Wey, M.-Y. (2007) Simultaneous removal of VOC and NO by activated carbon impregnated with transition metal catalysts in combustion flue gas Fuel Processing Technology 88 (6): 557–567.

Maeland, A.J. and Skjeltorp, A.T. (2001). Hydrogen storage in carbon material. US Patent 6290753.

Mackenzie, J.A., Tennant, M.F., Mazyck, D.W. (2005) Tailored GAC for the effective control of 2-methylisoborneol. Journal American Water Works Association 97 (6): 76–87.

Malik PK. (2003).Dye removal from wastewater using activated carbon developed a case studyof Acid Yellow 36. Dyes and Pigments; 56: 239–249.

Maroto-Valer M. M., Dranca I., Clifford D., Lupascu T., Nasta R., Leon y Leon C. A. (2006). Thermal regeneration of activated carbons saturated with ortho- and meta-chlorophenols. Thermochimica Acta 444: 148–156.

Matatov-Meytal, Y.I., Sheintuch M., Shter G.E., Grader G.S. (1997). Optimal temperatures for catalytic regeneration of activated carbon. PII: S0008-6223, 00103-6.

Matatov-Meytal Y., Sheintuch M. (2000). Catalytic regeneration of chloroorganics-saturated activated carbons using hydrodechlorination. Ind. Eng. Chem. Res., 39: 18–23.

Mizuta K., Matsumoto T., Y. Hatate, Nishihara K., Nakanishi T., (2004) Removal of nitrate-nitrogen from drinking water using bamboo powder charcoal, Bioresource Technology, 95: 255–257.

Mohanty K., Das D., Biswas M. N., (2008), Treatment of phenolic wastewater in a novel multi-stage external loop airlift reactor using activated carbon Separation and Purification Technology, 58 (3): 311–319.

Mohammadi T., Esmaeelifar A., (2005), Wastewater treatment of a vegetable oil factory by a hybrid ultrafiltration-activated carbon process, Journal of Membrane Science, 254 (1–2): 129–137.

Moreno-Castilla C., Rivera-Utrilla J., Joly, J.P. López-Ramón M.V., Ferro-García M.A., Carrasco-Marín F. (1995). Thermal regeneration of an activated carbon exhausted with different substituted phenols. Carbon, 33 (10): 1417–1423.

Mukherjee S., Kumar S., Misra A. K., Fan M., (2007), Removal of phenols from water environment by activated carbon, bagasse ash and wood charcoal Chemical Engineering Journal, 129 (1–3): 133–142.

Nakagawa K., Mukai S. R., Suzuki T., Tamon H. (2003) Gas adsorption on activated carbons from PET mixtures with a metal salt Carbon, 41(4) Pages 823–831.

Nakagawa K, Namba A, Mukai SR, Tamon H, Ariyadejwanich P, Tanthapanichakoon W. (2004); Adsorption of phenol and reactive dye from aqueous solution on activated carbons derived from solid wastes. Water Research 38: 1791–1798.

Namasivayam C, Kavitha D. (2002).Removal of Congo red from water by adsorption onto activated carbon prepared from coir pith, an agricultural solid waste. Dyes and Pigments; 54: 47–58.

Namasivayam, C., Sangeetha, D., Gunasekaran, R., (2007), Process Safety and Environmental Protection, 85 (2 B): 181–184.

Namasivayam C, Kumar MD, Selvi K, Begum RA, Vanathi T, Yamuna RT. (2001), Waste' coir pith – a potential biomass for the treatment of dyeing wastewaters. Biomass and Bioenergy; 21: 477–483.

Nishino H, Suzuki M, Hirota H, (1983).US Patent 4409125.

Ormad M.P., Miguel N., Claver A., Matesanz J.M., Ovelleiro J.L., (2007), Pesticides removal in the process of drinking water production, Chemosphere, doi:10.1016/j.chemosphere.2007.10.006.

Okada K, Yamamoto N, Kameshima Y, Yasumori A. (2003) Adsorption properties of activated carbon from waste newspaper prepared by chemical and physical activation. Journal of Colloid and Interface Science; 262: 179–193.

Otero M, Rozada F, Calvo LF, Garcia, AI, Moran A. (2003) Kinetic and equilibrium modelling of the methylene blue removal from solution by adsorbent materials produced from sewage sludges. Dyes and Pigments; 57: 55–65.

Park S.J., Kim Y.M., (2005), Adsorption behaviors of heavy metal ions onto electrochemically oxidized activated carbon fibers, Materials Science and Engineering A, 391(1–2): 121–123.

Passe-Coutrin N., Altenor S., Cossement D., Jean-Marius C. and Gaspard S., (2008) Comparison of parameters calculated from the BET and Freundlich isotherms obtained by nitrogen adsorption on activated carbons: A new method for calculating the specific surface area, Microporous and Mesoporous Materials, doi:10.1016/j.micromeso.2007.08.032.

Pelech R., Milchert E., Wróblewska A. (2005). Desorption of chloroorganic compounds from a bed of activated carbon. Journal of Colloid and Interface Science, 285: 518–524.

Phan N. H., Rio S., Faur C., Le Coq L., Le Cloirec P., Nguyen T.H., (2006), Production of fibrous activated carbons from natural cellulose (jute, coconut) fibers for water treatment applications, Carbon, 44 (12): 2569–2577.

Qu X., Zheng J., Zhang Y., (2007), Catalytic ozonation of phenolic wastewater with activated carbon fiber in a fluid bed reactor Journal of Colloid and Interface Science, 309 (2): 429–434.

Rangel-Mendez, J.R., Cannon, F.S. (2005) Improved activated carbon by thermal treatment in methane and steam: Physicochemical influences on MIB sorption capacity Carbon 43 (3): 467–479.

Rajeshwarisivaraj R, Sivakumar S, Senthilkumar P, Subburam V. (2001) Carbon from cassava peel, an agricultural waste, as an adsorbent in the removal of dyes and metal ions from aqueous solution. Bioresource Technology; 80: 233–235.

Reckhow, D. A., Singer, P. C., Malcolm R. M., (1990), Chlorination of humic materials: Byproduct formation and chemical interpretations, Environmental Science and Technology; 24: 1655–1664.

Rio S, Le Coq L, Faur C, Le Cloirec P. (2006) Production of porous carbonaceous adsorbent from physical activation of sewage sludge: Application to wastewater treatment Water Science and Technology; 53: 237–244.

Robinson KK (1990), US Patent 4954469.

Ross, D.K., (2006). Hydrogen storage: The major technological barrier to the development of hydrogen fuel cell cars. Vacuum, 80: 1084–1089.

Rozada F, Otero M, Parra JB, Moran A, Garcia, AI. (2005). Producing adsorbents from sewage sludge and discarded tyres: Characterization and utilization for the removal of pollutants from water. Chemical Engineering Journal; 114: 161–169.

Sakintuna B., Yurum Y., Cetunkaya S., (2004), Evolution of Carbon Microstructures during the Pyrolysis of Turkish Elbistan Lignite in the Temperature Range 700–1000°C Energy Fuels 18: 883–888.

Slaney A.J., Bhamidimarri R., (1998), Adsorption of pentachlorophenol (PCP) by activated carbon in fixed beds: Application of homogeneous surface diffusion model, Water Science and Technology, 38 (7): 227–235.

Shaalan,H. F., Montaser Y. Ghaly, Joseph Y. Farah, (2007), Techno economic evaluation for the treatment of pesticide industry effluents using membrane schemes, Desalination; 204 (1–3): 265–276.

Shin H.C., Park J.W., K. Park, Song H.C., (2002), Removal characteristics of trace compounds of landfill gas by activated carbon adsorption, Environmental Pollution, 119 (2): 227–236.

Singh K. P., Malik A., Sinha S., Ojha P., (2008), Liquid-phase adsorption of phenols using activated carbons derived from agricultural waste material, Journal of Hazardous Materials; 150 (3): 626–641.

Sirianuntapiboon S., Ungkaprasatcha O., (2007), Removal of Pb^{2+} and Ni^{2+} by bio-sludge in sequencing batch reactor (SBR) and granular activated carbon-SBR (GAC-SBR) systems, Bioresource Technology, 98 (14): 2749–2757.

Starvin A.M. and Rao T. P., (2004), Removal and recovery of mercury(II) from hazardous wastes using 1-(2-thiazolylazo)-2-naphthol functionalized activated carbon as solid phase extractant, Journal of Hazardous Materials, 113, (1–3) 75–79.

Takagi, H., Hatori, H., Soneda, Y., Yoshizawa, N. and Yamada Y. (2004). Adsorptive hydrogen storage in carbon and porous materials. Materials Science and Engineering, B108: 143–147.

Thinakaran N., Baskaralingam P., Pulikesi M., Panneerselvam P., Sivanesan S., (2008), Removal of Acid Violet 17 from aqueous solutions by adsorption onto activated carbon prepared from sunflower seed hull, Journal of Hazardous Materials, 151 (2–3) 316–322.

Thomas, K.M. Hydrogen adsorption and storage on porous materials 2007 Catalysis Today 120 (3–4): 389–398.

Tsai WT, Chang CY, Lin MC, Chien SF, Sun HF, Hsieh MF. (2001) Adsorption of acid dye onto activated carbons prepared from agricultural waste bagasse by ZnCl2 activation. Chemosphere; 45: 51–58

Tseng R, Wu F, Juang R. (2003) Liquid phase adsorption of dyes and phenols using pinewood-based activated carbons. Carbon; 41: 487–495.

Valix M, Cheung WH, McKay G. (2004) Roles of textural and surface chemical properties of activated carbon in adsorption of acid blue dye. Chemosphere; 56: 493–501.

Valderrama C., Cortina J.L., Farran A., Gamisans X., Lao C., (2007), Kinetics of sorption of polyaromatic hydrocarbons onto granular activated carbon and Macronet hyper-cross-linked polymers (MN200), Journal of Colloid and Interface Science, 310 (1): 35–46.

Vazquez I., Rodrıguez-Iglesias J., Maranon ECastrillon L ., Alvarez. M., (2007), Removal of residual phenols from coke wastewater by adsorption, Journal of Hazardous Materials, 147, (1–2): 395–400.

Wang S., Zhu Z.H., Coomes A., Haghseresht F., Lu G.Q., (2005), The physical and surface chemical characteristics of activated carbons and the adsorption of methylene blue from wastewater, Journal of Colloid and Interface Science, 284 (2): 440–446.

Wasik, E., Bohdziewicz, J., Blaszczyk, M., (2001) Removal of nitrates from ground water by a hybrid process of biological denitrification and microfiltration membrane. Proc. Biochem. 37: 57–64.

Weber Jr. W.J., Voice T.C., Jodellah A., (1983) Adsorption of humic substances: the effects of heterogeneity and system characteristics. J. Am. Water Works Assoc.; **75**: 612–618.

Wibowo N., Setyadhi L., Wibowo D., Setiawan J., Ismadji S., (2007), Adsorption of benzene and toluene from aqueous solutions onto activated carbon and its acid and heat treated forms: Influence of surface chemistry on adsorption, Journal of Hazardous Materials, 146 (1–2): 237–242.

Wu, F.C., Tseng, R.L., Juang, R.S. (1999) Preparation of activated carbons from bamboo and their adsorption abilities for dyes and phenol Journal of Environmental Science and Health – Part A Toxic/Hazardous Substances and Environmental Engineering 34 (9): 1753–1775.

Xu, W.C., Takahashi, K., Matsuo, Y., Hattori, Y., Kumagai, M., Ishiyama, S., Kaneko, K. and Iijimad, S. 2007. Investigation of hydrogen storage capacity of various carbon materials. International Journal of Hydrogen Energy, 32: 2504–2512.

Zabaniotou A., Stavropoulos G., Skoulou V., (2008), Activated carbon from olive kernels in a two-stage process: Industrial improvement, Bioresource Technology, 99 (2): 320–326.

Zhang H. (2002). Regeneration of exhausted activated carbons by electrochemical method. Chemical Engineering Journal, 85: 81–85.

Zhou, M.H., Lei, L.C. (2006) Electrochemical regeneration of activated carbon loaded with p-nitrophenol in a fluidized electrochemical reactor. Electrochimica Acta 51 (21): 4489–4496.

Utilization of Recycled Tires in Hot Mix Asphalt

Serji Amirkhanian, Khaldoun Shatanawi & Carl Thodesen
Clemson University, SC, USA

SUMMARY: The aim of the current chapter was to assess the utilization of recycled tires in hot mix asphalt. The blending of ground recycled tires with asphalt binder has been shown in both field and laboratory evaluations to significantly improve the performance of asphalt binders. Specifically, the high temperature performance grade of the asphalt binder is seen to increase significantly with the addition of crumb rubber. Numerous field evaluations have shown that the use of rubberized asphalt, while more expensive, leads to improved pavements requiring less maintenance. From an environmental standpoint, the benefits of rubberized asphalt are also significant. Within the scientific community, there is widespread agreement that rubberized asphalt produces superior performing asphalt compared to conventional binder. As such, the future of paving materials lies with polymer modified asphalt as these produce more durable pavements, which in the long run require less maintenance thus costing less. Future research in this field might involve the incorporation of various recycled materials and other technologies, doing so will provide superior pavements which are also more sustainable.

1 INTRODUCTION

In 1999, lightning struck a tire dump in Westley, California; the resulting smoke plume impacted nearby farming communities and caused widespread concern of potential health affects from exposure to the smoke emissions (Environmental Protection Agency, 2007). The tire fire produced large quantities of pyrolitic oil which flowed off the slope and into the drainage of a nearby stream. The pyrolitic oil was also ignited and caused significant smoke emissions on the ground due to the raging oil fire. Local and state agencies were unable to respond to the oil and tire fires, thus requiring the EPA regional coordinator to intervene using the Oil Pollution Act of 1990. The tire fire lasted for 30 days and the EPA response costs were estimated to be $3.5 million (Environmental Protection Agency, 2007).

While tire fires are infrequent, they cause a serious concern to public safety as well as being expensive to remedy. This danger becomes more apparent as it is estimated that approximately 300 million scrap tires are generated annually in the United States (Rubber Manufacturers Association, 2006). This places the generation of scrap tires at approximately one tire per person per year. The number of open landfills in the United States has been in a steady decline, more than 75% of all landfills have been closed within the past 18 years (Environmental Protection Agency, 2006). As such landfilling can no longer be considered a suitable, or sustainable, disposal practice.

Farris reports that rubber has been one of the most useful materials of the modern era. It helped create the industrial revolution, and was prized for its strength, elasticity, and wear-resistance in industry. However, due to its popularity, rubber also represents one of the most difficult recycling problems ever encountered (Farris, 2003).

American Society for Testing and Materials (ASTM) D6114 defines asphalt as "a dark brown to black cementitious material of solid or semisolid consistency at ambient temperatures in which the predominant constituents are bitumens which occur in nature as such or are obtained as a residue by refining petroleum." (ASTM, 2001a) The first modern asphalt road was built in Paris in 1858; this was followed by the first US road which was built in New Jersey in 1870. The reason for the

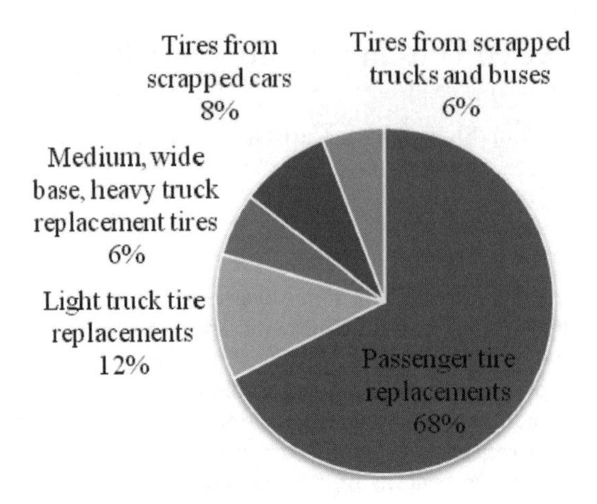

Figure 1. Scrap tire generation sources (RMA, 2006).

adoption of asphalt as a construction material lies in its engineering properties. It is a thermoplastic material which forms a strong cement with excellent adhesive properties. In addition to being a durable and water proof material, asphalt is also resistant to most acids, alkalis, and salts.

Due to the many excellent engineering properties attributed to asphalt, this material has become the principal construction material for the construction of roads and highways. However, as the volume of cars has increased so too has the stress on the roads. These stresses have led to increased road deterioration and the subsequent need to develop new materials to withstand the increased loading. Polymer Modified Asphalts (PMAs) have been introduced as a solution to this issue; these binders incorporate the use of polymers thus leading to increased viscosities at high pavement temperatures. These polymers have also been shown to increase ductility at low pavement temperatures and thus reducing the tendency for thermal cracking. These effects combine for an extended pavement life, and therefore an improved pavement material (Ruan et al., 2003).

The aim of the current chapter was to assess the utilization of recycled tires in hot mix asphalt. Specifically, by identifying the properties and benefits of crumb rubber modified asphalt, the importance of the crumb rubber grinding procedure, and the environmental impact of using crumb rubber modified binder.

2 CRUMB RUBBER MODIFIED BINDER (CRM)

The process of using crumb rubber, derived from the shredding of scrap tires, in asphalt was first developed in the 1960's as a method to improve the use of asphalt for surface treatments. In the 1970's, crumb rubber as a modifier started to be utilized in the production of hot mix asphalt producing what is known today as rubberized asphalt (Arizona Department of Transportation, 1989). Crumb rubber modified asphalt pavements have been widely used due to their proven benefits in reducing pavement performance cracking (Way, 2000; Dantas Neto et al., 2003; Amirkhanian, 2003; Lee et al., 2006a; Xiao et al., 2007). The addition of crumb rubber modifier (CRM) to asphalt has also been shown to be an effective method of increasing the performance grade of the asphalt binder, improving the high temperature properties, decreasing susceptibility to permanent deformation, and providing resistance to reflective cracking (Way, 2000; Dantas Neto et al., 2003; Amirkhanian, 2003; Lee et al., 2006a; Xiao et al., 2007). It also provides an environmentally friendly option for the disposal of the scrap tires (Takallou, 1991). As seen in Figure 1, there are many sources contributing to the production of scrap tires.

(a)

(b)

Figure 2. SEM micrographs of (a) Cryogenically and (b) Ambient Ground CRM at 30x magnification.

The two main procedures used to design and construct asphalt rubber mixtures are the "wet" and "dry" processes. In the wet process, finely ground crumb rubber (5 to 20% of rubber by total weight of the binder) is added to the asphalt cement. The dry process involves using crumb rubber (approximately 3% of total mixture weight) as a portion of the aggregate. The wet process is today the more popular method of crumb rubber modification, reported advantages of using this procedure include: Longer overall pavement life; Increased crack resistance; Increased rut resistance; Reduced oxidation (slower aging); Reduced maintenance needs/cost; Reduced traffic noise and increased skid resistance.

While standard Hot Mix Asphalt (HMA) production, paving, and compaction equipment may be utilized, experience has shown that the wet process, in general requires: Higher initial cost (compared to virgin binder); Experienced contractors in early stages of agency implementation; Use of blending unit or coordination with supplier for pre-blended binder; Agitated binder storage tank for rubber-modified liquid binder at HMA plant; and slightly higher production temperatures.

2.1 *Crumb rubber production*

Crumb rubber is produced by grinding scrap tires into fine particles, thus resulting in tire rubber with a granular consistency. The grinding procedure is typically done by ambient grinding of the scrap tires or by subjecting the scrap tire to extreme low temperatures through cryogenic freezing and subsequently shattering them. The effect of the grinding has been shown to produce crumb rubber particles of differing morphologies. From Figure 2, it can be seen that the cryogenically ground crumb rubber particles exhibit more of a crystalline morphology when compared to the ambient ground crumb rubber particles. Research has also shown that the differences in morphology account for a significant difference in surface areas, where ambient ground crumb rubber typically exhibits surface areas approximately 2.5 times greater than cryogenic ground crumb rubber (Putman, 2005; Putman et al., 2006).

2.2 *Crumb rubber gradation*

The gradation of the crumb rubber refers to the distribution of the various particle sizes contained within the given crumb rubber sample. Therefore, the gradation is an indication of the "coarseness" of "fineness" of the sample. The test method used to determine the gradation of the crumb rubber is specified by ASTM D5644 (ASTM, 2001b).

An example of the rubber gradation usually used in asphalt pavements are those of South Carolina and Arizona Departments of Transportation (SCDOT and ADOT, respectively) presented in Table 1. Here it can be seen that the gradation is presented in the form of an upper and lower limit. In order for a sample to be considered passing the DOT requirements it is necessary for the experimental gradation to fall within the upper and lower specification. Generally, the gradation of the material is specified upon ordering the material from the manufacturer, however, QC/QA procedures dictate that further tests must be performed to ensure the uniformity of the material.

From Table 1, it can be seen that the SCDOT materials tends to be finer and more uniform in nature, while the ADOT specified material tends to exhibit a wider range of sizes. As discussed

Table 1. Crumb rubber gradations for (a) ADOT and (b) SCDOT.

(a)

Sieve Number	No. 10	No. 16	No. 30	No. 50	No. 200
Opening size (mm)	2.000	1.190	0.600	0.300	0.075
Upper Specification (% passing)	100%	100%	100%	45%	5%
Lower Specification (% passing)	100%	65%	20%	0%	0%

(b)

Sieve Number	No. 20	No. 40	No. 80	No. 100
Opening size (mm)	0.850	0.425	0.180	0.150
Upper Specification (% passing)	100%	100%	50%	30%
Lower Specification (% passing)	100%	85%	10%	5%

in the laboratory studies section of this chapter, the gradation has been shown to exhibit differing effects on the modified binder properties.

2.3 *Asphalt rubber wet process interaction*

The nature of the asphalt rubber interaction in the wet process occurs through two principal mechanisms: particle swelling and degradation (this may include depolymerization and degradation) (Zanzotto & Kennepohl, 1996; Abdelrahman & Carpenter, 1999). The reaction between rubber and asphalt binder occurring during the "wet" blending process is not a chemical one (Heitzman, 1992), rather it is an absorption of aromatic oils form the asphalt cement into the polymer chains (key components of the CRM). During the blending process, the rubber particles are not melted in the asphalt cement, instead the rubber particles tend to swell two to three times due to absorption at high temperature of the asphalts oily phase (Abdelrahman, 2006).

Abdelrahman states the swelling experienced by the rubber is a function of time and temperature. As seen in Figure 3, there is a clear relationship between binder viscosity and variations in the particle size, it is also apparent form this figure that as time goes by, the particle will experience swelling, depolymerization, and finally contribute to a more homogeneous mixture (Abdelrahman, 2006).

The temperature and time at which the asphalt and rubber are blended are particularly important, as these affect the interaction process (Green & Tolonen, 1977). Specifically, as the temperature increases so too do the rate and extent of swelling of the rubber particles. This is especially true for fine rubbers, these tend to swell and depolymerize faster thus exhibiting a greater effect on the liquid phase of the binder. Coarse crumb rubber particles tend to have more effect on the binder matrix, but less effect on the liquid phase. Generally, liquid phase modifications tend to be more stable than matrix modifications (Abdelrahman & Carpenter, 1999; Abdelrahman, 2006).

2.4 *States currently using CRM asphalt*

A number of states have recognized the environmental and engineering advantages that asphalt rubber provides. States currently using asphalt rubber include: Arizona, California, Florida,

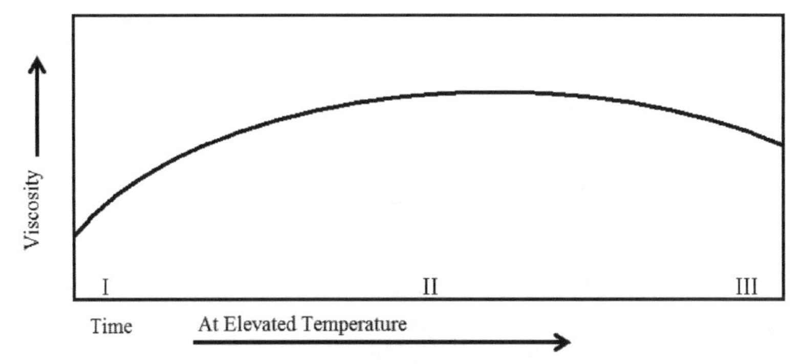

(a) Change of Binder Viscosity over Time at Elevate Temperature

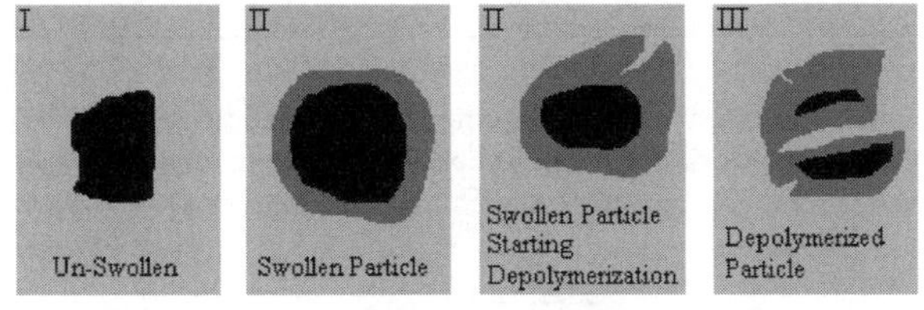

(b) Change of Particle Size over Time at Elevated Temperature

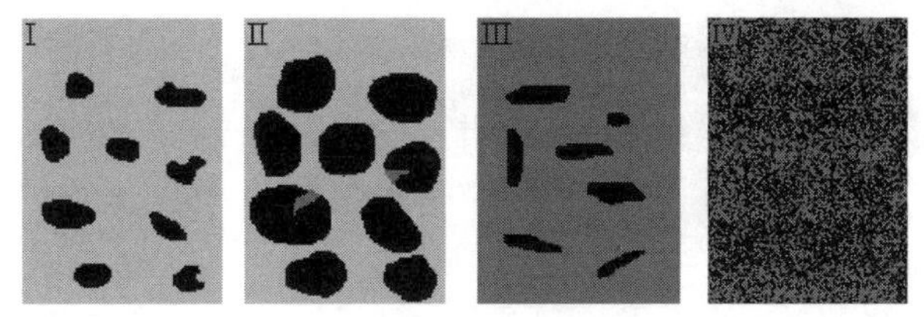

(c) Change of Binder Matrix over Time at Elevated Temperature

Figure 3. Progression of the asphalt-rubber interaction at elevated temperature (Abdelrahman, 2006).

South Carolina, Rhode Island, and Texas (Table 2). Of these, Rhode Island is the only state to use chemically-modified crumb rubber (CMCR) in asphalt.

As seen in Table 2, many states have different specifications for various rubberized asphalt mixtures. Several states utilize open graded friction courses (OGFCs) for many reasons. As seen in Figure 4, OGFC refers to a type of asphalt pavement which consists of an HMA which is designed to have a large number of air voids, thus permitting water to easily drain through the surface of the mixture. The advantages of such pavements are due to the rapid removal of storm water, and the consequent decreased threat of hydroplaning and increased skid resistance. Similarly, the flexible paving industry has identified rubberized asphalt as a suitable material for rubber-modified surface courses (RM SCs), dense-graded friction courses (DGFCs), gap graded friction courses (GGFCs), and rubber-modified open-graded friction courses (RM OGFC). Typically OGFCs are composed of approximately 93% aggregate, 7% asphalt binder, and a small amount of fibers, where a range of 12% to 20% rubber by weight of virgin binder is commonly used depending on the requirements.

2.5 *Field mixing procedure*

Asphalts may be modified with crumb rubber in situ; this method of modification is commonly called blending. Field blending of the asphalt binder is accomplished with equipment similar to the

Table 2. Types of asphalt rubber and uses in various states.

State	Applications used	% Rubber by weight of binder	Crumb rubber particle size
Arizona	GGFC and OGFC	20%	1.18 mm (#16 mesh)
California	GGFC and OGFC	18%–22%	1.18 mm (#16 mesh)
Florida	DGFC, OGFC, and SAMI (ARMI)	5%, 12%, and 20%	850 μm (#20 mesh) 425 μm (#40 mesh) and 210 μm (#80 mesh)
Rhode Island	OGFC and SAM	7% and 20%	850 μm (#20 mesh) and 210 μm (#80 mesh)
South Carolina	DGFC, OGFC, and SAMI	10%, 12%, and 20%	425 μm (#40 mesh)
Texas	DGFC, GGFC, OGFC and SAM	5% or 15%+	1.18 mm (#16 mesh) and minus #16 mesh

Key:
GGFC = Gap Graded Friction Course SAMI = Stress Absorbing Membrane Interlayer
OGFC = Open Graded Friction Course ARMI = Asphalt Rubber Membrane Interlayer
DGFC = Dense Graded Friction Course SAM = Stress Absorbing Membrane

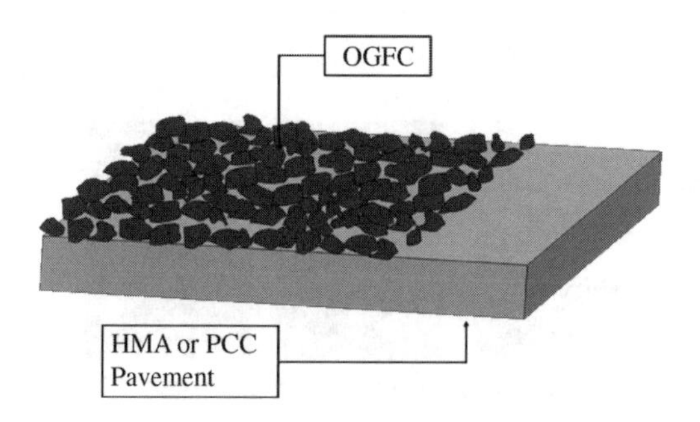

Figure 4. Open Graded Friction Course (California Department of Transportation, 2003).

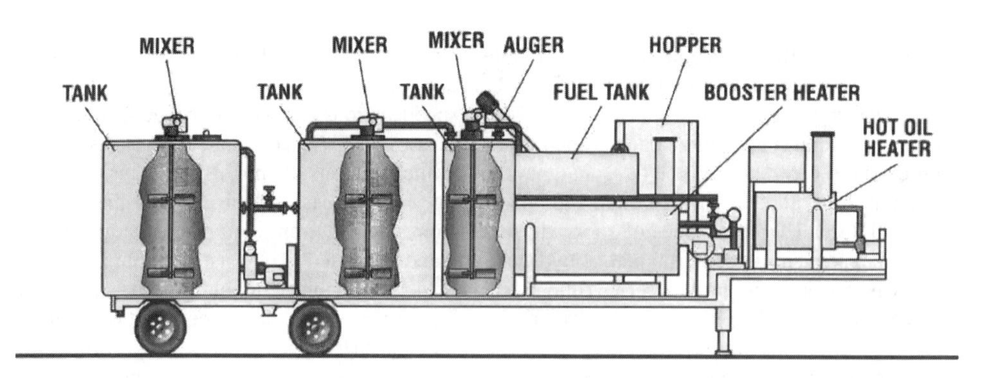

Figure 5. Components of a batch system for blending asphalt and crumb rubber (Courtesy of Heatec).

field blending unit shown in Figure 5. Blending of the binder with the crumb rubber is initiated by pumping the virgin asphalt binder (typically PG 64-22) into the booster heater and elevating it to the proper temperature. Next, the crumb rubber modifier (CRM) is transferred to the rubber hopper from 900 kg "super sacks" which open from the bottom. The appropriate amount of crumb rubber to be added is determined before hand and added accordingly. Due to the drop in binder temperature upon addition of the rubber, it is recommended that the virgin binder be preheated beyond 177°C (350°F). Failure to do so may result in a lower blending temperature, and consequently produce inappropriate mixing.

Blending is usually performed at a binder temperature of 177°C (350°F); this has been shown by laboratory studies to be the optimum blending temperature without causing excess oxidation. During blending, the rubber auger is set to the appropriate speed and the required binder temperature maintained. The rubber and virgin asphalt binder are initially combined in the mixing tank, upon completion of mixing the rubberized asphalt then overflows into one of the two larger agitated holding tanks. The rubberized asphalt is then stored until the mix is ready to be utilized on the paving surface.

3 PROPERTIES OF CRUMB RUBBER MODIFIED ASPHALT

In this section a number of common asphalt pavement distresses are described, the benefits of utilizing rubberized asphalt as a way of mitigating these phenomena are also discussed. Some of the distresses covered include reflective cracking, longitudinal cracking, permanent deformation, and thermal cracking. The use of rubberized asphalt as a noise mitigation tool is also introduced.

3.1 *Field applications and properties*

3.1.1 *Reflective cracking*

The occurrence of reflective cracking in flexible pavements is due to discontinuities in the underlying layers which are transferred through the HMA surface due to movement at the crack (Roberts et al., 1996). Roberts et al. report that possible causes contributing to reflective cracking include: Cracks or joints in an underlying concrete pavement; Low temperature cracks in the old HMA surface; Block cracks induced by the old HMA surface or those induced by subgrade soil cracking due to shrinkage whether stabilized or not; Longitudinal cracks in the old surface; or fatigue cracks in the old surface.

As with all cracking in asphalt pavements, reflective cracking is a serious problem which if untreated will contribute to the failure of the pavement. Pavement cracking will lead to more serious distresses due to the increased permeation of water and the decreased strength of the pavement following initial crack propagation. For these reasons it is of the utmost importance that pavements should be designed against cracking.

Rubberized asphalt provides a solution for both preemptively designing against cracking and also as a rehabilitation option. Figure 6 shows a photograph of a study performed in Arizona where one direction of old concrete pavement highway was overlaid with rubberized asphalt and the other using conventional asphalt. It can clearly be seen that the section overlaid with conventional asphalt was prone to the repetitive cracking commonly associated with reflective cracking, while the rubberized asphalt section did not experience any cracking on the main highway. The direction of highway utilizing rubberized asphalt only experienced reflective cracking on the shoulder; this was the part of the pavement which was paved with conventional asphalt too.

3.1.2 *SAM and SAMI*

The use of stress absorbing membrane interlayers (SAMIs) is one method of retarding reflective cracking; this procedure involves coating the surface requiring rehabilitation with a thin layer of rubberized asphalt. The purpose of this layer is to coat the existing layer with a very flexible, soft layer of rubberized asphalt which will not readily transmit the stress produced by the horizontal movements of the underlying layer to the HMA layer (Roberts et al., 1996). The soft, strain absorbing interlayer is expected to absorb high strains without cracking, thus preventing crack propagation of the crack tip from the old surface to the HMA overlay.

As seen in Figure 7, a SAMI will typically consist of a layer of rubber-modified asphalt binder, followed by a layer of crushed stone, topped by a layer of a new HMA. Typically, the rubberized

Figure 6. I-40 Arizona after 8 Years Performance Conventional Overlay on Left, Rubber on Right (Courtesy of ADOT).

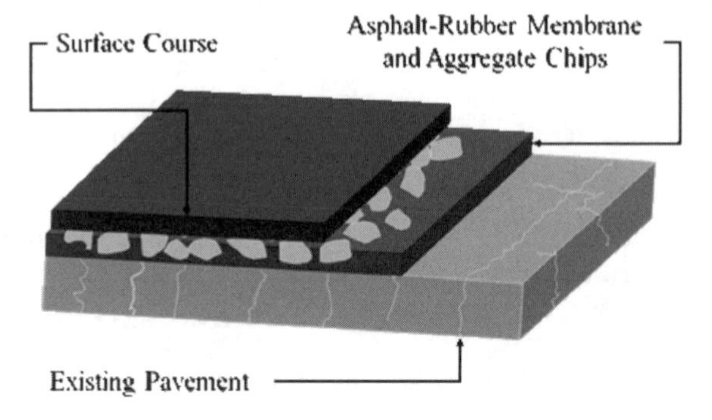

Figure 7. Stress Absorbing Membrane Interlayer (SAMI) (California Department of Transportation, 2003).

asphalt used in SAMI application will contain approximately 20% to 25% rubber by weight of virgin binder. Using this concentration of rubber in the SAMI equates to approximately 725 to 1,115 tires per lane-mile with a conventional HMA overlay. In the event an asphalt pavement was rehabilitated with a rubberized asphalt overlay and a SAMI, it is estimated that between 1,075 to 2,975 tires per lane-mile will be utilized.

As seen in Figure 8, field applications have confirmed the effectiveness of SAMI and SAM at combating the damaging effects of reflective cracking. Specifically, the photograph of the case study in Lubbock, Texas illustrates the improvement in physical appearance, rideability, and consequentially also the reduced maintenance of pavements incorporating a SAM or SAMI.

3.1.3 *Longitudinal cracking*

Longitudinal cracking is a phenomenon whereby cracking occurs predominantly parallel to the centerline of the pavement. These tend to occur at the joint between adjacent lanes of asphalt

Figure 8. Lubbock Texas (a) 1985 before SAM application (b) 2000, after 15 years performance (Courtesy of TXDOT).

Figure 9. Florida SR-16 after 8 Years Performance, (a) with 10% crumb rubber, and (b) control section (Courtesy of FDOT).

mixture or at the edges of the wheel paths in a rutted pavement (Roberts et al., 1996). Such distresses are damaging to the asphalt pavement as they tend to lead to raveling of HMA mixture adjacent to the longitudinal crack. This in turn tends to make the crack widen and ultimately accelerate the damage to the overall structure.

As seen in Figure 9, the use of rubberized asphalt has been documented to reduce the amount of longitudinal cracking taking place in pavements. In this study two test sections were placed along Florida's SR-16 highway, one test section was modified with 10% crumb rubber, while the control

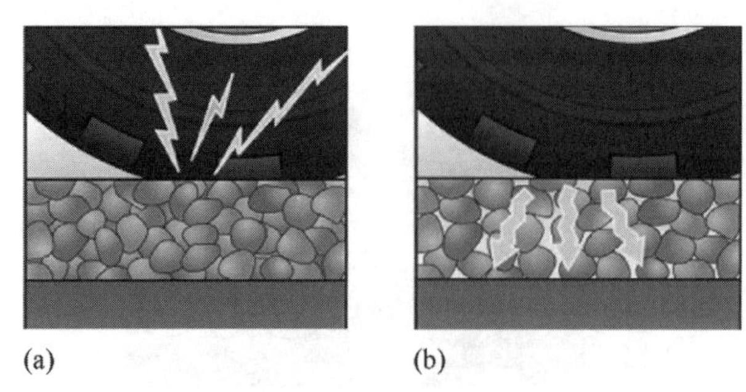

Figure 10. Noise emission in (a) a dense pavement and (b) in a rubberized asphalt porous pavement (Courtesy of Hokkaido Regional Development Bureau).

strip was made using only conventional HMA. It can clearly be seen in Figure 9 that the control strip exhibited more longitudinal cracking than the strip containing 10% crumb rubber.

Reducing pavement cracking is of the utmost importance to the pavement's health, as cracking leads to more severe problems. Therefore, many state agencies suggest rehabilitating the cracks during their early stages; such rehabilitations tend to provide more cost effective solutions in the long term as well as presenting better quality roads in the short term.

3.1.4 *Noise reduction*

Noise is defined as unwanted and or excessive sound. At a minimum, it can cause discomfort, and it has the potential to cause severe physical and psychological damages. Historically, traffic noise has been identified as a continuous environmental problem that affects the quality of human life. The earliest problem of traffic noise documented to human kind goes back to 44 BC (Schafer, 1994).

For many reasons (e.g., world population), both the number and size of vehicles have increased. Even though newer vehicles are designed to produce lower noise levels, the increase in vehicle density has kept traffic noise as a major setback in the improvement of life quality.

The major problem with the current techniques, such as wall barriers, used to control and reduce traffic noise is that the cost of one mile of sound wall barrier is estimated to range from 1 to 5 million dollars (Hanson et al., 2004). Also, most wall barriers tend to increase the level of noise for the highway user, since these barriers are of very high acoustic reflectivity (above 95%) (Anon, 1974).

Field applications have shown a significant reduction in noise levels when rubberized asphalt has been used as a paving material. The exact cause of the crumb rubber particle on noise reduction has not been identified yet; however, many researchers believe that the increased air voids present in rubberized asphalt may contribute to the reduction in noise emissions (Figure 10). One such example is the Superstition Freeway project in Phoenix, Arizona. As part of the reconstruction and widening of a 10-mile section of the road, the Arizona Department of Transportation (ADOT) placed a one-inch, rubberized open-graded friction course on top of the concrete pavement. As a result, the noise reported generated by traffic was reduced significantly by more than 5 decibels.

The Sacramento County Department of Environmental Review and Assessment performed a six year study on crumb rubber modified asphalts which ended in 1999. This study concluded that the crumb rubber modified asphalt pavement continued to decrease the traffic noise emissions even after six years (California Department of Transportation, 2003).

Despite the different techniques currently applied to control noise pollution, most of the present practices only seem to hide the problem from a certain group of people (residents near the roadway) rather than eliminate the problem from its source. The efforts of both the automotive and tire industries have met the consumer demand for quieter products; however, highway agencies have not been able to make a quieter product without a considerable cost increase. Therefore, the rubberized asphalt might offer an alternative solution to this problem.

Table 3. SHRP binder tests and purpose.

Equipment	Purpose	Performance parameter
Rolling Thin Film Oven (RTFO)	Simulate binder aging during HMA production and construction	Resistance to aging during construction
Pressure Aging Vessel (PAV)	Simulate binder aging during HMA service life	Resistance to aging during service life
Rotational Viscometer (RV)	Measure binder properties at high construction temperatures	Handling and pumping
Dynamic Shear Rheometer (DSR)	Measure binder properties at high and intermediate service temperatures	Resistance to permanent deformation and fatigue cracking
Bending Beam Rheometer (BBR)	Measure binder properties at low service temperatures	Resistance to thermal cracking
Direct tension tester (DTT)	Measure binder properties at low service temperatures	Resistance to thermal cracking

3.1.5 *Resistance to thermal cracking*

Thermal cracking of asphalt pavements occurs due transverse shrinkage cracking in the HMA layer due to low temperatures. This type of cracking is nonload associated, and occurs when rapid cooling rates and low temperatures develop tensile stresses due to shrinkage. Transverse cracking will occur if these stresses exceed the fracture strength of the HMA pavement layer (Roberts et al., 1996). As seen in the laboratory evaluation section for thermal cracking, some research has been done in an effort to evaluate the effects of crumb rubber modification on the cold weather properties of rubberized asphalt.

3.2 *Laboratory properties*

Following the completion of the Strategic Highway Research Program (SHRP), a new system of classifying asphalt materials was introduced in the United States. The procedure was called the Superpave (Superior Performing Pavements) system, and was designed for use in conjunction with increasingly complex tests and specifications (Table 3). Numerous studies have also been performed on rubberized asphalt mixtures using these testing methods. A brief summary is included in the following section.

3.2.1 *Handling and pumping*

AASHTO T 316 is the commonly used SHRP procedure for evaluating asphalt binder viscosity (AASHTO, 2006a). Achieving asphalt viscosity requirements is of utmost importance for ease of pumping as asphalt is generally stored in asphalt plants at temperatures between 149°C and 177°C depending on the grade or viscosity. However, fulfilling these requirements becomes more difficult with the increasing viscosity due to modification by crumb rubber (Stroup-Gardiner et al., 1993) as well as the specifications established by SHRP indicating that asphalt viscosity should not exceed 3.0 Pa.s at 135°C (Asphalt Institute, 2003).

Studies have shown that ambient CRM binder viscosity is typically higher than that of cryogenic CRM binder (Thodesen et al., 2008). As crumb rubber concentration increased the effect of the crumb rubber is more significant. While particle size (0.18–0.85 mm) has been shown to play a small role in dictating the binder viscosity; generally no significant differences were present between coarser (0.85 and 0.425 mm) ambient and cryogenic CRM binder viscosities. This study also indicated that when asphalt was modified with the finest crumb rubber (0.18 mm) the ambient CRM binder produced statistically significant higher viscosities than the cryogenic CRM binder (Thodesen et al., 2008).

In another study conducted by Lougheed and Papagiannakis, it was concluded that viscosity increases in rubberized asphalt occur due to the amount of aromatic oil absorption and rubber particle swelling. This effect was seen to increase with the increase in crumb rubber concentration, therefore, the more the crumb rubber was added, the greater the viscosity increase (Lougheed & Pappagiannakis, 1996).

All these factors are of interest when blending rubberized asphalt; this is because the increased viscosity of the modified binder requires higher temperatures during aggregate-binder mixing. Therefore, it is important that the appropriate amount of crumb rubber be added to the asphalt in order to ensure the most cost effective mix possible.

3.2.2 *Resistance to permanent deformation*

Permanent deformation, or rutting, is the phenomenon in which repeated loading of the asphalt surface causes the progressive movement of the pavement materials. Plastic flow or consolidations of the underlying material are known causes of this phenomenon. The complex shear modulus (G*) and phase angle (δ) are indicators of rutting tendency in the pavement (G*/sin δ) at high temperatures and of fatigue cracking (G*sin δ) at medium range temperatures. AASHTO T 315 provides specifications and procedures for obtaining experimental values of the complex shear modulus and phase angle using the Dynamic Shear Rheometer (DSR) (AASHTO, 2006b).

While increases in the crumb rubber concentration consistently produced higher G*/sin δ values, particle size (0.18–0.85 mm) was not seen to play a statistically significant effect on the rutting susceptibility of the CRM binder. Analysis of the data using Fishers LSD analysis indicated that ambient CRM binders produced significantly higher failure temperatures than cryogenic CRM binders (Thodesen et al., 2008).

In a study conducted by Putman and Amirkhanian, it was suggested that the reason for the increases in failure temperature were due to the physical interaction between the rubber particle and the binder. In this study, the effects of crumb rubber on asphalt binder were separated in two distinct elements: the interaction effect (IE) and the particle effect (PE). The IE is the effect of the lighter fractions of the binder diffusing into the CRM particles. While, the PE was identified as the effect of the CRM particles acting as filler in the binder. The findings of the study suggest that increases in the rut resistance of rubberized asphalt are due in large part to the crumb rubber particle acting as filler. This study also concluded that the properties of the base binder are very important in determining the rutting resistance of a binder (Putman & Amirkhanian, 2006).

3.2.3 *Resistance to thermal cracking*

AASHTO T 313 specifies the use of the Bending Beam Rheometer (BBR) to determine cold weather properties of asphalt binder (AASHTO, 2006c). This test is used to evaluate the binder's susceptibility to thermal cracking; this form of cracking occurs when the binder experiences a rapid temperature gradient at cold temperatures (Roberts et al., 1996). The creep stiffness (S(t)) and the rate of creep stiffness change (m-value) are the parameters used in determining the resistance to thermal cracking.

One study found that variations in particle size (0.18–0.85 mm) produced statistically significant effects on the low end failure temperature of the binder, as particle fineness increased so too did the resistance to thermal cracking. Ambient and cryogenic grinding procedures were seen to produce statistically similar CRM binders with respect to thermal cracking (Thodesen et al., 2008).

In another study, Bahia and Davis evaluated the impact of crumb rubber types and contents on the rheological properties of asphalt binders. This study concluded that the impact of rubber content (2–20%) was independent of the rubber source, but, was a linear function of the of the rubber content. Furthermore, the research concluded that for every 1% increase in rubber content a 4% decrease in stiffness was noted. Similarly to the high temperature properties, this trend is highly dependent on the base binder source (Bahia & Davies, 1995).

The findings from the Bahia and Davies study were confirmed in an independent study conducted by Gopal et al. This study concluded that the impact of crumb rubber on the m-value is highly dependent on the properties of the base binder. However, this study also concluded that in some limited cases, the use of certain combinations of crumb rubber size and content can either improve or adversely affect the low temperature performance grade of the asphalt binder. The researchers concluded that only when carefully designed and evaluated should crumb rubber be used to improve the low temperature properties of asphalt binder (Gopal et al., 2002).

3.2.4 *Compaction temperature*

Compaction temperature is an issue when dealing with rubberized binder; this is the case as the increased viscosity which characterizes rubberized asphalt typically requires higher compaction temperatures than conventional binders. Compaction temperature for conventional mixes is defined as the range of temperatures where the unaged asphalt binder has a kinematic viscosity of $280 \pm 30\,\mathrm{mm^2/s}$.

One study has shown that for specimens compacted using the Superpave gyratory compactor (SGC), the air voids contents of rubberized asphalt mixes (10 and 15% by weight of binder) tended to decrease as compaction temperature increased between 116 and 173°C. For the same study, control samples were prepared using an SBS modified binder and a virgin binder. The researchers found that for these mixtures the difference in air void contents was insignificant as the temperature was varied between 116 and 173°C (Lee et al., 2006b).

These findings indicate that compaction temperature is an issue which needs to be paid close attention during the preparation of rubberized asphalt mixtures. This is especially the case as the asphalt mix design will typically be based on the air void content of the compacted sample, therefore when conducting a mix design, it is essential that the appropriate compaction temperature be identified.

4 FUTURE APPLICATIONS OF RUBBERIZED ASPHALT

Today, many researchers are conducting research projects on other possibilities for rubberized asphalt; two applications gaining research interest are the application of rubberized asphalt with Reclaimed Asphalt Pavement (RAP) and with warm asphalt technology.

4.1 *Rubberized asphalt and RAP*

The removal of asphalt pavement is a routine task today when resurfacing and widening projects are undertaken are to be laid, the resulting removed material is today widely used in the production of new mixtures. The introduction of RAP into the paving industry has been a popular one as it has provided paving materials with considerable savings in material, money, and energy.

Research has shown that increasing concentrations of RAP produced rubberized asphalt mixtures with higher stiffness and Indirect Tensile Strength (ITS) values. These findings suggest higher stabilities, while also demonstrating improved rutting resistance over conventional mixes. Furthermore, during this study a good workability of the mix was observed by the researchers (Xiao et al., 2007).

4.2 *Warm mix rubberized asphalt*

Given the increasing concerns with global warming and increasing emissions, an effort is being made by the asphalt industry to lower its emissions by reducing the mixing and compaction temperatures of asphalt concrete (Figure 11). Warm mix asphalt (WMA) is a new technology which is rapidly gaining popularity due to the reduced energy requirements involved during the production of HMA. This technology is particularly applicable to rubberized asphalt as one of the main concerns with modified binders is the increased compaction temperatures associated with them. Typically WMA additives work by either reducing the viscosity binder or by improving the workability of the binder. The integration of such additives usually results in significant reductions of heat required to work the asphalt.

Figure 11. Comparison of emissions between warm and hot mix asphalt.

One study which evaluated the high temperature binder properties of rubberized asphalt blends with warm mix additives concluded that the addition of warm asphalt additives was shown to significantly affect the viscosities of the binders as well as improving the rutting resistance of the evaluated binders (Akisetty et al., 2007). To date, the incorporation of these two technologies together remains relatively untested, however, given the current trend of developing more and more sustainable technologies, this integration of technologies seems an interesting prospect for the future.

5 CONCLUDING REMARKS

The blending of ground recycled tires with asphalt binder has been shown in both field and laboratory evaluations to significantly improve the performance of asphalt binders. Specifically, the high temperature performance grade of the asphalt binder is seen to increase significantly with the addition of crumb rubber. Increases in performance grade are indicative of improved pavement quality at elevated temperatures, specifically in regards to rut resistance. Cold weather properties are also seen to be affected by the addition of crumb rubber, binder stiffness generally decreases as the crumb rubber concentration increases.

Numerous field evaluations have shown that the use of rubberized asphalt, while more expensive, leads to improved pavements requiring less maintenance. As such, rubberized asphalt mixes may actually be considered a more economically attractive option as they offer the benefits of conventional PMAs while offering a lower life cycle cost. State highway agencies are today less concerned with building new roads and more concerned with maintaining the existing ones. Therefore, applications such as SAM and SAMI are particularly attractive as they offer an effective method of pavement rehabilitation.

From an environmental standpoint, the benefits of rubberized asphalt are also significant. Given the current volume of scrap tires being generated, as well as legislation forbidding disposal of tires in landfills, there is a need for innovative approaches to this issue. Within the scientific community, there is widespread agreement that rubberized asphalt produces superior performing asphalt compared to conventional binder. As such, the future of paving materials lies with polymer modified asphalt as these produce more durable pavements, which in the long run require less maintenance thus costing less.

Future research in this field might involve the incorporation of various recycled materials and other technologies, doing so will provide superior pavements which are also more sustainable. An example of such a pavement might be one incorporating RAP, crumb rubber, and WMA additives. This pavement would be less susceptible to rutting and other pavement distresses while requiring less virgin aggregate and less heat. Current research suggests that such a pavement is feasible; however, to date more research is still required.

REFERENCES

Abdelrahman, M. A., & Carpenter, S. H. (1999). Mechanism of the Interaction of Asphalt Cement with Crumb Rubber Modifier (CRM). Transportation Research Record 1661, pp. 106–113.

Abdelrahman, M. (2006). Controlling the Performance of Crumb Rubber Modifier (CRM) Binders through the Addition of Polymer Modifiers. Transportation Research Record 1962, pp. 64–70.

Akisetty, C. K., Lee, S. J., & Amirkhanian, S. (2007). High Temperature Properties of Rubberized Binders Containing Warm Asphalt Additives. Journal of Construction and Building Materials.

American Association of State Highway and Transportation Officials. (2006c). AASHTO T 313: Standard Test Method for Determining the Flexural Creep Stiffness of Asphalt Binder Using the Bending Beam. Washington DC: AASHTO.

American Association of State Highway Officials. (2006a). AASHTO T 316: Viscosity Determination of Asphalt Binder using Rotational Viscometer. Washington DC: AASHTO.

American Association of State Highway and Transportation Officials. (2006b). AASHTO T 315: Standard Method of Test for Determining the Rheological Properties of Asphalt Binder Using a Dynamic Shear Rheometer (DSR). Washington DC: AASHTO.

American Society of Testing and Materials (ASTM). (2001a). D 6114 Standard Specification for Asphalt Rubber Binder. In A. B. Standards, Road and Paving Materials: Vehicle Pavement Systems. West Conshohoken, PA: ASTM.

American Society of Testing and Materials (ASTM). (2001b). D 5644 Standard Test Methods for Rubber Compounding Materials-Determination of Particle Size Distribution of Recylced Vulcanizate Particulate Rubber. West Conshohoken, PA: ASTM.

Amirkhanian, S. N. (2003). Establishment of an Asphalt-Rubber Technology Service (ARTS). Proceedings of the Asphalt Rubber 2003 Conference, 2, pp. 577–588. Brasilia, Brazil.

Anon (1974). Bulletin of the Acoustical and Insulation Materials Association. Park Ridge, Illinois.

Arizona Department of Transportation. (1989). The History, Development, and Performance of Asphalt Rubber at ADOT. Phoenix, AZ: Arizona Department of Transportation.

Asphalt Institute. (2003). Performance Graded Asphalt Binder Specification and Testing, Superpave Series No.1 (SP-1). Lexington, KY: Asphalt Institute.

Bahia, H., & Davies, R. (1995). Role of Crumb Rubber Content and Type in Changing Critical Properties of Asphalt Binders. Journal of the Association of Asphalt Paving Technologists, pp. 130–162.

California Department of Transportation. (2003). Asphalt Rubber Usage Guide. Sacramento, CA.

Dantas Neto, S. A., Farias, M. M., Pais, J. C., Pereira, A., & Picado Santos, L. (2003). Behavior of Asphalt-Rubber Hot Mixes Obtained with High Crumb Rubber Contents. Proceedings of the Asphalt Rubber 2003 Conference, 2, pp. 147–158. Brasilia, Brazil.

Environmental Protection Agency. (2007, September 6). Management of Scrap Tires. Retrieved January 28, 2008, from Tire Fires: http://www.epa.gov/garbage/tires/fires.htm

Environmental Protection Agency. (2006). Municipal Solid Waste Generation, Recycling, and Disposal in the United States: Facts and Figures for 2006. Washington, DC: Environmental Protection Agency.

Farris, R. (2003). UMass Magazine Online. Retrieved January 28, 2008, from One Giant Molecule: http://umassamherstmagazine.com/Winter_2003/One_Giant_Molecule_408.html

Gopal, V. T., Sebaaly, P. E., & Epps, J. (2002). Effect of Crumb Rubber Particle Size and Content on the Low Temperature Rheological Properties of Binders. Transportation Research Board. Washington, DC: Transportation Research Board.

Green, E., & Tolonen, W. (1977). The Chemical and Physical Properties of Asphalt Rubber Mixtures, Part 1 Basic Material Behaviour. Phoenix, AZ: Arizona Department of Transportation.

Hanson, D., James, R. S., & NeSmith, C. (2004). Tire/Pavement Noise Study. Auburn, AL: NCAT.

Heitzman, M. A. (1992). State of the Practice – Design and Construction of Asphalt Paving Materials with Crumb Rubber Modifier. Washington D.C.: Federal Highway Administration.

Lee, S. J., Amirkhanian, S., & Shatanawi, K. (2006a). Effects of Crumb Rubber on Aging of Asphalt Binders. Asphalt Rubber Conference, pp. 779–795. Palm Springs, CA.

Lee, S., Amirkhanian, S., Thodesen, C., & Shatanawi, K. (2006b). Effect of Compaction Temperatures on Rubber Asphalt Mixes. Asphalt Rubber Conference. Palm Springs, CA.

Lougheed, T. J., & Pappagiannakis, A. T. (1996). Viscosity Characteristics of Rubber Modified Asphalts. Journal of Materials in Civil Engineering, pp. 153–156.

Putman, B. J. (2005). Quantification of the Effects of Crumb Rubber in CRM Binders. PhD Dissertation, Clemson University, Department of Civil Engineering, Clemson, SC.

Putman, B. J., & Amirkhanian, S. N. (2006). Crumb Rubber Modification of Binders: Interaction and Particle Effects. Proceedings of the Asphalt Rubber 2006 Conference, 3, pp. 655–677. Palm Springs, CA.

Roberts, F. L., Kandhal, P. S., Brown, E. R., Lee, D., & Kennedy, T. (1996). Hot Mix Asphalt Materials Design and Construction. Lanham, MD, Md: NAPA Educational Fund.

Ruan, Y., Davison, R. R., & Glover, C. J. (2003). Oxidation and Viscosity Hardening of Polymer-Modified Asphalts. Energy and Fuels, pp. 991–998.

Rubber Manufacturers Association (RMA). (2006). Scrap Tire Markets in the United States. Washington DC: Rubber Manufacturers Association.

Schafer, R. (1994). The Soundscape. Rochester, VT: Destiny Books.

Takallou, B. (1991). Recycling Tires in Rubber Asphalt Paving Yields Cost Disposal Benefits. Elastomerics, pp. 19–24.

Thodesen, C., Shatanawi, K., Amirkhanian, S., & Putman, B. (2008). Effects of Crumb Rubber Properties on CRM Binder Performance. Transoportation Research Arena 2008. Ljubljana, Slovenia: Transportation Research Arena.

Way, G. B. (2000). OGFC Meets CRM Where the Rubber meets the Rubber 12 Years of Durable Success. Asphalt Rubber 2000, pp. 15–31. Lisbon, Portugal.

Xiao, F., Amirkhanian, S. N., & Juang, C. H. (2007). Rutting Resistance of Rubberized Asphalt Concrete Pavements Containing Reclaimed Asphalt Pavement Mixtures. Journal of Materials in Civil Engineering, 19 (6), pp. 475–483.

Zanzotto, L., & Kennepohl, G. (1996). Development of Rubber and Asphalt Binders by Depolymerization and Devulcanization of Scrap Tires in Asphalt. Transportation Research Record 1530, pp. 51–58.

Management, Disposal, Pathogen Reduction and Potential Uses of Sewage Sludge

Guadalupe de la Rosa, Elizabeth Reynel-Avila, Gustavo Cruz-Jiménez, Irene Cano-Aguilera & Francisco Martínez-González
Universidad de Guanajuato, México

Adrián Bonilla-Petriciolet
Instituto Tecnológico de Aguascalientes, México

SUMMARY: Disposal of sewage sludge is a major environmental problem in the world. An increase in population has also led to a significant increase in sewage sludge generation. For many years, landfill and agricultural use were common disposal practices. However, regulatory considerations are becoming stronger in many parts of the world. Several strategies have been investigated with the goal of providing new options for sewage sludge treatment, use and disposal. However, since these biosolids are rich in nutrients and organic matter, their utilization as soil conditioners appears to be one of the main sustainable options, especially for developing countries. The main goal of this paper is to provide a review of available strategies for management, disposal, pathogen reduction and potential uses of sewage sludge. A case study from Mexico is also presented on the reduction of *Salmonella* spp from sewage sludge generated at a Municipal Wastewater treatment plant and its potential use in soil conditioning.

1 INTRODUCTION

Treatment plants help to protect human and environmental health by reducing the amount of contaminants in wastewaters before the latter are returned to the surroundings (Fytili and Zabiotou, 2008). Wastewater handling processes generally involve three stages. The primary treatment consists of removing a substantial amount of the suspended solids from the used water, while secondary treatment involves the bio-oxidation of the remaining organic suspended and dissolved solids. Finally, the tertiary treatment processes are performed to increase the quality of the effluent. They include lime coagulation, flocculation, sedimentation, ammonia stripping, recarbonation, multimedia filtration, carbon adsorption and breakpoint chlorination. The wastes generated during primary, secondary and tertiary treatment processes are generally known as sewage sludge (Fytili and Zabiotou, 2008).

In wastewater treatment plants around the world, millions of tons of sewage sludge are generated every year (Babel and del Mundo, 2006; Laturnus *et al.*, 2007). The management and final disposal of this waste are crucial aspects, since they usually involve substantial costs and effort. As a consequence, the increase in sewage sludge generation is considered an important topic in the context of environmental protection (Domínguez *et al.*, 2006). It has been estimated that the average generation of these types of biosolids is around 40–60 g of dry matter per habitant per day for urban sewage plants (Otero *et al.*, 2003a; 2003b). In some countries, the production rate is significant, for instance, nearly 1 million m^3 per year of sewage sludge dry solids are generated in England, 4.2 millions m^3 per year in Switzerland and 50 millions m^3 per year in Germany (Midilli *et al.*, 2002).

Sludge may originate from several sources in a wastewater treatment plant. Therefore, sludge from different stages may possess specific characteristics varying from a liquid to a semisolid liquid. Generally, it is made of organic compounds, macro and micronutrients, non-essential trace metals, organic micro pollutants and microorganisms. It may also contain harmful substances such as detergents, salts, heavy metals, pesticides, toxic organics, among others (Zabaniotou and Theofilou, 2008; Fytili and Zabaniotou, 2008). It is of special relevance to mention that these residues are often contaminated with pathogens such as *Streptococci, Salmonella, Clostridia, Cryptosporididium, Giardia* and *Ascaris* (Blais *et al.*, 1992; Sahlström *et al.*, 2004; Pecson *et al.*, 2007; Graczyk *et al.*, 2008).

The presence and prevalence of toxic compounds and pathogens depends on several factors including: sewage origin, sewage treatment process and sludge treatment processes (Singh and Agrawal, 2008). Therefore, the physico-chemical characteristics of this material greatly depend on the plant type and its operation. As a consequence, an adequate strategy for proper management and disposal of sewage sludge should be related to its composition and properties.

The pathogen content is a crucial aspect that must be considered before any possible application or final disposal option, in order to avoid health risks for people and animals. For example, if sludge with pathogenic bacteria is spread on arable land, there is a risk of spreading diseases to people and animals. As indicated by other studies, the handling of sewage sludge is one of the most significant challenges in wastewater management (Fytili and Zabaniotou, 2008). There is a need to develop suitable strategies for pathogen reduction in sewage sludge. The main purpose of the current chapter was to provide a review of available strategies for pathogen reduction in sludge and its management and disposal.

2 PATHOGEN INCIDENCE IN SEWAGE SLUDGE

The use of sewage sludge in agriculture and others fields has increased in many countries. However, these practices should be associated with a prior knowledge of the pathogens and chemical constituents present in these residues. Many of the pathogens usually found in sludge are zoonotic bacteria, which are important in the eco-cycle. Several studies have reported the quantification and identification of these pathogens. For example, Kabrick and Jewell (1982) reported the relative populations of viruses, *Salmonella* spp., total and fecal coliforms, fecal streptococci and parasites in sewage sludge before and after digestion. Kothary *et al.* (1984) determined the levels of *Aspergillus fumigatus*, an opportunistic fungal pathogen, in compost and in air at a sewage sludge composting facility. In this study, up to a million colony forming units (CFU) were detected in 1 g of screened compost. Mininni and Santori (1987) suggested that negative effects can occur due to pathogens following sludge addition to soil. Epidemiological studies in the U.K. indicated that infections can arise, in particular from the presence of *Taenia saginata* and *Salmonella*.

In another study, Gaspard *et al.* (1997) investigated the parasitic contamination of helminths found in urban sludge. Parasitological analyses were performed on several samples collected from urban sludge, lagoon sediments and composts. The overall results showed that Nematode eggs (*Toxocara, Ascaris, Capillaria, Trichuris, Ascaridia, Enterobius*) represented 93.2% of total egg amount, whereas Cestode eggs (*Tenia, Hymenelopis*) were in the remaining 6.8%.

Many studies have documented the high frequencies of *Salmonella* in both, raw and digested sludge. In addition, the persistence of different serotypes of *Salmonella* has been reported (Sahlström *et al.*, 2004). Whether sewage sludge is being considered for disposal in landfills or for agricultural purposes, the determination of the presence of *Salmonella* spp. is extremely important since they display low dose infectivity rates. Besides, they can possibly be transferred from soil to animals and humans, or to fruits and vegetables through farming.

Salmonella spp. is responsible for gastroenteritis, enteric fevers and septicemia (Novinscak *et al.*, 2007). In a recent study, Sahlström *et al.* (2004) assessed the presence of bacterial pathogens in eight Swedish sewage treatment plants (STPs) with four different treatment methods. *Salmonella, Listeria monocytogenes, Campylobacter coli* and *jejuni, Escherichia coli* O157 were quantified. It was concluded that the sewage sludge produced in Swedish STPs did not comply with the regulations for unrestricted use in agriculture.

With respect to other pathogens, Al-Bachir *et al.* (2003) determined total microbial count and bacterial pathogens in samples of concentrated municipal sewage sludge stored for several months, with varying moisture content. The results indicated that in all tested samples, bacterial pathogens including *Enterobacter* sp., *Klebsiella* sp. and *Escherichia coli* were detected. On the other hand, Graczyk *et al.* (2008) reported the occurrence of *Cryptosporidium* and *Giardia* in sewage sludge and solid waste landfill leachate. High numbers of potentially viable, human-virulent species of *Cryptosporidium* and *Giardia* were reported in the analyzed wastes.

Finally, it is convenient to point out that there are few studies related to the quantification of pathogen content in sewage sludge compared to the information available for other pollutants such as heavy metals (Sahlström *et al.*, 2004).

3 TREATMENT METHODS FOR PATHOGEN REDUCTION IN SEWAGE SLUDGE

Sewage sludge treatment and disposal account for 40% to 45% of the capital and operating costs of a wastewater treatment plant. Depending on the sludge characteristics, this residue may represent a valuable resource. The main objective of sewage sludge treatment is its stabilization. This stabilization is a controlled decomposition of easily degradable organic matter that causes a significant reduction in volatile solids content, a change from an unpleasant smell to an earthy one, and pathogen reduction (Sahlström *et al.*, 2004).

Sewage sludge can be treated using several strategies that include mesophilic or thermophilic anaerobic digestion, chemical treatment, radiation, membrane processes and some combinations of these methods. All of them have advantages and disadvantages as well as differences in efficiencies for pathogen reduction. It should be pointed out that storage by itself is not recommended as a treatment method since pathogens can still be detected after one year (Gibbs *et al.*, 1997). The following sections are intended to provide a general description of some of the available strategies for pathogen reduction.

3.1 *Aerobic and anaerobic digestion*

Aerobic digesters can be operated in either a batch or on a continuous-flow condition. Aerobic digestion may be defined as the biological oxidation of organic sludge under aerobic conditions. Some advantages include fewer operational problems (thus, less laboratory control and daily maintenance are required), significant lower BOD concentrations in the supernatant liquor, and lower capital costs as compared to those for anaerobic digestions. On the other hand, disadvantages of aerobic as compared to anaerobic digestion include higher energy requirements (strong aeration and mixing are required), methane (an important useful by-product) is not produced, and the digested sludge has a lower solids content; consequently, the volume of sludge to be dewatered is significantly large (Borowski and Szopa, 2007).

Several investigators have reported that using aerobic digestion in addition to other treatment can significantly reduce pathogen content in sewage sludge. Kabrick and Jewell (1982) studied the effect of autoheated aerobic thermophilic digestion on the reduction of pathogen content in sewage sludge. The performance of this system was compared to that of conventional mesophilic anaerobic digestion. Both systems were full scale, continuously-fed facilities operated in parallel and used a feed sludge of thickened primary and waste-activated sludge. In this work, the relative populations of viruses, *Salmonella* spp., total and fecal coliforms, fecal streptococci and parasites were detected and compared before and after digestion. Their results showed that concentrations of viruses and *Salmonella* spp. in the aerobic effluent were below detectable limits in almost all samples. However, the anaerobic effluent contained detectable numbers of viruses and *Salmonella* spp. Bacterial indicator counts and parasite concentrations were less in the auto-heated digester effluent as compared to the effluent from the anaerobic digester. According to the authors, the simple auto-heated aerobic digestion process could be used to produce a virtually pathogen-free sludge at a cost comparable to that of conventional mesophilic anaerobic digestion.

Ponti *et al.* (1995) analyzed some process design features for an aerobic thermophilic sewage sludge pilot treatment plant. These authors indicated that the sludge hygienization is one of the objectives and benefits of the thermophilic treatment, where not only temperature but also the total solids content are important factors affecting inactivation of pathogens. Inactivation rates were increased by raising the temperature. In addition, the residual colony forming units (CFU) decreased when total solids content was reduced. These authors concluded that a continuous operation mode would not affect the quality of the hygienization.

The *autothermal thermophilic aerobic digestion* is an alternative system that may overcome the large operation volumes as well as energy consuming aeration devices required for the conventional aerobic digestion process. This technique is performed in biological systems that use aerobic and fermentative processes. Part of the energy obtained during the oxidation of organic matter is dissipated into the sludge as heat, which, if properly conserved, provides temperatures for a thermophilic operation (Borowski and Szopa, 2007). This system can be modified using an aerobic thermophilic pretreatment prior to the anaerobic digestion, which is known as the dual digestion process. The anaerobic digestion efficiency is enhanced by using the aerobic pretreatment as compared to the performance of a conventional single stage process. This modified treatment allows for a complete pathogen inactivation. Specifically, experimental results showed that the aerobic treatment may reduce the *Enterobacteriaceae* content below detectable limits (Borowski and Szopa, 2007).

Anaerobic digestion is also an appropriate technique for the treatment of sludge before final disposal, and is used worldwide as the oldest and most important process for sludge stabilization. Mesophilic anaerobic digestion of sewage sludge is more frequently used than thermophilic digestion. This process is defined as the utilization of microorganisms, in anoxygenic conditions, to stabilize the organic matter by its transformation into methane and other inorganic products including carbon dioxide. Some of the benefits this technique provides are that the biogas can be used as an energy source and the resulting digestion product can be employed as a soil conditioner (Castillo *et al.*, 2006). However, as a consequence of its high sensitivity and biogas production, it is important to note that this process needs large reaction volumes, gas collecting tanks and complex instrumentation (Borowski and Szopa, 2007). Taking into account this information, several strategies have been reported in the literature to improve the performance and operating conditions of this method. For example, the anaerobic digestion of sewage sludge in conventional reactors requires a hydraulic retention time of around 20 days. However, the study of Sanchez *et al.* (1995) showed that when using a fixed bed reactor this hydraulic retention time can be reduced to 3–7 days. Also, recirculation may increase the removal of organic matter in this system.

The efficiency of anaerobic digestion of activated sludge can be improved by using thermochemical or biological hydrolysis as a pretreatment (Park *et al.*, 2005). In fact, thermochemical pretreatment offers better results. On the other hand, the slow degradation rate of sewage sludge in anaerobic digesters is caused by the rate limiting step of sludge hydrolysis.

Tiehm *et al.* (1997) reported the effect of ultrasound pretreatment on sludge degradability using a frequency of 31 kHz and high acoustic intensities. They showed that ultrasound treatment resulted in raw sludge disintegration as was demonstrated by the increase in chemical oxygen demand (COD) in the supernatant and size reduction of sludge solids. Using ultrasound disintegration, a better degradability of raw sludge was achieved that permitted a substantial increase in throughput. Also, Hwang *et al.* (1997) evaluated the effectiveness of sewage pretreatment and its effect on the subsequent anaerobic digestion of waste activated sludge. They showed that microorganism cells in the sludge can be ruptured by a mechanical jet and smashed under pressurized conditions (5–50 bar). It was observed that high anaerobic sludge digestion efficiencies could be obtained according to the increase of microorganism cell rupture through mechanical pretreatment of this waste.

El-Hadj *et al.* (2007) reported that in many anaerobic digestion processes for sludge treatment, the hydrolysis of organic matter has been identified as the rate limiting step. These authors focused on the effect of ultrasonic pretreatment of raw sewage sludge before it was sent for mesophilic and thermophilic anaerobic digestion. They concluded that the use of pretreated sludge significantly improved the COD removal efficiency and biogas production in lab-scale anaerobic digesters when compared with the performance without pretreatment, especially under mesophilic conditions.

With respect to the anaerobic digestion process and its performance in pathogen reduction, Fukushi *et al.* (2003a; 2003b) attempted to decrease the number of *Salmonella* spp. using a simulated acid-phase anaerobic digester tested in a laboratory-scale batch experiment. They demonstrated that a reduction of *Salmonella* spp. can be achieved in a mixture of sludge and organic acid, simulating an acid digester of a two-phase anaerobic digestion process. Their results showed that a high concentration of organic acid at a pH value of 5.5–6.0, prevented the *Salmonella* spp. concentration from decreasing. In addition, these authors observed a complete destruction of *Salmonella* spp. within two days if the pH value was maintained below 5.5.

Gavala *et al.* (2003) studied the differences between mesophilic and thermophilic anaerobic digestion of sludge and the effect of the pre-treatment at 70°C on mesophilic and thermophilic anaerobic digestion of primary and secondary sludge. These authors claimed that thermal pre-treatment improves the reduction of pathogens, where the process could be developed at relatively low costs and low temperatures.

Song *et al.* (2004) studied and compared the performance of thermophilic and mesophilic co-phase anaerobic digestions using an exchange process between spatially separated mesophilic and thermophilic digesters with single-stage mesophilic and thermophilic anaerobic digestions. The results showed no difference in the reduction of total coliforms in the temperature co-phase system (98.5–99.6%) as compared to the one obtained in the single-stage thermophilic digestion. However, better results were obtained for the reduction of volatile solids and soluble chemical oxygen demand in the temperature co-phase anaerobic digestion system.

Recently, the treatment of domestic sewage has been the subject of a literature review reported by Aiyuk *et al.* (2006). The study was done in controlled environments having the anaerobic process and the upflow anaerobic sludge blanket (UASB) concept as the core, under natural high temperature conditions. According to these authors, the various stages involved in the UASB technology can eliminate a large fraction of the pathogens present in the raw wastewater, mainly through the pre-treatment sedimentation and ion exchange filtration.

Hybrid systems can be used to improve the reduction of pathogen contents in sewage sludge. One example of these systems is the scheme proposed by Carballa *et al.* (2007). These investigators studied the characteristics of digested sludge obtained from a combination of an oxidative pre-treatment with ozone followed by anaerobic digestion. The pathogens content after conventional and pre-ozonation treatment of sewage sludge were below the legal requirements. However, this goal was achieved at the expense of the deterioration of the dewatering properties of sludge when the ozone pre-treatment was applied. On the other hand, Ahn and Speece (2006) reported a novel fermentation procedure where a sludge blanket type configuration was adopted, in order to increase the hydrolysis/acidogenesis of the municipal primary sludge. This process was investigated under batch and semi-continuous conditions with varying pH and temperature. Their results demonstrated several advantages such as the production of pathogen-free stabilized solids, and excellent solids control.

In a recent effort, Shang *et al.* (2005) indicated that a membrane bioreactor (MBR) may serve as a pre-disinfection or disinfection unit, in addition to its solid/liquid separation and biological conversion functions, in order to produce high quality sewage effluents. They investigated the performance and factors affecting pathogen removal when a hollow-fiber membrane module submerged in an aeration tank and bacteriophage MS-2 as the indicator organism were used. Removal of the MS-2 phage was found to be promoted by physical filtration through the membrane. Biomass activity in the aeration tank and bio-filtration through the biofilm developed on the membrane surface. These results showed that membrane alone displayed poor virus removal but the overall removal increased substantially with the presence of biomass and the membrane-surface-attached biofilm.

Puchajda *et al.* (2006) managed to obtain pathogen standards using non-thermophilic acid digesters. It was proposed that the key mechanism responsible for fecal coliform inactivation was the presence of un-ionized volatile fatty acids. Lab-scale acid digesters were assembled and operated in a batch mode for 5 days at mesophilic (38°C) and low-mesophilic (21°C) temperatures and at different solids concentrations. The key factor recognized for successful pathogen inactivation was pH, which is also the main factor driving the shift in organic acids toward the un-ionized form. Compared to conventional mesophilic acid digestion, low-mesophilic acid digestion was effective in fecal coliform inactivation because the process maintained low pH values throughout the duration of the experiment, offered continuous release of organic acids, and showed high concentrations of organic acids in un-ionized form, including acetate, propionate, butyrate, and valerate.

Finally, Guzman *et al.* (2007) determined the presence and levels of pathogens and indicators in raw and treated sludge and compared their persistence after mesophilic and thermophilic treatments. Spores of sulphite-reducing clostridia were the most resistant micro-organisms.

3.2 *Radiation*

Radiation technology has been presented as a promising alternative for sludge treatment mainly due to its high efficiency in pathogen inactivation. In fact, the application of this technology in pilot plants started several decades ago (Lessel and Suess, 1977; Suess and Lessel, 1977). The most promising technologies and the effects of radiation on sludge characteristics have been discussed in a review reported by Wang and Wang (2007). Kawakami and Hashimoto (1977) showed that composting using radiation disinfected sewage sludge can be carried out under optimum conditions. This shortened the composting period, since it was not necessary to maintain high fermentation temperatures to reduce pathogen in sludge. This demonstrated the value of irradiated sludge in agronomic applications.

According to Ahlstrom (1977), the irradiation process does not increase the extractability and plant uptake of a broad range of nutrients and heavy metals from sludge-amended soils. However, it does eliminate the hazards associated with pathogen contamination when applying sludge to agricultural land. Irradiated sludge has also been evaluated as a supplemental foodstuff for cattle and sheep.

Watanabe and Takehisa (1977) reported that a suitable dose to reduce the coliforms to undetectable levels ranged from 0.3 to 0.5 Mrad in municipal sewage sludge cake. In addition, it was observed that no coliforms reappeared in 0.5 Mrad irradiated sludge cake during storage either at room temperature (6–16°C) or at 30°C. They reported that the adequate disinfection dose is considered to be 0.5 Mrad. *Pseudomonas cepacia* was a predominant bacterium in non-irradiated sludge cake. In a range of 0.5 to 0.7 Mrad, the residual flora consisted of *Bacillus* species. In addition, radioresistant *Deinococcus proteolyticus, Deinococcus radiodurans* and *Pseudomonas radiora* were isolated from sludge cake irradiated at dose levels of more than 1 Mrad.

Recently, Al-Bachir *et al.* (2003) studied the synergetic effect of gamma irradiation and moisture content on the decontamination of sewage sludge. Several samples of concentrated municipal sewage sludge, stored for different periods and with moisture contents, were studied to assess the effect of different gamma irradiation doses. After irradiation, total microbial count and bacterial pathogens were determined. The results indicated that in all tested sewage sludge samples, bacterial pathogens including *Enterobacter* sp., *Klebsiella* sp., *Salmonella* sp., and *Escherichia coli* were initially detected. All doses of gamma irradiation reduced the total counts of those microorganisms. These authors identified that the lowest lethal dose for tested bacterial pathogens was 5 kGy in air dried sewage sludge. Moreover, for wet sewage sludge having more than 40% moisture, the lethal dose was 1 kGy, where the cost per unit could be decreased.

Besides the effectiveness of this process, it should be noticed that the irradiation used to eliminate pathogens in sewage sludge may deleteriously affect properties of sludge organic matter (Wen *et al.*, 1997). As indicated by Wen *et al.* (1997), the gamma-irradiation may promote organic matter decomposition affecting the N and C concentrations. It is possible that compositional changes in irradiated sludge may affect its final application in agriculture.

Microwaves have also been found to be effective in destruction of pathogens in sewage sludge. Hong *et al.* (2004) investigated the mechanisms and roles of microwaves on fecal coliform destruction in sludge. They demonstrated cell membrane damage as microwave irradiation intensity and temperature increased. Above 60 ± 3°C, viable cells were rarely found when pure fecal coliforms were irradiated with microwaves. These authors concluded that microwave irradiation of sludge appears to be a viable and economical method for destructing pathogens and generating environmentally safe sludge.

3.3 *Chemical treatment*

For pathogen elimination purposes, the classical chemical approach is based on the increase of sewage sludge pH above 11 by means of the addition of lime or other compounds, with and without thermal treatment. In this process, lime has commonly been used to stabilize primary and excess activated sludge, temporarily preventing odors. If sufficient slaked lime is added to raise the pH above 11, almost all biological action ceases and a calcium carbonate precipitate is formed. Essentially all *E. coli* and *Salmonella typhosa* can be killed by high lime treatment. However, this treatment should be considered as a temporary sludge stabilization method, because it has been found that, when these biosolids are disposed in lagoons and if pH reduction occurs, a gradual increase in biological action occurs (Chale-Matsau, 2005). Marcinkowski (1985) analyzed the chemical inactivation of sludge where bacteriological and parasitological examinations of the sludge were carried out prior to, and following completion of, the inactivation process. The experiments indicated that sewage sludge deactivated with quicklime may be used for agricultural purposes.

During the last decade, other alternative chemical procedures have been proposed for sludge stabilization. Examples of these procedures are reported by Luca *et al.* (1996) and Boost and Poon (1998). A research study on the sanitary effectiveness in terms of pathogens (bacteria, fungi and helminth eggs) through biosolid stabilization was performed by de Luca *et al.* (1996). These authors used lime as a standard process compared to potassium ferrate (VI). Boost and Poon (1998) developed a modification of lime stabilization by the incorporation of pulverized fuel ash (PFA). These authors stated that this modified treatment was economic in terms of lime utilization and uses another waste product, PFA, which also requires disposal. In this study, several species of intestinal pathogen were considered for their ability to survive in raw sludge and then subjected to disinfection. They showed that a mixture of 60% sludge (600 g/kg): 40% CaO/PFA (44.44 g/kg/355.56 g/kg) with the ratio CaO:PFA 1:8, was able to prevent the growth of bacterial pathogens and maintain a pH above 11.0 for at least 7 days. These authors suggested that this modified sludge treatment is suitable for landfill.

Recently, Parmar *et al.* (2001) investigated the combined effects of enzyme, pH and temperature treatments for removal of pathogens from sewage sludge. A pH of 10 for 24-h at 23°C or a 3-h held at 50°C resulted in a 100% reduction of coliforms. However, a significant count of *Salmonella* species was still obtained. Additionally, pH adjustment to 12 held for 48-h at 23°C or pH adjustment to 10 or 12 and a 3-h holding time at 60°C was required to achieve a 100% reduction in *Salmonella* spp. Although a treatment with protease enzymes at 40°C with or without alkali treatment completely eliminated coliforms, *Salmonella* counts were reduced by two to three orders of magnitude. However, complete elimination of pathogens was observed when enzymatic treatment was carried out at 50°C. Application of alkaline protease, through its combined beneficial effects in pathogen reduction, solids reduction and improved solids settling, has potential as an effective procedure for sewage sludge treatment.

3.4 *Bioleaching*

Two main aspects limit the application of sewage sludge for agricultural purposes: the presence of potentially toxic concentrations of metals and the ineffective destruction of pathogenic microorganisms by conventional stabilization processes.

Regarding metals, bioleaching has been reported as a feasible option for the removal of these contaminants from sludge (Wong *et al.*, 2002). However, some efforts have been made to use bioleaching for both, metal and pathogen reduction. Blais *et al.* (1992) reported that in the bioleaching process, high sulfuric acid production ($pH < 2.5$) resulting from sulfur oxidation by indigenous thiobacilli, allowed for a considerable reduction in bacterial indicators in every examined sludge sample over a 5 day period. In addition, metal concentrations were reduced to levels compatible with recommended norms for intensive sludge agricultural use. Other contaminants including volatile solids were also reduced. These results indicate that this process can improve the stabilization of the digested sludge.

4 MANAGEMENT AND DISPOSAL OF SEWAGE SLUDGE

At the present time, there are three main options for the disposal of processed sludge: agricultural application as fertilizer, land fill and incineration. From these, the agricultural application has become the most widespread method of disposal. The basic methods used for disposal and management of sewage sludge such as landfill, land application and ocean dumping are becoming much less accepted because of the risk they represent to the environment and human health (Dominguez *et al.*, 2006). In order to avoid the possibility of damaging human health or the environment, wastes must be subjected to intermediate treatment, stabilized, encapsulated and made hygienic by removing pollutants (Shimaoka and Hanashima, 1996). The following section provides information on the available disposal strategies for sewage sludge.

4.1 *Ocean and lagoon dumping*

For several years, ocean dumping was a common way of disposing sewage sludge especially in countries with direct access to the sea. In the mid 1980s, an annual amount of 20 million tonnes of sewage sludge was dumped at sea every year (Laturnus *et al.*, 2007). Nowadays, few nations continue with this practice. Some reports have appeared on the effects on the environment caused by this disposal strategy (Carmody *et al.*, 1973; Adingra and Arfi, 1998; Friedman *et al.*, 2000). In fact, reports have been published on this in the context of policy, law and public perceptions (Kite-Powell *et al.*, 1998). As sea dumping of sewage sludge becomes more and more banned, other methods for its disposal have to be found.

4.2 *Composting*

Composting is the biological decomposition of sludge under conditions that allow the development of thermophilic temperatures resulting from biologically produced heat. Composting appears to be a viable alternative for sludge treatment since it reduces the load of pathogenic microorganisms, yielding a final product rich in organic matter and nutrients that can be used as a soil supplement for different horticulture, agricultural or land application purposes. Besides, it contributes to the suppression of certain plant diseases caused by several pathogens. However, this method displays several restrictions related to heavy metal concentration and moisture present in the sludge. Some authors affirm that it should not be used on crops eaten by humans, such as tomatoes, peppers, and similar vegetables (Novinscak *et al.*, 2007). According to Qiao and Ho (1997), composting of sewage sludge with bauxite refining residue (red mud) may overcome some of these problems. In fact, they studied how red mud affected the composting process. The results showed that the amendment with red mud improved the composting process by raising the temperature, removing moisture, and increasing the decomposition rate.

Traditional thermophillic composting is commonly adopted for treatment of organic wastes or for production of organic/natural fertilizers. A related technique, called vermicomposting (using earthworms to breakdown the organic wastes) is also becoming popular. These two techniques have their inherent advantages and disadvantages. Based on these techniques, Ndegwa and Thompson (2001) proposed an integrated approach to enhance the overall process and improve the products qualities. They investigated two strategies using the activated sewage sludge as substrate: pre-composting followed by vermicomposting and pre-vermicomposting followed by composting. Their results indicated that, a system that combines the two processes not only shortens stabilization time, but also improves the product quality. Combining the two systems resulted in a product that was more stable and consistent (homogenous), had less potential impact on the environment. For the compost–vermicomposting system, the product met the pathogen reduction requirements.

E. foetida was also used to produce manure of acceptable quality through vermicomposting. Gupta and Garg (2008) combined different amounts of primary sewage sludge and cow dung. The authors recommended a mixture of 30–40% of primary sewage sludge with cow dung to obtain good fertilized manure and earthworm growth rate.

Kowaljow and Mazzarino (2007) evaluated composted amendments of different origin (biosolids, municipal organic, inorganic wastes) for the restoration of soils burned by wildfires in the Patagonia, Argentina. They investigated chemical, biological and physical properties. Results imply that soil chemical and biological properties showed a high response to organic amendment addition. Inorganic fertilization enhanced higher plant cover than organic amendments, but did not contribute to soil restoration. Additionally, Andres *et al.* (2007) determined the effect of sewage sludge application to a degraded soil on the survival and growth of wild leguminous shrub species. Their results showed that the sludge application increased shrub growth and the spontaneous vegetation became quickly established in the amended plots.

4.3 *Combustion*

Sludge and other biomaterials are considered as source of renewable energy (Dominguez *et al.*, 2006). When the sludge is dry, it can produce heat amounts comparable to those obtained using low grade coal. Therefore, the heat generated by burning sewage sludge could be used to produce steam and electricity (CWMI, 1996; Fytili and Zabaniotou, 2008). According to Randall (1991), combustion or thermal treatment might represent a viable disposal technique for many of these waste streams because it can destroy hazardous organics and pathogens, and can significantly reduce mass and volume. The combustion systems used for the different waste streams are widely divergent because of the different physical characteristics of the materials. It is important to mention that advanced technology should be used for combustion; the release of some toxic substances (dioxins, mercury, and furans) must be controlled if using this option (Fytili and Zabaniotou, 2008).

4.4 *Gasification*

Fytili and Zabaniotou (2008) provided an appropriate definition for gasification: the thermal process during which carbonaceous content of sewage sludge is converted to combustible gas and ash in a net reducing atmosphere. Compared to incineration, gasification is more efficient and diminishes the emission of pollutants to the atmosphere.

In the future, the role of hydrogen may become more important, as some researchers suggest that the world's energy systems may undergo a transition to an era in which the main energy carries are hydrogen and electricity (Midilli *et al.*, 2002). Downdraft gasification of sewage sludge can be identified as a possible system for the production of hydrogen energy (Midilli *et al.*, 2002). The properties of sewage sludge that are known to influence the gasification are moisture content and chemical composition. Besides, the toxic organic compounds can be destroyed and the heavy metals can be fixed in the resultant solid ash (Midilli *et al.*, 2002). These authors have studied the hydrogen production potential from sewage sludge by applying a downdraft gasification technique. The gas obtained mainly consisted of hydrogen, nitrogen, carbon monoxide, carbon dioxide and methane. Around 10–11% was hydrogen which could be used for fuel cells.

Groß *et al.* (2007) reported two methods that use gasification with very promising results. These methods convert sewage sludge directly to energy with the advantage that they can be developed *on site*. The methods were called ETVS-Process (Entwässern, Trocknen, Vergasen, Strom erzeugen – dewatering, drying, gasification, electric power generation), and NTVS-Process (Niedertemperaturtrocknung, Vergasen, Strom erzeugen – low temperature drying, gasification, electric power generation). The NTVS-Process requires low temperature solar energy during the first step. The advantages of these systems are: Sewage sludge transportation is not necessary, since both processes are developed *on site*. There is a considerable mass reduction for the dewatering, drying and gasification process. In comparison to other current processes, the emission of CO_2 is lower. The only waste generated is the residual mineral ash, which is appropriate for landfill, and the ETVS-/NTVS-Process provides the energy to fulfill the needs of the Waste Water Treatment Plant.

4.5 *Pyrolysis*

Among the available alternatives for disposal of sewage sludge, pyrolysis has recently received significant attention since the operation conditions can be optimized to generate maximum production of char, oil or gas depending on the particular needs (Dominguez *et al.*, 2006). Pyrolysis is the process through which, organic substances are thermally decomposed in an oxygen-free atmosphere, at temperatures varying in the range of 300 and 900°C. This technique appears to be less pollutant than conventional methods. For example, Kaminsky and Kummer (1989) reported a study where a digested, thermally conditioned and dried sewage sludge was pyrolyzed in an indirectly heated fluidized bed reactor at temperatures ranging from 620°C to 750°C.

The products obtained from the pyrolysis process can be used as follow (Dominguez *et al.*, 2006): Char can be utilized as fuel, disposed when heavy metal concentration is high, or used as activated carbon to adsorb certain compounds; Oil can be employed as both fuel or as raw material for chemical production; and gas can be utilized as fuel. Particularly, hydrogen rich gas and syngas (synthetic gas, $H_2 + CO$) are becoming of great interest because of the former is an environmentally clean energy source and the latter can be converted to liquid fuels, which would help to reduce environmental pollution (Dominguez *et al.*, 2006).

Most of the research about gasification and pyrolysis of biomass normally utilizes fixed and fluidized bed reactors. It is known that this kind of wastes is poor receptor of microwave energy. Nevertheless, microwave-induced pyroylis is possible, if the raw material is mixed with an effective receptor (Menendez *et al.*, 2002). Dominguez *et al.* (2006) have investigated the use of microwave heating and compared the data with conventional heating. Their objective was to maximize the gas yield and to asses its quality as a fuel. The results of this study demonstrated that microwave ovens maximize the gas production. In contrast, the electrical ovens produced oil that could have environmental and toxicological impacts.

Different studies have been performed to maximize the yield of some pyrolysis products obtained using sewage sludge (Shen and Zhang, 2003; Kim and Parker, 2008). For example, Shen and Zhang (2003) investigated the pyrolysis of these biosolids under inert conditions in a fluidised-bed to determine the effects of temperature and gas residence time on the product distribution and composition with the purpose of maximizing oil yield. The temperature varied from 300 to 600°C and the gas residence time from 1.5 to 3.5 s. Three groups of products were obtained, a non-condensable gas (NCG) phase, a solid phase (char) and a liquid phase (oil). A maximum of 30% oil yield (wt% daf of sludge fed) was achieved at a pyrolysis temperature of 525°C and a gas residence time of 1.5 s; higher temperatures and longer gas residence times favored the formation of NCG, suggesting that secondary cracking reactions had occurred. Recently, Kim and Parker (2008) suggested that the pyrolysis is a promising way to produce bio-oil form sewage sludge. They performed a study using a laboratory scale-horizontal batch reactor. These authors established that temperature volatile solids were the most important factors affecting the yield of oil and char, however, sludge type also affected both results. Moreover, temperature and pre-treatment of sludge with acids, bases or catalysts (zeolite) did not improve the quantity of oil produced.

Hwang *et al.* (2007) compared the leaching of organic and inorganic compounds from the residues obtained from sewage sludge after pyrolysis and incineration treatments. The total organic carbon and metal concentrations in the pyrolysis residue was greater than the incineration ash; however, the column and batch leaching test performed showed a low organic matter and metal released. This study showed that pyrolysis could be recommended as an appropriate pretreatment for sewage sludge before landfilling, since the pyrolysis residues do not leachate organic and inorganic compounds in amounts that compromise the environment.

4.6 *Incineration*

Incineration consists of a dry combustion to produce an inert ash. The ash from sewage sludge is usually disposed of in a sanitary landfill. The fuel requirements depend on the fuel value of the sludge solids and the water content. Auxiliary fuel is required only during the startup of the incinerator. The most common types of incinerators used are the multiple-heart type and the fluidized-bed incinerator.

According to Fytili and Zabaniotou (2008), the advantages of incineration are reduction in sludge volume; decomposition of toxic organic pollutants; regeneration of heat energy, since it is comparable to brown coal; and odor elimination or reduction. However, it may cause air pollution and hence requires expensive treatment of the emissions.

4.7 *Low cost adsorbents*

One application that has received increasing attention during the last years is the utilization of sewage sludge as a low-cost alternative adsorbent for the removal of pollutants from wastewater. Several works have been reported in this field and some works are discussed in the next paragraphs.

Lister and Line (2001) performed the adsorption of Cd, Cu, Pb, and Zn ions from aqueous solution by sewage sludge and other wastes. They showed that sewage sludge was the most effective biosorbent of the waste products for all metal ions examined, adsorbing, for example, up to 39.3 mg g^{-1} of Pb at an initial concentration of 77.8 mg L^{-1}. A few years later, Otero *et al.* (2003a) investigated the dye binding capacity of adsorbents produced from sewage sludge generated by urban and agrofood industry wastewater treatment plants. Three types of biosolids were used, dried, pyrolyzed, and both chemically activated sewage sludge were used in single batch liquid-phase adsorption tests. Methylene blue was chosen as adsorbate because dyeing industry effluents represent an important environmental problem. The obtained results showed that the dried urban sewage sludge is the most efficient material for removing methylene blue dye from aqueous solution. In other study, Otero *et al.* (2003b) investigated the removal of crystal violet, indigo carmine and phenol using two different adsorbent particle sizes obtained from sewage sludge. Finally, these authors proposed that activated carbons made from sewage sludge show promise for the removal of organic pollutants from aqueous streams. Also, Annadurai *et al.* (2003) reported the application of microwave thermal treatment sludge in dye removal from water.

Activated sludge is one of the most abundant sources of microbial biomass (Choi and Yun, 2006). Certain types of microbial biomass can retain relatively high quantities of metals. Choi and Yun (2006) have tested capacities of metal loading for sewage sludge, anaerobically digested sludge, drinking water treatment plant sludge and leachate sludge. Cadmium was chosen as the sorbate because of its simple water chemistry (Choi and Yun, 2006). The results demonstrated that sewage sludge removed Cd most efficiently from aqueous solution (0.38 mmol/g), and showed the highest desorption efficiency (26.3%). This sorbent may have a potential use as high-value biosorbent of heavy metals.

Aksu and Gönen (2006) studied the simultaneous biosorption of phenol and chromium(VI) ions onto Mowital®B30H resin immobilized activated sludge using a binary mixture. In addition, the results were compared to those obtained using single phenol or chromium(VI) solutions in a continuous packed bed column. The equilibrium uptake (or column biosorption capacity) of each pollutant was determined by evaluating the breakthrough curves obtained at different input concentrations varying from 50 to 500 mg L^{-1} in single and binary systems. The maximum column biosorption capacities were of 9.0 mg g^{-1} and 18.5 mg g^{-1} for phenol and chromium(VI), respectively. On the other hand, adsorbents for the removal of hydrogen sulfide from moist air have been prepared from sewage sludge, waste oil sludge and their 50:50 mixture by pyrolysis at 650 and 950°C (Bandosz and Block, 2006). The catalytic performance of these adsorbents was evaluated in hydrogen sulfide reactive adsorption via a dynamic test. The adsorbents displayed a high capacity for hydrogen sulfide removal and high selectivity for its conversion to elemental sulfur.

Thawornchaisit and Pakulanon (2007) evaluated the potential application of dried sewage sludge as a biosorbent for removing phenol from aqueous solutions. They studied the influence of several factors such as pH, initial phenol and biosorbent concentration. The results of this study demonstrated that biosorption was strongly influenced by pH. This research showed that dried sewage sludge could be used as a low-cost biosorbent.

Taking into account that urban sewage sludge is carbonaceous in nature, this material may be considered a potential source for the production of activated carbon. Rozada *et al.* (2008) evaluated two materials produced from sewage sludge by pyrolysis of dried sewage sludge and by chemical activation of dried sewage sludge with $ZnCl_2$ followed by pyrolysis. The aim was to study the application of these materials for metal removal from water. Although the latter displayed higher capacity, both materials were able to adsorb Hg, Pb, Cu and Cr (III) in this order. The results indicated the potential application of these sewage sludge based adsorbents for the treatment of metal polluted effluents (Rozada *et al.*, 2008).

4.8 *Landfill and agriculture applications*

Landfill and land application have been suggested to be the most economical sludge disposal methods (Champagne, 2007; Singh and Agrawal, 2008). The simplicity of the practice for sewage treatment and disposal, the natural fertilization of vegetation derived from this practice, and water recycling and reuse were recognized as benefits in these early uses of land application of wastewaters (Babel and del Mundo, 2006). In developing countries, the disposal of these types of biosolids in open fields is a common practice, mainly due to the scarcity of appropriate disposal facilities (Babel and del Mundo, 2006). Some reviews related to the topic of land application of sewage sludge have been published in the literature during the last years. For example, Wang (1997) analyzed this topic for the case of China and concluded that proper land utilization of stabilized sewage sludge can make a positive contribution to agriculture, forestry, horticulture, and city development.

To apply this disposal strategy, it is necessary to know the waste composition as this alternative can result in serious problems and leaching of heavy metals and other contaminants to groundwater, surface water and soils (Babel and del Mundo, 2006). However, land application of sewage sludge appears to be reasonable, mostly because of its phosphorus content, nutrients and organic matter that can provide soil benefits (Lassen *et al.*, 1984). Compared to landfilling and incineration, utilization of sludge for agricultural use is considered as the best alternative for sludge disposal because it recycles both nutrients and organic matter (García-Delgado *et al.*, 2007). According to Mininni and Santori (1987), in the European Economic Community about 75% of sewage is treated and 40% of the residual sludges are used in agriculture producing significant savings in the use of fertilizers.

Depending on the composition, sewage sludge quality may vary. However, it is important to take into account that inorganic nitrogen as NO_3^{-1} and NH_4^+ becomes immediately available; organic nitrogen becomes available after mineralization, which depends on soil type and tends largely to diminish with time. The soluble forms of phosphorus are $H_2PO_4^-$ and HPO_4^{2-} and the percentage P-utilization can easily be predicted from the amount extracted by ammonium chloride. Potassium concentration in sewage sludge is frequently negligible. However, crop requirements for this element are high and often comparable to those for nitrogen. Organic matter mainly affects soil structure by increasing soil porosity, stability of aggregates and water retention; other properties affected by organic matter addition are pH and cation exchange capacity, which tend to increase.

Ingelmo *et al.* (1998) used different materials, including sewage sludge, to produce alternative substrates for ornamental plants and to improve the re-vegetation of a closed landfill. The cost of substrates was reduced and the quality of plants produced was not diminished. The authors reported a good quality and fast bushy authoconous vegetation covering of the landfill by using as treatment a superficial layer of dry sewage sludge. This result was similar to that obtained with standard and environmentally aggressive revegetating procedures. On the other hand, the findings of Sanchez-Monedero *et al.* (2004) indicated that land application caused an increase of both size and activity of soil microbial biomass and this was related to the degree of stabilization of the composting mixture. Sewage sludge stabilization through composting reduced the perturbance of the soil microbial biomass.

Stabilization of organic wastes before soil application is advisable for the lower perturbation of soil equilibrium status and the more efficient C mineralization. Hernandez-Apaolaza *et al.* (2005) indicated that mixtures of sewage sludge with others waste materials are suitable substrates in the production of ornamental plants. Finally, Ghini *et al.* (2007) evaluated the effect of sewage sludge on soil suppressiveness to the pathogens *Fusarium oxysporum* f. sp. *Lycopersici*, *Sclerotinia sclerotiorum* and *Ralstonia solanacearum* on tomato, *Sclerotium rolfsii* on bean, *Rhizoctonia solani* on radish, and *Pythium* spp. on cucumber. They reported that the effects of sewage sludge varied depending on the presence and nature of pathogen, methodology applied and on the time interval between the sewage sludge incorporation and soil sampling.

It is important to reiterate that sewage sludge also contain contaminants including metals, pathogens, and organic pollutants (Harrison *et al.*, 2006). In first instance, the sludge re-use in agriculture should be strongly associated with the characterization and quantification of pathogens in these wastes (Gaspard *et al.*, 1997). With respect to heavy metals, Mininni and Santori (1987) pointed out that the countries with more experience in sludge utilization reported in 1987 were France, Germany and the U.K. In these places, the disposal of wastes with high concentrations of heavy metals was allowed. Recently, there has been an increased concern because of the legal criteria for the heavy metal concentration in sewage sludge (Babel y del Mundo, 2006). In fact, Cd appears to be the most important contaminant because it can be accumulated from the soil by certain food plants (Dean and Suess, 1985). Therefore, heavy metal reduction before sludge application is a necessary step to achieve a more sustainable form of sludge management (Babel and del Mundo, 2006).

A study of various technologies for the extraction of heavy metals from sludge shows a broad range in metal extraction efficiencies, which can be attributed to differences in sludge composition, pretreatment of sludge, concentration and forms of metals. Specifically, for heavy metal removal from sludge, several strategies can be approached: a) Chemical extraction employing inorganic acids, organic acids, chelating agents and some inorganic chemicals (Lo and Chen, 1990; Veeken and Hamelers,1999; Babel and del Mundo, 2006); b) Bioleaching where microorganisms are applied to extract metals from ores through metabolic action of microorganisms (Tyagi *et al.*, 1988; Blais *et al.*, 1993;Chan *et al.*, 2003; Sreekrishnan *et al.*, 1993; Babel and del Mundo, 2006); c) Electrokinetics, which is based on electrokinetic phenomena that occurs when the soil/sludge is electrically charged with direct current. This extraction can be accomplished by electrodeposition, precipitation or ion exchange (Kim *et al.*, 2005; Wang *et al.*, 2005; Babel and del Mundo, 2006; Yuan and Weng, 2006); and d) Supercritial fluid extraction that is a liquid extraction process employing compressed fluids which are usually either liquids or gases, under supercritical conditions instead of normal solvents (Wai and Wang, 1997; Kersch *et al.*, 2004; Babel and del Mundo, 2006).

Recently, Harrison *et al.* (2006) have pointed out that available regulations generally require pathogen reduction and periodic monitoring for some metals prior to land application; however, there are no requirements for testing the presence of organic chemicals in sewage sludge, at least, in the U.S. Based on this fact, these authors have examined the peer-reviewed literature and official governmental reports. They found data for 516 organic compounds which were grouped into 15 classes. Concentrations were compared to EPA risk-based soil screening limits (SSLs) where available. For 6 of the 15 classes of chemicals identified, there were no SSLs. For the 79 reported chemicals which had SSLs, the maximum reported concentration of 86% exceeded at least one SSL. Eighty-three percent of the 516 chemicals were not on the EPA established list of priority pollutants and 80% were not on the EPA's list of target compounds. Thus, analyses targeting these lists will detect only a small fraction of the organic chemicals in sludges. Analysis of the reported data shows that more data has been collected for certain chemical classes such as pesticides, PAHs and PCBs than for others that may pose greater risk such as nitrosamines. Therefore, the results of this work reinforced the need for a survey of organic chemical contaminants in sewage sludges and for further assessment of the risks they pose. Thus, proper analysis of several compounds that are present in the sewage sludge must be performed. Specifically, it is convenient to quantify at least the content of pathogens, metals, halogenated and aromatic hydrocarbon solvents, volatile organics, chlorinated pesticides, petroleum hydrocarbons, phenols, surfactants, among others (Schnaak *et al.*, 1997; Bright and Healey, 2003; Harrison *et al.*, 2006).

The benefits and limitations of sewage sludge application as an organic fertilizer need to be investigated for various procedures of sludge treatment. Under this context, several studies have been performed to determine the most suitable type of sludge as organo-mineral fertilizer (Selivanovskaya *et al.*, 2001). Another potential way for sludge wastes utilization is to develop artificial soil based on these biosolids. Sewage sludge may contain significant amounts of nitrogen, phosphorous and trace elements. In Singapore, Stabnikova *et al.* (2005) found that subsoil mixed with sewage sludge could be a better medium for *I. aquatica* growth than with either ingredient alone. The addition of sewage sludge provided nutrients for plant growth. The weight of *I. aquatica* increased 12.6 and 11.5 times at 2% application rate of raw sewage sludge and dried sewage sludge respectively, in comparison with the weight of plants grown in artificial soil without addition of sludge. The artificial soil could be recommended for urban landscaping and gardening (Stabnikova *et al.*, 2005).

Recently, Rebah *et al.* (2007) indicated that wastewater sludge has also shown good potential for inoculant production as a growth medium and as a carrier (dehydrated sludge). Sludge usually contains nutrient elements at concentrations sufficient to sustain rhizobial growth. In some cases, growth conditions can be optimized by sludge pre-treatment or by the addition of nutrients. Inoculants produced in wastewater sludge are efficient for nodulation and nitrogen fixation with legumes as compared to standard inoculants. This new approach described in this review offers a safe environmental alternative for both waste treatment/disposal and inoculant production.

Verma *et al.* (2005) studied the feasibility of production of antagonistic *Trichoderma* sp. conidial spores using wastewater sludge as a raw material employing different suspended solids concentration (10–50 g/l) in shake flasks. This study successfully demonstrated the potential of wastewater sludge as a raw material for production of value added product, aiding in sludge management and proliferation of eco-friendly and economical biocontrol agents. In addition, wastewater sludges can be used as biopesticides, according to the study performed by Yezza *et al.* (2006). In this work, secondary sludge from three different wastewater treatment plants were used as raw materials for the production of *Bacillus thuringiensis* (Bt) based biopesticides in a pilot scale fermentor (100 L working volume). Their results showed that the secondary sludges are found to be suitable raw materials for high potency Bt biopesticide production.

Sewage sludge may also be used as soil amendment for metal stabilization. In this context, sewage sludge obtained from the Baltimore City wastewater treatment plant and composted with woodchips and sawdust at the Baltimore City composting (Orgro®) facility was used to reduce the hazard of Pb found in urban soils in Baltimore (Farfel *et al.*, 2005). High concentration of Fe and P found in biosolid composts reduced the bioavailability of Pb, and it was used in nine urban yards to prove its capability to reduce Pb exposure to residents. The Pb concentrations were above 800 mg Pb kg^{-1}. Soils were prepared before the treatment by rototilling at 20 cm depth. After that, the soils were treated with 6–8 cm depth of Orgro® biosolids compost, wich is rich in Fe and P. Then the soil was well mixed and rototilled. Kentucky bluegrass (*Poa pratensis*) was seeded in the place and it was well established in the yards. Different soil sampling was done during different stages of the experiment. Total Pb content and bioaccesible Pb was determined in every sample. Other areas near the site of investigation were used as controls. After one year, the grass grown in the treated soil was healthy and the bioaccesible Pb concentrations compared to pre-tillage were reduced to 64–67%. Changes in the total and bioaccesible Pb concentrations in the controls during the experiment were insignificant. This work showed the possibility of soil remediation in urban areas with high concentrations of Pb where mainly children could be at risk of Pb exposure.

The application of treated sewage sludge (biosolids) in semi-arid grass-lands has shown positive results (Wester *et al.*, 2003). For instance, in a 900-acres grassland and shrubland area in the northern Chihuahuan Desert, near Sierra Blanca, in Hudspeth County Texas, biosolids were applied at rates of 0, 3, 8, 15, and 40 dry tons per acre. This study started in 1992 and it was part of a ten-year research program with the goal of investigating the effects of biosolids applied at the Chihuahuan

Desert. The conclusions obtained from this study were: the biosolids helped in reducing soil erosion and water runoff; soil water infiltration was increased; the quality of the water was not affected; enhanced plant water use efficiency was observed; and an increase in forage production and quality was obtained.

In other study in the Chihuahuan Desert, the application of biosolids at rates of 0 (control), 7, 18 or 34 Mg ha^{-1} were studied (Jurado and Wester, 2001). The biosolids were applied to tobograss (*Hilaria mutica* (Buckl.) Benth). The tobograss herbage yield and total N concentration was monitored during the four-year study. The results showed that the application of biosolids was beneficial to this species.

4.9 *Biofuels*

Under the perspective of the Kyoto Protocol, countries all around the world have a compromise to reduce their greenhouse gas emissions. Ethanol blended gasolines have the potential to significantly contribute to this. Ethanol is an alternative fuel derived from biologically renewable resources and can be employed to replace octane enhancers and aromatic hydrocarbons. A potential source for low-cost ethanol production is to utilize lignocellulosic materials (crop residues, sawdust, woodchips, sludge, livestock manure). However, the challenges are generally associated with the low yield and the high cost of the hydrolysis process. (Champagne, 2007).

The only natural, renewable carbon resource large enough to be used as a substitute for fossil fuels is biomass, which includes all water and land-based organisms, vegetation and trees, as well as all dead and waste biomass such as municipal solid waste, sludge, agricultural residues and certain types of industrial waste (Champagne, 2007). Unlike fossil fuel deposits, biomass is considered a potentially sustainable renewable resource since a short period of time is required to replace what is used for bio-fuel production and energy generation. A good solution may be to subject biosolids or sludge to fermentative processes which could serve to stabilize the pathogens, provide the time necessary for the precipitation of toxic chemicals, and produce biogas. This option may be economically beneficial for municipalities.

In the literature, there are several studies related to the application of sewage sludge as fuel or to obtain biofuels (Yokoyama *et al.*, 1987; Boocock *et al.*, 1992; Dote *et al.*, 1992, Zabaniotou and Theofilou, 2008). Several years ago, Yokoyama *et al.* (1987) produced liquid fuels by means of the heating of sewage sludge under pressurized nitrogen over the temperature range 250–340°C in the presence of sodium carbonate. The yields and properties of the heavy oil produced depended strongly on the catalyst loading and reaction temperature. Liquid fuels having heating values of $\approx$33 MJ kg^{-1} were obtained in 50 wt% yields on an organic basis. Later, Boocock *et al.* (1992) studied the production of a low-nitrogen and low-sulphur substrate for pyrolysis to liquids by using the lipid fraction of a dried raw Atlanta sludge extracted with toluene. In recent times, Zabaniotou and Theofilou (2008) used sewage sludge as an alternative fuel at cement kilns. Because of cement plants burn fuel at 1400°C, the new sewage sludge-based fuel does not emit the harmful dioxin.

5 REDUCTION OF *SALMONELLA SPP* FROM SEWAGE SLUDGE GENERATED AT A MUNICIPAL WASTEWATER TREATMENT PLANT AND ITS POTENTIAL USE IN SOIL CONDITIONING: A CASE STUDY FROM MEXICO.

The reduction of *Salmonella* spp from sewage sludge using anaerobic digestion has been reported (Espinoza-Arzate, 2005; Hernández-Hernández, 2007). Chemical and biological characterization of sewage sludge was performed in order to determine nutrient, heavy metal and pathogens content. Subsequently, the material was subjected to anaerobic digestion to determine the effect on pathogen reduction. The main goal of this research was to evaluate the feasibility of using treated sludge as soil conditioner.

Table 1. Classification of biosolids according to the content of potentially toxic elements (NOM-004-SEMARNAT-2002).

	Classification	
Contaminant	Excellent mg kg^{-1} DW[a]	Good mg kg^{-1} DW[a]
As	41	75
Cd	39	85
Cr	1 200	3 000
Cu	1 500	4 300
Pb	300	840
Hg	17	57
Ni	420	420
Zn	2 800	7 500

[a]Dry weight

Table 2. Classification of biosolids according to their pathogen and parasite content (NOM-004-SEMARNAT-2002).

Classification	Biological indicator (MPN Fecal coliforms g^{-1} DW)	Pathogen (MPN[a] *Salmonella* spp g^{-1} DW)	Parasite (Helminth eggs g^{-1} DW)
A	<1 000	<3	<1
B	<1 000	<3	<10
C	<2 000 000	<300	<35

[a]Most probable number

5.1 *Generation of sludge at the water treatment plant*

Wastewater received at a wastewater treatment plant in Guanajuato (PTARG–planta de tratamiento de aguas residuales en Guanajuato-) originated from domestic and commercial activities. This plant displays a minimum capacity of 140 L s^{-1} and a maximum of 220 L s^{-1}. The sequence followed for wastewater treatment in this plant was as follows: coarse screening; sedimentation; fine screening; anaerobic biological treatment; aerobic biological treatment; clarification; disinfection using UV light, and sludge treatment through aerobic digestion for 30 days. The sludge obtained in this plant was stabilized through aerobic digestion with a maximum retention time of 30 days. Subsequently, biosolids were concentrated and dehydrated using press filters. The daily production of sludge in this plant ranged from 2 to 4 metric tons per day (Espinoza-Arzate, 2005), which were completely disposed of in a landfill.

In Mexico, the Norma Oficial Mexicana-004-SEMARNAT-2002 (NOM-004-SEMARNAT-2002) provides the specifications and values for maximum contaminant levels in sludge and other biosolids for their use and final waste disposal with the aim of protecting the environment and human health. Tables 1 and 2 present the classification of biosolids according to their potentially toxic elements (PTE) content and their pathogen and parasite content, respectively. Table 3 displays the allowed uses of biosolids according to their characteristics.

Sewage sludge generated at the PTARG usually displays large concentrations of pathogens; thus, a subsequent use in this form is not possible. Consequently, final disposal represents a big problem due to the large volumes generated and the risk they represent. Studies aimed at using

Table 3. Allowed uses of biosolids according to their characteristics (NOM-004-SEMARNAT-2002).

Type	Class	Potential use
Excellent	A	(a) Urban use. Allowed direct contact with the public during its application (b) Those established for Classes B and C
Excellent or good	B	(a) Urban use. Direct contact with the public during its application not allowed (b) Those established for Class C
Excellent or good	C	(a) Use in forests (b) Soil conditioning (c) Agriculture

Table 4. Proper conditions for biosolids sampling.

Parameter	Temperature (°C)	Maximum storage time
Fecal coliforms and *Salmonella* spp	4	48 h
Helminth eggs	4	30 d
As, Cd, Cu, Ni, Pb, Zn	4	180 d
Hg	4	13 d
Total solids	4	24 h

sewage sludge in agriculture are well documented (Celis *et al.*, 2006; Rebah *et al.*, 2007; Singh and Agrawal, 2008; Johansson *et al.*, 2008). However, since chemical and biological composition of sludge will depend on several variables, the characterization of these materials is extremely important before any application is intended.

Guanajuato in central Mexico is located in an area where agricultural practices are common. Thus, the objective of this study was to perform de chemical analysis of sewage sludge generated at a PTARG in order to determine if they contain the nutrients needed in a fertilizer, as well as their contaminant content. Moreover, biological characterization was carried out to obtain information about the potential risk associated with the presence and content of pathogens and parasites. Finally, sewage sludge was subjected to anaerobic digestion to determine the effect on pathogen reduction to levels that comply with the NOM-004-SEMARNAT-2002.

5.2 *Methodology*

The sewage sludge used in this study was obtained from a wastewater treatment plant located in Guanajuato in central México. This plant mainly receives domestic and commercial discharges. The first part of this investigation was aimed at quantifying fecal coliforms, *Salmonella* spp, helminth eggs, nutrients and heavy metals so as to an adequate classification could be achieved.

Composite samples were obtained at different times of the year of 2005. For this purpose, several grab samples were taken and then combined to form the composite sample for every month from February to June of 2005. This was done so that representative data could be obtained. Samples were stored for subsequent analyses. Temperature and maximum storage time for proper sampling are given in Table 4.

Every analysis performed in this work was done according to the Mexican regulation given in NOM-004-SEMARNAT-2002.

Table 5. Most probable number of several parasites and pathogens in sewage sludge samples obtained at a wastewater treatment plant in Guanajuato from February to May.

Month	Fecal coliforms (MPN g^{-1} TS)	*Salmonella* spp (MPN g^{-1} TS)	Helminth eggs (# eggs per 2 g DW)
February	4.6×10^6	2.4×10^4	5
March	7.5×10^5	4.3×10^4	3
April	2.4×10^6	2.4×10^4	6
May	1.1×10^7	4.2×10^4	3
June	1.1×10^7	2.4×10^5	3

Table 6. Concentrations of potential toxic elements in sewage sludge samples obtained at a wastewater treatment plant in Guanajuato from February to May[a].

	PTE concentration (mg kg^{-1} DW)					
Month	As	Cr	Cu	Pb	Ni	Zn
February	21.8	40.0	20.4	57.4	29.0	79.0
March	27.6	53.0	24.0	72.0	31.3	82.5
April	23.8	37.5	176.0	254.1	ND[b]	470.0
May	20.0	14.0	154.0	224.0	ND	420.0
June	28.5	24.7	244.3	265.0	ND	544.4

[a] Cd was not detected in samples
[b] Non detected

5.2.1 *Anaerobic digestion for reduction of Salmonella spp*
A series of anaerobic bioreactors were constructed using 250-mL round bottom flasks connected to a bottle containing 4% NaOH for biogas recovery. The composition in the bioreactors was as follows: 28 g PTARG sewage sludge and 20 mL of growth medium containing the following nutrients at the concentrations given in parenthesis: $FeCl_2.4H_2O$ (2000 mg L^{-1}), H_3BO_3 (50 mg L^{-1}), $ZnCl_2$ (50 mg L^{-1}), $CuCl_2$ (38 mg L^{-1}), $MnCl_2.4H_2O$ (500 mg L^{-1}), $(NH_4)6Mo_7O_{24}.4H_2O$ (50 mg L^{-1}), $AlCl_3.6H_2O$ (90 mg L^{-1}), $CoCl_2.6H_2O$ (2000 mg L^{-1}), $NiCl_2.6H_2O$ (142 mg L^{-1}), $Na_2SeO.5H_2O$ (164 mg L^{-1}). Bioreactors were maintained at a temperature of 37°C. Biogas production and *Salmonella* spp quantification were performed every three days for 48 days.

5.3 *Results and discussion*

In Mexico, these types of biosolids should comply with the regulation specified in the NOM-004-SEMARNAT-2002. Table 5 presents the results for the quantification of several pathogen and parasite components of the sewage sludge obtained at a PTAR in Guanajuato. As is shown, except for the month of March, the values obtained for fecal coliforms surpass the numbers established by the Mexican regulation (Table 2). Thus, it is not possible to classify these residues as Class A, B or C. Those residues collected in March can be classified as Class C. Additionally, the most probable number (MPNs) for *Salmonella* spp indicates that the sludge contains an enormous amount of these pathogens, not complying with the Mexican regulation. On the other hand, according to the results for helminth eggs quantification showed that the residues can be classified as Class B.

Table 6 displays the quantification of contaminants. No Cd was detected in the samples. Additionally, Ni was only present in samples obtained in February and March; however, the concentration of this metal was low as compared to the values given in Table 2. On the other hand, concentrations of As remained in the range of 20 to 28.5 mg kg^{-1} DW, while those for Cr varied between 14.0 and 53.0 mg kg^{-1} DW. Cu, Pb, and Zn showed interesting differences; in April, May, and

Table 7. Content of N, P, and Organic Matter in sewage sludge samples obtained at a PTARG from February to May.

Month	Ammonium N (%)	Total N (%)	Total P (as % P_2O_5)	Extractable P ($g\,kg^{-1}$ DW)	O.M[a] (%)
February	0.3	5.3	16.7	3.6	14.4
March	1.4	7.8	19.4	6.9	14.9
April	2.9	10.0	18.8	2.4	13.9
May	1.2	36.6	19.3	17	15.4
June	0.7	10.3	20.1	16	12.3

[a]Organic Matter

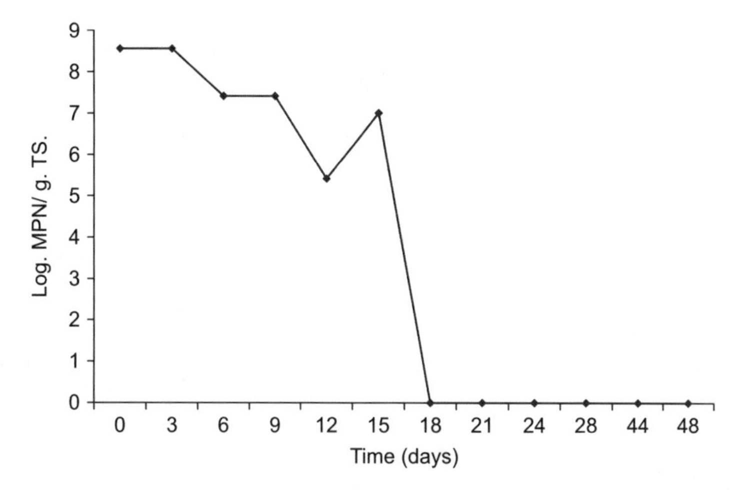

Figure 1. Fecal coliform decay in an anaerobic reactor containing sewage sludge from a wastewater treatment plant in Guanajuato. Data is given as Log (Most probable number).

June, concentrations of these metals where considerably higher than those found in February and March. However, comparing the results with the values given by the Mexican regulation, none of the elements exceed the limits. Thus, according to the concentrations of metals, sewage sludge from this PTARG can be classified as type Excellent.

Table 7 shows the results for N, P and organic matter content. As displayed, ammoniacal nitrogen is only a small fraction of total N. Similarly, extractable P represents a small amount of the total P. Since extractable P represents P soluble in water, and in consequence the one usable for plants, it might be necessary to add P to obtain the required concentrations. The concentration of organic matter indicates that, according to NOM-021-RECNAT-2000, these materials can be classified as a volcanic soil with high organic matter content. Thus, the dehydrated sludge might be able to provide organic matter for weathered soils regeneration. These results should be taken into account if these residues are intended to be used as soil fertilizers.

5.3.1 *Reduction of pathogens in the sewage sludge through anaerobic digestion*

Anaerobic bioreactors were maintained at 37°C for 48 days. Fecal coliforms and *Salmonella* spp quantifications, as well as biogas production were performed every three days. Figure 1 presents the content of fecal coliforms, Figure 2 displays the content of *Salmonella* spp through time, and Figure 3 shows the cumulative production of biogas. As results show, after 18 days, the presence of fecal coliforms was reduced to zero (Figure 1). At this time, the MPN for *Salmonella* spp was of

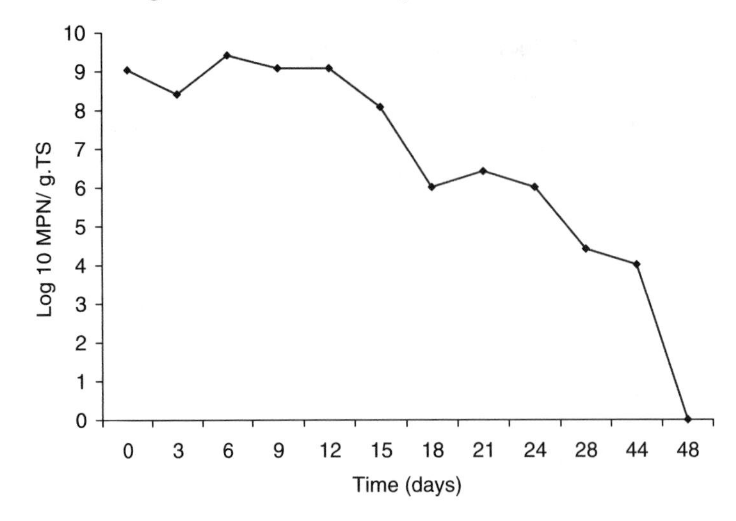

Figure 2. *Salmonella* spp decay in an anaerobic reactor containing sewage sludge from a wastewater treatment plant in Guanajuato.

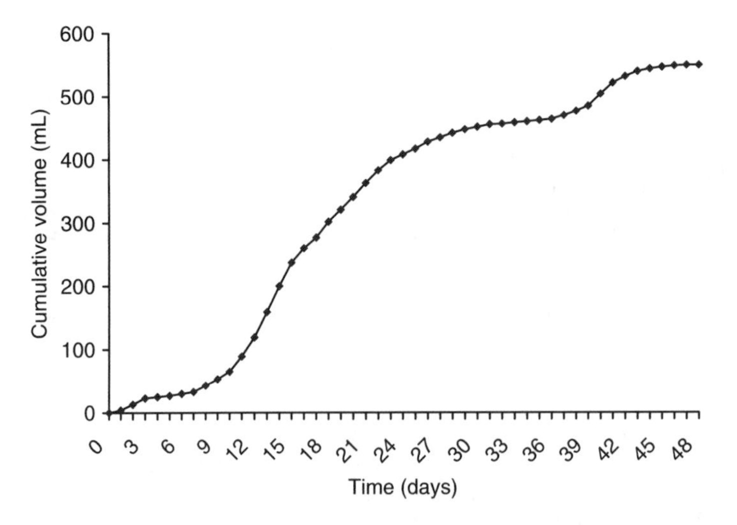

Figure 3. Cumulative biogas production in an anaerobic reactor containing sewage sludge from a wastewater treatment plant in Guanajuato.

1×10^6. This amount still exceeded the Mexican regulation. However, as shown in Figure 2, after 48 days this number was reduced to zero. Helminth eggs quantification showed a result of 10 eggs per g total solids (TS). Thus, according to the data obtained in this research, the biosolids can be classified as Type B.

Since metal content allows the classification of these materials as type excellent and after 48 day of treatment pathogens content complies with the Mexican regulation to be considered biosolids type B, it is possible to propose the use of the digested sludge for use in agriculture.

The total volume of biogas production was 550 mL. Therefore, the biogas production capacity of the sewage sludge obtained at this PTARG is about 20 mg g^{-1}. This biogas could be use for the generation of the energy needed to maintain the temperature in the anaerobic bioreactors, contributing to the sustainability of the system.

The nutritional and biological characteristics of these biosolids make them good candidates for use in agriculture. However, the characterization and quantification of toxic organic pollutants should be performed in order to ensure no other pollutants are being added to the soils. Thus, future research for this specific study should include appropriate analyses of organic contaminants content as well as the determination of long term transport of inorganic and organic sludge components in the environment.

6 CONCLUDING REMARKS

Several alternatives for proper management, disposal, pathogen reduction and the potential uses of sewage sludge have been presented. As this review shows, the identification, quantification, and reduction of potentially toxic elements, organic compounds, and pathogens are crucial aspects of sewage sludge management. With respect to pathogen reduction, sewage sludge can be subjected to anaerobic digestions, chemical treatment, radiation, and membrane filtration as single processes. However, combinations of these techniques have reported good results. Their utilization in agriculture appears to be the most accessible strategy mainly in developing countries. Recently, special attention has been given to determine their potential as a source of renewable energy. In fact, in several parts of the world, this strategy is already being exploited. Whether used in agriculture, as a source of energy or utilized for other purposes, efforts are still needed to monitor potential harmful effects on human and environmental health.

ACKNOWLEDGMENTS

The authors acknowledge the Universidad de Guanajuato, Instituto Tecnológico de Aguascalientes, CONCYTEG and CONACYT.

REFERENCES

Adingra, A.A., Arfi, R. 1998. Organic and bacterial pollution in the Ebrié lagoon, côte d'Ivoire. *Marine Pollution Bulletin* 36(9), 689–695.

Andrés F., Walter I., Tenorio J.L. 2007. Revegetation of abandoned agricultural land amended with biosolids. *Science of The Total Environment* 378(1–2), 81–83.

Ahn Y.H., Speece R.E. 2006. Elutriated acid fermentation of municipal primary sludge. *Water Research* 40(11), 2210–2220.

Aiyuk S., Forrez I., Lieven K., Haandel A., Verstraete W. 2006. Anaerobic and complementary treatment of domestic sewage in regions with hot climates—A review. *Bioresource Technology* 97(17), 2225–2241.

Aksu Z., Gönen F. 2006. Binary biosorption of phenol and chromium(VI) onto immobilized activated sludge in a packed bed: Prediction of kinetic parameters and breakthrough curves. *Separation and Purification Technology* 49(3), 205–216.

Al-Bachir M., Al-Adawi M. A., Shamma M. 2003. Synergetic effect of gamma irradiation and moisture content on decontamination of sewage sludge. *Bioresource Technology* 90(2), 139–143.

Ahlstrom S.B. 1977. Irradiation of municipal sludge for agricultural use. *Radiation Physics and Chemistry* 25(1–3), 1–10.

Annadurai G., Juang R.S., Yen P.S., Lee D.J. 2003. Use of thermally treated waste biological sludge as dye absorbent. *Advances in Environmental Research* 7(3), 739–744.

Babel S., del Mundo D. 2006. Heavy metal removal from contaminated sludge for land application: A review. *Waste Management* 26, 988–1004.

Bandosz T.J., Block K. 2006. Effect of pyrolysis temperature and time on catalytic performance of sewage sludge/industrial sludge-based composite adsorbents. *Applied Catalysis B. Environmental* 67(1–2), 77–85.

Blais J.F., Tyagi R.D., Auclair J.C., Lavoie M.C. 1992. Indicator bacteria reduction in sewage sludge by a metal bioleaching process. *Water Research* 26(4), 487–495.

Blais J.F., Tyagi R.D., Auclair J.C. 1993. Bioleaching of metals from sewage sludge: Effects of temperature. *Water Research* 27(1), 111–120.

Boocock D.G.B., Konar S.K., Leung A., Ly L.D. 1992. Fuels and chemicals from sewage sludge 1. The solvent extraction and composition of a lipid from a raw sewage sludge. Fuel 71(11), 1283–1289.

Boost M. V., Poon C. S. 1998. The effect of a modified method of lime-stabilisation sewage treatment on enteric pathogens. *Environment International* 24(7), 783–788.

Borowski S., Szopa J.S. 2007. Experiences with the dual digestion of municipal sewage sludge. *Bioresource Technology* 98(6), 1199–1207.

Bright D.A., Healey N. 2003. Contaminant risks from biosolids land application: Contemporary organic contaminant levels in digested sewage sludge from five treatment plants in Greater Vancouver, British Columbia. *Environmental Pollution* 126(1), 39–49.

Carballa M., Manterota G., Larrea L., Ternes T., Omil F.,Lema J.M. 2007. Influence of ozone pre-treatment on sludge anaerobic digestion: Removal of pharmaceutical and personal care products. *Chemosphere* 67(7), 1444–1452.

Carmody D.J., Pearce J.B., Yasso W.E. 1973. Trace metals in sediments of New York Bight. *Marine Pollution Bulletin* 4(9), 132–135.

Castillo E.F., Cristancho D.E., Arellano A.V. 2006. Study of the operational conditions for anaerobic digestion of urban solid wastes. *Waste Management* 26(5), 546–556.

Celis, J., Sandoval, M., Zagal, E., Briones, M. 2006. Effect of sewage sludge and salmon wastes applied to a Patagonian soil on lettuce (*Lactuca sativa* L.) germination. *Revista de la Ciencia del Suelo y Nutrición Vegetal* 6(3), 13–25.

Champagne P. 2007. Feasibility of producing bio-ethanol from waste residues: A Canadian perspective. *Resources, Conservation and Recycling* 50(3), 211–230.

Chale-Matsau, J.R.B. 2005. Persistance of human pathogens in a crop grown from sewage sludge treated soil. Ph,D. Thesis. *University of Pretoria, Pretoria.*

Chan L.C., Gu X.Y., Wong J.W.C. 2003. Comparison of bioleaching of heavy metals from sewage sludge using iron- and sulfur-oxidizing bacteria. *Advances in Environmental Research* 7(3), 603–607.

Choi S.B., Yun Y. 2006. Biosorption of cadmium by various types of dried sludge: An equilibrium study and investigation of mechanisms. *Journal of Hazardous Materials* 138(2–16), 378–383.

CWMI (Cornell Waste Management Institute), Center for the Environment. The Beneficial Uses of Biosolids/Sludge. Fact Sheet 6. (2008) Ithaca New York. Accesed on: Feb 26, 2008. http://cwmi.css.cornell.edu/Sludge/Beneficial.pdf

Dean R.B., Suess M.J.1985. The risk to health of chemicals in sewage sludge applied to land. *Waste Management & Research* 3(3), 251–278.

Dominguez A., Menendez J.A., Inguanzo M., Pís J.J. 2006. Production of bio-fuels by high temperature pyrolysis of sewage sludge using conventional and microwave heating. *Bioresource Technology* 97, 1185–1193.

Dote Y., Hayashi T., Suzuki A., Ogi T. 1992. Analysis of oil derived from liquefaction of sewage sludge. *Fuel* 71(9), 1071–1073.

El-Hadj T.B., Dosta J., Márquez-Serrano R., Mata-Álvarez J. 2007. Effect of ultrasound pretreatment in mesophilic and thermophilic anaerobic digestion with emphasis on naphthalene and pyrene removal. *Water Research* 41(1), 87–94.

Espinoza-Arzate, M.C. 2005. Valoración del lodo generado en una planta de tratamiento de aguas residuals de Guanajuato Centro. Tesis, Q.F.B. Universidad de Guanajuato.

Farfel M.R., Orlova A.O., Chaney R.L., Lees P.S.J., Rohde C., Ashley P.J. 2005. Biosolids compost amendment for reducing soil lead hazards: a pilot study of Orgro®amendment and grass seeding in urban yards. *Science of the Total Environment* 340, 81–95.

Friedman G.M., Mukhopadhyay P.K., Moch A., Ahmed M. 2000. Waters and organic-rich waste near dumping grounds in the New York Bight. *International Journal of Coal Geology* 43(1–4), 325–355.

Fukushi K., Babel S., Burakrai S. 2003a. Survival of *Salmonella* spp. in a simulated acid-phase anaerobic digester treating sewage sludge. *Bioresource Technology* 86(1), 53–57.

Fukushi K., Babel S., Burakrai S. 2003b. Survival of *Salmonella* spp. in a simulated acid-phase anaerobic digester treating sewage sludge. *Bioresource Technology* 86(2), 177–181.

Fytili D., Zabaniotou A. 2008. Utilization of sewage sludge in EU application of old and new methods – A review. *Renewable & Sustainable Energy Reviews* 12, 116–140.

García-Delgado M., Rodríguez-Cruz M.S., Lorenzo L.F., Arienzo M., Sánchez-Martín M.J. 2007. Seasonal and time variability of heavy metal content and of its chemical forms in sewage sludges from different wastewater treatment plants. *Science of The Total Environment* 382(1), 82–92.

Gavala H.N., Yenal U., Skiadas I.V., Westermann P., Ahring B.K. 2003. Mesophilic and thermophilic anaerobic digestion of primary and secondary sludge. Effect of pre-treatment at elevated temperature. *Water Research* 37(19), 4561–4572.

Gaspard P., Wiart J., Schwartzbrod J. 1997. Parasitological contamination of urban sludge used for agricultural purposes. *Waste Management & Research* 15(4), 429–436.
Ghini R., Rodrigues F., Bettiol W., Gatti I.M., Holanda A. 2007. Effect of sewage sludge on suppressiveness to soil-borne plant pathogens. *Soil Biology and Biochemistry* 39(11), 2797–2805.
Gibbs R.A., Hu C.J., Ho G.E., Unkovich I. 1997. Regrowth of faecal coliforms and *Salmonella* in stored biosolids and soil amended with biosolids. *Water Research and Technology* 35(11–12), 269–275.
Graczyk T.K., Kacprzak M., Neczaj E., Tamang L., Graczyk H., Lucy F.E., Girouard A.S. 2008. Ocurrence of Cryptosporidium and Giardia in sewage sludge and solid waste landfill leachate and quantitative comparative analysis of sanitization treatments on pathogen inactivation. *Environmental Research* 106(1), 27–33.
Groß B, Eder C., Grziwa P., Horst J., Kimmerle K. 2007. Energy recovery from sewage sludge by means of fluidised bed gasification. Waste Water Management. *In Press*. doi:10.1016/j.wasman.2007.08.016
Gupta R, Garg V.K. 2008. Stabilization of primary sewage sludge during vermicomposting. *Journal of Hazardous Materials* 153(3), 1023–1030.
Guzman C., Jofre J., Montemayor M., Lucena F. 2007. Occurrence and levels of indicator and selected pathogens in different sludges and biosolids. *Journal of Applied Microbiology* 103(6), 2420–2429.
Harrison E.Z., Oakes S.R., Hysell M., Hay A. 2006. Organic chemicals in sewage sludges. *Science of the Total Environment* 367(2–3), 481–497.
Hernández-Apaolaza L., Gascó A.M., Gascó J.M., Guerrero F. 2005. Reuse of waste materials as growing media for ornamental plants. *Bioresource Technology* 96(1), 125–131.
Hernández-Hernández, E. 2007. Eliminación de *Salmonella* spp de lodos residuales de una planta de tratamiento de aguas residuales del estado de Guanajuato para su posible utilización como fertilizantes. Tesis. Maestro en Ciencias del Agua. Universidad de Guanajuato.
Hong S., Park J.K., Lee Y.O. 2004. Mechanisms of microwave irradiation involved in the destruction of fecal coliforms from biosolids. *Water Research* 38(6), 1615–1625.
Hwang K., Shin E., Choi H. 1997. A mechanical pretreatment of waste activated sludge for improvement of anaerobic digestion system. *Water Science and Technology* 36(12), 111–116.
Hwang I.H., Ouchi Y., Matsuto T. 2007. Characteristics of leachate from pyrolysis residue of sewage sludge. *Chemosphere* 68, 1913–1919.
Ingelmo F., Canet R., Ibáñez M.A., Pomares F., García J. 1998. Use of MSW compost, dried sewage sludge and other wastes as partial substitutes for peat and soil. *Bioresource Technology* 63(2), 123–129.
Johansson, K., Perzon, M., Fröling, M., Mossakowska, A., Svanström, M. 2008. Sewage sludge handling with phosphorus utilization – life cycle assessment of four alternatives. *Journal of Cleaner Production* 16(1) 135–151.
Jurado P., Wester D.B. 2001. Effects of biosolids on tobograss growth in the Chihuahuan Desert. *Journal of Range Management* 54(1), 89–95.
Kabrick R.M., Jewell W.J. 1982. Fate of pathogens in thermophilic aerobic sludge digestion. *Water Research* 16(6), 1051–1060.
Kaminsky W., Kummer A.B. 1989. Fluidized bed pyrolysis of digested sewage sludge. *Journal of Analytical and Applied Pyrolysis* 16(1), 27–35.
Kawakami W., Hashimoto S. 1977. Enhanced compositing of radiation disinfected sewage sludge. *Radiation Physics and Chemistry* 24(1), 29–40.
Kersch C., Peretó S., Woerlee G.F., Witkamp G.J. 2004. Leachability of metals from fly ash: leaching tests before and after extraction with supercritical CO_2 and extractants. *Hidrometallurgy* 72(1–2), 119–127.
Kim W., Kim S., Kim K. 2005. Enhanced electrokinetic extraction of heavy metals from soils assisted by ion exchange membranes. *Journal of Hazardous Materials* 118(1–3), 93–102.
Kim Y., Parker W. 2008. A technical and economic evaluation of the pyrolysis of sewage sludge for the production of bio-oil. *Bioresource Technology* 99(5), 1409–1416.
Kite-Powell H.L., Hoagland P., Jin D. 1998. Policy, law, and public opposition: the prospects for abyssal ocean waste disposal in the United States. *Journal of Marine Systems* 14(3–4), 377–396.
Kothary M.H., Chase T.Jr., Macmillan J.D. 1984. Levels of aspergillus fumigatus in air and in compost at a sewage sludge composting site. *Environmental Pollution Series A, Ecological and Biological* 34(1), 1–14.
Kowaljow E., Mazzarino M.J. 2007. Soil restoration in semiarid Patagonia: Chemical and biological response to different compost quality. *Soil Biology and Biochemistry* 39(7), 1580–1588.
Lassen R., Tjell J.C., Hansen J. 1984. Phosphorus recovery from sewage for agriculture. *Waste Management & Research* 2(4), 369–378.
Laturnus F., Arnold K., Gron C. 2007. Organic contaminants from sewage sludge applied to agricultural soils. False alarm regarding possible problems for food safety? *Environmental Science and Pollution Research* 14, 53–60.

Lessel T., Suess A. 1977. Ten year experience in operation of a sewage sludge treatment plant using gamma irradiation. *Radiation Physics and Chemistry* 24(1), 3–16.

Lister S.K., Line M.A. 2001. Potential utilisation of sewage sludge and paper mill waste for biosorption of metals from polluted waterways. *Bioresource Technology* 79(1), 35–39.

Lo K.S.L., Chen Y.H. 1990. Extracting heavy metals from municipal and industrial sludges. *The Science of The Total Environment* 90, 99–116.

Luca S.J., Idle C.N., Chao A.C. 1996. Quality improvement of biosolids by ferrate(VI) oxidation of offensive odour compounds. *Water Science and Technology* 33(3), 119–130.

Marcinkowski T. 1985. Decontamination of sewage sludges with quicklime. *Waste Management & Research* 3(1), 55–64.

Menendez J.A., Inguanzo M., Pís J.J. 2002. Microwave-induced pyrolysis of sewage sludge. *Water Research* 36(13), 3261–3264.

Midilli A., Dogru M., Akay G., Howarth C.R. 2002. Hydrogen production from sewage sludge via a fixed bed gasifier product gas. *International Journal of Hydrogen Energy* 27(10), 1035–1041.

Mininni G., Santori M. 1987. Problems and perspectives of sludge utilization in agriculture. *Agriculture, Ecosystems & Environment* 18(4), 291–311.

Ndegwa P.M., Thompson S.A. 2001. Integrating composting and vermicomposting in the treatment and bioconversion of biosolids. *Bioresource Technology* 76(2), 107–112.

NOM-021-RECNAT-2000. Norma Oficial Mexicana. Especificaciones de fertilidad, salinidad y clasificación de suelos, estudio, muestreo y análisis. Diario Oficial de la Federación, Diciembre 2002.

NOM-004-SEMARNAT-2002. Norma Oficial Mexicana. Protección ambiental, lodos y biosólidos, especificaciones y límites máximos permisibles de contaminantes para su aprovechamiento y disposición final. Diario Oficial de la Federación, Agosto 2003.

Novinscak A., Surette C., Filion M. 2007. Quantification of *Salmonella* spp. in composted biosolids using a TaqMan qP CR assay. *Journal of Microbiological Methods* 70(1), 119–126.

Otero M., Rozada F., Calvo F., García A.I., Morán A. 2003a. Kinetic and equilibrium modeling of the methylene blue removal from solution by adsorbent materials produced from sewage sludges. *Biochemical Engineering Journal* 15, 59–68.

Otero M., Rozada F., Calvo L.F., García A.I., Morán A. 2003b. Elimination of organic water pollutants using adsorbents obtained from sewage sludge. *Dyes and Pigments* 57(1), 55–65.

Park C., Lee C., Kim S., Chen Y., Chase H.A. 2005. Upgrading of anaerobic digestion by incorporating two different hydrolysis processes. *Journal of Bioscience and Bioengineering* 100(2), 164–167.

Parmar N., Singh A., Ward O.P. 2001. Characterization of the combined effects of enzyme, pH and temperature treatments for removal of pathogens from sewage sludge. *World Journal of Microbiology and Biotechnology* 17(2), 169–172.

Pecson B.M., Barrios J.A., Jiménez B.E., Nelson K.L. 2007. The effects of temperature, pH, and ammonia concentration on the inactivation of *Ascaris* eggs in sewage sludge. *Water Research*, 41(13), 2893–2902.

Ponti C., Sonnleitner B., Fiechter A. 1995 . Aerobic thermophilic treatment of sewage sludge at pilot plant scale. 2. Technical solutions and process design. *Journal of Biotechnology* 38(2), 183–192.

Puchajda B., Oleszkiewicz J., Sparling R., Reimers R. 2006. Low-Temperature Inactivation of Fecal Coliforms in Sludge Digestion. *Water Environment Research* 78(7), 680–685.

Qiao L., Ho G. 1997. The effects of clay amendment on composting of digested sludge. *Water Research* 31(5), 1056–1064.

Randall Seeker W.M. 1991. Twenty-Third Symposium International on Combustion, Waste combustion. *Symposium on Combustion* 23(1), 867–885.

Rebah F.B., Prévost D., Yezza A., Tyagi R.D. 2007. Agro-industrial waste materials and wastewater sludge for rhizobial inoculant production: A review. *Bioresource Technology* 98(18), 3535–3546.

Rozada F., Otero M., Morán A., García A.I. 2008. Adsorption of heavy metals onto sewage sludge-derived materials. *Bioresource Technology In Press*.

Sahlström L., Aspan A., Bagge E., Danielson-Tham M., Albihn A. 2004. Bacterial pathogen incidences in sludge from Swedish sewage treatment plants. *Water Resource* 38(8), 1989–1994.

Sanchez E., Montalvo S., Travieso L., Rodríguez X. 1995. Anaerobic digestión of sewage sludge in an anaerobic fixed bed digester. *Biomass and Bioenergy* 9(6), 493–495.

Sánchez-Monedero M.A., Mondini C., Nobili M., Leita L. Roig A. 2004. Land application of biosolids. Soil response to different stabilization degree of the treated organic matter. *Waste Management* 24(4), 325–332.

Schnaak W., Küchler Th., Kujawa M., Henschel K.P, Süßenbach D., Donau R. 1997. Organic contaminants in sewage sludge and their ecotoxicological significance in the agricultural utilization of sewage sludge. *Chemosphere* 35(1–2), 5–11.

Selivanovskaya S.Y., Latypova V.Z., Kiyamova S.N., Alimova F.K. 2001. Use of microbial parameters to assess treatment methods of municipal sewage sludge applied to grey forest soils of Tatarstan. *Agriculture, Ecosystems & Environment* 86(2), 145–153.

Shang C., Wong H.M., Chen G. 2005. Bacteriophage MS-2 removal by submerged membrane bioreactor. *Water Research* 39(17), 4211–4219.

Shen L., Zhang D. 2003. An experimental study of oil recovery from sewage sludge by low-temperature pyrolysis in a fluidised-bed. *Fuel* 82(4), 465–472.

Shimaoka T., Hanashima M. 1996. Behavior of stabilized fly ashes in solid waste landfills. *Waste Management* 6(5–6), 545–554.

Singh R.P., Agrawal M. 2008. Potential benefits and risks of land application of sewage sludge. *Waste Management* 28(2), 347–358.

Song Y., Kwon S., Woo J. 2004. Mesophilic and thermophilic temperature co-phase anaerobic digestion compared with single-stage mesophilic- and thermophilic digestion of sewage sludge. *Water Research* 38(7), 1653–1662.

Sreekrishnan T.R., Tyagi R.D., Blais J.F., Campbell G.C. 1993. Kinetics of heavy metal bioleaching from sewage sludge—I. Effects of process parameters. *Water Research* 27(11), 1641–1651.

Stabnikova O., Goh W., Ding H., Tay J., Wang J. 2005. The use of sewage sludge and horticultural waste to develop artificial soil for plant cultivation in Singapore. *Bioresource Technology* 96(9), 1073–1080.

Suess A., Lessel T. 1977. Radiation treatment of sewage sludge – experience with an operating pilot plant. *Radiation Physics and Chemistry* 9(1–3), 353–370.

Thawornchaisit U., Pakulanon K. 2007. Application of dried sewage sludge as phenol biosorbent. *Bioresource Technology* 98(1), 140–144.

Tiehm A., Nickel K., Neis U. 1997. The use of ultrasound to accelerate the anaerobic digestion of sewage sludge. *Water Science and Technology* 36(11), 121–128.

Tyagi R.D., Couillard D., Tran F. 1988. Heavy metals removal from anaerobically digested sludge by chemical and microbiological methods. *Environmental Pollution* 50(4), 295–316.

Veeken A.H.M., Hamelers H.V.M. 1999. Removal of heavy metals from sewage sludge by extraction with organic acids. *Water Science and Technology* 40(1), 129–136.

Verma M., Brar S.K., Tyagi R.D., Valéro J.R., Surampalli R.Y. 2005. Wastewater sludge as a potential raw material for antagonistic fungus (*Trichoderma* sp.): Role of pre-treatment and solids concentration. *Water Research* 39(15), 3587–3596.

Wai C.M., Wang S. 1997. Supercritical fluid extraction: metals as complexes. *Journal of Chromatography A* 785(1–2), 369–383.

Wang M. 1997. Land application of sewage sludge in China. *Science of The Total Environment* 197(1–3), 149–160.

Wang J., Zhang D., Stabnikova O., Tay J. 2005. Evaluation of electrokinetic removal of heavy metals from sewage sludge. *Journal of Hazardous Materials* 124(1–3), 139–146.

Wang J., Wang J. 2007. Application of radiation technology to sewage sludge processing: A review. *Journal of Hazardous Materials* 143(1–2), 2–7.

Watanabe H., Takehisa M. 1977. Disinfection of sewage sludge cake by gamma-irradiation. *Radiation Physics and Chemistry* 24(1), 41–54.

Wen G., Voroney R.P., Winter J.P., Bates T.E. 1997. Effects of irradiation on sludge organic carbon and nitrogen mineralization. *Soil Biology and Biochemistry* 29(9–10), 1363–1370.

Wester D.B., Sosebee R.E., Zartman R. E., Fish E.B., Villalobos J.C. 2003. Biosolids in a Chihuahuan Desert Ecosystem. *Rangelands* 25(4), 27–32.

Wong, J.W.C., Xiang, L., Chan, L.C. 2002. pH Requirement for the Bioleaching of Heavy Metals from Anaerobically Digested Wastewater Sludge. *Water, Air & Soil Pollution* 138(1–4), 23–35.

Yezza A., Tyagi R.D., Valéro J.R., Surampalli R.Y. 2006. Bioconversion of industrial wastewater and wastewater sludge into *Bacillus thuringiensis* based biopesticides in pilot fermentor. *Bioresource Technology* 97(15), 1850–1857.

Yokoyama S., Suzuki A., Murakami M., Ogi T. 1987. Liquid fuel production from sewage sludge by catalytic conversion using sodium carbonate. *Fuel* 66(8), 1150–1155.

Yuan C., Weng C. 2006. Electrokinetic enhancement removal of heavy metals from industrial wastewater sludge. *Chemosphere* 65(1), 88–96.

Zabaniotou A., Theofilou C. 2008. Green energy at cement kiln in Cyprus – Use of sewage sludge as a conventional fuel substitute. *Renewable and Sustainable Energy Reviews* 12(2), 531–541.

Membrane Fouling and Cleaning in Treatment of Contaminated Water

Mattheus F. A. Goosen
Alfaisal University, Riyadh, KSA

Sulaiman K. S. Al-Obaidani & Hilal Al-Hinai
Department of Mechanical and Industrial Engineering, Sultan Qaboos University, Oman

Shyam. Sablani
Biological Systems Engineering, Washington State University, Pullman, Washington, USA

Y. Taniguchi & H. Okamura
Water Re-use Promotion Center, Japan

SUMMARY: The primary aim of this chapter was to review the mechanisms of membrane fouling and chemical cleaning methods as they relate to treatment of contaminated water. There is a need is to increase the understanding of membrane fouling phenomena, preventive means and membrane cleaning processes as it applies to the clean-up and desalination of oil contaminated seawater.

1 INTRODUCTION

Fiscal development and population growth has put increasing pressure on the world's limited fresh water resources (Goosen et al., 2004). In order to lessen this problem, desalination processes have been developed to obtain fresh water from the earth's vast supply of seawater. A major concern, however, is the location of a significant fraction of the world's desalination capacity in coastal areas of oil producing countries, such as in the Arabian Gulf (Al-Sajwani, 1998). Oil spills could have catastrophic effects on seawater desalination capacity in these regions.

The two most successful commercial water desalination techniques involve thermal and membrane separation methods (Al Obeidani et al., 2008, Hu and Scott, 2007, Al-Desouki and Ettoney, 2001). Thermal separation processes include multistage flash (MSF), multi-effect evaporation (MEE)/multi-effect distillation (MED), vapor compression (VC) and solar desalination. Membrane separation processes include reverses osmosis (RO) and electro-dialysis (ED). Reverses osmosis desalination is becoming increasingly more competitive with thermal desalination processes. Contamination of seawater is a problem due to oil spills, and dissolved and suspended microbial, chemical, and solid matter. This leads to fouling in membrane separation processes such as RO and also requires different cleaning techniques as a means of managing the problem.

The application of membrane filtration pretreatment for seawater reverses osmosis desalination plants enables clean water to be fed into the reverses osmosis modules on a stable basis even in the case when seawater is severely polluted and contains, for example, large amounts of organic material or oil (Al Obeidani et al., 2008, Goosen et al., 2004, Li et al., 2006, Taniguchi, 1997, Ahmed et al., 2003). Membrane life time and permeate (i.e. fresh water) fluxes are primarily affected by the phenomena of concentration polarization (i.e. solute build-up) and fouling (e.g. microbial adhesion, gel layer formation and solute adhesion) at the membrane surface (Sablani et al., 2001, Yang et al., 2007). This problem can be reduced by introducing a membrane pretreatment stage like microfiltration (MF) or ultrafiltration (UF).

Very low velocities over the membrane surface lead to increased concentration polarization and fouling, resulting in a rapid flux decrease and the need for frequent cleaning (Goosen et al., 2004).

Some foulants can be removed by hydraulic means such as filter backwashing. Most foulants can be removed by chemical means. Chemical cleaning is an integral part of a membrane process operation. It has a profound impact on the performance and economics of the process. Some manufacturers actually supply proprietary cleaners while others recommend commercial chemicals (Kaiya et al., 2000). The whole chemical cleaning process is still not well understood. Factors that need to be considered include; effectiveness of the process in removing the fouling layer; economics of the process (i.e. minimizing usage of chemicals); and the cleaning time.

Most of the produced oil is transported by sea. This exposes the marine environment to the dangers of oil contamination. A significant amount of pollution of the world's oceans is caused by shipping accidents and dumping from offshore oil and gas platforms (IMO/UNEP, 2002). A total of 360 million US gallons were spilled over a 15 year period in the Arabian Gulf region alone as reported by the Regional Organization for the Protection of the Marine Environment (Marine Emergency Mutual Aid Centre, 2003). This is significant since this region has the world's largest desalination capacity. Furthermore, oil spills can also harm marine life in various different ways; by poisoning after ingestion, by direct contact and by destroying habitats. Although the world's oceans are large, no oil can be spilled without affecting the ecosystem. Air and ocean currents can also transport pollutants for thousands of kilometers. Oil spills, therefore, affect more than just isolated locations (IMO/UNEP, 2002).

Conventional seawater desalination plants cannot operate if the feed seawater contains contaminants such as oil. There is a need for new pretreatment technologies. Micro-filtration, for example, could be used to remove oil prior to RO desalination in order to allow for continuous operation of a desalination plant (Taniguchi, 1997, Benito et al., 2002).

The primary aim of this chapter was to review mechanisms of membrane fouling and chemical cleaning techniques as they relate to treatment of contaminated water. The long term objective of our research is to increase the understanding of membrane fouling phenomena, preventive means and membrane cleaning processes as it applies to the clean-up and desalination of oil contaminated seawater as well as wastewater reuse.

2 MEMBRANE FOULING MECHANISMS

One of the most important problems facing the users of membrane technology is a decline in membrane performance due to fouling and concentration polarization (Goosen et al., 2004, Ridgeway et al., 1984) Applications of membrane filtration have expanded significantly in such fields as water treatment, food processing and medicine (Chebbi, 2001, Brehant et al., 2002, Strathman (check the spelling), 2001). Membrane fouling influences the economic viability of the separation process, whether it involves desalination or food processing. Accordingly, understanding and minimization of membrane fouling are crucial for the effective use of this technology (Hu and Scott, 2007, Al-Desouki and Ettoney, 2001). Factors such as flow conditions, pretreatment, membrane properties, water quality, cleaning agents, and cleaning performance are important for membrane fouling reduction and for membrane cleaning (Kaiya et al., 2000).

Fouling of membranes by chemical and biological contaminates transported in the feed can significantly reduce the energy efficiency and cost effectiveness of membrane applications (Goosen et al., 2004, Hu and Scott, 2007). This accumulation of contaminates is associated with an active decrease in permeate flux and mineral rejection of the membranes. A clear distinction must be made between concentration polarization and membrane fouling. Concentration polarization is the development of a concentrated gradient of the retained components near the membrane surface (Figure 1A). It is a function of the hydrodynamic conditions in the membrane system and is independent of the physical properties of the membrane.

The membrane pore size and porosity are not directly affected by the concentration polarization (Goosen et al., 2004). Fouling, on the other hand, is the deposition of material on the membrane surface or in its pores (i.e. solute adsorption), leading to a change in the membrane behavior

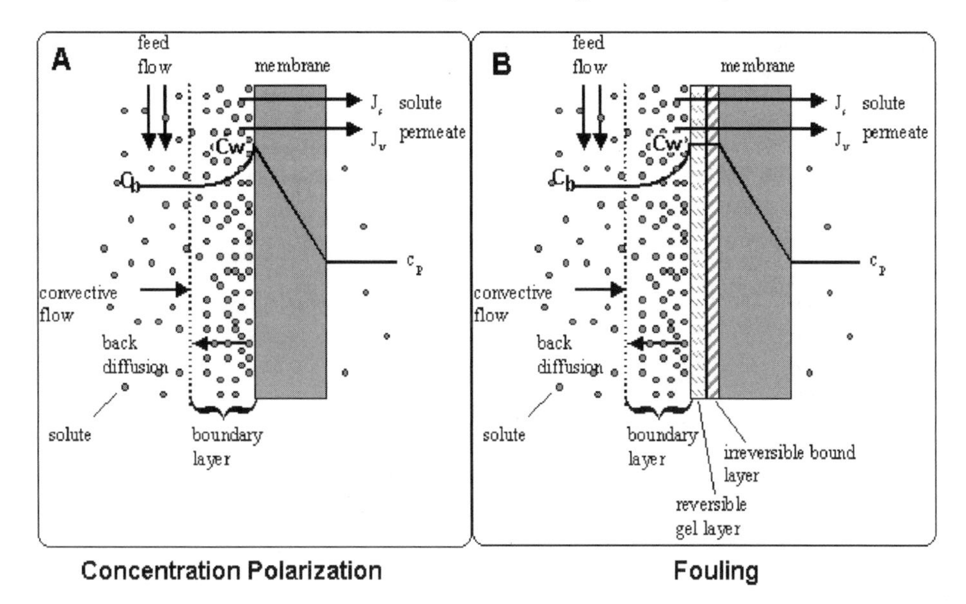

Figure 1. Concentration Polarization (A) and Fouling (B) at the Membrane Surface (Goosen et al., 2004).

(Figure 1B). Fouling must also be distinguished from membrane compaction. The later is the compression of the membrane structure under the transmembrane pressure, causing a decrease in membrane permeability (Marshall et al., 1993).

The nature of the adsorbed solute layer on the membrane surface depends on the type of fouling material present, whether it is, for example, microbial cells or oil droplets. This layer can be influenced by hydrodynamic conditions. Fouling within the membrane structure (i.e. pore plugging or pore narrowing), on the other hand, results in a change in the apparent pore size, pore size distribution and pore density of the membrane. It is likely that fouling in the immediate vicinity of the pore entrance dominates this behavior. In addition, the fouling mechanism is affected by the operating parameters such as the cross-flow velocity and fouling materials concentration (Goosen et al., 2004, Marshall et al., 1993).

The effects of feed properties, the membrane material and the processing variables on membrane fouling are summarized in Table 1. There are four major categories of membrane foulants: dissolved solids, suspended solids, biological organisms and non-biological organics (Stephen and Kronmiller, 1995).

Dissolved solids, which are scale-forming materials such as calcium and barium, may precipitate in the brine stream as their concentrations increase on the membrane surface. Suspended solids such as iron, aluminum or silica, on the other hand, stay floating in solution through a process of repulsion by a double layer of charge. The charge-repulsion characteristics of suspended solids also stabilize particulates such as carbon fines which may inadvertently leak from mixed media or carbon filters. Suspended solids tend to agglomerate and settle onto the membrane surface when concentrated past the point of their charge related stability.

In contrast, biological foulants such as bacteria, fungus, and algae have a tendency to be present in low concentrations. They literally grow into massive quantities that effectively block flow through the membrane surface. Non-biological organic foulants such as oil and humic acids, are also carbon-based chemical structures but are not living organisms. Studies have shown that natural organic matter is a major foulant (Cho et al., 1998).

Nikolova and Islam (1998) reported concentration polarization in the absence of gel layer formation using a lab scale ultrafiltration unit equipped with a tubular membrane. The solute adsorption described by Nikolova and Islam (1998) is reversible. The transition between this type

Table 1. Factors Affecting Membrane Fouling [Adapted from Marshall et al., 1993].

Factor	Effects
Feed properties	
Concentration	Decreases permeate flux, increases irreversible fouling in surface fouling and increases the rate of fouling when internal fouling dominates
pH and ionic strength	Affect the amount of fouling materials deposited and permeate flux
Component interactions	Affect the membrane retention
Prefiltration and removal of aggregates	Improves the permeate flux and decreases fouling rate
Membrane materials effects	
Hydrophobicity	Influences the deposition of fouling materials since they adsorb less to hydrophilic (wetable) than hydrophobic (non wetable) membranes
Charge effects	Can enhance permeate flux and reduce fouling if operated with membrane of similar charge to the fouling materials
Surface roughness	Increasing surface roughness may increase the tendency for fouling material to adsorb but this will reduce the competence of the dynamic membrane
Porosity and pore size distribution	Membrane fouling changes the pore size distribution and pore density which in turn affects permeate flow, component retention and membrane selectivity
Pore size	Membrane fouling is more severe with increasing pore size
Membrane consistency	Greater membrane consistency and tighter pore size are required to improve the flux and reduce the fouling rate
Effects of processing variables	
Transmembrane pressure	Increasing the transmembrane pressure in the low pressure range (<4 bar) will increase the permeate flux but it may also increase the fouling rate
Cross-flow velocity	Increasing the cross-flow velocity generally results in improving the permeate flux and reducing the fouling rate
Backflushing	Backflush pulses will blow the particles off the membrane surface which results in improving the permeate flux and reducing the rate of fouling
Temperature	The permeate flux will be enhanced and the fouling rate will be reduced as the temperature increases

of adsorption and irreversible fouling is crucial to determining the strategy for improved membrane performance and for understanding the threshold values for which optimal flux and rejection can be maintained. In a very thorough study, Chen et al. (1997) reported on the dynamic transition from concentration polarization to cake (i.e. gel-layer) formation for membrane filtration of colloidal silica. Once a critical flux, J_{crit}, was exceeded, the colloids in the polarized layer formed a consolidated cake structure that was slow to depolarize and which reduced the flux. This study showed that by controlling the flux below J_{crit}, the polarization layer may form and solute adsorption may occur but it is reversible and responds quickly to any changes in convection. This paper is a very valuable source of information for membrane plant operators, particularly in coastal areas which are susceptible to oil spills. By operating just below J_{crit} they can maximize the flux while at the same time reducing the frequency of membrane cleaning. More detailed explanations on the mechanisms of membrane fouling can be found in a paper by Goosen et al. (2004).

Attenuated Total Reflection (ATR) Fourier Transform Infrared (FTIR) spectroscopy can provide insight into the chemical nature of deposits, such as oil films, on the membrane surface [Howe et al., 2002]. The spectra of the foulants can be distinguished from the spectra of the membrane material.

ATR/FTIR can also indicate the presence of inorganic foulants as well as the ratio of inorganic to organic foulants. This could give investigators the ability to differentiate, for example, between oil foulants and calcium deposits. Identification of specific species deposited onto membrane surfaces can be carried out using Matrix Assisted Laser Desorption Ionization Mass Spectroscopy (MALDI-MS) [Chan et al., 2002]. Deposits on a membrane surface, before and after cleaning, can be analyzed using Scanning Electron Microscopy (SEM) in combination with Energy Dispersive X-ray (EDX) combined with a micro analysis system permitting quantitative determination of elements (Lindau and Jonsson, 1994).

In the development of strategies to improve operating conditions, non-destructive, real time observation techniques to detect and monitor fouling during liquid separation processes are of great importance. In a recommended paper by Li et al. (2002) ultrasonic time-domain reflectometry (UTDR) was used to measure organic fouling, in real time, during ultrafiltration with polysulphone (PS) membranes. The feed solution was a paper-mill effluent, which contained breakdown products of lignin or lignosulphonate, from a wastewater treatment plant. Experimental results showed that the ultrasonic signal response can be used to monitor fouling-layer formation and growth on the membrane in real-time. Traditional flux measurements and analysis of the membrane surface by microscopy corroborated the UTDR results. Furthermore, the differential signal developed indicated the state and progress of the fouling layer and gave warning of advanced fouling during operation. This research could be applied by desalination plants in coastal regions that suffer from oil spills.

3 USE OF MEMBRANES FOR OIL REMOVAL FROM CONTAMINATED WATER

Oil-in-water emulsions are one of the main pollutants from industrial and domestic sewage. Oily water in inland water ways and coastal zones is a serious environmental issue which needs to be addressed (Mohammadi et al., 2003). There are several techniques which can be employed for cleaning oily water; chemical emulsification, pH adjustment, gravity settling, centrifugal settling and heating treatment. Membranes may also be employed for treating such water (Hu and Scott, 2007, Yang et al., 2007, Li et al., 2006, and Lange et al., 1999).

In a related study on oil separation from contaminated water using reverse osmosis membranes, Mohammadi et al. (2003) showed a relationship between flux decline and oil concentration in the feed. A similar study was performed by Hodgkiess et al. (2001) on desalination performance of reverse osmosis membrane units under conditions where the feed water was contaminated with crude oil and fuel spillages. Their work examined the behavior of polyamide seawater and brackish water membranes using feed water containing NaCl at 2,000–35,000 mg/L concentrations. The effect of a range of contaminants on reverse osmosis membrane performance was assessed by comparing the water flux and salt rejection of membrane samples before and after their exposure to oil-based media. This is a recommended paper for those working on treatment of oil contaminated water.

In a study which may prove of use in the desalination of oil contaminated seawater, Zwijneenberg et al. (2005) employed membrane pervaporation for the recovery of oil from produced water. The latter is oil contaminated salty water from oil wells that comes out along with the oil. This type of work provides useful information on membrane fouling by oil films and the associated clean up process. However, the separation process was solar driven which may make large scale application in seawater desalination difficult.

The ultrafiltration separation behavior of oil-in-water emulsion of metalworking fluids was assessed by Bekassy-Molnar et al. (2002). The effect of different polymeric membrane materials, oil concentration in feed, transmembrane pressure, feed temperature and other process parameters on the permeate flux and oil rejection was measured. In addition, Hilal et al. (2002) provided a study of the washing cycle optimization of an ultrafiltration plant used to treat waste metalworking fluids. They assessed the interactions between oily foulants, membrane surface and oil emulsion.

However, it was unclear whether these results could be extrapolated to oil contaminated seawater due to the high metal ion concentrations.

Lindau and Jonsson (1994) considered the influence of different types of cleaning agents on polysulphone ultrafiltration membranes which had been used to treat oily waste water from a car washing station. They showed that after cleaning with an acidic agent, the pure water flux increased. However, the flux decreased continuously with time. In contrast, the flux, after membrane cleaning with an alkaline agent, was lower but time-independent. No studies have been reported in the literature, as far as the authors are aware, on the removal of oil from seawater using polyethylene microfiltration membranes, nor on the chemical cleaning of these types of membranes.

4 FACTORS THAT REDUCE MEMBRANE FOULING

Aspects which must be well thought-out in fouling reduction include flow conditions, feed pretreatment, and membrane properties. Both the module design and the circulation velocity over the membrane surface are important. Too low velocities, for example, lead to increased concentration polarization and fouling, resulting in a rapid flux decrease and the need for frequent cleaning (Goosen et al., 2004).

In an interesting study by Jenkins and Tanner (1998), they confirmed the results of Flemming and Schaule (1988), on chemically modified membranes that reduced microbial adhesion. One type was classified as a polyamide, the other utilized a new chemistry that formed a polyamide-urea barrier (i.e. surface) layer. The latter composite membrane proved superior in reverse osmosis operation similar to that of the polyetherurea membrane of Flemming and Schaule (1988), including rejection of certain dissolved species and fouling-resistance. These results suggested that the presence of urea groups in the membrane reduces microbial adhesion, perhaps through charge repulsion. Scientists should therefore be able to minimize microbial adhesion by controlling the surface chemistry of polymer membranes, through, for example, the inclusion of urea groups. Perhaps similar membranes could be developed which would reduce adhesion of oil droplets. This is one area where further research is needed.

Coagulation, to remove turbidity from water by the addition of cationic compounds, is a commonly used method (Choksuchart, 2002). Flocculators can also be used to remove suspended solids, organics and phosphorus from wastewater. They produce uniform microflocs, which can be removed by cross flow microfiltration (Chapman et al., 2002). Flocculated particles can form a highly porous filtration cake on a membrane surface. This will help inhibit fouling on the membrane by preventing the deposition of particles and therefore reducing the number of membrane cleaning cycles (Nguyen and Ripperger, 2002). There is a scarcity in the literature, as far as the authors are aware, on the effectiveness of these additives in removing oil from contaminated water.

Redondo (1999) employed chemical modification of a membrane surface was used in combination with spacers and periodic applications of bioacids. This paper, however, was short on specifics (e.g. details of chemical modification of aromatic polyamides membrane surface), and therefore not very useful to those looking for insights into membrane fouling.

When considering removal of oil from contaminated water, it is also important to consider other membrane properties such as electrical charge, and whether the membrane is hydrophilic or hydrophobic (Marshall et al., 1993). Until now these latter properties, though of importance for fouling, have seldom been considered.

5 TECHNIQUES FOR MEMBRANE CLEANING

While membranes used for the treatment of paints or for the production of desalinated water need to be cleaned less frequently (cleaning interval being a couple of months or more), membranes used in the food industry are generally cleaned at least once a day. Fouling of membranes, as well as the presence and growth of microorganisms, necessitate regular cleaning and disinfection

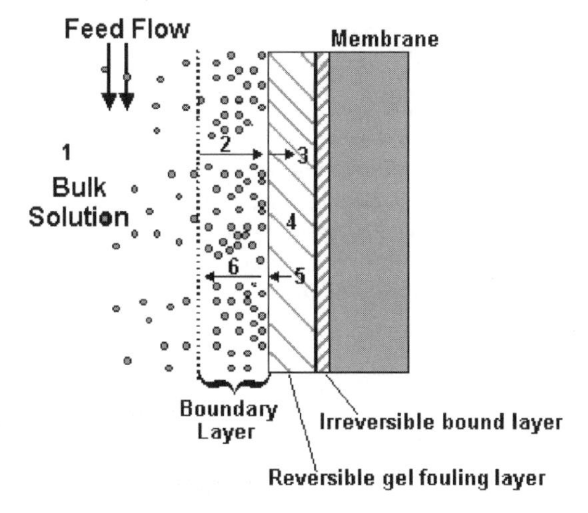

Figure 2. Chemical Cleaning Process at Membrane Surface: 1) Bulk reactions 2) Transport of detergent to interface 3) Transport of detergent into foulant layer 4) Cleaning reactions in fouling layer 5) Transport of cleaning reaction products back to interface 6) Transport of product to bulk solution. (Al Obeidani et al., 2008).

cycles (Ridgeway et al., 1984). Depending on the type of liquid processed, fouling and deposition can be of different types (Luss, 1990). Examples are: mineral, fat, protein, carbohydrate, and surface active agents (e.g. antifoams). Due to the small amounts of components deposited, it is often difficult to find analysis methods capable of determining the type of agents responsible for fouling (Marshall et al., 1993). Some foulants can be removed by hydraulic means such as filter backwashing or air scrubbing. Most foulants, including oil films, can be removed by chemical means. Chemical cleaning is an integral part of a membrane process operation. It has a profound impact on the performance and economics of the processes. Currently, chemical cleaning is based on recommendations from membrane manufacturers. Some of them supply proprietary cleaners while others use commercial chemicals (Kaiya et al., 2000).

The main goals of membrane cleaning are to restore the original permeate flow rate by removal of a flux-inhibiting fouling layer from the membrane surface, and to remove precipitated small molecular weight substances (e.g. salts, carbohydrates) from within the membrane structure. Chemical cleaning has great potential in cleaning of oil contaminated membranes.

The primary feature of a chemical cleaning process is a heterogeneous reaction between the detergent solution and the fouled layer (Tran-Ha and Wiley, 1998). The cleaning reaction can be divided into six stages (Figure 2):

- *Bulk reaction of detergents*. If the degree of fouling is high, a lot of the cleaning agent is consumed in the bulk solution and is not available for cleaning. It is thus important that as much foulant as possible is rinsed out with water before adding the cleaning agent.
- *Transport of detergent to the fouled surface*. The mass transport of detergents through the boundary layer.
- *Transport into the fouled layer*. The detergent can be transported into the fouled layer by capillarity or molecular diffusion. Due to their lower surface tension, surface-active agents have the ability to penetrate through crevices and due to their adsorption characteristics, they can be adsorbed onto the surface, thereby weakening the bond between the foulant and the membrane.
- *Cleaning reactions*. The processes which take place when the cleaning agents come into contact with the fouled layer can be subdivided into physicochemical transformations and chemical

reactions. These reactions help to overcome the cohesion forces between foulant particles and adhesion forces between fouling particles and the membrane surface.
- *Transport of cleaning reaction products back to the interface.* This is a diffusion process.
- *Transport of products to the bulk solution.* The cleaning reaction products are transferred through the boundary layer due to concentration gradients (back diffusion) or turbulence.

In order to obtain a good mechanical cleaning effect, the circulation flow rate should be higher and the pressure lower than those used during normal operation. Under these conditions the compressible fouled layer is relaxed (i.e. less compaction) and not as much able to withstand shear stresses (Tragardh, 1989). This would mean perhaps that an adsorbed oil film would be more easily removed under these conditions.

Characterization of the fouled membranes by ATR/FTIR before and after cleaning can also give a very good indication whether the cleaning process is effective or not (Tran-Ha and Wiley, 1998, and Belfer et al., 1999). The effectiveness of a disinfectant can be checked by microbiological tests. Such tests are unfortunately not very rapid.

The quality of the water used for rinsing and cleaning the membrane is important. The presence of iron, silica and calcium, for example, can lead to deposits which are difficult or impossible to remove (Tran-Ha, 1998, Tragardh, 1989). Chemicals used as cleaning agents should loosen and dissolve the foulant, keep the foulant in dispersion and solution, avoid spacer fouling, not attack the membrane (and other parts of the system), and disinfect all wetted surfaces (Tragardh, 1989). Besides the cleaning ability of a detergent, there are other important factors such as the ease with which it can be dispensed and rinsed away, its chemical stability during use, and cost and safety.

The choice of cleaning agents and cleaning conditions depends not only on the type of components deposited, but also on the resistance of the membrane, the module and the rest of the equipment. The chemical and thermal resistances of membranes vary greatly, depending on which polymers were used in the membrane. Polysulphone membranes, for example, are resistant in the pH range 1–13 and up to 80°C. They are quite resistant to oxidizing agents such as hypochlorite (Luss, 1990). On the other hand, cellulose acetate membranes can be used only at temperatures up to about 30–40°C and in the pH range 3–8.

Microfiltration mineral membranes composed of α-alumina are corrosion resistant and can be cleaned with sodium hypochlorite or sodium hydroxide as well as nitric acid, while ultrafiltration membranes (γ-alumina) have limited resistance to corrosion (Koseoglu, 1999). Phosphoric acid cannot be used for the cleaning of alumina membranes. Polyamide membranes are very sensitive to oxidizing agents and hence disinfectants such as hypochlorite and hydrogen peroxide cannot be used. Inorganic membranes made of zirconium oxide have very good thermal chemical resistance. The cleaning effect on polysulphone membranes varies depending on how the membranes are manufactured. In general, cleaning cycles affect the membrane and the membrane life more than normal running (Kaiya et al., 2000, and Tragardh, 1989). These studies have implications for cleaning of membranes that have been used for desalination of oil contaminated water. It is important to remember that while the cleaning chemicals may be effective in removing the fouling oil film, they could also damage the membrane over the long term.

6 CONCLUDING REMARKS

It is important to understand the nature of membrane fouling and how to reduce it, in order to improve membrane performance. Chemical cleaning, for example, can be applied to restore membrane capacity. There is a need to determine what cleaning chemicals work best without damaging the membrane. Additional work also needs to be done to find out what happens to the fouling resistance of chemically modified membranes over the long term.

The fouling process is a complex mechanism where the physico-chemical properties of the membrane, the type of cells, the quality of the feed water, the type of solute molecules, and the operating conditions all play a role.

Gaining a better understanding of membrane desalination of oil contaminated seawater is a major challenge facing both scientists as well as plant operators in many parts of the world. It is a challenge that we should be well able to meet.

REFERENCES

Ahmed A., Ismail S. and Bhatia S. 2003. Water Recycling from Palm Oil Mill Effluent (POME) using Membrane Technology. *Desalination*, 157, 87–95.

Al-Desouki H. and Ettoney H. M. 2001. Desalination Fundamentals, an Intensive Short Course, *MEDRC*, Sultan Qaboos University, Muscat, 20–24 January.

Al Obaidani, S., Al Hinai, H., Goosen, M. F. A., Sablani, S., Taniguchi, Y. and Okamura, H. 2008. Membrane Fouling and Cleaning in Treatment of Contaminated Water: A Critical Review. *Desalination*, (revised paper submitted July 2007).

Al-Sajwani, Taher M. 1998. The Desalination Plants of Oman: Past, Present and Future. *Desalination*, 120. 53–59.

Baird S. and Hayhoe D. 1993. Oil Spills. *Energy Fact Sheet, Originally Published by the Energy Educators of Ontario.*

Bekassy-Molnar, Xianguo Hu and Vatai G. 2002. Study of Ultrafiltration Behaviour of Emulsified Metalworking Fluids. *Desalination*, 149, 191–197.

Belfer S., Gilron J. and Kedem O. 1999. Characterization of Commercial RO and UF Modified and Fouled Membranes by Means of ATR/FTIR. *Desalination*, 124, 175–180.

Benito J. M., Rios G., Ortea E., Fernandez E., Cambiella A., Pazos C. and Coca J. Design and Construction of a Modular Pilot Plant for The Treatment of Oil-Containing Waste-Waters. *Desalination*, 147 (2002) 5–10.

Brehant A., Bonnelye V. and Perez M. 2002. Comparison of MF/UF Pretreatment with Conventional Filtration Prior to RO Membranes for Surface Seawater Desalination. *Desalination*, 144, 353–360.

Chan R., Chen V. and Bucknall M. P. 2002. Ultrafiltration of Protein Mixtures: Measurement of Apparent Critical Flux, Rejection Performance, and Identification of Protein Deposition. *Desalination*, 146, 83–90.

Chapman H., Vigneswaran S., Ngo H. H., Dyer S. and Ben Aim R. 2002. Pre-flocculation of Secondary Treated Wastewater in Enhancing the Performance of Microfiltration. *Desalination*, 146, 367–372.

Chebbi R. 2001. Viscous-Gravity Spreading of Oil in Water. *AIChE*, 47(2), 288–294.

Chen V., Fane A. G., Madaeni S. and Wenten G. 1997. Particle Deposition during Membrane Filtration of Colloids: Transition between Concentration Polarization and Cake Formation. *Journal of Membrane Science*, 125, 109–122.

Cho J., Amy G., Pellegrino J. and Yoon Y. 1998. Characterization of Clean and Natural Organic Matter (NOM) Fouled NF and UF Membranes, and Foulants Characterization. *Desalination*, 118, 101–108.

Choksuchart P., Heran M. and Grasmick A. 2002. Ultrafiltration Enhanced by Coagulation in an Immersed Membrane System. *Desalination*, 145, 265–272.

Flemming H.-C. and Schaule G. 1988. Biofouling of Membranes – A Microbiological Approach. *Desalination*, 70, 95–119.

Goosen M. F. A., Sablani S. S., Al-Hinai H., Al-Obeidani S., Al-Belushi R. and Jackson D. 2004. Fouling of Reverse Osmosis and Ultrafiltration Membranes: A Critical Review. *Separation Science and Technology*, 39(10), 2261–2298.

Hilal N., Busca G. and Atkin B. P. 2003. Optimization of Washing Cycle on Ultrafiltration Membranes used in Treatment of Metalworking Fluids. *Desalination*, 156, 199–207.

Hodgkiess T., Hanbury W. T., Law G. B. and Al-Ghasham T. Y. 2001. Effect of Hydrocarbon Contaminants on the Performance of RO Membranes. *Desalination*, 138, 283–289.

Howe K. J., Ishida K. P. and Clark M. M. 2002. Use of ATR/FTIR Spectrometry to Study Fouling of Microfiltration Membranes by Natural Waters. *Desalination*, 147, 251–255.

Hu, B. and Scott, K. 2007. Influence of Membrane Material and Corrugation and Process Conditions on Emulsion Microfiltration. *Journal of Membrane Science*, 294, 30–39.

IMO/UNEP Forum on Regional Arrangements for Co-operation in Combating Marine Pollution Incidents. 2002. London, 30 Sep.–2 Oct.

Jenkins M. and Tanner M. B. 1998. Operational Experience with a New Fouling Resistant Reverse Osmosis Membrane. *Desalination*, 119, 243–250.

Kaiya Y., Itoh Y., Takizawa S., Fujita K. and Tagawa T. 2000. Analysis of Organic Matter Causing Membrane Fouling in Drinking Water Treatment. *Water Science and Technology*, 41, 59–67.

Koseoglu S. S. 1999. Membrane Fouling and Cleaning in Food Processing, *Short course manual*, Texas A&M University System.

Lange R., Abilov F. A. and Orudjev A. G. 1999. Optimization of Oil-containing Wastewater Treatment Process. *Desalination*, 124, 225–229.

Li. H. -J., Cao Y. -M., Qin J. -J., Jie X. -M., Wang T. -H., Liu, J. -H. and Yuan, Q. 2006. Development and Characteristics of Anti-Fouling Cellulose Hollow-Fiber UF Membranes for Oil-Water Separation. *J. Membrane Science (?)*, 279, 328–335.

Li J., Sanderson R. D., Hallbauer D. K. and Hallbauer-Zadorozhnaya V. Y. 2002. Measurement and Modeling of Organic Deposition in Ultrafiltration by Ultrasonic Transfer Signals and Reflections. *Desalination*, 146, 177–185.

Lindau J. and Jonsson A. -S. 1994. Cleaning of Ultrafiltration Membranes after Treatment of Oily Waste Water. *Journal of Membrane Science*, 87, 71–78.

Luss G. 1990. Cleaning Membrane Systems in Process Industries, *Proceedings of the International Congress on Membranes and Membrane Processes,* Chicago, Illinois, 20–24 August.

Marine Emergency Mutual Aid Centre (MEMAC). 2003. Oil Spill Incidents in ROPME Sea Area: 1965–2002.

Marshall A. D., P. Munro A. and Tragardh G. 1993. The Effect of Protein Fouling in Microfiltration and Ultrafiltration on Permeate Flux, Protein Retention and Selectivity: A Literature Review. *Desalination*, 91, 65–108.

Mohammadi T., Kazemimoghadam M. and Saadabadi M. 2003. Modeling of Membrane Fouling and Flux Decline in Reverse Osmosis During Separation of Oil in Water Emulsions. *Desalination*, 157, 369–375.

Nguyen M. T., Ripperger S. 2002. Investigation on the Effect of Flocculants on the Filtration Behavior in Microfiltration of Fine Particles. *Desalination*, 147, 37–42.

Nikolova, J. D. and Islam, M. A. 1998. Contribution of adsorbed layer resistance to the flux-decline in an ultrafiltration process, *Journal of Membrane Science*, 146, 105–111.

Redondo J. A. 1999. Improve RO System Performance and Reduce Operating Cost with FILMTEC Fouling Resistant (FR) Elements, *Desalination*, 126, 249–259.

Ridgway H. F., Justice C. A., Whittaker C., Argo D. and Olson B. 1984. Biofilm Fouling of RO Membranes – Its Nature and Effect on Treatment of Water for Reuse. *Research and Technology Journal*, 94–102.

Sablani, S. S., Goosen, M. F. A., Al-Belushi, R., and Wilf, M. 2001. Concentration Polarization in Ultrafiltration and Reverse Osmosis: A Critical Review. *Desalination*, 141, 269–289.

Stephen R. and Kronmiller D. 1995. Membrane Cleaning Under the Microscope, Successful Cleaning Means Knowing the Foulant. Retrieved May 2008 from http://www.pwtinc.com/science.shtml?urlMenu1=1&urlSection=science&urlPage=TechBulletins...

Strathmann H. 2001. Membrane Separation Processes: Current Relevance and Future Opportunities. *AlChE Journal*, 47, 1077–1078.

Taniguchi Y. 1997. An Overview of Pretreatment Technology for Reverse Osmosis Desalination Plants in Japan. *Desalination*, 110, 21–36.

Tragardh G. 1989. Membrane Cleaning. *Desalination*, 71, 325–335.

Tran-Ha M. H. and Wiley D. E. 1998. The Relationship between Membrane Cleaning Efficiency and Water Quality. *Journal of Membrane Science*, 145, 99–110.

Yang, Y., Zhang, H., Wang, P., Zheng, Q. and Li, J. 2007. The Influence of Nano-sized TiO_2 Fillers on the Morphologies and Properties of PSF UF Membrane. *Journal of Membrane Science*, 288, 231–238.

Zwijnenberg, H. T., Koops, G. H. and Wessling, M. 2005. Solar Driven Membrane Pervaporation for Desalination Processes. *Journal Membrane Science*, 250, 235–246.

Assessment of Sustainability in Water Resources Management: A Case Study from the Dominican Republic

J.R. Pérez-Durán
Advisor on Water Resources, Instituto Nacional de Recursos Hidráulicos (INDRHI) – National Institute for Water Resources, Dominican Republic; EMR (Engineering, Management and Risk) Group

SUMMARY. This chapter presents an assessment of the water resources management system of a case study region, the Dominican Republic in the Caribbean, with a particular focus on sustainability. After a brief presentation on water resources management concepts, a methodology proposed for the assessment is developed and applied to country specific situation and issues of water resources. The "Building Blocks" assessment methodology, consisting of elements of the management functions or tools for water resources viewed in a simple graphic manner, helps to organize the assessment and is feasible to use in a participatory exercise due to its visual value and its potential to generate and guide discussions. Problems and issues related to water in the case study country are analyzed with reference to the tools or functions of water management as the evaluation unfolds. Results show that the water sector requires reform, which should be wisely based on the development of organization, planning and control tools for adequate administration of that resource in order to solve problems in water stressed regions of the country, identifying priority actions, and define strategies to solve current water quantity and quality issues, and to prevent aggravation of them in the future. Areas for improvement in water resources management are suggested.

1 INTRODUCTION

Concerns and interests on water resources issues are growing due to declining conditions in water quality, availability and climate variability. These have negative impacts on health and have diminished the potential for development, as well as increased suffering due to losses and damages caused by extreme hydrological events. Inherent to water resources is its geographic and seasonal variability and uncertainty, characteristics which are to be taken into account when water is to be managed for social and economic growth, while preserving the environment and natural patrimony. Water resources management has evolved from a partial focus or sectorial approach, with isolated actions in each particular sector making use of water, to a comprehensive view of a process which integrates all uses and conceives water itself as part of an ecosystem which is to be valued, preserved and used for the benefit of all. (Al Radif, 1999 and Jewit, 2001).

Given the not so optimistic view of the water situation in many countries, with forecasted worsening of shortages and growing conflicts (WMO/IDB, 1996, Biswas, 2004 and Al Radif, 1999), the capacity to make rational use of water and supply to all users in a coordinated and sustainable manner has been questioned, recognizing that solutions to water problems depend no only on water availability, but also on factors related to water management (Biswas, 2004). This is the case of the Dominican Republic, a developing country where an assessment of water resources management systems has been performed with the purpose of obtaining an insight concerning priorities and actions necessary for water conservation and to support improvement of living conditions of the poor and for sustaining economic activities, without threats to water bodies and the environment. To achieve this task of evaluating the situation of the water sector in the Dominican Republic, it has been thought both useful and necessary to develop a methodology that would guide the assessment in a simple, yet thorough way, brief and easy to understand, but without sacrificing neither conceptual nor technical depths of analysis.

Possibilities for improvement of living conditions, development of agriculture, tourism and many other important economic activities, are strongly dependent on the sustainability of water resources in the Dominican Republic. Management of water resources has to play a vital role in securing continued and sustainable use of water. Reforms and modernization of water resources planning and management are urgently needed. Different issues related to water resources management are briefly described trying to derive useful lessons from them and showing the priorities to be addressed. To convey a stimulating message of hope and potential for improvement of water resources management, some concrete experiences are highlighted and dealt with some detail, with refreshing lessons on participatory management and decentralization.

2 CHALLENGES AND TASKS FOR WATER MANAGEMENT

The importance of water is being increasingly highlighted as we gradually approach the limits of availability, or have probably trespassed it in some places, having passed from a safety to an unsafe area where this finite and vulnerable resource is becoming either unavailable (scarce or too polluted) or too difficult to fetch (too costly to access). While world population tripled during the twentieth Century, water consumption increased by a factor next to seven. It has been forecasted that population living in water stressed countries will increase from one third to two thirds (Global Water Partnership, 2000). Although access to water is considered a basic element for human life and a fundamental right, 1,100 persons do not have such access and 2,600 millions do not have proper sanitation conditions. An estimate of 1.8 million children die each year as a direct consequence of diarrhea, intestinal diseases and other diseases caused by unclean water or insufficient sanitation. Dysentery on its own is the second largest cause of deaths in infants (UNDP, 2006 and 2007). When adults are added to these tragic statistics, the figure rises to 5 million people dying annually from waterborne diseases, which is 10 times more than the amount of persons dying in wars all around the globe. Such dantean picture should be enough to motivate firm decisions to manage water more rationally and preserve its quality.

In developed countries diseases related to poor water quality and inappropriate wastewater disposal account for illness and loss of productivity equivalent to 2% of GDP **(Op. Cit, UNDP)**. Inversely, poor countries which invest in having better access to water and supply and sanitation, have a better economy (SIWI, 2005). Situations in the aforementioned scenarios are far from being all inclusive in relation to water problems. Extreme hydrological events like floods and droughts pose other great and menacing problems. Concerns on water pollution, irrational use of water, inefficient operation of existing infrastructure, climate change, institutional weaknesses and economic incapacities also add to the list.

3 WHAT IS WATER RESOURCES MANAGEMENT?

Water resources management is called to address several issues and problems. Where to get the required amount of water is in itself a challenge due to its evasive nature. The difficulty to identify and measure water has earned water a well deserved reputation as a "*fugitive resource*" (FAO, 1995 and UNDP, 2006). For its dynamic, changing and vulnerable nature, as well as its unequal distribution in time and space, it can be said that water resources management is therefore the management of variability and uncertainty.

To understand the dimension and challenge of variability in water management, it is useful to illustrate this issue considering that in Latin America, a richly water endowed region occupying 15% of the earth's surface, we have 30% of global precipitation and 13,120 million m^3/year flowing in surface water, equivalent to 33% of total water runoff in the world, for only 10% of the population on the planet. However, in this region 60% of the population live concentrated in 20% of its territory, where only 5% of all available water is (WMO, 1996). The per-capita estimate for renewable water resources in Latin America is 28,000 m^3/year, which is superior to the world average, more than four times the amount of water in the Central American and Caribbean region with an annual value of 6,890 m^3/hab, almost ten times the amount for the Greater Antilles of the Caribbean (Cuba, Haiti, Jamaica, Puerto Rico and the Dominican Republic), where average per-capita availability is 2,804 m^3/year, and ninety times the availability of Barbados with 313 m^3/hab/year. It is obvious that the regional value of available water and per-capita figure for the Latin American region disguise an uneven spatial distribution of water, which is well depicted by the fact that only one river, the Amazon, flows with 53% of all available water in the region (Ringler, Claudia et. al, 2000., and WMO/IDB, 1996). Spatial variability of water is also observed within one country, where territorial, demographic and water availability elements do not necessarily match.

Limited availability, whether seasonal or permanent, is a physical problem of water called scarcity. Growing population, rising economy and increase of industrial activities generate more demand and consumption of water. These factors, along with pollution of water bodies, increase with time and tend to diminish availability in such way that what was once a good source of water for a given population and for supporting economic activities in that area, becomes insufficient and inadequate. Another problem is the means to withdraw or divert and conduct water to the point of demand and distribute it among users. Not having the infrastructure or technology to carry and get water to where it is needed or not having the money to pay for this infrastructure or being deprived of it, which results in lack of access to water. The first case is a "physical scarcity of water" (probably also biochemical when considering water quality), while the second is what has been called the "economic scarcity of water". Other aspects like limited capacity of organizations, inappropriate institutional arrangements, inefficient infrastructure, low tariffs, losses and inefficient use of water and difficulties in procuring needed investment, plus other ingredients like lack of political will, are other dimensions to the water problems, which is not certainly limited to scarcity. At the Third World Water Forum (2003) it was recognized that water crisis is mainly a crisis of management. Water management is equivalent to administration of conflicts between human beings and amongst humans with their surroundings. A water management system is created to avoid and solve these conflicts (Dourojeanni, 2001).

With such diversity and magnitude of problems one has to agree that water management is an analytical and creative activity (Setti et al., 2002) oriented to the formulation and application of principles, policies, for establishing norms, structuring administrative systems and taking decisions for the use, control and protection of water resources.

For the management of water it is absolutely necessary to have a management model, where policy statements, norms and standards, organizations, administrative instruments (organizational system, licensing and permitting for water use, cadastre or registers systems, financial systems – water tariffs and investment schemes -, information systems, education, basin and national water planning, pollution control, vigilance and compliance mechanisms, among others), technology and infrastructure are oriented for optimizing the use of water, while at the same time preserving the quality and environmental value of water bodies, making water available in desired time and quantity for the use of present and future generations. Preparing plans, programming actions and investments, as well as implementing those policies, plans and actions are the activities of water resources management. To put it in business administration or management terms, water resources management requires applying organizations, planning, control functions to satisfy the needs of different uses, promoting rational use of water and optimizing objectives of social and economic development as well as conservation of quality and quantity of water.

It is also worthwhile to make a point in differentiating traditional water resources engineering and water resources management. While the first is an activity of specialized civil engineers, who provide technical and engineering solutions to the physical problems, water resources management is not exclusively limited to these specialist, and can involve several disciplines, trades and actors, including users, community organizations, farmers, businessmen, industries, journalists, government officials and politicians, working, ideally together or at least on same scenario, trying to find social, environmental and economic solutions to the problems and challenges that the world and communities face. The growing attendance of diverse crowds to World Water Forums, and the acknowledgement of the importance of water in several ministerial and presidential summits is a sign that water has become everybody's business. Prince Willem of Orange declared at the *World Summit on Sustainable Development* (WSSD) celebrated in Johannesburg in 2002, that from the previous WSSD in Río de Janeiro (1992) to Johannesburg, water climbed to the top of the agenda (IWMI, 2002).

More fashionable and "sustainable" terms like Integrated Water Resources Management (IWRM by its acronym), have become popular and conceive all users of water, including environmental demand, seeking a proper balance that would yield both economic benefits (profits) and social growth, including the poor, while not endangering the natural patrimony and securing conditions for it to continue to supply its valuable services for the enjoyment of life. As written in more a thoughtful definition, IWRM has been defined as "a process which promotes development and coordinated management of water, soil and related resources to have the results of maximizing economy and social wellbeing in a equitable manner without compromising the sustainability of vital or essential ecosystem" (GWP, 2000).

The trend in water management has shifted from a single purpose goal like agricultural development or hydropower development to a multipurpose development, where natural conservation and social equity lead to a wider vision and a more comprehensive and coherent approach of water resources management. Conceptual goals of IWRM is oriented to building a common platform in which all sectors using water would link their interest to a multisectoral coordination of water allocation and subordinate their proposals for intervention within a global context (ICWS, 2002). IWRM has promoted that the focus changes from exploitation of water resources to conservation and rational use, and from management of supply to demand management. The emphasis before was more in investing to build new canals and dams to supply water to more towns and irrigation fields. The aim now is to take into account all the demands, consider water availability and prepare an optimized plan for its use and conservation, weighing the effects of all proposed actions or requests for diversion of water and any discharge into water bodies in the context of the hydrographic basin's capacity to respond to that demand without harming the environment nor threatening natural resources.

Another important difference is that before there was an almost absolute participation of the government in all water management aspects, leaving to passive users a limited voice in this matter. Government was seen as responsible for performing all the tasks while now it is desired to have stakeholders become involved.

One key concept in IWRM's framework is the level of functions, calling for separation of roles and conceiving interventions of main water actors into three distinct areas of involvement. The Constitutional function has to do with developing and updating legislation, norms and standards, defining strategy and formulating policy and strategic plans. The Organizational function is where operational tasks of regulatory responsibilities are implemented, by an agency responsible for applying the norms, strategy, policy and strategic plans. The Operational function is the third level where water services are provided.

4 ASSESSMENT OF WATER MANAGEMENT SYSTEMS

The assessment of sustainability of water resources management has to be addressed in terms of how do the strategies, policies, norms and standards, strategic plans, institutions in the water sector and administrative instruments, which were previously mentioned as having to do with water resources management, help towards this aimed equilibrium of sustainable use of water resources. The key question is: How sustainable or unsustainable itself is the system called to preserve sustainability of the use of water for development for social and economic growth and preservation of water, as well as other associated natural resources and the environment? Considering that this is an important question to answer, an interest arises in being able to evaluate how the management model is leading, or misleading, to attain a proper use of water and tapping the benefits derived from its use, without harming the environment and guaranteeing water resources capacity to continuously sustain life and support development now and in the future. Water has been defined as a finite and renewable resource. In a way, with rapidly growing demand, scarcity of water in many places and contamination of water resources, what is pretended to be achieved by sustainable management of water resources is to monitor conditions pertaining to both attributes, finiteness and renewal, and where possible impose or promote discipline in the use of water so as not to approach limits of finiteness too soon and to better understand the natural processes of the water cycle not hampering this process of renewal, safeguarding quality and environmental conditions. Furthermore the term "sustainability", where appropriate, would be particularly dealt with in consideration of its relationship to environmental and natural resources base, but would generally imply technical, financial, administrative as well as the environmental dimensions.

To achieve this purpose of evaluating the capacity of a water resources management system, a methodology is proposed based on the elements or components necessary for water management. Since these elements are all considered essential, we opt for "baptizing" this methodology as the "*Building Blocks*" for assessment of water resources management systems. The aim is not to measure and evaluate water resources productive potential or aspects related to water service, but rather to evaluate the management system. Results and performance of the applied management scheme and quality of water services are off course indicative of the sustainability of the administrative model. In order to understand the capacity of a system to respond to an objective like that of managing water properly, identifying the components of water management and the criteria and evaluation terms for judging of the soundness of their application must be set forth.

Garduño (2001) proposes that the administration of water rights calls for implementation tools: planning models; guidelines and procedures for the filing, processing, granting and control of water abstraction and waste discharge permits; information system to systematically safeguard, retrieve and release all documents involved in each application, databases and follow-up; capacity building (training and enabling working environment, competitive salaries and rational promotion approach); and communication to enhance public awareness and education.

While the above set of tools is more oriented to evaluating the success of implementing and enforcing water law, it provides an interesting suggestion towards the core elements or "building blocks" of water resources management, which requires off course more than "good" water law, as it has already been wisely considered in the previous paragraph with tools related to education and communication.

From the works of Trewatha and Newport, (1982), and Robles and Alcérreca, (2000), management is defined as a set activities directed for the planning of human, technological an economic resources and organizing structures with task assignments. Comparatively, the assessment of the quality of the water management system has to take into account how human, technological, economic, financial, organizational and other resources, are arranged to achieve sustainable use of water. The proposed methodology is not aimed at awarding a grade for development of water resources management system for a country, but rather oriented towards revealing issues and challenges to be addressed and overcome. The objectives of this methodology are to identify and qualitatively evaluate performance of different components of water resources management; identify the relationships among them; find out which instruments or tools are in place and which others need to be developed; and to raise awareness among decision makers of the need for reorientation, organization and renewal of water resources management targets.

Another source for identifying components of a water management systems is a water plan, whether at the national or at the basin scale. Commonly proposed components, activities or products of such plans:

- Management and dissemination of information on water and climate
- Participatory management: decentralization and delegation of responsibilities to users and community organizations.
- Demand management: plans and technology for improvement in water use efficiency.
- Risk management accounting for extreme hydrologic events, floods and droughts.

- Water allocation: regulation of the availability of water for users, and administration of water rights.
- River basin planning and formulation of projects for multipurpose use of water resources.
- Strategy and policy formulation and institutional arrangement for water resources management.
- Financing value of water and economic, water accounts.
- Characterization of water bodies and watersheds.
- Integrated water management.
- Natural resources and environmental protection and water conservation.
- Education and communication.
- Indicative plans for sectorial uses of water.

From the above, the administration instruments or components for water resources management that can be considered in a water resources management system are: basin planning, risk management, education, information, and monitoring, institutional and organizational arrangement (water regulation agencies, service providers). Some other instruments of water management systems, not necessarily included as specific components of water plans are research, vigilance and inspection, information, licensing, permitting and awarding of rights for water use, system, registration and cadastre of water users, financing water management (tariff structure), norms and standards.

By selecting, arranging and "shuffling" some of the elements in both sets above, we finally have the wall made of Building blocks as shown in Table 1. Individual blocks are organized in layers or components (strategy and policy, legal, institutional, planning, monitoring, environment and natural resources, water allocation and quality control, financing and risk management). The order of layers is important; conceiving that foundational layers should come before other developments in water resources management. In a group exercise or round of questionnaires it is evident that some discussion is to concentrate on whether or not these proposed set of components of a water management system are all the layers or all the elements necessary or if there are some still missing. It is also to be expected that arguing can be raised about the justification to include a given "block" or disagreement as to which level or layer it should occupy. New blocks or elements could be added and some could even be discarded. In fact, for generating participatory input to the assessment, it is advisable to have the participants define the levels (components) and the blocks, and later on, determine best location of each block.

Several considerations are basic to determine criteria and scale of evaluation, having in mind that the tool is proposed for a qualitative assessment. Color, numeric or combined scales can be used, but a simple numeric scale is proposed, assigning a number from 0 to 5 for each block, and final score in each layer as addition of all blocks in that layer. Quality, functionality or efficacy of water management instruments can be graded according to the following: 0 = Absent (inexistent); 1 = Obsolete or ineffective; 2 = Incipient concepts and actions, isolated experiences and with questions on sustainability; 3 = Existing initiatives and proposals to improve; 4= Satisfactory or promising experiences and good potential for more improvement; and 5 = Well developed and improving. Some warning as to the use of number ranking is that the objective is not to award a grade and that interpretation of results should only be made in general terms, not pretending to accurately measure the level of development of a water resources management system, but rather to identify missing or weak elements or components and the relationship between them. Results can be of help to set priorities for development.

Premises for interpretation and application of this methodology are: a-) all elements or "blocks" are necessary for water management; b-) Blocks will have a score, indicating in each case the degree to which that specific element in water resources management has been attained or satisfactorily developed, notwithstanding that there is always room for improvement; c-) if a numeric scale is adopted, instead color scale (filled or empty blocks), the higher the score awarded, the more successful that specific element stands in assessment; d-) the more "blocks" of the "wall" with good scores, the nearer to success in assessment is the water resources management system; e-) without one or several of these blocks (zero or low score) the wall would be unstable and water resources management system would be judged to be unstable or not so sustainable; f-) the more

Table 1. Building blocks of water resources management.

RISK	Early warning system	coordination with Civil defense and other emergency institutions	Flood control action plan	Droughts	Risk mapping	Zoning protection
FINANCING	Investment programmes	Tariffs structure	Cost recovery	Incentives	Private participation	Water accounts
WATER ALLOCATION and QUALITY CONTROL	Allocation mechanisms (award, licensing and permitting)	Demand management	conflict resolution mechanisms	Quantity and quality integation	Compliance	water reserves
ENVIRONMENT and NATURAL RESOURCES	Norms and procedure	Control and vigilance	Impact assessment requirements	Ecological flows or demand	Environmental management plans	Watershed management plans
MONITORING	Climate stations	Hydrometric network	water quality measurement	groundwater	processing and analysis	disemination and public access
PLANNING	Inventory and assessment of water	Water Balance plans	national, state or regional planning	Basin planning	Indicative sectorial plans	Project optimization
INSTITUTIONAL	Role definition	Institutional arrangement	Descentralization and users organizations	Public participation administration of	Education and communication law and regulation	Research and development
LEGAL	drafting law	law	cadaster registry	water rights	law enforcement	
STRATEGY AND POLICY	Strategy	Policy formulation	Principles and Norms	Policy statement	Priorities	Policy evaluation

"empty blocks" (zero score) or "weak blocks", the more unsustainable the water management system would be; g-) the order of layers (rows) is important with regards to steps or stages that have to be followed or be ordered – institutional, legal and policy aspects should be at the foundation and influencing result in instability of the system; h-) there is not an indispensable place for layers, since it is possible that in one country institutional progress or monitoring capacities has taken place without an explicit policy statement; i-) size of blocks do not take into account relative importance with respects to other elements and all blocks are drawn of equal size.

In visual terms empty blocks, half filled blocks, scanty areas of the wall, or unsupported "bricks", would all reveal challenges, weaknesses or instability of wall. This situation would be characteristic of an unsustainable water management system. If upper layers, in comparison to lower layers, would be with higher score or having more filled blocks, this will also be considered not to foster stability or sustainability. More precisely said, this reveals that some tools for water resources management have been developed and are used without having a clear vision as to goals and that there are uncoordinated efforts in the water sector. Some developments, like technological progress, having for example an early warning system, and are sometimes partially achieved prior to having well developed organizational capacities or water allocation mechanisms.

As a graphic tool it is considered valid for a participatory exercise. It would be recommended to have knowledgeable or informed participants doing the exercise, but it should not be a requisite to demand great expertise. It would be very useful to generate and stimulate discussions. Guided discussion for group work would lead to consensus on color or numeric score to assign to each element or instrument. Voting and averaging votes is one alternative when strong disagreement does not allow the group to reach a consensus, but this situation should be avoided, since it would reveal confusion among participants regarding the concepts of water resources management. Prior explanation of the purpose of the exercise and on elements and components required for water resources management system would help to diminish potential for such lack of consensus. Informed participants would rely less on detailed explanations before being ready to embark on the use of this simple assessment tool.

To explain the use of the methodology, a concrete case, that of the Dominican Republic, which is best known to the author, will be used with the understanding that it is far from being a model on water resources management. It does provide however, an excellent opportunity to see how real water management is put in place or is not adequately addressed in a developing country. It is intended that examples, both good and bad ones, are pedagogically useful to explain how the boxes are filled or ranked. Concrete and practical cases having a potential for conveying positive messages and clear lessons on water resources management practice have been selected. To give it both substance and shape, general background information and more detailed information on specific issues having some relationship to sustainability or to water management are presented. This however, is not a diagnosis or country assessment exercise.

One advantage with the methodology is that the main subjects of assessment are not the institutions. Instead of being an evaluation of the water sector (the actors), it is more focused on the system and the tools comprising that system, which are necessary for the scheme to work effectively.

5 BACKGROUND INFORMATION ON THE DOMINICAN REPUBLIC

The Dominican Republic is located in the island of Quisqueya, called Hispaniola by Spaniards who settled after 1492, being this the second largest island of the Caribbean (77,914 Km^2). Its territory, occupying the eastern side of the island shared with Haiti, is 48,670 km^2. It has 1,389 km of coast with 70% of cities above 10,000 habitants and 75% of industry and most of tourism. Population is 8.5 millions and density of population is 169.31 hab/Km^2. Human development Index is 0.738 which would have the country in position 97. Average schooling is 5.4 years of education and analphabetism is 10.4%. Contribution to the GDP of agriculture is 11.67% (include agriculture, livestocks, fisheries and forest), crops 4.85% and water and electricity sector 1.82%. Weather is tropical and the island is in the route of hurricanes which frequently strike the country (needs references for all these data)

Water resources are a key factor that has supported the expansion of agriculture of the Dominican Republic, an activity that generates 30% employment and contributes to the economy with 12% of the GDP. Water is also a relief in energy generation with hydropower plants, which though small in capacity, alleviate dependence on imported oil, gas and carbon. Furthermore, tourism which accounts for 30% of all income has been based on natural richness, with particular preference to coastal environmental and natural attractions. There are signs of serious water problems threatening sustainability of development and justified concerns of possibilities of water conditions becoming aggravated if current tendencies in the water use efficiency, soil erosion, pollution of rivers and reservoirs, vulnerability to floods are not attended.

Flora is characterized by a high biodiversity with more than 5,500 species and high endemism (could you define % of endemism?). Such richness must be preserved. Both flora and fauna are affected by land use patterns, extensive cattle raising, pollution by agrochemicals, poor irrigation and drainage practices, deforestation and degradation of environment by pollution of rivers and coastal waters, due to discharge of untreated wastewater.

Payment balance is greatly impacted by the need to import oil used for generation of electricity and fuel. The country does not enjoy a stable electricity supply and several episodes of crisis have made it clear that it is vulnerable. It would be good for the country to expand the hydropower capacity which will be both for improvement of the energy deficit, reducing dependency on imported oil and alleviate the economy.

The uncontrolled growth of water demand, the progressive deterioration of watersheds and pollution of water and degradation of water bodies, with many unsatisfied segments of the population without access to water supply or with only limited access to it, is a risk and challenge for development and is threatening the natural patrimony of the country. On the other hand it is necessary to define strateges for investment in hydraulic and sanitary works to expand coverage of service and make a more efficient operation of existing hydraulic and sanitary systems, improving services in harmony with environment, making a rational use of water, identifying reserves within a growth plan.

Surface water resources potential for the Dominican Republic is estimated to be 19,000 million m^3/year, while potential for groundwater is has been estimated as 2,500 million m^3/year. Rainfall pattern is influenced by eastern winds and average precipitation is 1,450 mm/year, with low value of 422 mm/year and maximum average value of 2,305 mm/year. About 65% of the population has access to water supply services and only 21% have access to sewage systems. Irrigation area is 301,700 hectares, mostly irrigated by gravity by 89,317 farmers. There area 34 dams, 20 of which have a storage capacity above 1 million m^3, for a total storage volume of 1,500 millions m^3 and an installed hydropower capacity of 524.5 MW, annual generation being 1,340 GW – hr/year, which represents 17% of national demand and consumption of energy. These data, and wherever there are data with technical numeric information, needs references.

Infrastructural development, including dams and irrigation systems, has been prioritized in the 3 major basins, the Yaque del Norte, Yaque del Sur and Yuna river basins (Figure 1). As interpreted from Table 2) Yuna is the region with the highest proportion of irrigated land with 15.31 Ha/Km^2 (surface of irrigation system/area of watershed or group of basins) followed by Yaque del Norte with 12.70 Ha/km^2 and Yaque del Sur with 8.17 Ha/km^2.

6 WATER SUPPLY AND SANITATION

Access to the services of water supply systems is 79% of total population. Accumulated design capacity of all water supply systems in the country is 32 m^3/s. Half of the population in the rural area (50%) do not have access to water supply systems. As much as 6,000 communities and 1.565 millions are reported to be without water supply systems. Urban population that is served by water supply systems is 83% of urban dwellers. These figures are to be judged optimistic due to the fact

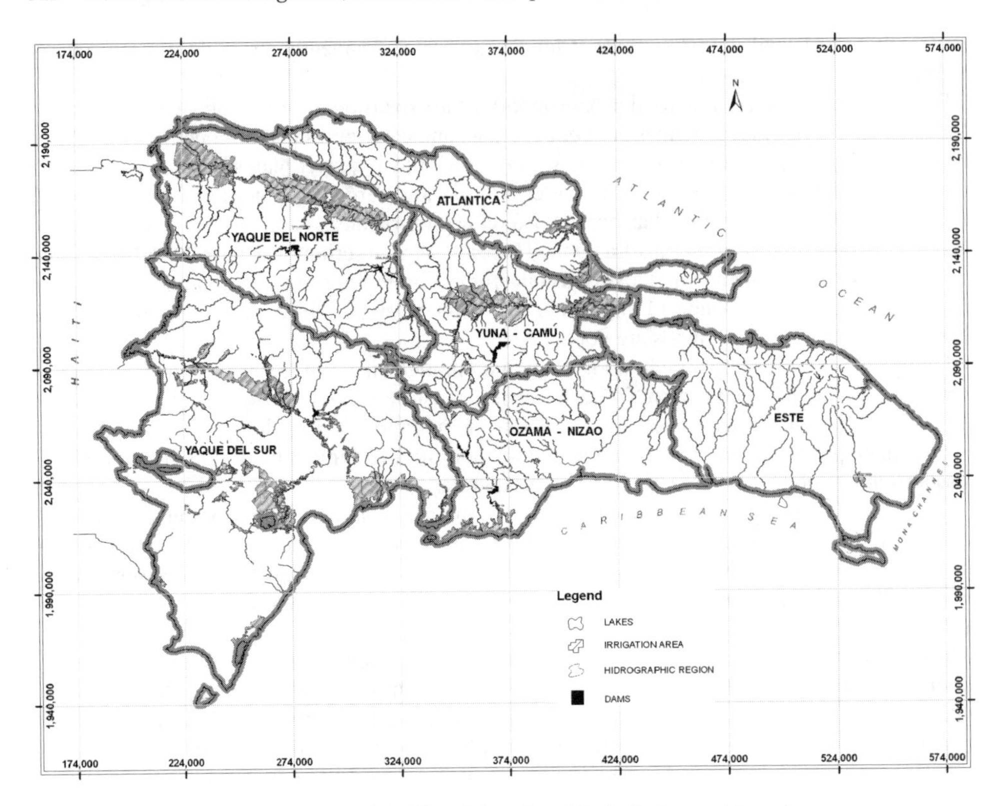

Figure 1. Irrigation areas and reservoirs of Dominican Republic by hydrographic regions.

Table 2. Infrastructural development by regions.

		Dams				Irrigation	
Region	area (km^2)	quantity	Storage capacity (million m^3)	Installed capacity (MW)	Generation (GW-hr/year)	Area (Ha)	(Ha/km^2)
Yaque del Norte	7,788.85	7	818.19	166.00	377.25	98,886.15	12.70
Yuna	5,258.38	4	517.96	68.10	282.90	80,505.09	15.31
Yaque del Sur	15,211.99	2	478.00	19.40	73.80	124,353.41	8.17
Ozama Nizao	6,221.93	4	360.30	204.00	401.11	18,167.74	2.92
Este	8,120.22	0	0.00	0.00	0.00	6,850.25	0.84
Atlántica	5,060.52	0	0.00	0.00	0.00	18,840.27	3.72
Total	47,661.89	17	2,174.45	457.50	1,135.06	347,602.91	7.29

that it includes access to water within 500 m and includes systems which have serious deficiencies in their treatment of water and quality control (CYE, 2003). Service is discontinuous and it is common in many areas that water is not available at the tap during 24 hours. In some areas people receive water only a few days during the week. An estimated 90% of population served by aqueducts have their own (in-house) regulation deposits, which is an indication of that the water supply systems are not dependable.

Almost 30% of water supply aqueducts do not have chlorination treatment. The average index of "potability" (percentage of samples with presence of total coliforms in relation to all samples analyzed) is below 80%. Water corporations of Santo Domingo and Santiago, two largest cities in the country, report higher values for this index (90 to 95%). It is estimated that as much as 60% of the population drink bottled water, which is a booming business with dramatic increase in consumption during the last 10 years. This is due to a combination of growing concern on water pollution, interest on improving healthy habits and a sense of social promotion. In any case, people generally distrust quality of water served by water utilities.

There is a big gap between access to water supply and sewage systems. Only 21% of population has access to sewage systems. The use of latrine and domestic wastewater disposal into septic wells are the sanitary conditions for 74.6% of the population, while 4.4% have do not have access to basic sanitation. This means a significant danger of pollution of groundwater. In some cases wastewater is discharged to the sea, being a source of pollution in beaches. Less than 50% of collected wastewater receives any treatment. Existing water treatment plants operate with serious deficiencies.

The Instituto Nacional de Aguas Potables y Alcantarillado (INAPA), created in 1962, is the State regulator of water supply system and sewage systems in the country, with responsibilities to design, build and operate these systems in all the country except for five municipalities which have created their own corporations, public or state owned, to manage water supply and sewage services (Santo Domingo in 1973, Santiago in 1977, Puerto Plata in 1997, Moca in 1997 and La Romana in 1998). Sewage services are very limited even in the capital city of Santo Domingo. Performance of the water corporations has been rather weak. Cost recovery is too low and subsidies from government support their operation. Only CORAASAN in Santiago reports above 80% of fee collection. Tariffs is low, there are great losses due to low efficiency of the pipe network (60%) to which illegal connections accounts for above 60%.

High inefficiency values are typical in Santo Domingo, were design capacity of all intakes from dams, rivers and ground water, to serve a population of 2.3 millions, is 16.00 m^3/s. This would mean that average water supply is nearly 600 l/d per person, which is more than twice recommended design value for water supply systems in developed countries. It is true that discontinuous service, leaks, and low flow – low pressure operation of network can prove this figure wrong (false), but real waste of water does provoke excess volume of water to run through drains.

INAPA has legal competence for regulation of the rest of corporations, but does not exercise this right. Following this model of detachment of INAPA, eleven other cities have proposed creation of their own corporation with the idea that in this way they will be able to have better services. The National Congress approved the creation of three new such corporations, but these systems are still under control of INAPA for operation.

INAPA currently operates 363 aqueducts, 255 of them serving 1,587,766 inhabitants from 764 rural communities and 108 systems covering 347 towns and cities with a joint population of 2,679,756 (INAPA, 2008). The intake for these aqueducts are 105 by gravity, 224 by pumping and 34 have both gravity and pumps. This factor makes operations costly and/or vulnerable to power failures, which are common. INAPA has currently in operation 111 water treatment plants for purification of water. Only 47 of them are reported to be operating satisfactorily or regularly. There are 28 sewage systems in the area served by INAPA and 15 wastewater treatment plants.

INAPA operates 133 small rural water supply systems. Since 1997 it has been training community organizations to become involved in administration and operation of these systems promoting decentralization. Some have stronger participation and are showing improvement in cost recovery. There are neraly 150 communities operating their own systems, which were built with aid from foundations or development agencies, usually channeled through local non- governmental organizations.

Case studies by Pérez and Segura for INDRHI (2006a, 2006b as well as more recent data from INAPA (2008) in several small water supply systems, have been analyzed examining organizational, financial and operational aspects, trying to extract useful lessons for water supply and sanitation reform and design of rural water supply programs. The main findings from this analysis are:

- When communities are organized and trained, they are able to operate the systems – some were able to expand the service to new areas and improve level of service, transforming and upgrading the systems from public taps to in-house taps after commissioned;
- Strengthening efforts of transfer process and existing community organizations should focus on organizational leadership, administrative and accounting systems, cadastre updating, measurement of flows, control devices and demand management arrangement, training of technical staff responsible for administration and operation, improved management of information (consumption, costs), benchmarking and quality certification, keeping stock and inventory of spare parts and tariffs structure based on consumption – organizational support and educational efforts are key topics and part of the strategy towards sustainability;
- Subsidy is still an important issue in some of them on items such as investment for infrastructure, energy to run pumps, replacement of equipment, but much less than bigger systems in cities;
- The smaller the community payment and fee collection is better;
- Timely and precise invoicing and fee collection mechanisms are key factors to success.
- Use of appropriate technology is important – traditional and conventional water supply systems are too costly solutions – designers are to be trained to design smaller systems (solar energy pumps for small communities are a good solution).
- Promotion strategy and stronger technical assistance should be provided by government agency. (please, reduce all this information and try to be accurate, brief and clear to comply with space)

Achieving water supply targets and millennium development goals is not a technical issue, but rather a financial and a social problem. A project law for the reform and modernization of the water supply and sanitation sector was drafted and has been subject of discussion at the National Congress since 2001. This proposal allows for private investment and awarding of license for private companies to operate water supply systems. The creation of two new entities, if passed by Congress, one acting as rector of the sector and the other would regulate operators. The future of INAPA was not well addressed in this project law, but ideas for its decentralization into 8 regional water utilities companies have been discussed. Even if by provisions of such legal reform and novelties it is made possible to attract private investment, it is clear that with without access to water and the great lack of sewage services, lagging way behind, the government would still need to have a strong participation in construction of water supply and sewage systems.

The water utility corporation of Santo Domingo (CAASD) has engaged private companies to provide commercial management services. Two companies invested in installation of meters and where made responsible for invoicing, fee collection, clients accounting, updating cadastre, incorporating new clients and client – point of contact services for specific areas in the city. In the areas where these companies work, invoicing is done on the basis of consumption (volume of water). Usual practice was at CAASD, and still is for the rest of nation, based on a fix sum of money (bill) per month. Assuming that a family used 30 m^3/month, applied tariff was RD$4.00/$m^3$ and fees were RD$120/month. Income generation has increased from RD$19 million to RD$60 millions per month, with no significant increase in tariffs. Residents in Santo Domingo are starting to save water to cut down the bills, by correction of leaks and control of valves, and CAASD reports important savings in water (Table 3).

7 IRRIGATION

Irrigation contributes to 43.77% of total agricultural production (INDRHI, 2006a and 2006b) in weight (tons) and 88.48% in value (Dominican pesos). Irrigation area expanded from 54,000 hectares in 1973 to 278 hectares in 1995 and it is currently 301,700 hectares, which is about 50% of the potentially irrigable land, taking onto account slopes, soil characteristics and water availability. This increment has not been accompanied by improvement in efficiency of use of water in irrigation

Table 3. Rural Water Supply Systems (reference).

No.	Community	Families	Population	Technical and Financial assitance	Invoicing (RD$/month)	Tariff (RD$/month) per family	Fee Collection (RD$/month)	Fee Collection (%)
1	Vuelta Grande	55	213	CARE, FUNDASUR, LEMBA, Bosque Seco.	1,100	20	1,000	92%
2	Cabeza de Toro	742	4,452	Visión Mundial, PNUD, FUNDASUR, LEMBA, Bosque Seco.	14,840	20	4,500	30%
3	El Granado	97	485	FUNDASUR, USAID	1,335	15	1,200	92%

systems, which is estimate be 25% as average value of combined efficiency in distribution network and field irrigation done by gravity. Irrigation is by far the largest user of water with a demand of 6,429.86 million m^3/year (reference). The cumulative growth of water demand for irrigation, municipalities and industries, including tourism, along with operational losses of water, create great pressure on water, which is evident in some regions exhibiting water deficits and where level of conflicts among users have increased.

Rice, a high water demanding crop, is the main crop in irrigation systems, occupying an area with 144,692 hectares, which is 47.96% of total area covered by irrigation systems. Rice, which is irrigated by inundation of fields, requires 1,572.99 million m^3/year of water (net requirement). Other important crops, as shown in Table 3, are plantain and banana which grow and are cultivated in an area of 43,115 hectares, vegetables with 21,252 hectares, sugar cane with 15,905 hectares (more cane goes under non irrigated agriculture) and beans with 10, 964 hectares.

Important results in water savings, reduction of production costs and use of less fertilizer in rice crop were achieved by Project PROTECAR, the Spanish acronyms for Technological Improvement Project in Irrigated Agriculture. This initiative was financed from 2001 to 2005, by the Japan International Cooperation Agency and the Dominican Government. PROMTECAR experiences offer an alternative with significant achievements reached through training of farmers and motivation to change traditional irrigation habits and adopt both water management and crop management techniques (Table 4).

Water distribution was significantly improved and programmed according to irrigation area and stage of crop requirements. Turns of shifts were organized from tails of the system, guaranteeing of service throughout crop cycle, which resulted in reduction of conflict in water allocation. Less water, less operational costs and less labour (8 hours, before 3 days) are successful exhibits of this Project. Reducing 65% of production costs and increased productivity are stimulating results.

Most outstanding results are that water savings of 44% and 24% reduction on the use of fertilizers. Water consumption was reduced from 25,000 m^3/Ha to 14,000 m^3/Ha. If this would be projected for the entire surface area of irrigation systems servicing rice crop, 692.08 million m^3/year, water that could be made available for other crops or other uses. This value is more than water supply demand for population in the whole country (679.86 million m^3/year) and such savings (692.08 million m^3/year) all by itself would be almost enough to cope with projected increase in water demand in the next 45 years, for the period 2005 to 2050 (704.22 million m^3/year), and along with present water demand be able to satisfy water demand for population in the year 2050 (1,384.08 million m^3/year). This was achieved within a typical gravity irrigation system fed by a canal. On better use of fertilizers, quantity of total nitrogen dropped form 124 Kg/Ha to 94 Kg/Ha. (INDRHI 2006a, 2006b)

The National Institute for Water Resources (INDRHI) is responsible for the management of irrigation systems and as provided by law has a expressed mandate to organize water users and

Table 4. Irrigation Net Water demand by major Crops (reference).

	Region (million m^3/year)							% Vol crop water Requirement/ Total volume for Irrigation
Crop	Yaque del norte	Atlantica	Yuna	Este	Ozama	Yaque Del sur	Total	
Rice	978.44	52.15	533.39	4.65	4.36	–	1,572.99	31.54
Musace as	626.52	5.30	–	0.04	75.11	747.01	1,453.98	29.16
Legums	13.99	0.17	–	0.05	0.95	161.64	176.79	3.55
Vegetables	19.03	0.15	6.69	0.14	4.02	73.95	103.99	2.09
Tuberculus	10.17	0.38	0.31	0.00	1.20	39.16	51.21	1.03
Grass	215.01	0.38	–	–	14.88	327.23	557.50	11.18
Sugar cane	–	–	–	–	13.39	793.32	806.71	16.18
Others	112.66	0.98	10.21	0.16	10.63	128.95	263.59	5.29
							–	
Total	1,975.82	59.50	550.60	5.04	124.54	2,271.25	4,986.76	100.00
% Vol crop water Requirement/ Total volume for Irrigation	39.62	1.19	11.04	0.10	2.50	45.55	100.00	

have them become involved in cleaning and maintaining canals. INDRHI has zonal delegation in 10 Irrigation Districts, but since 1987 has been transferring the administration, operation and maintenance of irrigation systems to the Water Users Associations (WUA). There are 89,317 farmers organized into 178 associations of irrigators, which are in turn organized into 28 Irrigation Boards and three independent associations, covering the totality of irrigated area as of 2006, having in this completed an era in irrigation management transfer, which should be consolidated with additional targets. The irrigation management and transfer program, promoted and conducted by INDRHI, has helped in the creation of WUA and capacity building of these organizations. The WUA have shown significant improvement and positive signs of maturity and development. They decide and establish their own tariffs and collect fees without intervention from INDRHI, making use of funds according to their annual operational plan in operation and maintenance of canals.

WUA has contributed to the elimination of gate delinquency (breaking of locks in gates at irrigation canals and bribery of government appointed personnel at canal gates to access more water than agreed or authorized, were typical situations in the past), to the improvement of operational efficiency in irrigation systems, to improvement of water distribution and irrigation programming, to a better handling of seasonal shortages of water, to reduce conflict among users and tensions between farmers and INDRHI, and to increase cost recovery. WUA seems to be more effective than INDRHI in the collection fee (average increase from 17% to 72%).

Chalas (2007) has proposed to measure impact of WUA's management of irrigation systems by analyzing the soil use index, the evolution of water depth applied in irrigation, and usage efficiency. Improvement in efficiency in the use of water depends on several aspects, some of them being irrigation technology, physical conditions of infrastructure, regulation mechanisms and tariffs. Maturity and stage of development of WUA's also has great influence. Yet from the study of the 3 first mentioned variables, it can be seen some WUA's are achieving good results in terms of amount of water used. In some of the irrigation systems area has increased and they are using less water. From the study of Chalas (Op. Cit) some illustrative data can be arranged to show that less water (water depth in mm) per hectares is being used (Table 5). Such is the case of the Azua Valley WUA, where the main channel operated with 16.00 m^3/s and now runs with 7 to 10 m^3/s, serving the same area. At the Fernando Valerio WUA the irrigation water depth per unit area relationship has been reduced in 88.96%. However, this improvement cannot be overstretched as an evidence

Table 5. Use of Water per Unit Area in selected Water Users Associations.

WUA	Initial year	Reduction in Water Depth (mm)	% of Reduction in Water Depth	Area increase (Ha)	% Increase in Area	Water depth/area (mm/Ha) initial	depth/area (mm/Ha) 2006	Difference in water depth/area (mm/Ha)	% Reduction water depth/area
Fernando Valerio	1992	2,238.10	59.82	21,044.10	260.63	0.46	0.05	0.41	88.86
YSURA	2001	−984.20	−43.24	1,120.69	13.13	0.27	0.34	−0.07	−26.61
UFE	1994	646.90	3.50	1,042.59	41.31	7.32	5.00	2.32	31.71
Mao	1992	−532.20	−22.77	1,304.86	15.60	0.28	0.30	−0.02	−6.21
Nizao – Najayo	1992	1,154.70	18.19	281.58	31.92	7.20	4.46	2.73	37.99
J. J. Puello	2001	−852.30	−109.99	167.70	2.06	0.10	0.20	−0.10	−105.74
Total		1,671.00		24,961.52		0.93	0.53	0.40	43.53

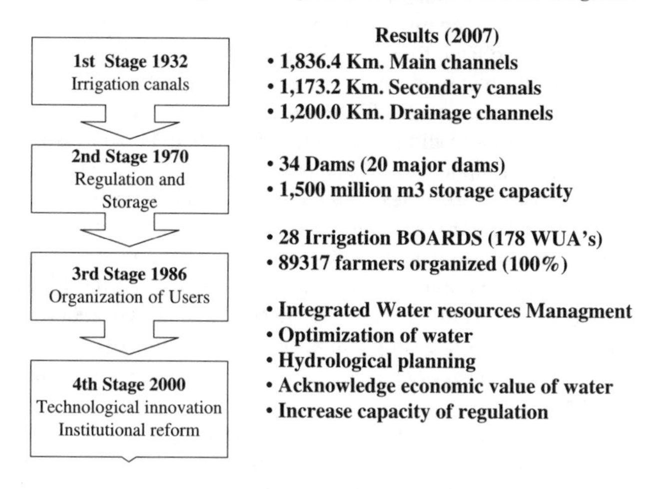

Figure 2. Development of irrigation and water resources management.

of proper management, since reduction of the use of water consumption is sometimes a result of water shortages due to different water allocation than originally foreseen and increase in number of users for example. In the case of Fernando Valerio, soil salinity problems as indicated bellow, are a probable sign of using less water than necessary. If such levels of salinity are confirmed, more quantity of water, more than crop requirements, is necessary for leaching of salts from soil.

Development of irrigation and water resources management is a three stages process (Figure 2). The first was the building of canals which took place since the late 1890's but was specially promoted after creation of the National Irrigation Service in 1932. The second stage is the era of dam development which took of in 1970. In the third stage there is a change in the pattern, shifting from infrastructural development to human and water management development with a milestone in decentralization. This stage, originating in 1987 with the creation of the first two water user associations in Azua and Santiago, is the stage of irrigation management transfer, which has been something INDRHI has done consistently and progressively, even with political changes in administration. A fourth stage is to follow in the 21st century, demanding institutional innovation, implementation of integrated water resources management, consolidation of irrigation management transfer, improvement of efficiency in the use of water, better use of technology, hydrological planning and further improvement in cost recovery and new financial schemes for the water services and infrastructural development.

Aspects to improve are operation of irrigation systems, which needs to be optimized so as to program water distribution as a function of the water demand from crops, installation of measuring and control devices, improve efficiency in irrigation systems and on- farm water management. Administrative and financial reform is needed to carry out development of accounting and management system, build more on more human resources and administrative talent, develop business and commercial culture of the WUA's, make better use of technology for handling cadastre, production and market information systems, introduce economic incentives to promote increase in water use efficiency, change tariff structure to correlate, where possible, invoicing to water consumption, design financial mechanism so that farmers have better access to credit for improving irrigation practice, introduce users certification and auditing instruments, create benchmarking to evaluate progress of WUA's, and optimize operational expending and increase investment among others.

Heavy duty equipment from INDRHI borrowed by the WUA's for their cleaning and maintenance of canals was one key component of subsidy, which has been reduced in some WUAs which bought their own machinery. Even so, there is opportunity for improvements so that WUAs be more self-sustainable with regard to operations and services. Another area where there still is a strong subsidy from the government is the cost of energy to run pumps for irrigation areas, bills which are paid by INDRHI. WUA's should also have to pay INDRHI for the water they use, and although have conceive this in their strategic planning, and even think of introducing fees for environmental services, this does not take place as yet. Tariffs structure need to be addressed, being still too low.

Use of technology and research would be assets to achieve better water resources management in irrigation systems. Increased support to WUA's is needed through effective extension services, oriented at augmentation in land use index, productivity and profitability of agricultural activity, modernized measuring and control structures, improvement of irrigation programming, clean production and environmental management plans.

More strengthening effort is required to consolidate the WUA's in order to reach higher levels of performance. A survey conducetd by CITAR-INDRHI (2008a and 2008b) evaluated WUAs performance. It revealed that WUAs need improvements in formulation of projects, public outreach, strategic planning, and assessment of organizational, administrative, logistic, and operational and financial outcomes and plans.

Other issues of concern to be resolved are water allocation and rights, urban growth, water quality, waste management and soil degradation. Competition from the water supply sector in some areas, having priority in water allocation defined by law, is depriving farmers and WUAs of pre-existing rights with no provisions for compensation.. This has been the case of the Rincón Reservoir where increased water demand for the city of San Francisco in the Yuna basin was committed for the water supply system. Likewise is the case of the Santana Diversion in the Yaque del Sur lower basin, where a new water supply for three provinces was built and affected the water balance in an already water stressed region. Urban growth, progressively expanding into areas under the influence of irrigation systems, has been acquiring lands where agriculture is no longer the use of that land, but rather housing projects. Some water quality problems occur in towns where a canal is passing through or nearby and has become the dumping site for both solid waste and wastewater. Use of agrochemicals is known to affect drainage water which is reutilized as well as groundwater. Salinity in soils caused by irrigation of land with poor drainage conditions in arid and semiarid zones or sub-humid areas irrigated with water with salt content. Such problem was identified in 1993 at the Fernando Valerio irrigation system with a surface area of 19,291.67 hectares. Evaluation rendered that 7,944.98 hectares were affected by salinity and speed of degradation was 200.63 Ha/year. This problem is estimated to be now affecting 10,553.19 hectares, which is 55% of (INDRHI, 2006) with light, moderate or strong salinity problems, but again more research is required for better and more precise assessment of these problems.

8 TOURISM

Tourism contributes to the economy with 7% of GDP and almost 30% of hard currency income (US Dollars and Euros), employment, geographically diversifying job offer, creating opportunity for service and construction of hotels and related infrastructure in the periphery of hotels. Tourism has grown from 5,800 beds in 1980 to 165,571 hotel rooms in 2007, with over three millions tourists visiting the Country each year. Development of tourism, mostly based on the sun – sea – sand recipe, occurs in fragile ecosystems and some areas have been cleared to make space for buildings violating the legal requirement of buffer zone area of 60 meters. Construction of hotels and golf courses have sometimes been done without consideration of environmental and natural values, cutting mangrove, desiccating wetlands and destroying coral reefs with great impact on marine biodiversity. Potential for growth in tourism will depend on water quality, including guarantee of water supply, clean beaches and well managed protected areas (World Bank, 2004). Diminishing water quality of rivers and beaches are a result of poor environmental conditions which is a hazard for tourism stability and potential in Puerto Plata. In the eastern tip of the island, the Bávaro and Punta Cana area, environmental degradation will restrict tourism in the future due to unsustainable water resources management. Overexploitation of groundwater has caused salt intrusion within 20 to 50 km inland. High water consumption of pools, gardens, laundry services and 30 golf courses in the country (350 m^3/year in an 18 holes golf course, where fertilizers are also applied) makes tourism a high water consumer, above demand for industry. Inappropriate discharge of wastewater and solid waste disposals are other challenges which demand specific planning and actions to be addressed in policy and strategy formulation.

Water supply and sanitation services are essential for hotels. Due to poor public water supply services or to location of tourist attractions, hotel developers and operators prefer to depend on their own water supply systems, usually by making wells. Overexploitation of groundwater resources however has been a problem and to avoid saline intrusion at the coast, they have had to drill other wells further from the coast, moving away from brackish water sources. Some hotel establishments are not a good example on wastewater discharge, but have been more and more assuming responsibilities to control pollution, some have been building their wastewater treatment plants, so as to protect rivers and sea. Bayahibe, in La Romana, is working on obtaining the "Blue Flag" award or category, meaning that it is a clean and safe beach, having as well environmental education activities. This is a good signal to join other 40 countries were blue flag is working.

Other demand management initiatives have started to be implemented in hotels, installing water saving devices in their plumbing and garden irrigation systems. Some hotels, as in other countries are inviting guests to contribute in reducing water and energy consumption on such specific habits as less demanding laundry services by les frequent change of blankets on beds and towels in bathrooms. For the more environmental and nature exploring preferences of the eco-tourist, water resources management is also important, since they value clean rivers and forested scenarios for health, mental or spiritually uplifting.

9 WATER BALANCE

Water balance is the accounting or inventory of the hydrologic cycle. Among the influx components of water balance that have to do with availability are natural elements like precipitation and related runoff, groundwater flows and storage in natural lakes or artificial reservoirs and return flow from wastewater that is added back to runoff. Efflux components in the hydrologic cycle are naturally occurring "losses" like evaporation and evapo-transpiration and natural stream requirements. Other efflux components, related to the use of water, are man made abstractions from rivers, lakes, reservoirs and wells for use in domestic water supply, municipalities, industries, livestock production and agriculture.

There are different techniques and approaches to water balance, some relying on comparison between runoff estimates to atmospheric variables like water vapor flux (evaporation) and transport

Table 6. Precipitation and Evapo-transpiration averages by regions in the Dominican Republic.

Region	Precipitation (P) or average rainfall (mm/year)	Evapotranspiration (ETP) average (mm/year)	P – ETP (mm/year)
Yaque Del Norte	1,263.00	1,379.28	−116.28
Atlántica	1,783.10	1,402.33	380.77
Yuna	1,641.40	1,339.11	302.29
Este	1,534.70	1,297.88	236.82
Ozama-Nizao	1,605.50	1,390.05	215.45
Yaque Del Sur	1,028.40	1,517.67	−489.27
Average	1,373.74		

for basins with big areas, continental or global scales, while others employ soil water budget model or a "bucket model" handling precipitation, potential evapotranspiration and soil water-holding capacity. A more commonly used method is the surface water balance study when precipitation and runoff measurements are available and evaporation can be reasonably estimated. For point from precipitation and available runoff stations of interest in the basin where there are no measurement stations, estimates of runoff from precipitation and available runoff stations, by more simple means or by more elaborate calculation using precipitation – runoff models can also be used (Reed et al., 1997).

Employing a hydrologic planning terminology, a water balance is the quantity of water available in a given location or region once water demands of different sectors have been met. It is an arithmetic net result of subtracting uses from inflows. This type of water balance, or comparison of total annual volume of surface water and groundwater available to total volume demanded for consumptive use in given regions or basins, is simply the difference between available water and demand and is a general – on average – description of the situation of the water availability in a given basin or region.

The water balance based on climate balances, uses an equation derived from the mass conservation principle ($Q = P - ETP$) (INDRHI 2006a and 2006b) (Table 6). The surface water balance as calculated in 1994, had some geographic limitations, since it is a general vision interpreting inflow as mean discharge at exit point of basin and outflow as accumulated water demands, the latter occurring at different specific points throughout the basin, not considered in their spatial distribution. Other inaccuracies requiring adjustments are that stored water in reservoirs is not taken into account and the fact that return flow from wastewater is also part of measured discharge at downstream – extreme – point. Average discharge values would grow progressively downstream in the basin unless abstractions take place along the river. To have a real water balance at the intake for a water supply system or an irrigation system, discharge value at that point would have to be determined.

Estimates for water demand for municipalities are based on population statistics and per-capita water consumption, distinguishing the urban from the rural area. Ecological demand has been estimated as excess water flows in rivers exceeding the 95% occurrence, either calculated from percentile values or from flow duration curves. Industrial demand comes from amount and types of industries with corresponding water demand estimated for each one of them at their location. Water demand for tourism has been estimated on the basis of existing locations of hotels, number of rooms and level of occupancy throughout the year and the water consumption estimates per tourist. Records of existing livestock population have been examined along with estimates of water demand to raise different types of animals to arrive at values for water demand for cattle (88.33% of total), sheep, pigs, goats, poultry, rabbits and horses. Water demand for irrigation in each irrigation district takes into account cultivated area, the crop demand month by month, and efficiency in the use of water (a global efficiency which includes efficiency of channels, application and varies from

Table 7. Water demand by region and sector in 2005.

	Water Demand 2005 by Region (million m^3/year)							
Sector	Yaque del Norte	Atlántica	Yuna	Este	Ozama – Nizao	Yaque del Sur	Total	Percentage of total demand (%)
Water supply	107.08	42.69	81.63	64.14	312.67	71.65	679.86	5.85
Irrigation	2,380.73	78.05	882.53	7.11	190.03	2,891.41	6,429.86	55.30
Livestock	86.74	84.28	93.17	99.95	85.73	88.36	538.23	4.63
Ecological	216.38	323.64	991.39	420.38	587.40	1,136.41	3,675.60	31.61
Industrial	40.65	15.99	30.61	24.96	119.60	27.29	259.10	2.23
Tourism	2.15	13.20	0.82	22.07	4.82	0.65	43.71	0.38
Total	**2,833.73**	**557.85**	**2,080.15**	**638.61**	**1,300.25**	**4,215.77**	**11,626.36**	**100.00**

Table 8. Projection of water demand from 2005 to 2025.

	Water Demand (million m^3/year)				
Sector	2005	2010	2015	2020	2025
Water Supply	679.86	760.76	843.8	928.5	1,013.08
Irrigation	6,429.85	6,429.85	6,429.85	6,429.85	6,429.85
Livestock	538.24	835.8	1,133.35	1,430.91	1,728.47
Ecological	3,675.60	3,675.60	3,675.60	3,675.60	3,675.60
Industrial	259.1	586.07	659.88	716.8	793.02
Tourism	43.71	94.29	124.8	165.98	221.57
SUB-TOTAL	**11,626.36**	**12,382.37**	**12,867.28**	**13,347.64**	**13,861.59**

Table 9. Projected water demand for 2025.

	Water Demand (million m^3/year)						
Region	Water Supply	Irrigation	Livestock	Ecological	Industrial	Tourism	Total
Yaque del Norte	158.66	2,380.73	285.12	216.38	148.79	1.90	3,191.58
Atlántica	63.31	78.05	282.65	323.64	56.57	10.02	814.24
Yuna	123.51	882.53	291.54	991.39	58.33	0.94	2,348.24
Este	97.76	7.11	298.32	420.38	47.71	197.50	1,068.78
Ozama-Nizao	461.30	190.03	284.11	587.40	429.77	10.68	1,963.29
Yaque del Sur	108.53	2,891.41	286.73	1,136.41	51.85	0.53	4,475.46
Total	**1,013.07**	**6,429.86**	**1,728.47**	**3,675.60**	**793.02**	**221.57**	**13,861.59**

10% to 32% according to conditions of irrigation systems and type of irrigation practiced in each one of them).

Estimates from the updated water balance (INDRHI – GRUSAMAR, 2007) providing information on water demand by sector and region are shown in Table 7. Results from projection of water demand by sector and regions, from 2005 to 2025, are shown in Tables 8 and 9, each with their respective growth expectation, except for irrigation where it is supposed that savings due to efforts of investment for improvement of infrastructure, modern technology for on – farm irrigation, and

Table 10. Water Availability in surface flow, groundwater, storage in reservoirs and reuse.

Region	Available Surface water (million m^3/ year)	Potential groundwater (million m^3/year)	Storage (million m^3/year)			Reused water (million m^3/year)		
			Reservoirs (Dams)	Small reservoirs	Total	Irrigation	Water Supply	Total
Yaque Del Norte	2,905.46	181.00	786.18	0.53	786.71	463.27	85.66	548.93
Atlántica	4,634.73	216.00	0.00	0.00	0.00	9.11	34.15	43.26
Yuna	3,600.96	236.00	550.33	0.05	550.38	83.98	65.3	149.28
Este	3,125.95	758.00	0.00	0.00	0.00	5.03	51.31	56.34
Ozama-Nizao	4,459.08	457.00	314.41	0.05	314.46	14.03	250.14	264.17
Yaque Del Sur	4,771.51	621.00	426.54	0.42	426.96	442.73	57.32	500.05
Total	**23,497.69**	**2,469.00**	**2,077.46**	**1.05**	**2,078.51**	**1,018.15**	**543.88**	**1,562.03**

improvement of operational techniques in irrigation systems will keep demand at same level even with limited expansion of irrigated area.

Available surface water is 23,497.69 million m^3/year and has been quantified from available records of water flows (discharges in m^3/s) from the existing network of 85 flow measurement stations distributed in the different basins of the country. Specific discharge values, calculated as a type of average of m^3/s per km^2 of surface area of basin for each basin, combined with precipitation records are used to define discharge values at given points and for the basins with limited measurement. In the case of groundwater, base flow, aquifers recharge values and potential for groundwater extraction are considered, having as a result that the total available water from aquifers is 2,469 million m^3/s, while recharge value is 4,161 million m^3/year. Reservoir storage capacity in 34 existing dams is 2,077 million m^3. Table 10 show available water data from updated water balance by INDRHI – GRUSAMAR (2007). Table 11 compares water availability with population to estimate per capita availability of water with average flows and with a firm guarantee of supply, considering this to be estimated as 80% probability of occurrence of flows records.

While the Country's general average figures is 2,628 million m^3/hab/year, it is evident from data in the above tables that the Ozama – Nizao region, where the major city of Santo Domingo with high concentration of population is located, exhibits stressful signs of shortages, with a per capita availability of 1,251 m^3/hab/year counting on average flows. This would be classified as a "water stress" condition applying the Falkenmark – Widstrand Index. If a firm guarantee condition would be considered, an average country value of 960.73 m^3/hab/year would have it classified as a "chronic water problem" and the 391.53 m^3/hab/year for the Ozama – Nizao region would mean an absolute scarcity. This already existing critical situation for some regions can still become considerably worse when water demand increase in the future.

Water balance figures for the Dominican Republic are shown in Tables 12 and 13 for the year 2005 and for projected increase in demand for 2025. Applying a scarcity index or degree of pressure index suggested by OMM/UNESCO (OMM No. 857) the country faces a strong pressure on water as shown in the tables for each region.

These results have not taken into account stored water in reservoirs, so there probably is a 2,078.51 million m^3/year margin depending on runoff regimes nor reuse of water, if properly treated, of 1,562.73 million m^3/year. Another favorable margin would come from excess flows during great floods, which are normally not measured in existing stations and consequently not appropriately taken into account. Nevertheless, the situation is critical for the Yaque del Norte and Yaque del Sur basins, where the 97.53% and 88.35% of the total available water, respectively, is already being used. WMO degree of pressure index of WMO (considers strong water pressure above 40%) should

Table 11. Per capita water availability.

No.	Region	Availability (Million m^3/year)	Potential for Groudwater (million m^3/year)	Total surface + groundwater available (million m^3/year)	% of Total available water	Population	% of population	Avalilability per capita (m^3/hab/year)	Firm water availability – 80% occurrence (million m^3/year)	Per capita availability with firm gurantee of supply
1	Yaque del Norte	2,905.46	181.00	3,086.46	11.89	1,478,113	14.96	2,088	789.00	533.79
2	Atlántica	4,634.73	216.00	4,850.73	18.68	661,581	6.69	7,332	1,245.00	1,881.86
3	Yuna	3,600.96	236.00	3,836.96	14.78	1,579,036	15.98	2,430	1,850.00	1,171.60
4	Este	3,125.95	758.00	3,883.95	14.96	919,613	9.31	4,223	1,712.00	1,861.65
5	Ozama-Nizao	4,459.08	457.00	4,916.08	18.93	3,930,708	39.78	1,251	1,539.00	391.53
6	Yaque del Sur	4,771.51	621.00	5,392.51	20.77	1,313,040	13.29	4,107	2,359.00	1,796.59
	Totals/ Average	**23,497.69**	**2,469.00**	**25,966.69**	**100.00**	**9,882,091**	**100.00**	2,628	9,494.00	960.73

Table 12. Water balance for 2005.

Region	Availability (million m^3/year)	Demand (million m^3/year)	Water Balance (Offer – Demand in million m^3/year)	Demand/Availability	
				Percentage (%)	Degree of Pressure
Yaque Del Norte	3,086.46	2,833.72	252.74	91.81	Strong
Atlántica	4,850.73	557.84	4,292.89	11.50	Moderate
Yuna	3,836.96	2,080.15	1,756.81	54.21	Strong
Este	3,883.95	638.61	3,245.34	16.44	Moderate
Ozama-Nizao	4,916.08	1,300.26	3,615.82	26.45	Medium
Yaque Del Sur	5,392.51	4,215.77	1,176.74	78.18	Strong
Total	**25,966.69**	**11,626.35**	**14,340.34**	**44.77**	**Strong**

Table 13. Water balance for 2025.

REGION	Availability (million m^3/year)	Demand (million m^3/year)	Water Balance (Offer – Demand in million m^3/year)	Demand / Availability	
				Percentage (%)	Degree of Pressure
Yaque Del Norte	3,086.46	3,191.58	−105.12	103.41	Strong
Atlántica	4,850.73	814.24	4,036.49	16.79	Moderate
Yuna	3,836.96	2,348.24	1,488.72	61.20	Strong
Este	3,883.95	1068.78	2,815.17	27.52	Medium
Ozama-Nizao	4,916.08	1,963.29	2,952.79	39.94	Moderate
Yaque Del Sur	5,392.51	4,475.46	917.05	82.99	Strong
Total	**25,966.69**	**13,861.59**	**12,105.10**	**53.38**	**Strong**

consider classifying this situation as critical rather than strong. Forecast in the Yaque del Norte region, very important for agricultural production, with high population and high water demand for irrigation (second largest in both) will be of deficit. Another region with water problems is Yaque del Sur, a poverty stricken region and therefore important for poverty alleviation policies, also having the largest demand of water for irrigation (2,891.41 million m^3/year), which represents 44.97% of total water demand for irrigation, and the second lowest water supply demand for population. Other regions that have favorable conditions according to the above results and tables, are also facing problems, like the Eastern region where most of the water for domestic, agricultural and tourism comes from wells and have absolutely no storage capacity, since there is no dam in that region. In the case of tourism, with all management of tourist attractions and facilities is concentrated at the coast, and projected to require in the year 2025 as much as 197.50 million m^3/year, or 89.13% of total water demand for this sector (221.57 million m^3/year), over exploitation of aquifers is degrading groundwater and making it difficult.

What will likely happen in the future, without adequate water allocation mechanisms, no demand management tradition, inefficient use of existing water infrastructure and proper water resources planning, is that water needs for ecological and agricultural purpose will suffer. As pointed out before such episodes are already taking place in certain locations and impact on natural resources and environment, with disregard of ecological flows and no control of pollution in water bodies, is putting at risk sustainability of many economic activities and even safe drinking water and its supply for population.

It is absolutely necessary to modernize irrigation infrastructure, improve efficiency in the use of water in all sectors, invest in better technology, implement demand management policies and techniques, strengthen institutions for regulation of water resources, develop research capabilities, water information management systems and plan for the future. Even with moderate increase in efficiency Storage capacity in reservoirs can mean an important safety margin. Countries like the United States of America and Australia have 5,000 to 7,000 m^3 of water stored per inhabitant in reservoirs, while Ethiopia only has 70 m^3 per person (references). In the Dominican Republic, increasing this value from actual 210.20 m^3 per person (only 17.9% of total available water is from regulated sources) and provide that environmental and social assessments and feasibility studies are carefully done, can mean an important difference and improvement of water security and avoid precarious conditions.

More studies are necessary in order to acquire more knowledge in the availability and needs of water. Wise investments and capacity building for hydrologic network, establishing modern measurement and control structures and devices of flows in water supply and irrigation systems, environmental education and water education for the population, and training of human resources, in particular in hydrology and better use of information technology (hydroinformatics) are advised. A revision of investment policy in infrastructure should be addressed, not to promote more investment in creating more capacity of water intake to cope with demand increase in existing towns and cities, as well as for new or expanded irrigation systems, without improvement on efficiency in distribution network, reduction of unaccounted water and illegal users and better tariff structure. Water allocation according to a basing planning will be useful to avoid improvising projects after project.

10 ENVIRONMENTAL AND WATER QUALITY ASPECTS

Some have compared evaluation of flow of water in a basin and its quality in relationship to other natural resources and environmental conditions in the watershed, with blood test on human bodies which serve to diagnose health conditions and detect health problems as if rivers were arteries and veins of the basin. Among many natural resources and environmental aspects that can be reflected in water courses there are physical, chemical and biological pollutions of water due to discharge of wastewater and solid waste, deforestation and erosion problems and inappropriate handling of mining residuals. Important aspects like evaluation and sanitary inspection for innocuity and acceptability of water supply and examination of physical, biological and chemical characteristics of water are basic for pollution control of water supply systems. Water quality monitoring and vigilance are essential tasks for adequate water resources management.

Laboratory facilities and monitoring programs are provide by INDRHI and other institutions. This facilities serve for quality and pollution testing. INDRHI's water quality monitoring has been carried out with irregularity for major river basins like Yaque del Norte, Yaque del Sur, Ozama and Higuamo. Measurement network has been well conceived and properly designed, but better logistics and improved operational capabilities are needed to carry out measurement and analysis campaigns more regularly. Initiatives to install floating gauges with transmission capacities via satellite for measurement of water quality in reservoirs and fixed river stations have not been successful. These would prove to be quite useful in treatment plants for water purification in water supply systems, since treatment can be adapted according to specific conditions of water coming from reservoirs or measured at rivers upstream and optimization of quality and treatment costs would be more probable. Analytical capacity of existing facilities and programs have been limited to basic parameters and management of information has also been limited to specialists engaged in the programs with access to the public somewhat restricted.

Proper assessment of water quality faces obstacles due to great voids in information and lack of identification of pollution sources, making it difficult to develop, calibrate and run predictive water quality models (SEMARN, 2001). This in turn, also makes difficult the evaluation of alternatives for the management of domestic and industrial effluents and for non-point pollution generated by wastewater runoff from agricultural lands and urban areas. Such stage of development in water quality control would improve both certainty and precision and achieve a better understanding of the effects of possible measures to counteract or lessen impact of pollution on health, environment and water quality.

Regarding norms they are considered updated and complete, although some recommendations from World Health Organization guidelines to include tests for fecal coliform (thermal-resistant) to evaluate risks on health due to transmission of intestinal diseases should be integrated, as well as some inorganic parameters. Laboratory capacities should improve in tests for inorganic parameters with significant health implications (nitrates and fluorides) and parameters like pesticides trihalomethane and heavy metals (INDRHI – GRUSAMAR, 2007). Revising norms for each specific use of water and development of indicators are necessary actions.

Potability index (percentage of samples with negative results for presence of coliforms) in water supply systems, analyzed during the 1992 – 2002 period, shows an increase from 1992 to 1994 but dropped to 77.6% in 1996. Last official figure in 2004 is 73.6%, which is far from the 95% target established in norms for human consumption. Some water supply systems are off course doing better than others with index values varying from 40.0% to 96.0%. Only 61.4% of water supply systems apply chlorination with adequate disinfection elements for purification processes and the situation for rural water supply systems is poorer with only 47.4% of existing systems doing disinfection by application of chlorine (Abreu, 2004). Water utility companies (state owned) do have infrastructure with 143 treatment plants, but with serious operational deficiencies and poor quality control.

The incidence of diarrhea has been studied in comparison to an index of quality of water supply service, considering population served by water supply system (easy access to water), access to sanitary – sewage -service (adequate excreta disposal) and water quality (percentages of water supply systems with chlorination systems). Relationship between the prevalence of diarrhea diseases in demographic and health survey of 2002 (ENDESA 2002) for children under five years of age and this indicator of quality of service are compared (Abreu, 2002). Results indicate that where more than 40% of population use water from aqueducts for drinking and quality index are lowest, diarrhea incidence is above 16% in children under five years. Where a greater percentage of the population is using bottled water for drinking, the incidence of diarrhea is less. People are gradually and increasingly changing their drinking water habits with a clear preference for bottled water. Percentage of population using bottled water for drinking varies from region to region, with a range of 16.2 to 77.9%. Ironically, where the quality of water service index is higher, people are depending less on the water from tap for drinking and do not used it for consumption. This manifestation of distrust in water quality from aqueducts is due to poor performance of water supply services and increased awareness of health implications, as well as new life patterns related to interest in a better social standard.

Monitoring water quality parameters like dissolved oxygen is important due to the fact that living organisms depend on oxygen to sustain their metabolic processes which produce energy for growth and reproduction. Bacteriological and microorganisms control tests are also important, particularly to control discharge of untreated water from urban wastewater and drainage into rivers, so as to avoid bacteriological pollution from fecal origin and transmission of infectious diseases, making water unsafe even for direct contact with body and submergence.

Microbian organisms are indicators of fecal pollution. Fecal coliforms are evidence of pollution from urban wastewater discharged into rivers, while total coliforms are indicators of insufficient treatment process of water. Another usual test is for *Pseudomona aeroginosa*, a pathogenic germ that can affect children, old aged and weak individuals. Inorganic components nitrites and nitrides, ions part of the nitrogen cycle, could be indicators of pollution due to inorganic fertilizers and sodium nitrite from food conservatives. Concentration of nitrates in water might be due to drainage water from agricultural lands and drainage runoff from open garbage disposal in open sites.

At reservoirs, monitoring water quality parameters is important so as to know the trophic level (conditions of eutrophication, a natural process of nutrient enrichment in water bodies, having as main indicators nitrogen and phosphorus). Systematic limnology studies (physical, chemical and biological characteristics of rivers and freshwater lakes) are recommended.

Main problems in the Dominican Republic related to environmental degradation and pollution of rivers and aquifers are due to solid waste and wastewater disposal from municipalities, agricultural lands and industries, and erosion and sediment transport in rivers due to watershed problems. Main pollution sources in Ozama, Yuna, and Yaque del Norte rivers are related to metallurgical , food elaboration, textiles, alcoholic and non-alcoholic beverages industries.

In the Yaque del Norte river basin (area 7,070 km^2 and length 207 Km) chemical characteristic of water at the upstream of the Tavera Dam is good (meet the standards?) for main uses. Microbiological pollution however is restrictive of use for recreation and direct contact. It is important to measure pesticides and agrochemicals due to intense cultivation of vegetables and flowers in Jarabacoa, to monitor risks of these elements to be transported with drainage runoff. Downstream of the Tavera Dam, (refer to Figure 1) an intake of major irrigation system (UFE), shows low salinity and sodium. Electro-conductivity, chlorines, total dissolved solids are dramatically incremented along the Las Charcas (160 μS/cm) – Otra Banda (220 μS/cm) – Quinigua (756 μS/cm) – Navarrete (704 μS/cm) – Jinamagao (796 μS/cm) reach with a clear effect of discharge of wastewater from the city of Santiago and withdrawals for irrigation system becoming a factor for increasing concentration of pollutants. Salinity of water is caused by agricultural drainage and discharge of wastewater from urban areas in Santiago with measurement at drainage flows between 1,000–1,350 μS/cm. Due to water from rivers Ámina (mean flow 10 m^3/s) and Mao (mean discharge 22 m^3/s), which are among the most important tributaries of the Yaque del Norte, water quality parameters improve downstream, but still keeping relatively high electro-conductivity values between San Rafael station (633 μS/cm) and Montecristi (503 μS/cm). Total dissolved solids show a similar behavior along the basin. Dissolved oxygen is above required level of 5 mg/l, all throughout – in all stations – the basin, which means that the Yaque del Norte river has favorable conditions for aquatic life. Favorable conditions are also found for nutrient content (nitrogen and phosphorus). These data needs technical references.

An important microbiological pollution has been detected in the Yaque del Norte River upstream from the Tavera reservoir due to wastewater from town of Jarabacoa and other communities in its vicinity. Total coliforms increase from 60 to 4,300 NMP/100ml at La Cienaga de Manabao station to 24,000 NMP/100 ml at Jarabacoa and above 110,000 NMP/100 ml at Jimenoa river in Hato Viejo. Measurement in other tributaries like Bao at La Placeta station has 9,300 NMP/100 ml and Jagua rivera t Higüero station (9,300 NMP/100 ml) and highest concentration at Jánico river above 110,000 NMP/100 ml due to discharge of domestic wastewater into the river and lixiviates from solid waste. Downstream of the dam the situation is better until the city of Santiago with colifroms concentration rising aguan above 110,000 NMP/100 ml due to wastewater discharge. At Navarrete is again 21,000 NMP/100 ml, and Jinamagao above 110,000 NMP/100 ml keeping this level of pollution until it reaches Montecristi. Health risks especially when no disinfection of water is carried out are high.

Vertical stratification profiles of Electro-Conductivity and Total Dissolved Solids variations with depth have been analyzed at the Tavera and Bao reservoirs in the Yaque del Norte river basin, with significant variation of approximately 5 μS/cm. Termoclines varied continuosly from water surface to bottom between 24 and 21 degress celcius. Oxyclines (variation of Dissolved oxygen with depth) showed a different behaviour, with the Tavera reservoir showing progressively diminishing values form surface to bottom, while at the Bao reservoir there is clear jump of the oxycline at a depth of 5 meters, where it reaches anoxia levels and anaerobic conditions were indicated by concentration of sulphur and characteristic odour of methanogenesis. This reservoir is used for water supply regulation for the city of Santiago, second largest in the country. The limit of photic zone (transparency of water) was less than 2 meters (INDRHI – GRUSAMAR 2007).

Urban and industrial discharges of the city of Santo Domingo, since colonial times located in the estuary of the Ozama river (area 2,686 km^2 and length 148 Km) contributes to organic matter and a low concentration of dissolved oxygen, less than 1 mg/l in fresh water area and area of mixed fresh water and saline water. Topographical conditions which influence backwater effects and DBO_5 concentrations above the levels of for protection of aquatic life, generates septic processes. Dissolved oxygen is higher in upper watershed (5 mg/l) and careful study of salt wedge behavior was necessary for design of new intake facilities for the Water supply system of Santo Domingo. The Ozama river, due to dissolved oxygen, nutrients loads, heavy metals and other parameters which are indicators of pollution, is considered an ecosystem in crisis (González and Gutiérrez, 2000). It is the major polluter of littoral waters providing high quantities of organic matter, greases, hydrocarburs, heavy metals and micro – organisms. All stations along the Ozama river have higher values than allowed by norms for recreation and conservation of aquatic life.

As can be expected, microbiological pollution is greater in the upper part of the Ozama basin with is an evidence of human intervention (revealed by analysis of coliforms and total coliforms, biological indicators of discharge of fecal matter into water bodies which was registered in all tests for all stations. In the estuary presence of heavy metals can be explained by activities related to the harbour, loading and unloading of ships, and abandoned wrecks.

Haina river, bordering the western side of the city of Santo Domingo, also has a complicated scenario in terms of water quality, affected by activities of several industries, including of the biggest port of the country and the oil refinery. Surface dissolved oxygen (4.4 mg/l) is bellow the limit for aquatic life protection in all stations. It is this basin that we find a fourth ranking for a world record for lead pollution, which is affecting a community of nearly 3,000, specially children. Source of such high lead concentration is a car battery recycling factory, whose residuals make the name of the town, "*El Paraiso de Dios*" (God's Paradise), a blasphemy.

The impact of mining activities (Gold, Nickel, Ferro-nickel, Aluminum and Bauxite) at the Yuna River basin (area 5,235.63 Km^2) has raised concerns on possible presence of cyanide, trophic level of reservoir in Hatillo, and other forms of pollution in streams flowing through mining areas of the Yuna river basin. This has motivated studies with results confirming acidity of water, which is characteristic of lixiviation process of soils in areas of open mining and high concentration of dissolved salts, evidenced by electro-conductivity values above 1000 μS/cm, and higher concentrations than allowed of heavy metals (Nickel, Iron). Anoxia conditions were detected as of 10 meters depth at the reservoir. Low concentration of cyanides were found, which is dangerous for animals and humans (0.05 mg/l CN- bellow allowable Standard for drinking water, irrigation).

Another concern in the Yuna basin is salinity problems in agricultural land in the lower part of the basin near the coast, due to inflow of salt water intrusion via natural and man made drainage courses. Test results indicated medium salinity levels for the Limón del Yuna irrigation area, making still possible to use this water for irrigation, provided there is a moderate degree of leaching and consideration of possibilities to promote plants which tolerate some level of salts. Electro-conductivity and dissolved solids are within admissible values. Systematic measurements of salinity are necessary to analyze penetration of salt wedge into streams and how this can affect losses in rice crops (Figure 3).

One other issue that demand attention is the sediment transport of the Yuna river discharging into sea at the Samaná bay, where this is affecting marine flora and fauna, threatening biodiversity, fishing and tourism.

In the San Juan River of the Yaque del Sur basin (area 4,800 km^2 and length 183 Km), dissolved salts indicated by high concentration of electro-conductivity has to be monitored since it can affect potential of productivity of irrigated crops, due to poor drainage of agricultural lands.

Dissolved oxygen values , which is one of the most valuable pollution tests due to the fact that living organisms depend on oxygen to sustain their metabolic processes which produces energy for growth and reproduction, has bellow 5 mg/l in some areas for San Juan river, but generally sufficient to promote fish and in the rest of the basin.

In given specific points of the basins untreated water from urban wastewater and drainage is causing bacteriological pollution from fecal origin which is dangerous for human consumption, direct contact with body and submergence. Results showed values of fecal coliforms between 2,300 and above 110,000 NMP/100 ml in all stations, which is an important sign of microbiological pollution. The river receives urban discharge when passing through city of San Juan and increases organic matter and microbiological pollution and lixiviates of garbage deposits on river banks have an important effect. Monitoring is to be carried out also during low flows – dry periods – when pollution conditions might be worse. Use of agrochemicals and pesticides is also to be monitored.

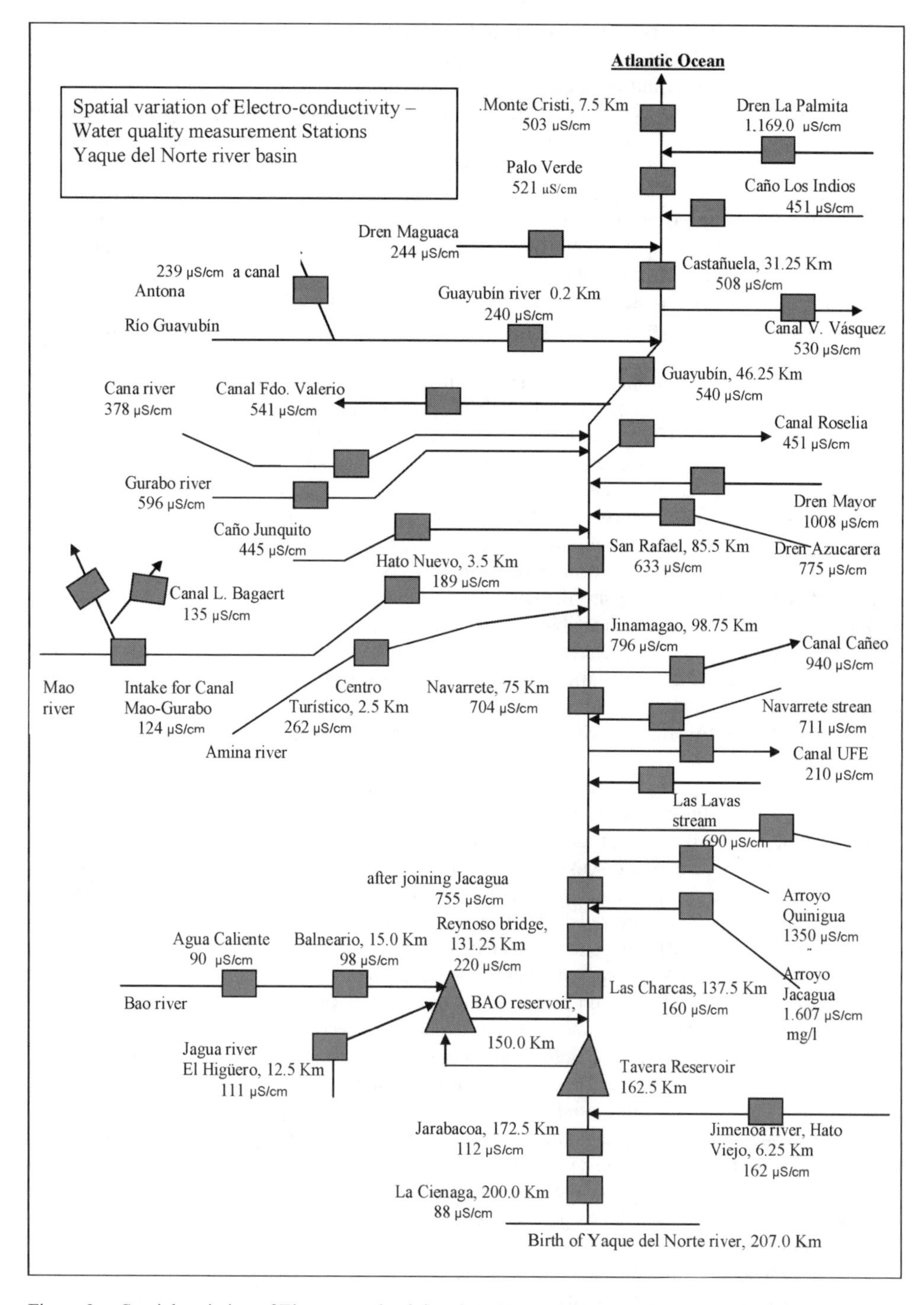

Figure 3. Spatial variation of Electro-conductivity along Yaque del Norte River (INDRHI, 2007).

Groundwater quality has been measured in 395 points in the southern part of the country, during 4 campaigns (1997 to 2000) with analysis of major ions. Samples from 40 wells used for water supply was made determining physical, chemical and microbiological analysis (total and fecal coliforms, enterococos and aerobios mesofilos), monitoring main indicator of pollution due to liquid of biological origin. Consistent presence of fecal coliforms, enterococus and estafilococus, which according to the norms should be absent, has been detected with quantities more or less of concern and some pseudomonas were found in some simples. More studies are required but unfortunately little follow up has been given to these stations. During 2001 to 2003, other 260 points for groundwater quality were studied in Northern part of the country in a one year testing for chemical and physical characteristics, including presence of heavy metals and pesticides. Results indicate that 29% of samples had laboratory results above allowable standard for human consumptions in different parameters (calcium, magnesium, sodium, chlorides, sulfates, ammonium, hardness and total solids).

Although results of bacteriological analysis are only considered to be preliminary and of seasonal effects, again coliform presence of NMP/100 ml reaches was detected and some stations had values above 1,100 NMP/100 ml. Pollution of groundwater due to Nitrogen possibly coming from agricultural activity or from wastewater from domestic sources and industry, have been found in some of the samples (5%) of samples in the Cibao Valley, Constanza and Cordillera Septentrional, with values above 45 mg/l, registered. Pollution due to pest control products is one of greatest concerns. Mineral and medicinal potential of springs was also explored.

Mineral waters and thermal patrimony of spring waters was confirmed in some sites with recommendations for deeper analysis of their chemical properties. Possible commercial value or exploitation of thermal waters is feasible especially for medicinal and cosmetic purpose.

Deforestation and soil use conflicts, manifested also in erosion and consequent sedimentation, are among main degrading factors affecting watersheds. Results from satellite imagery technology indicate that there is an alarming loss of vegetable cover, from 14.1% in 1980 to 27.5% in 1996, according to the USDA/Michigan State University cooperation programme for inventory and use of vegetable cover in the Dominican Republic (LandSat). Most studies confirmed deforestation with figures for rates of loss of forest cover up to 26,400 ha/year and a progressive desertification process from western border extending to the eastern side.

Some recent studies however question that traditional evaluation and mention increase of forest cover (ABT, 2002). While some discussions are currently going on for confirmation of this favorable change in trend, some probable causes are to be recognized including efforts to increase protected area (from 9 locations which covered 4.2% of national territory in 1980, to 19 areas covering 11.2% of country surface in the 1990 decade, to 70 areas covering 19.5% of the territory with the passing of environmental and natural resources law in the year 2000, to 86 areas covering 22% of land with provisions of a specific law on protected areas – SEMARN, 2004). Implementation of this regulation is a real challenge and objectives biodiversity conservation need to be clarified. Other possible reasons for a slow down on deforestation or stabilization of forested areas are reduction of migratory – subsistence – agriculture in upper parts of the watershed and increased migration from mountain areas to urban dwellings and costal areas. Most of the watershed management projects, due to small scale, no baseline studies and/or no monitoring system nor impact measurement, as well as prevailing unavailable information, cannot exhibit concrete impact on reduction of erosion and pollution (INDRHI, 2006a), nevertheless having good impacts promoting organizational capacity building at community level, achieving general educational goals regarding conservation of natural resources and attaining important rural – infrastructural – development objectives.

Estimates for erosion rates do maintain an increase however, with figures varying from 20 to 500 Ton/Ha per year for loss of soils. Many reports are loaded with estimates but there is little support for most of them. Calculations for Yaque del Sur river basin resulted in estimates of 184 Tm/Ha-year. Nagle (2005) using radioactive isotopes (cesium 137), estimated plots in hill slope erosion in 14 sites of the Nizao river basin, determining erosion rates in the 1963 to 2005 period, varying from 6.7 to 59 Tm/Ha-year and average 27 Tm/Ha-year. Applying Universal equation for loss of soils for the Nizao watershed would yield a higher value of 125 Tm/Ha-year. If accumulated sedimentation

Table 14. Reservoir Sedimentation.

No.	Reservoir	Initial Storage capacity (million m^3)	Current Storage Capacity (million m^3)	Volumen of Accumulated Sediments (million m^3)	Percentage of Reduction in Storage capacity (%)
1	TAVERA (Feb 1993)	173	137.1	35.9	20.75
2	VALDESIA (2006)	186	137.1	48.9	26.29
3	SABANETA (May 1999)	76.5	63.1	13.4	17.52
4	SABANA YEGUA (Dec. 1992)	479.9	422.3	57.6	12.00
5	RINCON (Dec. 1993)	74.5	60.1	14.4	19.33
6	HATILLO (Apr. 1994)	441	375.3	65.7	14.90
7	AGUACATE	4.3	1.2	3.1	72.09
8	BAO (Apr 1994)	244	150.7	93.3	38.24

in 3 reservoirs in this watershed is considered average erosion rate would be 41.6 Tm/Ha-year, above Nagle's estimate but less than other estimates (INDRHI, 2006a). Deforestation and migratory – low productivity – agriculture in mountain slopes have been blamed for most of the erosion, but recent studies point out that earth movement for construction of roads in hilly areas is thought to be among important contributing elements. The relationship between soil erosion in the watershed and reservoir sedimentation is not so clear, but sedimentation of reservoirs is definitively fast paced.

Reservoir sedimentation is critical in at least 6 dams, where level of sediments is dangerously reaching level of invert of intake structures or conduits for water supply, irrigation or powerhouse (hydropower), becoming a menace for operational functionality and diminishing socioeconomic benefits of these dams due to premature silting. Factors contributing to high sedimentation rates are high slopes of rivers, frequent floods, deforestation and absence of watershed management – soil conservation – projects upstream of dams. Forecasts of sedimentation and provision for accumulated sediments within dead storage when dams were designed, has been dwarfed by reality. Table 14 show accumulated sediments in 8 reservoirs, calculated by last batimetric survey, being evident that these surveys should be updated and become a regular routine and after extraordinary floods provoked with hurricane.

Accumulated silt is of concern at Bao, Aguacate and Valdesia, with 38.24%, 72.09% and 26.24% of volume is occupied by sediments respectively. Dead storage space is completely (100%) filled at Tavera dam, 60.8% filled at Valdesia and 57.6% filled in Sabaneta. Sediments have occupied 32.9% of useful storage volume in the Bao reservoir, and more than 15% of useful storage in Tavera, Hatillo and Rincón. Figures in Table 1 for other dams, except for Sabaneta and Valdesia, do not take into account hurricane Georges in 1998. Such high rates of sedimentation and buried intake structures within reservoir, motivated dredging at Valdesia dam in Nizao river to be carried out in 1984, but with limited results.

The Nizao river, a 132 Km long river flowing from elevation of 2,800 m.a.s.l to sea level, with a catchment area of 1,036.02 Km^2, is regulated by 4 dams Jigüey (167 million m^3), Aguacate (4.3 million m^3) Valdesia (186 million m^3) and Las Barias (6.1 million m^3). The first two having hydropower as single purpose (combined generation of 203.96 GW-Hr/year) came into operation in 1991, while Valdesia which is a multipurpose dam and Las Barias dam for irrigation, were both commissioned in 1975. Valdesia is the source of water supply for the city of Santo Domingo (6 m^3/s), provides water for irrigation systems (14,000 ha in Nizao – Najayo and Baní areas) and generates electricity (62.75 GW-hr/year). In Valdesia bottom outlet used for releasing water for irrigation and for bottom flush operation is buried under 14.4 m of sediments (INDRHI, 2007). Intake for penstock to powerhouse, which could be the next victim, is only 3.75 m above actual sediment level. Similar trends are observed at the Tavera and the Rincón dams, with sediment level are already 3 to 4 meters above the invert for water supply intake. Rincon and Sabana Yegua

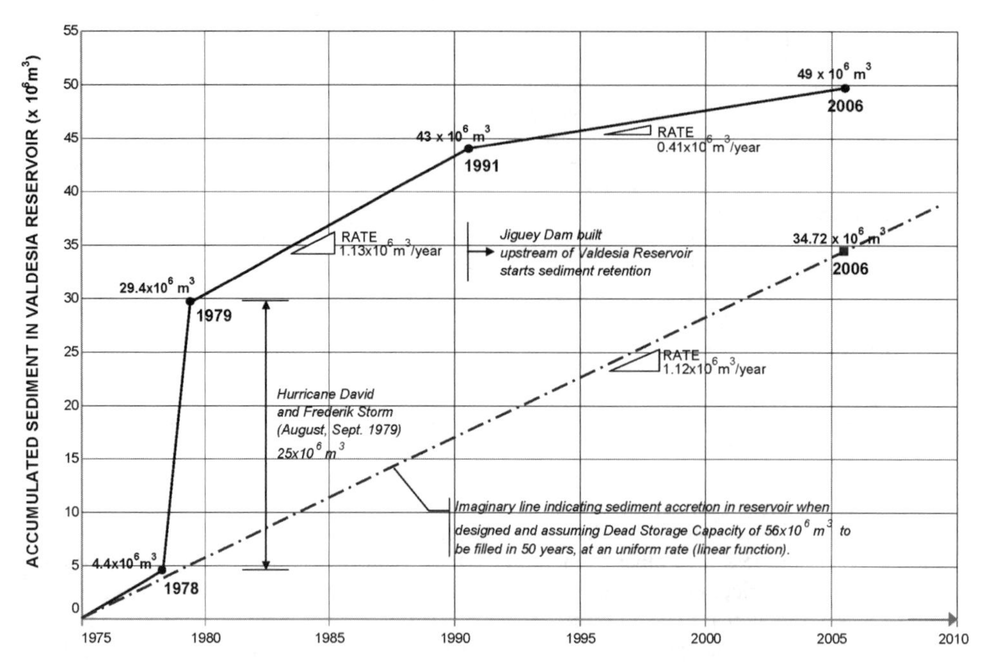

Figure 4. Evolution of Sediment Accumulations at the Valdesia Reservoir in Nizao River.

have sediments 5 to 8 meters above bottom outlets. If sediment wedge would move downstream, blocking of these intakes is possible.

When estimating year averages of sedimentation rates it should be considered that one single event like Hurricane David in 1979, a one in a hundred year flood in Nizao river, can fill for several years. It has been estimated that this sole extreme event (hurricane david was followed by FrederickStorm in less than a week, August – September of 1979), brought 23 to 25 million m^3 of sediments into the reservoir. In only 4 years of operation, Valdesia dam had 18% of reservoir volume occupied by accumulated sediments (Jiménez and Farias, 2002). After building of Jigüey dam, located upstream of Valdesia, sediment rates at Valdesia have decreased from 1.63 to 0.59 million m^3/year, considering input of hurricane (INDRHI, 2007).

Using data from 5 surveys of batimetry made by INDRHI at the Valdesia reservoir in the Nizao river, during 1975 to 2006, a graph of accumulation of sediment has been constructed (INDRHI, 2007). This has been drawn in upper "curve" of Figure 4 which is the real sediment trapped or accreted at the reservoir. Rates of sediment accumulation vary and there are reasons for it. For example, the Jigüey dam started retention of sediment as of 1991, and this influences a slower pace of silting of reservoir in the down stream dam of Valdesia due to sediment trapped upstream in the Jigüey reservoir. Just for the sake of comparison, if it is assumed that estimated volume of 56 million m^3 of dead storage contemplated in the design of the Valdesia dam, were to be filled in a 50 years period, as superimposed on the real sediment in reservoir measured by surveys traced by the author in a lower line of Figure 5, a clear illustration provides lessons of accelerated rate of reservoir sedimentation prematurely filling reservoir, of the great effect that a flood has, of the positive effect in sediment control that upstream retention structure can have, of the limitations of existing methods to estimate silting of reservoirs and of the lack of knowledge of its relationship with watershed erosion. One would like to suppose that deceleration of sediment accumulation rate is possible when sediment control measures are applied. Research on process of sedimentation trapped by a reservoir is necessary for better forecasting capacity in hydrologic design of reservoirs

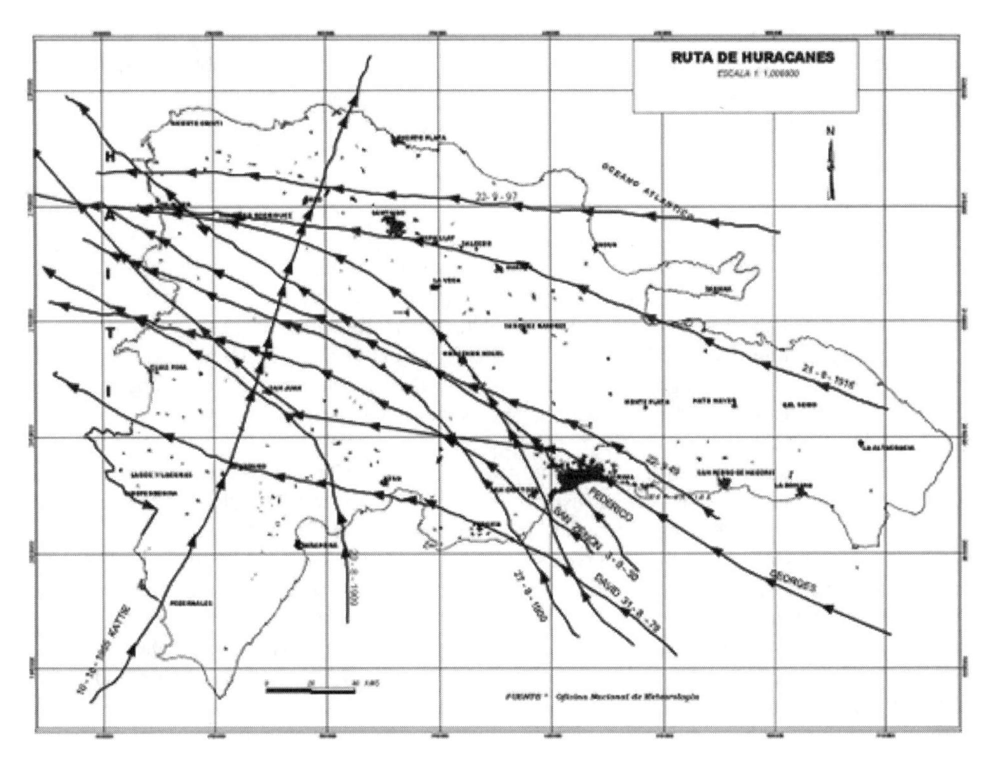

Figure 5. Routes of major Storms and Hurricanes in Dominican Republic.
Source: Oficina Nacional de Meteorología (ONAMET).

and more precise assessment of risks and impact of sedimentation in analysis of economic revenues of dams and of watershed protection projects.

Sediment extraction has also been planned and designed, recommending dredging of reservoir after analysis of several options because objective could be achieved in a shorter time and in a more effective manner than sediment removal by flushing, routing and by-pass. Sediment control via watershed protection (soil conservation practices) and sediment retention dams are sustainable solutions, which were judged to be discarded for high costs and limited effectiveness (Jiménez and Farias, 2002).

Rivers discharging waters with high sediment content and high concentration of fecal coliforms and other pollutants are affecting estuaries, beaches and associated ecosystems. There are 196 beaches with area 433 Km^2, which make up 95% of tourist attractions in the country. Other natural assets in a 1,389 Km of coast are 25 locations of sand dunes (60.4 Km), coral reefs (133 Km^2 up to 10 m depth, 610 Km^2 up to 30 m depth and 1,350 Km^2 in obscure zone more than 30 m depth), 325.3 Km^2 of mangrove, 55 locations of marine prairie (Geraldez, 2002) and two santuaries for marine mammals, including famous humpback-whales. Study by González et al. (2004) revealed that 29% of beaches have higher values for total coliforms than allowed by norms for skin contact (1,000 NMP/100 ml). In the Northwestern coastal reach (Montecristi – Puerto Plata), out of 19 beaches affected, 4 have values of total coliform in the order of 2,400 NMP/100 ml, and 6 have values between 60,000 and 110,000 NMP/100 ml.

Costal and marine resources are impacted by significant environmental pressure due to construction, excessive fishing and tourism. Growing population (70% of cities above 10,000 inhabitants are located at the coast), economic development, industrial development (75% of heavy industries located near the coast), sedimentation, lack of sewage systems and treatment plants and other means of land pollution due to mining, agriculture, industry, shipping activity and tourism continue to press beaches, coral reefs, mangrove, lagoons, estuaries and fisheries. Main problems now is excessive production of solid waste and inadequate handling of them depositing in open with problems of wash away and decomposition polluting surface and groundwater bodies.

11 FLOODS

Significant damages are caused by tropical storms and hurricanes in the Dominican Republic. Being hit and smitten by these extreme atmospheric phenomena is something out the control of humans, since the Hispaniola Island is in the path of hurricanes traveling through the Atlantic Ocean and originating in it north tropical area next to Africa's northern coast. The Dominican Republic is affected by one moderate to high intensity meteorological event every two years by one and by no less than four hurricanes in a decade. Figure 5 shows the routes of major storms and hurricanes in the Dominican Republic

Tropical storms and hurricanes have slashing wind speeds (150 miles per hour during Georges in 1998) and abundant rainfall, which usually results in floods and land slides. Intensive rain, which can come from a more regular as well, in turn causes greater flows than can be contained in normal river cross section, overflowing banks and passing on to flows in the flood plane. The flood magnitude is measured or evaluated in terms of total amount of rain during the meteorological event (mm), maximum amount of rain in 24 hours (mm/hr), maximum rainfall intensity (mm/hr), the evolution of discharges as seen in a hydrograph and peak discharges (m^3/s), and when available, the area flooded of a flood map. If there is a dam in river basin hydrologist are keen of elaborating and examining the in – flowing and out – flowing hydrographs (two curves) and peak discharges derived from this graphs, representing peak flow entering the reservoir and maximum discharge, if any, over or through the spillway. During some of the greatest flood events experienced in the country, David (1979) and Georges (1998) peak discharge values between 3,500 to 8,000 m^3/s have been estimated in rivers or calculated in hydrographs flowing into reservoirs. Highest discharge through a spillway has been in the order of 5,500 m^3/s. It is clear from this magnitude and frequency of events that flood prevention should be of highest priority in the country.

The impact of a flood is measured in terms of losses and damages. Losses include those aspects that cannot be recovered, like the unfortunate loss of human life, which though impossible to value, has an incalculable value for the country and in particular to families affected. Social inequalities are revealed or undressed by floods, and people suffer calamities loosing their loved ones, being wounded or harmed, loosing their houses or having to abandon it temporarily to looking for shelter (4% of the population during Georges in 1998) and becoming ill due to rise in diseases and epidemic episodes after floods suffer calamities. Costs of hospital emergencies and humanitarian aid (food and medicine distribution, sustaining camps for refuge) continue to add to damages. There are official records indicating remarkably related increased incidence of malaria after hurricanes David (1979) and Georges (1998) and of a leptospirosis outbreak killing 40 persons as one of the aftermaths of Noel tropical storm in October 2007. Damages also include affected roads, houses, and loss of productivity in different economic activities. There are direct damages as loss in crops and indirect damages include production income lost or not able to obtain due to direct damages, diminished income during recovery period and expenses for recovery itself. Cost of repairing a road (building back damaged stretch), costs of providing an alternative route while road is restored to normal or previous conditions, cost on additional miles or kilometers, more fuel consumed and extra time to drive through a rougher roadway or with a tougher and slower traffic are among possible consequences of floods. Health, education, communications, tourism, fishing, livestock production, irrigation, water supply (pumps and treatment plants out of service), forest and environmental sector, each have their own damages. There are also economic effects of disasters which contemplate macroeconomic implications like slow down of economic dynamics, exports reduction, increased government expending for relief, reprogramming debts payments and related fiscal impacts.

Table 15. Impact and dimension of risks of Storms and Hurricanes in Dominican Republic.

Name of Event	Date	Number of deaths	Estimated Direct Damages (USD Millions)
San Zenón	03 September 1930	4,500	15.00
Flora	October 1963	400	60.00
Inés	29 September 1966	70	10.00
Beulah	11 September 1967	N.A.	N.A.
David	31 August 1979	2,000	829.00
George	22 Sepetember 1998	283	1,337.00
Jeanne	September 2004	23	341.27

In the case of tourism, the ECLAC report (2004) after Jeanne in September 2004, highlights that the biggest lessons to learn was the damages related to hotel establishments with tourist in the eastern region of the country having to climb to roof tops to keep alive, were due to building of hotel infrastructure in the wrong place without proper consideration of the functioning on natural ecosystems, as it happened with the Bávaro lagoon and mangrove systems (ECLAC, 2004). There were environmental losses, costs for restoring water movement in the lagoon system, infrastructural damages, unforeseen expenses to face emergency and a probably well deserved drop in hotel occupancy. Drying of wetlands and destruction of mangrove areas are among sadly common criminal practices of some developers affecting fragile ecosystems and diminishing capacity of natural conditions to act as a buffer zone for storm and hurricane induced waves or surge tides, which can aggravate and prolong risks (ECLAC, 2004).

It has been estimated that different types and magnitudes of natural disasters have an annual estimated costs for the Caribbean of USD 1,500 millions and almost 6,000 lives (ECLAC 1998). Rodríguez (2005) considers losses and damages in selected extreme hydrologic events in the Dominican Republic are shown in Table 15.

Some determinant factors exposing or building vulnerability, are natural conditions as in the case of geographic location of the country and climate typical of the Caribbean, while others are "man-made". When natural factors meet pre-existing vulnerability caused by anthropogenic and risk augmenting factors like deforestation and erosion of watersheds, which cause a different response in the rain – runoff regimes with more "sudden" flows and floods (flash floods), inappropriate reservoir operation, execution of infrastructure projects (roads, bridges, canals, buildings) with no hydrologic, hydraulic nor environmental considerations, unplanned urban development, and conflicting land use, which is well illustrated by occupation of flood plain areas (for dwelling or crop growing) and the inappropriate use of soils in river basins, vulnerability is increased, making floods more dangerous.

An example of risky human settlement growth is the case of *Jimaní*, a town at the border with Haiti, right at the closure of the Soliette river basin, shared by Haiti and the Dominican Republic. Soliette river, runs from a height of 1,670 m.a.s.l. on Haitian ground and discharges into the Enriquillo lake, which is at the Dominican Republic's lowest elevation (−40 m.a.s.l.). The floods of the 24th of May of 2004 are one of the most tragic episodes in the history of the country and this time it was not caused by a hurricane, but a low pressure center combined with stationary front and a tropical disturbance, bringing 227 mm of rain in 12 hours (248 mm in 24 hours) measured downstream in Jimaní. Rain should have been higher upstream as it is in fact reported and recorded as 462 mm measured in 24 hours by a pluviograph in *Forest des Pins* on Haitian territory (Brath, et al, DISTART – UNIBO, 2007). This amount of rain, which has been estimated as the one in 100 to 200 years rain intensity in this catchment area of 165.78 Km^2, and a peak flow of hydrograph of 2,270 m^3/s according to reconstructed hydrology of the event,

(a)

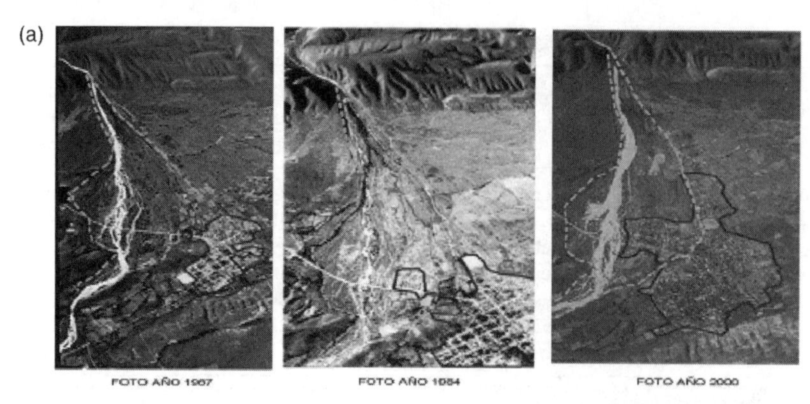

Figure 6-a. Evolution of Growth of the town of Jimaní, Dominican Republic, and flood area of the 24th of May 2004.

affected specially the most populated towns of the basin in *Fond Verrettes* in Haiti and *Jimaní* in the Dominican Republic. The loss officially recorded was 397 dead bodies found, plus 272 missing and 291 homes destroyed, 71 houses damaged and 300 families that were awaked without a house on the Dominican side. Human loss in Haiti surpassed these dramatic numbers (1,059 dead and 414 missing). This was a flash flood moving great quantities of sediments that surprised people in Jimaní who were sleeping at 1:00 a.m. when stroked by an alluvial phenomenon. This devastating force swept away with people and homes and also caused longer anguish to survivors suffering scarceness of water supply and electricity services and affected roads for several day after. One key lesson here was that even relatively moderate *climatic perturbations* have major consequences.

In trying to identify other factors apart form extreme hydrologic conditions, that contributed to build this horrifying situation, Pérez (2004) analyzed the growth pattern of the town of Jimaní, with available aerial photographs from different dates (1967, 1984 and 2000), from which it was easy to see evolution or spatial variation of the area of the town during a 33 years period and comparing with the area of river's floodplain during the night of may 2004, whose mark was clearly seen at the site and superimposed on the photographs. The results show, as seen in Figure 6, that the torn part of the town had been gradually growing into the floodplain.

The Italian cooperation (cooperazione Italiana allo Sviluppo – Istituto Italo Latino Americano – IILA) provided technical assistance by Universitá di Bologna for hydrologic and hydraulic study, training of engineers from both countries and drafting an action plan with structural and non-structural solutions to reduce hydrologic risks. The findings of the study highlight that high hydrologic vulnerability is due to: i-) intense deforestation that has flagellated Haitian territory (and also some part of valley area in Dominican side) notably augmenting propensity to disaster; ii-) expansion of unplanned or inadequately planned housing areas invaded the river's floodplain; and iii-) technical weaknesses by local and national institutions incapable of providing territorial safety. Technical support and human capacity building was precisely a main objective of a project that brought together institutions from both nations (INDRHI from the Dominican Republic and Ministère de l'Agriculture, des Ressources Naturelles et du Développement Rural from Haiti) to find a solution to a common problem. This hopeful cooperation strategy has great potential for transboundary water management and can be an example to other basins and to address other water management issues.

These type of situations amid the high frequency of tropical storms and hurricanes, with such sad records of losses and damages, require a well conceived set of actions to identify vulnerable areas, reduce risks and stimulate citizens, communities, local authorities, industries, business and financial sectors and high level government officials for preparedness. A flood management

(b)

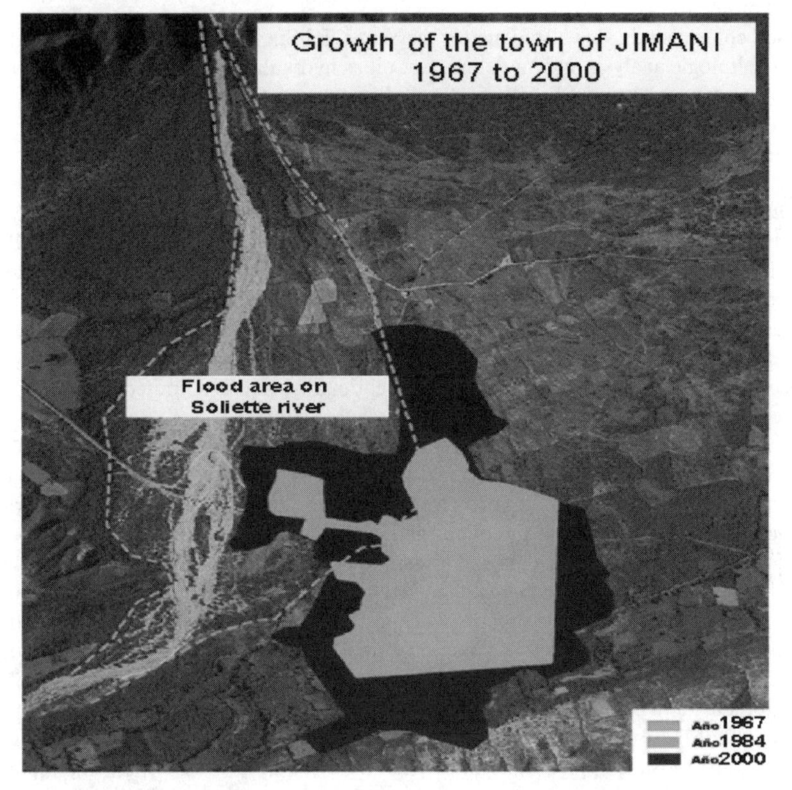

Figure 6-b,c. Evolution of Growth of the town of Jimaní, Dominican Republic, and flood area of the 24th of May 2004.

strategy and plan are urgent needs and these would have to consider studies, monitoring, warning mechanisms, investment for defense infrastructure and improvement of watershed or catchment, conditions, as well as pertinent norms and regulations.

The most complete flood control study in the Dominican Republic has been done for the Yaque del Norte river basin and was financed by the World Bank, resulting in a flood map for different magnitudes of floods defined by return periods (25, 50, 100 and 500 years). One pitfall of the study is that its flood map is a static picture, for floods of given magnitudes, not being able to use it for predictive for purposes of determining flooded areas for any other different discharge value, rather than guessing through interpolation of curves. Ideally, when rainfall estimates are forecasted, the watershed response to that rain could be worked out by virtue of hydrologic models that can supply estimates for discharge, which in turn can be processed, via hydraulic simulation, to provide water levels and flood maps. Another application could be for the operation of the Tavera dam on the Yaque del Norte basin, upstream of the city of Santiago, related to impact of discharge of spillway, which in fact has demanded serious decisions to be made without these advantageous tools. In this case there is a calibrated hydrologic model, but due to the fact that technology transfer was absent in this project and the use of the specific model developed and calibrated for the watershed is based on a software unknown to INDRHI's staff and of reserved rights by consultant providing the service (study), it is not possible to use it. In any case the flood maps for the fixed return periods can serve for planning and possibly for promoting actions to protect vulnerable areas and people at risk. This map was handed to authorities and several organizations in the basin. The study also contains a proposal with a proper balance of structural and non – structural actions and projects for flood protection.

A flood control study was also done for the Yaque del Sur basin, but flood maps were not elaborated. Hydrologic analysis and modeling, as well as hydraulic simulation of different magnitudes of discharges were also done. Flood control studies were performed for the cities of Bonao and La Vega in the Yuna river and Santo Domingo in the Ozama river (USACE, 2003), properly identifying risk zones by flood maps and also offering structural solutions.

Structural solutions for flood control like building dams to regulate river flows have been applied. The *Hatillo* dam in the Yuna River and the *Sabana Yegua* dam in the Yaque del Sur river, the two largest reservoirs (volume), are examples of dams with flood control purposes. INDRHI is frequently involved in building levees and small dikes of relatively small lengths to protect crop lands, roads, houses and buildings, as well as in fluvial works for river training, but also in relatively short distances. As to preventive Non structural solutions, including watershed management projects, displacement of people living in river flood plane, sirens alert systems, and early warning system have been planned, but applications are been exceptionally tried to a limited scale and with various degree of success.

INDRHI installed in 2005 an early warning system originally composed of 120 stations, some being there were pluviometric stations measuring rain intensity, others were stations to measure water depths, others were climate stations for temperature, relative humidity atmospheric pressure, solar radiation, wind speed and direction, and a few were tide gauge. This network allows for stations recording of hydrologic data and transmission in real time via GOES satellite and NEDIS approval, to two receiving stations, which meant a significant advancement for the country. More effort is required to enhance data processing capacities and to provide better access and sharing of this information to interested parties, including civil defense and meteorological service, as well as the general public. The fate of the stations is of great concern due to vandalism, theft, logistical support for adequate maintenance and some other operational difficulties. Currently there are not more than 40 of these stations transmitting. Financial, organizational and technical sustainability have been put to test. Doppler station has been installed by private tourist developers in Punta Cana in the eastern tip of the island, more for aviation purposes, but obviously of great use for rain forecasting, being still necessary to link this to INDRHI's routine and processing. Another such radar is said to be necessary in the extreme northwestern part o the country to cover the whole island and servicing Haiti as well.

Rodríguez (2005) has suggested several actions to improve flood management, including definitions of a national plan and sectors planning to prevent flood damages, strengthening of early warning system, human resources training, community education and a more participatory approach, optimizing operation of reservoirs, zoning and insurance for infrastructure and to protect crop production for farmers, regulations and requirements of flood risks assessment for infrastructure projects and revision and upgrading of construction codes so that roads, bridges, hospital buildings, water supply systems, tourist development, would be more securely designed and located to avoid risks and withstand effects of floods, and that these infrastructure do not increase risk nor unfavorably modify ecosystems due to lack of understanding or "voluntary ignorance" of fluvial dynamics and environmental conditions. Furthermore he proposes well articulated integration and more effective involvement of concerned agencies, including the army in the prevention and response to natural hydrologic disasters, since water and floods are topics of national security and should be a major defense issue.

One of the greatest challenges would be to induce and regulate zoning. Planning and enforcing territorial marshalling requires search for a solution with legal schemes, financial mechanisms, control and enforcement. Jimaní's insecurity was revealed by intense rainfall and a valid question is if vulnerability of other towns are to be known after a tragic flood event. Flood insurance, lending and mortgaging incentives from banks for safely located properties could possibly be useful tools, but would require demonstration of impacts and financial convenience for the insurance and credit institutions. Alternate restrictions for unsafely located and designed buildings could serve as a negative reinforcement for pedagogic stimulus. Flood insurance of estate infrastructure does not exists and though there was a project law drafted to make this compulsory, feasibility of such requirements is yet to be deeply analyzed and discussed. An attempt to introduce crop insurance in 1984 was short lived with closing of the agricultural insurance company (ADACA: Aseguradora Dominicana Agropecuaria) created mostly with state investment. Another investment of the state gave birth in 2002 to a new company (AGRODOSA: Aseguradora Agropecuaria Dominicana), offering insurance for a few crops (rice, plantain, banana and beans) in specified areas of the country, including coverage for losses due to floods, droughts, tornados and pests. Making reference to models of other countries as the FCIC (Federal Crop Insurance Corporation) in the United States of America for example, it has been suggested that instead of providing insurance directly, the State's involvement should be more a program of reinsurance and subsidy of insurance prime obtained through private companies.

12 WATER AND ECONOMICS

Water contributes to social and economic development in several ways. It is beyond doubt, that investments in water supply and sewage systems improve health and hygiene. Water is essential for food security also. Agricultural production for example is only possible in certain areas when water is provided for irrigation of crops. Irrigation, which fosters augmented yields or increased productivity, and consequently more profitability of agricultural production, is a catalyst for improvement of economic standard of farmers and their families, also stimulating general economy of rural areas by way of promoting increased opportunities of employment and other associated and diverse income generating activities.

In spite of the centrality of water to development, national scale indicators of economy, macroeconomics, development and poverty alleviation policies, do not reflect real contribution of water. This statement is not exclusive to the Dominican Republic, since the Water and Poverty Initiative (WPI), launched and supported by Asian Development Bank in February 2002 to explore the linkages between poverty and water security, elaborated interesting conclusions from shared experiences and discussions during 3rd World Water Forum in 2003 in Japan, emphasizing that: "A more complete understanding of the relationship between water security and poverty reduction is needed to improve the management of water resources and the delivery of water services"; Existing national policies for poverty reduction and development do not reflect potential contribution of water to poverty reduction". From one of the key sessions organized by WPI during the 3rd World Water Forum, we extract: "While only one of the millennium development goals mentions water specifically, all are impacted by access to water in one way or another. For example, halving the worldwide hunger by 2015 heavily depends on better water management". While better water management on its own is insufficient to alleviate poverty, none of MDGs can be achieved without better water management" (ThirdWWF 2003).

There is a strong need in the Dominican Republic for conciliation of the national development and poverty reduction policies and strategies, as well as government priority statements, to the investment programme of the government and to the country assistance strategies from development and donor agencies, as well as some multilateral banks, regarding real water problems and needs of the country and potential contribution of water to social wellbeing and economy. Water, which had a higher priority in investment before, is poorly recognized in the current sector preference agenda for international cooperation and financing.

The poor are the most afflicted segment of population during hurricane and floods disasters, as well as other water related dangers like droughts and pollution. Risk and vulnerability of the poorer is often exposed by these extreme events. They also tend to throw their lots in farming on river flood planes and hills. The first is subject to frequent flooding and the second involves cutting forest to plant, reap and later abandon plots as eroded and diminished soil fertility areas in a survival and predatory pattern of agriculture in mountain slopes. Small scale fishers are more impacted by biodiversity and environmental degradation affecting fishing potential and creating economic instability of their business. Better water management can make a key contribution to poverty reduction. This is recognized in internationally agreed targets to halve the proportion of people without access to drinking water and improved sanitation by 2015.

But water is also a key input to many industrial and other larger economic activities and thus water benefits both the rich and the poor.

To bridge the knowledge gap and attain a better understanding of relationships between water management and poverty, WPI has recommended (ThirdWWF, 2003) the development of effective indicators and monitoring systems to assess progress in realizing water-poverty targets. There is also a need for major advocacy programs to increase political and public awareness and support for pro-poor water management. The concrete suggestion is to "argue the case more effectively" for creating awareness among population, and especially politicians and economists, usually leading decisions in policy formulation and investment development.

To initiate the road that leads to that "better understanding" between water and wealth or poverty, the author has made an exercise analyzing the relationship between poverty and water availability. The first variable, poverty, was assumed to be represented by the percentage of poor population by provinces, using data from the Poverty Report elaborated by the National Planning Office (ONAPLAN, 2005). Water availability is thought to be well represented by the rainfall minus evapo-transpiration, accepting this to be some form of "effective rain" with data from the recently updated Water Balance (INDRHI – GRUSAMAR 2007).

This discussion should be expanded to take into account water infrastructural development and economy, since Yaque del Norte and Yaque del Sur are among the regions having more irrigation systems per surface area (Table 2) and have important percentages of their watersheds under dry conditions. This has to be interpreted as confirming justification of investment policy priority in irrigation to the drier areas, but measurement and analysis of impact of investment in irrigation as related to poverty alleviation and economic growth in those areas is a still pending issue. Another needed explanation is why the Yuna river basin, with largest proportion of area of irrigation systems per surface of catchment area, while being the rainiest region. Economic effectiveness and justification of seasonal supplementary irrigation to rice fields, main rice growing area in the country, would have to be clarified.

While there is a good, inversely proportional match of rain and poverty in the southwestern and northwestern regions, some eastern provinces however, like Hato Mayor and El Seybo, which are amongst the provinces with highest percentages of poor population (as much as 68.5% of families are considered poor), have abundant rain. In those provinces, topographical conditions, lack of infrastructure (economic poverty of water) and lack of reservoirs to regulate flow, which is true for the whole eastern region, can be probable factors of influence in great poverty exhibited in those two provinces. Abundance of water nor water scarcity cannot on its own explain richness nor poverty, since so many other aspects are involved in a situation that is still elusive to analytical reasoning. This exercise is only a limited approximation, linking only two variables, to try to identify possible relationships of certain factors involved in a much more complex reality.

A sector analysis of the economy indicates that tourism, commerce, construction, communications are the rising stars of the Dominican economy, while sectors like agriculture and manufacturing have been declining in their contribution to GDP. Agricultural production has been dropping from 25% of GDP in the 1960s to 13.4% of GDP in 1993 and to 11.5% of GDP in 2005. This includes livestock, forests, fishing and crops. Crop agricultural production was 7.6% of GDP in 1991 and 4.3% of GDP in 2005. Agriculture still keeps an important place when it comes to employment, with 14.6% of economically active population working in agricultural activities. For the rural area agriculture is on of the largest employer sector, representing 29% of employment.

Energy, hydropower plants contributed a record high 1,879 GW-hr/year, which is 18.7% of total energy generated, but average annual generation is about 1,200 GW-hr/year.

According to figures from the Central Bank, the relative importance in the economy of the water sector is indicated by its aggregate value in relation to total GDP, which represented 0.07% of GDP in 1995 and 0.04% in 2005. The contribution of water to the economy has been underestimated (INDRHI – GRUSAMAR, 2007). ADEAGUA, the Dominican association of private companies dedicated to the business of purifying and bottling water, have declared investments in the order of RD$4,000 millions pesos (approx. USD 126 millions). Surveys by National Statistics Office (ONE, 2006) revealed that 51% of total population are using bottled water for drinking and average expenses in buying those bottles of water is 30.98% of the amount of their water bills. This figure could be higher, since the cost of a 5 gallons bottle of purified water is at least 40% of a monthly water bill. It is not clear either if the figures from the Central Bank consider invoicing of water utilities companies, apparently out of their equation for aggregate value.

Other economic issues related to water management should be addressed. One of them is profitability of water supply services. Economic inefficiency is definitively a disgraceful characteristic of the water supply and sanitation sector. Tariffs in water and sanitation sector are low. The relationship between production of purified water over amount of water finally invoiced to consumers is high, which means that a good proportion of treated water is either lost or stolen. According to INDRHI – GRUSAMAR (2007) and assessment reports of the water supply and sanitation sector, the water utility company in Santo Domingo (CAASD) produced 378.605 million m^3 of water in 2005, but only invoiced 123.462 million m^3 (only 32.6%). Amount of total collected fees by CAASD in that year were 73.7% of total value of invoices and 41.94% of their operational costs and expenditures to purify, distribute water, maintain the water supply system and cover administrative and commercial management expenses. The difference is assumed by the government. Production costs (operational costs – in pesos – over production volume – m^3) is RD$ 3.63/m^3 (approximately USD 0.11/m^3). In spite that savings in expending is possible and justified, tariffs for water service, about RD$ 6 /m^3 (approximately USD 0.18/m^3) for a residence (house), would indicate that tariff structure is adequate to cover O & M expenses. The gaps are in inefficacy of invoicing, unaccounted water due to illegal connections and loss in the network, and poor fee collection. Operational costs per volume of water invoiced is RD$ 11.54/m^3 (approximately USD 0.34/m^3). Final annual fee collection income over costs of water service would make the price of water RD$ 15.11/m^3 (approximately USD 0.45/m^3).

A Fixed tariffs criteria for tariff structure in water and sanitation services are applied by INAPA and the most of the other five state owned water utilities, each having their own structure but generally taking into account the type and category of users (residential, commercial, industry, hotels, government institutions and poor communities), the location of establishment, type of building, the number of taps and estimates for volume of consumption. In part of Santo Domingo (CAASD) and Santiago (CORAASAN), there are residents which are charged by volume of consumption according to metered services and invoicing, having one tariff (RD$6/$m^3$ for a house) up to about 32 m^3 per month per house and an additional and increased tariff (133.33% higher) for consumption in excess of that basic volume. Some charge additional amounts for sewage service, but only CORAASAN does provides that service. They also charge for wells installed by home, business and industry owners, and the tariff is higher, the rationale apparently being that these users would be out of control as to the volume they consume.

In the case of irrigation, it is estimated that 60 to 80 percent of total annual agricultural production came from production in the irrigated areas during 1991 to 2004 (INDRHI, 2006a, 2006b, 2007). Importance of water used for irrigation however, is poorly recognized. Table 16 shows the statistics from the *Banco Agrícola de la República Dominicana*, a state owned bank providing financial services for promotion of agricultural production. For the analysis of loan applications, this bank

Table 16. Production Costs and Cost of Water of Irrigated Crops.

		Costs per tarea (RD$ Pesos / Tarea), 15.85 tareas = 1 Hectare									Cost of water		
No.	Name of Crop	Agricultural services (RD$/Ta)	Inputs (RD$/Ta)	Labor (RD$/Ta)	Subtotal (RD$/Ta)	Subtotal + unforseen 10% (RD$/Ta)	Finantial charges (RD$Ta)	percenrage of Finantial charges (%)	Finantial charges	Total, including unforseen and financial costs (RD$/Ta)	(RD$/Ta)	% of inputs	% of total production cost
1	Potato	855.67	7,562.20	3,383.34	11,801.21	12,981.33	778.88	6.00%	2,336.64	13,760.21	10.67	0.32	0.08
2	Sweet potato	658.50	347.10	1,596.88	2,602.48	2,862.73	257.65	9.00%	515.29	3,120.38	16.00	1.00	0.51
3	Banana	802.00	1,850.10	2,779.58	5,431.68	5,974.85	1,075.47	18.00%	1,075.47	7,050.32	32.00	1.15	0.45
4	Platain	711.00	1,474.50	2,723.32	4,908.82	5,399.70	971.95	18.00%	971.95	6,371.65	32.00	1.18	0.50
5	Cassava	769.00	519.10	2,029.46	3,317.56	3,649.32	656.88	18.00%	656.88	4,306.19	32.00	1.58	0.74
6	Red Beans	731.50	967.20	1,145.84	2,844.54	3,128.99	140.80	4.50%	563.22	3,269.79	8.00	0.70	0.24
7	Onion	633.33	5,179.40	4,137.64	9,950.37	10,945.41	985.09	9.00%	1,970.17	11,930.50	13.33	0.32	0.11
8	Garlic	841.00	11,500.63	4,091.58	16,433.21	18,076.53	1,626.89	9.00%	3,253.78	19,703.42	16.00	0.39	0.08
9	Pepper	630.67	4,712.81	1,934.52	7,278.00	8,005.80	480.35	6.00%	1,441.04	8,486.15	10.67	0.55	0.13
10	Rice	1,228.50	1,591.40	727.50	3,547.40	3,902.14	351.74	9.01%	702.39	4,253.88	23.50	3.23	0.55
11	Rice transplant	1,228.50	1,589.33	1,052.41	3,870.24	4,257.26	383.15	9.00%	766.31	4,640.41	23.50	2.23	0.51

uses data and analysis of production costs, which consider water as an input. As can be seen adduced importance of water is denied by the facts. Cost of water represents less than 4% of costs of all inputs and less than 1% of total production costs in all 11 crops considered.

Water tariff for irrigation is based on surface area and tariff structure differentiates between rice, pasture and other crops (vegetables and fruits, including plantain and banana) and take into account size of land in two categories, less than 10 hectares or above 10 hectares). Each WUA fix its own tariff, based on budget or total O & M costs per year. For the year 2007 tariffs for rice varied from RD$56 to RD$108 per tarea per year (approximately USD 26.50 to 51.10 per hectare per year), while tariffs for other crops were RD$34 to RD$63 per tarea per year (approximately USD16.09 to 29.81 per hectare per year). A study of CITAR related to tariffs for irrigation service (Ramírez and Chalas, 2008) analyzing tariffs of 11 WUAs covering an area of 163,168 Hectares (53.2 of total irrigated area in the country), translated water tariff in price per surface area (RD$/Tarea) is to its equivalent volumetric tariff by means of analyzing crop water requirements minus effective precipitation and multiplied by a coefficient to consider efficiency of the irrigation system in each area under study. Applying these concepts, tariffs for irrigation service in rice is calculated as RD$ 0.08 to RD$ 0.19 per cubic meter (approximately USD 0.0024 to USD 0.0057 per cubic meter) and that of other crops as RD$ 0.07 to RD$ 0.24/m^3 (approximately USD 0.0021 to USD 00.72/m^3). Average results seem to show that a similar price per volume of water is charged for rice (RD$0.11/$m^3$) as for other crops (RD$0.10/m^3). Water consumption in paddy fields will make the difference. When considering costs of O & M of both the WUA and INDRHI's Irrigation Districts, whose costs are not considered in the tariff of the WUAs, in those same 11 areas of selected WUAs for the study, average equivalent volumetric tariff would be RD$ 0.24/m^3, which exceeds 118 to 140 percent higher than the volumetric tariff calculated for rice (RD$0.11/$m^3$) and for other crops (RD$0.10/$m^3$) respectively. Implied average government subsidy, according to the study, is so that for every cubic meter of water delivered for irrigation of crops, INDRHI covers two thirds of the cost. One of the heavy loads borne by INDRHI is cost of electricity for operating pumps. More research is recommended to fine tune estimates of real costs structure of operating irrigation systems and to accurately determine composition of subsidy.

In terms of productivity the irrigation sector also needs to find better justification for investments. While it would still be true to say that irrigation increases yield in production, better outcomes would be expected in difference between rain fed agriculture and "irrigated" agriculture.

There is not a clear vision as to who should charge for water from rivers and wells. It is common that industries extracting water from wells and deriving waters from rivers, are approached by personnel of INAPA, the water supply institute and other water corporations (water utilities) of municipal jurisdiction, charged and invoiced in search for more income and better economic balance in this institutions. The water utilities nor the hydropower generation company (Empresa Generadora de Electricidad Dominicana – EGEHID) do not pay for using water either, which is defined, as well as other natural resources, as a state property by law.

INDRHI used to charge farmers for water served via irrigation systems, but after transferring all of the irrigation systems to water users associations (WUAs), these corporate organizations do not pay for water, although some have conceived to pay INDRHI in the future and even pay for "environmental services". Even when the farmers used to pay INDRHI, price for water was so low that accumulated income only made 5% of INDRHI's budget, the rest coming from government allocation in national budget.

Water fess of water utilities is significantly low in comparison to production costs (purifying and distributing water to house taps). In other words, the tax collected from all citizens is the major source for government expending and investment in providing water services. Some agencies promote users participation in investment in water supply facilities in small scale projects in the rural area and although the sense of appropriation that is derived from this is invaluable for sustainability and absolutely important, an interesting argument about why the government builds water supply systems for large cities without any requisites to financially better-off citizens, since investment is never a part of water tariffs. Again, the subject of water tariffs and justifications for sunk costs of investment and tax payment, and who benefits more from government paternalism, deserves some attention.

The legal provisions requires that people benefiting from government investments in canals are responsible for costs of operation and maintenance of infrastructure. Water titles are tied to land, making land-title holding in an irrigated area important. The effect of irrigation canals to economic appreciation of land is another economic benefit. In the early part of the ninety seventies, a legal scheme was designed to support agrarian reforms policies, in particular assigning land to the landless. This law, still kept ruling through successive amendments, means that when government builds canals that pass through properties of area above 100 hectares, landowners should pay for such benefit in kind, as much as half of their land, or in cash. This was an instrument in implementation of redistribution of land, which is now questioned because of effectiveness of in kind payment versus what would mean a cash contribution of landowners and the rights of the government to impose such payments.

Discussions on impact of large projects are also necessary. Dams generating electricity transmit their input into the national grid and poor communities, some having been displaced from what is now the impounded reservoir site, do not have a stable electricity services or have no electricity at all, nor have any other form of compensation, apart from having received a house and a small size land or having been paid for the land they had. It is paradoxical that remotely located and marginal economy communities in areas of hydropower projects have been frequently neglected or excluded from electrical power supply or services and only see the booming business around it and richness derived thereof, pass through their villages without none to non significant benefits. People living in the big cities are more benefited than poor villagers. One benchmarking experience that might be a model for other hydropower projects is that of "*Los Toros*", a small community located besides an existing water canal where a 9.2 MW hydropower plant (no reservoir) was built with funds from European Union.

Discussions and search for references on the topic of participatory decision making and stakeholder participation can find an interesting example in the *Los Toros Fund* (LTF). Approval and disbursement for development projects financed by the LTF are decided by a Provincial Commission heading the Fund, created by a Decree from the Executive Power in 2005 (485 – of September 2005), and composed by 13 members, with 8 of them from the private sector, including a representative from the Los Toros community, leaders of WUAs, two clergymen (a priest and a pastor), business sector and mutual cooperative representatives. Nine of the members live in the Province of Azua, which is an important factor since they are permanently in close contact with demands of communities. The other 5 members, including the Governor of Azua (appointed by the President), are government institutions related to the Project.

LTF is not a remedy fund, since there were no damages nor loss by the population, and displacement was not really an issue, with only two houses relocated due to construction. There was no resistance from the community and the LTF started after hydropower plant was built. Construction site was an existing canal built in the early 1980s and required no storage reservoir. It is indeed a good example for dam projects and can serve as a model to establish regional development funds, according to compensation policy principles and as a monetary benefit sharing mechanism where part of monetary flows generated by operation of dams is distributed to affected or concerned communities (UNEP, 2007), not necessarily exclusive of affected people. In the case of the LTF all, not part of, of the net income or benefit, is benefiting the whole province of Azua, not only the Los Toros and surrounding communities. The LTF is also considered to be an alternative model to think about when speaking of environmental services.

The LTF has a protocol to guide decisions, but lacking more standardized processes has not impeded the Provincial Commission of the LTF from working out a process for project selection as of December of 2005, holding community meetings and hearing the representatives of organizations from the communities in the different municipalities of the Province of Azua. The members of the Commission prefer to avoid bureaucratic formalities due to urgent needs in the province and are inclined to smaller projects with a reasonable geographic distribution of investment. Promoting replicable projects and fund generating activities or recovery of allocations from the LTF are among the ideas discussed for establishing future selection criteria.

One important area of improvement of the LTF is establishing formal project selection instruments and processes with norms and standards clearly known by the communities and potential applicants. Enhancing communication and public knowledge and acceptance of projects are elements that will benefit the consolidation of the LTF. Most urgent need is to formulate a development plan to guide decisions and programme investments. It is important to decide on a model for management of funds, leaving some room for the LTF to operate as a competitive fund, where project approval and investments decision will be based on demand. Selection criteria should be clear to all as well as information requirements (benefits and costs details). Improvement of transparency and accountability instruments of the agencies entrusted with the re-distribution of benefits is a key factor for successful implementation of LTF's objectives. Accounts receivable from CDEEE, which has been slow paying its dues to the fund, is one issue to resolve, having accumulated more than RD$300 millions in debts. Some pressure was and is necessary for regulating these payments. Ensuring compliance and strengthening the enforcement mechanism, which is a financial agreement and has a 15 year horizon (2015), and possibilities of becoming a permanent mechanism is a matter of discussion. The LTF is definitively creates jurisprudence in project financing. Profitability is by far outstanding. Energy generated from January 2001 to October 2007 (322,402.86 MW-hr) had a net income of RD$675,749,306.20, which is twice the cost of the hydropower project, more than "recovered" in 7 years.

Financing projects is one important issue to resolve, in particular as to the procurement of funds. Traditional sources of money like multilateral development banks, under "soft" conditions or terms (low interest rates and longer grace and repayment periods), are not longer financing dams or have serious concerns about evaluations to accept this type of projects and are only recently thinking of renewing their long ago faded support to the agricultural sector. The government's recourse has been to exploit the willingness of commercial banks to provide funds at higher interest rates, shorter grace period and shorter repayment horizon. Concession of roads has already been accepted and applied in two projects, but concessions for hydropower plants and water utilities is prohibited by law and will require more congressional discussions.

Discussion is going about environmental payment. In Yaque del Norte river basin and Yaque del Sur, some non-government organizations (NGOs) are promoting and trying to replicate experiences from Costa Rica of agreements made by the hydropower company and water utilities with individual farmers or residents in the upper watershed, for a careful exploitation of natural resources, so that they observe and practice conservation measures to either increase vegetation cover on their land and reduce soil erosion, or simply leave the land like it is. While some believe that this is a good model, others argue that a national system, not based on individual and voluntary agreements, but rather a compulsory scheme, except for tariff structure which should be negotiated between corporate parties, the water agency and hydropower supply company (only one state institution of national jurisdiction operating all existing dams), and transferring part of collected fee for water to the natural resources and environmental protection agency (Secretariat of Environment and Natural Resources) to finance watershed conservation projects for example, would be better. The water agency (INDRHI) has advocated for this type of model and thinks that the aforementioned *Los Toros* experience is a good argument of this other type of developmental approach. One other argument by the water agency is that this type of funds could also serve for sustainability of weather service and hydro-measurement network (installing, operating and maintaining stations), which is fundamental to measure the impact of conservation practices applied, but are in reality often neglected items in national budget allocation with consequent inefficiencies.

In any case, need for development and implantation of a reasonable model for financing water and soil conservation activities is necessary and would make good business sense for reduction of reservoir sedimentation, which is costly to resolve. Pertaining to analysis of benefits of sediment control activities is the consideration of costs to be avoided and functionality of dams. Costs of sediment extraction in Valdesia and Aguacate in the Nizao river, to recover bottom outlet capacity and prolong useful life of reservoirs, previously mentioned, is estimated as USD 16.814 millions. Notwithstanding, feasibility analysis indicate that in a 40 year period, with capital investment and operational costs of USD 28.79 millions (present value for costs of sediment control activities at a discount rate of 3 %), there would be a gross profit estimated as USD 581 millions. Considering annual benefits from generation of hydropower in the 3 dams in this one basin (Jigüey, Aguacate and Valdesia in the Nizao river basin) as USD 21.973 millions and benefit for irrigation USD 3.40 million/year. Costs for control of sedimentation would only be 5% of estimated benefits (INDRHI, 2007).

13 WATER AND EDUCATION

Every single person on earth uses water and the behavior of every individual, each organization, community, institution and industry, can affect rivers, aquifers, lakes and coastal waters. In a varied array of aspects and forms that goes from the quantity of water used, either at home, at the farm, at the office, or in a hotel, to energy consumption patterns, garbage and wastewater disposal habits and facilities, and even eating preferences, the values and behaviour of every person can make a difference in the growth or abatement of pressure on water resources or water degradation.

Water is everybody's business. The need for water conservation has intensified the call to integrate, in a wide spectrum, government's specialized institutions with responsibilities on water management or water services and other branches of national authority, local authorities, organized groups, communities, as well as the private, the industrial and financial sectors. Development of a new way of thinking on the importance of water and the need to have better institutional arrangement, improved legal frameworks and more stakeholder involvement for a more sustainable use of water, has required decades of talks, lengthy and unfinished debates, hundreds of seminars and conferences, and tons of papers written and exchanged among specialists of different fields, including water resources managers and planners, hydraulic engineers, irrigation specialists, environmental engineers, some sociologists, economist and even a few experts on water law. To translate experience and accumulated knowledge on water problems and concerns to the population, the common person who uses water at home or at work, will demand a special effort to bring about desired results of rational use of water. Putting water on the agenda of politicians, economists, youth, teachers and school children, cannot be postponed.

Search for sustainable solutions to the problems has to focus on the social and cultural dimensions of the water problems and the contribution of water users themselves towards a solution. Economic and technological inputs for solutions necessarily need to be envisioned with complementarities of education and cultural values, if decisive change in trends and durable effects of sustainable use of water resources are to be achieved. Participation and decentralization are strategic to integrated water resources management, but people need to be well informed and prepared to participate effectively.

In an effort to highlight the importance of water, raise awareness and motivate social action to face the present or future critical situation on water, an educational program was launched by INDRHI with collaboration of Secretariats (equivalent of ministries) of Education and Health, as well as water utilities, in the year 1997. The program is called *Programa Cultura del Agua* – water culture – (PCA), a common name in Latin America for such activities, and had as a reference model for its initial formulation and actions (Texas Water Watch program, 1997).

The PCA started actions with the "*Vigilantes de la Calidad de Agua*" (water quality guardians) sub-programme, whose main objective is to involve communities in the defense and preservation of water bodies in their surrounding territory. Awareness of conditions of water and identification of real or potential sources of contamination, is achieved by means of training community leaders and monitoring of water quality. The villages are and given an appropriate technology water quality kit for measurement of temperature, PH, electro-conductivity, dissolved oxygen, chlorine, nitrates and nitrides. Other parameters can be added to this kit. INDRHI worked with universities (Universidad Central del Este, Instituto Tecnológico de Santo Domingo) and NGOs (Asociación Para el Desarrollo, Inc, in Santiago, FUNDEJANICO and others) in approximately 80 rural communities located in the Yaque del Norte, Yaque del Sur, Higuamo and Soco river basins and established a network of Vigilantes del Agua (voluntary service) in several villages. Follow up on results of measurements and refills for the measuring kit is poor. Continued support of the programme is not guaranteed, but financing for new projects have been arranged to restart the VCAs.

One interesting project of PCA was the integration of water within the school syllabus, not seing water as a separate subject, but within each subject. With assistance of an expert from the German service for social and technical cooperation (DED), guidelines for teachers were developed with concepts and topics and classrooms and outdoor activities for each grade in primary and secondary school. The Secretariat of Education collaborated in this effort and guidelines were developed with intensive exchange between educators, environmentalists and graphic designers. Unfortunately, these materials, carefully conceived and designed, have not been printed nor applied at the classrooms.

Most identifiable strategy has been to focus on primary and secondary school population. Having children at school age is a key element, since, according to educational experts and psychologists, value and cultural learning is most effective during 4 to 6 years of age.

Another assistance of an expert provided by DED was the development of the *Saprobian* index to measure water quality, which is based on the quantity and type of living organisms – macro invertebrates – present in water (INDRHI, 2006a and 2006b). The Universidad Central del Este (UCE) in San Pedro de Macorís, also collaborated in calibration and local application of this model and a manual was developed for identifying representative species and determining water quality via a simple biological analysis demanding only a net and a minifying glass. Implementation of this biological test however, has been limited to initial stages of validation of methodology.

A recent initiative under the PCA is the "*Sala del Agua*" or Water Expo Hall created in 2006, which is a permanent exhibition receiving up to 12,000 visitors each year, that already deserved a recognition award from the regional Office for Latin America and the Caribbean of the International Hydrologic Programme (IHP) of UNESCO in February 2007. Children from schools or clubs are able to walk through five modules of water education guided by trained staff, raising their awareness on water problems, becoming acquainted with concepts on water (like the watershed and physical properties of water), learning about types of water measurement equipments, being informed through interactive methods of water pollution and floods, expressing their views on the centrality of water for development, needs for conservation and concerns on mismanagement of water at home and pollution of rivers, through game playing, writing poems, songs and sketches, drawing and painting, and finally making individual and collective solemn commitments to preserve water. Surprisingly, adults, university students and even graduate students that have also visited the Water Expo hall have been amused and enthusiastically engaged during the tour, which is adapted to their level by the guide.

"*Agua Móvil*" (water mobile), is a new sub-programme of the PCA started in 2007, having 12 trucks equipped with modern audio visuals technology, tents and chairs to accommodate their audience, visiting communities, organizations and schools, spreading the gospel on water conservation. Talks and group dynamics, are usually stimulated with a video – documentary on one of different topics of water and even fragments of popular films with scenes that touch water related issues. The activities of Agua Móvil are oriented to socialization and animation of community members towards water conservation. Proper staff selection, adequate training for the trainers and keeping high motivation in staff are among fundamental factors for success and expansion of this new subprogramme that seeks to make "water conservation converts".

Camps meetings of children, youth and nature explorers clubs are also among activities and strategies used by PCA. A recent pilot project was performed during 4 moths, working with the daughters and sons of farmers, members of an irrigation water users association (WUA) in Pedernales, a province in the southwestern extreme end of the country, bordering Haiti, where water, nature conservation and agricultural concepts were combined with drilling exercises, wilderness survival and camping techniques, covering a set of requirements in a different modality of explorers clubs classes or scouting levels. This experience, which had as one of its objective to get the next generation of farmers motivated with sustainable farming, should be further evaluated, developed and promoted. Explorers clubs and scouting activities are a good opportunity to use recreational active learning for water related issues.

Under the motto "*No more future than the one we build*", PCA has also worked with catholic churches and other Christian fellowship groups, in spiritual retreats of young parishioners. A one weekend programme covering topics of leadership and moral values is carried out, with the topics of water conservation neatly woven around the other learning objectives.

The water utility corporation of Santo Domingo (CAASD) also has an educational programme for children, focused on saving water at home and threats of unclean water. The educational department of CAASD makes a balanced use of videos and field trips to a water purification plant as learning experiences.

An interesting experience in the implementation of the *Sandwatch Project* in the Dominican Republic is reported by the Dominican National Commission to UNESCO. Through the network of associated schools of UNESCO and involvement of ministries of education, environment and culture, an national aquarium, provision of equipment for measurements, manual on the use of tools, training for teachers and working with students on monitoring environmental quality of beaches.

Among the PCA's most usual topics and messages are forest and watershed preservation and its relationship with flow of water, scarcity and pollution of water sources, water conservation, hygiene and prevention of diseases, and other general environmental and natural resources topics. Some field visits to dams have been done, but due to logistical and budgetary reasons, these have been too few. A few specific subjects should be developed and added to the programme, as special topics and concepts. Among this vulnerability to floods and a better understanding of downstream communities of dam safety issues.

Tools and strategies have been employed with varied degree of development and effectiveness. Use of printed graphical materials like posters, pamphlets and flyers, has been poorly developed and designed. Messages in flyers and brochures are not clear and have an unbalanced combination of words, too many, and pictures, too little and few. Utilization of mass communication means shows varied domain of production and consequently quality of products and effectiveness are varied. Radio and television have not been used by educational programmes. Sur Futuro, an NGO working with watershed conservation projects, did produce interesting, creative and attractive posters in the year 2003, officially declared as the national year of water. Some private companies and Sur Futuro had a few good television advertisement, but in general, the potential of television publicity campaign are yet to be tapped. The matter of costs involved in marketing water conservation is one factor probably restricting a better use of TV and radio.

One thing that should be clarified is the difference between an educational programme and a publicity campaign, whose messages are transmitted via posters and radio or television. Sometimes confusion and emotion over a "nice" slogan or isolated inspiring phrase invades and pervades educational objectives and degenerate in poor performance educational programmes and waste of efforts and resources. The other pervasive thing is that some private companies only use water or nature conservation as a means to create a good corporative image of environmental concern, and some of these companies are among main polluters of rivers discharging untreated water into it.

The production of local documentaries and films is one worthy project to embark on, since most materials of PCA for example, are videos and documentaries produced in other countries. The Secretariat of Environment and Natural Resources has produced some dealing with ecological and other natural aspects, not directly associated with water issues.

Music and folklore have not been exploited in educational campaigns, but experiences from other Caribbean countries like Barbados, where traditional festivals have taken motives of water in costumes, games and popular songs (Parsram, 2007), should further stimulate this idea. One critique about educational programmes is that the staff tend to try do everything. Proper space should be made for important contributions that talented musicians, visual artists, graphical designers, publicists, actors and cinematographers, can make to the water conservation advocacy and cultural movement. Music for example, has interesting characteristics that it does not ask for permission to get into your brains. If we are speaking of "water culture", we should let the cultural experts speak. Perez (2004) and Perez et al., (2004) have recommended that very popular and gifted artists or sports stars, who are very influential in the youths, can become allies to the water conservation campaigns. It is known that some have wisely joined to campaigns to promote aids or cancer prevention, combating abortion, illegal migration and other purposes, due to the strong public opinion influence they naturally exert. Having them on the water conservation movement would serve to promote and amplify awareness of citizens and society.

Some actions have been limited to scope and horizon of specific projects, not having continuity nor permanency. Main challenges of educational programmes are sustainability, creativeness and innovation, consistency and coherency. Sporadic and isolated messages and ephemeral support for educational projects are risks to. Avoid and overcome. There are voids in analysis of impact of promotional campaigns and room for improvement in the evaluation of teaching methods, and lack of measurement on how attitude of people towards water is changing. Exchange of experiences among different programmes is acsrse opportunities can and needs to grow. Use of educational and pedagogical tools and strategies are incipient degree of development, not specifically oriented or designed for community awareness.

Members of WUAs, youths and university students, not always so aware or concerned as one could suppose, are among important target groups. Institutional capacity building is necessary at rural level so that local organizations are able to develop their own programmes and seek funds for it, provided that technical assistance and some initial financial support is agreed with the water agencies or water utilities. Fostering and sponsoring of educational programmes is an important issue for sustainability.

The content and scope of University programmes for undergraduate studies in civil engineering should be revised. Water related subjects like hydraulic engineering and sanitary engineering, are probably focusing on creating knowledge and skills to solve the conventionalities of design for big cities and big structures like dams, not so environmentally and socially oriented and leaving aside certain issues that the students are more like to encounter in the professional careers. As for the rest of careers and programmes, although some environmental ingredients have been added to their requirements, more can be done to have the water issues permeating through course contents.

Training of staff involved in educational programmes should be planed and organized. Cooperation from publicity and marketing companies, radio and television stations should be sought to provide courses, workshops and internships. Other aspects to improve are publication and visibility of results. Net work of water educators and publicists, is one idea to promote for exchange of experiences. Research on education, psychology and sociology, are necessary to verify the feasibility and effectiveness of methodologies and analyze cultural and behavioral aspects and how educational strategies are able to impact communities. Topics like gender, culture are also to be studied and explored.

14 INSTITUTIONAL ARRANGEMENTS

The institutional arrangement to face the enormous challenges of timely and wisely satisfaction of society's diverse demands of water in appropriate quantities and of good quality is a central issue. Possibilities of successful implementation of an integrated approach to water resources management, or at least well coordinated, will depend, to a great extend, on the legal and institutional framework. The absence or insufficiency of capable institutions responsible for integrated water resources management has been recognized as one of the root causes of deficient water resources management (Lamoree et al., 2005). Traditional institutional schemes and past legal provisions have not helped to solve the problems. New dogmatic formulas of "in style" organizational set up for the water sector on integrated water resources management are not necessarily feasible in a given country. There is no universal solution to institutional arrangement and in each country there is an "*institutional ecosystem*" (Raast, 2005) which must be known in order to design the most suitable and effective framework, analyzing the institutional, social and political context within national reality. Even more important is that the institutions of the water sector have the capacity to implement the normative, regulatory or operational tasks assigned to them. Having the most adequate institutional framework to put integrated water resources management into practice is only half of what it takes for the story to have a happy end. There is an equally difficult question on how to bring about necessary changes in legal – institutional arrangements (Lamoree et al., 2005).

A few of the key components to consider for the assessment or design of the institutional framework are: policy and strategy formulation process and implementation capacities, water allocation and conflict resolution mechanisms, development of planning tools, legal and regulatory instruments, a monitoring network (water quality and quantity), water quality management, environmental assessment (norms, plans, control and vigilance), financial mechanisms, risks management capabilities and participatory approach.

In the Dominican Republic the existing institutional framework for water resources management is characterized by fragmentation of responsibilities among agencies, poorly developed regulatory framework, weak implementation of the very few existing regulatory mechanisms and meager enforcement capacity. Policy for the water sector is not explicitly defined no concrete effort to formulate it can be recalled. Information is disperse or unavailable and access to it is hazardous. The state agencies have been instruments for considerable investment for infrastructural development, but they have had poor performance and have been less successful in operation of water supply systems, water purification plants, sewage treatment plants and irrigation systems. As said before tariff structure, fee collection and economic performance are weak.

The "Instituto Nacional de Recursos Hidráulicos" (INDRHI: national institute for water resources) was created by Law No. 6 of 1965, having as main functions the planning for hydraulic development, performing hydrologic studies and operation of hydrologic network, management of irrigation systems with the participation of users and the construction of water infrastructure, including dams and irrigation systems. The "Secretaría de Medio Ambiente y Recursos Naturales" (SEMARN: secretariat of environment and natural resources) was created by Law No. 64 – 2000 in August of 2000, having several mandates related to water resources, most relevant ones related to groundwater management, surface and groundwater assessment in basins, water pollution control, approving or determining wastewater disposal locations, approving reuse of wastewater, establishing restrictions on the use of water and approving the plans and projects of INDRHI. SEMARN will take time to build capacity to be able implement these and for the time being water resources assessment and groundwater management capacities remain with INDRHI.

Most important effect of this Environmental and Natural Resources Law are a change in the Law 487 of 1969 (and subsidiary law – bylaw No. 2889 of 1977) for regulation of the use of groundwater, whereby SEMARN takes a role previously assigned to INDRHI of managing groundwater, permitting for drilling of wells and other control tasks, for which INDRHI is still better prepared. One positive development brought about by Law No. 64 – 2000 is the requisite of environmental impact studies for projects, including dams, irrigation systems, roads, housing complexes, and many others, having SEMARN the responsibility to rule on this process and to award licenses and permits. SEMARN has gradually developed norms and standards for environmental protection, including control of pollution. The creation in 2004 of an environmental court where to ventilate claims, examine lawsuits and prosecute trespassers of environmental offenses and deal with breach of the law, is also a significant achievement. Vigilance and enforcement capacities however are still under development and incipient. Getting lawyers acquainted with environmental and natural resources issues and having environmentalist become familiar with provisions of law is an important bridge to be built in order to get across in a two ways road.

It would seem that INDRHI is managing surface water and SEMARN has control of groundwater, but neither fulfills these tasks as such. Except for water users associations and those corporate users (water supply companies) deriving water from dams, INDRHI has little control of abstractions from surface waters or use of water. Informality reigns in matters related to legal entitlement to water and there is weakness in the enforcement of procedures to obtain or maintain rights to use water. INDRHI did develop a "hydro-agricultural" information system on a GIS platform to manage production statistics and use of water in the irrigation systems, but the potential of this tool in water management has not been tapped.

Existing water law (Law 5852 – 1962 on Domain of inland waters and distribution of public waters) requires that before proceeding to study or design any civil work on the river bed or before allocating a demand of water for a new user, the implementing agency would need to authorize a permit based on a prior application for the use of water. Permitting procedures are rarely enforced. Users like water corporations expand existing (operating) water system and build new ones without requesting a permit or right to use water or informing neither INDRHI nor SEMARN about it. Projects are not seen in the context of the water availability in the basin and on a project by project basis, frequently improvising and compromising future options.

The requirements of environmental studies and an environmental license has been, in some cases, an opportunity to have INDRHI, in coordination with SEMARN, approving planned use of water in a project or make observations about a project next to a reservoir site or too close to a river. These corporations as well as the hydropower company (state owned) plan, design and build their plants or systems, which sometimes include transfer of water from one basin to another, without an assessment of water reserves and existing water rights in the context of the basin. It also happens that the Secretariat (ministry) of Public Works builds bridges or new roads without consideration of legally established permitting and not taking into account hydrologic and hydraulic aspects.

Before the year 2000 INDRHI was under the secretariat (ministry) of agriculture and is now under SEMARN, which means a change and having INDRHI more involved or concentrate on the conservation of water, instead of being engaged with operational activities of managing irrigation systems. A careful design of that change process is necessary. INDRHI was the sole water agency, but is now legally defined as maximum authority of hydraulic infrastructure. Provisions of Law 64 – 2000 are not sufficiently detailed as to clarify relationship with INDRHI and with existing water law of 1962 (Law No. 5852 – 1962), nor have adequately taken into account INDRHI's role in hydrologic network, water quality monitoring and its capacity to carry out regulatory tasks. A need to harmonize the roles of SEMARN and INDRHI and the environmental and natural resources law with the water law is a priority. There are several issues where SEMARN and INDRHI need to reconcile their thoughts so as to have a common agreement in matters like river basin organizations, environmental payment services and most important the tasks which INDRHI is to perform as a regulator of water resources and its relationship to the sub-secretariat of Soil and water of SEMARN.

The Environmental and Natural Resources Law itself orders an updating of water law and a project law for a new water law was drafted and submitted to Congress in 2003. This proposal for a new water law is inspired in principles of integrated water resource management and the establishment of management instruments such as the river basin planning, water resources national master plan, licensing for the use of water and discharge of waste into water bodies, risk management, financial mechanisms, water information systems, education and research. This draft for the new water law proposes the separation of functions (normative, regulatory and operational), having SEMARN concentrate on policy, strategy, norms, standards and legal issues (constitutional function), while INDRHI would perform regulatory activities or the organizational function (hydrologic network, water assessments, water allocation, administration of procedures to award water rights and others). At the operational function levels the project law recognizes the role of water utilities (water supply corporations, irrigation – waters users associations). The draft also proposes that INDRHI is responsible for the subject of management of multipurpose dams.

Rodríguez (2006), analyzing the legal reforms initiatives and examining the structure, capacity and efficiency of the process, state that the irrigation management transfer process was ignore and suggests that the consolidation of this process be carefully considered and that consultation of this proposal with WUAs is an important step and recommending not to underestimate the capacity of farmers to contribute to this project law.

One objection posed to INDRHI's claims to be the regulator of water resources is that this it represent the main water use sectors, irrigation, and that if INDRHI is to be responsible for regulation of water resources, no operational tasks should be assigned to INDRHI. This, to many is to be both "the judge and party (plaintiff or defendant)". One other argument is that INDRHI should not be engaged in the construction of dams or irrigation systems if it is to regulate water resources, calling for a pure dedication to water regulation, since having both functions will always make infrastructural development be a priority within INDRHI and commitment to regulatory or conservation tasks is not guaranteed. The call for such an aseptic concept of water resources management is probably beyond practical convenience.

The water supply and sanitation sector has the "Instituto Nacional de Aguas Potables y Alcantarillados" (INAPA: national institute for water supply and sewage systems), created by Law No. 5994 de 1962, as the institution of national jurisdiction and mandate for building, operating and maintenance of water supply and sewage systems. INAPA should be a regulator of existing water supply corporations in Santo Domingo (Corporación de Acueductos y Alcantarillados de Santo Domingo – CAASD – created by Law 498 of 1973), Santiago (Corporación del Acueducto y Alcantarillado de Santiago – CORAASAN – created by Law 582 of 1977), Moca (Corporación del Acueducto y Alcantarillado de Moca – CORAAMOCA – created by Law 89 of 1997), Puerto Plata (Corporación del Acueducto y Alcantarillado de Puerto Plata – CORAAPPLATA – created by Law 142 of 1997) and La Romana (Corporación del Acueducto y Alcantarillado de La Romana – COAAROM – created by Law 385 of 1999). These corporations in practice act as independent operators with no regulation. New project laws have been submitted to the Congress for the creation of new corporations and some have been passed but not implemented. This, rather than a decentralization process, is a detachment not related to a strategy of better performance and is due to individual and independent interests of different municipalities not necessarily aiming at operational and economic efficiencies.

Following a study of the water supply and sanitation sector, financed through a non – reimbursable technical cooperation by the Interamerican Development Bank, a reform process was proposed and a law for this specific sector was drafted and submitted to Congress in 2001. Items in this proposal include the creation 2 new institutions a normative and planning entity and a regulating agency, allowing private operators to become involved in providing water supply and sanitation services through different kinds of contracts (BOT, BOO and others).

Rodríguez (2006 Op cit) considers that following subjects are to be revised in the law proposal: the future of INAPA is not properly addressed; links between with proposal for a new water law with missing links on topics like request of water rights, payment of operators for use of water and some provisions for environmental services; the risk of creating two new institutions for regulation of the sector and time for them to build competence are disregarded; investments in small towns where the private companies will not be interested in investing due to small size of the market and continued need of the government for infrastructural development has not been taken into account. A loan operation approved by Congress was tied to approval of the project law for the water supply sector and proposed reforms. Convenience of this loan driven approach to reform should be analyzed.

In spite of the fact the project law remains as a proposal, CAASD in Santo Domingo made contracts with 2 private companies to do the commercial tasks in its name (install meters and do invoicing and fee collections). Investments in meters and administrative costs are to be covered by collected fees. It is thought that good results have been achieved both in terms of savings in water and fee collection, especially considering that the government did not have to invest to buy or install the meters. Information on results however is not available and it is strongly recommended that a communication policy be established to overcome such stealth and for demand management measures to gain more acceptance.

Institutional aspects of hydropower generation are legally defined in the General Law on Electricity (Law No. 121 of 2001). The state owned "Empresa Generadora de Hidroelectricidad Dominicana" (EGEHID), created by this law as part of the holding of "Corporación Dominicana de Empresas Eléctricas Estatales" (CDEEE) operates all hydropower plants. The Superintendence of Electricity is to regulate private thermal power generators (gas, charcoal, oil), but the law defines exclusive right of the government to operate hydropower plants. In 2007 the Congress passed an Incentives Law for renewable energy allowing private companies to invest and operate small hydropower plants or solar or wind energy plants, up to 5 MW of installed capacity with several incentives, including some specific tax exemptions. Future discussion for updating of water law should consider issues as tariffs for the use of water in hydropower, project planning requirements for water allocation.

Concerning hydropower and water resources management one question to address with a clear and concise answer is: Who owns the dams and who is to operate it? Law 125 of 2001 creating EGEHID states that ownership of hydropower house is for EGEHID / CDEEE, while previous laws of 1975 state that ownership of dams are transferred to INDRHI. Except for two dams built by CDEEE, INDRHI has been all the rest of dams in the country, including dams whose main or sole purpose is hydropower generation and has also built most of small hydropower plant not requiring reservoirs. Whether the different components of the dam can be separated having the dam's body (earth or concrete structure), spillways, bottom outlets, and other civil works belong to INDRHI and only the penstock and powerhouse belonging to EGEHID / CDEEE is an interesting discussion.

Regarding reservoir operation, in practice EGEHID / CDEEE operates all gates and valves, including spillway gates for controlled spillways, Some tension often arises between the interest of farmers who prefer to save water for dry periods and the generation interests to generate all electricity possible to cover the deficit of electricity input into the national grid, specially when some thermal generators, consuming imported oil and other derivatives, are out of service. Strict adherence to priority of the use of water (water supply, livestock, irrigation, hydropower) is not always applied, having as apparent reasoning for temporal and emergency deviation of priorities the crisis of energy sector of the country, mostly depending on more costly (operational costs) thermal generation, and comparative economics of hydropower with respect to other productive uses of water, with higher prices for electricity and the non-consumptive use argument of hydropower sector, sometimes outweighing decisions over the irrigation sector due to its low or indirect economic return of the use of water. Inversely, during alert of hurricanes or tropical storms, decisions on safety measures would normally tend to suggest doing early preventive "discharges" to make available volume to effectively damp possible flood wave coming into the reservoir, while interests of electricity generation would try to prolong decisions so as not to "throw out" water through a spillway or bottom outlet. This situation of independently maximizing benefits for any specific does create serious conflicts in water management (Lamoree, 2005).

The Commission of the House of Representatives now studying both the water law proposal and the project law for the water supply an sanitation sector, is of the opinion that the Water Law should be considered and passed first. This is in line with thinking that "*intersectorial water allocation issues should be resolved before intrasectorial reforms are undertaken in the water use sectors*", but in reality experiences in Latin America show that in some countries a few water supply reforms which even include some levels of privatization, have been advanced before having a clear framework for the intersecorial water resources management level (Lamoree et al., 2005).

Within existing legal framework there are good elements that have not been enforced or seldom applied or neglected. Not having a new water law, though necessary to introduce concepts and instruments of integrated water resources management, should not be an obstacle to implement decentralization, improve cost recovery in water supply and irrigation systems, agree on water tariffs to be paid by corporative water users pay for water they are consuming of using, make water resources basin planning a standard for water allocation of new projects, define a national policy for water resources and develop and information system.

Findings of a study for implementation of integrated water resources management (ICWS, 2002) summarized characteristic of the Dominican water sector in this way: Principles of integrated water resources management are not well understood and separation of roles is unknown to many officials of the water sector; INDRHI has mixed functions performing organizational tasks (manager of water resources) and at the same time represents a water use sector (irrigation); no strategic inter-sectorial planning of water resources management takes place; water use planning at the operational level takes place through the "*Comité de Operación de Embalses*" (reservoir operation committee), and ad – hoc committee presided by INDRHI where other water use sectors participate (water supply and sanitation, hydropower); conflict resolution mechanism are poorly developed; there is no systematic water quality monitoring; there is no system of pollution licenses; there is no mechanisms for water policy and strategic water development plans; river basin authorities are non-existent; the system of water use right is not well developed; there is no operational enforcement mechanisms; there is no water resources regulation at basin level; and there are several on – going initiatives of reform process and proposals mainly productive sector oriented (water supply, irrigation and hydropower), but here is lack of coherence between various reform process.

Six years later, the principles of integrated water resources management are more familiar and except for planning, the rest of findings of 2002 are still valid. The existing institutional and legal frame work was designed with independent objectives and interests (sectorial approach) and the legal reforms initiatives remain much the same. Unfortunately, when the different sectorial law projects are discussed and compared, the institutional aspects catch all the attention, with representatives of each water use sector actively advocating for their share of power, with discussions across the table on who builds (infrastructural development), who controls awarding of bigger price contracts, and in essence negotiating which piece of the cake everyone is getting. Matters related to water rights, water allocation, river basin planning, pollution control, environmental protection, deserve much less attention of contestants.

In practical terms the only real action of water allocation is within the reservoir operation committee (ROC), having representatives of main water use sectors (INAPA, CAASD, CDEEE, INDRHI and ONAMET - meteorological office – and for hurricane season and specifically for hurricane alerts the committee is expanded) meeting every month to decide on volumes of water for each sector and at each dam. INDRHI and CDEEE play leading roles while the rest of institutions have a low profile (Lamoree, 2005). The proposal for the new water law considers the establishment of one ROC at each dam or reservoir to increase decision making participatory process involving more stakeholders dealing with more regional issues, rather than only one ROC at national level. The aforementioned assessment (ICWS, 2002) considers the ROC as rudimentary and authority oriented not conflict preventive.

Reservoir operation has to improve, developing operational rules curves, with a better link to early warning pluviometric and flow measurement stations and using well calibrated hydrologic and hydraulic simulation models for incoming flood and effects of discharge downstream respectively. The operation is now based on levels of the reservoir as indicated in the existing manuals, which are more reactive rather than proactive and predictive. The only legal base of the ROC is an agreement between CDEEE and INDRHI signed in 1974 with amendments in 1980, 1988 and 1991, having as objectives the "programming, implementation and control of reservoir operation to meet different demands (water supply, industry, irrigation, energy and ecological uses) and maximize energy production without affecting other users and minimized damages during extreme events – scarcity and abundance of waters" (text of the agreement). It also provides some negotiating ground for the fixing of tariffs, an issue never discussed.

Irrigation management transfer stands out as a positive example of decentralization and participatory solutions for management of water services, in this case irrigation. Transferred responsibilities to WUAs by INDRHI including administrative and operational tasks (the property is not transferred), have been performed without requiring a sectorial law on irrigation, since provisions of existing water law (Law No. 5852 of 1962) do foster the involvement of what was called "*irrigation societies*" in the cleaning and maintenance of irrigation and drainage canals; and INDRHI had a clear vision and a strong decision to make progress in delegating power to farmers in the irrigation systems, that showed continuity through out political changes and turnovers along two decades. Consolidation of this process is necessary and promoting WUAs is worthy of support. Innovation and strengthening in commercial aspects are priorities for the reinvigorating this process.

One interesting development is the formulation of the National Water Resources Plan (NWRP) carried by INDRHI since 2005, with technical assistance of the Spanish Government. The objectives of the studies for the NWRP are to diagnose the water situation of the country (updated water balance already available), identify priority actions and define an action plan and pertinent projects to solve current problems concerning water quantity and water quality, as well as to avoid future complications in the water sector, contemplating rational and sustainable use of water. It is expected that the NWRP will become an instrument for programming investment and solving problems affecting or threatening socioeconomic development of the country. One remarkable experience during the formulation of the plans has been the integration of community representatives with a participatory approach in the planning process having stakeholders identify main problems and issues within a watershed or a region and also rank those problems and perform problem analysis by means of several methodologies including GOPP (Goal Oriented Project Planning) and others, in order to identify measures and actions that can turn out into projects. The first time in the country that involvement of stakeholders goes this far and the first time planning is reaching this level. The participatory experience has been successful and outcomes of regional workshops and group exercises show that stakeholders do know the problems and interrelationships between those problems and possible alternatives for solutions. Hopes of continued participation of stakeholders at other stages of the NWRP, indicative sector plans, basin and/or regional perspective of water resources planning and project optimization based on water availability in the basins are high.

15 RESULTS OF ASSESSMENT

Having reflected on the issues described above on the country specific situation, one can proceed to examine the water resources management systems by making use of the proposed building blocks methodology. It must be taken into consideration that specific situations related to water stress, pollution detected in rivers, sedimentation of reservoirs, are not to be measured and that assessment concentrates on the management systems, rather than physical situation. Assembling and organizing available information on water problems is required. Nevertheless, existing problems are a reflection of weaknesses of the management system or aspects that could be addressed with improved approaches and tools to make water resources management components or elements, whether it be planning, monitoring or water allocation mechanisms, more sustainable. Having this in mind, the building blocks were individually evaluated and a color or numeric score was assigned to them. Using the more simple coloured blocks alternative as a first attempt or to initiate the scoring process provides a general picture of missing elements and facilitates a good start for the exercise, being able to identify the elements with extreme scores (inexistent or well developed) and making it easier to pass on to a numeric score, once it is decided which elements would have a 0 (absent), 4 (satisfactory or promising experiences and good potential for more improvement) or 5 (Well developed and improving) score.

The results of the coloured blocks approach are shown in Table 17, where we immediately realize that the important foundational layer of "Strategy and Policy" have voids; as well as those of water

Table 17. Results of Assessment – Coloured Building Blocks.

RISK	Early warning system	coordination with Civil defense and other emergency institutions	Flood control action plan	Droughts	Risk mapping	Zoning protection
FINANCING	Investment programmes	Tariffs structure	Cost recovery	Incentives	Private participation	Water accounts
WATER ALLOCATION and QUALITY CONTROL	Allocation mechanisms (award, licensing and permitting)	Demand management	conflict resolution mechanisms	Quantity and quality integation	Compliance	water reserves
ENVIRONMENT and NATURAL RESOURCES	Norms and procedure	Control and vigilance	Impact assessment requirements	Ecological flows or demand	Environmental management plans	Watershed management plans
MONITORING	Climate stations	Hydrometric network	water quality measurement	groundwater	processing and analysis	diseminaiton and public access
PLANNING	Inventory and asessment of water	Water Balance	national, state or regional planning	Basin planning	Indicative sectorial plans	Project optimization
INSTITUTIONAL	Role definition	Institutional arrangement	Descentralization and users organizations	Public participation	Education and communication	Research and development
LEGAL	drafting law	law	cadaster registry	administration of water rights	law and regulation	law enforcement
STRATEGY AND POLICY	Strategy	Policy formulation	Principles and Norms	Policy statement	Priorities	Policy evaluation

Table 18. Results of Assessment – Building Blocks with Numeric Scale

Components	Elements						Total (out of 30)	%
RISK	Early warning system 3	coordination with Civil defense and other emergency institutions 3	Flood control action plan 3	Droughts 1	Risk mapping 2	Zoning protection 0	12	40.00
FINANCING	Investment programmes 2	Tariffs structure 2	Cost recovery 1	Incentives 1	Private participation 3	Water accounts 1	10	33. 33
WATER ALLOCATION and QUALITY CONTROL	Allocation mechanisms (award, licensing and permitting) 2	Demand management 2	conflict resolution mechanisms 2	Quantity and quality integation 0	Compliance 0	water reserves 0	6	20.00
ENVIRONMENT and NATURAL RESOURCES	Norms and procedure 4	Control and vigilance 0	Impact assessment requirements 4	Ecological flows or demand 0	Environmental management plans 2	Watershed management plans 2	12	40. 00
MONITORING	Climate stations 4	Hydrometric network 3	water quality measurement 2	groundwater 2	processing and analysis 3	diseminaiton and public access 3	17	56. 67
PLANNING	Inventory and asessment of water 3	Water Balance 4	national, state or regional planning 2	Basin planning 1	Indicative sectorial plans 3	Project optimization 0	13	43. 33
INSTITUTIONAL	Role definition 2	Institutional arrangement 2	Descentralization and users organizations 4	Public participation 2	Education and communication 3	Research and development 2	15	50. 00
LEGAL	drafting law 3	law 1	cadaster registry 3	administration of water rights 0	law and regulation 2	law enforcement 0	9	30.00
STRATEGY AND POLICY	Strategy 0	Policy formulation 0	Principles and Norms 3	Policy statement 0	Priorities 3	Policy evaluation 0	6	20.00

allocation and quality control. These components should be priorities in any initiative for the improvement or reform of the water management system. Other priority components should be the environment and natural resources and financing. This graphic assessment also suggests an order for development of the water resources management system, being relatively easy to visualize the stages and process to follow. First items on the agenda should be strategy and policy formulation to arrive at an explicit policy statement. Some recent initiatives like the establishment of a research centre at INDRHI and current national water plan under way have made the picture better, but it is to be recognized that future periodic assessment is to prove if these elements can sustain achieved progress or make more advancements.

The graph results in Table 17 also show strengths or achievements in elements like water balance, due to recent update, decentralization and users organizations, due to progress and success in irrigation management transfer, and development and application of legal requirements on environmental impact assessment. These success stories can also provide clues as to the strategies for improvement in other elements.

When using a numeral scale it is then possible to assign an assessment score to all the elements or blocks and an accumulated score for the component (layer). Results are shown in Table 18, with a final percentage in each layer. Now, after those "light gray" blocks have been identified, one can proceed to assign them a more precise numeric evaluation score. The numbers or percentages should not be interpreted as approving or failure grades, but rather as revealing an order of magnitude of efforts needed to improve water resources management. It is evident that lowest ranking components are, again, "Strategy and Policy" (20%) and "Water Allocation and Quality Control" (20%), followed by "Legal" (30%), "Financing" (33%), "Environmental" (40%) and "Risks" (40%) components.

Elements such as Law, Cost Recovery, Incentives, Water Accounts and Droughts, are all considered being obsolete or inefficient, and should therefore be acknowledged as main items to be prioritized in an action plan for the modernization of the water resources management system. In the case of the "Monitoring" component, where there is more tradition and some modernization efforts have been made with luxury technology like "real time network of an early warning system", the score could have been higher. Difficulties with maintenance of hydrologic network and the early warning system, irregular water quality (surface and groundwater) and groundwater monitoring, make the final score to be lower than could have been expected, meaning that advances made are not sustainable or not being able to use them more effectively. It is believed that hydrologic monitoring has been debilitating in the last years and one underlying factor, as well as for most of the other components of the water management system, is lack of human resources. Renewal of and "cultivating" talented staff is a priority. Lastly, development of systematic management of hydrologic information is vital to water resources management.

16 CONCLUDING REMARKS

Problems and concerns of water resources in the country case study are related to signs of water stress, contamination of rivers and coastal waters with microbiological pollution, overexploitation of aquifers, poor quality of water for aquatic life in river reaches passing through cities, fast paced reservoir sedimentation and impact of uncontrolled urban growth and extreme hydrological events. Lack of sanitation services, insufficient wastewater infrastructure and irresponsible industrial and municipal discharge of solid and liquid wastes into rivers, are reasons for pollution and health problems. Low efficiency in the use of water and distribution network, losses of water due to non-accountability of users, subsidy dependent water services and poor cost recovery by service operators are characteristics of water quantity mismanagement. All these physical and economic water problems are related to the management system. It is necessary to allocate water having in consideration water availability, to improve monitoring and control of water pollution and adequately integrate or reconcile infrastructural and tourism developments. The absence of a policy statement and formulation process and water allocation mechanisms need to be solved.

Important topics to put on the agenda for improvement of the institutional framework are: weak control and weaker enforcement capacities, lack of rules to allocate water and informality on water rights. The existing institutional and legal frame work was designed with independent objectives and interests. Harmonizing reforms and working towards a common vision is recommended for any progress to be made in improvement of the water resources management system. There are signs of hope in water management with progress made in irrigation management transfer, educational programs on water conservation and environmental education, water balance update and the national water plan initiative.

The building blocks assessment tool, whose objectives are to identify missing elements of the water management system and evaluate the capacity of a country to face water challenges, can be useful for rapid appraisal exercises of water management capacities and is ideal for participatory assessment. Further testing and practical applications in other countries is relevant for its development and refinement. Results of the building block methodology indicate that strategy – policy formulation and water allocation mechanisms are top priorities to work on for the development of water resources management systems.

REFERENCES

Abreu, R.U. 2004. Informe Situación Actual de la Salud Ambiental Infantil en la Rep. Dominicana. Presentación Taller Definición Lineamientos Plan Nacional de Salud Ambiental Infantil, Santo Domingo, Julio, 2004.

Al Radif, Adil. Integrated water resources management (IWRM): an approach to face the challenges of the next century and to avert future crisis. (Conference on Desalination and the Environment, Las Palmas, Gran Canaria, November 9–12, 1999). Desalination 124, 1999 pp. 145–153.

Biswas, Asit K. Integrated Water Resources Management: A Reassessment. A Water Forum Contribution. International Water Resources Association. Water International, Volume 29, Number 2, June 2002. pp. 248–256.

Brath, A., Brandimarte, L., Castellearin, A., Di Baldassarre, G. DISTART – Universitá di Bologna. 2006. Proyecto de capacitación para la defensa Hidrogeológica del Territorio en la República Dominicana y Haití. Relazione Conclusiva. December 2006. pp. 1–4, 79–85.

Centro de Investigación en Tecnología de Agua para Riego (CITAR) – INDRHI. January 2008a. Estudio Diagnóstico de Desempeño de Juntas de Regantes 2007. pp. 1–22.

Centro de Investigación en Tecnología de Agua para Riego (CITAR) – INDRHI, Research Report – Calidad de las Aguas cuenca Ozama – Isabela, Santo Domingo (August 2007).

Centro de Investigación en Tecnología de Agua para Riego (CITAR) – INDRHI, Research Report AERH/08-08 – Análisis de las Tarifas y el Costo del Agua en Áreas bajo Riego en la República Dominicana (March 2008b). Santo Domingo. pp. 1–3, 11–16.

Chalas, J.R. 2007. Impacto de las Organizaciones de Usuarios en las áreas bajo riego de la República Dominicana *(Análisis Preliminar). July 2007. Centro de Investigación en Tecnología de Agua para Riego (CITAR) Instituto Nnacional de Recursos Hidráulicos (INDRHI. Santo Domingo, Dominican Republic.* (this seems not to be a bibliographic reference)

CYE Consult s.n.c. Estudio de Identificación y Formulación del Plan Indicativo de Trabajo del Sector Agua del 9no FED September 2003. (this seems not to be a bibliographic reference)

Dourojeanni, Axel 2001. Water management at the river basin level: challenges in Latin America. (Economic Commision for Latin America and the Caribbean, LC/L.1583—P, agosto de 2001, *Serie Recursos Naturales e Infraestructura N^o* 29)

ECLAC (Economic Commision for Latin America and the Caribbean) quoting Jovel and Zapata, 1993 in ECLAC's report on Damages evaluation caused by Hurricane Georges, 1998). (this seems not to be a bibliographic reference)

ECLAC (Economic Commision for Latin America and the Caribbean), Secretariado Técnico de la Presidencia, Programa de naciones Unidas para el desarrollo (PNUD). Octubre 2004. (this seems not to be a bibliographic reference)

Food and Agricultural Organization (FAO) of the United Nations. 1995. Water sector policy review and strategy formulation, A general framework, FAO Land and Water Bulletin No. 3, United Nations Development Programme (UNDP), World Bank and FAO, Rome, Italy, ISBN 92-5-103714-0. pp. 1–3.

Foro de Ministros de Medio Ambiente de Iberoamérica I, celebrado en La Toja, Reino de España, 21 y 22 de septiembre de 2001. (this seems not to be a bibliographic reference)

Garduño, H. 2001. Water Rights administration, experience, issues and guideline, Development Law Service of the Food and Agricultural Organization (FAO) of the United Nations, FAO Legislative Study No. 70, Rome, Italy, pp. 1–3.

Global Water Partnership (GWP) September 2000. Integrated Water Resources Management, TEC Background Papers No. 4, GWP, Stockholm, Sweden. ISSN: 1403-5324, ISBN: 91-631-0058-4. pp. 7.

González, G., Gladis, R. and Valdez, N. Calidad de las Aguas de las Playas de la República Dominicana. Santo Domingo, Secretaría de estado de medio Ambiente y Recursos Naturales, 2004. (this seems not to be a bibliographic reference)

Instituto Nacional de Recursos Hidráulicos (INDRHI)–OTS Corp, for Inter American Development Bank, Diagnóstico, Componente de Validación de Opciones y Prioridades, Análisis de factibilidad y Diseño del Programa de Manejo de Cuencas y Zonas Costeras, Santo Domingo. March, 2006. pp. 20, 94.

Instituto Nacional de Recursos Hidráulicos (INDRHI) 2007 Water Management Consultants (WMC), Análisis de la Rehabilitación de los Embalses y Represas de valdesia y Aguacate, Cuenca del Río Nizao, República Dominicana for Inter American Development Bank. pp. ii, xii, 5–12.

INDRHI, El INDRHI en el Desarrollo Nacional. 2006b. Editora Taller, Santo Domingo, Rep. Dominicana. pp. 107, 123.

International Centre for Water Studies (ICWS). May 2002. Elaboration of an Institutional Framework for Integrated Water Resources Management – Final Report to the Inter-American Development Bank, Amersfoort, The Netherlands.

International Water Management Institute (IWMI 2002). Final Report of the World Summit on Sustainable Development. 28th of August – 3rd of September. Johannesburg. ISBN: 92-9090-510-7. pp. 5, 7, 15, 16, 17, 21.

Jewit, Gram. Can integrated Water Resources Management sustain the provision of ecosystem goods and services. 2nd WARFSA/WaterNet Symposium: Integrated Water Resources Management: Theory, Practice, Cases; Cape Town, 30–31 October 2001.

Jiménez, O. and Farias, H., Problemática de la Sedimentación en el Embalse de Valdesia. Report for CDEEE, Santo Domingo, 2002, pp. 1–4, 27–29.

Lamoree, G. Bem, García, Luís, Pérez Raúl, Castro, Edmundo. Methodology for the Assessment of Institutional Frameworks for Water Resources Management, Experiences from Latin America. Water International Vol. 30, Number 3. September 2005. pp. 283–293. International Water Resources Associations (IWRA).

Nagle, Gregory. 2005. Los Efectos de un huracán sobre la pérdida de suelos en parcelas cultivadas en una cuenca tropical montañisa. 16, Ithaca, N. Y. Cornell University. (this seems not to be a bibliographic reference)

Oficina Nacional de Planificación (ONAPLAN), Informe General sobre Focalización de la Pobreza en República Dominicana, July 2005. (this seems not to be a bibliographic reference)

Oficina Nacional de Estadísticas (ONE), ENHOGAR 2006, Informe de Avances de Resultado, diciembre 2006 (this seems not to be a bibliographic reference)

Parsram, Kemraj, Educación sobre Conservación de Agua. Foro Nacional del Agua, February 2007. Santo Domingo. pp. 120–130. (this is not a bibliographic reference)

Pérez, R., González, E., Sánchez, M., Suero, M., Martínez, E. and Tavárez, V. 2004. (INDRHI). Inundaciones Jimaní–Río Soliette. Diagnóstico del Problema. June 2004. (this seems not to be a bibliographic reference)

Pérez, R. 2004. Programa Iberoamericano de Educación sobre el Agua. IV Conferencia General de Directores Generales Iberoamericanos del Agua (CODIA). Santo Domingo, March 2004. (this seems not to be a bibliographic reference)

Raast, Walter. 2005. Institutional design for water resources management. I Seminar on Integrated Transboundary Water Management. Lima, 2005. (this seems not to be a bibliographic reference)

Reed, S., Maidment, D. and Patoux, J. 1997. Spatial Water Balance of Texas, Center for Research in Water Resources University of Texas at Austin June 1997, pp. 1–4.

Ringler, Claudia, Rosegrant, M.W. and Paisner, M.S. Irrigation and Water resources in Latin America and the Caribbean: Challenges and Strategies. International Food Policy Research Institute, EPTD Discussion Paper No. 64. June 2000. pp. 11–17.

Robles, G. and Alcérreca, C. 2000. Administración, un enfoque interdisciplinario, Prentice Hall – Persons Educación de México, Mexico. ISBN: 968-444-421-4, p. 31.

Rodríguez, F. 2005. Floods in the Dominican Republic: Lessons Learnt. 1st Army Conference on Relief Operations in case of Disasters. Santo Domingo. 2005. (this seems not to be a bibliographic reference)

Rodríguez, F. 2006. Challenges of Legal Reform of the Water Sector in the Dominican Republic. World Bank session on assessment of water management systems 4th World Water Forum, Mexico, March 2006. (this seems not to be a bibliographic reference)

Secretaría de Estado de Medio Ambiente y Recursos Naturales (SEMARN), Atlas de los Recursos Naturales de la República Dominicana, Santo Domingo, 2004. (this seems not to be a bibliographic reference)

Secretaría de Estado de Medio Ambiente y Recursos Naturales (SEMARN) 2001, "Diagnóstico Económico Fiscal" (this seems not to be a bibliographic reference)

Setti, Arnaldo Augusto; Werneck Lima, Jorge Enoch Furquim; Goretti de Miranda Chaves, Adriana; de Castro Pereira, Isabella. 2002. Introdução ao gerenciamento de Recursos hídricos, Agencia Nacional de Energia Elétrica (ANEEL), Agencia Nacional de Águas, Organizao Meteorológica Mundial, Brasília, Brasil, pp. 90–93.

Stockholm International Water Management Institute (SIWI) and World Health Organization (WHO). Making Water a part of Economic Development: The economic benefits of improved water management and services. 2005. (this seems not to be a bibliographic reference)

Trewatha, R.L. and M. Gene Newport, Management, Business Publications, Inc., Plano Texas, 1982 ISBN 0-256-02713-7, pp. 5–8.

United Nations Development Program (UNDP) 2005 República Dominicana. Informe Nacional de Desarrollo Humano, Hacia una inserción mundial incluyente y renovada. 2005. ISBN 99934-55-81-4. pp. 86, 87.

United Nations Environmental Program (UNEP), Dams and Development: A compendium on Relevant practices for Improved Decision Making on Dams and their alternatives. Nairobi, 2007. ISBN 978-92-2816-3. pp. 11, 69–79.

United Nations Development Program (UNDP) 2006. Repot on Human Development, 2006, Beyond Scarcity, Power, Poverty and the World Water Crisis, Summary pp. 5, 41.

WMO/IDB (World Meteorological Organization and Inter-American Development Bank), 1996. Water Resources Assessment and management strategies in Latin America and the Caribbean. Proceedings of the WMO/IDB Conference, san Josè, Costa Rica, May 6–11.

World Meteorological Organization (WMO) and Inter-American Development Bank, 1996. (this seems not to be a bibliographic reference)

Third World Water Forum–Session report water, the MDGs and PRSPs Mainstreaming poverty reduction in water management. 19 march 2003. 3rd World Water Forum, Japan.

Author Index

Subject index